The Chora of Metaponto 3
Archaeological Field Survey
Bradano to Basento

Vol. II

The Chora of Metaponto 3
Archaeological Field Survey
Bradano to Basento

Vol. II

Edited by Joseph Coleman Carter and Alberto Prieto

Institute of Classical Archaeology
Packard Humanities Institute
University of Texas Press, Austin

Printed in China

First edition, 2011

Requests for permission to reproduce material from this work should be sent to:
Permissions
University of Texas Press
P.O. Box 7819
Austin, TX 78713-7819
www.utexas.edu/utpress/about/bpermission.html

♾ The paper used in this book meets the minimum requirements of
ANSI/NISO Z39.48-1992 (R1997) (Permanence of Paper).

The chora of Metaponto 3 / edited by Joseph Coleman Carter and Alberto Prieto. — 1st ed.
p. cm.
Includes bibliographical references and index.
ISBN 978-0-292-72678-9 (cloth : alk. paper)
1. Metapontum (Extinct city). 2. Metapontum (Extinct city)—Rural life and customs. 3. Agricultural colonies—Italy—Metapontum (Extinct city). 4. Greeks—Italy—Metapontum (Extinct city). 5. Excavations (Archaeology)—Italy—Metapontum (Extinct city). 6. Metaponto Region (Italy)—Antiquities. I. Carter, Joseph Coleman. II. Prieto, Alberto. III. Title: Chora of Metaponto three.
DG70.M52C56 2011
937'.773—dc22

2011004131
978-0-292-73481-4 (E-book)

For reasons of economy and speed, this volume has been printed from camera-ready copy furnished by ICA, which assumes full responsibility for its content.

To the memory of Dinu Adamesteanu,
to Antonio De Siena,
to students of Metaponto, past, present, and future.

Special Acknowledgment
These volumes and others in this series could not have been conceived,
much less carried through to publication, without the generous financial support of the
Packard Humanities Institute.
The Director and staff of the Institute of Classical Archaeology and its collaborators
express their profound gratitude.

Contents

Volume I

Volume II

Volume III

Volume IV Atlas

16
Introduction

Joseph Coleman Carter

Figure 16.1 A vision of the Metapontine landscape in the late 18th c., image by Claude Louis Chastelet. In the foreground, the Tavole Palatine on the Bradano River, 3 km from the ancient city, recently identified as a temple of Hera. (After Saint-Non, *Voyage Pittoresque ou Description des Royaumes de Naples et de Sicile* [1781–1786])

The original motivation in 1981 for the ICA survey was the desire to discover a Greek landscape as fully as possible, while it was still feasible (*Fig. 16.1*). Much of the primary documentation supplied by the survey is reproduced in Volumes I and IV. The full set of data will be available for search, query, and download on the book's companion website, including the information associated with the pottery assemblages (fabrics, classes, forms, and date ranges) for all sites. The spatial component is another essential and integral part of this study. Detailed maps are provided throughout the text, and they include basic topography, site locations, georeferenced features (such as the division lines) as well as the results of the geomorphological study of the region. A selection of the most relevant maps has been reproduced in Volume IV at a variety of scales (1:10,000; 1:25,000; 1:75,000). Online the reader can access these spatial data interactively.

Volumes II and III address how the survey results relate to evidence from other sources—excavation both in the chora and the city, aerial photography, special studies, and ancient texts—and attempt to weave this large body of information into a coherent narrative of the polis from its inception through its decline.

Chronological Scope

A major goal for the survey, in the early years especially, was to provide a spatial and temporal context for excavated sites in the chora. In the early 1980s, these consisted mainly of Greek farmhouses, necropoleis, and sanctuaries. Since then, the range has

Figure 16.2 Geologist Robert Folk (left) and survey team leader Cesare D'Annibale (right) at site 19 (outside the survey transects) in 1981, the first year of the survey. (PD)

expanded to include an important Neolithic village at Pantanello, a Late Roman farmhouse at San Biagio, a Medieval settlement, and others. By themselves, these could give some indication of the overall chronology of the chora and of the types of sites that existed in certain periods, and provide useful typologies of materials, but little about the overall pattern of settlement and its historical fluctuations. This last was to be the unique role of the survey at Metaponto.

From the beginning, the Metaponto survey aimed to document all periods of occupation—from the prehistoric to the Post-Medieval—with the same degree of thoroughness (*Fig. 16.2*). Though the primary focus was on the Greek and Roman periods, it would have been extremely short-sighted to ignore what came before and after. In this way, a balanced picture of a single geographic area over the long arc of human occupation could emerge; moreover, the originality of the Greek settlement—as well as its debts to previous occupants and its legacy to successors—could be objectively evaluated.

The six hundred years between the late 7th and late 1st centuries BC—the period of the independent polis and the two centuries following—left the widest and most indelible imprint on the landscape. Nearly 600 sites were documented in the 27.2-km^2 area surveyed between the Bradano and Basento from 1981 to 2001 (*Fig. 16.4*). Of these, 565 (95%) had ceramic material dateable to the period of the Greek polis;[1] of the over 29,000 fragments of pottery from all survey sites, some 25,000 document the Greek chora.

The pre-Greek occupation of the Metapontino was extensive, especially in the Neolithic and Bronze Ages. The pottery remains of this period number approximately 1,200 fragments; the stone implements, over 800. They come from a total of 136 sites (*Fig. 16.3*).

This publication was originally to have included a full treatment of the prehistoric period. For several reasons, primarily the incomplete state of the study of the material, the decision was made not to delay the publication of the rest and to devote a separate volume to the prehistoric.[2] In order to provide a continuous narrative of the human presence in the Metapontino, a brief treatment of the major trends of the prehistoric period appears here as Chapter 17. It is supplemented by individual site entries in the Gazetteer and a large-scale map with prehistoric sites in Volume IV.

The evidence for the Roman Imperial, Late Roman, Medieval, and Post-Medieval periods is modest compared to that of the Greek period. Approximately 1,800 ceramic fragments from 37 sites belong to the Roman period, and about 1,100 from 21 sites can be assigned to the Medieval. Four chapters (32–35) in

[1] A similar number were added to the record between the Basento and Cavone, which will be the subject of a future study.

[2] It will include the important Neolithic settlement excavated at Pantanello.

Volume III are devoted to the Roman and Medieval periods and complete the picture of human occupation of the Metapontino.

Survey, Excavation, and Aerial Photography

Even readers receptive to survey methods in archaeology may have doubts about the nature of the sites described by the survey. Chapter 18, which discusses the types of sites and materials from them, and Chapters 28–31, which describe the excavations of several survey sites, attempt to show that the surface remains are reliable indicators of what lies below. Further corroboration is found in the fact that the ceramic assemblages from survey sites identified as farmhouses, necropoleis, and sanctuaries are consistent among themselves and comparable to assemblages from excavated examples of these types. It is fortunate that work on the farmhouses, as well as several others excavated by the Soprintendenza, has been going on simultaneously with the survey publication,[3] thus making it possible to compare a wide range of information from excavated and surveyed sites.

Figure 16.3 Assemblage from site 81N, in Cogno del Prete in the Central Plateau, identified as a necropolis with evidence of Neolithic activity, and recorded in 1981, the first year of the survey. Note the polished basalt scraper (FS 81.2033).

Figure 16.4 The survey team in formation in the chora of Metaponto, July 2001. (AP)

Methodology

The methods employed to reconstruct the settlement patterns in the study area (the surveyed portion of the chora) are described in Chapter 19. The six centuries—four of the Greek polis and the two immediately following—have been divided into twelve 50-year periods (*Fig. 16.5*), and the study area into localities and regions. The chronology of the sites is based on an original analysis of the fine wares, principally the black-gloss pottery, developed in the early stages of research on the artifacts.

The approach and the dating methods employed throughout were, from the outset, designed to be flexible. The goal was ultimately not to provide a final statement, but to enable the reader to continue the study and modify it as the field advances, as it is bound to do. Not only advances in artifact studies but also in survey methods, new and improved technologies, plus future discoveries will all change and enrich the picture presented here.

Chora and Asty

Despite being considered among the best-known ancient Greek countrysides, many questions remain about the chora of Metaponto. This volume challenges or significantly modifies currently held ideas—some by others and even some expressed in earlier articles and studies by the authors. It also provides more details and, in some cases, new and different answers to some of the questions that long ago stimulated interest in the subject. When were the chora—and the asty—first settled? What were the roles of the early rural sanctuaries, and how did they relate to the religion of the city? What, if any, aspects of the life of the pre-Greek population survived in the chora? What was the social and economic status of the rural population vis-à-vis the urban? And the perennial question about landed property, which was the basis of citizenship and gauge of wealth in the Greek polis:

[3] The Soprintendenza generously permitted us to study excavations carried out in 1969 and 1979.

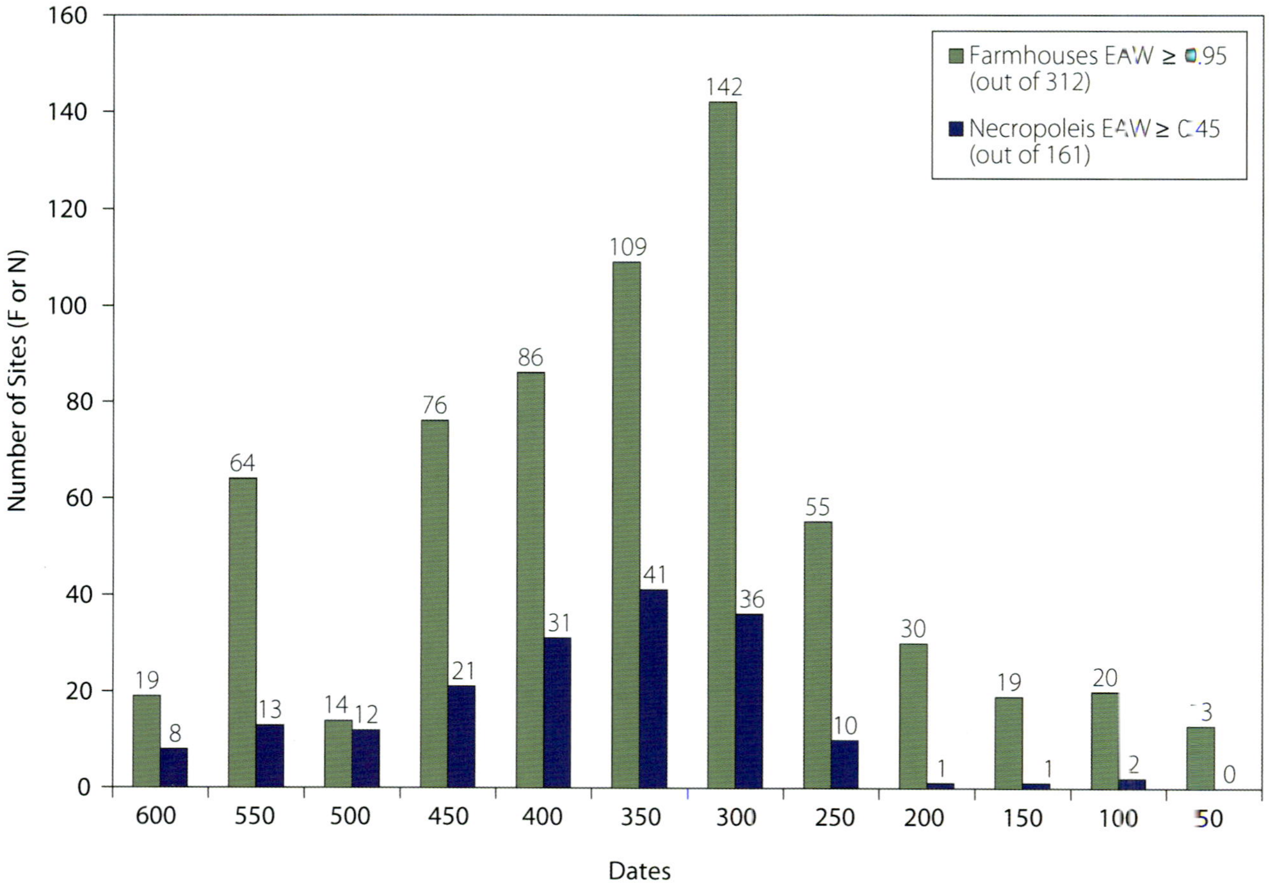

Figure 16.5 Counts of significant farmhouses and necropoleis by 50-year period showing increased and decreased settlement activity in the study area over the course of six centuries.

when and how was it first divided, distributed, and redistributed? These are some, but by no means all, of the questions which the evidence of the survey brings us closer to answering.

The chora, our immediate subject, is only half of the story of the polis of Metaponto. It cannot be separated from the asty (urban center), and the one cannot be fully understood without the other. Thanks to the inspired and enormously productive work of Dinu Adamesteanu, Antonio De Siena, and Dieter Mertens, a great deal is known about the asty. The results of their excavations over the last half-century are, regrettably, not widely known outside a circle of specialists. Full publication has long been anticipated and much desired, but up to now there have only been short accounts and articles on specific aspects in a variety of journals, some of which are not widely available in libraries outside of Italy. The parallel evolution of chora and asty is a major theme of this study, thanks to the generosity of the above-mentioned colleagues, and it is graphically illustrated by the series of maps covering the twelve periods (625–25 BC) which are the heart of Volume IV. Naturally, I take full responsibility for any errors of fact or interpretation that appear here.

Historical Overview

The earliest years of an apoikia, a new settlement, far from home, are almost always the most obscure. Metaponto is no exception. The survey results reveal that the earliest settlements in the chora are practically contemporaneous with that on the site of the future urban center (Ch. 20). When the pattern of farmhouses and necropoleis of the first century of settlement (625–525 BC) emerged, it was a revelation. Most of the settlers had stuck together in what we have termed Early Village Clusters. Five of them lie in our study area, with individual dwellings separated by regular intervals of approximately 600 or 300 m, close to springs, and facing each other across natural hollows along the major river valleys, or strung out at regular intervals along

Figure 16.6 View of the Central Plateau in 1982, from farmhouse site 257F in Cozzo del Prete.

the valley edges or tributary streams. Contemporary necropoleis and tombs besides the farmhouses are proof that some of these dwellings were permanent residences in the countryside at a considerable distance (6–11 km) from the coast and the nascent city center.

Three of the Early Village Clusters include structures that have been identified as sanctuaries because of the exceptionally large concentrations of associated votive material: terracotta figurines of divinities or related subjects and miniature votive cups. The villages appear to have been at least to some degree autonomous, with members bound together, in some cases, it could be argued, by a common cult and by communal eating and drinking rituals. The prime concern of these settlers was to establish themselves and their communities on choice land soon, if not immediately, after their arrival on the Ionian coast of Italy between the Bradano and Basento Rivers.

The villages grew rapidly and Metaponto in the mid-6th century experienced its first period of great prosperity (Ch. 21). The source of the wealth, as the coinage of Metaponto at this time and henceforth proudly proclaims, was grain. The agricultural potential of a range of crops, described in Chapter 19, has been documented by paleobotanical studies in several areas of the chora, including the Central Plateau (*Fig. 16.6*) There are indications of speculative development on virgin land; the arable Central Plateau between the two rivers was opened to settlement and smaller new villages grew up there throughout the 6th century. For a century, settlement there was largely restricted to a strip crossing it from north to south. The sanctuaries of the Early Village Clusters enjoyed their heyday then. By the end of the 6th century, the quantity of votive objects deposited there had fallen off dramatically, and thereafter these sites are barely distinguishable from farmhouses. The prosperity of the chora in the mid-6th century BC made possible a great building program in the asty. The modest shrines of the first settlement were replaced by two immense Doric temples, and, in the agora, the largest structure of its kind in the Greek world, a place of public assembly, Ekklesiasterion I, arose, with a seating capacity of approximately 8,000. These structures were the most visible aspects of an orthogonal city plan that gave direction to all future growth. The hand of a tyrant has been seen behind the changes of the asty. At the same time, this plan, it will be argued, had its complement in a first orthogonal division of the chora.

By the end of the 6th century, problems had begun to emerge (Ch. 22). The period **500** (525–475 BC) was an unsettled one for the poleis of Magna Grecia, which had grown to be very wealthy over the course of the 6th century BC. Many were wracked by internal dissent and internecine struggle as new settlers were pitted against the old, and there was widespread reaction to the considerable political power of the Pythagorean groups in many of the poleis. The most prosperous of all, Achaian Sybaris, was destroyed by a coalition led by Achaian Croton. The documentary sources that are sparse for Magna Grecia are particularly so for Metaponto, which is not mentioned in the context of these tumultuous events, but could hardly have escaped unscathed. The number of occupied farmhouses in the chora fell precipitously. Building in the city and its immediate environs ceased in about 520 BC, after the completion of the extra-urban Temple of Hera, the so-called Tavole Palatine (*Fig. 16.7*). This build-

Figure 16.7 Wheat field, Tavole Palatine (Temple of Hera, late 6th c. BC) in background. (CR)

ing is well known in the archaeological literature and the only nearly complete Greek temple still standing along the Ionian coast. The severity of the recession in the chora is hard to judge. The necropoleis showed undiminished activity. It may be that those who left the land sought shelter in the asty, but until the excavation of the residential quarter is undertaken—the biggest single lacuna in the archaeological record of Metaponto—that must remain an open question.

The revival of the chora and the asty in the 5th century BC (Ch. 23) seems to have been simultaneous and it initiated one of the most creative periods in the history of the polis. The number of farmhouses surpassed that of the flourishing mid-6th century, a large proportion of them being ex novo constructions on previously unoccupied sites. The chora of the first half of the 5th century, beginning about 480 BC, was the scene of one of the largest public works projects that the Greeks ever undertook of which any significant remains survive. The drainage of the chora, carried out by the construction of an extensive network of drainage canals that also divided the Central Plateau and other areas into parallel strips measuring approximately 210 m wide, flanked in many cases by roads, was completed by mid-century, celebrated by games and commemorated by a coin bearing the image of the river god Acheloos. Evidence of this process—the division lines that appear in the aerial photos and on the ground—is treated at greater length in Volume III in Chapters 28–31. The survey identified more of these lines and confirmed their double-function as drainage canals or roads (or both), and as highly visible boundary markers delineating properties. They are flanked by new necropoleis in this period.[4]

The new farmhouses are concentrated in the areas where the lines are most dense and well preserved, and the conclusion was drawn that a redistribution of land took place in the Central Plateau in the mid-5th

[4] A similar system, though oriented differently, was completed at the same time in the Basento-Cavone half of the chora.

century BC. Such radical acts were fairly commonplace in the Greek world in the 5th and 4th centuries, and they are frequently mentioned by the historical writers. The evidence from the chora of Metaponto is by far the fullest documentation of such a redistribution, but it is not alone. There is also the case of Kamarina in Sicily, where, though the archaeological evidence is slight, the textual evidence is unambiguous. *Bonifiche*, or land reclamation projects, frequently including a reorganization of the chora, were not an uncommon event in the poleis of Magna Grecia, Sicily, and mainland Greece in this period, as examples will show.

At the same time, the appearance of the monumental center in the city was altered decisively. Ekklesiasterion II, a much refined and embellished meeting place, replaced its crumbling predecessor on the same site. It was now clearly intended for the ekklesia, the assembly of citizens. An oracular sanctuary, described by Herodotos, arose in the agora and marked a new direction in religion.[5] The 5th century BC witnessed revival and reform in both chora and asty.

The growth of the chora continued, but slowed. Looking only at the city in the late 5th and early years of the 4th century BC (Ch. 24), it would appear to have been a period of decline. When Thucydides records Metaponto's participation in the Sicilian Expedition,[6] he also mentions stasis—the conflict between the better-off citizens and the demos (smaller farmers artisans, and the property-less)—that would have afflicted Metaponto, like most Greek cities.

At this point there is nearly complete silence in the ancient sources, and we must rely entirely on archaeology and numismatics. Already by the early 4th century BC, the temples of previous periods were not receiving needed repairs: Ekklesiasterion II was damaged and material from it was recycled in tombs. This had no noticeable repercussions in the chora, though this period in its history stands out for its quiescence in contrast to those that preceded and followed.

Change began in the middle of the 4th century (Ch. 25). The elevated number of necropoleis that began to be used for the first time in this period would indicate that new families and possibly new groups were moving into the chora. The sudden growth that extended from the mid-4th century to its end resulted in the maximum expansion of the chora, when both in population and in wealth it reached its zenith. The period 375–325 BC was again marked by what has been interpreted as a second redistribution of land, much more widespread than the first, in precisely the same area of the Central Plateau. Many new farmhouses were built in close proximity to the division lines. For the first time in this period, all over the Central Plateau, farmhouses appeared ex novo on sites never before occupied. They were arranged in clusters, and, like those of previous periods, they displayed a relatively uniform and regular spacing between components. Scarciullo, a large, flat area like the Central Plateau, ideal for growing grain, was opened up for the first time, and it was divided regularly into nearly equal-sized plots. In the last quarter of the 4th century BC and first quarter of the 3rd, material prosperity reached an unprecedented level, and it appears to have been widely shared throughout the Metapontino.

Investigations of the necropoleis of the second half of the 4th century indicate an influx of new groups, and the evidence of burial customs suggests that some, at least, were non-Greeks, though clearly familiar with Greek customs. The co-option or enfranchisement of local non-Greek populations, which began in Magna Grecia soon after the Greeks arrived, is also documented in Sicily and in northern Greece in this period. The archaeological evidence presented here for Metaponto, though capable of being read in different ways, is unusually full.

Whatever the composition of the population of the chora and the asty, there is renewed vitality that recalls two earlier eras: the mid-6th century BC and the middle of the 5th. It is manifested in the asty in the creation of a new center with a theater of startlingly innovative design that looks forward to the first Roman stone theaters two centuries later, and a stoa that frames the eastern side of the vast agora and anticipates the practices of the Hellenistic East. These buildings are at the cutting edge of architectural design, well in advance of similar developments on the mainland and in Asia Minor. Neither of these projects, unfortunately, was ever completely finished, though both were in operation in the last years of the 4th century.

There was military and political pressure from outside, from potentially hostile Lucanians and from the ambitious foreign generals. Called to southern Italy by hegemonic Taras to combat the external menace, the generals could and did turn on those

[5] 4.15.

[6] 7. 33.4–5; 57.11.

Figure 16.8 Abandoned farmhouse (Masseria Castellano) near Cozzo Parlante, beyond the study area to the west, well into the interior; Montescaglioso in background. (CDA)

whom they were there to protect. Metaponto, as we learn from Diodorus Siculus, one of the fuller sources for Metaponto in this period, fell victim to the ambitions of the Spartan general Kleonymos.[7] If the terms he imposed on Metaponto were honored, the strain on the city, the urban elite, and especially on the wealthier inhabitants of the chora would have been great indeed.

The early decades of the 3rd century must have been onerous, and they were punctuated at the end by the arrival on the scene of Pyrrhos of Epiros and the Romans. The epic struggle between Greeks and Romans, first joined at nearby Herakleia, changed the course of history for the poleis of Magna Grecia, including Metaponto. They never recovered. Metaponto, which had weathered other storms, began a steep downward trajectory that is reflected in the sharp fall in the number of farmhouses and necropoleis in the chora (Ch. 26). In the asty it is marked by the imposition over the grid plan of the city of an intrusive structure, known as the castrum, which, though it made use of the street grid, had nothing to do with the previous settlement. The structure, highly similar to ones erected by the Romans all along the Tyrrhenian coast for defense against the Carthaginians, is amazingly close in design to that at Pyrgi, except that it is much larger. The castrum was connected by the main east-west plateia of the asty to a newly created ship basin and to the chora, ever a source of grain

[7] Diodorus Siculus 20.104.3.

and other provisions useful to an army. The evidence strongly suggests that the Romans were in control of the asty from the mid-3rd century BC onwards. There are some slight indications that the chora, on the other hand, may have been largely in Lucanian hands by this time: major sanctuaries of the chora at San Biagio and Pantanello lost their religious character and were built over by farmhouses. The new population of the chora was either ignorant of the sacred places of their predecessors, or saw no reason to respect them.

The decline of the population inhabiting farmhouses in the chora accelerated. In the first half of this period, Metaponto was a base for Hannibal, who occupied what was left of the asty from 211–207 BC. The Roman garrison of the castrum had been betrayed by the Metapontines, or at least by some elements of the population, eager to rid themselves of the Romans. Retribution would have been expected, and it came. When we measure the population of the chora again, it had fallen tremendously (Ch. 27). There were only 20 farmhouses in the study area, as before, compared to the 142 at the height of Metaponto's prosperity. They do not cluster together, and in most cases they are separated by relatively great distances. Many were completely ex novo. There had been virtually no new farmhouses in the chora since 300. Now a third of the farmhouses were constructions on sites never before occupied.

The territories of poleis that did not support Romans against Hannibal—or did not support them enough—were frequently divided into small plots for the colonists that Rome sent out. Alternatively, these areas became *ager publicus populi Romani*, public land that could be leased out to private citizens, who were often well-connected politically. The ancient sources are silent on the fate of the territory of Metaponto, but the wide spacing between farmhouses, isolated in contrast to previous periods, strongly suggests that it became *ager publicus* and eventually the private property of a privileged few. Interestingly, the areas surrounding farmhouses, and presumably under their control, are quite large, but not larger, it would seem, than the 500 *jugera* limit that Roman agrarian law set as the maximum that an individual could rent from the state and eventually own. After the failure of the Gracchan reforms, the spaces between farmhouses were considerably larger on the whole.

Both excavation and survey have revealed that a number of kilns were active in the chora at this time. One of them at Pantanello was producing amphorae, pottery, and roof tiles on an industrial scale, presumably for large agricultural establishments in the region, or for export. By this time, the settlement on the site of the asty had been reduced to a shadow of its former self, and it was reduced further still, during the occupation of Metaponto by Spartacus in 73–71 BC. He used it as his base and a rallying point for the numerous slaves employed in the surroundings, either in farming or transhumant pastoralism. Studies of animal and limited plant remains indicate that Metaponto was engaged in both activities during the last two centuries BC. Using documentary sources exclusively, Arnold Toynbee argued a half-century ago that southern Italy after the Second Punic War consisted entirely of latifundia populated by slaves and sheep.[8] He was convinced that these events had created the conditions in southern Italy—poor peasants, wealthy nobles, malaria, and inertia—that he witnessed as a student in the early 20th century (*Fig. 16.8*). The results of the survey and other research in the chora of Metaponto offer an alternative vision, much richer and more variegated, one that is, however, still far from complete.

This marks the end of the period covered by this volume. The following chapters present the data and the methods used to reach the conclusions summarized here.

[8] Toynbee 1965 II, Chs. VII and IX especially.

17
The Prehistory of the Chora

Joseph Coleman Carter

The chora of Metaponto has long been the most frequently cited example of a dispersed Greek settlement, with hundreds of farmhouses and other types of sites scattered throughout its agricultural territory. Before the ICA survey began in 1981, it was believed that the Greeks had introduced this settlement pattern to the countryside in southern Italy.

The results of the field survey—approximately 1,200 ceramic fragments and 840 lithics—have created a very different picture of the Metapontino in the prehistoric periods than existed before (see *Map 4*, Vol. IV). Before the survey began in 1981, only two prehistoric sites were known in the study area, the area covered by the survey: the unexcavated Bronze Age settlements at San Marco[1] and San Salvatore (*Fig. 17.1* and *Table 17.1*). A third, close to the study area, was Saldone, with Bronze Age material. In no case were there sufficient elements to characterize these sites. The remnants of a protohistoric Iron Age village were discovered in 1969 on the plateau of Cozzo Presepe (see *Fig 17.13*) just outside the northeast corner of the original Bradano-Basento transect.[2] The majority of the previously known pre- and protohistoric sites are located to the south and west of the study area. These include the important and fully published Iron Age necropoleis of Incoronata indigena and San Teodoro,[3] which lie in the Basento-Cavone transect (*Fig. 17.1*). The extremely important Bronze Age site at Termitito to the south of the Cavone (*Fig. 17.1*) was discovered and excavated beginning in the early 1980s. At that time, the contemporary Bronze Age site of San Vito di Pisticci,[4] which was included in the survey, was first noted. Both these sites have produced considerable quantities of Mycenaean-style pottery and attracted wide attention, yet apart from preliminary reports they remain thus far unpublished.[5] The Iron Age village beside an area of the extensive necropolis at Incoronata indigena was excavated in the same decade.[6]

[1] S. Bianco 1978.

[2] *Cozzo Presepe*, 285–317.

[3] Chiartano 1983; 1994; 1996.

[4] Site 29 (outside the study area): De Siena and Bianco 1982.

[5] De Siena 1996, 166–75.

Between 1981 and 2001, in the relatively small space of the study area, the survey recorded a total of 55 sites with Neolithic material, including pottery and evidence of stone working (*Fig. 17.2* and Gazetteer, Vol. IV). Thirteen of them were identified as "domestic structures" and five as "nucleated settlements." The remaining sites yielded evidence of Neolithic or suspected Neolithic activity. The clearly identifiable Bronze Age sites (*Fig. 17.5*) are fewer, with a total of 16. There are eight possible domestic structures, four possible nucleated settlements, and four areas of Bronze Age activity.

At another 65 sites (*Fig. 17.6*) there were signs of prehistoric activity that could not be assigned to a particular period. In most cases the evidence consisted of difficult to date stone manufactures, and at six sites there were only undateable fragments of impasto pottery. This brings the total of sites with prehistoric material of all periods to 136, a density of 3.7 sites per km^2, spread over four or five millennia.

This is probably not the whole picture. Small fragments of impasto pottery and lithics are not as obvious in the field as the fired and decorated pottery typical of a Greek or Roman site. The survey team's ability to spot isolated fragments of impasto pottery was very good, but it is a fact that much of the prehistoric material collected came from sites that also had historic material. There is thus a high correlation between prehistoric sites and those settled by the Greeks that may be to some degree an artifact of the process.

The Neolithic and Eneolithic in the Metapontino

It is a fact that the fertile land and abundant sources of water, which consistently attracted settlement from the period of the developing polis in the late 7th century BC onwards, also appealed to the early agricul-

[6] De Siena 1996, 175–95.

Figure 17.1 Map of known prehistoric and protohistoric sites in the Metapontino and vicinity. (JT)

turists of the Neolithic and Eneolithic periods (6th to late 3rd millennia BC), when the forests were virgin and game abundant. The precise conditions that attracted the early Neolithic farmers to the Metapontino cannot, however, automatically be assumed to have been those that the Greeks knew. In the 6th millennium, there had been widespread climatic change in the transition between the Boreal and Atlantic climatic periods, which saw the establishment of moist conditions with forests of deciduous oak. During the sub-Boreal (3000

Name-Location	Chronology	Site type	Bibliography
Conca d'Oro	IX–VIII c.	Necropolis	S. Bianco 1998, 26–28
Cozzo Presepe	IX–VII c.	Settlement	Morel 1970a; Adamesteanu 1974a, 77–78; *Cozzo Presepe* 210–14, 285–317
Craco	VII c.	Settlement	De Siena 2001b, 20–21
Difesa San Biagio, Montescaglioso	Bronze Age, X–VII c.	Indigenous settlement, necropolis	Canosa 1986, 171–75; S. Bianco 1986, 19
Ferrandina	VII c.	Settlement	De Siena 2001b, 20–21
Fiumicello, site 20	Neolithic, Iron Age	Necropolis	See field survey 1981 notes
Incoronata greca	VIII–late VII c.	Protocolonial settlement	Orlandini 1976; Carter 1978; 2006, 55–74; *Incoronata*, 1–5; Orlandini 1999; De Siena 1996
Incoronata indigena	X–late VIII c.	Settlement, necropolis	De Siena 1996, 165–66, 175–90; S. Bianco 1998, 16; Chiartano 1983; 1994; 1996
Lama San Nicola	VII c.	Habitation	Nava 2002, 739–40; De Siena 2002, 31–32
Masseria Cardillo	VIII c.	Settlement	Adamesteanu 1974a, 66
Metaponto, Andrisani	VII c.	Settlement	De Siena 1986; 1990; 2001, 21
Metaponto, Castrum	VII c.	Settlement	L. Giardino 1978a
Montescaglioso	Bronze Age–Medieval	Settlement	De Siena 2001, 20–21; S. Bianco 1986, 22
Pantanello	Eneolithic	Settlement	Ingravallo 1981; S. Bianco 1983b; Carter 1984, 469–70
Pisticci	VIII–VI c.	Settlement	Adamesteanu 1974a, 137–44; Adamesteanu and Vatin 1976, 116; De Siena 2001, 20–21
Pizzica	Neolithic/Bronze Age	Habitation	Nava 2002, 736; Radi 1999, 64–65
Saldone	Neolithic, Bronze Age	Settlement?	S. Bianco 1986, 19–20; Radi 1999, 58; De Siena 1996, 163
San Biagio sanctuary	Neolithic, Iron Age	Prehistoric settlement	Adamesteanu 1974a, 56, 66; Cipolloni Sampò 1999, 125
San Biagio farmhouse	Neolithic	Farmhouse	D'Annibale in Lapadula (forthcoming)
San Leonarco	VIII c.	Indigenous settlement	Adamesteanu 1974a, 66–67
San Marco	Bronze Age, VIII c.	Indigenous settlement	Adamesteanu 1974a, 66; S. Bianco 1986, 19
Santa Maria d'Anglona	Middle Bronze Age–early VII c.	Indigenous settlement, necropolis	S. Bianco 1998, 19–22; De Siena 1996, 162
San Salvatore	Middle Bronze Age	Necropolis	S. Bianco 1998, 15; S. Bianco 1986, 19
San Salvatore, sito 29*	Neolithic/Eneolithic	Necropolis	Nava 2003a, 675
San Teodoro	IX–mid-VIII c.	Necropolis	Chiartano 1983; 1994; 1996; Nava 2003a; De Siena 1996
San Teodoro Vecchio, sito 68*	VII c.	Canal with retaining walls	Nava 2003a, 670–71
San Vito di Pisticci, site 29	Bronze Age	Settlement	Bianco and De Siena 1982; De Siena 1996, 162–66; S. Bianco 1983b, 134–36; De Siena 1986a
Tavole Palatine	Neolithic	Deposit	Lo Porto 1981a, 25–26
Termitito	Mid-VII c.	Farmhouse	De Siena 1996, 171–72
Termitito	Middle/Late Bronze Age–VII c.	Indigenous settlement	Bianco and De Siena 1982; De Siena 1996, 166–73
Tinchi	VIII–VII? c.	Indigenous settlement	Adamesteanu 1974a, 66
Valle Sorigliano	End IX–mid-VII c.	Necropolis	S. Bianco 1998, 22–24
Viggiano, site 57*	VII c.	Habitation	Nava 2002, 738; 2003a, 672; De Siena 2002, 30–31; 2003, 512; 2005, 450

Table 17.1 Known prehistoric and protohistoric sites in the Metapontino and vicinity. Light blue: in the Bradano-Basento area but outside the study area; grey: in the study area. Site numbers refer to the ICA survey. *Sites marked with an asterisk were investigated by the Soprintendenza and numbered according to its system.

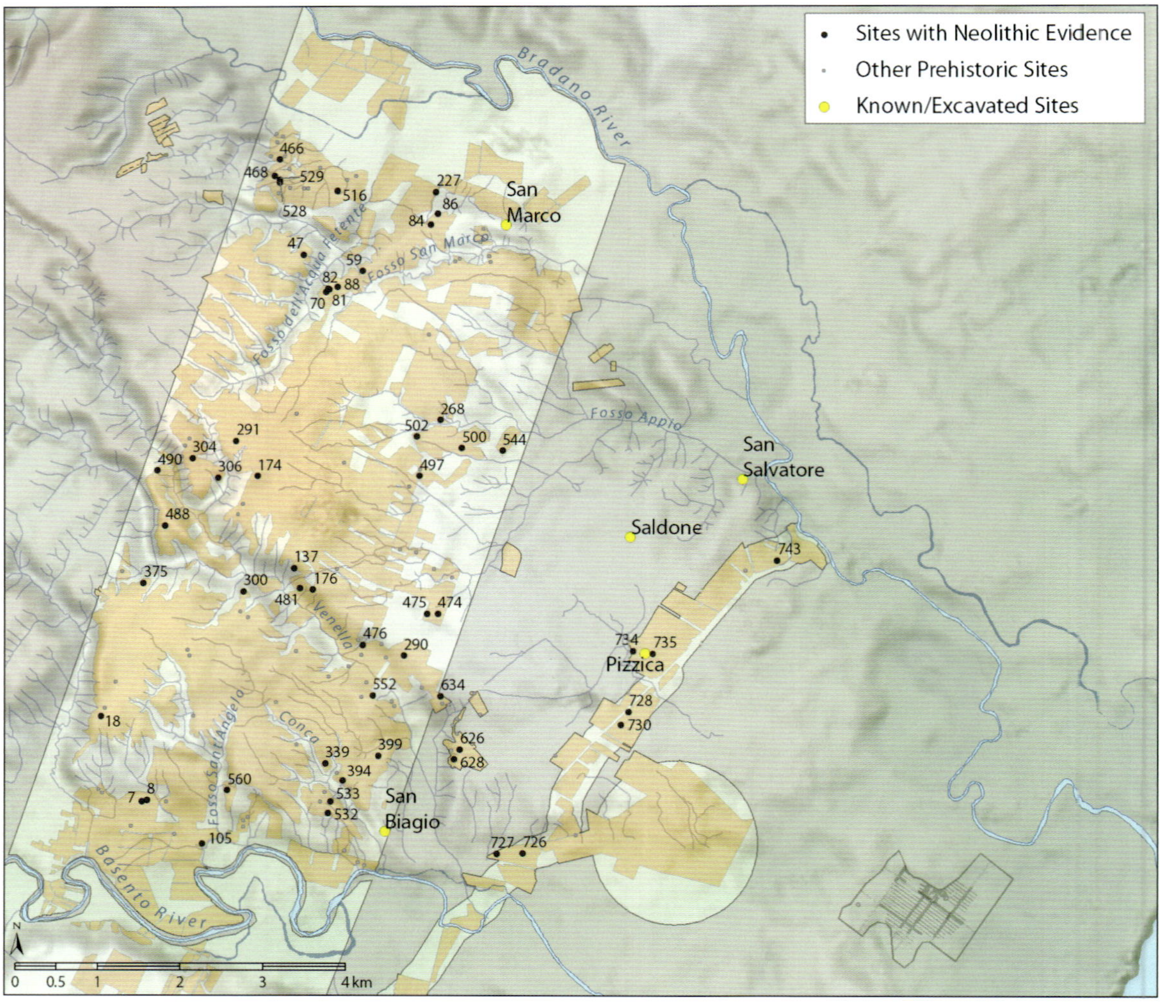

Figure 17.2 Map of sites with Neolithic evidence in the study area. (JT)

or 2500 BC) down to about 800 or 700 BC, precipitation decreased and the holm oak predominated in Mediterranean forests. Most of the Eneolithic and all of the Bronze Age coincided with the sub-Boreal. With the sub-Atlantic, beginning about 700–500 BC, precipitation increased and temperatures were cooler.[7] There were significant tectonic events, such as the positive bradyseism (uplift) that affected southern Italy as a whole. In the case of Metaponto, there was negative bradyseism (descent), which was probably one of the causes of a correspondingly sudden rise of the water table throughout the coastal area in the late 6th or early 5th century BC.[8]

Abbott (Ch. 2) has convincingly argued that the process of deforestation and plowing that began in the Metapontino in the Neolithic irremediably altered the landscape. The earliest major erosional episodes have been dated by C14 analysis to the 4th millennium BC. There was a period of stabilization in the Bronze Age, due in part to the lighter pressure on the land that a predominantly pastoral economy would have exerted. The alluvial build up in the river valleys is a constant factor from the Neolithic onward. It has blocked from view sites of all periods lying in or near the valley floors. The last great transformations of the valleys took place about a thousand years ago. Since then relatively minor flooding, as recently as 1959 and 2004, has deposited thick layers of alluvium on portions of the coastal plains.

The slopes have suffered less, though erosion has continued to modify their contours. The single biggest agent of change from the 1950s onward—in ever accelerating pace—has been agricultural development, which has bulldozed away hillsides and flattened

[7] Delano Smith 1979, 312–15.

[8] Boenzi et al. 1988, 1989; Abbott 1997, 202–03.

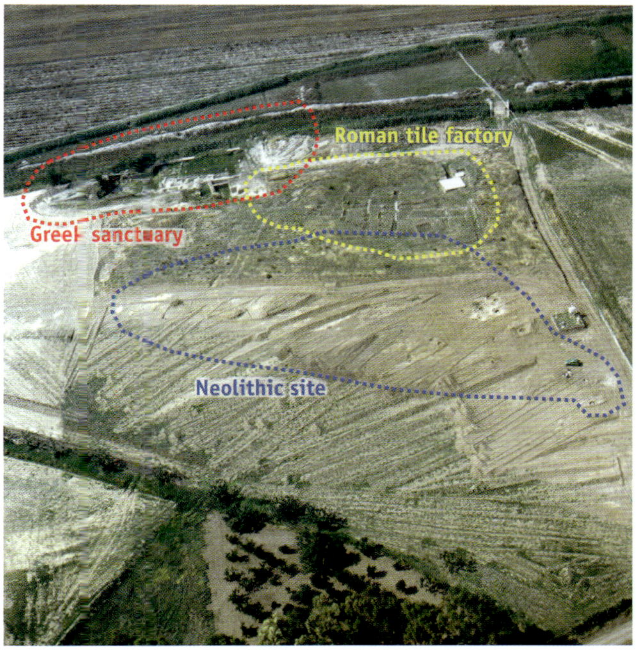

Figure 17.3 Aerial photo of Pantanello, looking east, with Greek sanctuary, Roman-period tile factory, and Neolithic village site in foreground. (ALC)

fields, destroying sites on the slopes and the marine terraces, with serious negative effects on this once relatively pristine Mediterranean landscape.

The earliest of the 55 sites with Neolithic material is site 300 in the upper reaches of the Venella valley, an area that also attracted Bronze Age settlers (*Fig. 17.2*). It is an Early Neolithic site with cardium-decorated Impressed Ware dateable to ca. 4500 BC.[9] The largest Neolithic site, site 227, located at San Marco on a low slope overlooking the Bradano, apparently had no subsequent occupation, though the San Marco area was a center of intensive settlement in the Bronze Age.[10] Sites 7 and 8, on a ridge overlooking the Basento valley at Sant'Angelo, were substantial Neolithic sites, with abundant pottery of the Middle Neolithic as well as polished stone tools and preserved daub (mud chinking).

A major Late Neolithic to Eneolithic site of the 3rd millennium BC was excavated at Pantanello[11] by the survey team led by Cesare D'Annibale (*Fig. 17.3*).[12] Situated on a low spur facing the Basento and the coast, its discovery came about quite by chance. There were a few fragments of impasto pottery on the surface in 1974 when a small area on the summit of the spur was excavated by a group from the University of Lecce.[13] Then, one morning in June 1983, the author observed a bulldozer scraping away the surface of the north side of the hill to prepare it for an experimental almond grove for the Azienda Sperimentale "Pantanello." A hurried visit revealed the presence of a large number of patches of grey soil distinctly visible against the yellow-brown virgin soil of the hillside. Closer inspection showed that these contained not only an abundance of impasto pottery (*Fig. 17.4*), but stone tools, animal bones, and daub from the wattle hut walls. The daub had impressions of seeds—impressions that, together with the carbonized seeds and abundant faunal remains, would become the primary evidence for agriculture and animal populations in this stretch of the Ionian coast in the Eneolithic period (ca. 3000–2500 BC).[14] Some 26 features—many identified as pits and several as hut floors—were subsequently excavated at Pantanello. One contained evidence for bronze working, among the earliest known examples in Italy.[15] Another nearby excavated site of this period in the Metapontino is a recently discovered burial at San Salvatore.[16]

Figure 17.4 Three impasto vessels from the Neolithic site Pantanello: *left*, one-handled cup (h 9 cm); *right*, ovoid jug with cylindrical neck (h 16 cm); *center foreground*, one-handled dish (h 3.5 cm).

[9] Site 300 was also that of Greek burials and a major Medieval site, one of the most important in the area; see Ch. 34, 1163–70.

[10] San Marco was lightly settled by Greek farmers, who preferred the opposite slope of the small valley with its springs.

[11] Cipolloni Sampò 1999, 69–70; Radi 1999, 64–65.

[12] S. Bianco 1983. This site is the subject, together with the prehistoric survey sites, of a future study.

[13] Ingravallo 1981; Carter 1984, 469–70.

[14] Carter 1984, 469–70; S. Bianco 1983, 10, 15; Cipolloni Sampò 1999, 68–69; Bökönyi and Gál 2010, 4–5.

[15] The results will form a forthcoming volume in this series.

[16] See Table 17.1, San Salvatore, sito 29. Nava 2003a, 675.

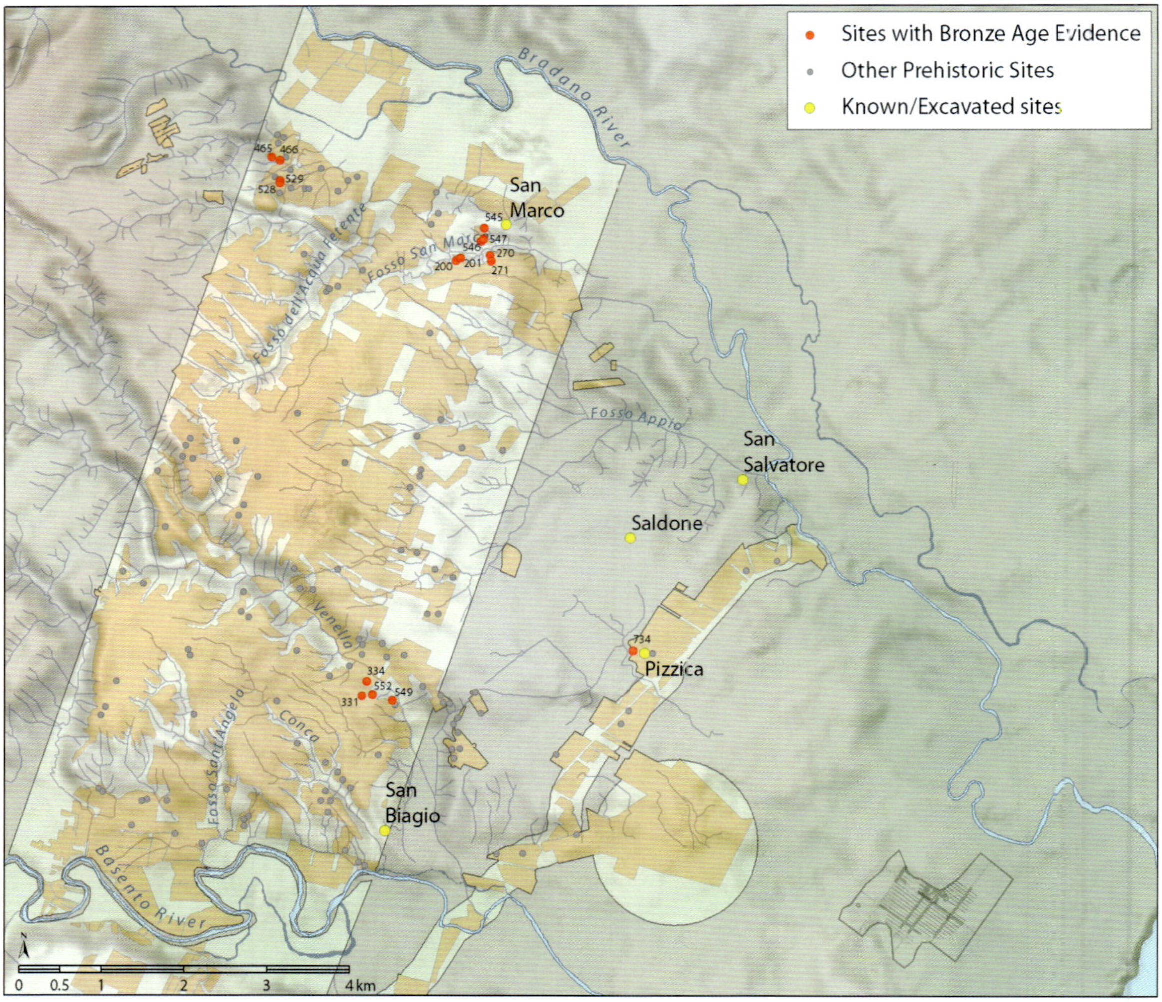

Figure 17.5 Map of sites with Bronze Age activity in the study area. (JT)

The reasons the occupants of the prehistoric and of subsequent periods, including the Medieval, chose this spot were probably the same: an abundant supply of water from a still extant spring, and good quality clay that was exploited in some, if not all, of these periods.[17] There were nearby forests that supplied wood for these operations as well as habitat for the animals, which are documented by the faunal remains from the Late Neolithic pits and the Greek and Roman occupations.[18]

The limitations of the evidence—in particular the difficulty of dating the impasto pottery found at many sites, the problem of site visibility, and the widespread and constant alluvial and erosional activity—are such that any description of the Neolithic settlement pattern is valid only in the broadest sense. The areas occupied by Neolithic settlements, or where possible Neolithic activity has been detected, are distributed principally along the river valleys. Giardinetto and San Marco on the Bradano, Sant'Angelo Grieco on the Basento, and the main tributary of the Basento in the survey transect, the Venella valley (in particular its north slope), all have excellent springs. The upper Venella, at the headwaters, is an area where Neolithic activity seems to have been widespread.[19] There was much prehistoric activity for the whole length of the north slope of the Venella. It is especially pronounced in several areas of the lower marine terraces, in Lago del Lupo and Cogno del Prete (possible Neolithic

[17] The Pantanello location also proved attractive to later settlers. The south slope of this small rise became the site of an extensive Greek rural sanctuary, known as the Pantanello sanctuary (late 7th to 3rd c. BC). Later, there was an extensive Roman industrial complex which manufactured pottery and tiles, and engaged in metallurgy. For these two important sites, a separate volume in this series is being prepared. Preliminary publications are Carter 1983a and 1994a.

[18] Bökönyi and Gál 2010, 41–48.

[19] They were also densely occupied by Greek settlements.

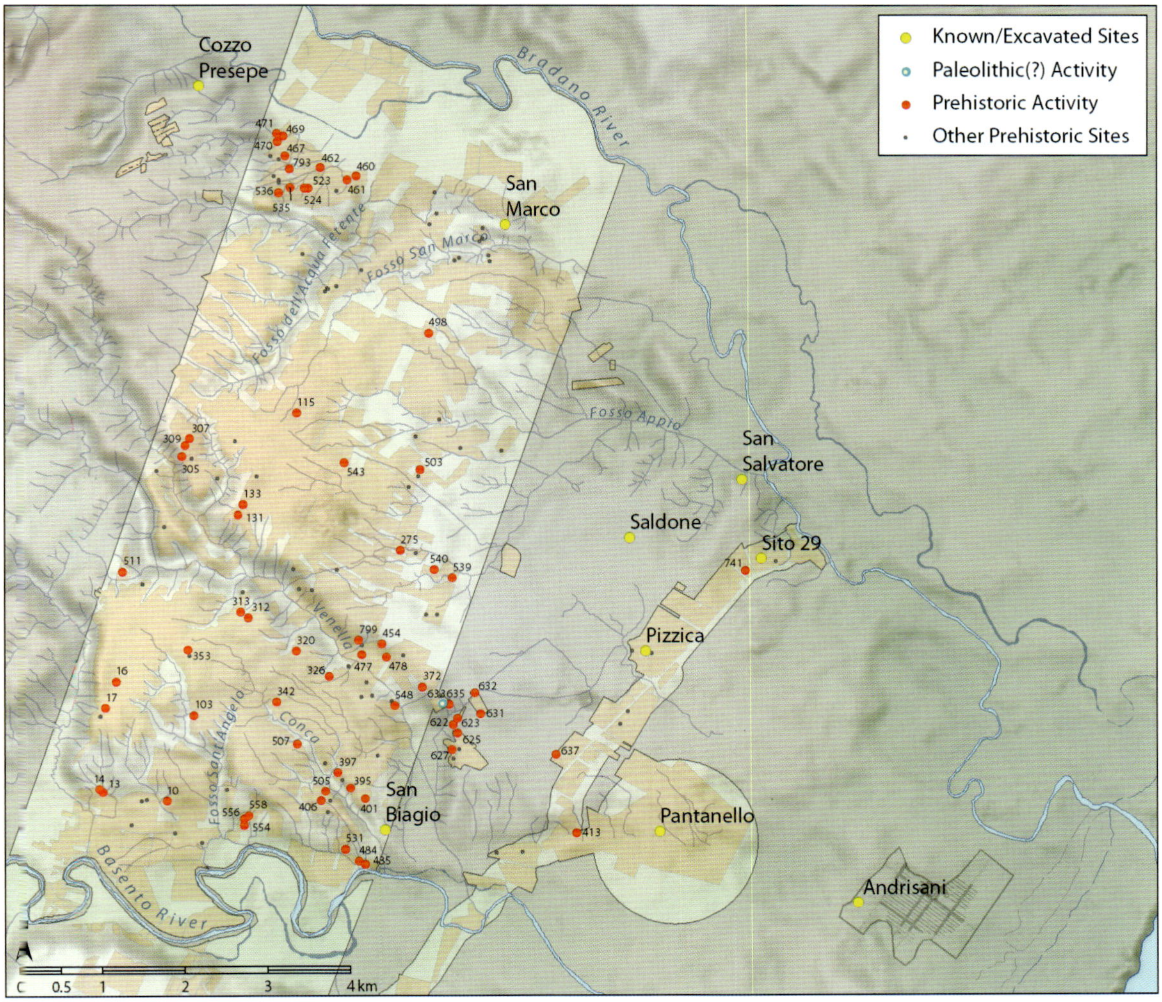

Figure 17.6 Map of sites with prehistoric activity in the study area. (JT)

domestic structures at sites 268, 500, 544; see the Gazetteer in Vol. IV) and in Pizzica where Greek settlement before the 5th century BC was limited (e.g. sites 734 and 735).

On the whole, it is reasonably clear that Neolithic sites were chosen because of their proximity to fertile valley floors and at the same time to good springs, which come out along the valley slopes (see Ch. 1). Their elevation varies between 40 and 60 m above sea level (masl). The highest and also the oldest, site 300, was located on a point just over 100 masl and about 55 m above the fertile bottomland where two smaller streams join to form the Venella valley. This valley, protected and exceptionally well watered, is ideal for agriculture, as well as animal husbandry; in fact, it is still today a center of sheep raising and cheese production.

The Bronze Age

In contrast to the preceding period, the sites occupied during the Bronze Age (late 3rd millennium to 10th century BC) were much reduced in number and their distribution across the study area more limited (*Fig. 17.5*). There is obvious continuity in two areas with a concentration of substantial occupation sites: at Giardinetto (sites 466 and 529) and at San Marco with two adjacent groups of sites on the promontory. The Bronze Age occupation of San Marco has long been known,[20] but because of the episodic nature of the investigations it has not been possible to form a clear idea of the settlement and its evolution. The survey identified three groups of structures (sites 545, 546,

[20] S. Bianco 1981. The Greek period at San Marco is represented by a single burial (site 548) directly opposite site 270, across the Fosso San Marco. The site of San Marco is currently occupied by Masseria San Marco, an extensive modern farm and *agriturismo* that has compromised much of the archaeological site.

and 547) and two activity areas (sites 200 and 201) divided between two locales.

Just across the small valley is a third site, site 270, a Bronze Age domestic structure, and a nearby area of Bronze Age activity, site 271. Site 270 in the Greek period was occupied by one of the earliest rural sanctuaries in the chora. The presence on the same site of a Neolithic or generally prehistoric settlement and a Greek sanctuary was noted at Pantanello. The survey shows that this relationship included not just the well-known Greek sanctuary sites, such as the Temple of Hera on the Bradano (Tavole Palatine)[21] and the Sanctuary of Artemis at San Biagio,[22] but also previously unidentified sanctuaries. Site 397 in Avinella is a prehistoric activity area, and sites 334 and 477 in the Venella valleys were identified as sites of Bronze Age and prehistoric activity, respectively.[23] The common factor between these very different cultures in widely separated periods of occupation is the presence of water.[24]

The third broad area of Bronze Age occupation in the study area consists of a close-knit group of sites in the Venella valley. Sites 331, 552, and 549, on the south bank of the Venella at mid-course, were identified as possible Bronze Age domestic structures, separated from each other and from site 334, an area of Bronze Age activity, by several hundred meters. The fertile area of the Central Plateau that Neolithic farmhouses exploited was apparently ignored in this period. There are, however, several sites with Bronze Age material closer to the sea, at San Salvatore, Saldone,[25] and at San Biagio,[26] on the site of the future Sanctuary of Artemis. At site 734 in Pizzica, impasto pottery and stone tools document a probable Neolithic to Bronze Age habitation site, and recent excavations in Pizzica revealed a votive deposit and a multiple burial (in flexed position) of Middle Bronze Age date.[27]

The wider Bronze Age context, beyond the study area, includes the important Late Bronze Age phase, roughly the 13th–11th centuries BC, corresponding to the later part of the LH IIIB and IIIC phases of the Aegean Bronze Age. Material from this period appears at a number of sites along the Ionian coast, including San Vito (survey site 29)[28] located in the half of the chora to the south of the Basento. Its commanding position (approximately 125 masl) conforms very well to the models of Late Bronze Age settlement on the Ionian coast, proposed by Peroni and his collaborators for the Sybaritide,[29] and by De Siena for the Metapontino and surrounding areas.[30] Sites were chosen for their panoramic positions overlooking rich farmland, and were utilized as centers for the storage of agricultural produce. Emblematic are the huge wheel-made containers found at these and other large sites like Timmari in the Bradano watershed 50 km from the coast.

Unlike San Vito, the contemporary site of Termitito[31] on the south side of the Cavone valley was well-preserved and, indeed, lends weight to these views about the economic dominance of these sites. There has been, however, no systematic search for the lesser sites of this period and so the extent of their control remains in doubt. The principal feature and focus of activity at Termitito was a pit (A/80) which measures 10 m in diameter and is 4.5 m deep. The findings of pottery, a stove, and loom weights led the excavator to postulate a wooden floor dividing the upper storey for living and the lower for storage. Smaller pits around the main one would also have been used for storage. Signs of fire damage show that at some point settlement contracted, and was followed by a more modest phase of occupation.[32]

On a clear day, much of the Ionian coast towards Taras and the Salentine peninsula is visible from the elevated and restricted area of the Termitito plateau. This site, like San Vito, reflects contacts with the Aegean world, as evidenced by large quantities of Mycenaean-style pottery (*Fig. 17.7*)[33]—much of it probably

[21] Lo Porto 1981a; Nava 2001, 944.

[22] For the Middle Bronze Age at San Biagio, see Cipolloni Sampò 1999, 121; Nava 2000, 691.

[23] Site 334 became a Greek rural sanctuary, as well as the center of an Early Village Cluster in the early 6th c. BC.

[24] A prehistoric potsherd near a Greek spring sanctuary may have no significance at all, beyond a natural interest in the spring in both periods. Speculations about the continuity of religious observance from the prehistoric period suffer from an almost total absence of information about the context or of material unambiguously documenting religious practices before the Greeks in this area of Italy. Adamesteanu 1974a, 56; Torelli 1977.

[25] Cipolloni Sampò 1999, 125; S. Bianco 1991–1992, 515, Fig. 4.

[26] Cipolloni Sampò 1999, 125, Late Bronze Age with some proto-Villanovan.

[27] Nava 2002, 736.

[28] Bianco and De Siena 1982.

[29] Peroni and Trucco 1994, 831–79.

[30] Giardino and De Siena 1999a, 24–27.

[31] De Siena 1996, 166–73.

[32] Similar though smaller interred structures are known in the Metapontino at Iron Age Incoronata indigena in the 7th c. BC (De Siena 1996, 180–92), as well as in Incoronata greca (De Siena 1986d) and throughout areas of the Mediterranean and Black Sea subsequently occupied by Greeks. Cf. Carter 2006a, 55–68.

[33] De Siena 1986a, 31; cf. Peroni and Trucco 1994, 845–48.

Figure 17.7 Mycenaean-style kylix from Termitito (h 11.5 cm). (CR, courtesy of the Soprintendenza)

locally produced—mixed with sub-Apennine Bronze Age material. It may have continued in existence from the latest phase of the Bronze Age (*Bronzo finale*) into the early Iron Age, documented by proto-Villanovan and Italo-Protogeometric pottery.[34] The first half of the 7th century BC saw the destruction of a defensive rampart, or *agger*, and an increase in the size of the settlement with semi-interred huts,[35] similar to those found elsewhere in the Metapontino at this time, for example at Lama San Nicola[36] and most prominently at Incoronata greca.[37]

The Iron Age (9th–8th Centuries BC)

The Neolithic and Bronze Age presence throughout the study area is well documented by the survey. As noted in the introduction to this volume, they will be the object of a separate study. The Iron Age is practically invisible in the study area. Evidence for this period is concentrated at several large sites, Incoronata and Cozzo Presepe which border on the study area, but lie outside it. The research on these sites by the Soprintendenza and its collaborators, including ICA, is summarized here.

Incoronata is a very large site consisting of a main area covering a broad plateau known as "Incoronata indigena." A full study of a portion of the relatively small hilltop known as "Incoronata greca" will be the subject of a future study. This summary treatment of the Iron Age (9th and 8th centuries BC), and of the 7th century, or protohistoric period, when contacts with Greeks intensify, aims to provide chronological continuity.

The abandonment of these large centers along the Ionian coast in the 11th and 10th centuries BC is accompanied by a fall off and eventual cessation of contacts outside of Italy, and a decline in the quality of manufactured goods.[38] The appearance of proto-Villanovan pottery suggests the presence of different groups, but that cannot be confirmed as the necropoleis associated with Termitito and San Vito have not been found. De Siena observes that the same phenomena—defensively positioned sites and large storage vessels—occur both at continuously occupied sites and at more recent ones, and argues that there were no mass movements at the end of the Bronze Age, but rather internal developments resulting from the crisis that involved the major centers at the end of the Bronze Age. There was a recovery, De Siena thought, in the 10th and 9th centuries, but without significant changes. The Bronze Age sites seem to have been abandoned for new sites on the lower marine terraces closer to the broader valleys and coastal plains.

The broad terrace known as Incoronata, including San Teodoro, is the main site for this period corresponding to the 9th and 8th centuries BC in the Metapontino (*Fig. 17.8*). It overlooks the Basento from the south in the Basento-Cavone transect, and is set back 8 km from the coast in an extensive area of the flat, low-lying (approximately 60–65 masl) interior marine terraces. Defense, natural or man-made, seems not to have been as pressing a concern as it had been in the preceding period. The numerous springs at the base of the terraces provided a constant supply of water.[39]

Incoronata

The necropoleis of Incoronata indigena and San Teodoro. Large areas of Incoronata were occupied by extensive necropoleis (*Fig. 17.8*). Excavations began in 1970 and continued until 1987. This was a period before agricultural projects had seriously compromised the archaeological contexts. The excavations documented 532 burials in two areas: Incoronata indigena and San Teodoro.[40] They correspond in date to much of the Iron Age in this region, ranging from

[34] De Siena and Bianco 1982.

[35] See n. 31.

[36] Nava 2002, 739–40.

[37] The site was subsequently occupied by a Greek farmhouse and a kiln dating to the second half of the 7th c. BC, and a very large 2nd-c. BC–to–2nd-c. AD Roman villa. See Ch. 27, 912–13, and Carter 2006a, 135–36, Table 4.A with bibliography.

[38] De Siena 1999b, 219–21.

[39] Later in the mid-6th c. BC these became the sites of small sanctuaries. E.g., site 57: Nava 2003a, 672–73, Pl. 37.2.

[40] The necropoleis have been fully published in a long article by Chiartano (1983) and in three beautifully illustrated volumes which have regrettably not had a wide distribution: Chiartano 1994; 1996.

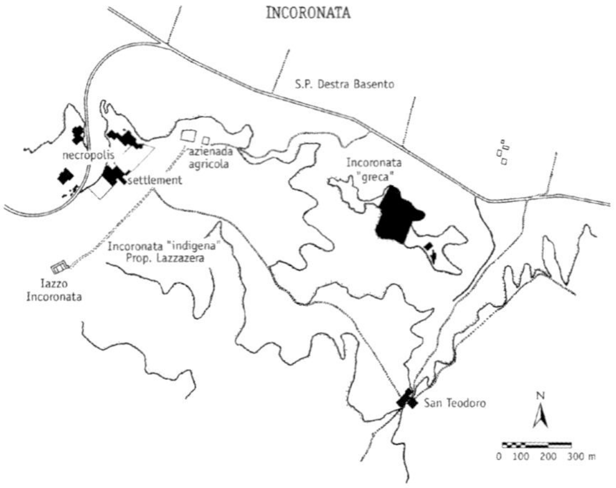

Figure 17.8 The sites of Incoronata greca, Incoronata indigena, and San Teodoro on the south side of the Basento River valley, 8 km from the Ionian coast. (CW)

Figure 17.9 From tomb 117, San Teodoro: *center*, Matt-painted two-handled biconical olla (h 23.5 cm); *left*, impasto bowl (h 4 cm); and *right*, small vessel (h 3 cm), placed in the mouth of the olla. (VM, Ancient Crossroads, ICA; courtesy of the Soprintendenza)

the last quarter of the 9th century to the mid-8th century BC. The mostly single burials were in the flexed position (*rannicchiati*). They were enclosed by rough sandstone slabs and river cobbles, used either separately or in combination, and were covered by a layer of cobbles. Infants were apparently not buried together with adults, but separately in ceramic containers near the dwelling,[41] though there are a few examples of mother and child burials.[42]

Both males and females were buried in the flexed position, with important gender-related differences: females, identified by their grave goods, for example, faced northwards, and males in the opposite direction. There were also qualitative and quantitative differences in the grave goods. The arrangement of the body in the grave is quite similar to that in all the surrounding Iron Age necropoleis (see *Table 17.1*). The flexed position is found in Neolithic and Bronze Age burials in the region, and occurs regularly in indigenous burials in the interior in the historical period. The grave goods typically consist of a large container for liquids (a *brocca* or a jug), and a small cup or dipper (*tazza*) in coarse impasto. Occasionally the burials include a large vessel of purified clay decorated with a geometric design in the Matt-painted style known as "South Italian Geometric"[43] (*Fig. 17.9*). The cup is often found in the mouth of the large vessel, and both are almost always placed at the feet of the deceased—a practice that is found in burials in the colonial chora.[44]

Females were laid to rest wearing bracelets and rings, in garments decorated with numerous bronze ornaments, including pendant rings, pectorals, and large spiral fibulae (*Figs. 17.10* and *17.11*). The loom weights and bobbins represent the deceased's sphere of activity. Males, by contrast, were buried with little, if any, personal ornament—at most, a simple pin with a twisting form to close the garment (*fibula ad arco serpeggiante*)—but almost always with a spear point (*Fig. 17.12*), and sometimes with a sword, a spear point, and a razor. The males with swords were a minority and presumably enjoyed a higher status than those with only spear points.[45] On the whole, the differences between burials were not so great, and the conclusion can reasonably be drawn that there was a high degree of equality among the population interred at San Teodoro and Incoronata.[46]

[41] De Siena 1990a, 76; Chiartano 1994, e.g. T.16.

[42] Chiartano 1994, 19.

[43] Chiartano 1994, 17, e.g. T. 147, Pl. 1; Yntema 1990a, 40–41, Fig. 20.

[44] E.g., many of the burials in the Pantanello necropolis that date from the early 6th c. BC onwards. J. Hall 1998, 580. At Pantanello the large container at the feet is a pelike or a storage amphora. The drinking vessel, often in the mouth of the large vessel, is a black-gloss skyphos. See Carter and Hall 1998, e.g. T. 87, 151, etc.

[45] See Chiartano 1996, 16–17, for a discussion of typical grave goods.

[46] De Siena 1990a, 77.

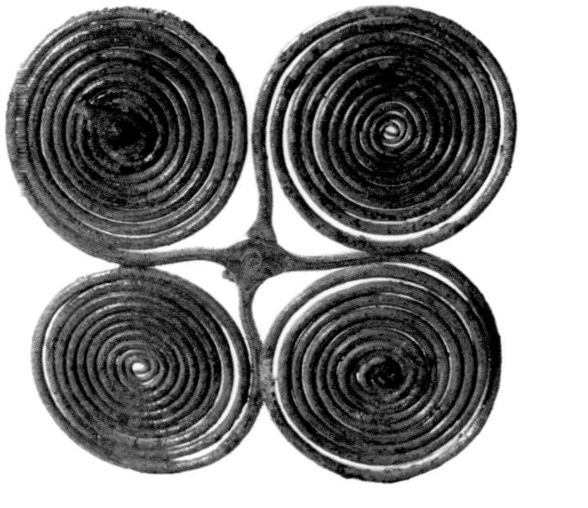

Figure 17.10 Bronze fibula with spirals (h 11 cm, max diam. spirals 4 cm). San Teodoro tomb 64. (VM, Ancient Crossroads, ICA; courtesy of the Soprintendenza)

The burials at Incoronata indigena and San Teodoro are organized in loose clusters, which are similar to, though perhaps not as distinct as, those of the huts in the excavated area of the Iron Age settlement.[47] For the settlement, family groupings of huts have been hypothesized.[48] The clusters of burials in all likelihood reflect family groups, too.[49] A cobble-paved surface running through both the Incoronata and San Teodoro necropoleis was first thought to divide the Iron Age burials.[50] The "road" has now been correctly recognized as the lowest level of cobble-lined drainage canals dug in the first half of the 5th century BC as part of the bonifica of the chora.[51]

The settlement at Incoronata indigena. A portion of the Iron Age settlement and a contiguous area of burials were excavated just to the north of the Incoronata indigena necropolis described above. This site is referred to as Incoronata Azienda Agricola to differentiate it from other areas of the Incoronata plateau. Approximately 40 burials and 30 features convincingly interpreted in most cases by De Siena as hut floors were discovered.[52] The ovoid pits of

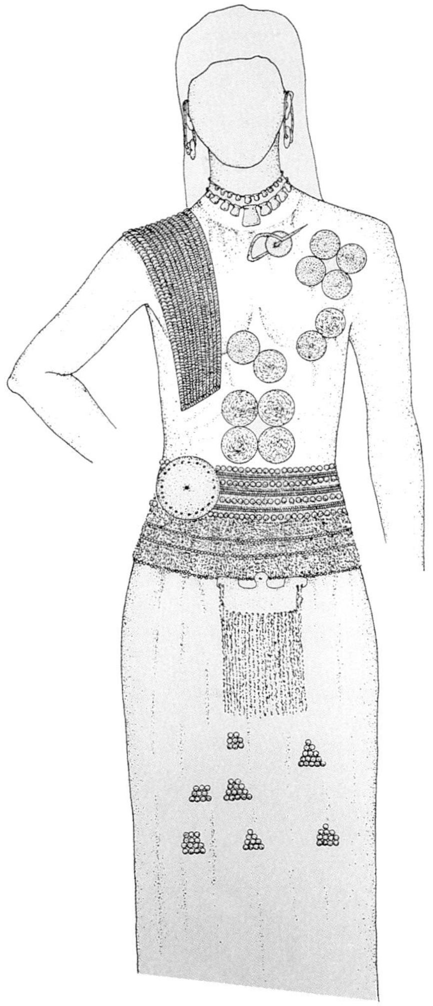

Figure 17.11 Reconstruction drawing of an ensemble of bronze and iron ornaments, fibulae, belt, and jewelry from the burial of a woman; dated to the 7th c. BC. Alianello, contrada Cazzaiola, T 316. (Courtesy of the Museo Nazionale Policoro and the Soprintendenza)

various depths (approximately 0.5 m and 1 m) were filled with daub from hut walls, the remains of small portable ovens, loom weights, fragments of impasto pottery as well as of vessels made with purified clay and decorated in the Matt-painted geometric style, and of very large pithoi or dolia similar to those used in the Bronze Age. The large vessels are evidence of specialized craftsmanship that seems to have carried over from the preceding period, as are the molds for casting bronze rings, examples of which were found in the burials.[53] The mix of objects reflects the domestic functions of the huts at the time of their abandonment.

[47] Chiartano 1994, 2: plans A and E. Possible clusters are T.302–307 and T. 365–70, 373–81, 383 on Plan A.

[48] De Siena 1990b, 221.

[49] Chiartano 1994, 18–19. The clusters which vary in size are suggestively similar to the nuclei of the Pantanello necropolis. Cf. Carter 1998e. So far there is only a preliminary study of the skeletal material included in Chiartano 1994, 39–40. A full study by R. and M. Henneberg is now in progress.

[50] It appeared to anticipate the road dividing burials corresponding to the Division Line Road in the Pantanello necropolis. Carter 1998b, 42–45.

[51] De Siena, pers. comm., May 2007. In fact, it was part of the same system of division line roads as that in the Pantanello necropolis.

[52] De Siena 1990a, 72–75; 1996, 180–92.

[53] De Siena 1990a, Pl. 7.

Figure 17.12 *Left*, bowl with zoomorphic handle (h 9 cm); *right*, bronze javelin point (l 13 cm). Tomb 125, San Teodoro. (VM, Ancient Crossroads, ICA; courtesy of the Soprintendenza)

The upper levels of the huts and the contents were truncated here and elsewhere on the plateau by intensive agricultural activity in recent times. What remains are the lowest levels and, in some but not all instances, traces of the holes for the posts that supported these semi-interred structures. In one case the postholes outline rectangular structures with gravel pavements. There were potsherds representing activity areas in front of a number of the huts. The regularity of these features has suggested to the excavator, De Siena, that the plan of the settlement was "quasi geometrical."[54]

The settlement can be dated on the basis of the pottery,[55] including the handmade impasto clay situlae used for the burial of infants, and a bronze fibula, to the mid-8th century BC.[56] To the north of the hut settlement on the opposite side of the 5th-century BC cobble-lined drainage canal is an area of burials of the same sort as those found elsewhere at Incoronata. It has been dated from the late 9th to the first half of the 8th century BC.[57] Some of the burials, therefore, are earlier than the settlement. Those corresponding to the last years of the settlement have not as yet been found.

Cozzo Presepe

Besides Incoronata, the only other location of a major Iron Age settlement near the study area was on the plateau known as Cozzo Presepe (*Fig. 17.13*). Little is known about its state in the 8th century BC.[58] This site, excavated in the late 1960s and early 1970s by L'École française de Rome, the British School in Rome, and in the mid-1990s by the Soprintendenza, occupied approximately 7 ha. It overlooks the valley of the Bradano and was in a good position to control its lower course. The Iron Age settlement occupied the most defensible part, the northern height (at an elevation of 125 masl), from the middle of the 8th century BC until sometime around 600 BC, as documented by the abundant Bradano Middle, Late, and Sub-Geometric Matt-painted pottery (e.g., *Fig. 17.21*).[59] As the British excavators admit, given the restricted area investigated, little more can be said about the settlement. It was apparently destroyed by fire and on the same spot arose a massive agger, or fortification wall, measuring 2.5 to 3 m thick, as the carefully documented stratigraphy and ceramic records indicate.[60] It was in use until the end of the 5th century BC.[61]

Iron Age Incoronata Greca

The most interesting of the late Iron Age settlements—from the point of view of the future Greek

---

54 E.g. the Bradano Middle Geometric olla (775–725 BC) found in pit A88: De Siena 1990a, Pl. 6.

55 De Siena 1996, 188.

56 De Siena 1990a, 75; 1996, 175.

57 De Siena 1990a, 76.

58 The first excavation by L'École française de Rome (Morel 1970a) adds little of substance to the picture of the indigenous site. For the British School publication, see *Cozzo Presepe*, 210–14, 317–85. The subsequent, more extensive work by the Soprintendenza has yet to be published. It focused on the possibly Lucanian village of the 4th c. BC. Nava 1997, 481–83.

59 The distinctive Matt-painted Iron Age pottery of southern Italy occurs with great frequency at settlement sites, but not in necropoleis. There is relatively little southern Italian Early Geometric (825–775 BC) from the large necropoleis of Incoronata indigena and San Teodoro discussed above. By contrast, Bradano Middle Geometric (775–725 BC) has been found in the indigenous centers of Gravina in Puglia Incoronata indigena, in abundance at Cozzo Presepe, and at Incoronata greca.

60 Macnamara concludes that the destruction of the native settlement (ca. 600 BC), which shows some Greek influence in the late 7th c., was caused by the Metapontines (*Cozzo Presepe* 221, 288). The construction of the defensive wall on the highest point of the promontory has been connected with the hostilities between the Oenotrians, Tarentines, and Metapontines, for which the source is the 5th-c. BC historian Antiochos of Syracuse, cited by Strabo (6.1.15) and quoted in its entirety at the end of this chapter. See *Cozzo Presepe*, 385. Among the pottery of Phase II (the destruction and later), there is at least one vessel, an ovoid aryballos, which should be dated in the second half of the 7th c. BC (*Cozzo Presepe*, 319, Fig. 107, No. 119). Several fragments of *a filetti* cups at Cozzo Presepe (323–25, Fig. 110, Nos. 148–52) are clearly of the same and even of slightly earlier date than the earliest forms of *a filetti* cups from the survey.

The possibility exists that the site, which appears to have been purely indigenous in the 8th c. BC, may actually later have been a mixed settlement or one at least in contact with the earliest colonists in the chora. It does not follow unequivocally that a "native settlement" was able to maintain its traditional Iron Age lifestyle for as much as half a century after the arrival of the Greeks; see *Cozzo Presepe*, 384.

61 At this point it was abandoned and a sanctuary with monumental architecture occupied this commanding point of land. A subsequent defensive wall was erected at the end of the 4th c. BC; it has been suggested that this was in reaction to a perceived threat of expansion by new waves of Lucanians moving from the interior. It is possible that the late 4th c. BC occupants of Cozzo Presepe were themselves Lucanians. A. De Siena, pers. comm., 1996 (see Ch. 25, below). Nava 1997, 481–83.

Figure 17.13 Aerial photo of Cozzo Presepe viewed from the northwest toward the chora. (ALC)

presence on this stretch of the Ionian coast—is that at Incoronata greca (*Figs. 17.8* and *17.15*). Despite extensive damage to the earlier structures by the 7th-century BC settlement, it is also the best preserved and most thoroughly studied.[62] This site occupies the entire surface (about 4.5 ha) of an isolated spur of the same extensive marine terrace where the two early Iron Age necropoleis and settlements lie nearly 1 km away on either side (*Fig. 17.8*).[63] The Iron Age settlement at Incoronata greca consisted of structures resembling in all essential respects those at Incoronata indigena.[64] A large number of the pits of varying dimension and depths, all of which appeared to Orlandini to be deposits for waste disposal, have now been identified by De Siena as the substructures of huts constructed over a wooden frame, supported by posts and covered with daub, as at Incoronata indigena. This type of structure was probably very similar to those wattle and daub huts of the Neolithic at Pantanello. De Siena's argument receives its clearest and best support from the excavations of the 7th-century BC huts at Andrisani on the coast.[65] These are, as we shall see, by far the largest and most elaborate of such huts with their interconnecting areas of different depths, and the remains of ovens, daub, and much pottery.

As the settlement and necropoleis on the broader plateau of Incoronata indigena and San Teodoro began to contract in the mid-8th century BC, the settlement of Incoronata greca grew until by the late 8th century BC it was the only one in the area. Shortly afterward it was destroyed, violently according to Orlandini, to be replaced by a purely Greek "emporion."[66] This interpretation has been much debated.[67]

The Iron Age in Survey: Indigenous Pottery

Remarkably, not a single piece of pottery that is clearly of Iron Age date was recovered by the survey in the study area. That is not to say that there was no one occupying this rather large area before the arrival of Greek settlers. There almost certainly was. Iron Age impasto ware, the crucial evidence, has successfully eluded both the surveyor and the excavator in the study area. Impasto is not easy to spot by itself and small bits of it are not diagnostic. Without a recognizable shape it is difficult, though perhaps not impossible, to distinguish impasto of any of the major prehistoric periods. None of the distinctive shapes present at Cozzo Presepe[68] or at Incoronata, just outside the study area to the north and south, were recognized among the survey materials.

Only a handful of fragments of wheel-made pottery of the thousands collected from the Bradano-Basento transects can be identified as indigenous or possibly indigenous forms. None of these can be securely dated to the 8th century BC and may well be later in date. The decoration, if there was any, has disappeared from the sherds collected at sites 104, 155, and 285. All of these sites were occupied by Greek farmhouses, two of which (155 and 285) are among the earliest in the chora, in the Sant'Angelo village cluster (see Ch. 20). One fragment of a rim of an olla

[62] The bibliography has been collected in the series *Incoronata* 1–6.

[63] The necropoleis corresponding to the indigenous settlers have not been found, though in the case of Incoronata greca its possible location has recently been identified.

[64] De Siena 1995, 192–94.

[65] *Greci sul Basento*, 199–212.

[66] Stea 1999.

[67] See discussion, Carter 2006a, Ch. 2.

[68] *Cozzo Presepe*, 314–17.

Figure 17.14 Indigenous olla rim from site 285 (rim diam. 18.5 cm). (CR)

from each has been documented (*Fig. 17.14*). Less certain are the bases from sites 336, 554, and 542 and the rim from site 247.

The general paucity of indigenous material in the study area confirms—at least for this area—the proposed model of an Iron Age population concentrated at few sites like Incoronata in proximity to good farm- and pasture lands. The most that may safely be inferred from these isolated fragments of indigenous pottery found on sites dated to the 6th century BC and later is that there was some activity that required that type of vessel, and perhaps an occupant of indigenous origin at these sites. Vessels with indigenous forms continued to be used well into the 6th century BC and later at indigenous sites in the interior and Greek sites in the chora.

Three clearly indigenous pieces turned up in the excavation of the sanctuary by the spring at Pantanello (*Fig. 17.17a–c*). The olla rim was found in a votive deposit by the spring in its earliest phase, the body fragment of a Daunian cup with handle attached came from the collecting basin beside the spring, and the fragment of a small, Matt-painted geometric vessel dates to ca. 700 BC, antedating the sanctuary. Such pottery has not been found except at Cozzo Presepe and Incoronata. Its appearance at the spring suggests, at the very least, that Pantanello was frequented in the century preceding the Greek settlement of the chora. It is probably no coincidence that the olla rims were found at three sites that were among the first to be occupied as the Greeks took possession of the chora. Depending on their date, the fragments of indigenous pottery may indicate a previous Iron Age presence, which was erased by the Greek sanctuary and farmhouse; if the pottery is of 7th-century BC date, it may indicate that

Figure 17.15 Aerial view of Incoronata greca from the northeast looking inland. (Courtesy P. Orlandini)

Figure 17.16 Pit B, Incoronata greca, partially exposed, excavation 1977. (JCC)

the occupants of the farmhouse were in contact with the indigenous population of the interior. Perhaps we can see here evidence of attempts by family members of mixed heritage to maintain in some fashion the culture of their ancestors. The possibility that the early settlers represented a fusion of Greeks and indigenous peoples is given some support by the persistence of the burial customs of the 7th century BC in following centuries.[69] The 6th-century BC burials in the chora include a number of the enchytrismos type,[70] similar to those of the small necropolis at Incoronata Azienda

[69] Carter 1998c, 64–66; Carter 1998f, 167–71; Carter and Hall 1998, 310 (T 299, 301).

[70] E.g. tomb 191 in the Pantanello necropolis: Carter and Hall 1998, 417. Cf. Nava 2003a, 675, also for the enchytrismos burial in San Salvatore on the Bradano side of the chora.

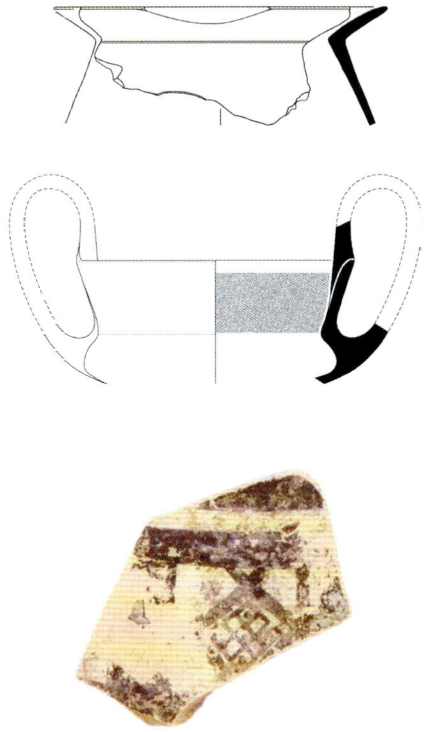
a
b
c

Figure 17.17 Fragments of indigenous pottery from the Pantanello sanctuary: a) olla rim from spring channel (rim diam. 9.5 cm); b) Daunian carinated cup from the collecting basin (rim diam. 12.1 cm); c) Matt-painted geometric fragment from the western side of the sanctuary (h 6.5 cm). (CR)

Agricola, with adults in the traditional flexed position of indigenous burials, and the children in Greek amphorae (see below). More difficult to date precisely are the semi-flexed burials in the area of the Pantanello necropolis.[71]

The Wider Context

The Iron Age burials of Incoronata indigena and San Teodoro are expressions of a well-defined culture, which was common along the Ionian coast of Italy in an area corresponding to that dominated later by the Greek settlers of Metaponto, Siris, and more generally, to contiguous areas north and south. Settlements near the necropoleis at San Leonardo below Pisticci[72] and Santa Maria d'Anglona[73] in the Siritide above Policoro were identified but not excavated.[74] Unfortunately, very few settlement sites have survived and been excavated, so that the definition of this culture is based almost exclusively on similar burial practices. The well-built stone tombs, the flexed position of the deceased, and the closely comparable grave goods for both males and females, are found, for example, at Valle Sorigliano in the Siritide[75] and at dozens of other excavated sites. This cultural koine has been associated by scholars with the "Oenotrians" or "Choni,"[76] names given by Greek writers to the peoples who occupied an extensive region of central southern Italy from the Ionian coast to the Tyrrhenian, before the arrival of immigrants from Achaia and elsewhere. The unity as an *ethnos* ascribed to the indigenous peoples may well have been a propagandistic invention of later sources, but there perhaps is an element of truth in the claim that their mythical ruler, Italos, whose ancestry was traced to Greece before the Trojan War, was responsible for the fundamental change from pastoralism to an economy based on agriculture.[77]

At the end of the 8th century BC, the occupants of the Metapontino were essentially domiciled at two sites: the native villages at Cozzo Presepe and Incoronata greca, and possibly the small satellite sites documented by the survey on the extended plateaus of greater Incoronata.[78] Santa Maria D'Anglona in the Siritide to the south was a comparable central place in the 7th century BC.

By the beginning of the 7th century BC, these native centers had begun to develop contacts with a wider world that would, over the course of that century, transform the settlement of the territory. In the Siritide the influence and contacts are reflected by prestige items in burials—the major source of information about the occupants of that territory—such as bronze vessels, "antenna" swords from the Tyrrhenian, amber from the Adriatic, and long swords and glass from the eastern Mediterranean.[79] With the foundation of Taras and Sybaris at the end of the 8th century, and of Siris in the mid-7th, Greek influence began to spread rapidly, affecting the development of this stretch of the Ionian coast in a decisive way for the next four or five centuries. Though it is rare, evidence of Greek influence was perceptible already in the 8th

[71] Carter 1998f, 167–71.
[72] Lo Porto 1969.
[73] Whitehouse and Whitehouse 1969, 34–75.
[74] S. Bianco 1998, 21.

[75] Frey 1998, 33–41.
[76] The earliest was Hekataios of Miletos, who lived in the early 5th c. BC.
[77] De Siena 1999b, 221. The Choni, a subgroup of Oenotrian stock, are more specifically identified with the Ionian coast.
[78] To be discussed in a future study devoted to the survey between the Basento and Cavone Rivers.
[79] S. Bianco 1996, 37–40; 2001, 807–12.

century. It consisted of fragments of imported Greek Geometric pottery from the excavations at Incoronata greca,[80] on the site of the future asty of Metaponto,[81] and out of context in the area of the Pantanello necropolis.[82] It may not be coincidental that all three of these areas later participated in varying degrees in the transformations of the 7th century BC.

The 7th Century BC

Incoronata Greca in the 7th Century BC

The 7th-century BC settlement which canceled the Iron Age village at Incoronata greca remained unique for fourteen years after its discovery in 1970. Together with that, in 1984, of the related settlement of mid-7th–century BC huts at Andrisani near the mouth of the Basento, it has significantly altered perceptions of the relationship between the Greek and indigenous worlds on this stretch of the Ionian coast at the outset of the colonial movement. Subsequent excavations in 2000 and 2001 at San Teodoro and on the valley floors below have appreciably broadened our understanding of the settlement at Incoronata greca and its relation to the site at the mouth of the Basento. The uniqueness of Incoronata greca in this respect is due, at least in part, to the excellent state of preservation of the site. Other sites, notably those in and around Policoro, where Siris[83] is thought to have been situated, have not been so well preserved, though evidence is accumulating that shows the hilltop at Policoro played a similar role in the development of the Siritide, with the difference that the settlement there seems to have been more widely dispersed than that at Incoronata.

The pace of discoveries affecting interpretations of early Greek contacts and colonization in this area of Italy has been so rapid that theories are outmoded or in need of serious revision even before they have become a part of the scientific literature. A case in point is the apparent contrast between the centralized settlement at Incoronata greca, and the dispersed one at Policoro. The discovery of comparable structures, coeval with Incoronata greca in the valley floor below nearby San Teodoro, suggests that more such sites may well lie buried within the thick alluvial deposits. The late-7th-century BC dwellings in the Bradano valley at Pantano, also discovered in 2001 (see below and Ch. 20), are another example. Incoronata greca may represent only the more elevated part of a much wider settlement dating to the second half of the century.

The possible roles that Incoronata greca played in the progressive Greek occupation of the coast are still not clear. Was it a Greek emporion[84] in an indigenous world in dissolution as Orlandini held? Or an indigenous center that responded to the external stimuli and developed internally a hybrid culture and economy in which the indigenous population was actively involved? De Siena and Giardino have emphasized the prominent role the indigenous population played in the growth of settlements along the coast and the retreat of purely indigenous sites into the interior.[85]

Figure 17.18 Imported Late proto-Corinthian kotyle from Pit B, Incoronata greca (h 10.0 cm). (CW)

How "Greek" was Incoronata greca in the 7th century BC? The answer depends on an understanding of the process by which this part of the peninsula became, in a fairly short time, predominantly Greek in culture. The presence of large quantities of imported Greek (*Fig. 17.18*) and locally produced pottery in a variety of Greek styles is unique, if only for its sheer volume, but the presence of the pottery alone is not by itself an accurate indication of the ethnicity of the occupants. The debate on this subject has been intense and so far inconclusive.[86] This is not the place to reopen it, since the primary focus of this study is not the ethnicity of the settlers in the various periods, but the manner in which they occupied the land. Important evidence with which the question might be partially resolved—an in-depth comparative analysis of the human remains—does not yet exist. Related necropoleis have not yet been explored[87]

[80] *Greci sul Basento* 126, cat. no. 55 and S. Leach (Davis), pers. comm., 1989.

[81] De Siena 1990a, 86–87.

[82] S. Savelli, pers. comm., 2008.

[83] The precise location and date of Siris are, however, still matters of diverging opinion among scholars.

[84] Most recently Stea 1999.

[85] Giardino and De Siena 1999a, 32–33.

[86] Carter 2006a, Ch. 2.

[87] The excavation of intact burials would help to clarify the question of the physical characteristics of the population, since comparative material

Figure 17.19 Rectangular structure with an interior space of 9.7 m^2. Mudbrick walls on a stone socle, 7th c. BC, Incoronata greca SW spur. (TH)

nized.[89] Some of the larger rectangular structures, however, were indeed houses and not necessarily for Greeks, if the evidence of the pottery in use there can be trusted. Some of the 7th-century BC pottery in use was clearly indigenous.[90]

The question of which Greeks arrived in or were in contact with the Metapontino in the late 8th and early 7th century BC has also been much discussed, as have the similarities between the pottery of Incoronata and that of the settlement at Policoro.[91] There are similar motifs in the locally made ceramics,

and comparative skeletal material from Greece itself has not been studied. Because the value of this sort of research has only slowly begun to be appreciated by archaeologists and historians, this is not likely to happen soon.

The archaeological evidence—above all, the pottery of the 7th century BC—clearly reflects close contact with Greeks, if not their actual presence on the site. The architectural appearance of the hilltop settlement at Incoronata changed in the 7th century BC. Beside the ovoid huts (*capanne*) constructed over cavities in the earth using wattle and daub,[88] there arose square or rectangular structures on stone socles with mudbrick walls and thatched roofs (*Fig. 17.19*). These were also semi-interred, that is, erected over rectangular cavities which in the tradition of the oval cavities served as storage places. At first they were interpreted as Greek *oikoi* (residences), but now their probable function as storehouses has been recog-

for example, the heraldically posed horses framing a tripod on large *dinoi* found at Policoro[92] and very similar ones from Incoronata.[93] Kolophon on the Asia Minor coast was, according to tradition, the origin of the colonists at Siris;[94] it has been argued that the pottery proves Ionian Greeks were also involved at Incoronata greca.[95] A fragment of a *dinos* with heraldic horses has, however, also turned up at Croton, the late-8th-century BC Achaian origin of which has not been questioned.[96]

A very large amount of the Incoronata greca and Policoro pottery reflects Greek influence from areas other than Asia Minor (*Fig. 17.20*). This distinctive pottery, referred to as "colonial," represented a considerable technical advance over the indigenous Matt-painted wares made on a slow wheel (*Fig. 17.21*) and the traditional handmade impasto. The forms are Greek, and Greeks were undoubtedly responsible for the introduction of the fast potter's wheel. Still, some-

from Metaponto does exist: Henneberg 1998; Carter 2006a, 80–83. If De Siena is correct in identifying the destroyed burials at San Teodoro as being part of that necropolis, then the Iron Age practice of burying in graves surrounded and covered by river cobbles would have persisted into the 7th c. BC, alongside alternative methods of disposal, such as the enchytrismos burials found in the small, 7th-c. BC burial ground at Incoronata Azienda Agricola.

88 All of these were originally interpreted by Orlandini as Greek waste pits, despite the large numbers of intact vessels, tools, daub, and plant remains found in them. See Carter 2006a, Ch. 2, for a recent summary of the debate, with bibliography.

89 Orlandini 1999, 201.

90 See, e.g., the "Rectangular House" in Carter 2006a, 73.

91 Orlandini 1999, 201–04.

92 Adamesteanu and Dilthey 1978, 519, Fig. 7.

93 *Greci sul Basento*, 154–55, 192, Nos. 96–7, Pls. 57–59.

94 Pugliese Carratelli 1999, 184–87.

95 The chance discovery in 1970 of a single piece of Late Wild Goat pottery imported from Asia Minor on the site prompted the first investigations at Incoronata greca. Adamesteanu 1974a, 72–73.

96 Sabbione 1984, 255, Pl. 32, 2.

Figure 17.20 Imitation proto-Corinthian cup from Pit B, Incoronata greca (h 11.5 cm, diam. 17.0 cm). (CW)

Figure 17.21 Upper part of Bradano Late Geometric olla with tent motif from Pit B, Incoronata greca (h 14.3 cm). (CW)

thing of the spirit of the indigenous potter survives in the large size of many vessels (*Fig. 17.22*), their flamboyant geometric decoration, the eclectic choice of shapes, and the original combination of motifs from the Ionian world (Samos), from the Peloponnese (Argos), from central Greece (Euboea), and from western Greece (Ithaca). These surely reflect the heterogeneity of the contacts, if not of the craftsmen who used them.

Significantly, this pottery is found nowhere in the study area, neither in the Bradano-Basento nor in the Basento-Cavone watersheds. The only other places in the Metapontino where it has been found are the settlement below San Teodoro and that at Andrisani (see above). Why was this locally produced pottery not found at other contemporary indigenous settlements, such as Cozzo Presepe (see above) or the smaller indigenous sites like Lama San Nicola?[97] Whatever trade contacts Incoronata may have had with these sites, they did not include pottery. The impressive numbers of commercial amphorae of the SOS type and pithoi from Corinth (*Fig. 17.23*) and Attica, from the islands of Rhodes and Mytilene and the coast of Asia Minor, attest the importance of trade in the economic life of Incoronata. The small *oikoi* were crammed full of pottery, mostly amphorae.[98] More limited numbers of these containers were recovered from the interred oval huts, such as Pit B on the southeastern spur.[99] Oil and wine, it would seem, were being stockpiled at Incoronata for export through the port at the mouth of the Basento. The fertile valleys and marine terraces surrounding it were ideal for growing cereals. These were being exploited from the hilltop sites as they surely had been in the past, but now a surplus was produced. Trade and internal consumption of the olive and grape, as well as cereals and animals (especially sheep and pigs),[100] must have formed the basis of the economy. The technology for turning these raw materials into marketable and transportable commodities would not have required the direct intervention of Greeks.[101] The exclusive focus on the Greek contributions, as De Siena has urged, makes the indigenous population into passive recipients of a superior culture from abroad. The Greeks were attracted by the success of the indigenous peoples. If Greeks lived in mixed settlements in the midst of an indigenous world, clearly they were there, at least at the beginning, at the pleasure of the native population.

The hilltop site at Incoronata greca was linked to the contemporaneous settlement at the mouth of the Basento. Despite the fact that only small areas of the coastal site have been excavated, the connection between the two cannot be doubted. Soundings in three separate areas, the largest of which is Andrisani, have indicated that the settlement covered an extensive but probably discontinuous area. The two fully excavated

[97] Nava 2002, 739–40; De Siena 2002, 30–31.

[98] *Incoronata* 5, 19–51.

[99] Carter 2004, 374–81.

[100] See Carter 2006a, 78–80. The paleobotanical evidence for ancient agriculture in the Metapontino from the Neolithic to Late Roman periods will be the subject of a future study. For evidence of animal husbandry at Incoronata greca, see Bökönyi and Gál 2010, 6, 10.

[101] Greek propaganda attributes to the Greeks the introduction of the grape and olive into Italy. The most that can be safely said is that Greeks were probably responsible for improving agricultural methods, and perhaps the plant and animal stocks; see Carter 2006a, 79–80.

Figure 17.22 Stamnos with linear "barbell" design in Colonial ware from Pit B, Incoronata greca (h 26.5 cm). (VM)

huts at Ancrisani,[102] though larger and more elaborate than any discovered at Incoronata, are of the same wattle-and-daub construction, containing the same sort of ceramics: indigenous Matt-painted ware, impasto, and large vessels in the colonial style, including fragments of *dinoi* familiar from Incoronata and Siris and perirrhanteria like those produced at Incoronata. These were found together in the pits, mixed with the daub and carbonized wood from the structure that stood over the cavities. The deposit in Hut A may contain some later material, possibly including some Early Corinthian pottery, which was entirely absent at Incoronata. Its kitchen area, with an oven and Incoronata-style pottery, was intact, as was the entire deposit of Hut B.[103] Seated at the mouth of a navigable river, the village was well situated to provide for the shipment of the agricultural products of the interior.[104] The case for its actually having played that role is strengthened by the discovery of commercial amphorae in these huts, including the SOS type.[105]

The discovery on the valley floor below San Teodoro of habitations contemporaneous with the last phase of Incoronata greca and of the settlement under Metaponto has added a new and important dimension to the history of the settlement of the Metapontino in the second half of the 7th century BC. The buildings excavated in the very limited area of Soprintendenza site 57[106] consisted of two adjacent one-room spaces sharing a common wall, each with an entrance to the north. Constructed of mudbrick on a stone socle, like the rectangular houses at Incoronata, they were dwellings used also for storage. There was no trace of a tile roof of the sort that became standard from the 6th century BC onward in the chora of Metaponto. The area of these spaces varies from 15 to 19 m^2. These were probably two distinct one-room habitations like those of the 7th-century BC Greek village of Amastuolo[107] in the chora of Taras, which arose, like those at Incoronata greca, over a native village. Other early colonial dwellings of the 8th and 7th centuries BC in Sicily, such as Megara Hyblaia,[108] and on the Black Sea at Berezan[109] in the 6th century BC are comparable. They were united by a continuous courtyard in front, which served as an outdoor extension of the two units. The contents of these dwellings included a carefully made olla, a purely indigenous shape, and

Figure 17.23 Imported Corinthian amphora (Koehler Type A) from Pit B, Incoronata greca (h 73.0 cm). (CW)

[102] De Siena 1986b, 140–51.

[103] De Siena 1986b, 141–45.

[104] An analogous function has been posited for an as yet undiscovered site at the mouth of the Sinni or Agri, in relation to the site on the hilltop at Policoro; see S. Bianco 2001, 809.

[105] De Siena 1986b, 144.

[106] De Siena 2002, 30–31.

[107] Maruggi 2002, 45–54, pls. 4–7; 1996a.

[108] *Mégara Hyblaea* I, 1.62, 1.60.

[109] Solovyov 1999, 59–62, Fig. 43.

numerous spools, attesting the importance of sheep raising and weaving. The possibility remains that the two spaces, though each has a door to the outside, are part of a single domestic unit. In this case, the room with the spools could have been the woman's quarters and the other used as space for eating and sleeping.[110] A proto-Corinthian-style pyxis lid and a large number of *a filetti* cup fragments place the probable date for the occupation of the houses in the third quarter of the 7th century BC.[111]

The sudden abandonment of the houses below San Teodoro at this time, according to the excavator, links their fate to that of Incoronata and the settlement at the mouth of the Basento. They would not have been far off a valley road connecting Incoronata with the coast. The model for the settlement of the Metapontino at this time must now be expanded to include dwellings on the valley floor near springs, built using Greek or Greek-inspired construction techniques. It is doubtful that this was an isolated case, and this pattern continued, as noted above, into the late 7th century BC in the Bradano valley. The site on the coast at Andrisani, that at Incoronata on the inland hilltop, and the dwellings at San Teodoro in the valley below were all destroyed in a conflagration at the same time, according to De Siena.[112] There were no signs of life again until the following century, when a small sanctuary occupied the height at Incoronata[113] and the residential quarter of the growing asty covered the huts beside the river.

Cultural Diversity in the 7th Century BC

There were evident signs of cultural diversity in burials at Incoronata, well before the end of the 7th century BC when Greek immigrants established themselves throughout the Metapontino. The isolated burial of an adult in supine position was discovered on the northern edge of the hut settlement in Incoronata indigena. The wooden coffin rested on a carefully made bed of cobbles and was covered by a single, large stone slab. If this burial is contemporary with those in the flexed position in the area, then the conclusion drawn by De Siena that this is a "stranger of elevated rank"[114] is certainly justified; in the absence of grave goods, however, the burial cannot be securely dated. The fact that it was surrounded by hut floors and far removed from other burials on all sides suggests that it was intrusive.

Nearby at Incoronata Azienda Agricola, De Siena excavated another smaller necropolis consisting of two dozen burials, and on the basis of these he has made a convincing case that culturally diverse or mixed groups had already begun to bury at Incoronata in the 7th century BC. The adults were all found in the flexed position typical of 8th-century BC inhumations, with no grave goods except for a single impasto vessel. The careless construction of the tombs contrasted sharply with that of the earlier period. More than half of the burials were of children in Greek storage amphorae, which, together with the few grave goods, date this necropolis to the second and third quarters of the 7th century.[115] They are hence contemporary with the last phase of the settlement at nearby Incoronata greca (see *Fig. 17.8*). The chronology of the enchytrismos burials depends on the container, as the grave goods are scarce. A single *a filetti* cup, two proto-Corinthian aryballoi, a local imitation of a proto-Corinthian geometric cup, and a hydria with banded decoration, like those from Incoronata greca and Andrisani on the coast on the site of the future asty of Metaponto are the exceptions. Parallels have been drawn with the Madonelle necropolis at Policoro,[116] some 20 km to the south, where such enchytrismos burials are well documented.[117] This is one of a number of similarities between the Metapontino and Siritide in the 7th century BC. The modesty of these arrangements led the excavator to speculate that the group who buried here was distinct from the preceding indigenous community, and of economically inferior, if not of subaltern, status.

The end of the mixed settlement of the chora that characterized the first half of the 7th century BC and the beginning of the pattern of dispersed farmhouses, densely distributed throughout the chora, are certainly connected with these events, though just how is not yet clear. A solution suggested by De Siena[118]

[110] De Siena 2002, 39 n. 15.

[111] The sequence of *a filetti* cup types has never been precisely defined. They cover the 7th c. BC; see De Siena 1986b, 141.

[112] De Siena 1986b, 156; Giardino and De Siena 1999b, 339; Nava 2002, 738–39.

[113] Carter 2006a, Ch. 2.

[114] De Siena 1990a, 78.

[115] Some 100 m to the north, beside the modern structures of the Azienda Agricola: De Siena 1990a, 80.

[116] De Siena 1990a, 80, 83.

[117] De Siena 1990a, 84.

[118] Giardino and De Siena 1999b, esp. 341–45.

has the virtue of resolving the problems raised by the contradicting versions of the founding of the apoikia as reported by Strabo (6.1.15) and the archaeological evidence for the violent end or hurried abandonment of these sites.

The Traditions about the Founding of Metaponto

The traditions about the founding of Metaponto are many and contradictory. Alfonso Mele has reduced them to three main themes: the indigenous origin, centered on the figure of Metabos, the heroic Achaian ancestors of the Pylians of Trojan War fame, and the destruction of the Achaian foundation myths.[119] Each of them is embedded in Strabo's brief sketch of the origin of Metaponto. The text of Strabo with a translation (H.L. Jones) is reproduced here.

> [15] Ἑξῆς δ' ἐστὶ Μεταπόντιον, εἰς ἣν ἀπὸ τοῦ ἐπινείου τῆς Ἡρακλείας εἰσὶ στάδιοι τετταράκοντα πρὸς τοῖς ἑκατόν. Πυλίων δὲ λέγεται κτίσμα τῶν ἐξ Ἰλίου πλευσάντων μετὰ Νέστορος, οὓς οὕτως ἀπὸ γεωργίας εὐτυχῆσαί φασιν, ὥστε θέρος χρυσοῦν ἐν Δελφοῖς ἀναθεῖναι. σημεῖον δὲ ποιοῦνται τῆς κτίσεως τὸν τῶν Νηληιδῶν ἐναγισμόν· ἠφανίσθη δ' ὑπὸ Σαυνιτῶν. Ἀντίοχος δέ φησιν ἐκλειφθέντα τὸν τόπον ἐποικῆσαι τῶν Ἀχαιῶν τινας μεταπεμφθέντας ὑπὸ τῶν ἐν Συβάρει Ἀχαιῶν, μεταπεμφθῆναι δὲ κατὰ μῖσος τὸ πρὸς Ταραντίνους τῶν Ἀχαιῶν, τῶν ἐκπεσόντων ἐκ τῆς Λακωνικῆς, ἵνα μὴ Ταραντῖνοι γειτνιῶντες ἐπιπηδήσαιεν τῷ τόπῳ. δυεῖν δ' οὐσῶν πόλεων, τοῦ δὲ Μεταποντίου ἐγγυτέρω τοῦ Τάραντος, πεισθῆναι τοὺς ἀφιγμένους ὑπὸ τῶν Συβαριτῶν τὸ Μεταπόντιον κατασχεῖν· τοῦτο μὲν γὰρ ἔχοντας ἕξειν καὶ τὴν Σιρῖτιν, εἰ δ' ἐπὶ τὴν Σιρῖτιν τράποιντο, προσθήσειν τοῖς Ταραντίνοις τὸ Μεταπόντιον ἐν πλευραῖς οὖσι. πολεμοῦντας δ' ὕστερον πρὸς τοὺς Ταραντίνους καὶ τοὺς ὑπερκειμένους Οἰνωτροὺς ἐπὶ μέρει διαλυθῆναι τῆς γῆς, ὅπερ γενέσθαι τῆς τότε Ἰταλίας ὅριον καὶ τῆς Ἰαπυγίας. ἐνταῦθα δὲ καὶ τὸν Μετάποντον μυθεύουσι καὶ τὴν Μελανίππην τὴν δεσμῶτιν καὶ τὸν ἐξ αὐτῆς Βοιωτόν. δοκεῖ δ' Ἀντίοχος τὴν πόλιν Μεταπόντιον εἰρῆσθαι πρότερον Μέταβον, παρωνομάσθαι δ' ὕστερον· τήν τε Μελανίππην οὐ πρὸς τοῦτον, ἀλλὰ πρὸς Δῖον κομισθῆναι ἐλέγχειν ἡρῷον τοῦ Μετάβου καὶ Ἄσιον τὸν ποιητὴν φήσαντα, ὅτι τὸν Βοιωτὸν
>
> Δίου ἐνὶ μεγάροις τέκεν εὐειδὴς Μελανίππη
>
> ὡς πρὸς ἐκεῖνον ἀχθεῖσαν τὴν Μελανίππην, οὐ πρὸς Μέταβον. οἰκιστὴς δὲ τοῦ Μεταποντίου Δαύλιος ὁ Κρίσης τύραννος γεγένηται τῆς περὶ Δελφούς, ὥς φησιν Ἔφορος. ἔστι δ' ἔτι καὶ οὗτος ὁ λόγος, ὡς ὁ πεμφθεὶς ὑπὸ τῶν Ἀχαιῶν ἐπὶ τὸν συνοικισμὸν Λεύκιππος εἴη, χρησάμενος δὲ παρὰ τῶν Ταραντίνων τὸν τόπον εἰς ἡμέραν καὶ νύκτα μὴ ἀποδοίη, μεθ' ἡμέραν μὲν λέγων πρὸς τοὺς ἀπαιτοῦντας, ὅτι καὶ εἰς τὴν ἐφεξῆς νύκτα αἰτήσαιτο καὶ λάβοι, νύκτωρ δ', ὅτι καὶ πρὸς τὴν ἑξῆς ἡμέραν.

> [15] Next in order comes Metapontium, which is one hundred and forty stadia from the naval station of Heracleia. It is said to have been founded by the Pylians who sailed from Troy with Nestor; and they so prospered from farming, it is said, that they dedicated a golden harvest at Delphi. And writers produce as a sign of its having been founded by the Pylians the sacrifice to the shades of the sons of Neleus. However, the city was wiped out by the Samnitae. According to Antiochus: Certain of the Achaeans were sent for by the Achaeans in Sybaris and re-settled the place, then forsaken, but they were summoned only because of a hatred which the Achaeans who had been banished from Laconia had for the Tarantini, in order that the neighbouring Tarantini might not pounce upon the place; there were two cities, but since, of the two, Metapontium was nearer to Taras, the newcomers were persuaded by the Sybarites to take Metapontium and hold it, for, if they held this, they would also hold the territory of Siris, whereas, if they turned to the territory of Siris, they would add Metapontium to the territory of the Tarantini, which latter was on the very flank of Metapontium; and when, later on, the Metapontians were at war with the Tarantini and the Oenotrians of the

[119] Mele 1998; cf. also Dunbabin 1948, 31–35.

interior, a reconciliation was effected in regard to a portion of the land—that portion, indeed, which marked the boundary between the Italy of that time and Iapygia. Here, too, the fabulous accounts place Metapontus, and also Melanippe the prisoner and her son Boeotus. In the opinion of Antiochus, the city Metapontium was first called Metabum and later on its name was slightly altered, and further, Melanippe was brought, not to Metabus, but to Dius, as is proved by a hero-temple of Metabus, and also by Asius the poet, when he says that Boeotus was brought forth "in the halls of Dius by shapely Melanippe," meaning that Melanippe was brought to Dius, not to Metabus. But, as Ephorus says, the coloniser of Metapontium was Daulius, the tyrant of the Crisa which is near Delphi. And there is this further account, that the man who was sent by the Achaeans to help colonise it was Leucippus, and that after procuring the use of the place from the Tarantini for only a day and night he would not give it back, replying by day to those who asked it back that he had asked and taken it for the next night also, and by night that he had taken and asked it also for the next day.[120]

Strabo 6.1.15

Strabo's account emphasizes continuity, in Achaian terms, with a significant Bronze Age founding of the colony by Nestor of Pylos, and its destruction at the hands of an indigenous people anachronistically referred to as "Samnites." Antiochos' more realistic account, as Giardino and De Siena see it, makes the impetus for the founding by Achaians sent for by Sybaris, an attempt to block the expansion to the south of Dorian Taras, founded in the late 8th century BC. An Achaian colony would also have subtracted fertile agricultural land from Ionian Siris, founded ca. 650 BC. The problem for the Achaians was the already existing settlements along the Basento valley and these were dealt with swiftly and violently. At the same time, the foundation of Achaian Metaponto created a *de facto* boundary along the Bradano, between lands in the sphere of Sybaris and those of Taras to the north.[121] Antiochos' version ascribes both political and economic motivations to the founding, and does nothing to support the idea of a very ancient Achaian claim on the land. This can be understood in light of the rivalry that existed, in Antiochos' time and earlier, between Achaian Sybaris and Metaponto and Dorian Taras for control of the rich land of Siris. The story reported by Strabo of the mythical origins of Metaponto in the Bronze Age had propagandistic value to the aristocratic families in their claims against the indigenous population and later waves of colonists settling in the chora. The mythical accounts of founders that Strabo mentions include Metapontus, Melanippe and her son Boeotos, Metabos, to whom a hero temple was built, Daulios of Crisa (near Delphi), and Leukippos. This crowded field of founders has been convincingly deconstructed by Mele and each is seen in the light of the historical currents of the period in which the particular figure was most likely introduced into the traditions of the founding.[122]

One of the points in the account of Antiochos that has caused perplexity and a divergent interpretation is the statement that the Achaians arrived in a land that had been abandoned and that there were "two cities." The explanation proposed by De Siena—that the two cities, Incoronata greca and the settlement under the future colony, were abandoned "or destroyed" and that the new colonists were persuaded to take the Metapontino because it was near to Taras—fits the archaeological evidence well. Indeed the site of Metaponto is closer to the territory of Taras, and the foundation of the apoikia there in effect laid claim to all the land to the south of the Bradano and north of the territory of Siris.

The presentation of the chronological development of the land between the Bradano and Basento resumes in Chapter 20 with a discussion of the early settlers of the Greek chora.

[120] Text and trans. from H.L. Jones 1983 (Loeb).

[121] Giardino and De Siena 1999b, 341–45.

[122] Mele 1998.

18
Types of Structures in the Chora

Alberto Prieto with Joseph Coleman Carter

> Τοῦτ' <ὀλίγον> Κλείτωνος ἐπαύλιον, ἥ τ' ὀλιγῶλαξ
> σπείρεσθαι, λιτός θ' ὁ σχεδὸν ἀμπελεών,
> τοῦτό τε ῥωπεῖον ὀλιγόξυλον· ἀλλ' ἐπὶ τούτοις
> Κλείτων ὀγδώκοντ' ἐξεπέρησ' ἔτεα.
>
> This is Clito's little cottage, this his little strip of land to sow, and the scanty vineyard hard by, this is his patch of brushwood, but here Clito passed eighty years.[1]
>
> Leonidas of Taras (3rd century BC)

Because of both human and natural history in this area, the localization of sites in the Metapontino has been fraught with much less uncertainty than in many other parts of the Greek world. More controversial—and essential to the interpretation of the survey results—is the identification of the types of sites. In this the investigators of this chora have again been fortunate. Because of the history of archaeological exploration since the 1960s, more is probably known of the sites in this Greek countryside than of all but a few. The surveyor has a wide variety of excavated sites with their materials as background to draw on for comparisons.[2] Still, in the determination of site type there is inevitably a degree of subjectivity, as the materials that appear in the surface scatter do not always give an unequivocal indication of what might lie beneath.

There are obvious dangers in interpreting the function and type of a site when using only the limited information of its surface artifact assemblage (Ch. 5). The tendency is to simplify and identify any large survey scatter as a "farmhouse" site.[3] The variety of possible structures that could be found in an ancient Greek countryside was in reality quite wide, as the documentary sources well illustrate. As Osborne cautions:

> Archaeology indicates the presence of buildings, but literary and epigraphic information is needed to interpret the use to which buildings were put ... The importance of the literary and epigraphic evidence here is that it warns us against expecting that the material remains will provide a full picture of the agricultural activity on any site. Many agricultural activities will leave no archaeological trace.[4]

With this salutary reminder, we will discuss the documentary evidence about rural structures and activities from non-archaeological sources—primarily Greek and Latin literature and epigraphic texts—as well as the archaeological evidence from the Greek world of excavated rural structures. Here the particular emphasis is on evidence from Italiote colonies and close neighbors of Metaponto, and, of course, from Metaponto itself. Non-Metapontine examples are drawn from a wide variety of sources and the types of structures described depend on local conditions, which in many cases are quite different from those of the Metapontino. There is, however, a commonality in agricultural practices that justifies considering all possibilities for each particular one.

Tombs and Necropoleis

Those who resided in the countryside were generally laid to rest there as well. Practical reasons prompted members of a farming family to bury their dead near their dwelling: it was presumably more economical and facilitated the traditional observances that had to be carried out periodically. From the mid-5th to the mid-3rd century BC, and probably even before, there were numerous roads that flanked properties in the chora of Metaponto (Ch. 28), traversing the landscape and eventually leading to the city. Burial beside such roads would have had a degree of public exposure, which seems to have been desirable though it certainly would have had less prestige than burial along thoroughfares closer to the city.

---

[1] Text and trans. W.R. Paton's Loeb edition of *The Greek Anthology* (1916–1918), 6.226.

[2] Carter 2006a, especially Ch. 4.

[3] Osanna 1992, 68.

[4] Osborne 1992, 21–22.

The act of burying one's ancestors on one's own land had great symbolic power. This practice is well illustrated in Athens, the Greek polis best documented in ancient written sources. According to a client of Demosthenes, tombs, among other things, "are to be found in most private pieces of land" (μνήματ' ἔχειν τινὰ καὶ τἄλλ' ἅπερ καὶ τοῖς πλείστοις χωρίοις συμβέβηκεν).[5] The defendant cited the presence of tombs as proof of the private status of his rural property in Attica. Although the tombs in question belonged, in fact, to the family of the previous proprietor, their presence helped clarify its status when the plaintiff challenged that the property was public.

Most burials were placed near the edges of properties, often with highly visible ceramic or stone markers, to reinforce the boundary lines. Since desecration of a burial was—and still is—considered a very serious matter, some citizens went so far as to bury their dead in property, both public and private, that did not belong to them in order to lay a permanent private claim.[6] A recent study emphasizes the sense of ownership or possession imparted by burial on private property, noting that families may very well have lived in nucleated settlements, but nonetheless buried their deceased on the land to reinforce their claim to it.[7]

In mainland Greece, the archaeological evidence for isolated rural burials and small rural necropoleis is scattered unevenly. Relatively large numbers of sporadic burials, both individual and collective, have been recognized in Boeotia and southern Attica in the same periods,[8] while burials in the southern Argolid in the Archaic to Hellenistic periods are few.[9] Rural burials throughout Magna Grecia and Sicily, however, are well documented in most chorai and periods. In the chora of Taras, discoveries of isolated tombs are common, as are small "family" necropoleis of 5 to 10 burials. Much larger and rarer necropoleis of up to 150 burials are also known.[10] Isolated tombs and small necropoleis, which begin to appear there in the 4th century BC, are common. Comparable casual discoveries have been made in the chora of Herakleia,[11] as well as in the chora of Croton.[12] At Poseidonia-Paestum, isolated rural tombs and necropoleis have been dated to the 6th century BC, several decades before the earliest isolated farmhouses.[13] Although most of the early rural burials in the countryside surrounding Poseidonia are indigenous or mixed, the presence of Archaic Greek sanctuaries strongly suggests that the chora was under Greek influence beginning in the 6th century BC, with implications for the makeup of the workforce in that period. By the 4th century BC, the occupants of the chora were either Greek or Hellenized *indigeni*, whose material culture was already thoroughly Greek.

Most of the burials in the chora beyond a certain distance from an urban center are rural by definition,[14] and the majority, surely, belonged to the inhabitants of the chora. There is some anthropological evidence from Pizzica, however, that people who were biologically closer to those dwelling in the city were also buried there (Ch. 30, 1113–14). The areas considered "suburban" or "peri-urban," approximately 2 km from the city walls and coastline toward the interior, contained the major urban necropoleis, such as that found at Crucinia[15] and the recently discovered necropolis at Torre di Mare with some of the earliest Metapontine burials.[16]

[5] Dem. 55.13–15, trans. A.T. Murray, Cambridge, Mass. 1939 (Loeb). See also Osborne 1985b, 17.

[6] E.g., the prohibition of burials recorded in the lease of the lands sacred to Dionysos in Herakleia Tablet I.137–138 (Arangio-Ruiz and Olivieri 1965, XIV.645).

[7] Snodgrass 1998, 40–41. Tombs were probably common landscape features in agrarian societies. According to the Roman surveyor-writer Siculus Flaccus, who lived in the 2nd c. AD, "men are accustomed to putting tombs at the edges of their fields" (*sepulchra in extremis finibus facere soliti sunt*). Thulin 1971, 104; Campbell 2000, 106–07. A brief directive on the roles of tombs in landscape organization (*De sepulchris*) was attributed to the emperor Tiberius (AD 14–37). Blume/Lachmann/Rudorff 1848, 271–72; Campbell 2000, 220–221. In a third Roman landscape text (*Ex libris Dolabellae*), the author advises the surveyor looking for an obscured field boundary in a dispute between neighbors to begin by locating a tomb near the field's edge; the missing boundary should be within 5 feet of either side of the tomb. See Blume/Lachmann/Rudorff 1848, 303; Campbell 2000, 222–23.

[8] See Snodgrass 1998, 38–41; Lohmann 1992, 33–35, 51; 1993, 123–26, 136, 184–94.

[9] *Greek Countryside*, 424–25 (site A16), 427 (A25), and 474–75 (C41).

[10] Alessio and Guzzo 1989–1990, 368. Many of the known rural Tarantine necropoleis are documented in Osanna 1992, 25–38, particularly nos. 5, 8–10, 13, 15, 18–19, 23, 28–33, 35–36, 45–47, 50, 52–53, 56–74; several others are documented in Cocchiaro 1981.

[11] Osanna 1992, 107–14 nos. 5, 8–13, 16–17, 22, 24–26.

[12] Osanna 1992, 191–99 nos. 2, 4–7, 12, 14, 18, 20, 23 28–31 33, 38. To date the survey of the chora of Croton has recorded a total of some 560 sites with over 70 necropoleis or isolated tombs. See Carter and D'Annibale 1985 and 1993. The Croton survey will be the subject of a future study.

[13] E. Greco 1979, 18–24; 2001, 186.

[14] See Snodgrass 1998.

[15] De Siena 1999b, 232–36.

[16] Nava 2005, 327–29.

Figure 18.1 *Left*, intact cist tomb, tomb 350, Pantanello necropolis; *right*, disturbed slabs from a cist tomb found at survey site 471N, Giardinetto, a large necropolis site.

Although Metapontine rural burials are found singly, groups are more typical, and range from single families to multiple or extended households. Thousands of burials have been discovered, and presumably thousands more await discovery.[17] Locations of unexcavated burials can generally be identified on the basis of the following criteria: small surface area (ranging from 4 to 100 m^2, depending on recent land use); concentrated and isolated artifact scatter; ceramic vessels with forms and decoration typically associated with funeral contexts and used as grave goods or tomb markers[18] (lekythoi, kraters, and pelikai and storage amphorae); human bone; and rarely, the ashy residue of cremation.

Metapontine rural burials are sometimes distinguished by the presence of the tomb materials themselves. Stone slabs in local conglomerate (*puddinga*) or limestone (calcarenita di Gravina), quarried from as far as 20 km away (see Ch. 1), were used for the construction of cist tombs (*Fig. 18.1*) and stone sarcophagi. Large, flat Corinthian-style pan tiles (*Fig. 18.2*) containing abundant flakes of bronze-colored biotite mica were apparently manufactured specifically for the construction of *a cappuccina* tombs, since they are rarely encountered at sites without burials. Tombs could be constructed from ordinary Laconian-style roofing tiles as well. The most basic form of burial container known in the Metapontino was a simple pit or *fossa* excavated in the earth, occasionally lined with cobbles or plaster.[19] The remains of single burials are also frequently found close to farmhouse scatters, usually within approximately 25 m. These are distinguished by some or all of the above-mentioned characteristics, as well as by a distinctly higher artifact density than the core farmhouse scatter.

When multiple indicators are lacking, and the only criterion is the composition of the ceramic assemblages, the identification of burials becomes more problematic. Those of the Pantanello, Saldone, and Sant'Angelo Vecchio necropoleis were characterized by a wide range of ceramic shapes and types, including oil vessels (primarily lekythoi), drinking vessels (cups, skyphoi, etc.), and large containers for liquids (amphorae, pelikai, etc.), in varying combinations. A significant proportion—approximately 20%—of the Pantanello burials contained either no grave goods at all or simply no ceramic artifacts. The number of vessels buried with the deceased, whether inside or outside of the tomb proper, was generally very low: 58% of burials contained three or fewer vessels, 46% two or fewer, and 25% only one. In only 23% of burials were four or more vessels found, and the maximum number of vessels recorded was nineteen (T 350).[20]

With the partial exceptions of figured vases (especially those of large dimensions) that were used

[17] For the pervasiveness of Greek burials in the Metapontino generally, see Carter 1998f and Uggeri 1969, 53. The sheer profusion of Greek burials in the Metapontino is perhaps not so unusual, but the laudable effort by the Soprintendenza to recover them is, and has added considerably to our understanding of Metapontine society. One of the most recently excavated necropoleis, identified by Uggeri as long ago as 1969, is located in the central Pizzica locality, roughly 1.5 km northeast of Pantanello and 3.5 km northwest of the asty, and contained 77 graves. See Nava 2001, 9–7–49. The widening of the coastal road SS106 has resulted in the discovery of a thousand or more burials between 2006 and 2008. See reports by De Siena (2007) and Osanna (2008).

[18] Carter 1998d 116–17; Snodgrass 1998, 38.

[19] On the various rural burial types in the Metapontino and their structural components, see Carter 1998c, 60–103. For the types of stone employed, see Carter 1998b, 24, fig. 38; Carter 1998c, 73; Folk 1998. On the tomb tiles, see also Folk 1983. During geological investigation in Calabria, Folk (1988) located a likely origin of the biotite in a diorite quarry on the eastern edge of the Sila massif, near Cerenzia. The biotite may have been added to strengthen the clay matrix without appreciably increasing its weight.

[20] J. Hall 1998, 563–65. On the use of larger vessels as grave markers, see Carter 1998d, 124–27.

Figure 18.2 *Left*, collapsed tile-covered tomb, tomb 151, at Pantanello necropolis: *right*, ceramic assemblage and pan tile fragments from survey site 471N, the same site illustrated in Fig. 18.1.

exclusively as grave markers, and certain classes of lekythoi found only in burials (see Chs. 8 and 29), identification of burials based solely on ceramic artifacts is challenging. The wide range of ceramic wares and decoration typically encountered in Metapontine necropoleis are also the wares and styles characteristic of domestic contexts. They include decorated fine wares (red figure, Gnathia style), black gloss pottery in a variety of shapes (cups, lebetes, lekythoi), banded ware (one-handler dippers, hydriai), undecorated coarse wares, miniature table and storage amphorae, and a wide range of cooking wares, especially pots or chytrai.[21] Among these, black-gloss, banded, and coarse wares predominate in the excavated necropoleis.[22] Terracotta figurines are occasionally found among the grave goods, especially in burials of women and children (see Ch. 13).[23]

Farmhouses

The recent emphasis placed by some scholars on defining universal models of settlement seriously overlooks important regional and local evidence from the Greek world.[24] Generalization about the Greeks being city-dwellers commuting to their farmhouses may apply to some areas, but the existence of the isolated Greek farmhouse, homestead, or country estate in other places is a fact, established by excavations in different, widely-separated parts of the Greek world, and supported by literary and epigraphic evidence. A classic paper on farmhouses by Pečírka[25] and subsequent work over the last three decades in the Aegean, western Mediterranean, and Black Sea[26] have added significant new examples of this widespread phenomenon. Interest in domestic rural and urban architecture has grown dramatically with publication of individual houses and of synthetic studies.[27] Nowhere has this interest been more intense than among archaeologists working in southern Italy, where investigation has embraced the coastal territories[28] and the indigenous interior as well.[29] Yet in some quarters of the Classical archaeological community there still exists a stubborn resistance to the idea that Greeks lived as farmers on the land; moreover, there is a reluctance to recognize the inherent variability in the form and function of farmhouses determined by the specific requirements, environmental conditions, and social organization of individual poleis spread over a very wide geographical area.

[21] On the presence of cooking vessels and storage amphorae in burials, see Carter 1998d.

[22] J. Hall 1998, 564.

[23] See Malone 1998.

[24] E.g., Foxhall 2001, 217–18.

[25] Pečírka 1973.

[26] For Olbia, see Buyskikh and Buyskikh 2001. For Chersonesos, see Saprykin 1994; Carter et al. 2000. For Kamarina, see Pelagatti 1980–1981; Di Stefano 2001.

[27] Nevett 1999.

[28] For Metaponto, see Carter 2006a, 134–50. For Herakleia, see L. Giardino 1996b; 2005. For Locri, see Barra Bagnasco 1996b. For Elea, see Cicala 2002.

[29] Russo Tagliente 1992, and in general, *Ricerche sulla casa*.

Occupation and Continuity

Isolated farmhouse structures, used as either temporary or permanent residences, as this study will show, were numerous in the chora of Metaponto from the 6th to 3rd centuries BC. The ceramic assemblages from the excavated Metapontine farmhouses document a continuous history of settlement in the chora. The earliest examples were occupied in the 7th and early 6th century BC at Termitito and Cogno del Pero respectively. The latest rural residences for which we have evidence date to the 4th century AD.[30] A farmhouse typically would be occupied for a century or more, presumably by several generations of the same family. Rural necropoleis usually contain two or three generations of a family, rarely more.[31] As discussed later (see Ch. 25), building sites continued to be attractive and would be reoccupied after periods of abandonment by new structures that reutilized materials from the earlier buildings.[32] Excavated examples of farmhouses of the 4th century BC—the date of the majority—are very similar in size,[33] measuring on average roughly 200 m^2 and rarely as much as 600 m^2.

Location and Orientation

Fundamental, of course, was ownership of the land, either through inheritance, rental or tenant agreements, or sale. Many farmers chose already established sites on their plots, where materials were available, such as fluvial cobbles, tiles—tile kilns from all periods have been found throughout the chora (see below)—perhaps surviving mudbricks, and even wood from the original structures on the site.[34] Other factors would have included the always essential proximity to water, and where possible, a favorable exposure to prevailing breezes.[35]

Excavated farmhouses in the Metapontino generally face southeast towards the asty and the Ionian Sea. Such an orientation was general practice in the Greek world, as observed by Xenophon: καὶ σύμπασαν δὲ τὴν οἰκίαν ἐπέδειξα αὐτῇ ὅτι πρὸς μεσημβρίαν ἀναπέπταται, ὥστε εὔδηλον εἶναι, ὅτι χειμῶνος μὲν εὐήλιός ἐστι, τοῦ δὲ θέρους εὔσκιος ("I showed her that the whole house fronts south, so that it is obvious that it is sunny in winter and shady in summer").[36] The author of the pseudo-Aristotelian *Oeconomica* also noted: καὶ πρὸς εὐημερίαν δὲ καὶ πρὸς ὑγίειαν δεῖ εἶναι, εὔπνουν μὲν τοῦ θέρους, εὐήλιον δὲ τοῦ χειμῶνος. εἴη δ' ἂν ἡ τοιαύτη κατάβορρος οὖσα καὶ μὴ ἰσοπλατής ("For well-being and health, again, the homestead should be airy in summer, and sunny in winter. A homestead possessing these qualities would be longer than it is deep; and its main front would face the south").[37]

To make the most of the natural topography and breezes, most Greek farmhouses in the chora of Metaponto were laid out parallel to the coastline, following the northwest-southeast contours of the marine terraces that step down to the sea.[38] This means that these farmhouses were aligned perpendicular to the northwest-southeast axis that bisected the chora.[39] Only sites constrained by other topographic situations, along the slopes and valley walls, were likely to face a different direction.

Materials

Construction techniques and plans are well documented by the excavation of over a dozen structures.[40] The materials employed in the construction of Greek rural structures at Metaponto and elsewhere—tiles, stones, mudbrick, and wood—all were conveniently

[30] De Siena 1996, 172, and Adamesteanu 1974a, 82–85. See below, Lapadula, 1145.

[31] Carter 1998e.

[32] For excavated examples, see Carter 2006a, 134–50, with bibliography. Also Adamesteanu 1974a, 82–87; 1970a, 235; 1968, 254; 1967b, 46; Adamesteanu and Vatin 1976, 115–16; Carter 1980a, 18–19; 1983b, 18; A. Bottini 1983a, 470–71; Nava 2002, 733–35.

[33] Adamesteanu 1967b, 46; Adamesteanu and Vatin 1976, 116. See also Uggeri 1969, 61. The largest excavated farmhouse, Fattoria Stefan in the Lago del Lupo locality, measured 35 x 20 m (Adamesteanu 1973, 56). The excavated farmhouse at Bufalara, which was occupied and abandoned in the 6th c., was slightly smaller at 20 x 15 m, suggesting that the size of farmhouses increased in the Classical and Hellenistic periods. These preserved socles are thus analogous to the houses excavated by Jones et al. and the dilapidated stone structures documented by Young (1956) and Lohmann (1992, 1993), cited (by, e.g., Pettegrew 2001, 204) as firmer proof of the existence of farmhouses and associated structures in Attica than more opaque surface artifact scatters.

[34] See Pettegrew 2001, 202.

[35] Recommended to Roman builders by Vitruvius 6.6.1–7.

[36] Xenophon, *Oeconomicus* 9.4–5, text and trans. E.R. Marchant 1923 (Loeb). Many excavated Greek farmhouses face south: e.g., Jones/Graham/Sackett 1973, 360; Jones/Sackett/Graham 1962, 77, 103–05 (with numerous published plans).

[37] Aristotle, *Oec.* 1345a.31–33, text and trans. G.C. Armstrong 1935 (Loeb). Kent (1948, 298–99) noted that the proportions and orientation of all of the modern farmhouses he observed on Delos and Rheneia conformed to this ancient tradition.

[38] See Ch. 1 on geological formations of the Italian peninsula.

[39] A 6th-c. farmhouse excavated in the chora of Herakleia was found to have its residential side facing north and its courtyard facing south. See Ghinatti 1982, 51–52.

[40] The most recent examples are described summarily in Nava 2002, 735, and for the chora north of the Bradano, Schojer 2001, 124–25. The Greek and Roman Metapontine farmhouses excavated by ICA will be published in detail. See Amore 1992–1993.

Figure 18.3 Scatter of fluvial cobbles, site 224. (CDA)

Figure 18.4 Wall base constructed from fluvial cobble, excavation of Fattoria Fabrizio. (JCC)

at hand or easily portable.[41] Materials can be a useful indicator of the date of a farmhouse. In the earlier periods of the settlement of the chora, cut stone was literally beyond the reach of the farming population. Quality stone was not available in the chora. The quarrying and shipping of good stone was expensive and generally reserved for use in monumental public structures, such as sanctuaries and temples, and the tombs of the wealthy.[42]

Blocks of good quality stone began to appear in foundations of farmhouses in the Late Archaic and Early Classical periods, reflecting a general rise in prosperity among a larger section of the population at that time. Imported stone could be found in private dwellings and in the sarcophagi of the rural, as well as urban, necropoleis.[43] The architectural stones employed in rural structures from the 2nd century BC onward are, in some cases at least, the spoils of dilapidated rural sanctuaries and the like, a sign of the general transformation of rural life.[44]

As noted above, materials from Archaic period dwellings were often reused in subsequent structures on the same site. For example, fragments of roof tiles from earlier buildings were incorporated as leveling material in foundation courses from the 5th century BC onward.[45] The socle, uniformly 40–50 cm wide, was composed of dry-packed fluvial cobbles (*Figs. 18.3–4*), preferred over the abundant marine cobbles presumably for their relatively greater sphericity, making them better packing material. Walls were raised in mudbrick formed from the ubiquitous Plio-Pleistocene marine clays, usually available in the immediate environs.[46]

[41] Carter 2006a, 137–38; Pettegrew 2001, 196.

[42] Carter 1998c, 69–89.

[43] Carter 2006a, 213. Evident also in the building history of the asty, where there was much building but on a more moderate scale than in the mid-6th c. BC.

[44] E.g., the triglyph from the Temple of Hera (Tavole Palatine) and numerous other stone architectural elements recovered in the Roman rural structure (a transhumance station?) in contrada Casa Teresa (Lissi Caronna 1998–1999, 178, 185, 197; Mertens 1974, 214), the several stone architectural elements dating from the Archaic through Hellenistic periods found at the Pantanello sanctuary (Carter 1983c, 23), the Doric stone capital at the 4th-c. BC farmhouse in contrada Sant'Angelo Grieco (Carter 1983b, 18), and the several decorated Archaic stone architectural elements re-used in a farmhouse dated to the 2nd–1st c. BC in contrada Pantano north of the Bradano (Schojer 2001, 124–25; 2002, 83–84), part of a structure of clear Metapontine origin. On the lack of good local building stone, the depredation of stone structures, and the reuse of stone blocks in the Metapontino generally, see Adamesteanu 1970a, 234; 1974a, 85–87; 1973, 56; L. Giardino 1982, 158; De Siena 1999b, 233.

[45] Adamesteanu 1970a, 235; 1974, 85; Uggeri 1969, 62.

[46] Carter 2006a, 136; Adamesteanu 1973, 56; 1974a, 82–85; Adamesteanu and Vatin 1976, 115; Uggeri 1969, 61–62; Amore 1992–1993, 61. This remained the primary construction technique of the Roman Metapontino, too, through the 4th c. AD. To judge from the frequent appearance of Roman structures on or near the remains of Greek farmhouses and monumental buildings, the Romans, like the Metapontines in the 4th and 3rd c. BC, viewed ruined structures as sources of raw materials for new structures. See Carter 1980a, 22–25; 1983a, 439; 1983b, 17–18; Lissi Caronna 1998–1999, 195; also Adamesteanu 1973, 57; Adamesteanu and Vatin 1976, 116. The same stone, mudbrick, timber, and tile construction technique was recommended by Xenophon (*Mem.*

Figure 18.5 Excavation at survey site 1F, July 1981, shortly after the survey began. The site proved to be that of a Greek farmhouse, known as Sant'Angelo Grieco, occupied and periodically modified over a long period from the 6th to 3rd c. BC and expanded in the 2nd and 1st c. BC. (JCC)

Figure 18.6 Detail, with the fallen tile roof in place, of the excavation of the 4th- and 3rd-c. BC farmhouse known as Fattoria Stefan, one of the largest and best preserved of the excavated examples of a rural residence. It was excavated in 1969. (Courtesy of the Soprintendenza)

Once abandoned, the wooden roof beams of a farmhouse inevitably rotted, causing the tile roof to collapse onto the beaten earth floor. After direct exposure to rain, the mudbrick walls melted, covering the cobble socle and collapsed roof in a heap of clay (*Figs. 18.5–7*). The countryside around Metaponto preserves the remains of even quite recently occupied agricultural structures that are not so different in plan from the ancient.[47]

The masses of stones, mudbrick, and roof tiles produced by the plowing of ancient farmhouse foundations were clearly visible in early aerial photographs of the archaeological zone as light or dark spots contrasting strongly with the uniform grey tones of the cultivated fields. On the ground, the stones are recognizable (*Fig. 18.3*), even when the landowner has placed them in a heap to one side of his field; the yellow-gray color of the mudbrick stands out against the red-brown (*terra vergine*) soil.[48]

In the chora of Metaponto, as noted, a considerable number of rural structures from the Greek period have been excavated. These sites were chosen for investigation because of the suspected good condition of the buried buildings. Reports indicated that deep plowing had not occurred and surface scatters were light. Deep plowing now seems comparatively benign in contrast to the years of landscape alteration that have reduced some of the best examples to smears of fluvial cobbles, tiles, clay, and domestic ceramics.[49] The window of opportunity for future research grows inexorably smaller with each passing year. A major desideratum is good examples of farmhouses of the 6th and 5th centuries BC. One hopes it is not too late.

Contents

Identification of excavated rural structures as residences has been based on two criteria: their architectural form and their contents. Because the first cannot be ascertained through surveys, what objects might be expected from the scatter of an unexcavated site

3.1.7) and appears to have been a general Greek architectural tradition in regions abundant in the materials in question: see Jones/Sackett/Graham 1962, 77, 106 (with bibliography). The excavated farmhouse in the Cogno del Pero locality near the northwestern edge of the chora of Metaponto, however, is the only known exception to the rule. Its socle was built in the 6th c. BC of marine cobbles retrieved in the immediate vicinity rather than fluvial cobbles, and it was never rebuilt (Adamesteanu 1973, 54; 1974a, 82–85).

47 Carter 2006a, 142, 187 n. 32.

48 Adamesteanu 1973, 52.

49 This stage of destruction was reached in Attica even earlier. See Pečírka 1973, 136–37.

Class	Fattoria Fabrizio*	Survey farm average, 300**
Fine ware	0%	< 1%
Black gloss	18%	39%
Banded ware	7%	6%
Plain ware	40%	40%
Cooking ware	27%	11%
Amphorae	5%	2%
Miniatures	1%	< 1%
Terracotta figurines	1%	1%

Table 18.1 A comparison of ware class percentages from Fattoria Fabrizio and the average for 24 surveyed farmhouses of similar date. (*Based on provisional figures, E. Lanza Catti, pers. comm., Aug. 2009; **Average of 24 farmhouses with best date **300**. Cf. Appendix C18.)

Figure 18.7 A room in Fattoria Fabrizio excavated in 1980. The tile roof had fallen, covering the floor and the objects illustrated in Figs. 18.8–18.11. (JCC)

that would indicate that it was not only a center of agricultural production, but also a home?

The contents of excavated farmhouses in the chora of Metaponto are currently the object of detailed study that will be presented in separate volumes in the series on the chora devoted to the farmhouses. The survey team and the authors of the volumes, however, had the excavation inventories at their disposition. A preview of the results from the typical small, single-family farmhouse, Fattoria Fabrizio, shows the principal categories of material (*Table 18.1* and *Figs. 18.8–11*). The percentages of the principal pottery classes correspond broadly to the averages of sites in the survey identified in the field as farmhouse sites.[50] Although it is not conclusive proof, the assemblages of farmhouses in the survey are very similar in many points—not just to Fattoria Fabrizio, but to other excavated farmhouses (see *Appendices C19b, C21,* and *C23b*).

Signs of year-round habitation are unmistakable in the large quantities and broad ranges of ceramics present at the Metapontine farmhouse sites. Excavation at Fattoria Fabrizio turned up black-gloss pottery (*Fig. 18.8a*) used for consumption of food and drink on ceremonial occasions, as well as banded and undecorated table wares for everyday use (*Fig. 18.8c*). Amphorae and pithoi were used for storing foodstuffs, while a range of vessels—bowls, lekanai, mortars, casseroles, pots, and pans—were utilized for preparing, cooking, and serving meals (*Fig. 18.8b, d,* and *18.11*).

Other familiar farmhouse items include metal objects such as a bronze strainer (see below *Fig. 18.12*) and tools, loom weights (*Fig. 18.8e*), terracotta figurines (*Fig. 18.9*), and miniature votive vessels (*Fig. 18.10a, b*); occasional finds, such as iron tools, stone saddle querns, and stone press blocks, are further evidence of domestic activities at these sites.[51] A very typical find in the chora of Metaponto and throughout this area of Italy is the loom weight, the only surviving evidence for weaving (see Ch. 14) and an indirect indicator of pastoralism and animal husbandry. These activities continued into the period of Roman domination in southern Italy.[52] Loom weights are also found as votives in sanctuaries both rural and urban, but not in burials.[53] The terracotta figurines and miniature votive vessels are signs of cultic activity within the domestic setting, indicating prolonged, if not permanent, occupation; they are also found in burials and at sanctuary sites.[54]

Some artifacts in surface scatters are evidence of the habitation and abandonment phase, others are residual, from earlier periods of occupation. The sheer mass of varied ceramic fragments recovered on most farmhouse sites—black-gloss pottery, coarse and cooking wares, as well as the frequent fragments of

[50] Carter 2006a, 33–38, fig. 1.28. The comparison of survey assemblages with those from excavated farmhouses is discussed at length in Ch. 25.

[51] See, e.g., Adamesteanu 1974a, 88.

[52] L. Giardino 2005, 405, and Ch. 27, 914.

[53] See De Siena 1993.

[54] E.g., the relief plaque recovered from Fattoria Fabrizio (Carter 1980a, 20; 2006a, Fig. 4.5). Single votive miniatures have been recovered at the Fabrizio and Stefan houses in the Lago del Lupo locality. Complete studies of several farmhouses excavated by ICA and the Soprintendenza are forthcoming.

a
black-gloss dish
[FF-80-185PL]
0 2 cm

b
cooking pot
[FF-80-162L]
0 2 cm

c
undecorated two-handled jug
[FF-80-1P]
0 4 cm

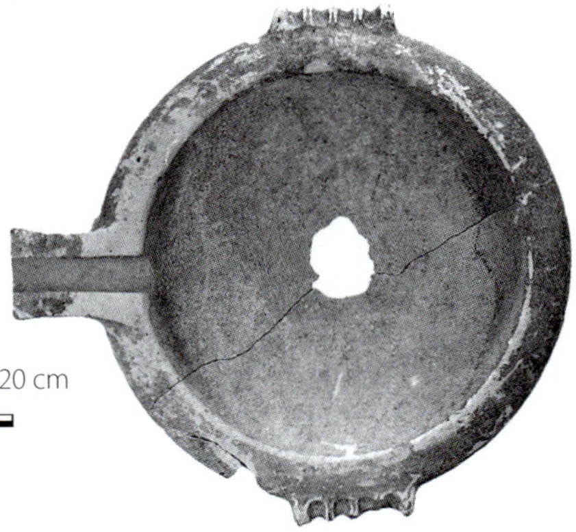

d
mortar
[FF-80-251S]
0 20 cm

e
terracotta
loom weight (inscription: "IN")
[FF-80-195T]
0 5 cm

Figure 8.8 Objects from the excavation of Fattoria Fabrizio: a) black-gloss dish, b) cooking pot, c) undecorated two-handled jug, d) mortar, found undisturbed in situ (see Figs. 18.7 and 18.11), e) pyramidal loom weight, 6th c. BC, with incised letters.

Figure 18.9 Fragment of votive terracotta plaque representing a young woman with a sheep on her shoulder from Fattoria Fabrizio, most likely located just inside the western half of Room III or outside the external wall before it was smashed into fragments that were then scattered well into the area of "Room VII."

a
undecorated krateriskos
[FF-80-57P]
0 2 cm

b
banded ware one-handler
[FF-80-189PA]
0 2 cm

Figure 18. 10 Miniature votive krater and banded ware one-handler cup from Fattoria Fabrizio.

pithoi and amphorae and countless tile fragments—suggests that some families may have abandoned their houses without removing the most portable, immediately useful (*Fig. 18.11*), or even most expensive items, and that scavenging was relatively rare.[55] The reasons for a hurried departure may have been fear and uncertainty in the troubled period of the 4th and 3rd centuries BC, but other explanations are possible as well. The evidence of life and death in the chora in any case proves that the isolated farmhouse was a reality at Metaponto.[56]

The rare documentary sources, inventories from farmhouses widely scattered in time and place from all over the Greek world, are an independent source

[55] See Pettegrew 2001, 202–03.
[56] And not as some have claimed "the conceptual child of settlement archaeology." Pettegrew 2001, 190.

Figure 18.11 Cooking pot (chytra) found under collapsed roof tiles at Fattoria Fabrizio [FF-80-166P]. (JCC)

of evidence of rural life. Most are from mainland Greece, but they should be at least broadly applicable to the Metapontine situation. The published inventories of the Attic and Thasian properties of the Hermakopidai,[57] seized by the Athenian state in 415/414 BC to be sold at auction, list numerous items that could arguably have been found on any Greek farmhouse. Among the dry foodstuffs were almonds, coriander, barley, several varieties of millet, wheat, lentils, sesame, figs, and bitter vetch; among the moist items were olives, olive oil, grapes, wine, and vinegar. Chaff and husks were also present, presumably to be used as fodder for livestock. Storage of many of these substances would have required ceramic vessels. Certain dry foodstuffs, such as the cereals, could have been conserved in sacks or baskets. Many are listed as the contents of various ceramic vessels, particularly amphorae, stamnoi, and pithoi. Among the numerous metal and wood agricultural tools were bands for binding sheaves, hoes and mattocks, pruning-hooks, pitchforks, harrows, winnowing-shovels, drills and augers, and vine stakes or props; even a millstone, clearly a valuable item, was destined to be auctioned. Empty, perhaps decorative or domestic vessels included Chian, Eretrian, and Panathenaic amphorae, hydriai, jars, and kraters (some of the latter may have been metallic, though usually such objects are described explicitly as of bronze, etc.).[58] Several kneading troughs, both ceramic and stone, were also listed, as well as stone troughs which may have been employed in pressing.[59] And there was much more generic material indicative of heavy labor (pulleys, planks, spare handles, beams, etc.).

All of this material, whether perishable or not, would have required significant storage space. Some of this space may have been made available within the confines of the residential structures on the various farms, while more space may have been created by constructing dedicated storage sheds (see below) or annexes to the main farmhouse. Such material and artifacts could have been reasonably expected at a functioning Greek farmhouse of the early Hellenistic period, when many of the Metapontine farmhouses were active. Some of them, those made of ceramic and stone at least, should be present in the debris scatter of an abandoned and plowed farmhouse even after two and a half millennia.[60]

Other Agricultural Structures

Thirty-five sites listed in the Gazetteer have been designated "Other Agricultural Structures" because, in the judgment of the survey team, they did not meet the minimum requirements, either in terms of size or composition of the scatter, for identification as farmhouses. The other possible functions that nonresidential structures may have served are suggested by the documentary evidence and more securely identified structures on rural sites in other chorai.

Animal and Produce Sheds

The variability in the spread, density, and the contents of artifact scatters in survey results from within regions and from different regions of the Greek world has recently prompted reconsideration of the range of rural structures present in the ancient countryside. Storage sheds and animal pens, among others, are recognized site types either used in conjunction with, or in lieu of, temporary or permanent residences.[61]

[57] Pritchett 1953; 1956; Arangio-Ruiz and Olivieri 1965, 421–430.

[58] Arangio-Ruiz and Olivieri 1965, I^3.422 (Pritchett Stele II) lines 18–21, 41–60. A 4th–3rd-c. BC inscription from Tenos (Arangio-Ruiz and Olivieri 1965, XII/5.872.54–55, 94) recording acts of sale of various rural and urban properties in three cases specifies that the ceramic items contained within the structures on the properties were to belong to the new owners. Presumably the ceramic items in question consisted mostly of fine-gloss and plain table wares, as well as cooking, food-processing, and storage wares.

[59] Arangio-Ruiz and Olivieri 1965, I^3.422 (Pritchett Stele II) lines 9-12, 32-36, 118-119, 189, 270-73, 306; Arangio-Ruiz and Olivieri 1965, I^3.425 (Pritchett Stele V) line 31. See Foxhall 1993, 198; 1997, 261.

[60] Kent 1948.

[61] Pettegrew 2001, 192. Whitelaw's (1994, 168–69) ethnoarchaeological evidence demonstrates that, in modern Aegean contexts at least, rural structures formed a continuum as much as a simple set; one type of structure could be transformed into another through addition or subtraction of space.

There is significant literary and epigraphic evidence for the existence of these small, non-residential rural structures. The author of the pseudo-Aristotelian *Oeconomica* summarizes a farmer's concerns about non-residential structures:

> Οἰκίαν δὲ πρός τε τὰ κτήματα ἀποβλέποντα κατασκευαστέον καὶ πρὸς ὑγίειαν καὶ πρὸς εὐημερίαν αὐτῶν· λέγω δὲ κτήματα μέν, οἷον καρποῖς καὶ ἐσθῆτι ποία συμφέρει, καὶ τῶν καρπῶν ποία ξηροῖς καὶ ποία ὑγροῖς, καὶ τῶν ἄλλων κτημάτων ποία ἐμψύχοις καὶ ποία ἀψύχοις καὶ δούλοις κα ἐλευθέροις, καὶ γυναιξὶ καὶ ἀνδράσι καὶ ξένοις καὶ ἀστοῖς.
>
> In constructing a homestead, we have to provide for the stock which it is to shelter, and for its health and well-being. Providing for the stock involves questions such as these: What type of building is best for the storage of crops and of clothing? How are we to store the dry crops, and how the moist ones? Of the other stock, how is the living to be housed, and how the dead? and what accommodation are we to make for slaves and free, for women and men, for foreigners and fellow-citizens?[62]
>
> Aristotle, *Oec.* 1345a24–30

The Herakleia Tablets contain a wealth of evidence that may be especially useful in elucidating the Metapontine situation, given the proximity of the two poleis. In Tablet I.138–144, among stipulations to which the lessees of the lands sacred to Dionysos were to adhere, is a provision for the construction of three spaces by each lessee of the first three plots:

> οἰκοδομησῆται δὲ καὶ οἰκίαν ἐν τοῖς χώροις τούτοις, βοῶνα, μυχόν, ἀχύριον· τὸν μὲν βοῶνα τὸ μὲν μᾶκος ϝίκατι καὶ δυῶν ποδῶν, τὸ δὲ εὖρος *h*οκτὼ καὶ δέκα ποδῶν, τὸν δὲ ἀχύριον μὴ μεῖον τὸ μὲν μᾶκος *h*οκτὼ καὶ δέκα ποδῶν, τὸ δὲ εὖρος πέντε καὶ δέκα ποδῶν, τὸν δὲ μυχὸν πέντε καὶ δέκα ποδῶν παντᾶι. Ταῦτα δὲ παρεξόντι οἰκοδομημένα καὶ στεγόμενα καὶ τεθυρωμένα ἐν τοῖς χρόνοις ἐν *h*οῖς καὶ τὰ δένδρεα δεῖ πεφυτευκῆμεν· αἰ δὲ μή, κατεδικάσθεν πὰρ μὲν τὸν βοῶνα ϝὲξ μνᾶς ἀργυρίω, πὰρ δὲ τὸν ἀχύριον τετόρας μνᾶς ἀργυρίω, πὰρ δὲ τὸν μυχὸν τρῖς μνᾶς ἀργυρίω.
>
> And he will build on these lands a structure, (consisting of) a cow shed, a storage shed, and a chaff shed: the cow shed (must be) 22 feet long and 18 feet wide, the chaff shed not less than 18 feet long and 15 feet wide, and the storage shed 15 feet on all sides. They will present these structures built, roofed, and equipped with doors by the time in which the trees must be planted; if not, they are to be fined 6 minae of silver for the cow shed, 4 minae for the chaff shed, and 3 minae for the storage shed.[63]
>
> Herakleia Tablet I.138–144

The wording of the text is at first glance ambiguous: four structures are named (οἰκία, βοών, μυχός, and αχύριος), but the dimensions of only three are specified. The absence of a specified fine for failure to build the οἰκία, however, indicates that only three spaces were intended, probably to be built side-by-side as a single entity[64]—the οἰκία—for convenience or to reduce the cost to the lessees. By constructing the three as a single unit, four exterior walls could be spared, as well as a certain amount of timber and roof tiles. This arrangement also reduced the loss of surface area in the property to a minimum.

Instructions are not given as to where within the sacred plots the structures are to be built. Given the care and expense invested in recovering and redefining the sacred properties, the high value of the crops to be planted there, the variability of the property's quality (alternating arable, non-arable, and scrub thicket, I.18-20ff.), and the general profusion of rules and regulations in the contract, it is odd that the Herakleian authorities should have failed to specify or restrict particular locations for the construction of permanent structures by forbidding construction in the arable areas. Since these three units would likely have served future lessees, the demos and its agents might reasonably have been expected to take an interest in their location.

The structures were all specific to agricultural functions consonant with Herakleia's well-known agrarian orientation.[65] The cow shed was clearly a

[62] Text and trans. G.C. Armstrong 1935 (Loeb).

[63] Text from Arangio-Ruiz and Olivieri 1965, XIV.645. Trans. A. Prieto.

[64] As interpreted by E. Greco (2001, 191–93). Osborne (1985a, 124) misreads the passage as referring to four structures, rather than three. M. Brunet (1992, 41–42) has demonstrated the same meaning in the single instance of οἰκία among the Delian lease inscriptions.

[65] In an interesting coincidence, many of the more than 500 modern non-residential agricultural structures ("field shelters") documented on

house for the four-legged "tractors" used to plow the fields. The chaff shed was presumably a space for the storage of straw, used for bedding or possibly hay, to be fed to the oxen housed next door. The storage shed might have contained the produce from the plots or agricultural tools employed in farm activities.

Structure	Greek Feet	Meters
Herakleia Tablet: cow shed	22 x 18	6.5 x 5.4
Herakleia Tablet: chaff shed	18 x 15	5.4 x 4.5
Herakleia Tablet: storage shed	15 x 15	4.5 x 4.5
Typical Metapontine farmhouse	51 x 51	15.0 x 15.0
Dema House, Attica	85 x 58	25.0 x 17.0
Vari House, Attica	60 x 46	17.6 x 13.7

Table 18.2 Comparison of rural structure sizes.

Equivalent structures of similar dimensions were doubtlessly present in the chora of Metaponto, located 20 km to the north (*Table 18.2*). A comparison of the size of these structures with farmhouses of comparable date from Metaponto and elsewhere makes clear that they were too small to ever be confused with a farmhouse. They may well be represented in the archaeological record by the so-called Greek Agricultural Structures (GAS), identified in the Metaponto survey. They are characterized by their small dimensions and sparse pottery assemblages.

The Dema house in Attica is comparable in size to the average excavated Metapontine farmhouse, while the Vari house is notably smaller.[66] The Herakleian cow shed, in turn, is only a quarter of the size of these structures, and the chaff shed and storage shed cover even less area. They were likely built using the same technique and materials employed in the farmhouses. Cattle sheds may well have also been an integral part of some farmhouses as appears to be the case with Fattoria Stefan at Metaponto.[67] The lessees were to equip these structures with doors and roofs, making them secure and fully functional, and thus the remains of storage facilities like these would appear roughly similar to the remains of a farmhouse: a scatter of stones, tiles, and clay. Some farmers may have housed surplus containers, cooking or table wares (for meals in the field?) in their sheds; in this case fragments of pithoi, amphorae, and plain table vessels may be expected in the artifact scatter.

Herakleia Tablet I.71 mentions cheese-pressing facilities (τυρεΐας) built on low rises in the Akiris (modern Agri) River valley. Such locations make sense: milk is highly sensitive to heat and relatively difficult to transport, requiring metal or ceramic pails or skins; also, the coagulant agents were readily available from the flocks themselves and native flora. Dairy-processing would have taken place preferably in shaded areas in close proximity to pastures or folds—as is done in southern Italy today—to minimize transport time.[68] The presence of these facilities presupposes the raising of sheep and goats in the chora of Herakleia. In view of the large quantities of sheep and goat remains on every excavated site,[69] the chora of Metaponto was almost certain to have had extensive flocks and therefore may have been equipped with dairy-processing facilities as well.

Cheese is a convenient method of converting a considerable part of the constituents of milk into a product that will keep well, is less bulky as it contains much less water, is of high nutritive value, and is both palatable and readily digestible.[70] It could be preserved even longer by storing it in containers filled with brine, much like the Greek feta consumed today.[71] Much of the equipment used in cheese production was likely to have been portable and made of perishable materials: skins for transporting the milk, or woven fiber strainers for separating the curd from the whey,[72] wooden molds and shelves[73] for storage of hard cheese. Equipment that might have survived should include large vessels, perhaps pithoi, for conservation of briny cheeses and ceramic or bronze strainers such as those found at the Fattoria Fabrizio farmhouse (*Fig. 18.12*) and also in a burial in the Pantanello necropolis.[74]

the Aegean island of Keos are sheds for animals and storage sheds for straw and equipment (Whitelaw 1994, 168).

66 Adamesteanu 1967b, 46; for the Dema house, see Jones/Sackett/Graham 1962, 76; for the Vari house see Jones/Graham/Sackett 1973, 362. Cf. Carter 2006a, 146, Table 4.C.

67 Carter 2006a, 144.

68 Isager and Skydsgaard 1992, 91; J.G. Davis 1965, 1:3-6.

69 Carter 2006a, Ch. 1; Bökönyi and Gál 2010, Ch. 1.

70 J.G. Davis 1965, 1.3.

71 Tamime/Dalgleish/Banks 1991, 11–12.

72 As in Homer, *Od.* 9.247.

73 Carter 2006a, 138–40.

74 Prohászka 1998, 819.

Figure 13.12 Fragment of a bronze cheese grater from the excavation of Fattoria Fabrizio [FF-80-211M].

The Herakleia Tablet (II.44–51 and 75–80) mentions two generic structures (οἰκίαι) as landmarks in the chora sacred to Athena. One of the duties of the Heraklcian boundary magistrates was the clearing of paths to these structures, presumably to facilitate access by the lessees.[75] These were likely simple sheds or shelters rather than houses, containing a single space (unlike the complex three-space οἰκία of Tablet I), constructed with the typical stone socle, mudbrick walls, and tile roof. Other lease inscriptions from the Classical period mention equivalent structures, often with the provision that the lessee maintain the integrity of the roof and walls.[76] An οἰκία may have contained a few tools and ceramic containers such as coarse-ware jugs, amphorae, and pithoi.

Demosthenes lumps all the structures on Phainippos' large estate at Kytheros into a single generic category, calling them simply οἰκήματα,[77] in some of which were stored "grain, wine, and other things."[78] This contrasts with Herakleia Tablet I, which is explicit about the separation of functions for structures and provides detail on specific agricultural activities.

The inventories pertaining to the estates of Delian Apollo list a series of rural structures, many of which correspond to those mentioned in the Herakleia Tablets. A cow shed (βούστασις) and a chaff shed (ἀχυρών) are listed, as well as a granary (σιτοβολών), a pithos shed (πιθών), a millhouse (μυλών), a sheep-pen (προβατών), and a "cookhouse" (ἰπνών, perhaps a simple bread oven, or a separate kitchen for slaves).[79]

Four to eight sites in the southern Argolid were identified as possible storage structures of the Late Classical and/or Hellenistic periods and were described as follows:

> Field buildings are similar to farmsteads in some respects, though perhaps smaller in size. Architectural remains may be present, but it is assumed that the assemblage of artifacts will contain a smaller range of 'domestic' types than those of the other categories of settlements. . . . Field buildings . . . are very small and isolated sites with only a very limited range of artifact types. These field buildings may be the remains of seasonal houses, storerooms, or animal folds.[80]

Some uncertainty exists regarding the precise number because of either ceramic chronology or site size and structural remains. All were classifiable among the smallest sites documented according to the criteria set forth by the investigators (i.e., area less than 0.3 ha); in the same period a minimum of 64 and a maximum of 87 farmsteads were identified.[81] These figures produce a minimum storage structure/farmstead ratio of 1:8, and a maximum of 1:22, both indicating that storage structures were relatively rare in the southern Argolid in pre-Roman periods.

Paleobotanical and palynological evidence from the sanctuary at Pantanello in the Metapontino documents the presence of the Mediterranean quartet—olive, vine, cereals, and legumes—in the chora from the earliest phase of its Greek occupation in the late 7th century BC onwards.[82] The Herakleia Tablets prove that olives, vines, and barley were still being successfully cultivated on a large scale on the floodplains of the neighboring chora to the south during

[75] The same generic term appears in an Amphictyonic account of leases of Apolline estates on Delos dated ca. 432 BC, recovered at Athens (Arangio-Ruiz and Olivieri 1965, I2 377.16; see also Kent 1948, 258–59). There is no information on the nature of these structures.

[76] E.g., Arangio-Ruiz and Olivieri 1965, XII/5.568.11-13 (Keos, 5th–4th c. BC), XII/7.62.25-26 (Amorgos, 4th c. BC), II-III2/1.1241.17–18 (Attica, 300/299 BC); II-III2/2.2498.22–24 (Attica, 321/320 BC).

[77] Dem. 42.2, 6, 8, 19, 26, 30.

[78] Dem. 42.19, 30.

[79] Kent 1948, 292–95.

[80] *Greek Countryside*, 249–50.

[81] The storage sites are B65–66, B71–72, B80, B98, F55, and G14. *Greek Countryside*, 251–52.

[82] See Carter 2006a, 23; 1990, 419–22; 1987, 183, 191; Costantini 1983; Sullivan 1983. The grape seeds from the Pantanello sanctuary's collecting basin, dated to the late 4th c. BC, were recently subjected to an original morphometric analysis using wild and cultivated varieties for comparison, with the unequivocal result that they were cultivated (Costantini and Biasini Costantini 2003, 9–10). However, palynological evidence for the presence of the vine at the Pantanello sanctuary in the very late 4th and early 3rd c. BC is so diminished as to suggest that cultivation of the vine was very limited, and perhaps even eradicated by the mid-3rd c. BC, as perhaps generally in the Metapontino. The presence of olive was substantial by the mid-4th c. BC but then diminished rapidly by the end of that century, disappearing from the pollen record almost entirely by the mid-3rd c. BC. See Carter 1990, 420–22; 1987, 183, 191–93.

the 4th century and beginning of the 3rd century BC. The farmers would have needed specialized pressing and crushing equipment to realize the full value of their produce. Evidence for the tools and spaces used to process the crops is remarkable for its almost complete absence in the surface finds from the survey, and its rarity among the finds from excavated farmhouses at Metaponto.

Iron sickles and plowshares, indispensable tools of cereal culture, have been recovered from excavations of both farmhouses and sanctuaries.[83] Many of the implements involved in handling the grain, sacks or baskets, and other characteristic furnishings, such as ladders and lofts for keeping dry foodstuffs elevated and dry,[84] were surely wooden and therefore unlikely to have left tangible traces in the archaeological record. Threshing floors are particularly ephemeral, silos were often buried, and granaries made especially for the purpose are rarities before the Roman period. Their absence can be explained by the perishability of the material from which these tools would be constructed or their intrinsic value as raw materials to be reused.[85] All three, however, have been identified in the archaeological record at other Greek sites, but surprisingly none of them so far at Metaponto.

Threshing Floors

In the passage where he mentions the οἰκήματα on Phainippos's estate Demosthenes also mentions two threshing floors (ἁλῶνες).[86] Each one is at least a plethron in extent (probably the linear measure of the perimeter rather than the area), a quantity and size apparently considered excessive and therefore indicative of the owner's wealth and the vastness of his holdings. Although Demosthenes does not describe these in any detail, they were likely constructed of packed stones.[87] Several apparently ancient threshing floors, measuring as much as approximately 20 m in diameter, were documented in the Sounion district of southern Attica by J. Young in the 1930s and 1940s. They were defined by a ring of stones and a packed stone surface, and one was partially terraced into the bedrock.[88] Several more, reaching up to 22 m in diameter, were identified by Lohmann in the *Athene deme* in southern Attica.[89] Given the abundance of cobbles and clay at Metaponto, and the loose, sandy/gravelly nature of the surface on the terraces, it is likely that threshing-floors at Metapontine farmhouses were constructed of cobbles packed in a clay matrix. Judging from the Attic evidence, threshing floors were typically sited near the tops of hills and on the edges of minor eminences where the wind required to separate grain and chaff would have been unobstructed.[90]

Mills

On the grounds of the Masseria del Concio in the chora of Herakleia were found the remains of two curious, cylindrical structures only 3 m apart. Judging from the incomplete extant remains, both were built of squared stone blocks. The preserved diameters are different (ca. 13 and 10 m), and the larger western structure had a second, inner stone ring of ca. 3 m in diameter. Within and around this site were recovered a terracotta figurine fragment and several terracotta antefix fragments, suggesting a cult place.[91] Greco has recently interpreted one of these structures as a mill, composed of a small reinforced circular housing for the millstone and a wider circular wall to accommodate and restrain the poor creature (a donkey, or more likely, an ox) that turned the mill. The other structure, he suggests, was a granary.[92]

The survey in the chora of Metaponto turned up a relative abundance of saddle querns—some 35 complete or fragmentary examples—beyond the numerous examples recovered in excavated farmhouses (*Fig. 18.13*).[93] This is evidence that milling in the Metapontino in the Archaic and Classical periods, as in other parts of the Mediterranean, generally took place

[83] Carter 1987, Figs. 263–67.

[84] Columella, *De re rustica* 1.6.9-10. See also Varro, *Res rusticae* 1.57.

[85] Grape processing required wicker baskets, wooden pressing boards, and ceramic jars. Treading vats of wood or ceramic also played their roles, but the ceramics would have survived though no distinctive ceramic vessel exclusively used for wine production has been found in any of the excavated farmhouses or in any of the assemblages from the sites in the survey. See Isager and Skydsgaard 1992 56–57; Vandermersch 1994, 35–36.

[86] Dem. 42.6.

[87] Packed earth is another common construction technique for threshing floors in Mediterranean societies (Burford 1993, 117). Columella (*De re rustica* 1.6.23–24) recommended stone threshing floors because the grains separate from the hulls more easily and rapidly when the treading surface does not give way beneath the weight of the treading animals, which would allow the grains to be rammed into the ground. On threshing floors generally, see Isager and Skydsgaard 1992, 53–55.

[88] Young 1956, 124–26, 140.

[89] Lohmann (1992, 39, 42, 48) notes that such large threshing floors were probably worked by oxen, perhaps drawing sledges.

[90] See the numerous site maps in Young 1956 and Lohmann 1992.

[91] Osanna (1992, 107) presents a history of these structures.

[92] E. Greco 1995, 463–65.

[93] Adamesteanu 1973, 56.

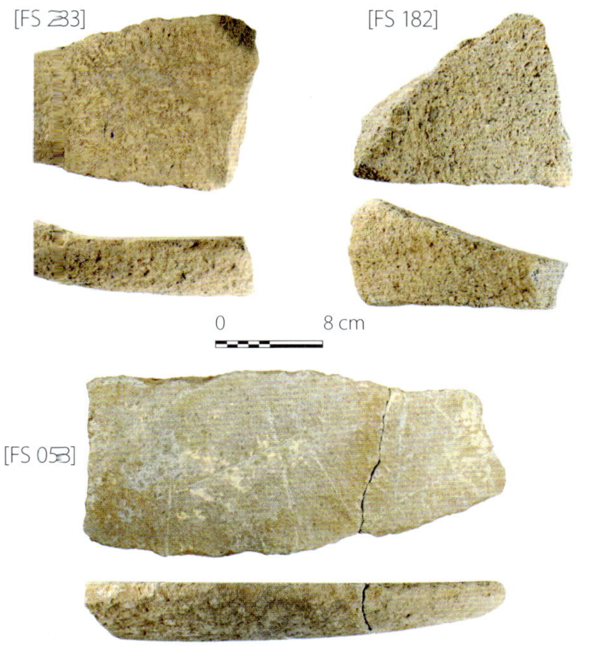

Figure 13.13 Three saddle quern fragments, viewed frontally and in profile from the field survey.

in the individual household, which would have prepared its flour throughout the year as needed.[94] Grain could also have been ground in mortars, or at least hulled in preparation for grinding on a saddle quern.[95] There is no evidence from the chora of Metaponto for the existence of hopper-rubber mills actuated by long handles, many examples of which were recovered at Olynthos.[96] The presence of mills serving crowded population centers, and their absence on isolated farmsites makes sense, as a mill represented a significant economic investment. They might well turn up in the yet-unexcavated urban center of the city.

The form of the saddle querns changed little over time. Most of those documented in the chora of Metaponto were found at sites occupied in multiple periods (e.g., sites 82, 182, 413, 521), appearing at Pantanello as early as the Neolithic period. Its earliest documented use in the Greek chora perhaps is at the exclusively Late Archaic site 523. The position of the saddle quern as the dominant flour-milling tool throughout Metapontine history is corroborated by sites 42, 83, and 360—all occupied in the late Classical through early Hellenistic periods—and site 95. Saddle querns and mortars were versatile tools used also to grind legumes and other hard food.

Most of the saddle querns noted in the survey were made of volcanic stones such as basalt. Such stone is not found in the immediate vicinity of Metaponto but was imported from the Monte Vulture region of northern Basilicata and as far off as the area around Vesuvius.

Silos

Silos cut in the rock, with associated pottery of the Hellenistic period and of the 1st and 2nd centuries AD, were noted in the chora of Herakleia (Piano del Forno at Trisaia). It was thought that they might indicate the location of the port of Siris.[97] Structures interpreted as grain-storing silos or sheds have been amply documented in the chora of Emporion in northeastern Spain in the 5th and 4th centuries BC. They are typically associated with population centers and are occasionally found in tight groups of as many as a dozen. Palynological and paleobotanical records have been cited as proof of the predominance of cereals in the agriculture of Emporion. Taken together, the evidence points to an intensive cereal culture oriented toward export.[98]

Metaponto's heavy investment in cereal culture[99] would have required storage facilities, but no freestanding silo, either above ground or below, has ever been identified. In the extensive excavations of the urban center, no central repository occupying the agora or adjacent areas has been postulated, such as the granary of Hieron II at Morgantina.[100] The ancient port of Metaponto was a transit point for great quantities of grain, but the best evidence, coming from amphorae containing carbonized cereals, found in a magazine in the port of Metaponto, dates to the Late Roman period.[101]

[94] Foxhall 1993, 193. On the saddle quern in Greek grain processing see Moritz 1958, 36–41; Amouretti 1986, 138–40.

[95] Moritz 1958, 23–28, 42–61, 103–21; Amouretti 1986, 135–37, 140–47. Moritz dated the invention of the lever mill to at least the 6th c. BC, on the basis of hopper-rubber mills discovered at sites occupied in Archaic times, such as Kaulonia, south of Croton (Orsi 1916, 730–31). Many of these sites, however, were occupied for centuries afterwards or reoccupied sporadically, so that an Archaic date cannot be assumed. Pieces of both hopper-rubber and rotary mills were recovered in Laconia, along with saddle querns (*Laconia* II, 177–79), though the dating is at variance with the arguments presented here. Fragments of hopper-rubber mills also appear at Late Classical/Hellenistic sites in the southern Argolid (*Greek Countryside*, 448 for site B33, 449 for site B36).

[96] Foxhall 2007, 152.

[97] Quilici 1967, 123–140, sites 57–67.

[98] Plana-Mallart 2001, 550; 1994, 31. Athens appears to have been one of the major trading partners.

[99] Carter 2006a; Costantini and Biasini Costantini 2003.

[100] Deussen 1994.

[101] L. Giardino 1983.

Figure 18.14 Press or millstone found at site 42.

Figure 18.15 Remains of a buried pithos found during the excavation of the 3rd–4th-c. AD farmhouse site at San Biagio.

Presses and Pithoi

Grape processing required wicker baskets, wooden pressing boards, and ceramic jars. Treading vats of wood or ceramic also played their roles.[102] The ceramics would have survived, though no distinctive ceramic vessel exclusively used for the wine production has been found in any of the excavated farmhouses or in any of the assemblages from the survey sites.

A grape or olive press was discovered in the Late Republican villa at Termitito in the chora of Herakleia.[103] At Metaponto survey site 42, a round conglomerate stone, measuring 80 x 60 cm with a rounded boss in the center, was found (*Fig. 18.14*). This object, probably quarried in the vicinity of Montescaglioso, may have been part of a press, or arguably part of a Roman rotary grain mill of the 2nd–1st century BC.[104]

Structures characterized mostly by heavy storage wares, such as amphorae and pithoi, as well as ceramic roof tiles were likely storage sheds housing wine or oil. An Attic inscription from the late 4th century BC describes an unnamed rural structure apparently used for functions related to grape growing and wine making. It accommodated 80 buried pithoi and 4,000 wooden vine-props or stakes.[105] The nine intact and fire-damaged pithoi belonging to a leased property mentioned in a 3rd-century BC inscription from Keos may have been housed in a similar structure, though for the "country estates" of Chersonesos such a number of pithoi was not excessive and would have been comfortably housed in the central fortified tower of the residential-industrial farmhouse unit.[106]

[102] Isager and Skydsgaard 1992, 56–57; Vandermersch 1994, 35–36.

[103] De Siena and Giardino 2001, 156.

[104] Amouretti 1986, 144–47.

[105] Vanderpool/McCredie/Steinberg 1962, 54–55 no. 138.

[106] Carter 2006a, 152–53.

According to Columella, wine had to be kept below ground level in order to maintain a relatively even temperature.[107] Buried pithoi or floor levels below ground ought to be criteria for identifying such buildings. Metapontine colonial farmhouses had rooms with large numbers of amphorae and buried pithoi in the Late Roman period (*Fig. 18.15*).[108] It seems likely, therefore, that since no freestanding structure exclusively filled with wine-storage vessels has been identified in any excavation, wine was stored in the residence of the family farm.

The neighboring chora of Taras, unlike Metaponto, has not been surveyed intensively, but oil production has been documented at a Greek farmhouse near Grottaglie.[109]

Conclusions

Despite the documentary, iconographical and paleobotanical testimony, there is strikingly little archaeological evidence of the tools used to process these crops—grape or olive presses or millstones—in the Metaponto area.[110] It could be that much of the wine making and olive growing was intended for local consumption, and did not require these elaborate tools or numerous transport and storage vessels that are

[107] Columella, *De re rustica* 1.6.9.

[108] See Carter 2006a, Ch. 4 for Fattoria Fabrizio (4th c. BC) and San Biagio (4th c. AD). See also Nava 2002, 735–36.

[109] Andreassi 2001, 1027–28. Compare the recently discovered Archaic Greek oil facility at Klazomenai (Koparal and İplikçi 2004). On grape and olive presses as well as olive crushers generally, see Amouretti 1986, 162–75; Isager and Skydsgaard 1992, 60–65; Foxhall 1993; Frankel 1993; Foxhall 2007.

[110] E.g., grinding stones from the ancient farmhouse in contrada Portico near Montescaglioso on the north bank of the Bradano (date unknown—Adamesteanu 1974a, 88); traces of a grape or olive press in the late Republican villa at Termitito in the chora of Herakleia (De Siena and Giardino 2001, 156); and Metaponto survey site 42, mentioned above.

Figure 18.16 Neck and handles of a Metapontine Greco-Italic amphora from the excavations of the kiln deposit at Pantanello (h 24.3, w 30.2). This ware type was widely distributed in the chora of Metaponto from the late 4th to the 2nd c. BC, the probable date of this example. (BB)

not much in evidence in the assemblages from farmhouses. Amphorae and pithoi are found at every site, but their numbers do not suggest export industries. The fundamental importance of cereals, however, for the Metapontine economy is not in doubt. After harvesting and threshing, the grain would have been shipped and manufactured into flour at its ultimate destination. So it is not surprising that large mill stones are not in evidence. The modest saddle querns would have sufficed for home consumption, and they are ubiquitous in the chora.

Not enough of the Metapontine farmhouses have been excavated to be completely sure that larger establishments like Fattoria Stefan were not involved in wine making in the Early Hellenistic period. Discoveries of transport amphorae of the Greco-Italic type (*Fig. 18.16*), along with the kilns that produced them in the 3rd century BC and later, strongly indicate that wine and probably oil were produced for export at this time in the chora of Metaponto (see Ch. 12).[111] Still, the scarcity of presses and other equipment for the purpose of transforming the grape and olive requires a fuller explanation.[112]

Recognizable grape presses and related basins for fermentation, not to mention numerous amphorae and huge pithoi, are common fixtures of the large farmsteads of the Hellenistic period. As mentioned above, there are many examples in the chora of Chersonesos on the Black Sea. Here wine was a major export crop and wine making was organized on an industrial scale.[113] A widespread transformation of agriculture from the smaller self-sufficient family farm to the larger commercial agricultural ventures characterized the Hellenistic period throughout the Mediterranean world and the Black Sea basin.[114] This can be seen also at Metaponto, and it becomes particularly evident from the last half of the 4th century BC onwards, gaining momentum in the period of Roman domination of southern Italy. It is reflected in the diminution of the number of farms and a corresponding increase in the size and complexity of some of the farmhouse structures.

Reservoirs and Wells

Life in the countryside depended on having a reliable supply of water for cooking and small-scale irrigation, watering for livestock, as well as for human consumption. Reservoirs associated with farms, especially in mainland Greece, must have been far more numerous than the ancient documentary sources would indicate. An inscription from the late 4th century BC is a rare example, which records that a certain Euboean farmer, Chairephanes, contracted with the Eretrian government to drain a marshy lake (λίμνη) at his own expense in return for guaranteed usufruct of the land for a predetermined period and at a predetermined lease price.[115] The lake was to be drained primarily by subterranean canals leading to a constructed reservoir (δεξαμενή), no more than two stadia (1200 Greek feet, approximately 400 m) in perimeter and equipped with a sluice-gate (θύρα) to allow neighboring farmers access to the water reserve for irrigation.

The geological structure of Metaponto (Ch. 1) guaranteed numerous springs, so excavated examples of reservoirs are rarer and the collecting basin at the Pantanello sanctuary is exceptional, but far from unique. A similar structure was excavated in the Sanctuary of Artemis at San Biagio[116] where, as at

111 On the contents of Greco-Italic amphorae, see Vandermersch 1994, 86–87.

112 De Siena and Giardino (1994, 201, 209) have noted the peculiar absence of heavy agricultural processing equipment in the Late Republican period at Metaponto, a period marked by the manufacture of transport amphorae at various sites.

113 Carter 2006a, 151.

114 Carter 2006a, 153.

115 Arangio-Ruiz and Olivieri 1965, XII 9.191.

116 Adamesteanu 1974a, 55–64.

Figure 18.17 Discovery in 1974 of Archaic-period level and materials in the side of a recently cleared irrigation ditch, the future Pantanello sanctuary excavation.

Figure 18.18 Torso of a terracotta figurine of 6th-c. BC date representing a female divinity or offerant, found in the Archaic level in the side of the irrigation ditch.

Figure 18.19 The Pantanello sanctuary seen from the west, area of the Archaic structure and cobble pavement, 1976. In the foreground, the capital of a votive pillar or anta of a temenos wall. (JCC)

Pantanello, it would have provided water for man and beast.[117] At the site of the Pantanello sanctuary, a reservoir would never have been identified by the survey. There was no indication of this reservoir on the surface. It was buried by more than 1 m of alluvial deposits and would not have been discovered except that a modern irrigation ditch (*Fig 18.17*), widened in 1974, revealed a long layer of Archaic Greek pottery and a terracotta figurine of the early 6th century BC (*Fig 18.18*).

The first phase of the basin dates to the first half of the 6th century BC (*Figs. 18.19–21*). Measuring approximately 11 x 14 m, it collected the water from an all-weather spring and was bounded on three sides by a clay embankment. The fourth side on the west was formed by the temenos wall of an adjacent open-air sanctuary which channeled the spring and protected it from encroachment by animals. In one of two channels inside the temenos was a carefully cut rectangular pit intended for filling amphorae and other vessels, examples of which were found in situ. The collecting basin was paved with cobbles. Probably used by both humans and animals, it contained many fragments of vessels, including louteria, and votive figurines and plaques.[118]

Excavations occasioned by the 1999–2002 construction of oil, gas, and water conduits across the chora

[117] Carter 2006a, 163–73.

[118] Carter 1994a, 183–96; 2006a, 164–66.

Figure 18.20 Pantanello sanctuary, 6th-c. BC temenos wall, reutilized in the 4th-c. BC phase of the collecting basin.

Figure 18.21 Pantanello, cobble floor of 6th-c. BC sanctuary.

Figure 18.22 Semi-cylindrical terracotta well liners used as tomb cover, Pantanello necropolis tomb 258.

revealed a number of deeply buried sites. Among these were several cobble-lined basins, apparently reservoirs. They contained neatly arranged pottery of Late Archaic date (ca. 460 BC) and were suggested by the excavator, A. De Siena, as having a probable ritual function.[119]

The literature on wells and water supply and the Greek laws governing them in the asty and chora alike is extensive.[120] There is good reason to believe that wells were widely used in the chora at least as early as the late 4th century BC. The evidence is mostly indirect. Cylindrical well liners in terracotta found a secondary use as tombs in rural necropoleis like those at Pantanello and elsewhere.[121] Three half well-liners were placed end-to-end to create a vaulted structure for the deceased, always an adult, usually female (*Fig. 18.22*). There were 17 of these at Pantanello, and the grave goods in every case dated to the period 320-280 BC. The diameter of the liner was .55 m on the average. Only one example, in the Pantanello sanctuary, has actually been found in situ being used as a well-liner directly over the sacred spring (see *Fig. 25.22*).[122] Dug wells would have made living in areas far from streams or springs, such as Lago del Lupo, decidedly more practical and convenient, and their use may explain why some areas were densely occupied for the first time in the late 4th century BC.

Fortifications

Towers serving as lookouts, among other functions, were a familiar part of the landscape in many areas of the Greek world. The circular foundations of the mill in the chora of Herakleia (see Mills above) recall the stone towers, sometimes freestanding, characteristic of southern Attica, Thasos, and the Aegean basin.[123] As seen in the northern Black Sea region, a tower could also be an integral part of the farmhouse, used for defense and for storing agricultural produce, especially wine.[124] No circular tower, either freestanding or incorporated into a farmhouse, has been discovered in the chora of Metaponto, though a single example was provisionally identified as such.[125] It has further

119 Nava 2003a, 675–76.

120 Wikander and Wikander 2000.

121 See also Fig. 25.25, below.

122 Carter 1998c, 100–01. See also Fig. 25.22, below.

123 See Young 1956; Osborne 1986; also Arangio-Ruiz and Olivieri 1965, XII/5.872.49, 52 (Tenos, late 4th/early 3rd c. BC). The latter of these Tenian towers housed a pithos, perhaps a small cistern or large storage vat rather than a single ceramic vessel, and was roofed with tiles. See also Nowicka 1975.

124 On the Chersonesan farmhouses, see Pečírka 1973, 123–29; Dufkova and Pečírka 1970; Chtcheglov 1992; Saprykin 1994. Carter et al. 2000, 151–56.

125 Adamesteanu (1968, 254; 1973, 57) suggested that the central room of a farmhouse excavated on the Casamassima property (Lago del Lupo locality) was a tower because of the slightly wider socles defining it. See

been suggested that the farmhouse known as Fattoria Stefan had a rectangular tower, but there is no absolute proof that this was the case (see *Fig. 25.18*).

Another type of defense was the fortified town (*phrourion*). Thucydides (2.18) mentions one such example, Oenoe, located on the border between Attica and Boeotia. In the vicinity of Metaponto, there are several possible candidates for phrouria. These would have served in times of conflict as guard posts and forts protecting the chora and hinterland (*proschora*) of the asty. The only site near Metaponto that closely compares to the situation described by Thucydides is Cozzo Presepe, a fortified village on the edge of the chora, just beyond the northwest corner of the study area. Located on a high plateau that dominates the Bradano valley, it commands the view both inland and toward the sea. An indigenous village in the 8th century BC (see Ch. 17), it was surrounded first by a solidly built stone wall at the beginning of the 6th century, then by a wall made of tiles.[126] The 4th-century village, which the most recent excavator has suggested was not Greek but indigenous, was also fortified. It could very well have been part of a defensive system organized by the recently arrived Achaian colonists, but the case for a defensive role for 4th-century Cozzo Presepe is more doubtful.

Adamesteanu put forth a suggestion that two other indigenous settlements, at Pisticci and Pomarico Vecchio, may have, along with Cozzo Presepe, formed a defensive ring around the chora, but this idea has been questioned.[127] The fortification at Pomarico Vecchio, though Greek inspired, protected an indigenous center that operated independently of Metaponto.[128] The site of Bernalda has also been suggested as the site of a possible phrourion even though there is no archaeological evidence to support the assertion.[129]

In the chora of Taras to the north, the recently discovered 7th-century Greek settlement at Amastuolo that replaced an earlier indigenous one has been thought to have been a phrourion.[130] It is similar in many respects to fortified Cozzo Presepe, but also to Incoronata (see Ch. 17), an unfortified, mixed Greek and indigenous settlement that was not a defensive outpost of the Achaian colony and was destroyed either immediately before, or immediately after, the arrival of the new proprietors of the chora.[131]

For the urban center of Metaponto the earliest wall dates to the late 7th or early 6th century BC, and was also replaced by an imposing one in the 4th century BC.[132] By and large, however, both asty and chora were lightly defended, indicating that relations with their Greek and indigenous neighbors were generally good for most of their long history together.

Kilns

Ceramic kilns were a ubiquitous feature of the chora of Metaponto from an early period.[133] Pottery of every sort was a daily necessity. Amphorae and pithoi were required for transport and storage of agricultural products. Although thatch was still used to roof the huts and oikoi of mixed Greek and indigenous settlements at Incoronata greca, which ceased to be occupied about 630 BC, farmers began to cover their homes with tiles in the first wave of new settlers in the late 7th century BC. Locating kilns as near to the source of demand as possible was both convenient and economical, given the cost and difficulty of land transport. Local ceramic production was greatly facilitated by the presence of all-weather springs and sources of very good clay, especially along the Basento valley, where the majority of kilns excavated in the chora are located. Nine of the twelve kiln sites found by the survey have been dated to the Greek period (6th to the 3rd centuries BC). Of these, six overlook the Basento valley, two the Bradano, and only one was located in the interior (Lago del Lupo).[134] One of the two Late Hellenistic period sites (175–25 BC) was found along the Basento, and one in the interior (Demanio

also Carter 2006a, 144, on possible tower associations.

126 Morel 1970a.

127 Adamesteanu 1974a, 76, 91 passim; Tagliente and Lombardo 1985; Osanna 1992, 54.

128 *Pomarico Vecchio* I, 53–54.

129 E. Greco 2001, 176–78; Osanna 2001, 207–10; Guaitoli 2002, 238–40, who distinguishes these sites from phrouria by virtue of their larger size and more ostensibly residential character. For examples, see Osanna 1992, 16–17, 33–38 nos. 45, 70, and 72; Alessio and Guzzo 1989–1990, 368 (Monte S. Elia; cf. also 395 no. 175, a possible watchtower); Alessio 2002, 90–92, 108.

130 Maruggi 2002, 45–54; 1996a: 202–18. Some, such as Dell'Aglio (2002, 23–24) are skeptical of the agglomerated settlement ("village") as a pervasive type in the Archaic period at Taras.

131 Carter 2006a, 51–89.

132 De Siena 1998, 141–51 (see Ch. 20, Fortifications).

133 Carter 2006a, 150. Sites where kilns have been identified: 9, 12, 105, 213, 248, 299, 411, 521, 530 (Greek period) (see Ch. 21); 95, 105 (Roman period); 300 (Medieval period).

134 A 6th-c. BC pottery kiln with a load of B2-type Ionic cups was excavated at a site overlooking the Cavone valley at San Basilio by the Soprintendenza in the 1990s. A. Bottini 1994, 705–06, Pl. 50.

Figure 18.23 *Left*: modern plaster impression from ancient mold, *right*: head of a satyr, Sant'Angelo Vecchio. (CW)

Figure 18.24 Vitrified slag or kiln furniture from survey site 12F/K, a farmhouse with a probable kiln in Sant'Angelo. These fragments are highly visible in surface scatters and attest to ceramic production in the vicinity.

Campagnolo). The Medieval kiln site was deep in the interior (Campagnolo). Kilns were associated with houses from the earliest period. An excavated example is the farmhouse and kiln site at Termitito, just to the south of the Cavone in the chora of Herakleia dated to the late 7th century BC.[135]

In several instances kilns were located in close proximity to sanctuaries. The kiln near the sanctuary at San Biagio was engaged in the production of votive objects and terracotta figurines.[136] The deposit at Sant'Angelo Vecchio contained molds (*Fig. 18.23*), as well as votive plaques and statuettes, which would indicate that a kiln and perhaps a sanctuary were nearby, though neither has yet been located.[137] Extremely well-built kilns of the 2nd to 1st century BC have been found along, near, and even partially overlying previously abandoned Greek sites at Sant'Angelo Vecchio (twin kilns) and the sanctuary at Pantanello (paired large and small kilns). The reason for this overlap is presumably the natural resources—clay and a constant supply of water—that both types of sites would have required.[138]

Misfired pottery (wasters), kiln furniture (spacers),[139] and slag are common surface indicators of the presence of kilns (*Fig. 18.24*).[140] Besides these, other possible clues that a site may have been occupied by a kiln are *tuyères* (bellows spouts for metalwork) and special tiles used in the construction of the kiln itself.[141]

The ceramic kilns of the Archaic period and 4th century BC that have been excavated in the urban center of Metaponto consist of a circular firing chamber (approximately 1.3 to 1.7 m) and short *praefurnium*, all formed of clay.[142] Though reusable, they were expendable, and often replaced.[143] The kiln below the farmhouse at Sant'Angelo Grieco (survey site 1) was a much more solid construction (*Fig. 18.25*). Since no ceramic material is associated, it is possible that the kiln was of Greek date, but it might also have been in use during the latest phases of the occupation of the farmhouse in the 1st century BC to 1st century AD. The kilns at Pantanello and Sant'Angelo Vecchio are also solidly built and could be considered permanent. The firing chamber's perforated floor, made with special tiles and praefurnium, was constructed as a single unit from fired tiles and mortar. The larger Pantanello kiln (see *Fig. 27.11*) was carefully modified to improve its performance.[144]

135 See Carter 2006a, 150, 186 n. 18, for bibliography.

136 Adamesteanu 1974a, 60.

137 Edlund 1986.

138 Carter 2006a, 150.

139 Cracolici 2004a.

140 On the close relations in their origins, at least between ceramic manufacture and metallurgy—both types of pyrotechnology—see C. Giardino 1998, 7. This is a topic for further research and will be discussed in a subsequent volume.

141 Cuomo di Caprio 1992, 32.

142 D'Andria 1980a; 1980b.

143 Silvestrelli forthcoming a.

144 Carter 1983a, 430.

Figure 18.25 Excavation of an undated kiln below the Greek farmhouse 1F, Sant'Angelo Grieco.

Quarries

Despite the absence of good quality stone for building—that is, stone suitably resistant to the elements and capable of being cut or carved precisely—the lithic resources of the chora were nonetheless exploited, especially in the late 7th and first half of the 6th century BC. As noted in the discussion of farmhouse structures, cobbles were transported from river bottoms. Together with the irregular chunks of local conglomerate that underlie the whole chora, they were used for the foundation courses of rural dwellings. The same conglomerate was also quarried and used in structures of greater architectural pretension, including the foundation and perhaps part of the elevation of Oikos C,[145] the earliest surviving cult building in the asty, the temenos walls in the earliest phase of the Pantanello sanctuary (*Figs. 18.20* and *20.33*), and the structure around the spring at San Biagio. Local conglomerate was also used in the nucleus of aristocratic burials in the early-6th-century BC phase of the urban necropolis at Crucinia. With increasing prosperity the buildings in the city, like the peripteral temples and the ekklesiasterion of the mid-6th century BC, used limestone imported from sites in the interior beyond the chora and from the territory of Taras and its non-Greek neighbors. Conglomerate was again in use in the monumental tombs of the wealthy landed aristocracy all over the chora at Pantanello, San Biagio, and elsewhere. These date for the most part to the late 4th and 3rd centuries BC. The particularly unsettled political and military situation in this period may have hindered the commerce in stone, but it did not prevent its use in the new theater and stoa of the asty (Ch. 25).

A stone quarry, with slabs cut and ready for extraction, was discovered during excavation at Pizzica in 1999 (*Fig. 18.26*). This was located at the intersection of a major longitudinal division line road and a transverse road (see Ch. 28). The slabs measured 0.75 x 1.85 m, approximately the size used at the Pantanello sanctuary. Two others, sites 141 (*Fig. 18.27*) and 171, were found with partially finished slabs. The slabs at site 171 measure 1.00 x 1.25 m and 1.00 x 1.50 m These quarries almost certainly belong to the period of the Greek colony and probably to its early years.

In sum, procuring large amounts of stone for the Metapontino required relatively great effort and expense. The nearest large-scale sources were the enormous quarries in the highlands around modern Matera and Gravina and in the chora of Taras, both on the coast and on the Murge plateau.

Sanctuaries

A striking characteristic of the Greek colonial settlements along the Ionian coast of Italy is the distribution of sanctuaries throughout the countryside.[146] These vary in terms of size and importance, and in the chora of Metaponto were particularly plentiful. Roughly 20 sites have been identified as rural sanctuaries, with half of them the object of at least limited excavation. In terms of their architectural significance, they range

[145] Known as Temple C after its 5th-c. BC enlargement with lavish use of marble from Greece. Mertens 1999, 274–75.

[146] Genovese 1999; Cipriani 1996; R. Leone 1998; *Spazio del rito*, passim; Carter 1994a, 180–83; 2006a, 157–73.

Figure 18.26 Excavation at Pizzica of a quarry in an outcrop of local conglomerate (*puddinga*) stone, showing roughed out slabs (see Ch. 28).

Figure 18.27 Site 141Q, probable quarry with roughed out slabs.

from simple farmhouse-like structures to monumental constructions. The most impressive of these is the Doric peristyle Temple of Hera (Tavole Palatine), which was built around 520 BC and is located some 3 km from the city walls, overlooking the Bradano River (*Figs. 16.1, 18.28,* and *22.9*). It is a smaller counterpart of the great temples of the urban center, the Temples of Hera and Apollo,[147] and the foremost of the known examples of rural sanctuaries of more or less monumental proportions in the chora. Most of its two lateral colonnades are still standing in situ.

The second best-known Metapontine sanctuary is that of Artemis and Zeus Aglaos at San Biagio on the Basento. Its figured terracotta frieze, dating to the late 7th century BC, is the earliest architectural element from any site in the Greek chora.[148] There was a structure around the spring; only a few elements and some foundation courses attest to the temple that once stood on the terrace above. In the mid-5th century BC it had, like the Temple of Hera, elaborate terracotta revetments that imitated those of the urban peripteral temples. The votive objects included a remarkable series of 6th-century BC terracotta figures[149]—exotic even by south Italian standards—bronze, and coins reflecting contacts with some of the major poleis of Magna Grecia.

The sanctuary beside the spring at Pantanello, whose discovery was earlier mentioned, is the most fully excavated of the more architecturally ambitious rural sanctuaries yet found. It is the most useful for our purposes here because of its contents, which included a full range of pottery used in rituals and ritual repasts, as well as votive objects such as terracotta figurines and plaques. There were also an iron plowshare, fibulae, loom weights, and, most remarkably, preserved organic materials, such as immature olives, presumably offerings of first fruits, as well as the bones left over from animal sacrifices.[150]

The extensive site of Pantanello—the lower section alone covered some 2000 m^2 along the Basento where the valley joins the coastal plain—closely resembles San Biagio. With its sacred spring and additional collecting basin (see Reservoirs above),[151] its design and organization fully illustrate a distinctively hybrid rural sanctuary form that developed in Magna Grecia. The stoa beside the spring temenos was the only roofed structure in the Archaic phase of the sanctuary. A fine limestone capital from a votive monument (*Fig 18.30a*) was found on the cobble pavement of the 6th-century BC sanctuary.[152]

An oikos replaced the stoa in the mid-4th century BC. It was located just above and to the west of the

[147] Mertens 2006, 216–17, and Ch. 22, 742–43.

[148] Mertens-Horn 1992, 24.

[149] Olbrich 1979.

[150] Carter 2006a, 23–26; 163–73. An inscribed inventory from an Archaic sanctuary in the chora of Taras lists among the sanctuary's possessions: "16 iron (tools): hammers and pitchforks and hoes and six spits for meat, six cleavers and cereal measures and all the ceramic material necessary for the structure" (Lo Porto 1987, 42, 46; Alessio and Guzzo 1989–1990, 363).

[151] A common feature of the known indigenous sanctuaries, e.g., the sanctuary at Chiaromonte (Barra Bagnasco and Russo Tagliente 1996, 186–93).

[152] It is comparable to some of the aniconic rough stone (*argoi lithoi*) votives from the Sanctuary of Apollo in the asty. Adamesteanu 1970b; Doepner 2002; Carter 2006a, 202, fig. 5.7.

Figure 18.28 Temple of Hera (Tavole Palatine), late 6th c. BC (see Ch. 22). (CW)

spring. That is the most likely time when a small temple also rose on the slope above the lower sanctuaries. From this structure remain a foundation and numerous architectural elements, including a triglyph (*Fig 18.30b*),[153] fragmentary capitals, and column drums reutilized in the Roman tile factory that occupied the site in the 2nd century BC. This building, though less impressive, would have formed a pendant with the Temple of Hera (Tavole Palatine).

The monumental Doric capital (*Fig. 18.30c*) found at Sant'Angelo Grieco (survey site 1), some 13 km up the Basento valley, was in all likelihood part of a sacred building erected in the mid-5th century BC. Its location along the upper Basento may signal the furthest extensions inland of the chora proper.[154] It would have

Figure 18.29 Head of a votive figurine in the daedalic style (h 7.8 cm) ca. 630–620 BC from Saldone. Other examples in this style have been excavated at San Biagio in the chora and in the sanctuary of the asty. (CW)

occupied a position analogous to the sanctuary with altar at Cozzo Presepe (see Ch. 20, Fortifications) at the extreme northwest corner of the chora.

Within this roughly symmetrical pattern are distributed the smaller, less prepossessing sanctuaries. These can be identified by the presence of stone or terracotta architectural elements, such as those found at Incoronata, San Salvatore, and Pantanello, usually in combination with votive figurines (*Fig. 18.29*) or miniature vessels. The claims of other sites as sanctuaries are based almost exclusively on votive deposits, such as the sites at Sant'Angelo Vecchio and Crucinia-Favale. Some cases are more problematic and are based on reported evidence no longer available, such as at Avenia (San Vito), Pizzareddo, Acheloos, and Torre di Mare.[155] The last three sites, like Crucinia-Favale, lie just outside the city walls. This list will be expanded, one hopes, as archaeological exploration of the chora continues. Sites in arable farmland are at great risk, but those deeply buried in the alluvial plain, like Pantanello and Pantano on the north bank of the Bradano,[156] hold the promise of extraordinary preservation of both structures and their contents.

Five sites were identified as sanctuaries in the surveyed portion of the chora between the Bradano and the Basento: sites 397, 266, 270, 334, and 477. These were so classified because of the very high numbers either of terracotta votive figurines or of miniature votive vessels compared with other sites. There could be others among the sites provisionally designated as farmhouses. Site 266 could be a farmhouse as well as

153 The triglyph may date to the late 4th–early 3rd c. BC (D. Mertens, pers. comm., 1988).

154 A Sanctuary of Artemis was located at the very eastern edge of the chora of Taras (Osanna 1992, 13–14, 21, 31–32, Nos. 37 and 40). In a much-discussed study, de Polignac (1991) has treated at length the subject of the strategic location of major sanctuaries at important frontiers: e.g., the Sanctuary of Hera at the mouth of the Sele, on the border between the territory of Poseidonia and the area to the north under Etruscan influence in the 6th and 5th c. BC.

155 See Carter 1994a, for the meager bibliography on these tentative sanctuaries. For Pizzareddo, see Giardino and De Siena 1999b, 347.

156 Schojer 2001, 124–25; 2002, 83–84.

Figure 13.30 Architectural fragments: a) anta capital from Pantanello sanctuary (h 20 cm), 6th c. BC; b) triglyph from the kiln deposit, Pantanello sanctuary (h 88 cm), 3rd c. BC; c) Doric capital from Sant'Angelo Grieco (h 35 cm), mid-5th c. BC.

a production site for terracottas and ceramics and not a sanctuary at all. The assemblages of pottery from sanctuaries, as evidenced by known and excavated sanctuary sites, like Pantanello, are not so very different from those of known farmhouses. This is not surprising, as the preparation and consumption of food was a principal activity at both types of sites. The occupants of the farmhouses had their own domestic cults, as is clear from excavated farmhouses such as Fattoria Stefan, replete with votive figurines and vessels, though in limited numbers. It is not easy to judge from a surface collection whether the site was a residence or a place of worship and congregation—and perhaps some sites were all three.

The most common type of sanctuary of all perhaps was the altogether least prepossessing, a votive deposit beside or in a collecting basin, at the base of a marine terrace. It is the Pantanello sanctuary in embryonic form. At least one other example has been identified by De Siena but many others surely lie hidden under the alluvial deposits along the valley edges.[157]

The written evidence for rural sanctuaries in Magna Grecia is not great, but of all the structures of the Metapontine countryside, they alone caught the attention of the poets. Bacchylides immortalized the Sanctuary of Artemis at San Biagio.[158] Leonidas of Taras spoke for all the small rustic springs and their resident gods and goddesses:

Πέτρης ἐκ δισσῆς ψυχρὸν καταπάλμενον ὕδωρ,
  χαίροις, καὶ Νυμφέων ποιμενικὰ ξόανα,
πίστραι τε κρηνέων, καὶ ἐν ὕδασι κόσμια ταῦτα
  ὑμέων, ὦ κοῦραι, μυρία τεγγόμενα,
χαίρετ'· Ἀριστοκλέης δ' ὅδ' ὁδοιπόρος, ᾧπερ ἀπῶσα
  δίψαν βαψάμενος τοῦτο δίδωμι γέρας.

Acqua che fresca scorri dalla rupe
spaccata, salve! Salute, statuette
pastorali delle Ninfe, e voi conche
delle fonti! E salve ancora alle vostre
piccole immagini spruzzate d'acqua,
o fanciulle! Anch'io, il viandante Aristocle,
a voi offro questo corno
che ho immerso qui per togliermi la sete.[159]

Leonidas of Taras (3rd century BC)

157 Osanna 2008, 926
158 See below, Ch. 21, 708.
159 *The Greek Anthology* 9.326, text and translation Quasimodo 1969, 112–13.

19
An Approach to the Chora

Joseph Coleman Carter

The detailed examination of the changing settlement pattern of the chora over six centuries (late 7th–late 1st century BC) is the subject of the remaining chapters in this volume. It is important first to define the key aspects of this research: the term "chora" itself, the populations that inhabited the area, the geographic extent of the chora, and the natural environment and its economic potential. Explanation then follows of how the chora was subdivided for investigation (the study area, regions, and localities), the assemblages, and the procedures used to date sites and establish the thresholds for when sites are considered active and the basis on which their relative importance is evaluated. The reader will also be introduced to the various data tables used throughout later discussions. Some of these subjects are examined in previous chapters (for example, assemblages in Ch. 5). The aim here is to bring all methods and materials together to prepare the reader for the analysis of the historical development of settlement that begins in the next chapter.

Definition of "Chora"

The brief set of definitions that follows is intended to illustrate the various facets of the chora (χώρα) that the Greeks thought it worthwhile to discuss.[1] The term "chora" itself, as it is regularly used in the book, refers to the rural part of the polis that lies immediately outside the asty (the ancient urban space). Chora and asty are complementary parts of a single reality—a reality, however, which is far from uniform or static. The common meaning of the term chora refers to the agricultural land, on which the city depended for its livelihood, and which generated surpluses for trade and profit. This included fields (ἄγροι), the soil of which was suited for plow-based agriculture (ἄρουρα).

At Metaponto, the geomorphological units that contained the highest concentrations of such land were the coastal plain and river valleys—the richest land of all—and the marine terraces of the Central Plateau. This was the pedion (πεδίον), the heart of the chora, corresponding in Metaponto to the Central Plateau. This was the prime arable land, γῆ ἀρόσιμος. It was divided into plots, or kleroi (κλῆροι), which were the properties of individual citizens.

The public act of dividing, the dasmos (δασμός),[2] was closely connected with the creation of a new settlement or apoikia (ἀποικία). Like the polis, this was an organic group of households (οἶκοι), and lots were distributed to heads of families. The dasmos took place in unoccupied (ἔρημος χώρα) or conquered land. The redistribution of previously occupied land was referred to as γῆς ἀναδασμός, and of this there are two possible instances in the chora of Metaponto (Chs. 23 and 25). The possession of a kleros was a necessary condition for citizenship.

The chora also includes land for the growing of grapes and olives that was not necessarily cultivatable, and could be rough; this was referred to by ancient sources as γῆ ἀγροική. But because it was rough does not mean it was necessarily less productive, as Lepore notes. The sloping valley walls with their springs and typical vegetation—the macchia or maquis of the Mediterranean world, or the phrygana of the Black Sea[3]—were particularly suited to pasturing animals, and on the gentler slopes, growing olives. While the good plowland might be mostly privately owned by citizens (ἡ τῶν ἰδιωτῶν χώρα), some portions of the rougher land, particularly along its borders, may also have been owned in common (δημοσία χώρα).

Some portions of the chora were set aside for the gods (ἱερὰ χώρα). This land could be agricultural and was leased out to citizens. Such was the case of the sacred lands of Athena and Dionysos, as described in the bronze Herakleia Tablets. The text of these tablets provides a vivid portrait of the chora of Herakleia, the neighboring colony to the south, in the 4th century BC.[4]

[1] See fundamental studies by Asheri (1966) and Lepore (1968; 1973).

[2] Asheri 1966, 5–13.

[3] Carter 2006c, 31.

[4] Arangio-Ruiz and Olivieri 1965. Uguzzoni and Ghinatti 1968.

Outside the more settled areas of the chora and towards the interior lay expanses of land that were undivided, γῆ ἀδιαίρετος. At what point this became a sort of no man's land, between the Greek settlements along the coast and the indigenous settlements of the interior, is difficult to determine. Its extent, which varied much over time, is sometimes referred to as the eschatia (ἐσχατιά).[5] The pedion was thus the nucleus, and the eschatia, the periphery, with the latter being less secure against external threat.

The Populations of the Chora
Indigenous and Greek Inhabitants

The term chora is sometimes used to more loosely indicate not only the land immediately outside or adjacent to the asty, with its fields and cult places, but also the wider area, which included the eschatia. There the influence of the city was felt, though the laws of the polis may not have applied, and the inhabitants may well have been the indigenous populations of the interior. There are many examples of the co-option of the earliest occupants of the territory throughout Magna Grecia and Sicily. That of Sybaris, which included twenty-five villages and four local indigenous groups or ethne, is among the best documented.[6] Greek influence can be seen in most material aspects of life in the interior. Burial is a partial exception. Indigenous peoples—the Oenotrians in the present instance—continued to bury their dead in the contracted position. It is clear from the results of archaeological research in Basilicata over the last half-century that the influence of Metaponto on the material aspects of culture was felt deep in the interior already in the 6th century BC, soon after the arrival of the Greek settlers on the coast: for example, at the site of an indigenous farmhouse known as Pagliarone (see below, The Farther Chora).

The questions of the ethnic composition and social structure of the chora in particular are serious and complex issues that cannot be given the treatment they deserve in this volume. It will be, I hope, both useful and sufficient to state very briefly the basic facts and assumptions that are treated almost as givens in the chapters that follow.

The few documentary sources[7] state that the immigrant population consisted of Greeks, predominantly of Achaian origin.[8] This is reinforced by the evidence of early inscriptions in the Achaian alphabet,[9] which include that of the late-7th-century BC city wall of Metaponto.[10] The heterogeneous nature of the settlement is no doubt reflected in the unusually varied foundation myths associated with Metaponto.[11]

The sources are virtually silent on the constitution of the Metapontines and their social structure. That there was a landed aristocracy is evident from Bacchylides's ode to the athletic victor, Alexidamos (Ch. 21); this can also be deduced from the archaeological evidence of burial. Strains between oligarchic and democratic factions—a given in the Greek world—were noted early, for example, in Thucydides' brief mention of stasis at the end of the 5th century (see Ch. 24), and are evident in the rare documentary sources of the 4th century (see Ch. 25). There is neither documentary nor archaeological evidence for slavery, though there is evidence for it in Achaian Croton in the 4th century BC.[12] It seems unlikely to have existed at Metaponto before the period of Roman domination in the last two centuries BC.

The physical state and health of the chora and city populations are the subject of ongoing investigations. A first contribution is the study of the burial population in the Pantanello necropolis,[13] which has since been expanded to the urban population from the necropolis at Crucinia and the necropoleis that lined the coastal road from the Bradano to the Cavone, and to the native pre-Greek population of Incoronata.[14] Significant biological diversity between the inhabitants of the chora and those of the city or asty has been demonstrated.[15]

One result of these studies that is directly relevant to conditions in the chora was the discovery that malaria was probably widespread even in the earlier periods of the Greek settlement, and probably before. The population suffered synergistically from hereditary anemia, probably thalassemia.[16] Potentially more life threatening was the treponematosis,

[5] Tagliente and Lombardo 1985.

[6] Strabo 6.1.13; Lepore 1973, 22.

[7] Principally, Strabo 6.1.15 (cited in Ch. 17).

[8] Morgan and Hall 1996, 209–11, and 213.

[9] Dubois 2002, 81–133.

[10] De Siena 1998, 144–45, Fig. 2.

[11] Mele 1998.

[12] Spadea 1984, 144–49.

[13] Henneberg and Henneberg 1998.

[14] Subject of a future study.

[15] Henneberg 1998.

[16] Henneberg and Henneberg 1998, 528.

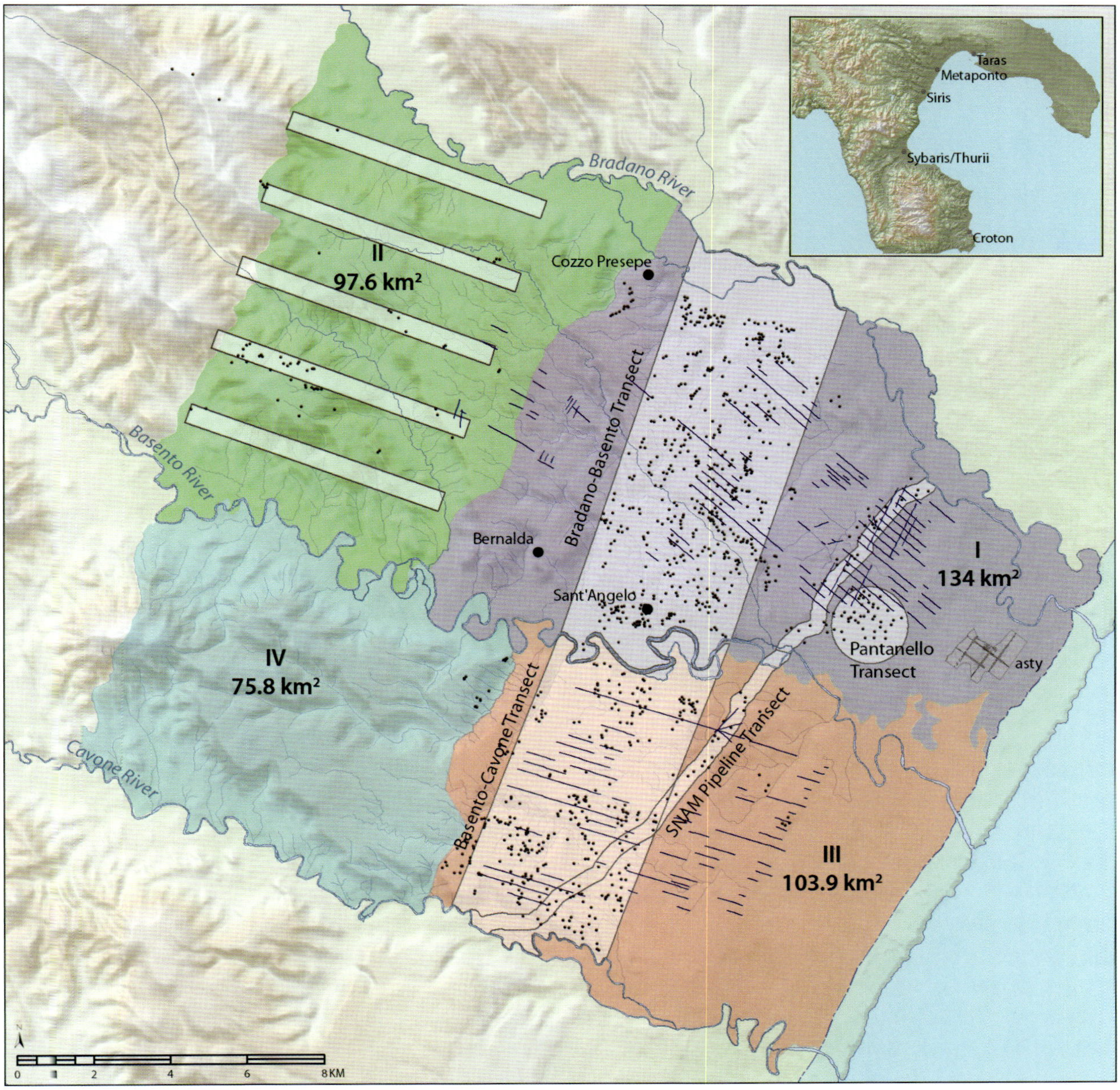

Figure 19.1 Map of the chora, showing division lines, transects, and the zones of the nearer chora (I and III) and the farther chora (II and IV). Inset shows neighboring poleis. (JT)

evidenced by sclerotic thickening of the bones.[17] The chora, in short, was not a particularly healthy place to live. In this respect, it differed little from others along this stretch of the Ionian coast.[18] Average lifespan, nonetheless, was generally longer than that of ancient urban populations, and comparable to some 19th-century peasant populations.[19]

[17] Henneberg and Henneberg 1998, 533.

[18] There is evidence for malaria and treponematosis in the territory of Siris/Herakleia, cited by Henneberg and Henneberg (1998, 528).

[19] Henneberg and Henneberg 1998, 511.

Neighboring Poleis

The expansion of the chora to the north was limited by the territory in the control of the Tarentines (*Fig. 19.1, inset*). Taras, a Dorian foundation of the late 8th century BC, was a little less than a century older than Metaponto. According to Antiochos, the 5th-century BC pro-Tarentine Syracusan historian cited by Strabo,[20] the Tarentines were on the flanks (ἐν πλευραῖς) of the Metapontines. He recounts

[20] Strabo 6.1.15. The entire passage is reproduced with translation at the end of Ch. 17.

that when the Metapontines were at war with the Oenotrians of the interior and the Tarentines, a reconciliation was effected regarding the portion of land that marked the boundary (ὅριον) between Italia and Iapygia. Italia was equated with Oenotria, its earlier name, and it included the Siritide (the chora of Siris) and the Metapontino (Μεταποντίνη). This disputed land was the area controlled by Metaponto and, though the date of the reconciliation is not specified, it was plausibly after the arrival of the Achaian colonists sent by Sybaris, but probably not immediately afterwards. The Metapontines had to have consolidated their hold enough to represent a real but still apparently manageable threat to their opponents. The territory of the Tarentines, Antiochos adds, was contiguous with Metaponto (συνεχὴς τῷ Μεταποντίῳ), and he specifies was "outside of Italia." He calls its inhabitants Iapyges.[21] This northern boundary of the chora of Metaponto was not just a boundary between neighboring rival Greek poleis, but also between different geographical areas inhabited by different native peoples.

Though rivers are frequently mentioned by Strabo as divisions between geographic areas, the Bradano for some reason is not. It has been assumed, not unreasonably, by modern commentators to have been the natural feature that signaled the boundary between the Metapontino and the Tarantino, between Oenotria-Italia and Iapygia-Apulia. The current course of the river as it nears the coast lies very close to the northern wall of the city, at one point within 100 m—too close to be a comfortable boundary between the land claimed by the two poleis. Even in light of Strabo's account of the motivation for the foundation (κτισμός)—to block Tarentine expansion into the south and add the territory of Siris[22]—it is still too close to the city. The ancient course, according to aerial photographic research, reaches the coast 2 km north of the site of the asty. This is a more credible limit of Metaponto's territorial claims, but it is probably not the real one, or at least was only a limit in the early years of the colony.

Archaeological evidence suggests that the territory under Metapontine control or influence extended farther to the north of the older course of the Bradano. The late-7th–early-6th-century BC houses and later sanctuary at Pantano, excavated in 2000–2001 in the path of the SNAM project (see SNAM Transect, 1999, below), are identical to other contemporary structures of the Metapontino, as are the burials in the Stornara necropolis 3 km distant. The tombs line the road that passed the extra-urban Temple of Hera (Tavole Palatine) on the right bank of the Bradano on its way to Taras,[23] some 40 km to the east. Not only are the materials included in the burials the same as those found in the urban and rural necropoleis of Metaponto on the opposite bank of the Bradano, the funerary rites are closely comparable.[24] Given that these sites are within several kilometers of the asty of Metaponto, it is plausible that they belonged to the agricultural chora, within the jurisdiction of the polis.

The Cavone River (ancient Akalandros), south of the Basento, has routinely been taken as marking the boundary between the Metapontino and Siritide, an ἀποικία of the middle of the 7th century BC. According to tradition, it was occupied by settlers from Ionian Kolophon on the eastern coast of the Aegean. The influence, if not the presence of Ionian Greeks like those who established themselves at Siris[25] has been seen at the site of Incoronata greca in the 7th century BC. Incoronata was destroyed about 630 BC, probably by the Achaians of Metaponto, who arrived at about this time on the shore between the Bradano and Basento. After Siris was conquered by a coalition of Achaian Greeks from Sybaris, Croton, and Metaponto in the first half of the 6th century BC (the dates proposed by scholars range from the early part to the mid-century[26]), there was no obstacle to the expansion of the chora of Metaponto to the Cavone and even beyond. In fact, there is a small but growing body of evidence—again, burials of the Metapontine type—to the south of the Cavone that indicates a Metapontine presence there at least as early as the early 5th century BC.[27]

[21] Strabo 6.1.4.

[22] Strabo 6.1.15.

[23] De Siena 2002, 36.

[24] Nuzzi 2008.

[25] About the date and the location of early Siris much ink has been spilled. See Lombardo 1998, 47–48, for the recent status of the debate; also Orlandini 1999. There seems to be general agreement now that the fortifications, huts, and extensive necropoleis at Policoro belonged to Siris, over and beside which the Dorian Herakleia (founded by Taras in 432 BC) grew up. The *Oxford Classical Dictionary* locates it (and erroneously, also Herakleia) at Nova Siris some 10 km to the south on the Sinni River. In a separate entry, Herakleia is correctly located at Policoro.

[26] Orlandini 1999, 207–10.

[27] Osanna 2008, 927.

The Extent of the Chora

The intensively surveyed Bradano-Basento transect is only a sample of the chora of Metaponto—a limited but nonetheless generous one. It is natural to want to draw inferences from this area to the chora as a whole. Thus, it will be useful to ask at this point how big the whole really was and how the study area (defined below) related to it. The answers to these questions will rely on approximation and inferences from archaeological data. There are no certainties, but the attempt is both necessary and useful.

The criteria that guided the original choice in 1981 of the area to be surveyed were essentially two: it had to be large enough to be a reliable sample, and it had to be representative of the natural setting, including all of the topography characteristic of the territory occupied by Greek settlers between the Bradano and Basento (*Fig. 19.1*). The western limit was first set in 1981 at the point that marked the furthest extent into the interior of the division lines (the clearest evidence of the division of the plowland of the Central Plateau and other smaller plateaus perhaps as early as the 6th century BC; see Ch. 28). This limit was arbitrary at the time the transect was laid out and it turned out that settlement, not necessarily by Greeks, continued much deeper inland.

In 1990, the survey was extended into the Basento-Cavone watershed, and the same criterion—the extent of the division lines, which in this case followed a different orientation (*Fig. 19.1*)—determined the expansion of the survey towards the interior. The western border of the original 1981 Bradano-Basento transect, and its extension in 1990 to the Cavone, was an arbitrary limit, although it did in fact correspond roughly to topographical features, principally the major drainages, and a cultural divide.

It soon became clear that the settlement continued toward the interior and proceeded inland as the terrain became more dissected and the area of cultivable terrain narrower. A distinction was therefore made between the best lands of the nearer chora and those of the farther chora. These divisions are represented graphically as four zones, which stretch inland approximately 27 km (*Fig. 19.1*). Zones I and III form the nearer chora (simply "chora" hereafter), and they cover an area between the Bradano and Cavone of roughly 240 km^2.

In 2002 the survey was extended inland to the point where there was no longer any surface evidence of the distinctive structures and pottery, particularly the black-gloss fine ware, which was in use throughout the period of the Greek polis. This point, reached in 2007,[28] is 27 km from the coast and it defines the western perimeter of the farther chora. The entire area occupied by Metapontine farm families, or culturally similar indigenous families, was 412.7 km^2. To put this in perspective, this area was about a sixth of the size of Attica, the Athenian chora, and about half the size of the combined "nearer" and "farther" chora of Chersonesos on the Black Sea.[29] Metaponto was a medium-sized polis, but next to Sybaris, Croton, or Syracuse in the West, it would have seemed relatively modest. It is on approximately the same scale as Poseidonia or Locri, but probably more prosperous than either of these on account of its agricultural productivity.

The Nearer Chora

The nearer chora is defined as the agricultural territory of the polis and the rural counterpart of the asty. This was the dwelling and final resting place of a sizeable number of the citizens—and no doubt some noncitizens, who wrung their livelihood from the earth and soon made Metaponto a flourishing place. Despite attempts at quantification, the nearer chora cannot be exactly defined spatially because it changed shape over time. It expanded and contracted in response to political events and economic conditions from its beginning in the late 7th century BC down to the 2nd century BC when Metaponto had lost its independence and the Greek chora likely became, at least temporarily, the property of the Roman state (*ager publicus populi Romani;* see Ch. 27).

The area of the chora could be defined as the entire area between the Bradano and Cavone at any point in time, minus the area of the asty. The considerable extent of the asty, some 156 ha, was set aside from the start in the late 7th century BC and was separated from the chora by the western wall.[30] Its surface area was never smaller, having been reserved apparently from the first years of the new polis, as a space for the homes of the citizens and the temples and sanctuaries of the gods. The Metapontines, as Mertens and De Siena have argued, showed foresight and unexpected organizational

[28] See *Ancient Territories 2006–2007*, 56.

[29] Chtcheglov 1992, 250.

[30] For the Archaic-period wall, see De Siena 1998, 141–51.

skills in the *ab initio* functional division of the space within the walls.[31] No burials of any period were found within its perimeter until the Roman era. In fact, the urban fabric expanded slowly, reaching its maximum extent only in the late 4th century BC.

The chora probably did not extend right up to the city's walls. There may well have been an area—a sort of *pomerium* or buffer—reserved from the start for burials and sanctuaries. This is inferred from the fact that the area immediately outside the city began to fill up with tombs, beginning in the first half of the 6th century BC. The burials of the earliest settlers, which should belong to the last quarter of the 7th century and early decades of the 6th, have, with a few notable exceptions, so far escaped detection in the areas close around the city, though some of that date have been revealed deeper in the chora through excavation (see Ch. 20) and by the ICA survey.[32] Thereafter, the area immediately outside the city became a place of burial. In the 5th and 4th centuries BC it was occupied by separate areas of densely spaced tombs. The necropoleis—Crucinia, Casa Ricotta, Casa Teresa, and Torre di Mare—form a band around the city that extends in all directions several kilometers, principally to the west, but also to the north and south.[33] There seems not to have been a single "urban necropolis" but a series of distinct larger or smaller clusters of tombs. As will soon become apparent, burial was widespread throughout the chora in all periods (see especially Chs. 24 and 25). The misleading distinction between "urban" and "rural" is to a large degree artificial. It is more about the distance from the city than about the status of the people buried there.[34] As studies of these necropoleis have shown, there are typical Metapontine burial rituals throughout the territory, irrespective of the location.

Farther from the asty, there are larger necropoleis, though these are separated by greater intervals and interspersed with farmhouses. Concentrations of burials are found along the central division line, L15, which was a major axis through the chora. These continue from the recently discovered site at proprietà Mutinati[35] (where L15 intersects with the Basento Road at a distance of 1.5 km from the city walls) up the valley of the Basento to the extensive necropoleis at Pizzica D'Onofrio and Pantanello (320 burials), and still further out at 3 to 3.5 km in proprietà Morlino.[36] The pattern of burial clusters at the intersection with division-line roads is now clear (cf. *Map 9*, Vol. IV).

The question of how far outside the asty the chora begins is further complicated by the fact that the earliest farmhouses within 3 or 4 km of the city do not appear before the 5th century BC, though surely the fields Close to the City were cultivated from the beginning. The absence of farmhouses would indicate instead that the farmers of these nearby plots lived in the asty and commuted daily to their fields.

Beyond the band of necropoleis that surrounded the asty, the absolute boundary of the chora on the east was the coastline, which has moved approximately 1 km outward since ancient times (*Fig. 19.1*). The prograding of the coastline and migration of the river channels was actively taking place before and throughout the four centuries of the Greek chora, and afterwards (see Ch. 2). Only very recently has the prograding begun to reverse.

The Farther Chora

The limits of the chora to the west and toward the interior are much more difficult to determine. All efforts to draw a line, as mentioned above, are, at best, approximations. From the beginning it was a moving frontier. The Metapontino was not a deserted land, an ἔρημος χώρα, when the Greeks began to arrive. The first Greek contacts with the indigenous inhabitants were made at the sites of Incoronata and Andrisani (Metaponto) at the end of the 8th and during the 7th century BC (see Ch. 17). They ended about 630 BC, as mentioned above, with the sudden abandonment of both sites and their possible violent destruction. On the site of Cozzo Presepe, some 12 km distant from the new asty, a strong defensive wall and Greek pottery in large quantities appear about 600 BC. The settlement directly overlay an Iron Age village, which, unlike Incoronata, showed little or no sign of earlier, direct Greek contacts.[37]

31 Mertens 2006, 49; De Siena 2002, 34.

32 The earliest excavated burials outside the city (in the Crucina necropolis) and in the chora (Pantanello necropolis) date to the first half of the 6th c. BC.

33 Rocchietti 2002, 17–22. See Pl. 2 for a plan of the necropoleis of Metaponto's asty based on published information. It will need to be drastically revised in light of recent discoveries; see Osanna 2008.

34 See Finley 1979, 40–41, on the lack of distinction in law between the asty and chora. The "urban" and "rural" dichotomy had no political implications.

35 Osanna 2008, 927.

36 Carter 1998b, Fig. 2.19

37 *Cozzo Presepe*, 210–21.

We have created what is an artificial division between the nearer and the farther chora along a line that corresponds with a deep drainage just to the west of Cozzo Presepe, and includes all of the divided chora (*Fig. 19.1*), but it is clearly impossible to say whether the Metapontine Greeks in the 6th century BC or later respected this arbitrary limit. It will be immediately evident in this study that Metapontine farmers moved into the interior at least as deeply as Cozzo Presepe on the Bradano and Sant'Angelo on the Basento side possibly as early as the late 7th century BC. The settlers had established homes and had begun to bury their dead there by this early date.

The area up to the major drainages between Sant'Angelo and modern Bernalda on the Basento, and just beyond the site of Cozzo Presepe on the Bradano, forms a compact whole with a large expanse of relatively flat marine terrace. Only one major drainage, the Venella, interrupts it, separating the localities of Scarciullo and Avinella in the southwestern corner of the transect from the main area of the marine terraces of the Central Plateau (*Fig. 19.1*). This area includes all of the division lines, now identified as drainage ditches, constructed in the first half of the 5th century BC (see Ch. 28). The area so defined also includes a high percentage of the marine terraces that are ideally suited to the plow.

The Moving Frontier

The land along the eastern edge of the farther chora is virtually indistinguishable from the adjacent western edge of the nearer chora with respect to topography, fertility, and water sources. There is no reason not to believe that it too was similarly settled. How completely under Metapontine control these further areas beyond the divided chora were, however, has been questioned,[38] and it has been suggested that the economy was, like that of the nearer chora, based on agriculture and not principally on pastoralism. The 6th-century BC indigenous cluster of farmhouses, organized and operating like their Greek neighbors, at Pagliarone (see above) only 18 km from the coast is a good example of the peaceful symbiosis of indigenous and Greek farmers in this period.[39] It lay just outside the nearer chora but within easy reach of Cozzo Presepe.

There is good evidence now that Metapontine influence, if not control, may have encompassed the farther chora as early as the 6th century BC. In 2005, the survey was extended and five narrow transects were defined, parallel to each other and perpendicular to the coastline (*Fig. 19.1*). Sites with all of the external signs—the very same, or similar, black-gloss pottery and the same construction materials, including roof tiles—characteristic of the colonial settlement of the nearer chora have been documented in these transects on steeper, less fertile, and more dissected slopes stretching deep into the interior. The limit of "Greek" settlement was identified in 2007, at Masseria San Felice near modern Pomarico. Up to this point Archaic fine wares and black-gloss pottery of later date appear on the sites.[40] Beyond this point the recognizable fine pottery is either prehistoric, indigenous, or of the Roman period and later.

The Economic Potential of the Chora

The Land

As this study will show, the chora of Metaponto was densely occupied from the 6th century BC until the 3rd, and less densely but continuously for the six or seven centuries after that. The greatest appeal to the early settlers was the agricultural potential of the fertile land that would, in the course of the 6th century, make it one of the most prosperous poleis in the Greek West. Also of interest were the resources of the interior, the woodland inhabited by a great variety of animals. In addition to the animal population to be found in the chora and eschatia, the Greeks imported new and improved species, superior to those of the earlier native population, and they were successful in improving them further through selective breeding. The fauna of the chora have been the subject of a recently published volume in this series.[41] The flora, both domesticated and wild, will be the subject of another study. To anticipate here the paleobotanical studies and summarize very briefly several of the more important conclusions of the archaeobotanical study will, I hope, enrich the discussion of the settlement pattern.

The study of the plant life of the chora depends heavily on the evidence from a single important and well-dated deposit from the extra-urban sanctuary at Pantanello.[42] This has been supplemented by a

[38] Whitehouse and Wilkins 1989, 108–11; Osanna 1999.

[39] Amore 1992–1993, 89–96.

[40] C. D'Annibale, pers. comm., July 2007; *Ancient Territories* 2006–2007, 63–65.

[41] Bökönyi and Gál 2010.

[42] The results of the study of the seeds and preserved woods and other plant remains by L. Costantini, Rome, and the analysis of the pollen

recent palynological study of the fill from a drainage canal of the 5th-century BC bonifica of the chora (see Chs. 23 and 28).[43] The sanctuary deposits are truly exceptional for the variety of plant remains—not just seeds, but also wood, leaves, and pollen—and for their excellent state of preservation. Furthermore they were excavated under controlled conditions, which has permitted precise dating of the material. The pollen from the drainage canal complements the picture presented by the Pantanello site. Not only is it possible to reconstruct the environment and appearance of the chora in a detailed way, the knowledge gained of the actual crops raised by farmers in successive periods is, so far, unique in the Greek world.

Among the trees that the earliest Greek settlers would have found in the humid areas of the chora (such as the Pantanello sanctuary and the drainage canal) were alder (*Alnus*), hazelnut (*Corylus*), and willow (*Salix*). The wood and objects made from maple, poplar, and fir (*Abies alba*) were preserved in the anaerobic environment of the collecting basin of the sanctuary. The fir, worked as thin, wide planks, was certainly imported, probably from the interior; a small stand of huge white fir near Potenza, and large ones in the Pollino, exist down to the present (see Ch. 27). The olive, identified in both pollen studies, and in the form of a tool found in the basin, was apparently valued not just for the fruit it produced but also as a useful wood. Not far from the locations of the samples existed a mosaic of maquis (macchia) consisting of phillyrea, pistacchia, and scrub oak. Forests of pine and deciduous oak stood, as they do today, on the edge of the chora, in the eschatia.

The cultivated plants consisted of cereals, including two types of wheat, spelt (*Triticum dicoccum*), bread wheat (*Triticum compactum*), and barley (*Hordeum*), as well as legumes in rich variety, with chickpea, fava bean, lentil, and pea. The constant presence of the legumes in both the pollen profiles and in the macrofossil record has strongly suggested that they played an important role in the human as well as the animal diet at Metaponto. Metapontine farmers likely knew to sow legumes, which are rich in nitrogen, to revitalize the fallow fields: vetch was prominent among the macrofossil remains. The discovery of alfalfa (*Medicago*) at Pantanello in a 4th-century context is its earliest known appearance in Italy. It was surely used to support the superior livestock (sheep, cattle, and horses) that were being imported and developed locally by Metapontine breeders.

The olive and the grape (*Vitis vinifera*) complete the list of important crops for which there is evidence in pollen record as well as among the preserved seeds. To fill out the picture of economic activity in the chora are the grazing indicators, *Centaurea* and *Plantago*. The abundant presence of the *Liguliflorae* family (sunflowers), especially the *Chichoroideae* (lettuce and dandelions), and the *Chenopodiaceae* (goosefoot) and *Amaranthaceae* (amaranth and tumbleweeds) testify to the existence of fallow or abandoned fields.

Given the well-dated contexts both from macrofossil remains and pollen, it is possible, as Costantini, Sullivan, and Mercuri have done, to infer major changes in the relative importance of different plant types, and therefore changes in the agriculture regimes in the chora during the six centuries covered by this study.[44] The elevated levels of the grazing indicators strongly suggest that animal husbandry was a major activity in the 6th century, and the study of faunal remains from Incoronata and Pantanello bears this out.[45] The cereals and grasses are also very prominent. Legumes were probably already planted as a crop in the 6th century BC.

The plants of the maquis provided forage for sheep and goats especially. The maquis plant species were relatively low in the 6th century BC, but increased as sheep and goats declined in importance in the 5th and 4th centuries BC. From the 6th century, the cultivation of cereals and plow-based agriculture played a major role in the economy, and oxen, the "tractors" of the ancient world, came to dominate the animal population.[46] The cultivated olive appeared already in the 6th century, the first period of the developed chora, but by the 5th century was an important crop, as both the Pantanello and Pizzica pollen samples show. The olive and grape require some year-round attention. While other explanations are possible, this may explain in part at least why permanent dwellings, which previously had been few in Pizzica and close to the asty, suddenly began to appear there in numbers from the 5th century BC onward.

by Donald Sullivan have been published in preliminary reports since 1982–83. See Carter 2006a, 23–29, for bibliography.

[43] Unpublished report submitted to ICA by A.M. Mercuri, University of Modena, 2009.

[44] Costantini 1983; Sullivan 1983; Mercuri 2009.

[45] Bökönyi and Gál 2010, 6–8.

[46] Carter 2006a, 30, Fig. 1.22.

To judge from the pollen, the agricultural economy of the Metapontino in the mid-4th century was reasonably diversified between cereals, olive, and the grape.[47] Both olive and grape reached maximum levels at this time. By the late 4th and early 3rd centuries, however, at the period of the maximum expansion of the rural population (see Ch. 25), the olive is less important, the grasses and cereals increase, and the legumes peak. Towards the end of the 4th century BC, there was a decisive shift toward grain crops, at the expense possibly of olive and grape cultivation and pasture. The over-emphasis on a single crop may have depleted the land, and, with other developments, hastened the decline of the chora. By the mid-3rd century, as the area under cultivation steadily decreased, legumes and grapes had dropped sharply and abandoned fields were overrun by weeds. In the last two centuries BC, the cereals decline, but the olive and grape seem to stage a recovery and grazing indicators remain steady.

This is the current picture of the development of agriculture and animal husbandry in the chora. For an in-depth treatment of the animal populations and their dynamics, there is now the volume by Bökönyi and Gál.[48] In light of this knowledge, the shifting patterns of the settlement of the chora acquire a plausible rationale and a deeper significance.

Navigability of the Basento and Bradano Rivers

In most periods of the year, the streams that flank the chora are unimpressive, slow, meandering creeks rather than rivers.[49] In the rainy season they become torrents and regularly go outside their banks and occasionally flood their valleys and the coastal plain (see Ch. 2 on the rivers). It is difficult to imagine that they were formerly broad, free-flowing rivers that were used for transportation inland for some distance from their mouths. There is, however, documentary evidence for this, not from our ancient sources, but from several authoritative Medieval and Renaissance geographers and writers. The highly educated and respected Islamic geographer Muhammad al-Idrisy, who worked in Sicily in the 12th century at the court of the Norman king Roger II, is our best source on the rivers. His maps were copied for centuries without revision. Of the Bradano he writes,

> "...(it) is a river that flows without mixing its waters with those of the Basento. It first appears as two branches which ... join at the level of Rocca S. Giuliano [near Matera] and take their name of Bradano and then flows among cultivated fields to the sea. On the banks of this river pines grow in abundance, which when cut are then transported to the sea on the river's current. Pitch and tar are extracted from them and exported in various lands."[50]

Several documents of the Angevin period attest to the navigability of the streams at least in their final stretches. One dated January 1271 recounts that Charles I of Anjou, on the occasion of his arrival in the realm of the king of France, urges that boats be readied on the Bradano and Basento for transporting men and supplies.[51] Leandro Alberti, writing in 1525, describes a series of places in the area, including Torre di Mare, a fortified center on the coast, which dates from Medieval times if not before (see Ch. 34). His description of the rivers indicates that they are already less negotiable because the winter floods knock down the "shaking bridges" and because of the ever more frequent raids by pirates.

Michele Lacava, writing at the end of the 19th century, speaks of the port of the ancient city, which he thought could be identified with the stagnant Lago di Santa Pelagina. He notes the remains of structures and claims that the mouth of the Bradano must have been a port, and likewise the mouth of the Basento.[52] These sources from the Medieval period to the present lend some weight to the idea, developed in Chapter 26, that the rivers were navigable in ancient times, and on occasion played a key role in Metaponto's history.

The Study Area

The study area is defined as the area of actual coverage by the survey, that is, the fields walked, plus a buffer of 50 m on all sides (*Fig. 19.2*). All of the areas covered in the transects are included in it, but it also includes some areas immediately outside the transects that were also surveyed. With its 50-m buffer, the study area measures 36.1 km^2. The area of the transects is 50.1 km^2, and the actual areas covered within them is 27.3 km^2 (*Table 19.1*).

[47] Carter 2006a, 28, Fig. 1.20.

[48] Bökönyi and Gál 2010.

[49] Research by E. Lapadula, pers. comm., 2009.

[50] Idrisi 1994.

[51] Filangieri 1950, 1,273–77, Vol. 9, 151.

[52] Lacava 1891, 94–95; cf. L. Giardino 1983 on the excavation of the port.

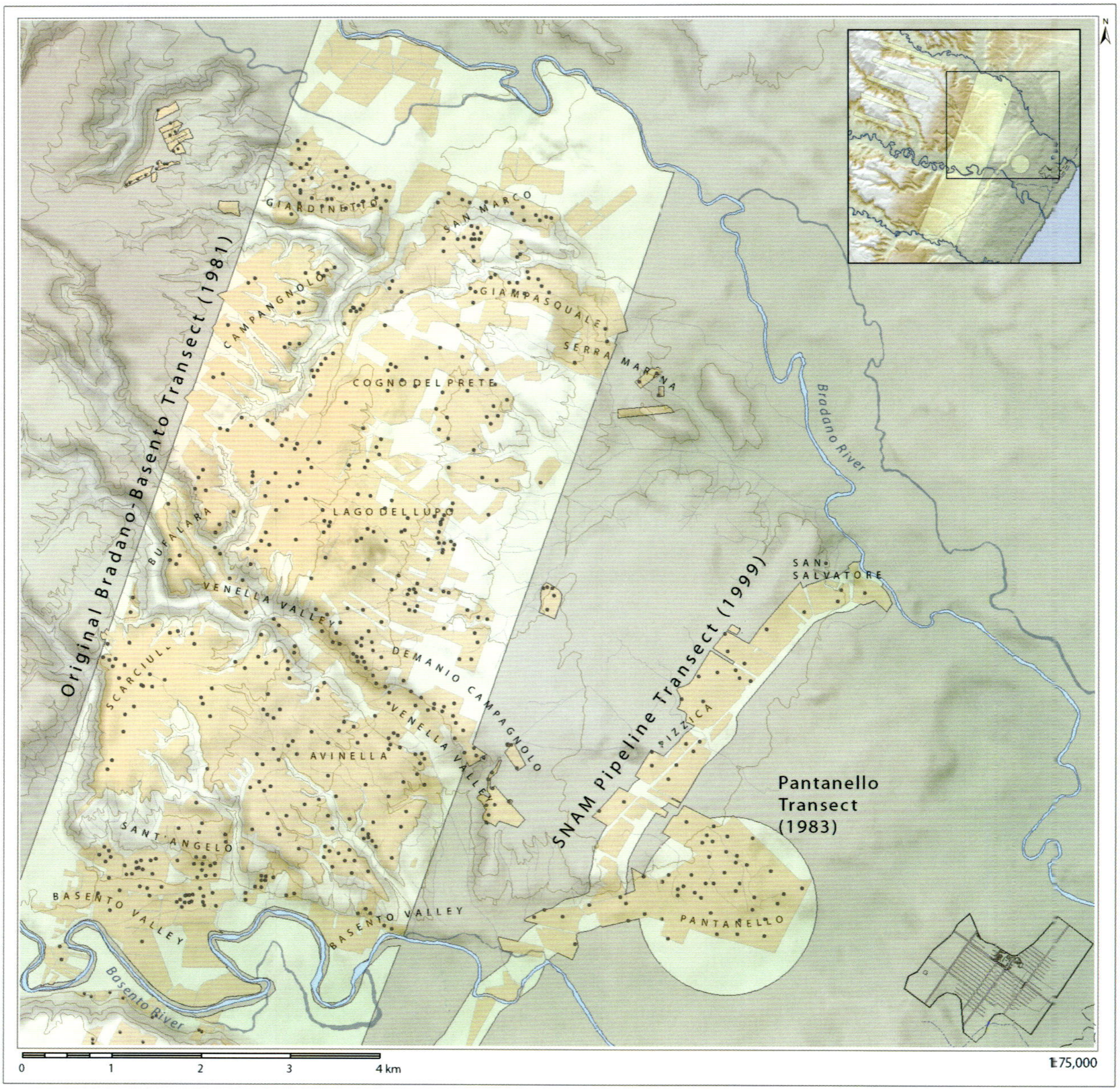

Figure 19.2 Map of the chora, showing the Bradano-Basento study area, with transects and localities. Inset shows entire chora. (JT)

The Transects

Bradano-Basento transect, 1981. The study area consists of three transects of varying size and shape (*Fig. 19.2*). They complement each other and provide a representative sample of the nearer chora as a whole (see Ch. 3). The original and largest transect, stretching between the modern courses of the Bradano and Basento Rivers (later extended to the Cavone), is bounded on the east and west by arbitrary north-south running parallel lines 4.2 km apart. The eastern boundary is on the average 7.25 km from the city walls of Metaponto. This transect samples all of the topography and the main geomorphological units typical of the nearer chora—river terraces, gently sloping marine terraces interrupted by drainages, and the slopes that connect the valley and drainages and the terraces. The units with the largest area are the marine terraces, which form the Central Plateau, and the smaller plateaus of Avinella and Scarciullo. The shading (tan) indicates fields that have been thoroughly covered by the survey team at least once, and in many cases twice or three times at intervals of one to ten years. The dates of first coverage and revisits are recorded in the Gazetteer (Vol. IV). A total of 524 sites was recorded in the original Bradano-Basento transect between 1981 and 2001, including prehistoric, Greek, Roman, and Medieval sites.

Pantanello transect, 1983. A circular transect with a diameter of 2 km was laid out so as to provide a context for four major excavation sites: the Greek sanctuary (1974–1982, 1990–1991); the Roman tile factory (1974–1976, 2008); a Neolithic village (1983) situated on a low spur overlooking the Basento River; and the extensive Pantanello necropoleis (1982–1986), located some 500 m to the southwest along the Basento Road, which connected the Basento valley, the coastal road, and the ancient city. A total of 43 sites was identified during the 1983 campaign. This transect comprises a low marine terrace, fluvial terraces, and a portion of coastal plain, the fourth major geomorphological unit of the Metapontino that is not represented in the two other transects.

The SNAM transect, 1999. This irregular-shaped transect, which followed the projected path of major oil and gas pipelines across the Metapontino, was designated in the summer of 1999 prior to the beginning of pipeline construction.[53] The transect is 17.3 km long and varies in width between 200 and 500 m.[54] In the portion lying between the Bradano and Basento, 28 sites were recorded. Excavations were conducting before and during the construction of the pipeline (see Ch. 28).

Survey Coverage, Bradano to Basento

The total area included within all three transects is 50.1 km^2 (*Table 19.1*). Of this a total of 27.3 km^2 or roughly 55% was completely and thoroughly examined by the survey team between 1981 and 2001. Revisits to the same areas in the transects, beginning in 1990, have confirmed the accuracy of the original observations,[55] and strengthened the conviction that few if any sites escaped notice in the areas surveyed. The remaining unsurveyed areas in the transects, 13.3 km^2, were either inaccessible for various reasons, or, because they lie deeply buried in river valleys, were soon discovered to be unproductive. As a result of the SNAM excavations by the Soprintendenza, we now know that the sites located there in considerable numbers are too deeply buried in most cases to have left any trace on the surface. The sites in the Bradano valley below Giardinetto were noted, but the surface material was not collected. Those in the Basento, especially farther upstream, were more in evidence.

Transect	Area (km^2)	Area covered	Study area 50-m buffer
Bradano-Basento	44.5	23.8	30.5
Pantanello	3.1	1.8	ca. 2.0
SNAM	2.5	1.6	ca. 3.6
Totals	**50.1**	**27.2**	**36.1**

Table 19.1 Basic statistics for the three transects.

The Study Area Subdivided

Localities. The discussion of the study area and its development has been made more manageable by subdividing it into smaller units. Instead of creating a new system with artificial boundaries, the traditional place names for the divisions of the Metapontino were adopted. The origin and etymology of the names, often either descriptive like "Giardinetto" (Little Garden) or "Pantanello" (Little Swamp) or colorful like "Lago del Lupo" (Wolf Lake) or "Cogno del Prete" (Priest's Knoll?), is obscure. The important thing is that they are still in use by the current population, a circumstance that has facilitated discussions with local informants. These place names were consistently recorded during the survey and are presented in the Gazetteer (Vol. IV). Their use has the advantage that it is independent of any other division, such as modern jurisdictional boundaries, or scientifically determined units such as those of the geomorphological and soil maps. The localities are indicated on most maps reproduced here.

The decision to use these traditional place names, however, resulted in some necessary modifications in the way areas are described. During the course of this study it became apparent that the locality names did not always coincide exactly with the groupings of sites that needed to be described as a unit. In only two cases—Giardinetto and Giampasquale—do all the sites fall into a single locality. For example, the Early Village Cluster Sant'Angelo includes sites from

[53] Nava 2001, 944–50. See full discussion of the SNAM transect and excavations in Chs. 3 (84–87) and 28–30. Additional pipeline and facilities were created in this corridor by the *Consorzio Bonifica di Bradano e Metaponto*, with the intent of modernizing the irrigation system in the Metapontino.

[54] A total of 66 sites were documented between the Bradano and Cavone by the survey team in 1999.

[55] The dates of the original visits and revisits are recorded in the Gazetteer (Vol. IV). Materials collected from the surface on each occasion are separately recorded there.

the traditional localities of Sant'Angelo and Basento valley. The solution adopted was to use one collective name—e.g., Sant'Angelo—as the name of the location for all sites in the cluster. Likewise for Avinella, which included many sites from Avinella and some from Venella. The Venella cluster was composed of sites from Venella, Avinella, and Demanio. These collective names are used in all the tables describing clusters.

Sites in a locality such as Sant'Angelo that are not included in a cluster, as well as those from localities with very few sites, have been combined in two groups with the designations AVS and DBSBS. The group AVS, which stands for Avinella, Venella, and Sant'Angelo, contains the sites in these localities that do not belong to the specific Avinella, Venella, and Sant'Angelo clusters. Widely separated, they are located in three different areas on the Basento side of the study area, relatively far from the ancient city or asty. There is a total of 34 AVS sites. The same applies to DBSBS, which designates 46 sites from five different areas—the Demanio Campagnolo, Basento valley, San Salvatore, Bradano valley, and Serra Marina localities—which lie Close to the City between the Basento and Bradano valleys. Without this reorganization the use of the basic tables would have been far more complicated.

Regions. At a certain point when the analysis of the settlement patterns in the study area was well advanced, description on a larger scale than that of localities was required. The division into four regions evolved naturally, and became essential for describing larger trends. The decision was made to adopt this division with the realization that (1) the areas nearer the city (ἡ πρὸς τῇ πόλει χώρα) had very different characteristics in the 5th century BC than before; (2) the five early clusters of sites along the major river valleys all behaved in much the same way; (3) the settlement of the area of the Central Plateau (το πεδίον) between the valleys had its own trajectory, distinct from that of the clusters; and finally (4) there was a sort of reciprocal relationship between areas Close to the City (especially the localities of Pizzica and Pantanello) in the SNAM and Pantanello transects, and those at the farthest extent of the study area (bordering on the ἐσχατιά), which were divided from the Central Plateau by deep drainages, the Venella and the Fosso dell'Acqua Fetente.

Once the decision had been made, tables tracing the settlement history were created to reflect the divisions in four regions:

Early Village Clusters (Giardinetto, Sant'Angelo, Avinella, Venella, Giampasquale);
The Central Plateau (Lago del Lupo, Cogno del Prete);
Far from the City (Campagnolo, Bufalara, Scarciullo, San Marco, AVS);
Close to the City (Pizzica, Pantanello, DBSBS).

All but the Early Village Clusters correspond to a continuous topographical area composed of neighboring localities. In a sense, the Early Village Clusters also represent a single, discontinuous geographic unit, that is, the edges of the great river valleys at the points where they are joined by drainages that reach into the main interior landmass that is composed of almost flat marine terraces.

Survey Coverage and Site Density by Region and Locality

The survey coverage was just about equally divided between the four regions (*Table 19.2*). In terms of settlement density, however, the regions differ greatly. Of the total sites from all periods, the Early Village Clusters have much higher concentrations of sites. The number of sites per km^2 (19.3) in any of the localities of the Early Village Clusters, with the exception of Giampasquale, is greater than that of all the localities in the other three regions.

Using surface area and densities of sites in surveyed areas, it is possible to make at least an approximate calculation of the total number of sites of all types and periods in the entire chora (nearer and farther). The total number of sites in the study area, which represents 18% of the total area of the nearer chora, is 595, of which 565 (95%) were either occupied or frequented in the Greek period.

The average density of sites using these numbers is 16 sites per km^2 for sites of all types and periods (to the nearest whole number) and 15 sites per km^2 for Greek sites of all types (again, to the nearest whole number). This is a density that is exceptionally high for surveys in the Greek world. The density of Greek sites is "five to ten times greater" than the total site density encountered in survey projects elsewhere in the Mediterranean.[56]

Assuming the same density of sites and the same ratio of Greek sites to sites of other periods throughout

[56] S M. Thompson 2004, 67, and Fig. 6.4.

Region	Locality	Survey coverage (km^2)	Survey coverage (% of total)	Total sites	Density (all sites)
Early Village Clusters	Giardinetto	0.91	3.4%	36	40
	Sant'Angelo	1.00	3.7%	43	43
	Avinella	2.02	7.5%	48	24
	lower	(1.27)	(4.7%)	(33)	(26)
	upper	(0.75)	(2.8%)	(15)	(20)
	Venella	1.57	5.8%	61	39
	east	(0.90)	(3.3%)	(40)	(44)
	west	(0.67)	(2.5%)	(21)	(31)
	Giampasquale	0.63	2.3%	14	22
	Subtotal	**6.13**	**22.7%**	**202**	**33**
Central Plateau	Lago Lupo	4.15	15.3%	95	23
	Cogno Prete	2.79	10.3%	47	17
	Subtotal	**6.94**	**25.6%**	**142**	**20**
Far from the City	Campagnolo	1.32	4.9%	16	12
	Bufalara	0.79	2.9%	13	16
	Scarciullo	2.45	9.0%	28	11
	San Marco	0.93	3.4%	19	20
	AVS*	1.74	6.4%	34	20
	Subtotal	**7.23**	**26.6%**	**110**	**15**
Close to the City	Pizzica	1.57	5.8%	21	13
	Pantanello	2.00	7.4%	44	22
	DBSBS**	3.28	12.1%	46	14
	Subtotal	**6.85**	**25.5%**	**111**	**16**
	Totals	**27.2**	**100.0**	**565**	**20.8**

Table 19.2 Survey coverage, total Greek sites, and density by region and locality. (*Avinella, Venella, and Sant'Angelo. **Demanio Campagnolo, Basento valley, Salvatore, Bradano valley, and Serra Marina.)

the whole of the nearer chora—both reasonable assumptions, borne out by the preliminary results in the nearer chora to the south of the Basento[57]—the total number of Greek sites of all types can be calculated to be 3,255, or 3,000 in round numbers. The nearer chora was the home of predominantly Metapontine farmers of Greek origin, although mixing with locals probably began immediately.[58] The density of sites in the farther chora is likely to have been far less and the proportion of native to Greek farmhouses much higher. These are very broad order of magnitude approximations and should be treated as such.

Types of Sites

The following identifications could be made for the 565 Greek sites in the study area (see also *Appendix C2*):

- 312 (55%) farmhouses;
- 161 (29%) necropoleis;
- 5 (1%) sanctuaries.

All other types comprise 7% of the total:

- 2 kilns;
- 3 quarries;
- 35 "Greek agrarian structures" (GAS)

For the remaining 47 sites (8%) a function could not be ascertained, and these are called "unidentified Greek features" (UGF). As mentioned in Chapter 3, determinations were made in the field on the basis of multiple criteria, such as the extent of the scatter, presence of building materials, pottery types, and other diagnostic materials. This information for each site is compiled in the Gazetteer (Vol. IV).

The various types of sites are not uniformly distributed over the study area. A total of 114 farmhouses (36% of the total) are located in the Early Village Clusters, the smallest in terms of area of the four regions, comprising just 22.7% of the total surface of the study area (*Table 19.2*). Sixty necropoleis or 37% of the total are also located in the clusters and 4 out of 5 of sites identified as rural sanctuaries are there.

There is a considerably higher ratio of necropoleis to farmhouses in the Early Village Clusters than in either the Central Plateau or in the localities Far from the City, but in those Close to the City there were almost as many burial sites as residences. These are the bare statistics. Their historical implications will begin to emerge in the remaining chapters in this volume.

The Artifacts

The surface collection of artifacts—above all the pottery—provides crucial dating evidence as well as information about the activities that took place at the survey sites. Between 2001 and 2007, some 29,000 fragments were studied and classified by fabric,

[57] For preliminary reports, see *Ancient Territories*, ICA's annual reports from 2000 to 2007. The Basento-Cavone half of the chora will be the subject of future study.

[58] Carter 1998f, 167–72. See Henneberg 1998 for the biological identity of the settlers in the chora.

	AFW	BG	BF	RF	Gna	BoB	XG	GW	BW	PW	Ck	Amp	Min	TC	Total
Total fragments	479	8,750	3	212	43	2	2	203	2,143	12,902	1,575	2,586	231	342	29,486
Total identified by form	479	5,460	3	172	27	2	2	41	1,786	7,921	582	565	112	123	17,160
Total dated	464	3,027	3	192	43	2	2	203	128	267	0	565	21	120	4,922

Table 19.3 Counts of fragments by ware, showing total number of fragments, those identifiable by form, and those by date.

form, type, subtype, function, and, where possible, date. This information resides in databases that are a fundamental part of the Metaponto GIS (see Ch. 4), which includes complete counts of all fragments. The pottery of the six centuries between 625 and 25 BC (*Table 19.3*) consists of nearly 10,000 fragments of fine wares, including a very limited number of fragments of imitation proto-Corinthian and Colonial wares of the 7th century BC, Archaic fine wares (AFW) such as Ionic cups, black-gloss (BG), black-figure (BF), and red-figure (RF) pottery, Gnathia style (Gna), a few fragments of black-on-buff (BoB), Xenon Group (XG), and a larger group of grey ware (GW).

By far the largest and most important of the fine wares for this study is the black gloss (BG), which numbers approximately 8,750 fragments and is the subject of an exhaustive exposition in Chapter 6. The general shape (open or closed) could be securely identified for some 5,460 fragments, of which the majority could be classified by specific form. Of these, 3,027 could further be assigned date ranges with a high degree of precision. 201 fragments of grey ware (GW) were collected.

Fragments from plain/banded wares used for storage and table service numbered 15,045. Undecorated plain wares (PW) are six times more common than the banded wares (BW), which have bands or stripes usually of red-brown slip. More than half the plain-ware vessels could be identified as to form or shape and be assigned a function. Besides the fine wares and plain wares, practically every site produced cooking wares, and many sites yielded one or more fragments of amphorae. Of the 1,575 fragments of cooking-ware vessels (Ck), 582 could be identified by form; of the 565 fragments of amphorae rims and bases (Amp), 336 could be assigned to a production center and a broad date range.

The miniature votive vessels (Min) in plain-ware fabric, as well as the terracotta figurines and votive plaques, provide valuable indications of cult life and religious beliefs at more than a hundred of the survey sites. The forms of these tiny vessels are known in almost half of the 231 examples. Of the 342 figurines and other objects in terracotta with figural decoration (TC), 123 could be identified and 120 assigned an approximate date range.

The 54 fragments of wheel-made lamps, which are included in the counts for each of their classes, and 119 examples of loom weights complete the range of objects in terracotta. The only large class of non-ceramic object are the grinding stones. A total of 45 fragments were found on 35 sites, most made of basalt and associated with farmhouses. They are the object of a detailed analysis that will appear later. Far rarer are the coins—9 examples in bronze from 7 sites—of which 3 come from outside the current study area. Each important class of artifact, with the exception of the grinding stones, is dealt with separately in Chapters 6–15.

Dating Methods and Data Presentation

Of the 565 sites in the study area belonging to six centuries between 625–25 BC, 371 (66%) could be dated with some degree of confidence. On the basis of the pottery, 269 of 312 farmhouses and 74 of the 161 necropoleis have been dated (*Appendix C2*). All five sanctuaries are securely dated. More than a third—18 out of 47—of the unidentified Greek features (UGF), but only 5 of 35 Greek agrarian structures (GAS) had enough associated dateable pottery to be assigned to a specific chronological period. Two quarries out of three, and both isolated kilns had some types of pottery characteristic of the period or other reasons for considering them part of the Greek chora, but not enough to provide a secure date. Kilns and burials or tombs associated so closely with the farmhouse scatter as to be considered inseparable components of it have been assumed to have existed during the period of the farmhouse's life unless there is evidence to the contrary. Seven of the 8 farmhouses with kilns, and 26 of the 30 farmhouses with burials beside them have been assigned a date.

Class	600	550	500	450	400	350	300	250	200	150	100	50
Black gloss (BG)*	104.70	304.02	74.52	237.88	295.52	350.00	871.01	194.38	93.25	33.47	4.22	0.74
Black figure (BF)	—	0.03	2.94	2.03	—	—	—	—	—	—	—	—
Red figure (RF)	—	—	0.01	4.42	50.54	91.76	39.96	0.19	—	—	—	—
Gnathia (Gna)	—	—	—	—	0.28	8.99	23.91	1.81	—	—	—	—
Black on buff (BoB)	—	—	—	—	0.02	3.09	0.89	—	—	—	—	—
Xenon Group (XG)	—	—	—	—	0.43	1.41	0.16	—	—	—	—	—
Figured totals	—	0.03	2.94	6.45	51.28	105.24	64.92	2.00	—	—	—	—
Grey ware (GW)	—	—	—	—	—	—	—	—	2.26	46.46	107.53	41.21
Subtotals	**104.70**	**304.05**	**77.46**	**244.33**	**346.80**	**455.24**	**935.92**	**196.39**	**95.51**	**79.92**	**111.75**	**41.95**
Total: 2,994												

Table 19.4 Counts of dateable fine-ware fragments by period. *Includes Archaic fine wares (AFW) for **600–500**.

Black Gloss and Date Bins

Black-gloss pottery (BG), most of which was locally produced in the kilns of Metaponto by potters in the city or the chora, was the fine ware most in use throughout the entire history of the chora. In the early years of the polis it was largely limited to vessels for drinking, primarily cups. The initial study of the black gloss, completed in 2002, was extensively revised and greatly enlarged in 2009–2010. The results appear as Chapter 6. The dates of the very earliest pottery have been modified and the consequences of this revision are discussed in Chapter 20. In the 5th and 4th centuries BC, the range of shapes expanded rapidly. Within functional groups, new forms emerged and older ones disappeared. Throughout the history of the chora's settlement, black-gloss pottery was a major component of every assemblage, be it farmhouse, necropolis, sanctuary, or other. As the number of sites increased and the range of shapes expanded over time, so did the quantities of this material (*Table 19.4*).

The period 625–175 BC and the following century and a half (175–25 BC) of Roman domination of Magna Grecia under the Republic have been divided into a dozen 50-year "date bins" (625–576 BC, 575–525 BC, etc.). For the sake of economy they are referred to throughout this text by their mid-dates (**600**, **550**, etc.), in bold font.

Equivalent Artifact Weight (EAW) and Appendix A

The Equivalent Artifact Weight, or EAW, is one of two key indices in this study. It is a measure of the quantity of black-gloss pottery at a site dateable to a particular 50-year span. Its values range from zero upward but only very rarely does it surpass 20. An EAW value of 0.95 indicates that there is the equivalent of one fragment of black-gloss pottery at the site in that particular period. The date ranges for black-gloss pottery are typically about a half-century, but may be more narrow. Only very rarely will the date range of a particular fragment fit precisely into a 50-year date bin. The EAW value is the sum of the fractional portions of pottery ranges that fall in a date bin. For example, if a fragment has a date range of 425 to 350 BC, it will add .66 to the EAW value of the site in **400** (425–375 BC) and .33 to the period **350** (375–325 BC). In Appendix A, located at the end of this volume, the sums are recorded to two decimal places. The dating procedures are discussed in detail in Chapter 4. Appendix A lists the sites in numerical order and contains the following information:

1. type of site;
2. traditional name of the locality;
3. geomorphological unit;
4. elevation in meters above sea level (masl);
5. number of dateable artifacts from the site (a gauge of the reliability of the date).

Then comes the core of the table: the EAW values for each site in every period from **600** to **50**. In the final column in Appendix A, the "best date" for the site has been recorded. This is reached by averaging the date ranges of all pottery from a site. It does not always correspond with the period of the highest EAW value for a site in cases where a site has comparable EAW values in more than one date bin (for example, site 1 in Sant'Angelo or site 395 in Avinella).

	600	550	500	450	400	350	300	250	200	150	100	50
Farmhouse	0.64	1.64	0.35	1.13	1.43	1.54	3.88	1.00	0.74	0.68	1.94	1.38
Necropolis	0.93	1.67	0.45	0.74	1.08	1.44	1.43	0.36	0.29	0.16	0.55	0.21
Sanctuary	4.74	16.73	3.76	4.52	1.38	1.26	5.42	1.45	N/A	N/A	N/A	N/A
Other	0.20	1.26	0.03	0.46	0.56	0.28	0.75	0.13	0.37	0.10	N/A	N/A

Table 19.5 Average Equivalent Artifact Weight (EAW) values for farmhouses, necropoleis, and sanctuaries by 50-year period.

Site type	Class	600	550	500	450	400	350	300	250	200	150	100	50
Farmhouse	BG*	69.08	192.02	46.75	186.53	259.83	302.39	786.01	179.58	89.29	32.49	4.20	0.74
	RF	—	—	0.01	2.01	17.94	48.26	22.71	0.01	—	—	—	—
	Gna	—	—	—	—	—	6.30	20.29	1.40	—	—	—	—
	BoB	—	—	—	—	0.02	1.09	0.89	—	—	—	—	—
	XG	—	—	—	—	0.29	0.69	0.01	—	—	—	—	—
	GW	—	—	—	—	—	—	—	—	2.24	46.03	106.44	40.78
Sanctuary	BG	23.71	83.67	18.79	22.60	5.38	5.58	21.56	4.34	1.18	0.11	—	—
	XG	—	—	—	—	0.14	0.72	0.14	—	—	—	—	—
Necropolis	BG	11.11	23.28	8.81	25.98	25.27	37.33	52.25	9.17	2.04	0.67	—	—
	BF	—	0.03	2.94	2.03	—	—	—	—	—	—	—	—
	RF	—	—	—	2.41	32.50	43.49	17.25	0.18	—	—	—	—
	Gna	—	—	—	—	0.28	2.69	3.62	0.41	—	—	—	—
	GW	—	—	—	—	—	—	—	—	0.02	0.43	1.09	0.43
GAS	BG	0.26	1.73	0.01	0.99	0.65	1.32	2.97	0.07	—	—	—	—
UGF	BG	0.53	3.31	0.16	1.80	4.40	3.38	8.22	1.23	0.74	0.20	0.02	—
Subtotals		104.70	304.05	77.46	244.33	346.80	455.24	935.92	196.39	95.51	79.92	111.75	41.95
Total: 2,994													

Table 19.6 Dated fine wares class totals by site type and period. *Includes Archaic fine wares (AFW) for **600–500**.

Variations in the Quantities of Fine Ware, 600–50

There is a considerable variation in the quantities of dateable fine wares in the six centuries under consideration (*Table 19.6*). Less than 5% of the total can be dated to the first period of the settlement in the study area in the original chronology (and even less in the revised chronology). Since the fine wares are the basis for dating all the sites, there is a strong correlation between the quantities of fine wares and the number of sites in particular 50-year periods. There are four periods when the average EAW value per farmhouse drops below the equivalent of one fragment—in **600**, **500**, **200**, and **150** (*Table 19.5*). Necropoleis on the average have more than a half a fragment (EAW ≥ 0.45) in all periods down to **250**. After this date the number of necropoleis with or without pottery dropped precipitously and did not recover. The long period of settlement contraction seen in the study area had begun (Ch. 26). The four Archaic sanctuaries have by far the highest average EAW values between **600** and **450**. The one later 4th-century BC site identified tentatively as a sanctuary, 266S, has only slightly more dateable fine wares than the average for all farmhouses in **300**.

The black-gloss fine ware predominates in all periods (*Table 19.6*) and the dating of farmhouses, in particular, depends on it. Red-figure pottery, however, plays an important role in dating necropoleis in the period **400** to **300**, when this class of pottery was produced in great quantities in the kerameikos of Metaponto.

When is a Site Considered Active?

A total of 207 out of 321 farmhouses identified in the study area have an EAW of 0.95 or greater in one or more periods between **600** and **50**. This is the subgroup,

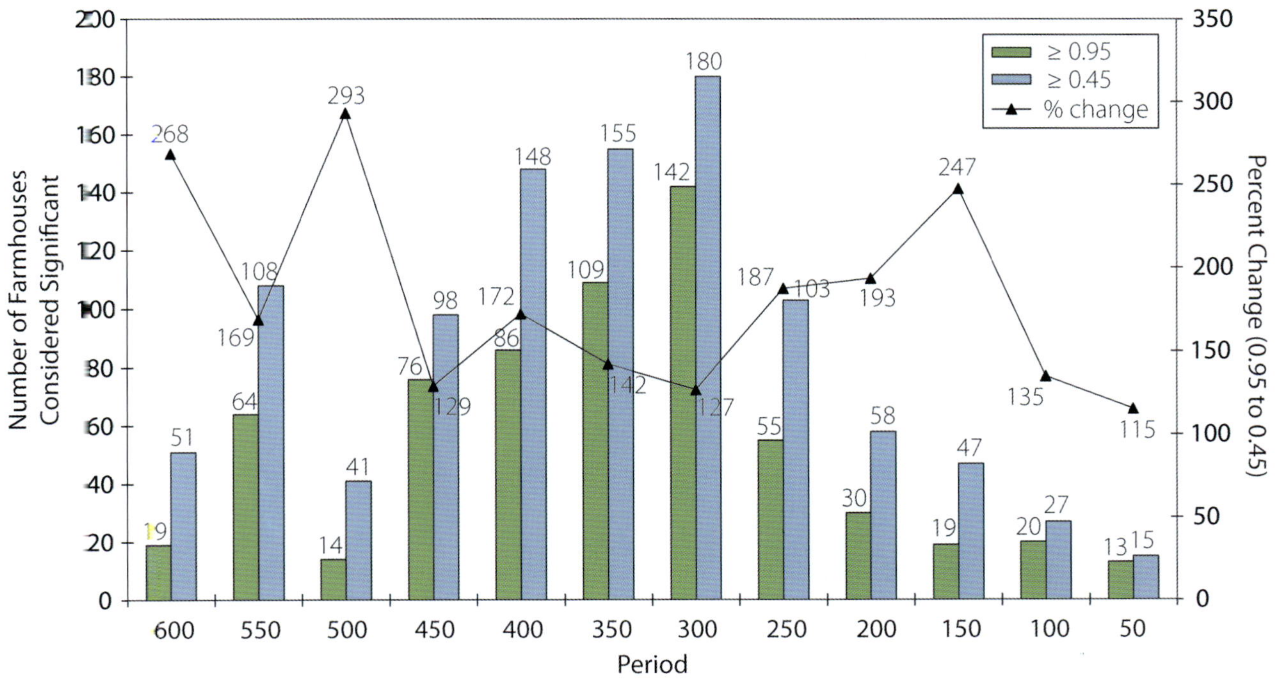

Figure 19.3 Comparison of the number of farmhouses above the EAW thresholds of 0.95 and 0.45 (second magnitude), **600–50**.

on which all conclusions about the changing pattern of settlement in the chora will essentially be based. We have chosen the threshold of EAW ≥ 0.95, or the equivalent of one fragment of black-gloss pottery. It is an arbitrary decision, but a necessary one in order to define unequivocally the precise point at which we say that a farmhouse was active and occupied. In most cases this does not mean that a single fragment with a date range that corresponds to the 50-year period was found on the site, but rather that there are several fragments whose date ranges partially fall within the period and whose sum is the equivalent of one fragment. There is, of course, the possibility that a site with a large number of black-gloss fragments with enough partial date ranges to fill several date bins will be considered active in all, when in fact it was really only active in the one in which the most black gloss was concentrated. With this method such a result is inevitable. The threshold could have been higher, which would have resulted in a sparsely populated chora; alternately a lower limit might have been chosen, and in fact, we have considered farmhouses with an EAW less than 0.95 but greater than 0.45 (the equivalent of half of a black-gloss fragment per 50-year period). There are only 55 farmhouses that surpass the 0.45 threshold but not the 0.95 level (*Appendix C2*). Farmhouses with an EAW value of 0.45 or greater are recognized as second magnitude. They figure more prominently in the first three periods of the chora. Black-gloss pottery, and pottery in general, is relatively scarce in the early years and the average EAW value for farmhouses (*Table 19.5*) is quite low—only 0.64 in **600** and 0.35 in **500**. Pottery of this period is nonetheless present and the study area was occupied in **600** or in **500**, but at a much lower level.

What effect would reducing the EAW threshold for farmhouses from 0.95 to 0.45 have? A larger percentage of farmhouses will have an EAW greater than 0.45 in some than in other periods. The differences between farmhouses with an EAW ≥ 0.95 and those with an EAW ≥ 0.45 from one period to the next are represented here in graphic form (*Fig. 19.3*). They are greatest in **600**, **500**, **250**, and **150**, all periods of rapid change. There are more second magnitude farmhouses relative to those meeting the 0.95 limit in these periods than in any others.

The effect of considering sites with this lower threshold in the chronological development of the farmhouse population is considerable, but the overall picture does not differ in the essentials. There are similar dips and peaks, but the differences were simply more exaggerated in the periods of rapid change.

In the case of necropoleis, a lower threshold (EAW ≥ 0.45) was chosen because these sites and the surface collections are much smaller. One consequence is an inevitable loss of chronological definition. A single fragment of black-gloss pottery could, for example, also date a necropolis to the successive periods, if its date range is equal for both. Of the 74 necropoleis that have any black-gloss pottery at all, 60 (81%) can be said to be active in one or more 50-year period. This compares closely with the farmhouses. Of the 269 farmhouses with pottery, 207 (77%) have EAW values of 0.95 or more (*Appendix C2*).

These choices of threshold values for including sites in the chronological description of the settlement pattern were arbitrary but reasonable, and they permitted a reconstruction of the development of the chora, which, as it turns out, compares closely with that traced for the ancient city by its excavators. The reader will judge whether these decisions and the resulting reconstruction are convincing or not.

Farmhouses and Necropoleis

Adopting the EAW thresholds of 0.95 for farmhouses and 0.45 for necropoleis results in the chronological distribution presented in Fig. 16.5 and Appendix C1a. These are the numbers used throughout this study, and this graph and table will be referred to repeatedly. They are grouped with several other basic tables and graphs in Appendix C.

Artifact Assemblages

The total assemblage of ceramics and other objects collected from each site is essential information for interpreting site function. The full evidence for the assemblage from each survey site resides in the assemblage tables. They consist of individual records of all classes of pottery, fine and utilitarian together, including amphorae and cooking ware. The artifact assemblages make it possible to determine not only the periods in which a site was occupied, but also much about the nature of the site, the activities of its occupants, and the relation of individual sites to others. The study of the assemblages is basic to understanding the nature of the settlement in successive periods and is one of the underlying strengths of this study. Sample assemblages representing the major types of sites—farmhouses, necropoleis, and sanctuaries—are illustrated in the discussion of the chora's historical development.

Assemblage Tables

Sample tables from nearly two hundred sites (179) are illustrated in Volume IV. They were chosen because the dates of their fine wares are concentrated narrowly and they represent periods of particular historical importance (**600–550**, **450**, **300**, and **100**). Arrangement is by ceramic class. Within each class, the fragments are described by form/type, followed by date range, function, and quantity. Each assemblage table is accompanied by two small summary tables, arranged by ceramic class and functional group (see below), with total counts and percentages. This basic information is used in all discussions of assemblages and in comparisons of assemblages from different periods.

The plain/banded wares (cf. Ch. 9) make up half or more of typical ceramic assemblages. A wide range of forms is represented, but by far the most common forms are the jug or oinochoe and the lekane, which alone make up half of these wares. Most were in some way involved in food preparation, storage, and service. The next largest single component of assemblages is the black-gloss pottery. Vessels used for drinking and serving liquids predominate in the early periods, but the range of shapes broadens considerably in the 5th and 4th centuries BC. The categories of cooking ware and transport amphora are major constituents of every assemblage, each contributing about 20% of the total.

Much of the pottery in a particular assemblage cannot be assigned to a specific period, as, for example, the plain/banded ware that typically constitute 30–40% of an assemblage. An assemblage often represents an accumulation from one 50-year period to the next on a site, which, in many cases, may have continued in use over a very long period of time. As a result, there will not appear to be very much difference in the composition of assemblages between farmhouses that were occupied for long periods. To describe the changes that take place over time—and changes there were—it would be desirable to assign most of the forms and functions of the undated pottery present to the period in which the majority of the dateable pottery is concentrated. In fact, this has been done for many sites that appear to have had a duration of a century or less, but this must be done with care. There are almost no sites with large scatters that do not have at least some pieces of pottery that belong to a much earlier or later period than that indicated by the majority of the dated pottery. In individual cases, pieces of undateable pottery may be wrongly assigned to a period, but the overarching goal is to recognize general patterns, not to date specific forms to specific periods. In most cases, it is fair to assume that the plain/banded and cooking wares were fairly conservative, and did not experience the relatively rapid evolution of forms observed in most of the fine wares.

Functional Groups

In addition to the formal description, where applicable a single probable function has been assigned to each piece of pottery or other object belonging to an assemblage: for example "drinking" has been assigned to cups, "food preparation" to mortars, and "ritual/votive" to terracotta figurines. These designations are based on the most probable use and certainly do not cover all the possible functions that a piece of pottery may have had. A jug, for example, will always be assigned the function "short-term storage," even though this type of vessel could have been used to serve wine ("drinking service"). With these limitations in mind, the function designations have been applied consistently to all pottery with an identifiable form. The analyses for the assemblages in each period will show that farmhouses, not surprisingly, have very similar percentages of functional groups. The same general sorts of activities were taking place at all of them. The larger assemblages have all of the functional groups and the smaller ones may be missing some of them, but the overall pattern is consistent. The composition of the farmhouse assemblages, however, does change over time. For example, the range of black-gloss pottery shapes and functions, as noted, increases over time, while others decrease and these changes are widespread. The assemblages of functional types, with some notable exceptions, demonstrate a remarkable degree of consistency, at least over time during the period of the polis, **600** to **200**. New forms appear, but the essential functions remain unchanged until the period of Roman control when the changes are significant (Ch. 27).

Very much the same result is apparent in the assemblages from the five rural sanctuaries and the more numerous necropoleis. Not surprisingly, these assemblages differ from those of the farmhouses and from each other in some of their constituent elements, with a greater or lesser emphasis on certain functions, but overall they are amazingly similar in many respects. The importance of this conclusion cannot be overstated. It validates both the methods adopted for the study of the pottery and supports the identification in the field of sites as farmhouses, sanctuaries, or necropoleis.

To assure statistical relevance, a chi-square test was applied to the analysis of assemblages by functional group (*Appendix D*). A number of farmhouses were compared that had dateable fine ware concentrated in four periods, **600–550**, **450**, **300**, and **100**. There were statistical differences in the percentages of functional groups from each of the farmhouses of a single period, but these were caused by only a few atypical dates. There were statistical differences between farmhouses and necropoleis of all periods, and between farmhouses and sanctuaries (see Ch. 21), as would be expected.

Significance and the Final Multiple Criteria Evaluation (FMCE)

Why FMCE?

The study of the pottery and other objects from the site assemblages had made it possible to see them in their historical contexts, to date them using the dateable fine wares, and to understand through the analysis of functional groups the basic activities that took place at the three major types of sites (farmhouses, necropoleis, and sanctuaries). A fundamental question about the nature of this settlement remained: was it as truly egalitarian as might appear at first glance, or were there differences in the material circumstances of the farmers who occupied the farmhouses, cultivated the fields, worshipped at the rural sanctuaries, and buried in the necropoleis of the chora? We can never know how much land the occupant of a farmhouse actually controlled, but if we assume that the land upon which the farmhouse was located represented a significant part of the farmhouse's property, is this a question the results of survey archaeology could illuminate?

Earlier attempts to quantify the amount of land controlled by the occupant of a farmhouse were not satisfactory largely because they involved too many assumptions that are unverifiable.[59] We decided early on in this study to develop a method of determining the relative importance of each site and we call it "significance," which makes use of no external sources or assumptions but relies strictly on the data from the sites themselves.

[59] Carter 1990, 425–30.

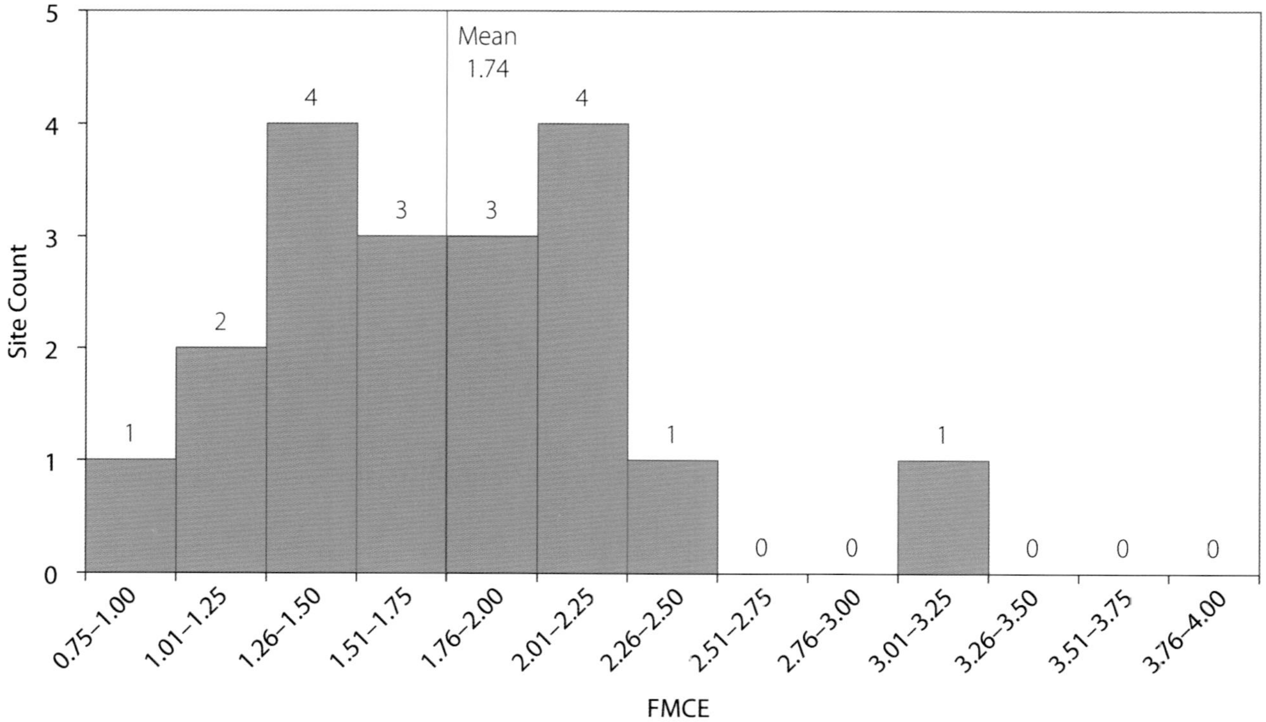

Figure 19.4 Farmhouse Final Multiple Criteria Evaluation (FMCE) distribution in **600**.

A weighting system (described in detail in Ch. 4), is used to evaluate all of the basic evidence recorded by the survey: the size of the scatter, the density of the ceramic roof tile scatter, and the quantities of dateable and undateable pottery. The result is the Final Multiple Criteria Evaluation (FMCE) coefficient. The computed FMCE is not an expression of the physical size of the site, only of its significance in relation to all the other sites in this area of the chora in a particular 50-year period.

One of the basic uncertainties of the survey process is that we can never know what percentage of the original assemblage of the site is represented by the finds collected from the surface. A small number of pottery fragments may actually represent a very large site. One of the advantages of the FMCE analysis is that a correction for this uncertainty is effectively modeled into it (see Ch. 4).

FMCE Tables, Appendix B

FMCE values were calculated for all farmhouses and sanctuaries where there is at least one period with the equivalent of one dated piece of pottery. Therefore all farmhouses and sanctuaries with an EAW ≥ 0.95 are considered significant in at least one period. These values are recorded in Appendix B at the end of this volume.

A graph of FMCE values for **600** shows a very even and almost normal distribution around a mean FMCE value of 1.74 (*Fig. 19.4*). Those for the succeeding periods differ from each other (see *Appendix D: Fig. D1, Table D4*). A chi-square test shows that these differences are significant, and thus that the distribution of values reflecting levels of material accumulation does change over time. This is a positive indication of social and economic change.

Necropoleis have been excluded from the FMCE analysis, in part because the ceramic remains are so sparse, with no dateable pottery from half of them, and in part because the pottery fragments on the surface of surveyed necropoleis, unlike the more deeply buried grave goods found through excavation, cannot reveal as much about the economic level and social status of the occupant. The five sanctuaries were included in the analysis because sanctuaries resemble farmhouses closely in all but one respect: the elevated number of votive objects that define them.

The FMCE calculation was possible only for a restricted number of farmhouses (those with EAW ≥ 0.95). Had more been known about the excluded farmhouses, they would surely have affected the conclusions. This fact should underline the fundamental importance of not considering individual sites, but only the broad patterns that emerge from both the EAW and FMCE calculations. The detailed analysis of the evidence and the rigor of quantitative analysis can only take us so far. In any case, the very nature of survey archaeology—partial preservation of sites and materials—should always warn us against too literal an interpretation of the results.

The Chora at a Glance
Distribution of Significant Farmhouses in Region and Locality, 600–50

The information in the Appendices regarding significance will appear many times throughout the rest of this volume, often imbedded in averages for specific areas or periods. The following four tables represent in numbers the development of the entire study area over the whole of its history.

Appendix C4, "Rates of Occupation for Significant Farmhouses in Localities 600–50," gives the number of occupied farmhouses in a period as a percentage of all farmhouse sites, past, present, and future, in a given locality.

Appendix C5, "Significant Farmhouses in Localities as a Percentage of the Total in the Study Area in Each Period 600–50," complements the previous one by indicating the relative importance of individual localities and their changes over time.

Appendix C6, "Ex novo: Newly Significant Farmhouses as a Percentage of the Total Significant Sites in Each Locality in All Periods 600–50,"shows the newly significant farmhouses as a percentage of the total number of significant farmhouses in each period. This table indicates the change and vitality in the chora highlighting newly created farmhouses on sites not previously utilized.

Appendix C7, "Average FMCE Values, 600–50," shows the average FMCE values of farmhouses from 600–50. Like the others limited to farmhouses, it shows the major trends in the development of all the localities and regions over six centuries. In this table the changing fortunes of the various regions and localities can be read in the numbers of occupied farmhouses and their FMCE averages.

Distribution of Significant Farmhouses in Geomorphic Zones

Appendices C11a and C11b provide a complementary view of the chora's development, based on its soil, topography, and the relation to significant farmhouses. The study of the geomorphology of the chora (Ch. 2) was crucial to the interpretation of the survey results, revealing why some areas could not be expected to produce visible surface remains. It is no less fundamental for an understanding of the relationship between the settlers and the land. The distribution of sites in the principal geomorphological units changes over time and reveals important shifts in the agricultural economy of the chora. The increasing importance of the Pleistocene marine terraces is evident (*Fig. 19.5*). This is the prime cereal growing terrain. As its importance increased, the relative importance of the alluvial terraces (T_0–T_4) in the river valleys decreased. Though the valleys contained the richest soils, they were also the most subject to periodic, sometimes disastrous flooding, a constant feature from the prehistoric period, accelerated by deforestation of the interior and intensive agriculture, down to the present day.

The valley walls were best adapted to grazing and, depending on the slope, also to viticulture and olive growing. As the studies of faunal and plant remains have shown, all of these activities were relatively more important in the early years and the later Roman phase. The soil of the mass movement deposit (*Fig. 19.5*), which is restricted to the Giardinetto locality, was perhaps the most productive of all. An unusually high percentage of sites was located in this aptly named "Little Garden" locality in the first four centuries of the polis.

The geomorphological unit as well as the elevation of each site is recorded in the EAW Table in Appendix A, together with the type of site and the number of dated artifacts recovered from it. The FMCE tables in Appendix B have this information and, in addition, the area of the artifact scatter. The geomorphological study has shown why the number of observed sites can only give a very approximate

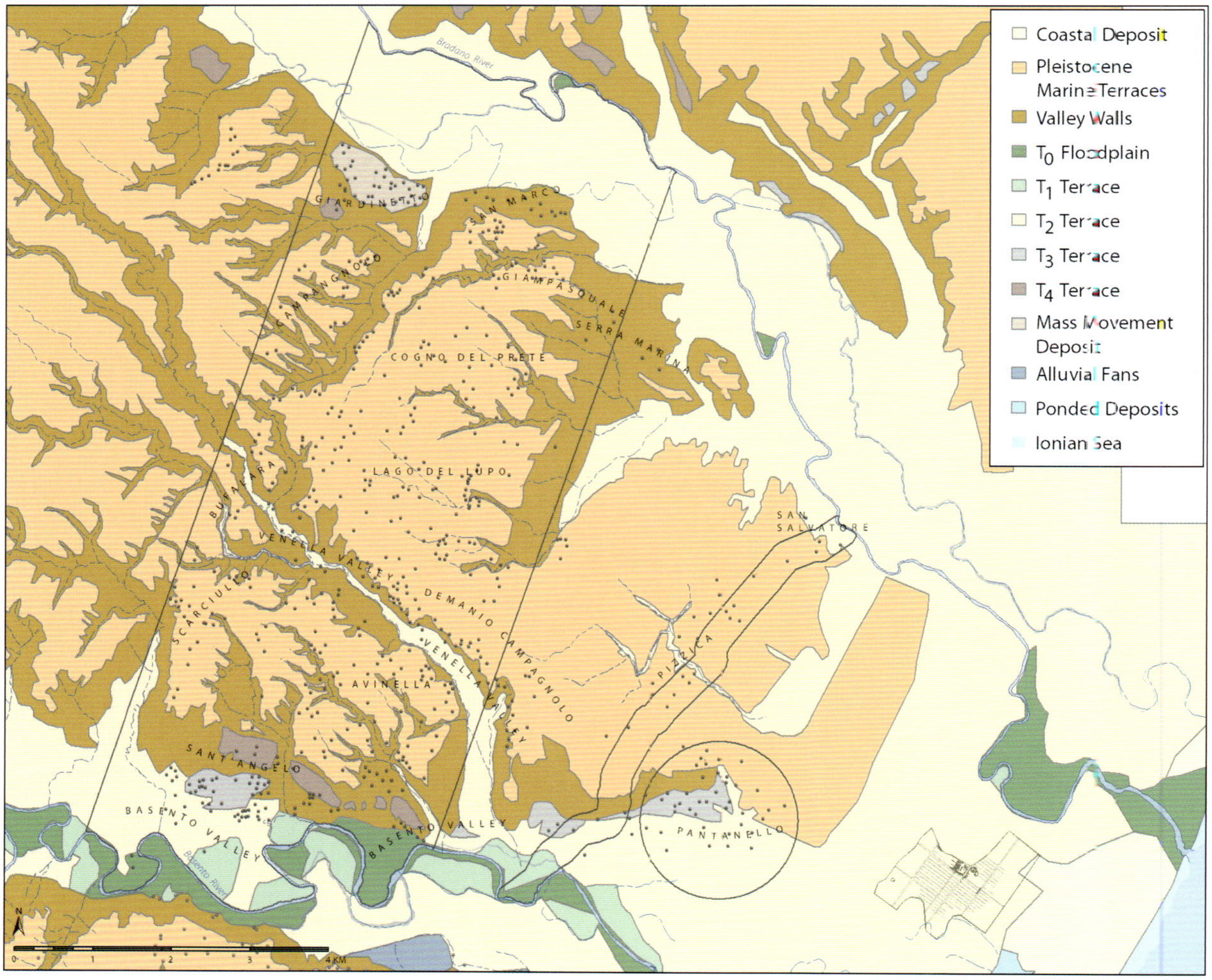

Figure 19.5 Geomorphological map. (JT, after JTA)

idea of the original population of the study area. Many sites were deeply buried by successive episodes of alluviation in the river valleys, and, as noted earlier, only recently has very limited excavation begun to reveal their presence. The geomorphological study has defined the areas of the basic units: the coastal plain, the five alluvial terraces (T_0–T_4), the valley walls, the marine terraces, and the mass movement deposit. This is the basis for an estimate that will be closer to the original number of sites. As is clear from the table of observed density in geomorphological units (*Appendices C11a* and *C11b*), there are about as many sites on the alluvial terraces T_3 and T_4, which are raised above the valley floors, as there are on terraces T_0 to T_2, which are in the valleys but occupy four times the area.

There is every reason to believe that the lower area should have been at least as densely occupied as the higher ground, but let us conservatively estimate that the site density in the valleys was only half of that of the high terraces and about equal to that of the Pleistocene marine terrace, at 16.3 sites per square kilometer. This would result in an original population of terraces T_0, T_1, and T_2 of 90 instead of 51 sites. The difference of 39 sites is considerable. Such a density would result in an overall count for the study area of 604 sites instead of 565, and a density of 16.4 sites per square kilometer (cf. Survey coverage and site density by region and localities above, where the buried sites are not considered, and Ch. 20).

Elevation of Sites throughout the Chora

There is a considerable range in the elevation of sites in the study area, from the low-lying coastal plain to the marine terraces Far from the City. This information was taken directly from the contour maps or the digital elevation model (DEM).[60]

An average elevation for farmhouses and necropoleis has been calculated for each of the 50-year periods (*Appendix C10*). Trends that reflect the developing settlement pattern are immediately apparent. The overall trend for farmhouses is from lower to higher with a major correction in **500** (see Ch. 22) and only minor variations down to **250**, after which dateable necropoleis have disappeared from the record and the character of the settlement changes fundamentally in the final three 50-year periods under Roman domination. This shift to higher ground generally reflects accelerating settlement in the area of the Central Plateau and other smaller areas of the elevated marine terraces. It indicates indirectly the continual importance of cereal growing as this is the terrain ideally suited to that crop. It may also reflect a partial abandonment of the river valleys, which had from the late 6th century BC become less attractive due to increased flooding (see Chs. 22–23).

The trend for necropoleis is just the opposite. Their average elevations appear to vary inversely with those of the farmhouses. Since many farmhouses have necropoleis associated with them, this at first would seem paradoxical. Instead it reflects an historical aspect of the settlement pattern. In the early periods, the majority of farmhouses with identifiable and dateable necropoleis were relatively far from the city at higher elevations. There were very few farmhouses close to the city before the 5th century BC. From that point onward, the settlement began to spread out, and all available land was densely occupied, including the coastal plain. As a consequence, the average elevation of farmhouses started to fall. Furthermore, there was a much higher ratio of necropoleis to farmhouses in those areas closer to the city, which only increased from the 5th century and throughout the 4th and early 3rd centuries BC.

Distances of farmhouses from the city and between nearest neighboring farmhouses and necropoleis

The changing spatial relationship of farmhouses and the urban center has been measured and summarized in a graph (*Appendix C9a*). This reveals fluctuations in the number of sites in three broad, 3-km-wide bands radiating out from the coastal site of the asty. The absence of observed sites close to the city and the preference for settling far from the center in the earliest period (**600**) is evident, as is the preference for the middle distance which included much of the principal grain growing region—the Central Plateau—in the periods of the greatest expansion of the chora (**350–300**) and of Roman domination (**150–50**).

Among the basic statistics to which reference will be made throughout this volume is the distance between farmhouses and nearest neighboring necropoleis (*Appendix C16*) and between farmhouses and their nearest neighbors, which gives an idea of the density of settlement in successive 50-year periods (*Appendix C12*).

This chapter has introduced the concepts, methods, and assumptions underlying the research in the chora. Basic statistics that apply to all periods from **600** to **50** are contained in tables and graphs and have been placed in the back of this volume in Appendix C, for the reader's convenience.

[60] See Chs. 3–4. There is a good correlation between the SRTM DEM and the georeferenced 1:10,000 scale ESACTA map series, on which survey sites were originally recorded in the field. GPS units were used to record site locations (easting, northing, and elevation) from 1999 onwards and those elevations also correlate closely with the SRTM DEM.

20
The Pioneers: Historical Development of the Chora, 600

Joseph Coleman Carter

ἥδε δ᾽ ὥστ᾽ ὄνου ῥάχις
ἕστηκεν ὕλης ἀγρίης ἐπιστεφής·
. . . .
οὐ γάρ τι καλὸς χῶρος οὐδ᾽ ἐφίμερος
οὐδ᾽ ἐρατός, οἷος ἀμφὶ Σίριος ῥοάς.

. . . while Thasos stands here like
the spine of a donkey, wreathed with unkempt forest.

It's not a beautiful or lovely place
or charming, like the Siris River lands.[1]
Archilochos of Paros (second half of 7th c. BC)

This chapter picks up where the narrative begun in Chapter 17 left off, with the Iron Age and 7th century BC. As noted earlier, no distinctively Iron Age forms were recognized among the impasto pottery collected in the Bradano-Basento watershed; most of the fragments of undecorated indigenous wares are probably not pre-Greek in date. Knowledge of the first half of the 7th century BC, a period crucial to the understanding of the early contacts of Greek and indigenous peoples, is based on Incoronata greca, and the other recent discoveries mentioned in Chapter 17, and conclusions about it must remain tentative until these excavations are fully published.[2]

In the second half of the 7th century BC, when the Greek settlers arrived in the territory and on the site of the future urban center of Metaponto, changes took place in the organization and functioning of the chora that stand in sharp contrast to those of all the preceding periods. The organization of the Metapontine chora, and that of the urban center, was practical and logical. It did not just casually evolve, but, as we shall see, was carefully planned from the beginning in response to immediate and pressing needs. The survival of the apoikia depended first and foremost on a rapid solution to the questions of how to distribute land in an extensive territory formerly the possession of others—and how then to manage it.[3] At the same time, there was the question of how to plan the future asty as a religious, commercial, and political center of the polis. There were few comparable experiences on which the early Metapontine settlers could draw. In Achaia, the original homeland of many settlers, the urban experience would arrive two centuries or more after the asty of Metaponto had been laid out.[4] In the discussion that follows, the full range of evidence from the chora will be brought to bear on the question of the city's historical development because, as the research of the past forty years has revealed, both are complementary parts of a whole, the polis.

Considerations about Evidence
The Partially Buried Landscape

The distribution of farmhouses, sanctuaries, and necropoleis active in the first period of the chora clearly demonstrates the attraction that the river valleys had for the earliest settlers of the chora (*Map 5*, Vol. IV; *Figs. 20.1* and *20.2*). This is particularly evident on the Basento side, where sites were found along the lower slopes of the valley, in proximity to good all-weather springs and to the Basento Road, a land route from the coast to the interior that soon became a major thoroughfare. Settlers were attracted to the tributaries of the Basento (the Conca and the Venella) as well as the Bradano (the Fosso San Marco). The Fosso Appio and the Fosso dell'Acqua Fetente played important roles in the settlement but are only partially included in the study area. The extreme alluvial activity in the Bradano may well have been responsible

[1] *Frag.* 18. Text from J.A. Moore, 1961, *Selections from the Greek elegiac, iambic, and lyric poets*, Harvard University Press. Trans. from M.L. West, 1994, *Greek Lyric Poetry*, Oxford University Press.

[2] ICA's excavation at Incoronata will be the subject of a separate study in the series.

[3] The tendency with recent Anglophone scholarship to downplay the importance of the founding, or eliminate it entirely, is illustrated by the recent articles by Morgan and Hall (1996) and Osborne (1998). Earlier studies, e.g. Dunbabin 1948 and Graham 1964, rely on the ancient sources and tend to overemphasize the connection between metropolis and apoikia. The body of archaeological evidence bearing on the early history of the polis has increased dramatically since the 1960s.

[4] According to Morgan and Hall (1996, 169), poleis are certainly attested in Achaia only in the 4th c. BC and the evidence linking Achaia and Achaian colonies is thin (213). Despite a substantial amount of archaeological evidence to the contrary, they deny the existence of an urban center at Metaponto before the mid-6th c. BC (209).

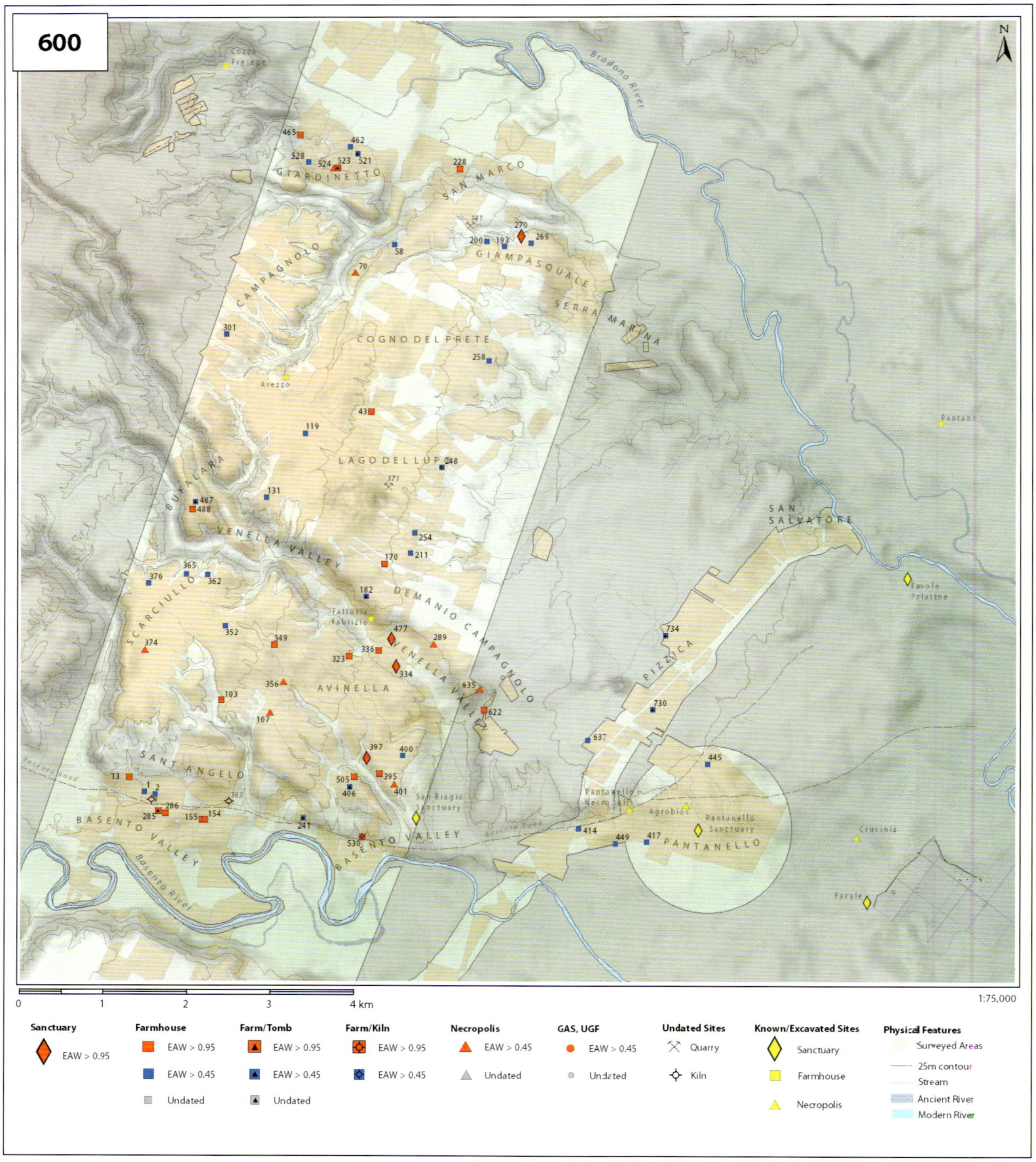

Figure 20.1 The study area in **600**. (JT)

for the absence of detectable sites on the lower slopes of its valley walls. There the sites are discernible at an elevation of about 30 m above sea level and higher, as at Giardinetto on the Bradano, and at slightly lower elevations on the alluvial terraces at Sant'Angelo on the Basento side.

Equally clear is the newly arrived settlers' avoidance of the interfluve, the interior area between the river valleys. This is perhaps to have been expected, since the wide and fertile valleys were a natural means of communication with the nascent urban center on the coast. It is striking that the earliest settlers were apparently not at all timid about forging deeply inland along the valleys far from the coast and near the

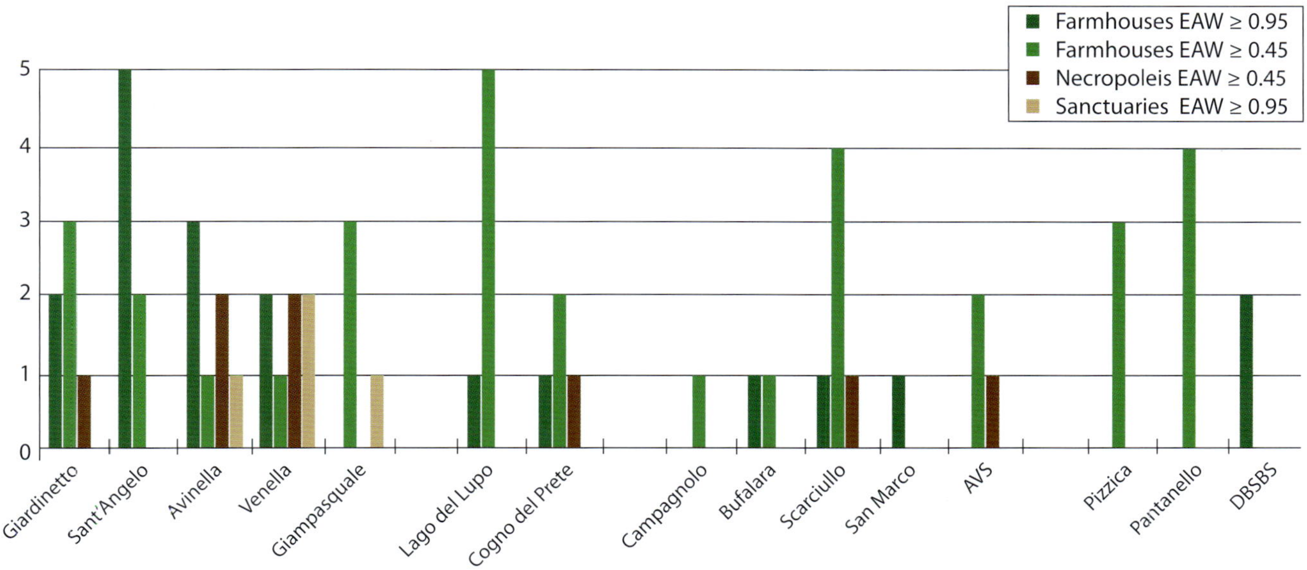

Figure 20.2 Distribution of farmhouse, necropolis, and sanctuary sites among localities, **600**. Totals: significant farmhouses, 19; second magnitude farmhouses, 34; necropoleis, 8; sanctuaries, 4.

earlier hilltop settlement of the indigenous population at Cozzo Presepe, which, probably not coincidentally, began to show evidence of a strong Greek influence, if not of an actual Greek presence, at about this time.[5] The fact that there is Early Corinthian pottery on that site, of course, is no indication that Greeks were present there in the late 7th century BC.

Less clear, however, is the extent and density of the valley settlements in this period. As the geomorphological study in Chapter 2 has shown, the ancient landscape in the river valleys has been buried, sometimes by as much as 3 m of alluvial sediment. The sites on the higher T_3 and T_4 terraces are unlikely to have been covered by these thick alluvial deposits; but the valley floor, at a level contemporaneous with Greek settlement elsewhere in the chora, is often well below the level on which modern agricultural activity in the valley occurs today. Sites located in the valleys have been invisible to the survey; however, during excavations in the path of the SNAM pipeline in 2000–2002, farmhouses and necropoleis were found all across the valley up to the rivers, where they were waterlogged and often well preserved.[6] Farmers apparently were not deterred by the possibility (which must have been frequently realized then as in more recent times) that the stream would rise over its banks and flood the surrounding countryside.

The attractions of the fertile land and nearby water were strong. Sanctuaries grew up beside springs at the valleys' edge (T_2), which fed the two rivers. Had not the ancient landscape near the rivers been masked by alluvial deposit, this area would probably have been discovered to be among the more densely occupied in the first century of the settlement.

At Pantano (*Fig. 20.1*), just north of the Bradano and outside of the Bradano-Basento transect,[7] the excavation of a group of square, one-room dwellings (*oikoi*) revealed what appears to be among the earliest evidence of permanent colonial residence in the chora. This and similar discoveries have altered the picture of the historical development of the Metapontino in several important respects. The structures at Pantano, it would seem, belong to groups of similar dwellings forming what could be described as a hamlet. These oikoi are comparable in form and dimension to the 7th-century ones discovered by De Siena in the Basento valley below San Teodoro (see Ch. 17). Both sets of excavated houses—those along the Bradano and those beside the Basento—were probably at least as old as the earliest surviving structures in the area of the temples and agora of the asty (Oikos C and the ikria). De Siena believed that the oikoi below San Teodoro were contemporary with the 7th-century BC huts (*capanne*) discovered under the street grid of the residential quarter of the asty.[8]

[5] *Cozzo Presepe*, 214–23, 331–43. See Ch. 17.

[6] De Siena, pers. comm., 2002.

[7] This is an area of the Metapontine chora under the administration of the Soprintendenza of Puglia (see Ch. 19). See Andreassi 2002, 800–01, pls. 85–6; Schojer 2002, 81–84.

[8] Nava 2002, 738, pls. 67.2, 68.1; De Siena 1986b.

Previously known early sites on the valley floor included the Pantanello sanctuary,[9] which lay beside an all-weather spring under more than 1 m of river-borne alluvium and slope wash. That site would have entirely escaped detection had it not been for a recently cleaned modern drainage canal that cut through it (see *Fig. 18.17*), revealing a consistent layer of Archaic pottery and a terracotta image of a female divinity (see *Fig. 18.18*).[10]

Dating the Earliest Sites in the Chora

The earliest sites in ancient settlements are usually few and almost always the most difficult to date precisely. Not only is there relatively little material to study, there is often also a wide range of possible dates for it. The chora of Metaponto is no exception. For example, one cannot date the locally-made fine wares of the second half of the 7th century and first half of the 6th century with the same precision—to the half-century or even decade—that is possible with the much greater production of the Metapontine kerameikos in the late 6th to early 3rd centuries (see Ch. 6).

The chronology of the black gloss and earlier fine wares employed in this study, and which supports the reconstruction of the chora's settlement in all periods, was adopted in 2002. In 2009, it was substantially revised by the authors of Chapter 6 in light of recent advances in the field. By the time the revisions to Chapter 6 had been made, the manuscript of Volume II was complete. The options were, essentially, two: to make extensive revisions to the early chapters of the book; or to compromise—to retain the earlier chronology and offer the additional data contained in Chapter 6 as well as an alternate interpretation of the earliest period. This second option was chosen because the dating of the succeeding periods is relatively unaffected and because, after all, the study of the relevant pottery—and thus the chronology—is still far from definitive.

The discipline of archaeology, by necessity, is based on relative chronologies that are constantly in flux as new material is discovered and as new methods of analysis are developed. Such variations in dating were anticipated when the analysis employed here was developed. Different historical reconstructions of the chora are, of course, possible, and two will be presented here (see below, The Chora in **600**: An Alternative View). The goal of this study is not a single reading of the data and their relationships, or even a single version of historical events. The quantitative analyses employed throughout aim to be rigorous so that the arguments are clear, but the ultimate goal is to trace the broad movements in the population of the chora.

In the reconstruction of the chora's development in the following pages, based on the original 2002 higher chronology, both significant and second-magnitude sites are included in the discussions of **600** and **550**.[11] The reason for the exceptional treatment of farmhouses in these early periods is that the quantities of Archaic fine wares—the sole component of the EAW calculation—are much smaller. Including both significant and second magnitude farmhouses helps to define the areas that were frequented and provides a fuller and more nuanced picture of the earliest settlements. In subsequent periods, only significant farmhouses will be discussed, although the second-magnitude sites continue to be shown on the maps.

The Chora in 600
Early Village Clusters

Groups of two, three, or four sites occupy areas along the major river valleys: at Giardinetto and Giampasquale along the Bradano, at Sant'Angelo and Avinella along the Basento, and in the mid-Venella valley (between its mouth at San Biagio and its headwaters in the interior). There would certainly have been many more such clusters outside the study area along the edges of all river valleys deep in the interior. Coverage, even within the study area, was far from complete. Site 530 along the Basento, for example, probably belonged to a large group of sites, but regrettably that cannot be known, as the area immediately surrounding it had been bulldozed before the survey began. A small percentage of undated farmhouses would have belonged to the early settlers, as well as some of the undated necropoleis that probably related to the dated farmhouses of this period. Together these would have added to the total number of sites in a cluster.

Giardinetto. Giardinetto is a theater-like hollow, which opens to the north and slopes down to the Bradano River. To the south, it rises to the level of the marine terraces, from which it is cut off by a deep

[9] Carter 1994a, 183–98; 2006a, 163–73.

[10] The discovery was made by Giuseppe Di Taranto, foreman of the ICA excavation team 1974–present (Carter 1974, 12).

[11] Farmhouses with an EAW value ≥ 0.95 are considered significant; those of the second magnitude have an EAW ≤ 0.95 but ≥ 0.45. In all periods, necropoleis with an EAW value ≥ 0.45 are considered active.

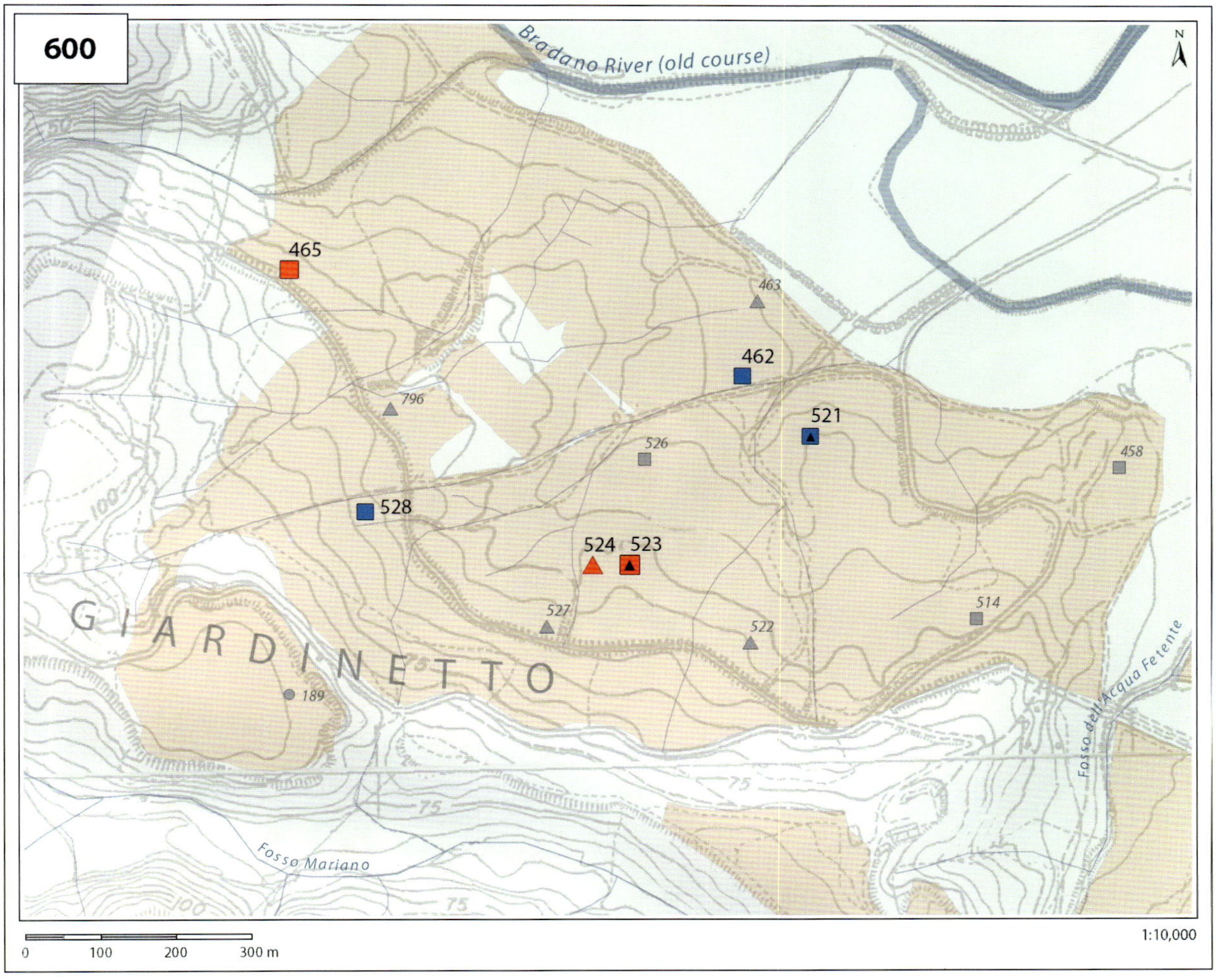

Figure 20.3 Giardinetto in **600**. See Fig. 20.1 for map key. (JT)

drainage that forms a major tributary to the Acqua Fetente (*Figs. 20.3–5; Table 20.1*). This self-contained area attracted some of the earliest settlers to the chora. Besides being supplied with water from two springs, it occupied a convenient middle ground between the rich soils of the river valley and the extensive high plains of the Central Plateau (Cogno del Prete and Lago del Lupo), and it is near enough to both to be easily reached on foot. The best lands of the Central Plateau are no more than an hour's walk away. The terraces stretch about 1.3 km east to west, parallel to the river valley, and slope down from south to north for a distance of about 0.5 km. Their elevation ranges from between 30 to 70 masl.

Site no.	Period			
	600	550	500	450
465F	1.4	3.7	0.7	3.8
462F	0.8	2.2		2.1
521F/T/K	0.5	0.5	0.2	7.4
523F/T	2.3	9.3	2.4	1.3
524N	0.6	1.4	0.6	2.4
528F	0.5	0.5		

Table 20.1 Giardinetto **600,** with EAW values rounded to the nearest tenth through **450.** Dark green = significant farmhouses or sanctuaries (EAW ≥ 0.95) and active necropoleis (EAW ≥ 0.45). Light green = second magnitude farmhouses (EAW ≥ 0.45, but ≤ 0.95).

Farmhouses. Sites 465F and 523F/T sit on a level terrace at the same elevation (50 masl), with commanding views of the surrounding countryside.[12] They are separated by just under 600 m, but are on different sides of a small, spring-fed stream. Site 528F is located between them at a slightly higher elevation (63 masl).

[12] During a visit to site 465F in October 2007 the soil was being removed from the site and it was being plowed down the hill.

Figure 20.4 Giardinetto, **600**. 3-D reconstruction using a satellite image of 2006 draped over a DEM, looking south. (ML)

Sites 521F/T/K and 462F are on a similar, but lower terrace downslope, roughly 300–400 m from the old course of the Bradano.

Giardinetto covers a surface area of only 0.91 km^2; the cluster covers a little less than half of it. There is no area of low density or open space—this is the only cluster in the study area where this occurs. This locality rapidly became one of the most densely packed in the chora, and only the Sant'Angelo and Venella settlements approach it in this respect. The particular character of the soil—it is a unique mass movement deposit (MMD)—and the presence of springs certainly contributed to this unusual situation.

Farmhouse 521F/T/K had a kiln. Several early sites, on the Basento side, were also associated with kilns, e.g. 12F/K at Sant'Angelo and, in the Basento valley, 530F/K, and the kilns may also have been functioning with the first generation of inhabitants at this site in the chora. It may well be the case that already from an early period, the clusters of early farmers were self-sufficient in this as in other respects. Although the production of Ionic cups in the chora has been documented by excavation,[13] as yet few traces of ceramic production have been found at sites—other than these farmhouses—before the later 6th century BC. Unfortunately it is not possible to date the kiln activity independently from the farmhouse, and 521F/T/K was also in use for two centuries after 450 BC. The kiln may have belonged to the later farmstead, or it may have operated continuously over the whole life of the farm. The basic requirements—water and a good source of clay—were present in all periods.

Necropoleis. The pairing of farmhouses and necropoleis is our strongest indication of permanent residence in the chora. The most obvious reason for associating a farmhouse and a necropolis is the distance between them. The typical distance between a farmhouse and related family burial grounds in the study area has been calculated at 50–170 m (*Table 20.2, Appendix C16*). While it is more likely that a necropolis was first used in the period to which the earliest recovered pottery belongs, there is still the possibility that it was already in use as a burial ground in an earlier phase of the chora. As excavations have shown, the earliest burials in the chora have few, if any, grave goods. The possible relationships between farmhouses and places of burial are crucial to interpreting the nature of a settlement. The farmhouse and related necropolis was a basic unit of the settlement from the beginning, but farmhouses could also have been occupied by families who buried elsewhere.

[13] A. Bottini 1994.

Figure 20.5 Giardinetto, looking northwest toward the Bradano valley and Montescaglioso. (JCC)

Farm	Period	Distance	Necro.	Period
465F	**600–550**, 500, **450–250**, 200	100 m	466N	**400–350**
		170 m	467N	(undated)
523F/T	**600–300**	50 m	524N	**600, 550, 500, 450**
		140 m	527N	(undated)
521F/T/K	600–550, **450–250**, 200	50 m	520N	**300**
528F	600–550, **350–250**	130 m	796N	(undated)
462F	600, **550, 450–400, 300**, 250	100 m	463N	(undated)
		100 m	791N	**400–350**

Table 20.2 Giarcinetto, farmhouses and distances to necropoleis. Bold dates indicates periods of significance (EAW of farmhouses is ≥ 0.95, and that of necropoleis is ≥ 0.45).

Farmhouse 521F/T/K had burials right beside it, as did 523F/T. Although the association is beyond question, the dates of the burials cannot be determined independently, and so doubt remains as to whether the earliest occupants buried on the site. The family at 523F/T may also have utilized necropolis 524N, only 50 m away to the west. This necropolis was in continuous use from **600** to **450** and is the most secure evidence that some early settlers of Giardinetto buried their dead in the village.

Farmhouse 462F may have used 463N, an undated necropolis 100 m away, or 796N, also undated and some 100 m downslope toward the river. Site 528F was 140 m from undated necropolis 796N. These are purely hypothetical pairings based on proximity, and without dates they must remain so.

Significance, land, and continuity. In **600**, there were two significant farmhouses (465F and 523F/T) and three of second magnitude (462F, 528F, and 521F/T/K). In **550**, the significant farmhouses increase to seven (with four second-magnitude farmhouses; see *Table 21.1*). Such a group of contemporaneous dwellings could be considered a hamlet or a very small κόμη.[14] As newcomers moved into the area, the size of the settlement increased as they further defined the arc, which arrangement gave all the farmhouses a clear view of their neighbors and the valley below (*Fig. 20.4*).

The raising of livestock was an important aspect of the Metapontine economy from the 6th century onward when new and improved breeds of livestock

[14] The Greek term for a self-contained group of houses in a rural setting is *kōme*, which is variously translated as "township," "village," or "hamlet." As an alternative, M.H. Hansen (pers. comm., March 2011) has suggested χώριον (see also M.H. Hansen 1995, 50). Hamlet is the word in English that best approximates the situation at Giardinetto—a small group of houses, especially one apparently without a church, or in this case, a sanctuary or obvious site for the community to congregate (see discussion on Villages in Ch. 21).

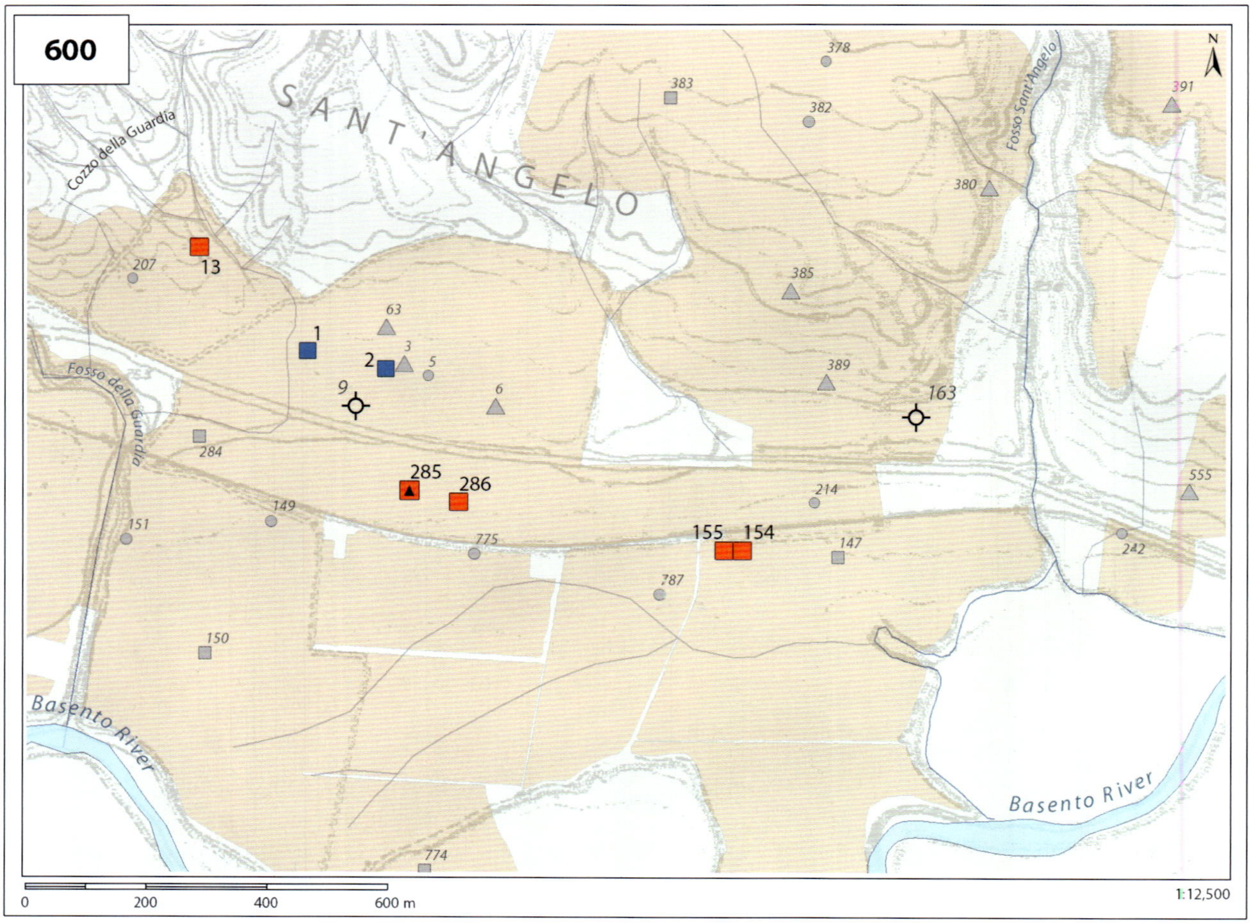

Figure 20.6 Sant'Angelo in **600**. See Fig. 20.1 for map key. (JT)

were introduced.[15] Giardinetto was well placed for grazing and watering herds. While the earlier inhabitants of this area of the chora could produce a surplus of crops on fields adjacent to their dwellings, this would not have been the case as the population grew. It would have been necessary for farmers of Giardinetto to move to other, vacant lands to be able to produce surpluses comparable to those possible in the preceding period. Besides the valley below the nascent hamlet, the obvious areas for expansion were the marine terraces to the south and the extensive, gently rolling hills of the Central Plateau. The rapid economic growth of the Metapontino in **550** would support such a scenario (see Ch. 21).

Sant'Angelo. In its location and topography, this area of the Basento valley is almost a mirror image of Giardinetto on the Bradano (*Figs. 20.6–8*; *Table 20.3*).[16] It lies below the marine terraces, protected by their heights and that of the oval-shaped hill known as Cozzo della Guardia to the north and northwest. To the south lies the Basento River. On the east the locality is bounded by the deep Fosso Sant'Angelo. At a distance of about 9 km from the coast, it was marginally closer to the early asty than Giardinetto (at 11 km).

The terrain consists of a series of gentle terraces that gradually merge with the valley (*Fig. 20.8*). An old bed of the Basento is within 500 m of the early sites. The terraces stretch 1.4 km from east to west, and slope down to the river valley for a distance of about 0.5 km. There is a degree of surface visibility of ancient sites in this alluvial plain that is not encountered in the Bradano valley, and sites dating to later periods have been detected close to the old river bank.[17]

[15] See Carter 2006a, 79; Bökönyi and Gál 2010, 17–29.

[16] This cluster has been named Sant'Angelo, though it contains sites from two localities: Sant'Angelo and the Basento valley.

[17] For example, site 105F/K, first significant in **400**.

Figure 20.7 Sant'Angelo, **600.** 3-D reconstruction using a satellite image of 2006 draped over a DEM, looking north. (ML)

This locality has the appearance and the feeling of being nestled, like Giardinetto, but without the focus and scooped out effect that makes the other setting unique. The terraces here are flat or slightly convex, and they are lower, ranging from about 30 to 20 masl. A shallow drainage divides the area in two, with each half occupied—as at Giardinetto—by farmhouses. The similarity of the two areas perhaps reveals something about the motivations of these early colonists. They may have been drawn to these cozy niches between the high terraces, which could be utilized for grazing and eventually for cereal cultivation, and the valley. The fertile, alluvial plains of the Basento, well supplied with water from springs, were apparently unoccupied at this time and were thus available for a variety of purposes, including pottery manufacture. A kiln (see *Fig. 18.25*), in fact, lies just below site 1F (*Fig. 20.10*), and both were excavated by ICA.[18] There is also an active spring.

Site no.	Period			
	600	550	500	450
13F	1.5	3.5	0.2	3.2
1F	0.6	1.4	0.1	3.3
2F	0.5	0.5		
285F/T	10.1	8.6	1.3	1.3
286F	1.0	4.0		0.4
155F	3.7	10.5	6.3	3.4
154F	2.8	4.2		0.8

Table 20.3 Sant'Angelo **600,** with EAW values rounded to the nearest tenth through **450.** Dark green = significant farmhouses or sanctuaries (EAW ≥ 0.95) and active necropoleis (EAW ≥ 0.45). Light green = second magnitude farmhouses (EAW ≥ 0.45, but ≤ 0.95).

Farmhouses. With the exception of the solitary 13F, the early farmhouses here appear to be grouped together: 285F/T and 286F along with 1F and 2F to the west of the drainage, and farmhouses 154F and 155F to the east. The pottery assemblage of site 285F/T (*Fig. 20.9*) which includes 188 fragments, 108 of which are identifiable as to form, is the largest from an early farmhouse. It is one of the few with a fragment of indigenous pottery (*Fig. 17.14*). The latter two are less than 400 m from the old course of the river, across the valley bottom. Both the first and second magnitude sites at Sant'Angelo follow the slope of the terraces in a linear pattern. This arrangement strongly indicates that they were placed close to the Basento Road, which led up the valley toward the interior. The impression of farmhouses lining the road is strength-

[18] Site 1F was occupied from **600** through the Early Imperial period. The site will be the subject of future study.

Figure 20.8 Sant'Angelo terraces and Basento valley at dusk. In foreground is Cesare D'Annibale, who led the survey team from 1981 to 1991, and from 2003 to present. (JCC)

Figure 20.9 A selection from a very large assemblage from site 285F/T in Sant'Angelo (significant **600, 550, 500, 450**). (MLG)

ened by the location of other sites of this period further down the valley in areas, which, unfortunately, were only selectively surveyed. The significant site 530F/K, much further down the valley, was probably located by the road or not far off it (see *Fig. 20.1*).

Necropoleis. One notable difference between the Sant'Angelo cluster and most others is the apparent scarcity of dateable necropoleis of any period near the farmhouses. Only at site 285F/T are burials closely associated. The pottery of the burials, accompanied by human bone, was inextricably mixed with that of the dwelling. While it is not possible to be absolutely certain about their date, we may surmise that burials from the first or second generation occupied the site.[19] The isolated necropoleis in this area, some of which were close to early farmhouses (e.g. 63N and 3N), were not dateable, and so possible associations are at best hypothetical.

Significance, land, and continuity. The three areas of significant farmhouses in **600** were separated by about 600 m—the same distance between the two in Giardinetto. The second magnitude pair of 1F (*Fig. 20.9*) and 2F rose halfway between 13F and another pair, 285F/T and 286F. Again, the regularity of this spacing recalls that at Giardinetto. The dating of farmhouses with EAW ≥ 0.95 indicates that the farmhouses were widely spaced, both at Giardinetto and Sant'Angelo in the earliest period. Each could have controlled an area of about 28 ha. Infilling would have begun in the next 50-year period, when the original plots were subdivided and the adjacent area available for cultivation greatly diminished. This is the most likely scenario. A lower second-magnitude threshold (EAW ≥ 0.45) for active farmhouses would suggest that the land surrounding farmhouses was limited from the beginning to about 7 ha.

The pairing of sites, a phenomenon evident in many areas of the chora in **550**, is already visible at Sant'Angelo a half-century earlier. On the evidence of site 285F/T, the most significant and the earliest farmhouse in the whole of the study area in **600**, it is tempting to conclude that the Sant'Angelo settlement was more precocious in its development than Giardinetto and the other clusters. Only five farmhouse—465F in Giardinetto, 228F in San Marco on the Bradano side, 155F in Sant'Angelo, 505F in Avinella, and 530F along the Basento—have FMCE values greater than 2.0 in **600** (*Appendix B*).

[19] The majority of the dateable pottery from the site is from the earliest period (EAW 10.09 in **600** and 8.64 in **550** with much lower values after that).

Figure 20.10 ICA excavation at survey site 1F, the Greek and Roman farmhouse at Sant'Angelo Grieco, July 1981, two weeks after it was recorded by the survey team. (CW)

Avinella. The Avinella cluster differs from both the Giardinetto and Sant'Angelo clusters. It was located along the Conca drainage, a smaller tributary of the Basento, parallel to and west of the wide valley of the Venella (*Figs. 20.11–15; Table 20.4*).[20] The Avinella cluster is inland and hidden away from the broad valley of the Basento, as is the Venella cluster. The settlers along the Conca gave up direct and easy communication with the coast and the asty. This narrow valley, however, offered an easy route for penetrating the interior and gaining access to the level expanses of the marine terraces. The soils of the marine terraces, lighter with more sand and less clay, are ideal for growing cereals. A similar choice of topographical setting but on a grander scale characterized the Venella cluster.

The valleys offered the settlers then, as they do shepherds today, shelter and an enclosure for their herds. Sheep and goats, as well as cattle, were an important component of the Metapontine economy, particularly in the Archaic period,[21] and required a constant, reliable source of water. The valley floors furnished lush grass, and the native scrub (*macchia*) characteristic of the sloping valley walls provided material for browsing goats. Thus, along the Conca valley drainage, the settlers were well situated to exploit both the agricultural and the grazing potential of the land, and they were even more conveniently located with respect to both than were the settlers at Giardinetto or Sant'Angelo (*Fig. 20.11*).

The long drainage with dramatic steep sides that passes below Masseria Sant'Angelo Vecchio to the east of the Sant'Angelo cluster is comparable in many ways (see *Fig. 20.1*), yet, apart from the Greek farmhouse and necropolis known as Sant'Angelo Vecchio, it attracted few settlers before the period of Roman domination in the 2nd century BC. Perhaps its steep slopes were too rugged, making it difficult to reach the valley bottom.

Farmhouses. Some early sites—395F and 505F along the lower Conca, and 349F in the upper valley—were placed deliberately at the very edges of the narrow valley. Each

[20] The Avinella cluster contains sites from the Avinella and Venella localities.

[21] Carter 2006a, 31.

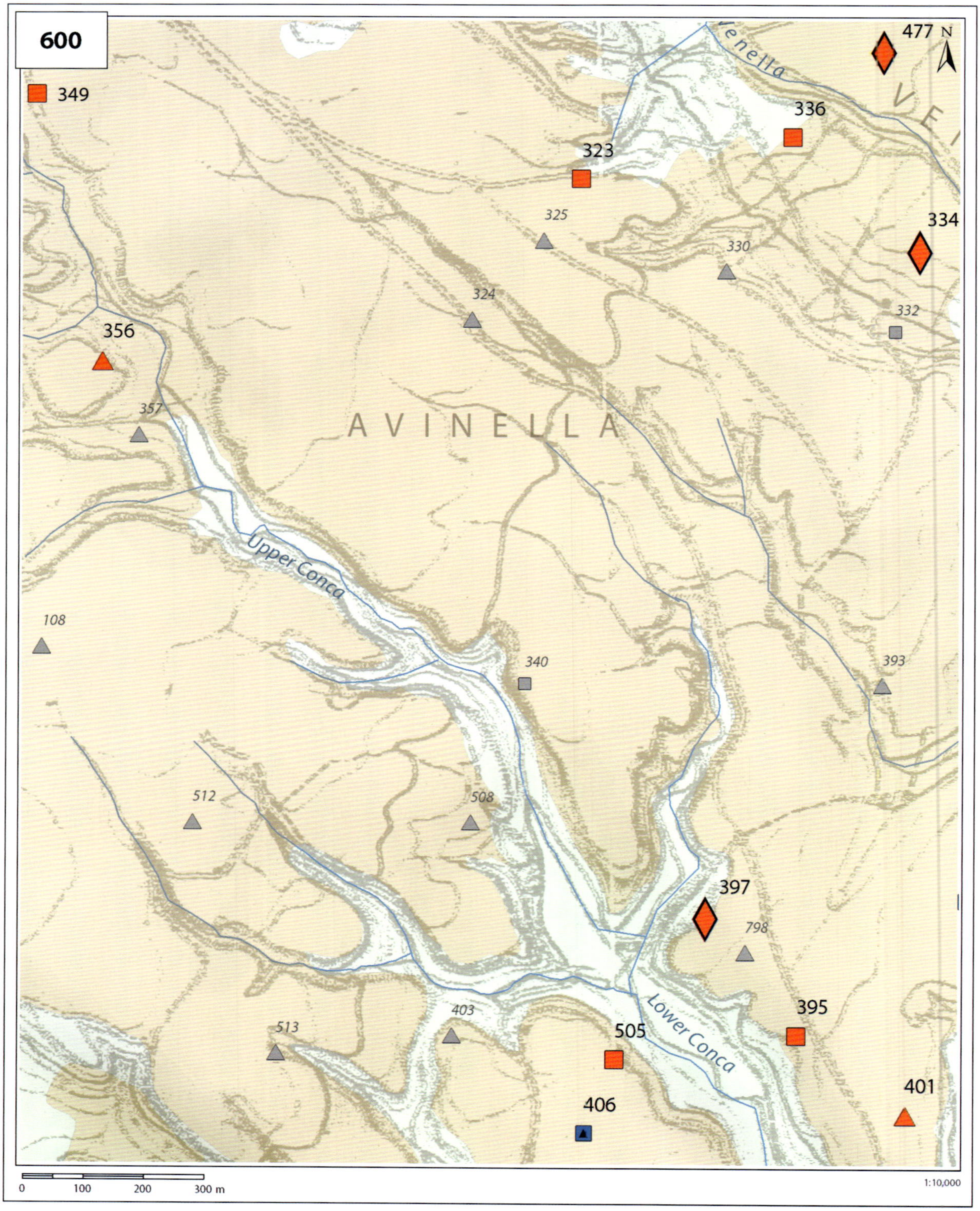

Figure 20.11 Avinella in **600**. See Fig. 20.1 for map key. (JT)

occupies its own point or headland (*Fig. 20.11*). These were attractive positions because they were at the edge of good, flat, fertile fields and near water. The stream lies below and springs issue from the valley walls at the interface between the sandy mantle of the marine terraces and the impermeable clays beneath.

Figure 20.12 Avinella, **600.** 3-D reconstruction using a satellite image of 2006 draped over a DEM, looking northwest. (ML)

A number of the early settlers chose a strategic point up the valley where three smaller tributaries join to form the larger stream. From here to site 349F near the head of the valley the distance is 1.5 km. The occupants of this distant farmhouse appear to have been fearless pioneers, independent-minded and not at all bothered by not having close neighbors; this may even have been a motivating factor. Such solitary beings are found on most frontiers. The closest farmhouse in **600** was ca. 1 km away on the Venella.

The area around the mouth of the valley where it joins the broad plain of the Basento could not be completely surveyed, but, if site 530F is any indication, this area was not ignored by the settlers. The 1 km wide space between sites 530F and 505F, however, shows no evidence of human occupation in **600** or at any time thereafter.

	Site no.	Period			
		600	**550**	**500**	**450**
Lower Avinella	397S	8.6	34.8	9.3	6.1
	395F	1.4	4.0	1.2	5.6
	505F	1.0	1.0		
	406F/T	0.8	2.2		0.4
Upper Avinella	356N	1.1	2.2	0.7	
	349F	1.0	3.7	0.3	
	107N	0.5	0.5		

Table 20.4 Avinella **600,** with EAW values rounded to the nearest tenth through **450.** Dark green = significant farmhouses or sanctuaries (EAW ≥ 0.95) and active necropoleis (EAW ≥ 0.45). Light green = second magnitude farmhouses (EAW ≥ 0.45, but ≤ 0.95).

Necropoleis. Although one would expect that the early settlers of the Conca, like those of Giardinetto, buried their dead in the vicinity of their dwellings, the distance separating farmhouses and tombs is considerably greater, and the relationship between them harder to establish. There are three active necropoleis in or near this cluster. Two (356N and 107N) lie along the edge of the upper section of the main stem and along a tributary. There is no obvious connection between the smaller necropolis 356N and a farmhouse. It seems unlikely that this well-dated early necropolis

Figure 20.13 View of the macchia-covered slopes of the Conca in Avinella. (JCC)

prominently perched on the valley edge was isolated, but the best candidate is a undateable structure, 359GAS (Greek agricultural structure), which could have been a shed, temporary shelter, or possibly a very poorly preserved small farmhouse, some 120 m to the west on the same plateau. Likewise, site 107N, at the end of a small tributary of the Conca, may have been associated with the undated 106F (*Map 5*, Vol. IV).

A large necropolis site, 401N, continuously in use during the whole period of occupancy of the chora, stands apart about midway between the Conca and the Venella. Though apparently isolated—it is 200 m from the nearest farmhouse (395F)—it was probably associated in some way with the major sanctuary at San Biagio. It lay on a line midway between the already famous cult place of Artemis and the small sanctuary at the "three forks". The distance to each sanctuary from 401N—roughly 500 m—is approximately the same as the distance between the Pantanello necropolis, which was also in use at this time, and the Pantanello sanctuary. In both cases, sanctuary and necropolis seem to have served a wider territory. It is tempting to conclude that the early farmers in Avinella sought to bury their dead at this communal, albeit more distant, burying ground.

Sanctuaries. The sanctuary site 397S occupies a central position in the Avinella cluster above the junction of the "three forks" (*Figs. 20.14* and *20.15*). Site 397S was identified as a sanctuary because of the very high concentration of miniature votive vessels and terracotta figurines (*Fig. 20.16*). The scatter, which covered a relatively small area, also contained an extraordinarily high proportion of dateable pottery, most of it collected during the first visit in 1982.[22] Four other sites in the study area identified as sanctuaries had similarly elevated numbers of miniatures and terracotta figurines. Three of these—334S and 477S in the Venella and 270S in Giampasquale—date to the earliest settlement of the chora, and are likewise at or near the center of their respective clusters.

Significance, land, and continuity. At the "three forks," the early sites face each other across the valley and there is a certain regularity in their spacing that suggests that limits on expansion were in force and respected. The distance between significant sites 397S and 395F is approximately 300 m; between 395F and 505F, 300 m; between 505F and 397S, 300 m. They thus form an equilateral triangle. Some farmhouses are at greater or lesser distances from each other. One second magnitude site, 406F/T, for example, is located in the center of the flat terrace to the south of the Conca and is only about 120 m from the significant site 505F on the edge of the ravine;

[22] The total counts of pottery fragments from three visits include: 174 black gloss, 27 banded ware, 111 plain ware, 11 cooking ware, 15 amphorae, along with 85 miniature votive vessels and 10 terracotta figurines (see Gazetteer, Vol. IV). The best preserved example from site 397S has been dated by R. Ammerman (Ch. 13) to the second quarter of the 6th c. BC, thus slightly later than the first use of this site. See **No. 210** in the catalog in Ch. 13.

Figure 20.14 Sanctuary site 397S above the "three forks" of the Conca in Avinella, 1982. (CDA)

Figure 20.15 Sanctuary site 397S in Avinella after the site had been converted to a vineyard, with Cesare D'Annibale, 2007. (JCC)

another second magnitude site (400F) lies considerably farther away from its neighbors. The distance of 200 to 300 m separating farmhouses in Avinella seems to be standard in the settlement of the chora in this period and was not much altered later on. The early settlement of the Conca and surroundings enjoyed a remarkable continuity throughout the whole of the history of the chora and into the Roman period.

Figure 20.16 Assemblage from sanctuary site 397S in Avinella. (MLG)

Venella. The Venella cluster is located halfway up the course of the Venella, along both sides of the stream. It consists of two sanctuaries and three farmhouses that can be confidently dated to 600 (*Figs. 20.17–19; Table 20.5*).[23] In most respects comparable to the Avinella cluster, Venella differs in one obvious way. While some sites occupy the valley's edge, others take full advantage of the much wider valley floor, and convenient access to water and well-watered pastureland. This cluster had easy access to vast expanses of relatively level, well-drained fields on either side of the valley, on the plateau to the south between the Venella and the Conca, and the far greater Central Plateau to the north between the Venella and the Bradano.

Farmhouses. One of the farmhouse sites, 336F, lay on the same middle terrace as sanctuary 334S (36 masl), while the other, site 323F, is on the edge of the plateau some 40 m higher. A distance of 350 m and a steep

[23] The Venella cluster contains sites from the Venella, Avinella, and Demanio Campagnolo localities.

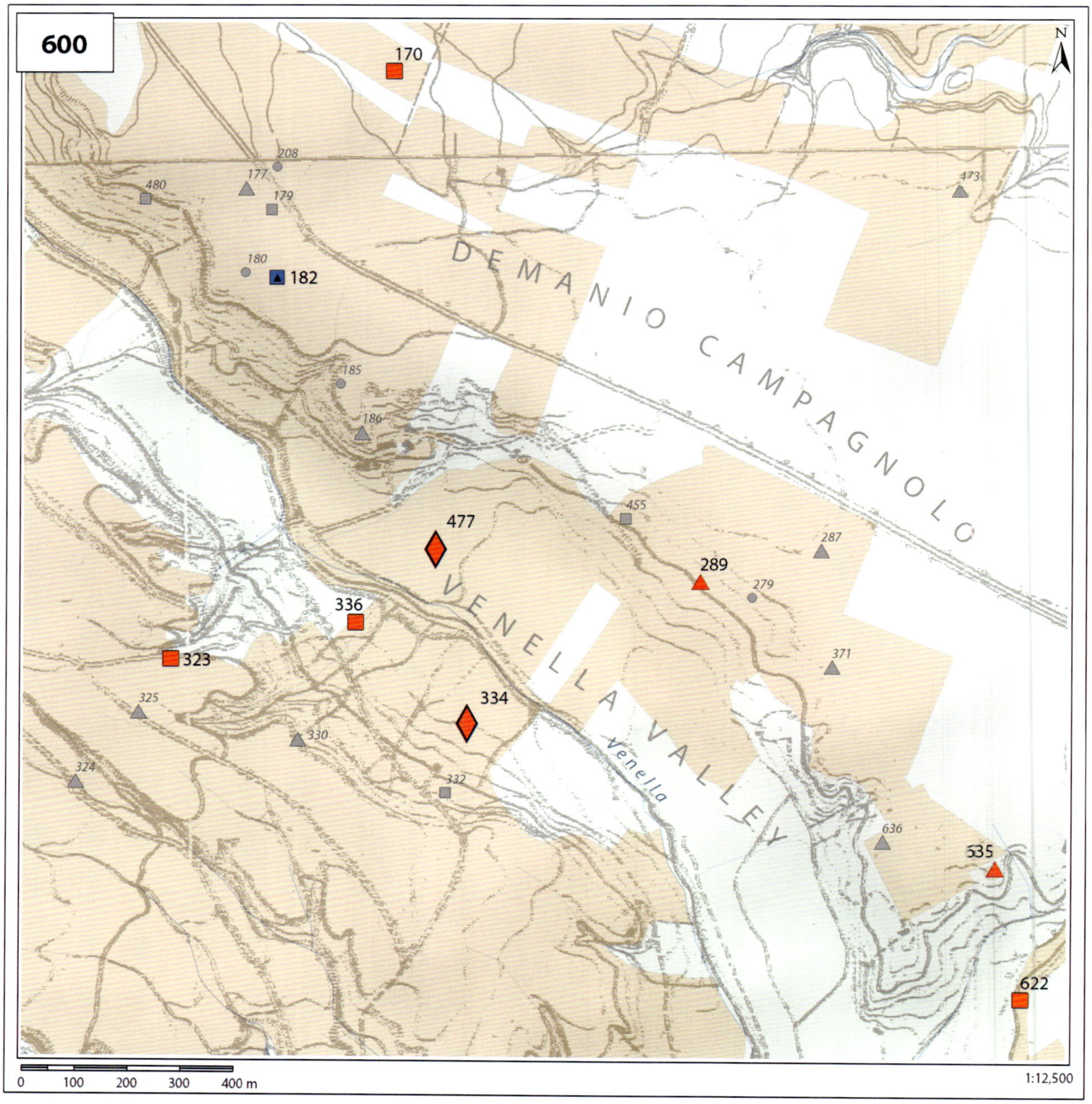

Figure 20.17 Venella in **600**. See Fig. 20.1 for map key. (JT)

slope separate the two farmhouses, and they are 550 m and 250 m from the sanctuary 334S, respectively. The life of these farmhouses during their period of greatest activity in the late 7th and 6th centuries BC parallels that of the sanctuaries.

Necropoleis. None of the earliest farmhouses in this area can be securely associated with a burial—the only possible pairings of farmhouse and necropolis are 323F on the western side of the Venella with the undated necropolis 325N. They are approximately 90 m apart. Necropolis 289N was frequented already in 600, though no farmhouse was in sight at that time in the area immediately surrounding it, which was thoroughly covered by the survey. The family using it may not yet have established a residence in the chora, or it may have been more distant than usual. There is, however, another possible connection with sanctuary 477S, which is approximately 500 m up the valley. This is the distance that separates necropolis 401N as well as the sanctuary at San Biagio and the Pantanello necropoleis from the Pantanello sanctuary. Both the

Figure 20.18 Venella, **600.** 3-D reconstruction using a satellite image of 2006 draped over a DEM, looking northwest. (ML)

necropolis and sanctuary in this case had a relatively short duration in the 6th century, whereas the Pantanello and San Biagio sanctuaries persisted into the 3rd century, as did their associated necropoleis.

All that can be said for certain about the date of the burials at 182F/T is that they belong to the period of the dwelling's occupation (600–400). In no instance can we be sure that the burials were those of the early settlers. Since the occupation of the site 182F/T apparently had no interruptions in this period, we may be reasonably sure that if not the first generation, then at least their descendants were laid to rest near the dwelling.

	Site no.	600	550	500	450
Venella west	334S	2.0	11.2	2.2	2.5
	336F	1.8	6.8	0.4	1.3
	323F	1.4	3.4	0.1	1.0
Venella east	477S	8.1	23.2	4.7	12.1
	289N	0.8	2.5	0.8	1.9
	182F/T	0.6	1.5	1.0	5.6

Table 20.5 Venella **600,** with EAW values rounded to the nearest tenth through **450.** Dark green = significant farmhouses or sanctuaries (EAW ≥ 0.95) and active necropoleis (EAW ≥ 0.45). Light green = second magnitude farmhouses (EAW ≥ 0.45, but ≤ 0.95).

Sanctuaries. On the west bank, one sanctuary, site 334S, occupied a terrace (36 masl) close to the valley floor. The other sanctuary, site 477S, was located 50 m east of the stream at the same elevation in a low-lying area, which is currently marshy valley floor because of a spring that feeds the Venella. Like many of the springs around which larger and smaller sanctuaries grew up (e.g. Pantanello and San Biagio), this one is still active. The two sanctuaries are comparable to each other in terms of the extent of the scatter, ceramic assemblages, and votive objects.

Significance, land, and continuity. Second magnitude site 182F/T, 600 m from 477S, pointed the way for future developments on the marine terrace overlooking the middle course of the Venella. This was a first tentative approach to an area that saw many settlers in the next half-century and beyond.

Giampasquale. The tributaries of the river valleys proved very attractive to the earliest settlers of the chora, as the settlements along the edges of the Conca and in the Venella illustrate. The broadest of the

Figure 20.19 View of the Venella valley from Fattoria Fabrizio, looking west. (AP)

tributaries on the Bradano, the Fosso Appio, lies outside of the study area and the survey transects (*Figs. 20.20–21*; *Table 20.6*). Its spacious and gently sloping terraces certainly drew settlers, as the results of earlier surveys in the 1960s show.[24] A systematic survey here was not undertaken by the ICA team because by the 1980s, this area had been much compromised by modern agricultural activity. Some of the sites located fifteen years previously could be recognized upon revisit only with difficulty.[25] One of the two tributaries that were surveyed on the north side of the chora is the torturous and steep-sided Fosso dell'Acqua Fetente. It cuts deeply through the extreme northwest corner of the 1981 survey transect and divides Giardinetto from San Marco and the Central Plateau (Cogno del Prete) to the east. This gorge was explored from its mouth on the Bradano to its source far to the southwest, where it nearly meets the Venella. Unlike the lower course of the Venella, it is not inviting. The other fork originates outside the survey transect to the southwest of Cozzo Presepe.

Site no.	600	550	500	450
270S	5.0	13.0	2.1	
269F	0.5	1.3	1.2	3.4
193F	0.5	0.5		0.7
200F	0.6	1.4		1.6

Table 20.6 Giampasquale **600,** with EAW values rounded to the nearest tenth through **450.** Dark green = significant farmhouses or sanctuaries (EAW ≥ 0.95) and active necropoleis (EAW ≥ 0.45). Light green = second magnitude farmhouses (EAW ≥ 0.45, but ≤ 0.95).

Between the Acqua Fetente and the Fosso Appio is a small tributary, the Fosso San Marco, which resembles the Conca topographically, though on a smaller scale. The south bank has been surveyed, but the north side at its mouth is dominated by a sprawling modern agricultural establishment, the Masseria San Marco. It was not possible to carry out an intensive, systematic survey on this side of the drainage, though earlier investigations there revealed a major Bronze Age presence on the promontory and surrounding areas (see Ch. 17).

Early settlers chose the south bank of this tributary, and probably the north as well, to settle in the first wave of colonization. The sanctuary, 270S, is the only significant site there in **600**.

Farmhouses. Three second magnitude farmhouses lay in close proximity: 269F, 193F, and 200F. The second magnitude farmhouses follow the valley's edge more or less closely at intervals of 200 to 300 m.

Sanctuaries. To judge from the ceramics (including early black-gloss pottery, a fragment of a 7th-century SOS amphora, and many miniature votive vessels), sanctuary 270S was one of the earliest sites in the study area. The sanctuary occupies its own little point jutting out into the valley. The site was thoroughly

[24] Uggeri 1969; unpublished Soprintendenza and French surveys. The French sites appear on the distribution maps in the Atlas (see map key).

[25] G. Di Taranto, pers. comm., 1996. Di Taranto participated in the early and later surveys.

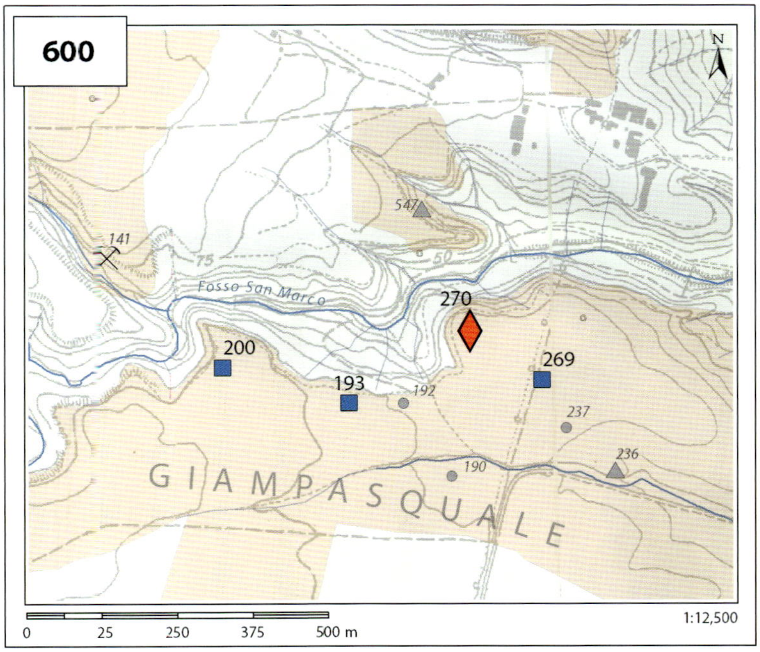

Figure 20.20 Giampasquale in **600**. See Fig. 20.1 for map key. (JT)

destroyed at some point after our 1982 visit; only a few fragments of pottery could be found on a revisit in 2007. The natural setting was defaced by an ill-conceived project to create flat terraces. It was subsequently abandoned and allowed to erode—an eyesore, a useless "field," now, lacking top soil and without any historical or agricultural value. Its fate is not unique.

Necropoleis. No necropoleis were documented in this area that can be associated confidently with the farmhouses or sanctuary. An undated necropolis, 236N, is 200 m southeast of 269F. It lies downslope, some 150 m distant at the head of a small secondary drainage. It may have served the family occupying 269F or have been associated with 237UGF, a structure that was so poorly preserved that it was unrecognizable as to type. An undated necropolis and a quarry occupied the north side of the valley opposite the sanctuary.

Significance, land, and continuity. The sanctuary 270S was the only significant site in this cluster in **600**. The second magnitude farmhouse 269F was only about 100 m away, and two others (193F and 200F) lined the valley's edge moving inland. They were closely spaced with less than 250 m separating them from each other and the sanctuary. The Giampasquale cluster, apart from the sanctuary, was less precocious than the others, and its future development was remarkably uneventful.

Single sites, 600. (*Fig. 20.1*) Of the sites in the study area that the early settlers occupied, only four appear to have been isolated. All were in areas that were intensively surveyed, making it highly likely that they were indeed alone in the landscape. Site 228F in San Marco crowned a promontory on the south side of the Fosso dell'Acqua Fetente, and was occupied for a century before being abandoned forever. The very large necropolis 70N, which was particularly active from **600** to **450**, is located nowhere near a contemporary farmhouse. It sits on a slope on the south side of the Acqua Fetente, halfway along its course, in an area devoid of contemporary structures. The farmhouse closest in date to the necropolis, 58F (second magnitude in **600**), was located 500 m to the east. While there may have been understandable motives in seeking a remote place to live and farm, it is not so easy to find a reason for locating a burial ground in such a remote area.

Both sites 228F and 70N indicate that the area of the Acqua Fetente was not ignored by the early settlers, though it was clearly not as attractive to them as the Venella. The gentler parts of the Venella were immediately settled, but at least one brave soul ventured much further up the valley to the point where it split into two forks. There a farmhouse was established in a position to observe the confluence and a good stretch of the valley. Farmhouse 488F was the first site in this area, and it was occupied continuously, except in **500**, when so many sites were temporarily or permanently abandoned (see Ch. 22). It was soon joined in the next half-century by a number of sites on the surrounding bluffs.

Site 103F in Scarciullo is perched on the east edge near the sources of the deeply-incised and many-branched drainage that enters the Basento at Masseria Sant'Angelo Vecchio, 1.5 km downstream. It was almost 1 km away from its nearest neighbor in **600**, farmhouse 346F on the upper reaches of the Conca. Like that site and 488F in the upper valley of the Venella, it had a commanding position on the plateau overlooking the confluence of two branches of the main stem of the Fosso Sant'Angelo.

Figure 20.21 Giampasquale, **600.** 3-D reconstruction using a satellite image of 2006 draped over a DEM, looking west. (ML)

The advantages of living in or beside the valleys, especially for farmers involved in animal husbandry, have been touched on above. The earliest settlers and their successors for the first three centuries down to **300** almost invariably resided where water was close at hand. That is what makes the choice of site 43F seem uncharacteristic. It is located in the center of the high plateau of the marine terraces far from any drainage, and from any obvious source of water. The site was occupied in **600** and **550**, and, after a pause in **500**, more or less continuously until the 1st century BC—a truly noteworthy record. Clearly the site had its attractions, located as it was in an undulating fertile plain. Nonetheless, it must have taken courage to go it alone in the early days.

Assemblages, 600

How did the early farmers in the chora live? What did they eat and drink and how did they prepare and serve meals? What other activities were performed in that basic unit, the farmhouse? These are very important questions and there is abundant evidence in the site assemblages.

But how to extract information about the earliest period? The problem of determining the typical pottery assemblage of the earliest farmhouses in the chora would seem almost unsolvable. Each assemblage consists of all the pottery collected from a surveyed site, but the fine wares indicate that this represents a build up over at least a century for the 19 farmhouses in **600**, and probably more (see *Appendix A* for the EAW values for these sites in **600** and all subsequent periods). Only one site, 285F/T, had more fine ware dateable to **600** than to **550**, according to the original chronology. Six others had about equal amounts, but half of these also had far more fine ware dateable to later periods. As will become apparent, the same sites were often occupied in two or more periods, frequently separated by a century or more. The remaining 12 farmhouses had more pottery from **550** than from **600**, often considerably more.

The best opportunity to discover the typical range of pottery in **600** comes from considering the two farmhouses 285F/T and 530F/K which had no significant occupation after **550** (these are also the oldest farmhouses in the revised chronology). The fine ware from these sites dates to **600** and **550** in approximately equal proportions, and the assemblages should contain at least an equal amount of the undated pottery used on the site in **600** and **550**. In terms of functional groups, these two sites both have roughly the same percentage of vessels associated with drinking (see full assemblage tables for both in Vol. IV). They also compare closely with ten other sites with fine wares concentrated in the periods **600** and **550** (see *Fig 21.19*). Site 530F/K, however, had a considerably higher than expected count of drinking vessels than the others, and 285F/T had proportionately far fewer "food-service" vessels (lekanai) (see chi-square 8, *Appendix D* and *Table D11*).[26]

The farmhouse assemblages can also be compared in terms of the relative percentages of the different classes of ceramics that occur at each—fine ware, plain/banded ware, cooking wares, and so forth, and a similar result is obtained (Ch. 21; *Fig. 21.20*). The farmhouses of these periods, including 285F/T and 530F/K, have comparable proportions of the different classes.[27]

We may cautiously conclude that the basic functions and classes of ceramic vessels were present at the earliest farmhouses in the chora, as exemplified by 285F/T and 530F/K, and that although the proportions changed over time, this basic unit existed at the beginning in much the same form in **600** as it did in **550** and in subsequent centuries.

Significance

The relative importance of farmhouses in **600**, as measured by their FMCE values, is recorded in Appendix B and the first of a series of twelve graphs in Appendix C (*Appendix C8,* **600**). The FMCE value makes possible a comparison of the significance or material importance of a farmhouse, as compared synchronically with all farmhouses of the same period, and diachronically with itself in all other periods (see Ch. 4).

The values are low in this first period in the history of the chora. Values are evenly distributed around a mean of 1.74, but there is one obvious outlier, site 285F/T, which is by far the most significant farmhouse in **600**. This is not a surprise in view of the relatively great quantity of dateable fine ware from the site. The overall impression of the distribution of values is that these were not great differences in the significance of farmhouses, and that this would indicate a more egalitarian society than that of later periods. The alluvial terraces were choice land, and the average FMCE value of the seven farmhouses on the T_2 terrace in this period (2.60) gives an indication that those who settled there were better off materially than those on higher ground. The average of the five farmhouses on T_3 and T_4 is 2.30, and the average of the terraces combined (2.48) is well above the average (2.14) for the study area as a whole in **600/550** (see *Appendix C7*). A comparison of the distribution of FMCE values in successive periods shows greater variation. The changes in overall distribution from one period to the next proves to be statistically significant (see *Appendix D*, chi-square 1, and *Table D4*).

Types of Sites in the Early Chora

Farmhouses. The surface indications of a farmhouse in general are building stones, river cobbles, pieces of sandstone, and a scatter of roof tiles and pottery. The scatter is usually more extensive than that of a necropolis or a sanctuary. The composition of the farmhouse assemblages is quite standard and they tend to remain that way, with minor modifications, down to the 2nd century BC, when major changes do occur (see Ch. 27, *Ware class proportions in* **100**).

What form of shelter existed for the rural households of the end of the late 7th and early 6th centuries BC? The answer comes from a small but growing number of excavated examples from the chora

Two basic floor plans have been recognized. The first is a square plan consisting of one (or possibly two) rooms facing a courtyard in the Basento valley below San Teodoro, discovered by De Siena during the SNAM excavations (see Ch. 17).[28] It is dateable by the pottery to the second half of the 7th century BC.

The second is a *pastas* plan with smaller rooms (usually three) in a row opening onto a long one (the pastas). An example of this type was found at Termitito in the chora of Siris to the south of the Cavone and dated also to the second half of the 7th cen-

[26] The test compares the counts in four combined categories—food preparation, which combines food preparation and cooking wares; storage, which combines short- and long-term storage vessels; and so forth—among 12 farmhouses. There is a statistically significant difference between farmhouse assemblages. The cross tabular analysis (*Table D11*), however, shows that there is a great deal of similarity among them. The difference between expected and observed values is not great. There are more drinking vessels than expected at site 530F (standardized residual of 2.1) and fewer than expected food service vessels at 285F/T (standardized residual of -1.9).

[27] The assemblages of **600** and **550** differ from those of later periods in their ware class proportions (chi-square 7, *Appendix D*).

[28] Nava 2002, 738, pl. 67.2.

Figure 20.22 The sacred spring at the Sanctuary of Artemis, San Biagio in the background. Modern chapel of San Biagio on the hilltop, looking northwest. (After Mertens 2006, Fig. 49)

Excavated examples of early burials in the chora are admittedly rare, and they are all quite modest. Substantial burial containers do not appear in the record before the middle of the 6th century, either in the chora or the close environs of the asty with the unique exception of an elite group in the necropolis of Crucinia just beyond the western wall.[33] This may explain in part why necropoleis of early date are not so obvious and may not have been identified in the survey. The excavated examples were, for the most part, simple graves with at most a wooden coffin and few if any grave goods. The earliest monumental tombs in the chora, monolithic sarcophagi, are not in evidence before the last decades of the 6th century BC.[34]

Sites were identified as necropoleis during the survey on the evidence of tomb tiles and fragments of sarcophagi, human bone, evidence of cremation, and their relatively small scatter size. The pottery assemblage from site 70N—the most representative example from this period—contained dateable fine wares, amphorae, and good quantities of coarse ware (see Assemblage Tables, Vol. IV).

tury BC. Both of these types have good parallels in Greece and the wider Greek world from the end of the 8th century BC onward.[29] The simple rectangular house—the more modest of the two—provided a living space of 14 to 20 m^2, though it could be smaller. Oikos G at Incoronata,[30] for example, had only 12.3 m^2 of floor space. The square, one-room houses below San Teodoro had footprints of 15 m^2 and 19 m^2.[31] The pastas houses at Termitito and at Cospite-Caserta (also in the chora of Siris) and at Policoro measure 62.5 and 102 m^2 respectively.[32]

Necropoleis. The early settlers of the chora very soon had to face the problem of where to dispose of their dead. Only eight sites identified as necropoleis by the survey have some probability of having been in use in the late 7th or early 6th century BC (*Appendix A, Table C14*). It is very possible that there were others, for example, those containing no dateable fine wares but fragments of Corinthian type A amphorae, which have a broad range between 700 and 450 BC (see *Table 21.7*).

Sanctuaries. The Sanctuary of Artemis on the Venella is among the very earliest securely dated sites in the chora. It is located beside a spring in a hollow on the west bank of the Venella at its mouth, just before it enters the Basento valley (*Fig. 20.22*). The terracotta revetment with a scene of the departure of a hero in a chariot drawn by winged horses, dated to the last quarter of the 7th century BC, is among the earliest of this type of architectural decoration in the Greek world (*Fig. 20.23*).[35] The site commands a broad secondary valley that was ideally suited to both agriculture and pastoralism—concerns that, to judge by the terracotta votive figurines, were reflected in the cult (*Fig. 20.24*).

[29] Fusaro 1982.

[30] See Carter 2006a, 64–67, and Tables 2.A–E, for information about plan divisions, orientation, and bibliography. The function of these Incoronata oikoi has been debated. They may simply have been storage depots. The paucity of the usual array of coarse wares used for living has been noted and it has called into question the interpretation of Incoronata greca as a whole. Incoronata will be the subject of future study.

[31] Cf., n. 28 above. These were probably separate houses, according to the excavator, A. De Siena (pers. comm., July 2007).

[32] M. Tagliente 1986, 131, Pl. 30.1-2. Carter 2006a, 135–49.

[33] De Siena 1999b, 233–35.

[34] Carter 1998f, 178.

[35] Mertens-Horn 1992, 6–45; Olbrich 1976 for terracotta figurines.

Figure 20.23 Terracotta frieze representing the departure of a warrior (Achilles?) for battle in a chariot drawn by winged horses, ca. 625 BC. (VM, ICA Ancient Crossroads; courtesy of the Soprintendenza)

Figure 20.24 (Right) Statuette from the Sanctuary of Artemis, San Biagio, first half of 6th c. BC. (VM, ICA Ancient Crossroads; courtesy of the Soprintendenza)

The sites identified as rural sanctuaries in the survey were in all probability no more substantial than the farmhouse structures that their material remains closely resemble. Their pottery assemblages differ little from those of nearby farmhouses. What sets them off from the farmhouses, as noted before, are the considerably larger quantities of votive vessels and terracotta figurines. The sites themselves are integral parts of the small rural communities at Avinella, Venella, and Giampasquale, where they occupied a central position, but were regularly spaced like the farmhouses themselves. They are near springs, like the farmhouses, but on the whole at higher elevations than major sanctuaries like San Biagio and Pantanello. Could these cult places be expanded versions of the household cults that existed in many farmhouses where a few terracotta figurines and votive miniatures are a rather common occurrence? Several observations would favor this interpretation. The sanctuaries are distributed among the small communities at Avinella, the Venella on both sides of the stream, and at Giampasquale. The sanctuary might, in fact, have been the home of the head of a family group residing together in the Early Village Cluster.[36]

A connection between rural cults and the families residing in the immediate neighborhood has been argued well in the case of San Biagio.[37] This may have been the case also for the sanctuaries recognized by the survey. As opposed to sites like San Biagio, which continued to grow in importance in later periods, these more modest sanctuaries do not continue much beyond the end of the 6th century, or continue with much reduced importance when some families appear to die out and there is an influx into the chora of new family groups.[38] The two extensively excavated rural sanctuaries in the chora of Metaponto at San Biagio[39] and Pantanello, and the less explored Sanctuary of Hera on the Bradano (Tavole Palatine), were already in existence by 600 BC.

Fortifications. The immigrant Greek settlers, according to Strabo (see Ch. 17), did not have a friendly reception. Fortifications for the early settlers are therefore remarkable for their nearly total absence. The Archaic wall of the city consisted of a single line defining the western perimeter of the asty. There is an inscription on the wall in the Achaian alphabet, which De Siena, the excavator, dates to the end of the

36 Parker 1996, 56–66, discusses the relationship of family and cult in Archaic Attica: "The cults controlled by a *genos* were in origin nothing other than the private observances of an aristocratic club."

37 De Siena, pers. comm., Aug. 2007. Cf. Montepaone 1986.

38 See discussion of extra-urban cults in Attica, focusing on Brauron, in Osborne 1985b, 154–82.

39 De Siena 2001a.

Figure 20.25 Plateau of Cozzo Presepe, with Montescaglioso on the site of a major indigenous center, visible in the distance. (JCC)

7th century BC.[40] In the chora the unique example of an early fort was located at Cozzo Presepe (*Fig. 20.25*; see Ch. 17) just outside the extreme northwest corner of the study area. The ethnic identity of the occupants of the one excavated structure of this period, the so-called Archaic house—whether Metapontine farmers, members of a garrison, or possibly indigenous farmers—is a matter of speculation.[41] The frontier between Greek and indigenous settlements in the interior appears to have been porous. The much-debated passage of Strabo (reproduced in its entirety at the end of Ch. 17), however, mentions a first settlement by the Metapontines that was wiped out by non-Greeks,[42] and a war between the Metapontines, the Oenotrians of the interior, and the Tarentines. These events would seem to offer the best pretext for constructing a fort, one whose role may as well have been anti-Tarentine as a protection against the native population of the interior.[43]

The Chora in 600: An Alternative View

The Revised Fine Ware Chronology

One of the main results of the 2009 study of fine ware was that a number of the fragments previously identified simply as *a filetti* cups were reclassified either as local imitations of proto-Corinthian or as Colonial ware (descending probably from the Corinthian tradition as represented earlier by the Thapsos cups), or as local versions of the Ionic cups (type B2 of probably eastern Greek inspiration). The imitation proto-Corinthian and Colonial wares have been dated to the late 7th and early 6th centuries BC; the B2 Ionic cups, to the period of 550–500 BC.

The imitation proto-Corinthian and Colonial wares are the only fine wares among the survey finds that were dated securely by the authors of Chapter 6 to the first period of the settlement of the chora, and they are very few. This and the date range for the Ionic cup, which is narrower and lower than that originally employed in 2002, had substantial implications for the dating of sites and the interpretation of the settlement patterns in the Archaic period, especially in **600**. The differences between the view of the chora based on the original chronology (*Map 5*, Vol. IV) and that using the revised chronology (*Fig. 20.26*) are immediately apparent.

The Earliest Chora

Only two farmhouses—285F/T and 530F/K—appear in both reconstructions, and both have abundant pottery from the next period, **550**. A single fragment of an imitation proto-Corinthian cup (*Fig. 20.27*), dated to the last decades of the 7th century BC, has been identified at the farmhouse 530F/K in Avinella on the Basento side of the study area. Six fragments of

[40] De Siena 1998, 141–51.

[41] *Cozzo Presepe*, 230–35.

[42] Musti 1988, 117–20; Guzzo 1983.

[43] *Cozzo Presepe*, 385. Osanna 1999, 278.

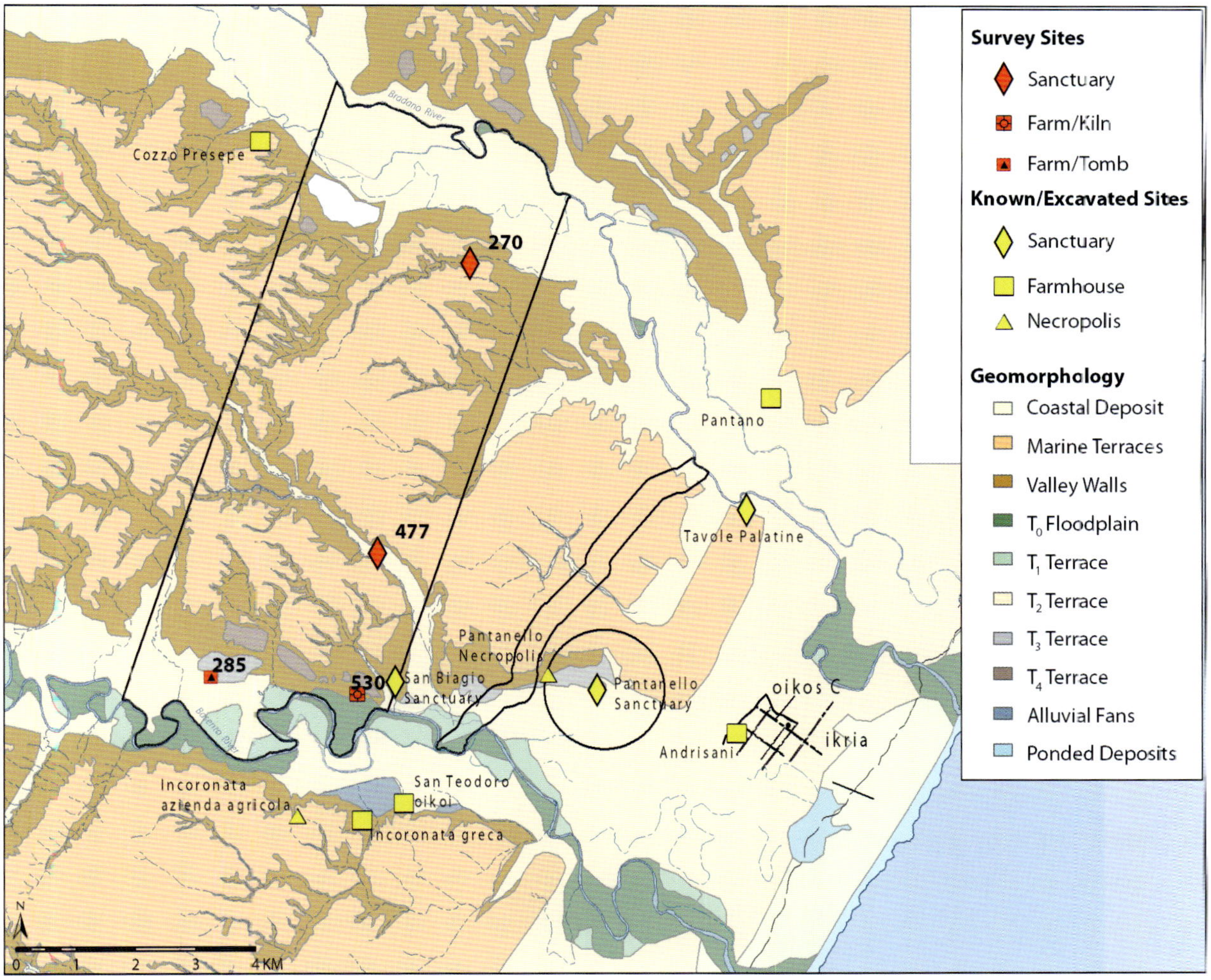

Figure 20.26 The chora in the late 7th and first half of the 6th c. BC, alternative chronology for survey sites, with other early sites laid over the geomorphological map (see Ch. 2, Fig. 2.2). (JT)

Colonial ware, the so-called *a filetti* cups (*Fig. 20.28*), argue for a late-7th- to early-6th-century BC date for farmhouse site 285F/T in the Sant'Angelo cluster. Pottery of the same type has been reported in the excavation in the urban sanctuary and under the ikria in the agora.

Sites identified as farmhouses are extremely rare in the alternative view of the chora in 600, but these two examples demonstrate that early settlement occurred inland, and, as the geomorphological map shows, along the edges of valleys. Like the early sanctuaries, these farmhouses were active at the same time as the earliest structures of religious and civic importance at the future asty. The contemporaneity of the settlement of the chora and development of the asty in the first period is certainly more obvious according to the original chronology, but is supported also in the alternative view. This is an important and far-reaching conclusion.

Both of these farmhouses were among the most significant in the chora throughout the 6th century, but they declined rapidly thereafter. Despite the small amount of dateable fine ware at these sites in 600 according to the revised chronology, there is no reason to doubt their continuity of function throughout the late 7th and 6th centuries. As noted above, site 285F/T also produced one of the extremely rare fragments of indigenous pottery, an olla of unknown date, that at least opens the possibility that the occupants of the early farmhouse may have included some of indigenous origin (see *Fig. 17.14*).

There are only two other survey sites with ceramic material of securely 7th-century BC date according to the revised chronology. Sites 477S and 270S were

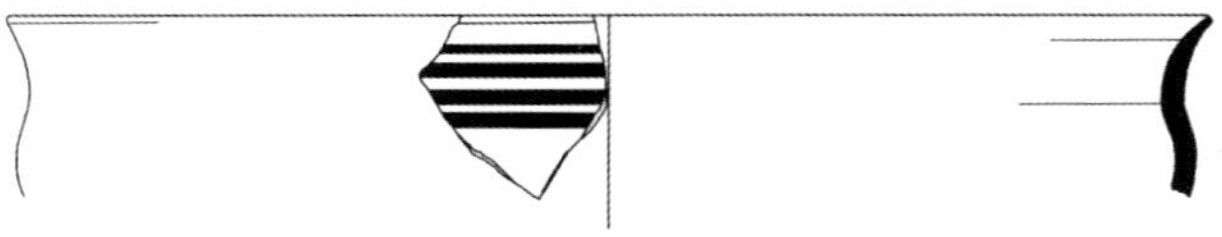

Figure 20.27 Imitation proto-Corinthian cup rim from site 530F/K. (Ch. 6, AFW No. 1)

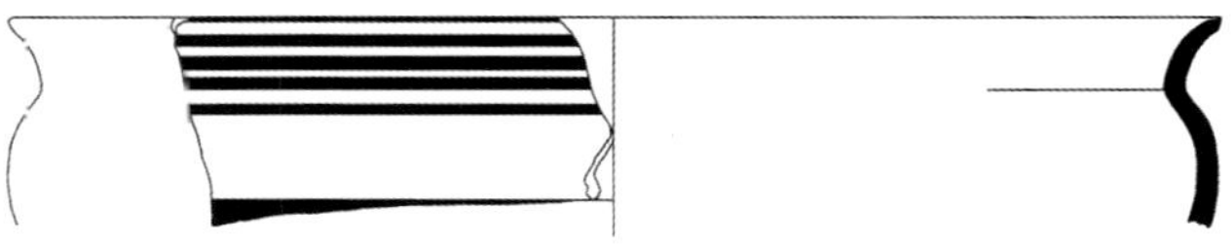

Figure 20.28 Colonial ware *a filetti* cup, vessel from site 285F/T. (Ch. 6, AFW No. 3)

both identified as sanctuaries because of large numbers of terracotta votive figurines dateable to the 6th century BC and contemporary miniature votive vessels. Each had one rim fragment of an imported SOS amphora in its assemblage (see Ch. 12, Nos. 1 and 2). These places were both frequented at this time, and though we have no early examples of votives as we do from the securely dated, late-7th-century sanctuary at San Biagio, it seems likely that religious activity was already taking place. Like the farmhouses, these sites are documented by much securely dated 6th-century BC pottery as well as terracotta figurines (Ch. 13, Nos. 229 and 230). A terracotta figurine of a standing female divinity, sub-daedalic in style, recovered by the survey team from site 397S shows that it was in use as a place of worship early in the 6th century (*Fig. 20.16*; Ch. 13, No. 210).

According to the revised chronology, the chora in the late 7th and first half of the 6th century BC would appear to have been practically uninhabited, with only four sites in the Bradano-Basento transect. There is, however, very good evidence that this was far from the case. It comes not from survey but from excavation, both of burials and of sanctuaries (*Table 20.7*).

The Pantanello Necropolis

The earliest burials in the area of the Pantanello necropolis (*Fig. 20.29*) are earthen graves (T 299 and T 301),[44] dated by C14 analysis of the bone. The probable dates are the 8th to 7th centuries BC and 7th to mid-6th centuries BC, respectively. The only grave good was a single iron pin from T 301, and both skeletons were in a slightly flexed position, strongly suggesting that the occupants were of non-Greek origin. A third undated burial, containing a decidedly more flexed skeleton, was discovered about 30 m to the northwest in what became nucleus 8 of the Pantanello necropolis. This burial was only about 10 m away from a pair of burials, both in the supine position, which had been placed in wooden coffins held together by iron nails and clamps (T 276 and T 277).[45] One burial is dated to ca. 575 BC by a single Corinthian quadrifoil aryballos (*Fig. 20.30*). In the same area lay a ceramic deposit (CD 105), in all likelihood a destroyed burial, containing Corinthian pottery of the first half of the 6th century BC.[46]

This cluster of burials, which would become one of the largest burial grounds of the Greek chora, was already in active use in the 7th and early 6th centuries BC. The close proximity of supine burials with Greek grave goods—plausibly of the early Greek settlers of the area—to burials following indigenous practice suggests continuity or coexistence.[47] A mixed settlement, or continued use of this burial area, is further suggested by two ceramic deposits (CD 30, CD 289), containing both indigenous and Greek pottery, including Corinthian A amphorae and Colonial ware or *a filetti* cups (CD 30) of the

Area	Locality	Site	Finds
Basento valley and tributaries	Sant'Angelo	285F/T	Colonial ware
	Avinella	397S	Terracotta figurines
	Basento valley	530F/K	Imitation proto-Corinthian pottery
	Venella	477S	SOS amphora
	Pantanello	Sanctuary	Colonial ware terracotta figurines
		Necropolis Crossroads	Corinthian pottery, C14 dating of bones
		Necropolis East	Enchytrismos burial
	*Incoronata greca	Sanctuary	Terracotta figurines, pottery
Bradano valley	Giampasquale	270S	SOS amphora
	San Salvatore	Site 29	Enchytrismos burial
	*Cozzo Presepe		Colonial ware
	*Pantano		Colonial ware?

Table 20.7 Some active sites in the late 7th or first half of the 6th c. BC (*outside of the study area).

[44] Carter 1998f, 168–69; Carter and Hall 1998, 301.

[45] Carter 1998f, 171; Carter and Hall 1998, 359, 361.

[46] Carter and Hall 1998, 306.

[47] Carter 1998f, 169–71.

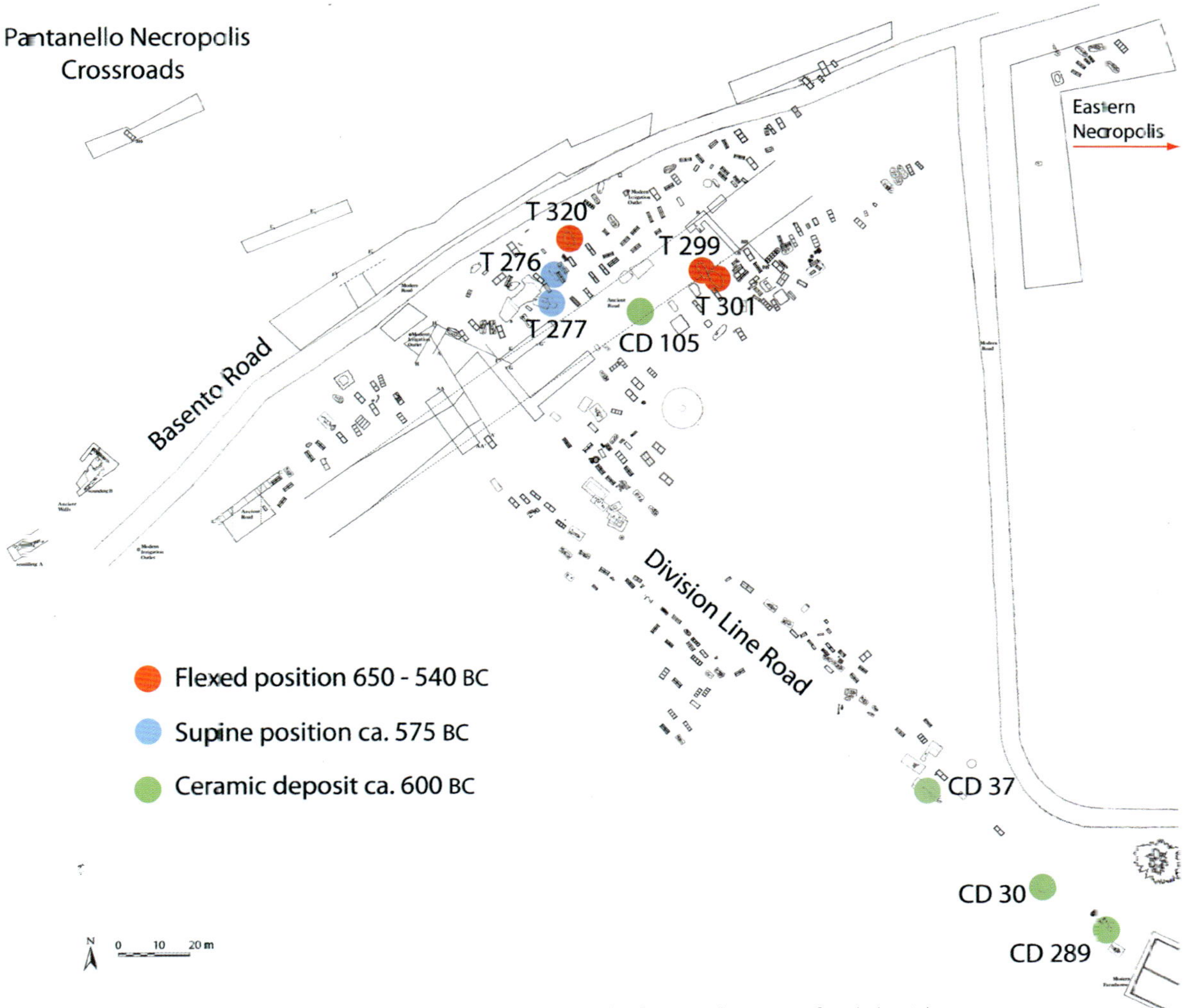

Figure 20.29 Plan of Pantanello necropolis showing location of early burials.

late 7th and early 6th centuries BC.[48]

Enchytrismos burials using pots as containers appeared already in the mid-7th century at Incoronata (see Ch. 17), by what seemed to the excavator to be a subaltern, possibly mixed population. The same practice reappeared in the Eastern necropolis at Pantanello (T 191) in the mid-6th century. The evidence for enchytrismos burials of this period continues to grow, with a recently excavated example from San Salvatore.[49] In the same area two more burials were excavated in 2009 (Soprintendenza tombs 144 and ex. 175, *Figs. 20.31* and *20.32*); these contained imported Corinthian pottery (like that used in the enchytrismos burials) of the first half of the 6th and perhaps even the late 7th century BC.[50]

There are several significant facts about these relatively early burials, including the supine ones, at Pantanello: they were deeply buried and hence the likelihood of their contents becoming surface finds was much reduced; the pottery used for burial in the first half of the 6th century in the chora seems to have been almost exclusively imported Corinthian or local imitations of the same; and their consistent and growing numbers are a clear indication that the chora was more densely occupied than the survey results would indicate when following the revised lower chronology for fine wares collected from the surface by the survey teams.

Excavated sanctuaries at San Biagio (ca. 625 BC to early 3rd century BC), Pantanello (ca. 600 BC to early 3rd century BC), and Incoronata (630–550 BC,

48 Carter and Toxey 1998a, 697 (*a filetti* cups), 706–07 (banded ware hydria). Morter and Leonard 1998 (Corinthian A amphorae).

49 Nava 2003, 675–76.

50 De Siena forthcoming. I am grateful to Antonio De Siena for the information about these and the photographic documentation (pers. comm., Oct. 2010). The hydria used in the San Salvatore enchytrismos burial closely resembles those from Pantanello, CD 289 (n. 49, above).

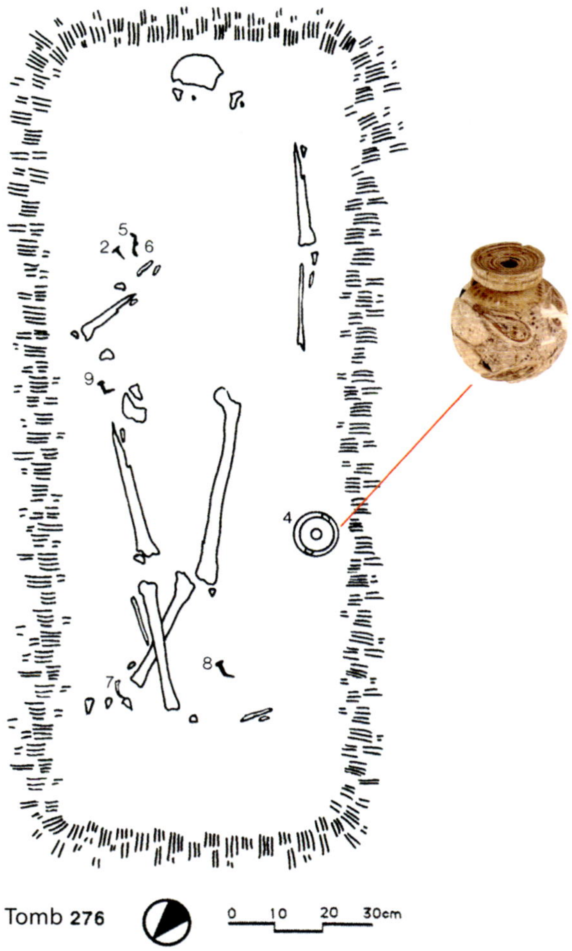

Figure 20.30 Pantanello necropolis tomb 276 with quadrifoil Corinthian aryballos, ca. 575 BC. (CW)

located outside the study area on the right bank of the Basento) were certainly among the earliest structures in the chora. By themselves, they might not necessarily indicate a substantial population resident in the chora before the mid-6th century, which, according to the revised chronology, would be the date of the first appearance of the B2 Ionic cups found in abundance in sites throughout the chora. But together with the evidence of the farmhouses (285F/T, 530F/K) and the excavated necropoleis discussed above, they do make a strong case for settlement in the first period, 625–575 BC. These dates are based on artifacts including: Middle and Late Corinthian pottery (*Figs. 20.30* and *20.31*), and possibly Early Corinthian as well (*Fig. 20.32*); locally made Colonial ware or *a filetti* cups (*Fig. 20.28*); and daedalic and sub-daedalic votive figurines. Corinthian pottery of the same sort, it should be noted, was found in large amounts in the excavation of Oikos C, which is among the first structures with substantial traces that survive in the asty.[51] All the sanctuaries have appeared in earlier scholarly literature,[52] and all three will, one hopes, soon receive the full publication they deserve.[53]

Like the early burials, these sanctuary sites were deeply buried, and were located at all weather springs, sites of crucial importance for the early settlers and their descendants. It is hardly surprising that they became places of worship and congregation. This is true of Pantanello on the Basento, and of other sanctuary sites of greater or lesser importance, like the tiny collecting basin with bird sacrifice in the Basento valley (see Ch. 21).[54] Most were deeply buried under layers of alluvial deposit,[55] and few if any traces were visible on the surface when archaeological investigation of the chora began in the 1960s. This fact about the early sanctuaries is particularly germane to the reconstruction of the chora in this or any period.

The Pantanello sanctuary, as noted in Ch. 18, was discovered by pure chance. Nothing was visible on the surface to indicate a structure that had grown in size and importance in the Archaic period and later.

The site was below the modern water table and hence partially submerged. In its earliest phase it consisted of a spring, open to the sky with two channels and a votive deposit with a complete terracotta half statuette and miniature votive vessels (*Fig. 20.33*). A fragmentary, locally-produced Middle Corinthian conical oinochoe was found by the mouth of the spring, and further downstream fragments of *a filetti* cups were recovered. Together they provide a terminus ante quem for the sanctuary's first use. In one of the channels there was a rectangular pit designed for filling amphorae or other large vessels; it contained fragments of a Corinthian A amphora and an Ionic cup of very great size. Perhaps already at this stage

[51] A. De Siena, pers. comm., Jul. 2010.

[52] San Biagio: Adamesteanu 1974a, 55–64; Olbrich 1979; and De Polignac 1991. Pantanello: Carter 1994a, 183–98; Incoronata: Carter 1994a, 175–76.

[53] The Pantanello sanctuary will be the subject of a future volume in this series.

[54] Osanna 2008, 926.

[55] The sanctuary at Incoronata is a possible exception. The fact that only votives and architectural revetments and roof tiles were found, with no trace of a structure, suggests that it may originally have been located elsewhere, and that for some reason these remains were later transported to the hilltop site.

Figure 20.31 Assemblage from a burial in San Salvatore, tomb ex 175. Included are six locally produced Metapontine-type pyxides with concentric circles, two locally produced Corinthian style kotylai, and an amphoriskos. The burial dates to the first half of the 6th c. BC. (Courtesy of the Soprintendenza)

Figure 20.32 Early Corinthian alabastron from tomb ex 175 in San Salvatore. The vessel dates about 600 BC, and is the earliest element among the grave goods. (Courtesy of the Soprintendenza)

the channel and the spring area had temenos walls constructed of conglomerate stone.

A further important finding from this excavation was that among the predominantly Greek materials were a half-dozen fragments of indigenous pottery (see *Figs. 17.14, 17.17*). This evidence adds a dimension to the reconstruction of the first period that hardly appears at all in the survey evidence: the presence, albeit limited, of the indigenous people in the chora at this time.

The discussion of evidence for the occupation of the chora contemporaneous with the growth of the future urban center, based largely on the evidence of excavated sites, is far from exhaustive. Enough has been said, however, to illustrate the inadequacy of survey evidence alone to give a balanced picture, especially as regards this early period.

A Synoptic View of the Chora in 600

Had an early traveler to the Greek West in the late 7th century BC, a contemporary of the poet Archilochos of Paros, visited the site of Metaponto and viewed its broad and fertile coastal plain, he would surely have agreed with the poet that no country is as fair or desirable or lovely as that around the banks of the Siris, or of its near neighbor, the Bradano. His first impression of the primitive settlement in a swampy area between the Bradano and the Basento would not have been so enthusiastic (see Asty, below).[56] There, on the dunes beside pools of stagnant water, stood huts of wattle and daub, or simple one-room adobe shelters. The cult buildings of wood and mudbrick were protected and adorned by terracotta friezes, before which the settlers sacrificed to the patron gods who had emigrated with them. Close by was the great open space where markets were held, and where the fresh immigrants congregated on wooden bleachers to discuss matters of pressing concern to the community and celebrate holy days with choral hymns and dancing.

Beyond lay the coastal plain and wide stretches of river valleys (*Figs. 20.34* and *20.35*), newly cultivated fields. Venturing inland along the major waterways and their tributaries, he would have seen where the majority of the new settlers resided. Their farmhouses

[56] But considering the origin of the settlers, the nascent asty was a true innovation on this shore. While this phenomenon was not unusual elsewhere in Greece, it was completely unknown in their native Achaia until the following century.

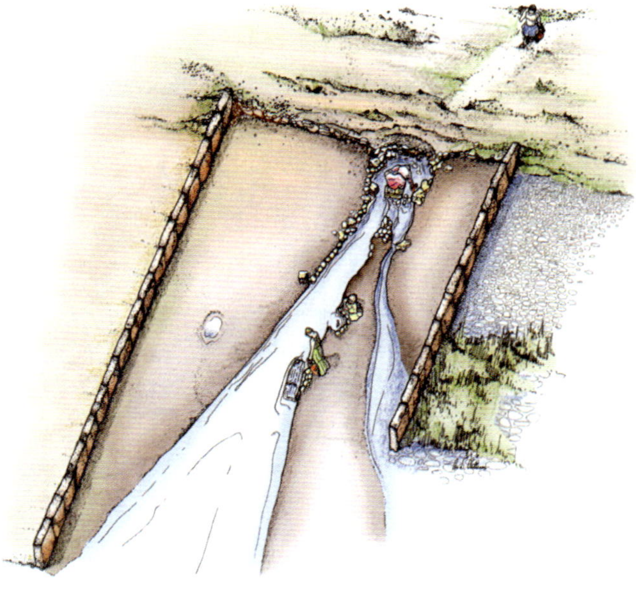

Figure 20.33 Reconstruction of spring sanctuary at Pantanello. (ALP/EH after Carter 2006a, Fig. 4.34)

were modest—one- or two-room mudbrick houses or wattle-and-daub huts—and situated in pairs or clusters along both banks. Burials, either near the farmhouses or in communal necropoleis, were likewise simple, with earthen graves and few if any goods. Our visitor might have been struck by the number of springs within easy reach of the farmhouses. Some springs were already the focus of simple sanctuaries, where worshippers brought votive offerings to a female divinity, and which may have acted as centers for the small communities. He might also have noted activity around the valley's abundant clay deposits, as villagers prepared to make pottery in a few scattered kilns next to farmhouses. The sight of these clusters might well have reminded him of similar Greek villages elsewhere, such as northwest Peloponnesos, where many of the original settlers originated, but the expanses of cultivable soil, like the abundant springs, would have recalled to his mind no place he had known before. Truly, this was Megale Hellas, and a place to make a new start.

At a distance of about 30 stades (6 km) from the coast, both wider and narrower valleys lead into the interior. Their bottom bands were covered with lush grass, and their edges with a thick growth of wild olive, evergreen oak, and pistachio,[57] offering diverse forage for cattle, sheep, goats, and swine. The livestock provided the pioneers with traction, wool, and meat. Gallery forests consisting of willow and alder lined the major valleys and forests of pine and deciduous oak and maple were found farther inland. Both were inhabited by fallow deer, roebuck, boar, and wild cat, which supplemented the diet of the early settlers. This wild game was especially important in these early years, before the fields had begun to produce a steady and secure supply of food.

Had our visitor given his curiosity full rein, he would have wandered up the river valley where pioneers had ventured fearlessly into the interior. Here they had begun to carve out new fields and to pasture their animals in rolling open plains. Approaching the limit of the farming settlement, some 60 stades (12 km) from the coast, he would have drawn near the fortified promontory (later known as Cozzo Presepe) that dominated the valley of the Bradano. Beyond this point were farms, similar to those closer to the coast, where the indigenous people lived. Their Greek neighbors, with whom they traded and intermarried, called them "Oenotrians."

As the next chapters will show, the asty and the chora underwent a dramatic transformation after 600. Still, the basic settlement pattern had been set already, along with the challenges that future Metapontines would face.

Despite the remarkable, constantly renewed fertility and great beauty of the land, this choice of location posed at least one serious disadvantage. The lower swampy areas harbored mosquito-borne diseases, and although the inhabitants might have been protected to some degree by hereditary anemia (most probably thalassemia),[58] the problem of excess water in both the asty and the chora would have to be addressed.

In this first round of settlement, there was still plenty of choice land available. Farmhouses, however, are separated by regular intervals of as little as about 300 m in the Sant'Angelo Early Village Cluster, and 600 m in Giardinetto. The rather precise pattern suggests that the sizes of the plots of land on which a farmhouse was to be located were determined beforehand. Spacing of around 300 m would correspond to a plot from 7 to 9 ha.

With the exception of Sant'Angelo, the clusters in the first period were consistent in their sparseness; crowding would begin to be a problem later (see *Fig.*

[57] Carter 2003, 508.

[58] Henneberg and Henneberg 1998, 527–37.

Figure 20.34 View across the Bradano valley from below Cozzo Presepe toward the north. (AP)

Figure 20.35 View from below modern Bernalda across the Basento valley to the south. (AP)

20.1). The total number of significant farmhouses dated to **600**, according to the original chronology for fine-ware pottery, is nineteen.[59] As mentioned in the introduction, the alluvial deposits covering the T_0–T_2 terraces may have made a good many sites invisible to the survey team. Had the density of farmhouses on these terraces equaled that found on the higher ones, there might have been many more farmhouses closer to the river in the study area.

Subsequent waves of settlers would face choices of where to live. One option would have been to fill in the village clusters, at first reducing the size of some

[59] If the revised chronology is the only criterion used and this pottery is the only source of information for the period, then the number drops to two. Evidence from excavations suggests that this estimate errs on the low side, and the true population may have been somewhere between these extremes.

of the earlier plots. From the 6th century on, however, the distance between farmhouses varies little. In fact, many clusters expanded outward in a linear fashion, adding new sites at regular intervals along roads or river valleys. They could also strike out for new areas on the marine terraces, as will occur in the next period, or risk it alone, like the pioneering families who bypassed the protected valleys to venture fearlessly into the unknown spaces of the interior. Such solitary outposts, far from the nearest neighbors, would persist for centuries.[60] The spirit of rugged individualism coexisted with that of the communities found at 6–12 km from the coast, beyond the limits that could be comfortably reached from the primitive asty.

The four sanctuaries discovered in the study area in this period would continue to be central to rural life down to the early 5th century BC. Among the first were the sanctuaries in Giampasquale and the one dedicated to Artemis and Zeus in the Venella valley. The cult building at San Biagio antedated the earliest known structures in the asty. It was decorated with a very early example of a continuous terracotta frieze. The sanctuary at San Biagio and others, such as Pantanello and the sanctuary of Hera on the Bradano (Tavole Palatine) grew in importance and became in later times, in the case of San Biagio, associated with a powerful family. The village that formed around the three forks of the Conca had a small rural sanctuary. This feature was not present in the villages along the main rivers, but is characteristic of those along the smaller tributaries, like the Conca and the Fosso San Marco, and the larger ones like the Venella, where two neighboring sanctuaries might have served twin villages.

From the beginning, necropoleis of varying size and duration were a constant feature in the Metapontine countryside. When all periods are considered, there are almost exactly half as many necropoleis as farmhouses. Because the earliest burials were, as we know from excavations, often simply earthen pits, with few or often no grave goods, they are probably underrepresented in the census of **600**. Nonetheless, the two farmhouses with associated burials and the eight nearly contemporaneous necropoleis demonstrate that residence in the countryside was already permanent in many cases; these were not seasonal occupants of the land who commuted from the asty. Many chose to bury their dead nearby, but there were still others who preferred to travel farther and bury at necropoleis associated with sanctuaries (e.g., at site 401N along the Venella and at Pantanello[61]).

Kilns producing pottery were also a perennial feature of the chora. Eleven sites with evidence of pottery kiln operation were identified by the survey, with at least one and possibly more belonging to this early period.[62] There is significant evidence to suggest that the early settlers felt the need to create their own distinctive ceramic vessels and had the skills to do so.

The presence of sanctuaries, necropoleis, and kilns indicates that the villages would have been quite autonomous. The early settlers were self-sufficient, and the indigenous population in the interior provided a ready market for manufactures and agricultural surplus. Demand in the following centuries would support various industries in both asty and chora. In spite of the young asty's unimpressive appearance, the stage was set for an explosion of economic growth over the next century and beyond.

The Asty in 600

The site of the asty was occupied by a village of huts well before the establishment of the consistent settlement in the chora in the late 7th and early 6th centuries BC. Two sets of huts (located in areas designated as proprietà Andrisani and proprietà Lazzazara; see Ch. 17) were excavated under the level of the street grid of the 6th-century BC asty.[63] The two areas may have belonged to separate clusters or to a single large village. The huts, contemporaneous with the latest phase of the village of Incoronata, were more elaborate and spacious, containing assemblages of mixed Greek and indigenous pottery similar to those at Incoronata. Both sites appear to have been destroyed at the same time, about 630 BC. Imported proto-Corinthian pottery, and locally produced vessels inspired by it,[64] were found in abundance at Incoronata and at the Andrisani site, but no Early Corinthian ware was found at Incoronata and only a few fragments of it at Andrisani, in possibly disturbed contexts.[65] The ceramic evidence from the lowest levels (in the areas destined as the central public spaces—the sanctuary of the principal gods and the agora) consists of imitation proto-Corinthian and Transitional pottery.[66] The

[60] Sites 43F and 103F.

[61] See *Necropoleis*.

[62] Site 530 F/K. See also A. Bottini 1994, 705–06, Pl. 50.

[63] De Siena 1986b, 139–40.

[64] *Incoronata* 3, 35–40.

[65] De Siena 1986b, 141.

[66] De Siena 1980, 97.

Figure 20.36 Aerial view of asty looking east, with sanctuary in foreground. From left to right, Temples D, B, A, and C. In the background, the agora with the ekklesiasterion/theater to the left. (After Mertens 2006, Fig. 273; courtesy Dieter Mertens)

near contemporaneity of the last phases of Incoronata and Andrisani has suggested a cause and effect relationship between the arrival of Achaian settlers and the sudden destruction of the earlier sites.

In support of De Siena's and Giardino's interpretation of Strabo's text[67] is the fact that the settlement of the last quarter of the 7th century on the site of the future asty gives every appearance of having been a clean start without any sign of continuity with what went before.[68] This also appears to be true, on the whole, of the settlement of the chora. Just as the earliest of excavated structures in the asty are in contact with virgin soil, so the sites of the clusters have almost no trace of Iron Age occupation[69]—a fact that is driven home by the apparently complete abandonment of Incoronata greca until the establishment of a small sanctuary there in the first half of the 6th century BC.[70]

[67] Strabo 6.1.15. See above, Ch. 17, 589–90.

[68] For a discussion of the stratigraphical relations of the hut village at Andrisani and the colonial levels overlying it, see De Siena 1986b, 141–42.

[69] A single fragment of an undecorated olla at farmhouse 285F/T is, up to now, a unique example of indigenous pottery from a farmhouse site (Fig. 17.14, above). For indigenous pottery from the Pantanello sanctuary, see Fig. 17.17.

[70] Carter 2006a, 74–78. Incoronata will be the subject of a future study.

The perimeter of the urban center was determined from the start in the late 7th century BC, according to its excavators, and was not increased or altered significantly until the Roman period. Such generous and farsighted advance planning is not commonplace in any era. In its current state, the site of the asty is a palimpsest of at least eight centuries of life (*Fig. 20.36*). It was protected on the north and south by the natural barriers, the Bradano and Basento. To the east it was bounded by the coast, and on the west, facing the interior, by a stone wall dating to the end of the 7th century or the beginning of the 6th century BC (see Fortifications, above). Its total extent of 156 ha[71] far exceeded the needs of the early settlers. Clearly the founders had great ambitions, but also a plan in mind which is all the more surprising considering the lack of obvious precedents. The precocious attempts of the early settlers at city planning resemble those at Megara Hyblaia and Syracuse more than any on the Greek mainland.

A persuasive case has been made by Mertens and De Siena that this future urban space had been organized from the beginning into three major areas of

[71] Mertens 1998, 125–26.

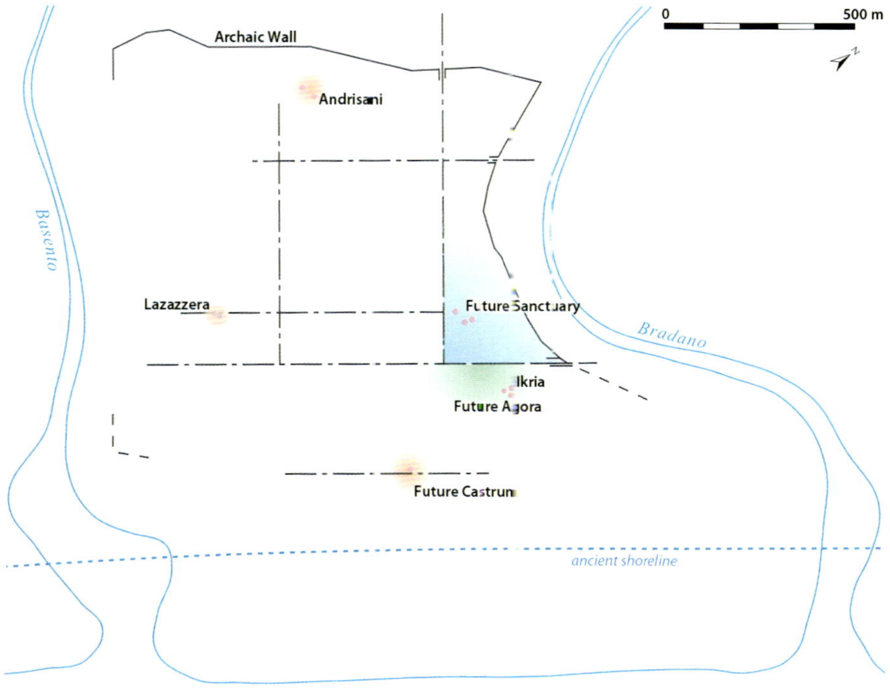

Figure 20.37 Plan of the site of the future asty. (After Mertens 2006, Fig. 47; courtesy Dieter Mertens)

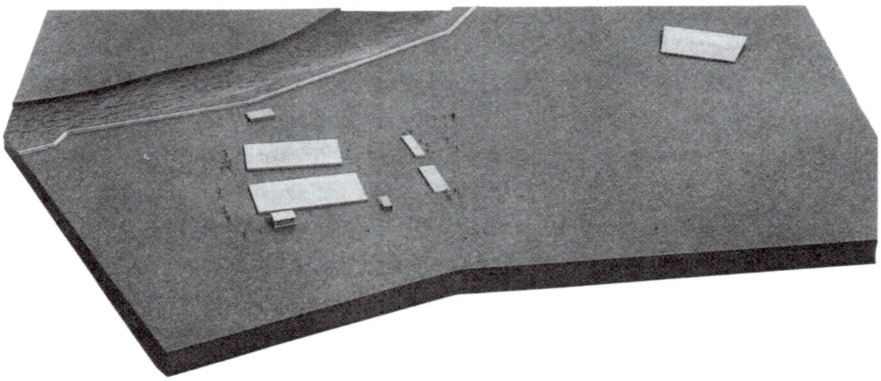

Figure 20.38 Model of sanctuary and agora in the first half of the 6th c. BC. (After Mertens 1999, 252, Fig. 1; courtesy Dieter Mertens)

activity (*Fig. 20.37*): the area sacred to the principal divinities of the city, the urban sanctuary; the place of public assembly, the agora; and the residential quarter. In the whole history of the excavation no signs were found that the destination of any of these areas changed during the lifetime of the independent polis. The earliest structures were wooden, as evidenced by deposits of charred wood found in the sanctuary and in the agora on sites later occupied by more substantial structures in stone (*Fig. 20.38*). The function of the structure later replaced by Ekklesiasterion I seems clear (see Ch. 21); it was a wooden bleacher (ikria) and it was associated with pottery said to date to the end of the 7th century.[72] Such a structure was known from literary sources to have stood in the Athenian Agora and its physical form was perhaps reflected on a fragment by the vase painter Sophilos,[73] active in the first quarter of the 6th century BC. The ikria of the Metapontines was a remarkable structure—as were its even more noteworthy successors—because it shows that among the first concerns of the settlers was not only to establish farms and create a new home for

[72] Mertens 1999, 250; 2006, 47.

[73] Arias/Hirmer/Shefton 1961, 39.

their gods, but also a place to gather as a community.

It is natural to wonder whence came the inspiration for the concept of such a settlement. The point has been made that it is easier to understand a planned asty in Sicily, because the majority of the settlers came from centers where planning either already had begun or was being contemplated. Still, the most detailed evidence of early planning comes not only from Sicily, but also Italy. Planning was "in the air" and Metaponto should be viewed in relation to other, unfortunately less fully documented, sites like nearby Siris and Sybaris. In contrast to Metaponto, early Siris in the late 8th to mid-7th century BC may have consisted of a number of clusters of semi-interred huts,[74] like the hut village at the mouth of the Basento that preceded the residential quarter of Metaponto. Sybaris also may have been dispersed in a similar way, but there is some evidence—the recent results of a study of the excavations of the late 60s and early 70s—that there was a regular plan with houses consisting of multiple rooms on rectangular stone socles, divided by streets, already in the mid-7th century BC.[75] The conjectured transformation of Siris and Sybaris from "hut villages" to planned settlements with city walls in the 7th century would have paralleled that of Megara Hyblaia in Sicily in the 8th century. The Early Archaic wall of Metaponto may have imitated or been in response to an earlier one on the acropolis of Siris Polieion.[76]

Figure 20.39 Frieze in terracotta from Oikos C in the asty (h 24.5 cm). A procession featuring female divinities in a cart drawn by hinnies, dated ca. 600 BC. (Courtesy of the Soprintendenza)

The evolution of the sanctuaries at Metaponto is among the best documented of any polis in Magna Grecia. Carbonized remains of wooden buildings beside the great mid-6th–century BC peripteral temples, the Temple to Hera (A2) and the Temple to Apollo Lykeios (B2), and in the vicinity of Temple D, are said to be associated with pottery of the end of the 7th century BC or the beginning of the 6th. The earliest votive object recovered was a bronze griffin head, dated to the late 7th century BC, belonging to a cauldron and tripod like those famously dedicated at Olympia. The rough stones known as *argoi lithoi*, some bearing inscriptions to Apollo with the name of dedicators, mirror the cult practices of the homeland.[77]

The small structure to the north of the future Temples of Hera and Apollo, known as Oikos C, was the earliest cult place to have left significant traces. It rose on foundations of local conglomerate quarried in the chora on which rested a socle of limestone (imported from neighboring sites in the interior) that probably supported a mudbrick wall. The building was decorated with a frieze representing a procession of divinities and other figures including a cart drawn by hinnies, as Bökönyi has observed (*Fig. 20.39*).[78] Fragments of a frieze from the same mold have been found at Siris Polieion and at Francavilla Maritima in the chora of Sybaris, the earliest Greek sanctuary known on this coast. It was implanted on a pre-existing indigenous cult place.

The Achaian settlers with no direct experience of urban life brought with them a familiarity with the sanctuaries of the chora and it is in the chora that they established their earliest places of worship—and probably their first substantial buildings. As at Sybaris, so at Metaponto, the earliest known dateable cult place—the shrine beside a spring at San Biagio—is in the chora, and it may even have antedated the cult

[74] Mertens 2006, 62, with an up-to-date bibliography. Siris covered an area of ca. 140 ha and its mudbrick defensive wall on the acropolis has been dated to the mid-7th c. BC.

[75] Carando 1999, 170, Fig. 3.

[76] Adamesteanu and Dilthey 1978, 522.

[77] Adamesteanu 1970b; Doepner 2002. The early dating of the *argoi lithoi*, based on epigraphical as well as stratigraphical arguments, is certain.

[78] S. Bökönyi, pers. comm., 1987.

Figure 20.40 View of the columns of the Temple of Hera (Tavole Palatine) on the banks of the Bradano from the northeast looking upstream. (After Mertens 2006, Fig. 51; courtesy Dieter Mertens)

place of Hera on the Bradano later occupied by the Doric temple known as Tavole Palatine (*Fig. 20.40*) as well as the sanctuary by a spring at Pantanello, though not by much. The proof of the 7th-century BC date for San Biagio comes from the structure itself. The terracotta frieze, representing a warrior departing for battle, is stylistically a quarter-century earlier than the processional frieze from Oikos C. They have been dated by Mertens-Horn to ca. 625 BC and ca. 600 BC respectively.[79]

Apart from the pre-existing hut village at Andrisani, little is known about the unexcavated residential quarter of the city. For now, the houses of the late 7th and early 6th centuries BC are a total blank. Were they of mudbrick on socles of stone like those at Siris Polieion,[80] or Strombi at Sybaris, both located outside the presumed limits of the city? Did the urban lots of early Metaponto reflect the influence of the division of the chora, as has been suggested for early Selinous?[81] As excavations in the chora have shown, the farmhouses Cogno del Pero and perhaps Fattoria Fabrizio,[82] with their parallel rows of rooms and apparent omission of a central courtyard, define a new and typically Metapontine type as early as the mid-6th century BC.

The topography of the site chosen by the arriving settlers dictated in fundamental ways the plan of the city as it developed. The sandy dunes became building sites; the stagnant pools of water had to be drained or avoided. From the first the Metapontines were challenged by the water and, according to De Siena and Mertens,[83] their responses were reflected in the arrangement of streets, which follow drainages, north and south toward the rivers. From the beginning, Plateia A (the major east-west road) was the artery that connected the heart of the asty—the public spaces, sanctuary, and agora—with the chora through the western gate (Porta Settembrini) in the earliest city wall. It has been suggested that experiences from the chora—e.g., the problems of drainage and dividing the available agricultural land into fair allotments—may possibly have anticipated solutions to these problems in the city, or they could have been contemporaneous and complementary.[84]

[79] Mertens-Horn 1992, 23–24, 50–57.

[80] M. Tagliente 1986, 131–32 (house in Cospito Caserta, see above).

[81] Mertens 2006, 174.

[82] Carter 2006a, 135. The earliest phase at Fattoria Fabrizio can now be securely dated to the mid-6th c. BC.

[83] De Siena 1998, 143–44; Mertens 1999, 248–49; 2006, 436.

[84] Mertens 1999, 249.

21
Golden Harvests: The Expansion of the Mid-6th Century

Joseph Coleman Carter

...οὕτως ἀπὸ γεωργίας εὐτυχῆσαί φασιν,
ὥστε θέρος χρυσοῦν ἐν Δελφοῖς ἀναθεῖναι.

...and they so prospered from farming, it is said, that they dedicated a golden harvest at Delphi.[1]

Strabo 6.1.15

Θέρος : Μετα[π]όντ̣[ιοι]
δεκάταν : τοπόλον[ι]

La moisson, les Métapontins en dîme à Apollon[2]

Dedication at Delphi

From its tentative beginnings in the late 7th and early 6th centuries BC, the settlement of the chora expanded dramatically in the middle years of the 6th. At this point it assumed the overall shape that it would maintain throughout the following three centuries in the life of the Greek polis. In retrospect, these were the years of its fullest independence and greatest economic success, celebrated famously by the "golden harvest" (θέρος χρυσοῦν), offered to Apollo at Delphi, which consisted of sheaves of grain made from solid gold.[3]

The number of farmhouses in the study area more than tripled, growing from 19 farmhouses in **600** to 64 in **550** (*Appendix C1a*). While the Early Village Clusters expanded significantly, there was also a decisive shift of settlement toward the interior, to areas that were previously only lightly occupied, if at all. This was the first of three periods of exceptional growth (the others being **450** and **350–300**). The prosperity of the Metapontines during this first agricultural boom was reflected in the great building projects in the urban center, and in the striking of silver incuse coinage about the middle of the 6th century BC,[4] which produced some of the most elegant and impressive coins ever minted by the Greeks (see *Fig. 22.10*).

As the coinage attests, the wealth was based primarily on cereal production, for which the relatively flat marine terraces of the interior were ideally suited (Ch. 19). Their importance can be seen by comparing the overall distribution of sites in the study area in 550 (*Map 6*, Vol. IV; *Figs. 21.1* and *21.2*) with that in 600 (*Figs. 20.1* and *20.2*). A strip of sites that runs NE–SW across the chora from the Venella at the south to the greatly expanded cluster at Giampasquale stands out. In this area there were only two significant sites in **600**; four significant farmhouses appear, where before there had been none. In the more remote interior, a single significant farmhouse at the headwaters of the Venella in **600** was joined by six more in the next half-century. The distribution of second magnitude farmhouses further indicates areas of interest.[5] The changes that were taking place in this period and the patterns of settlement that were emerging are best understood by looking in detail at each of the Early Village Clusters and the newly populated areas of the Central Plateau and the upper Venella valley.

The Chora in 550

Early Village Clusters

Giardinetto (*Fig. 21.3*). After an unexceptional start in **600**, this cluster entered a phase of rapid growth in **550**, and became one of the most consistently and densely populated of clusters.

[1] Text and trans. H.L. Jones 1983 (Loeb).

[2] Text and trans. Vatin 1991, 55, Figs. 11, 12. "Harvest, the Metapontines a tenth part (to Apollo)."

[3] Vatin 1991, 53–54, Fig. 10.

[4] The unique technique, whereby the relief design on the obverse repeated the intaglio on the reverse, was introduced by Metaponto and other poleis in Magna Grecia and Sicily, and was retained by Metaponto long after double-relief coins were in general use elsewhere in the Greek world.

[5] See Ch. 22, below. Although they appear on maps of all periods, second magnitude sites are particularly relevant to the discussions of the earliest periods in Chs. 20, 21, and 22. The counts of dateable pottery from **600** and **500** are very low compared to other periods down to 200 (*Table 19.4*). These sites have pottery dateable to the period, but not enough to be considered significant. Still, they indicate areas where some activity at a lower level may have been taking place. We cannot be sure that the portions of date ranges that add up to 0.45 may not belong to pottery that was deposited earlier or later than the period in question. For a discussion of dating methods, see Ch. 19, Black-gloss pottery and date bins. See also Ch. 4.

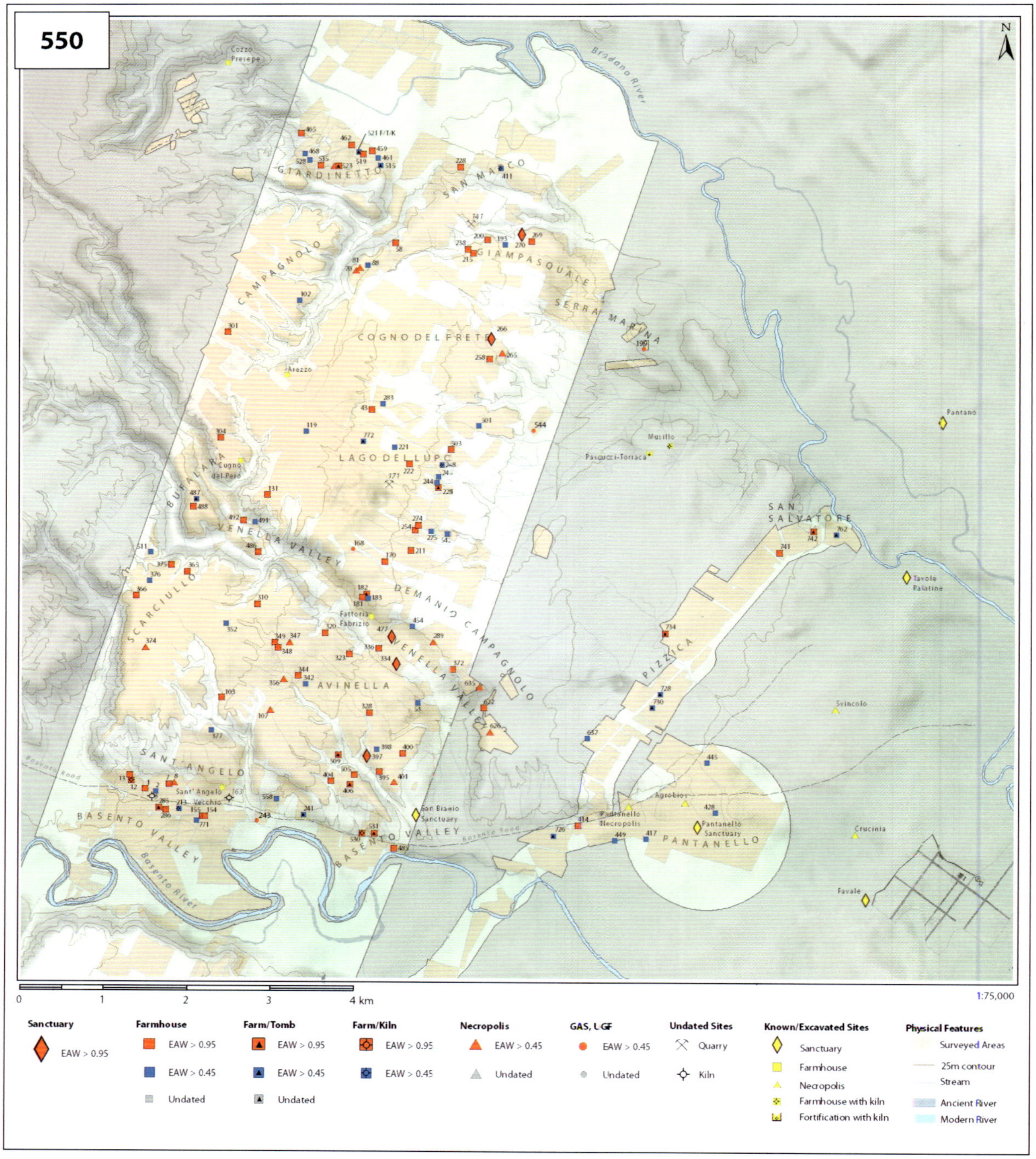

Figure 21.1 The study area in **550**. (JT)

Farmhouses. The number of farmhouses tripled in this period. To the earlier five farmhouses, six more ex novo farmhouses were added, for a total of eleven (*Table 21.1*). The two original significant farmhouses of **600** (523F/T and 465F) continue to be by far the most fully documented with dateable black-gloss pottery. These "first families" and their descendants, or successors on the site, continue to occupy the spot for the next three centuries in the case of site 465F. Of the farmhouses that came into existence for the first time in this half-century—surely in response to the agricultural boom—only 459F did not continue afterward: like many others here and elsewhere, it did not survive the crisis of **500**[6] and the site was abandoned forever.

6 See below, Ch. 22.

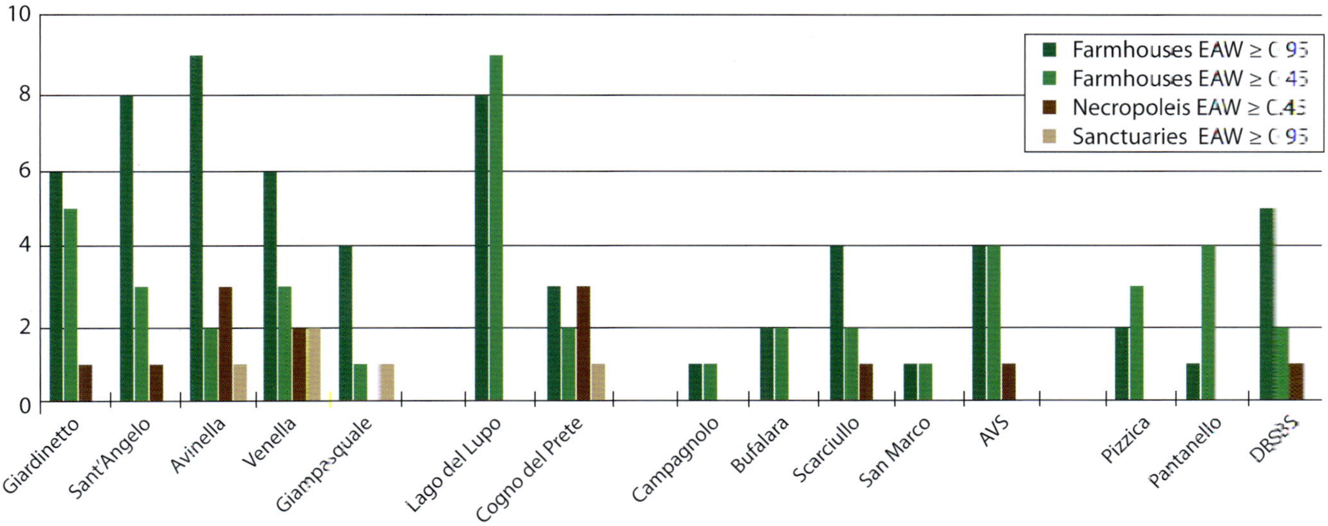

Figure 21.2 Distribution of farmhouse, necropolis, and sanctuary sites among localities, **550**. Totals: significant farmhouses, 64; second magnitude farmhouses, 44; necropoleis, 13; sanctuaries, 4.

On the western side of the locality, three new sites (535F, 468F, and 528F) filled in the spaces between the original two farmhouses. There is a steep drop-off from 535F to the valley below, where there were apparently no sites in this period. The new sites to the east, however, formed a second cluster within the larger Giardinetto group, on a series of gently sloping terraces that reach down to the floor of the Bradano valley (cf. *Figs. 20.3* and *20.4*).

The distance separating the two original farmhouses at Giardinetto in **600** was about 600 m, or approximately three stadia. In the subsequent period the distance separating nearest neighbors was much less but was still respectable. Considering only significant farmhouses, the distance separating them on the west side of the drainage varies from approximately 200 to 400 m. On the eastern side the farmhouses are much more crowded, with approximately 100 to 150 m between them. Their proximity suggests that they may be in some sense "paired."[7] Evident in **550**, here and elsewhere, is a tendency for two farmhouses to appear very close together, with 100 to 120 m between them: for example, 459F and 519F; 461F and 515F/T; 468F and 528F.

Necropoleis. No necropoleis and no more farmhouses with associated burials were added in **550**, despite the fact that the number of significant farmhouses rose from two to six. The ratio of places of burial to dwellings, as a result, fell sharply: two burial sites to six farmhouses (1:3). This ratio stayed fairly constant over time, despite the fact that over all periods there were half as many necropoleis as farmhouses (12:22). Only one necropolis in Giardinetto was dateable to this period. Site 524N was in continuous use from **600** to **450**, and the farmhouse with burials, 523F/T, from **600** to **300**. (To this period also belong two second magnitude farmhouses with burials, 515F/T and 521F/T/K, which do not become significant until a century later.) Though it is difficult to prove chronological relationships, in part because of the small amount of materials on the surface of burial sites, the proximity of the many undated or weakly dated necropoleis to farmhouses is, indeed, suggestive (*Table 20.2*). The average distance between a necropolis and its nearest farmhouse neighbor at Giardinetto is approximately 100 m, thus well within the range to be considered a potentially related burial. Very few farmhouses in Giardinetto are far removed from a necropolis.

Significance, land, and continuity. The two original farmhouses of **600** would have been the two most significant in this period, were it not for the newcomer site 459F, which existed only in this period. It has a higher FMCE value than the two that preceded it. The phenomenon of single period sites occurs elsewhere in the chora in this period, and in others where expansion is rapid, e.g., **450** and **300**. The average

[7] See discussion of pairs, below.

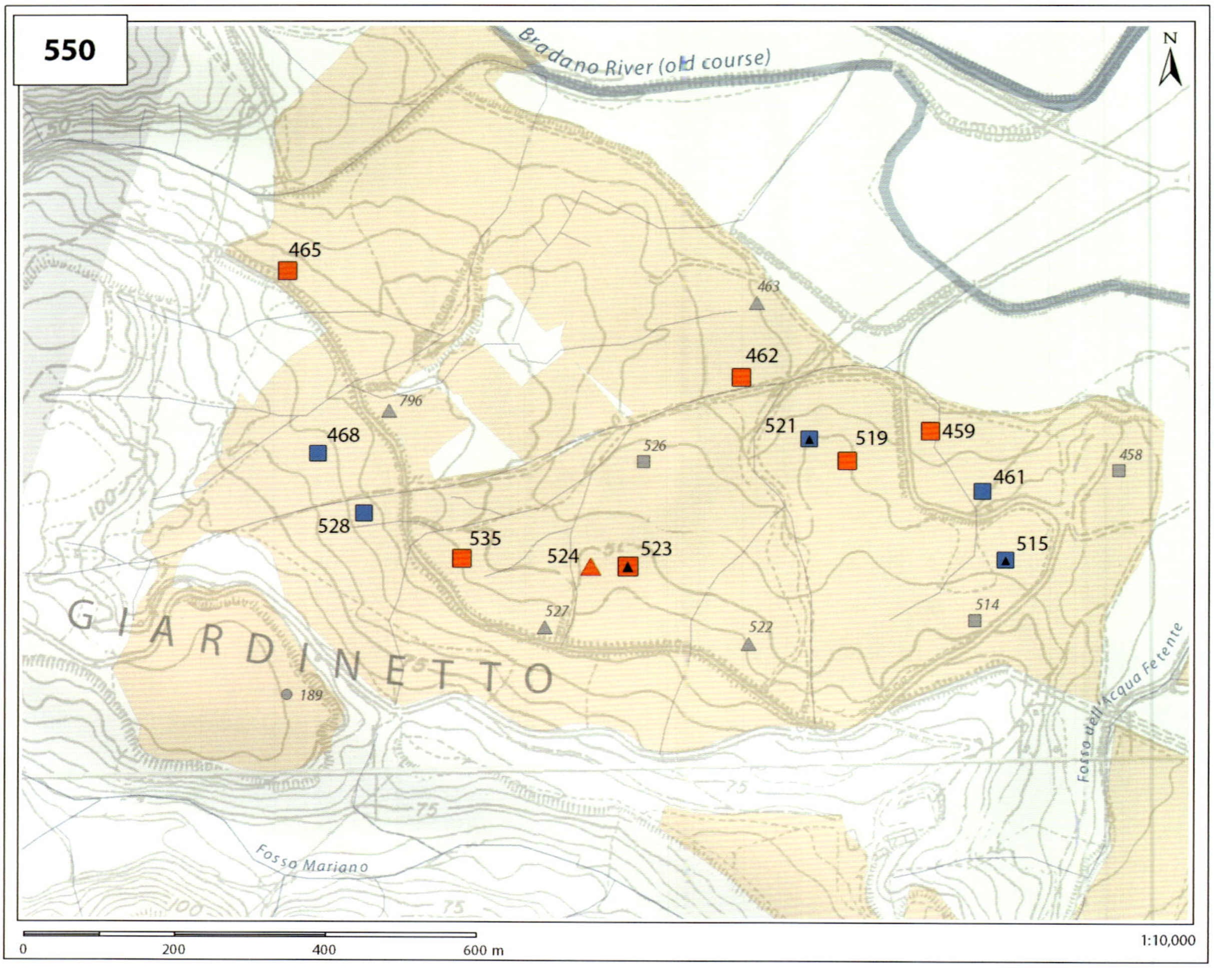

Figure 21.3 Giardinetto in **550**. See Fig. 21.1 for map key. (JT)

FMCE value for Giardinetto in **550** is 2.46, which is considerably higher than the average of 2.17 for the study area as a whole at this time (*Appendix C7*). Four farmhouses at Giardinetto are above this overall average and only two below it. The lower FMCE values (with the exception of 459F) are found on the east side of the cluster. This is the area of greatest crowding and the lowest density of necropoleis, which indicate permanent residence. Together these facts suggest that this area may have been occupied only seasonally by owners with permanent residences and necropoleis in the asty or elsewhere.

Sant'Angelo (*Figs. 21.4* **and** *21.5*). Sant'Angelo in **600** was the most precocious of the clusters, so it is not surprising that the rate of growth between **600** and **550** in this cluster, formed around three groups of early farmhouses, was somewhat slower than that at Giardinetto. There was a total of eleven farmhouses

Site no.	Period			
	600	550	500	450
465F	1.4	3.7	0.7	3.8
468F	0.1	0.9		
462F	0.8	2.2		2.1
459F	0.3	1.7		
521F/T/K	0.5	0.5	0.2	7.4
519F	0.3	1.6	0.2	1.8
523F/T	2.3	9.3	2.4	1.3
524N	0.6	1.4	0.6	2.4
535F	0.3	1.6	0.2	
528F	0.5	0.5		
461F	0.1	0.9		
515F/T	0.1	0.9	0.1	2.2

Table 21.1 Giardinetto **550,** with EAW values rounded to the nearest tenth through **450.** Dark green = significant farmhouses or sanctuaries (EAW ≥ 0.95) and active necropoleis (EAW ≥ 0.45). Light green = second magnitude farmhouses (EAW ≥ 0.45, but ≤ 0.95).

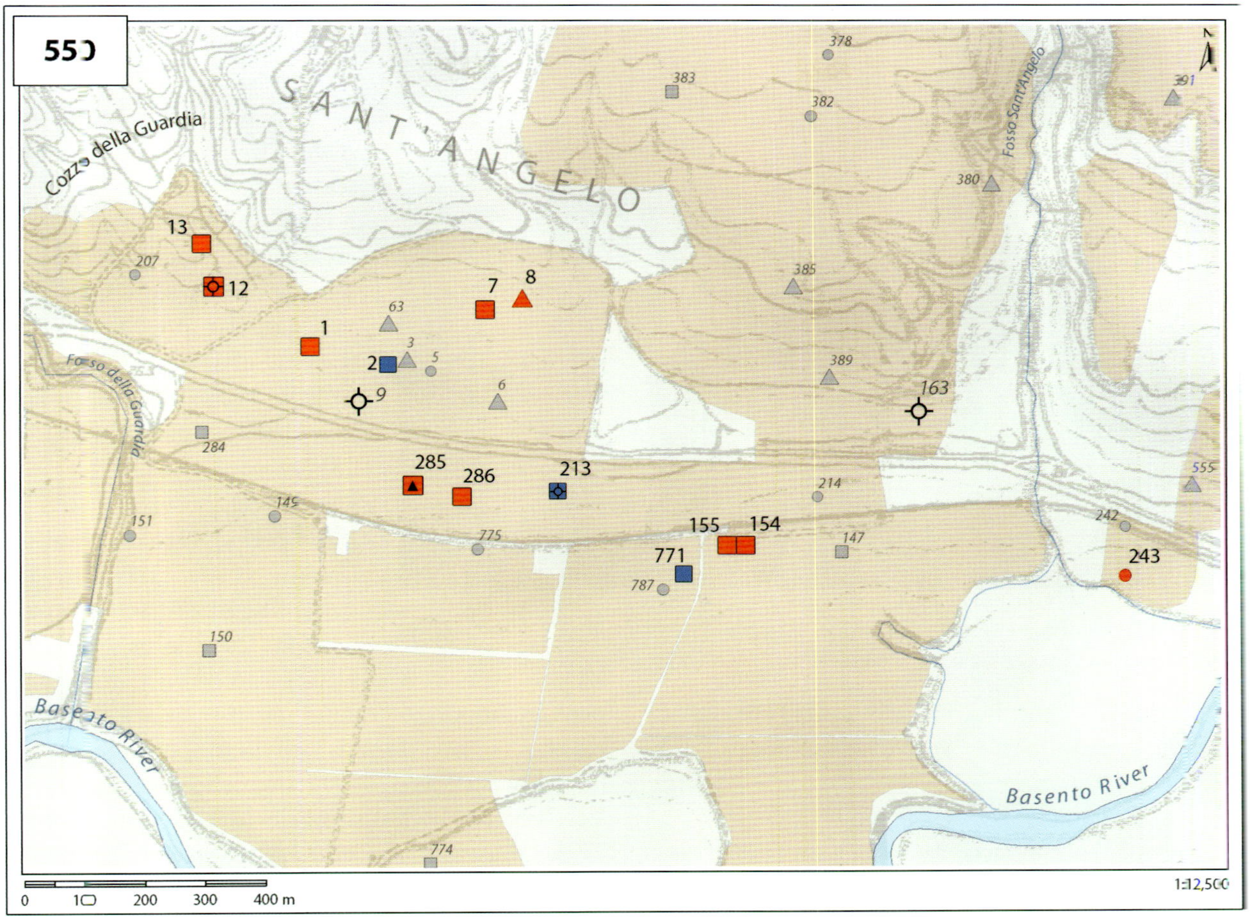

Figure 21.4 Sant'Angelo in **550**. See Fig. 21.1 for map key. (JT)

(eight significant, three second magnitude) in **550**, as compared with seven (five significant, two second magnitude) in **600** (*Appendix C3: Sant'Angelo*). Of the eleven, five (three first magnitude, two second magnitude) started ex novo in **550**. They are sites 12F/K and 7F (and 213F/K and 771F of second magnitude).

The elongated shape of this cluster and its extent were evident already in **600**. The steep slopes on the north and west and a very deep drainage on the east (the Fosso Sant'Angelo) embrace the settlement, which is divided by a single small drainage (compared to the pair of drainages in Giardinetto). The widening of the cluster occurred on its western side in analogy with the similar agglutination of new sites in Giardinetto. The difference is that at Sant'Angelo growth occurs along a road, whereas this does not appear to be the case at Giardinetto. In both cases this occurs on one side of the drainage. No new sites appear to the east of it.

Farmhouses. Three of the newer farmhouses (12F/K, 1F, and 7F) were located at the far western end and at higher elevations than the well established pairs of sites (285F/T and 286F; 155F and 154F). Nestled in a cleft below the Cozzo della Guardia with earlier site 13F was the new site 12F/K. To the immediate east there was also infilling, but it appears to have been carefully dispersed. Sites 1F, 7F, and 285F/T form an equilateral triangle, 300 m on each side, that is reminiscent of the one in Avinella in **600**, except that there the sites are divided by precipitous valley walls. This distance seems to be the preferred minimum wherever possible for separating sites in this period.

Sites 13F and the new site 12F/K were guaranteed a secure source of water from springs that regularly issue at the foot of heights along this shore of the Basento (*Fig. 21.5*). The early farmhouses were supplied by springs, by the Fosso della Guardia and the river, only 600 m to the south. The spring on the slope below site 1F was still active in 1981 when the site

Figure 21.5 Sant'Angelo in summer 2007. This view looking to the south across the Basento was taken from the site of farmhouse 1F, excavated as Sant'Angelo Grieco. (JCC)

was excavated, a month after being identified in the survey. Several miniature votive vessels were found around its mouth.

Necropoleis. Likely associated with new site 7F is a new necropolis, site 8N, located not more than 50 m away. A documented example of a contemporaneous farmhouse and necropolis nicely parallels 523F/T and 524N in Giardinetto. In **550**, there is one necropolis of this date and eight farmhouses—a ratio very close to that of Giardinetto in this period. There were, however, far fewer necropoleis in relation to farmhouses in the Sant'Angelo cluster than at Giardinetto (compare *Appendix C3: Giardinetto* and *Sant'Angelo*). Considering all periods, the ratio of 8 necropoleis, dated and undated, to 22 farmhouses is low (about 1:3, as compared to 1:2 at Giardinetto). Five out of eight of the necropoleis at Sant'Angelo are undated, a much higher ratio than at Giardinetto. If the undated ones are taken into account, there are some possible necropolis-farmhouse associations. Site 63N for example, though undated, is only 20–30 m from 2F, a second magnitude farmhouse in **600** and **550** which was not active in later periods. It is no more than 120 m away from site 1F, which was in use from **600** to **300**. Although only one necropolis is securely dated to this period, it marks an increase with respect to **600**, when the only burials associated with a farmhouse were those at site 285F/T.[8]

Kilns. The slag in the surface scatter at two of the new farmhouses provides evidence of ceramic production. It is hard to know in precisely which period this activity took place, but in the case of site 12F/K it is likely to have been **550**, the principal phase of the nearby farmhouse 13F according to the EAW values (*Appendix A*). The most likely periods for the operation of the kiln at site 213F/K are later, beginning in **450** and continuing to **250**. These kilns are the first signs of an activity that came to play an important role along this valley throughout the Greek period and well into the Roman. Kilns have been excavated at Pantanello and Sant'Angelo Vecchio. In the Sant'Angelo locality, just below site 1F, a kiln was found near the excavated farmhouse at Sant'Angelo Grieco (*Fig. 18.25*).

Significance, land, and continuity. As was the case at Giardinetto, the sites that were significant in **600** continued to be so in **550** and they are joined by three more (1F, 7F, and 12F/K). The average FMCE value for Sant'Angelo in **550** is 2.42, considerably higher than all other clusters including Giardinetto (2.29) and than the average for all significant sites in the study area in this period (2.14) (see *Appendix C7*). There are five above the average and three below. The low values are fewer (site 13F with FMCE of 1.53) and the high values both higher (155F, 3.00) and more evenly distributed than in Giardinetto, which suggests that this cluster may have been more successful in this period than the others, and that the success was more evenly shared by the members of the community. Its location, with good supplies of water, along a major artery (the Basento Road), and beside a navigable river, certainly must have contributed. Giardinetto, with the second highest average FMCE value, was also favored in these respects.

[8] These, of course, may well have belonged to later phases of the farmhouses which continued to be significant until **450**

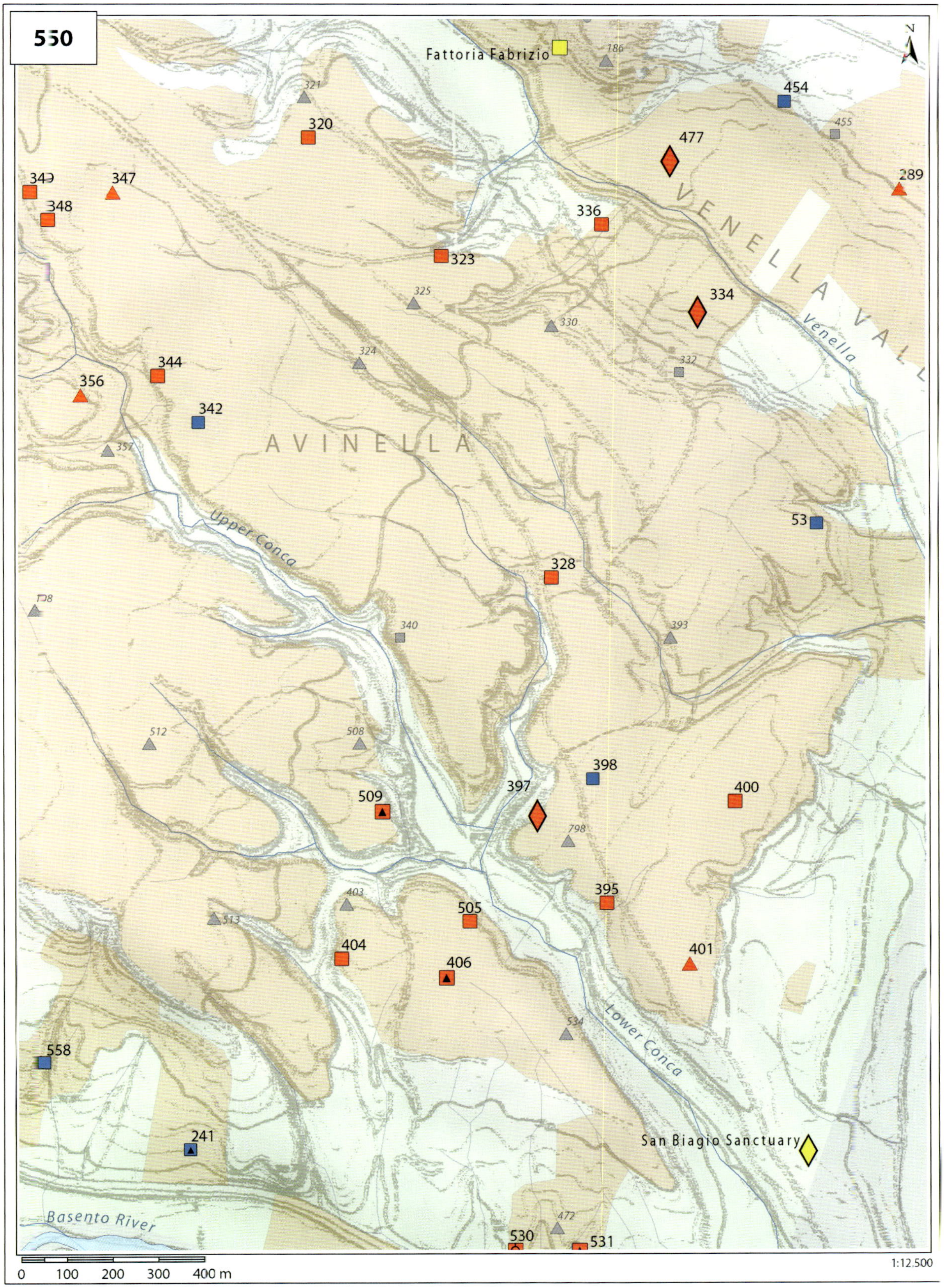

Figure 21.6 Avinella in **550**. See Fig. 21.1 for map key. (JT)

	Site no.	Period			
		600	550	500	450
Lower	397S	8.6	34.8	9.3	6.1
	395F	1.4	4.0	1.2	5.6
	404F	0.3	1.7	0.1	1.0
	505F	1.0	1.0		
	406F/T	0.8	2.2		0.4
	398F	0.1	0.7	0.1	
	509F/T	0.2	1.8		
Upper	342F	0.1	0.9	0.1	3.0
	344F	0.1	1.9		2.0
	356N	1.1	2.2	0.7	
	347N	0.1	0.9		0.4
	349F	1.0	3.7	0.3	
	348F	0.1	1.2	0.7	0.7
	107N	0.5	0.5		
	328F	0.3	1.7	0.6	1.5

Table 21.2 Avinella **550,** with EAW values rounded to the nearest tenth through **450.** Dark green = significant farmhouses or sanctuaries (EAW ≥ 0.95) and active necropoleis (EAW ≥ 0.45). Light green = second magnitude farmhouses (EAW ≥ 0.45, but ≤ 0.95).

Avinella (*Fig. 21.6*). The number of significant farmhouses in Avinella in the two areas initially settled in **600**—the lower and the upper Conca—increased dramatically from 3 to 9 in **550** (*Table 21.2*). The settlements at Giardinetto and Sant'Angelo were characterized by wide spaces with high visibility, where individual farm families were in visual contact with most, if not all, of their neighbors. Those village clusters with commanding views over the broad valleys of the Bradano and Basento are compact, but open to the wider world. They are in relatively early contact with the settlement on the coast. At Avinella, the settlers followed edges of narrow valleys and created separate settlements at the three forks of the Conca and 1.5 km further upstream. At the three forks the farmhouses are rather close together, within distances ranging from approximately 150 to 300 m, but separated by deep ravines. The upper Conca, though more sparsely settled, follows a similar pattern. There are two, possibly three, independent nodes, connected by branches of the stream.[9]

[9] The pattern is reminiscent of a more recent frontier, in the Appalachian Mountains, where hollows were occupied by successive waves of immigrants, who followed the valleys inland and maintained to a large extent their independence.

Farmhouses. In the lower reaches of the Conca, the number of significant farmhouses increased from two in **600** to five in **550**, with the addition of significant sites 404F and 509F/T along with newly significant 406F/T. There is also one new second magnitude site, 398F. In the upper reaches four new significant farmhouses, 344F, 328F, 349F, and 348F, appear where previously only one lone farm family had ventured (see *Fig. 20.11*; *Table 20.4*). All of the significant farmhouses in both areas continue and are more fully documented in this period with dateable black-gloss pottery.

These new sites in the lower Conca occupy spots that are more remote from the original core at the three forks. Most are along the edge of a valley (e.g., 404F, 509F/T, and 328F). Site 406F/T, on the other hand, sits in the interior of a plateau, the center of prime agricultural land that is perfectly adapted to plow-based agriculture and cereal cultivation. Further inland, along the main stem of the upper Conca, we find the new significant farmhouse 344F, and the second magnitude 342F. Site 328F sits alone at the head of the east branch. About 400 m farther up the Conca from 344F is new site 348F (see *Fig. 21.6*). The early settlers who occupied 349F were separated from the newcomers in 348F by less than 100 m. The two sites would seem to constitute a pair, like those that appeared in Sant'Angelo at this time.

The valleys, as noted earlier, offered good conditions for herding animals, and in the wide spacing of sites along the Conca and their progression into the interior we may recognize the search by the early settlers for new pasturelands on the plateaus, as well as the valleys. Those who located their farmhouses near the centers of the small plateaus between drainages, as we have suggested, may instead have been motivated by the growing demand for cereals and for flatter fields. Neither of these activities, herding nor agriculture, need have excluded the other and it well may be that the deep penetration of sites like 348F and 349F reflects the desire of the occupants to possess the fertile plains extending back from the edges of the Conca.

Necropoleis. The three necropoleis that were active in Avinella in **550** were all to be found in the upper reaches of the Conca. 347N, which came into use in

550, probably served both the earlier farmhouse 349F, which was significant in **600** and **550** but not later, and the newly significant 348F. The occupants of site 344F could have utilized the early necropolis 356N as burying grounds, though located 170 m away and on the opposite bank of the Conca. This cannot have been the case with the large necropolis 401N, which was truly isolated and remained so for four centuries.[10]

Sanctuaries. Though it was frequented already in **600**, sanctuary 397S had its heyday in **550**, after which the number of votive objects fell off precipitously. In this respect, it was like the other sanctuaries identified by the survey.

Significance, land, and continuity. Where sites are concentrated around the three forks, there seems to be a uniform measure of approximately 300 m separating both the older and the newer sites, as at Sant'Angelo. The spacing would have corresponded to an area of approximately 7.5 ha, with the farmhouse at its center. This amount of land is greater than the size that some scholars have suggested would be adequate for subsistence in mainland Greece, but less than that possessed by a wealthy farmer.[11] Newly added sites 404F and 509F are separated from the older 505F by this distance. They form a roughly equilateral triangle (bridging the west bank of the Conca), similar to that formed in **600** with the sanctuary 397S at the apex and the original two farmhouses, 395F and 505F, forming the base. This fact might indicate that there were constraints—perhaps a maximum distance of separation between farmhouses?—in planning new settlements, and that a process of allocation of land was already in place from the beginning.

The contrasts between the sites along the narrow valleys of Avinella and those that face the broad valleys of Bradano and Basento are not limited to topography and spacing, but include also the less immediately perceptible matter of site significance. The sites in Avinella are measurably less significant than those at Giardinetto or Sant'Angelo. Excluding the sanctuary 397S, whose FMCE in **550** is 2.8, the average FMCE value for the significant farmhouses of the lower Conca is 1.98 (*Appendix C7*). That for the upper Conca, 2.38, is even greater than the average of 2.29 at Giardinetto in **550**. The greater significance of the upper Conca with respect to the more permanent settlement lower down the valley would seem to be a consequence of the rapid expansion that was cut short by the crisis in the chora in the next half-century.[12]

	Site no.	Period			
		600	550	500	450
West	334S	2.0	11.2	2.2	2.5
	336F	1.8	6.8	0.4	1.3
	323F	1.4	3.4	0.1	1.0
	320F	0.1	1.2	0.8	5.1
	53F	0.1	0.9		
East	477S	8.1	23.2	4.7	12.1
	454F	0.1	0.9		0.4
	289N	0.8	2.5	0.8	1.9
	372F	0.3	1.7	1.0	0.9
	183F	0.1	0.9		0.9
	181F	0.4	2.5	0.1	
	182F/T	0.6	1.5	1.0	5.6
	400F	0.6	1.4		

Table 21.3 Venella **550**, with EAW values rounded to the nearest tenth through **450**. Dark green = significant farmhouses or sanctuaries (EAW ≥ 0.95) and active necropoleis (EAW ≥ 0.45). Light green = second magnitude farmhouses (EAW ≥ 0.45 but ≤ 0.95).

Venella (*Figs. 21.7* **and** *21.8*). From the first, the settlement on the Venella had two sanctuary sites, the more significant 477S on the east and 334S on the west (*Fig. 21.7*), facing each other across the stream. One of the characteristics of villages is that they often consist of groups who share the same cult (see Villages, below).[13] If a village had but a single cult, then there may have been two villages facing each other in the Venella. The sanctuaries are separated by the stream, but only by approximately 350 m. It would have been possible, surely, for a village to have had several places of worship.

The impression that east and west behaved somewhat differently from each other, however, is given substance by a sudden increase of sites on the east side (*Table 21.3*). Apart from the sanctuary 477S and ne-

[10] See below, Ch. 23, 760.

[11] E.g., Foxhall 2002. Hodkinson and Hodkinson (1981, 274, and n. 127 following Burford Cooper) suggest a farm of 18 ha for "wealthy households" and 4.5 ha for "ordinary hoplites."

[12] The higher value for this group is surely due to the fact that many of these sites were short-lived. For an explanation of the frequent high values of single-period sites, see Ch. 4, and n. 24, below.

[13] Osanna 2001, 205, with references.

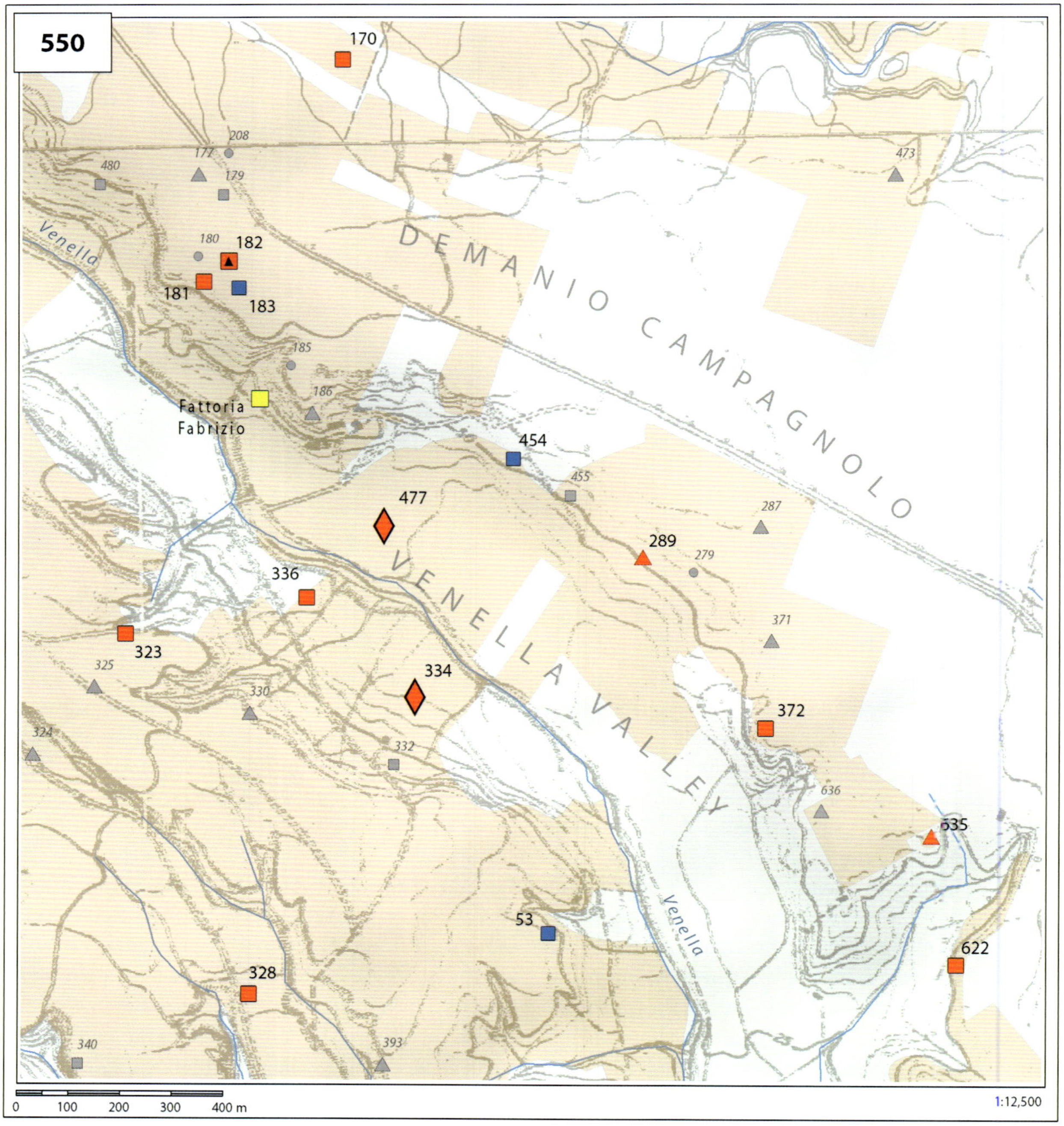

Figure 21.7 Venella in **550**. See Fig. 21.1 for map key. (JT)

cropolis 289N, the west bank was deserted in **600**. In **550**, four new farmhouses appeared on the east bank (see below) as opposed to only two new sites on the west bank.

The definition of this settlement in the middle course of the Venella as a village cluster is open to a more fundamental question. How extensive can a settlement be before it ceases to be a cluster? The Venella settlement may well have stretched up and down the valley almost without break. Site 372F, farthest to the southeast along the eastern valley slopes, for example, is close enough to its nearest neighbors to the south, the major Archaic necropolis 635N and the farmhouse 622F, for these sites to be considered part of a continuum.[14] It is quite conceivable—even very likely—that there was a major concentration of sites at

[14] That conclusion could not be confirmed because the survey coverage of the lower Venella valley was, unfortunately, incomplete.

the mouth of the Venella with the partially excavated sanctuary at San Biagio on the west side at its center (see *Fig. 21.1*). Site 622F is separated from the central Venella settlement by a deep drainage and it is equidistant—about 1.5 km as the crow flies—from the San Biagio sanctuary and site 477S. The San Biagio sanctuary is actually visible from site 622F while site 477S is not. Topography and visibility would favor the association of 622F with a hypothetical San Biagio cluster. The same arguments could apply to the probable association of 372F and 477S.

Farmhouses. Three first magnitude farmhouses, 181F, 182F/T, and 372F, sprang up ex novo in **550** on the east bank, along with two of the second magnitude, 183F and 454F. The excavated site of Fattoria Fabrizio, also on the east bank, was first occupied in **550**.[15] On the western bank, there was just one new significant farmhouse (320F; see *Map 6*, Vol. IV) and one second magnitude site (53F).

Whereas the earlier sites were at lower elevations within the valley (with the exception of site 323F), the new sites in **550** are all located along the valley's edge (for site elevations, see *Appendix A*). The new settlers had more commanding views and were more conveniently located with respect to the fertile plateaus of Demanio Campagnolo and Lago del Lupo. Except for one tight-knit group of sites (181F, 182F/T, and 183F) on the east bank above Fattoria Fabrizio, the farmhouses are spaced at increasing distances from each other, moving out from the two sanctuaries. The distance from each sanctuary to the nearest farmhouse is approximately 200 m, from that farmhouse to the next, 300–350 m, and so on increasing to a 400 m interval creating the impression of a centrifugal expansion along each bank in both directions (*Fig. 21.7*).

Necropoleis. There were two active necropoleis in the Venella cluster in **550**, 289N and 635N, both located on the eastern side of the stream, together with one farmhouse with burials, 182F/T. There were no necropoleis on the west side contemporary with the three significant farmhouses. Burials were being made at 289N from **600** to **450**, and at 635N from **600** to **500** (as well as **350** to **250**), but these two necropoleis are 400 m distant on either side from the nearest contemporary farmhouse, 372F. Both, however, are near the limit of the survey area of the lower Venella, and we may suppose that the farmhouses with which they were presumably associated lay outside it.

Figure 21.8 Sheep and goats grazing in the upper Venella valley. (JCC)

Sanctuaries. Both sanctuaries reach the peak of their importance in **550**. Both suffered in the crisis of **500**, but only 477S recovered strongly, though briefly, in **450**.

Significance, land, and continuity. There is a real difference in terms of significance between the settlement on the west bank of the Venella and that on the east. The FMCE values of both of the older farmhouses on the west (336F and 323F) increased dramatically—from very much below average values in **600** to comfortably above average in **550** (*Appendices B* and *C7*). Only the new farmhouse 320F lagged behind. On the east, however, the new farmhouses remained below average—in fact, sites 372F and 320F have the same value of 1.73, compared to the average FMCE value 2.14 for the study area as a whole in **550**. Two out of the three new farmhouses are of below average significance, the only exception being the single-period site 181F. How should this clear contrast be interpreted? In social and economic terms? Or rather as an indication of different stages of development? Such questions suggested by the survey required the fuller evidence supplied by the excavation of farmhouses such as Fattoria Fabrizio.

Lago del Lupo (*Figs. 21.9* and *21.10*). The momentous advance in the settlement of the chora in the second and third quarters of the 6th century BC was the occupation of the fertile terraces of the Central Plateau. The new farmhouses are restricted to a narrow strip

15 The site of this farmhouse, the subject of a separate volume in this series, was first occupied in the 6th c. BC. The occupation which has left the fullest record belongs to the 4th c.

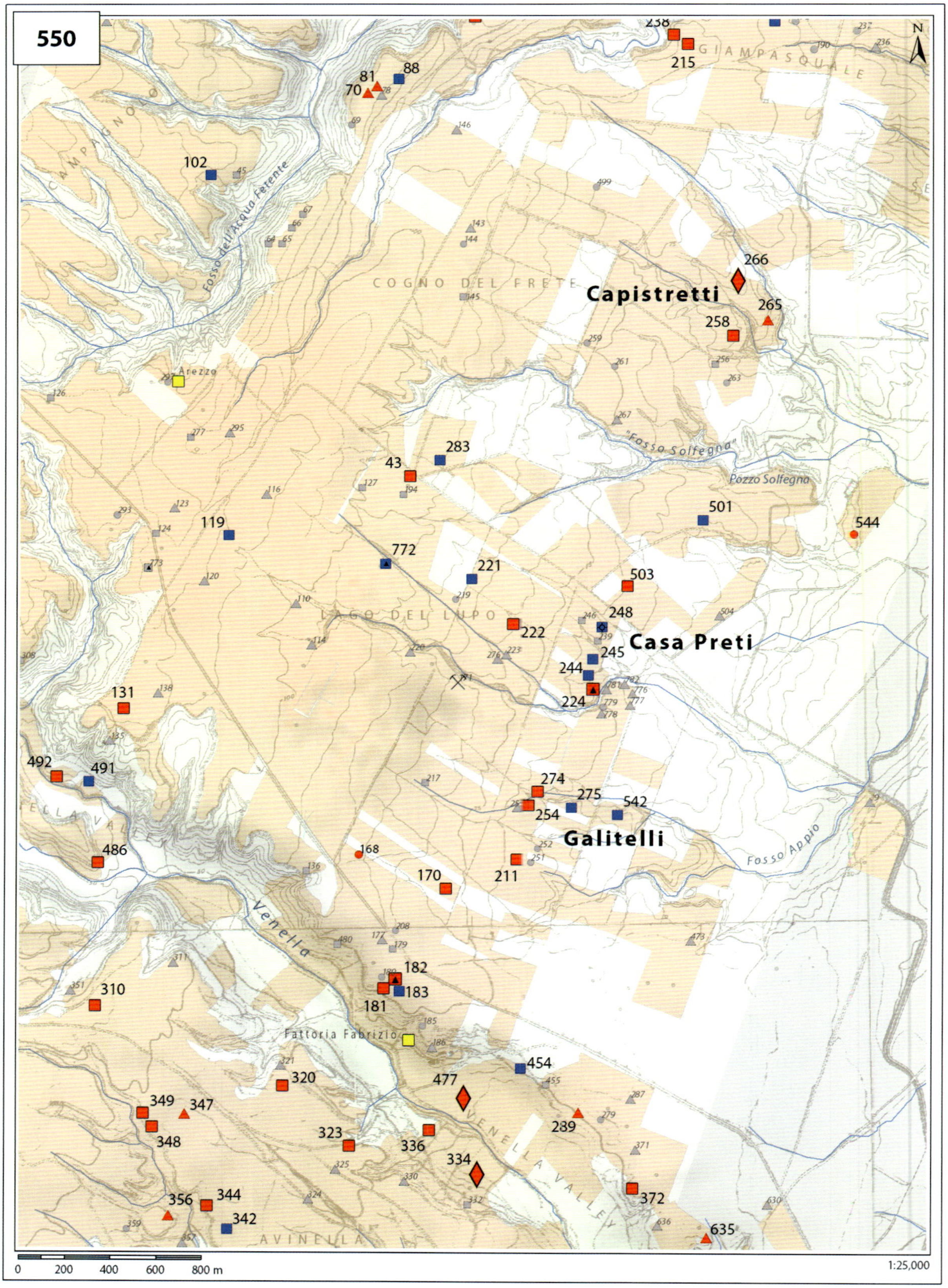

Figure 21.9 The Central Plateau in **550**. See Fig. 21.1 for map key. (JT)

Figure 21.10 Lago del Lupo and Cogno del Prete, **550**. 3-D reconstruction using a satellite image of 2006 draped over a DEM. (ML)

that at first glance seemed to be continuous, but on closer inspection was revealed to be a series of three clusters separated by distances between 500 m to 1.2 km. They are located within a corridor approximately 200 m wide, which, extended to the north, would have completely traversed the Central Plateau from the Venella to Giampasquale. In addition, there are two sites between east Venella and the southernmost cluster of the Central Plateau. Sites 170F and 211F lie in the corridor and are not as isolated as sites 222F and 43F. Most of these farmhouses are located in Lago del Lupo; the easternmost are in Cogno del Prete. These three clusters have been given the name of the known topographical feature nearest to each (*Table 21.4*):

Galitelli. This cluster is formed by sites 254F, 274F, 275F, and 542F and named after the modern farmhouse, Casa Galitelli.

Casa Preti. Sites 224F, 248F/K, 245F, and 244F form a rather tightly-knit group near the modern Casa Preti farmhouse (*Fig. 21.11*), from which the cluster takes its name. It may have included site 503F as an outlier. This cluster is located

	Site no.	Period			
		600	550	500	450
	170F	1.0	4.9	1.1	1.0
Galitelli	211F	0.6	1.7	0.6	0.4
	274F	0.3	2.0	0.8	1.4
	254F	0.9	3.1		1.0
	275F	0.1	0.9		
	542F	0.1	0.9		
Casa Preti	224F/T	0.3	1.7	0.1	1.0
	248F/K	0.5	0.5		
	245F	0.1	0.9		
	244F	0.1	0.9		
	503F	0.3	1.4	0.3	1.6
	222F		1.0		
	501F	0.1	0.9		
Capistretti	266S	0.1	1.4	0.5	2.0
	258F	0.8	2.2		
	265N	0.1	0.9	0.5	0.5

Table 21.4 Lago del Lupo **550,** with EAW values rounded to the nearest tenth through **450.** Dark green = significant farmhouses or sanctuaries (EAW ≥ 0.95) and active necropoleis (EAW ≥ 0.45). Light green = second magnitude farmhouses (EAW ≥ 0.45, but ≤ 0.95).

Figure 21.11 Casa Preti, site of the modern village of Spineto. (JCC)

near the modern village of Spineto.[16] The ancient village grew up in this spot because of the abundant springs in the fields around Spineto and persisted, after a hiatus in **500**, for several centuries.

Capistretti. The distance (1.2 km) from the Casa Preti cluster to the third, the Capistretti cluster, is three times that between it and the Galitelli cluster. This northernmost of the clusters on the high plateau is the least dense, consisting of just three sites: 258F, 265N, and a sanctuary and/or possible ceramic production center, 266S.[17] One km separates the southernmost cluster, Galitelli, from the Venella (sites 181F, 182F/T, and 183F) and the same distance divides Capistretti from Giampasquale on the Bradano side of the chora.

The uneven intervals between clusters could well be the result of locating them beside reliable and constant sources of water. The Galitelli cluster arose at the head of a small drainage that becomes a much larger one: the Fosso Appio (see *Fig. 21.1*), which traverses the chora from south to north to empty into the Bradano. The farmhouses of Casa Preti are nestled beside a bend in a larger spring-fed wet-weather stream that joins the Appio. The spacing between clusters would have been more regular and the intervals between them commensurable had a cluster also grown up along the drainage known as the Fosso Solfegna, which takes its name from the Pozzo Solfegna (or "Sulphur Well"), and is a major affluent of the Appio. Settlers may have avoided it because of the unpleasant odor, despite its medicinal properties.[18] Capistretti, the northernmost of the mid-6th-century BC clusters in the Central Plateau, lies close to another tributary of the Appio, and may have stretched farther north and east. Survey coverage in this area of the drainage was, unfortunately, very incomplete. It is certainly

[16] Spineto is one of the planned communities with small church, created in the early 1950s as part of the post-WWII bonifica and agrarian reform movement in southern Italy. The modern village was short-lived; the beneficiaries of the reform did not adapt the dispersed pattern of the ancient settlement and soon moved to town.

[17] The sanctuary would not have served that function until the late 4th c. BC, the date of the votive figurines recovered from the site (see Ch. 13, Cat. Nos. 87–103).

[18] There is no permanent settlement along this drainage until the period of Roman domination in the 2nd c. BC (site 502F; see below, Ch. 27, 896).

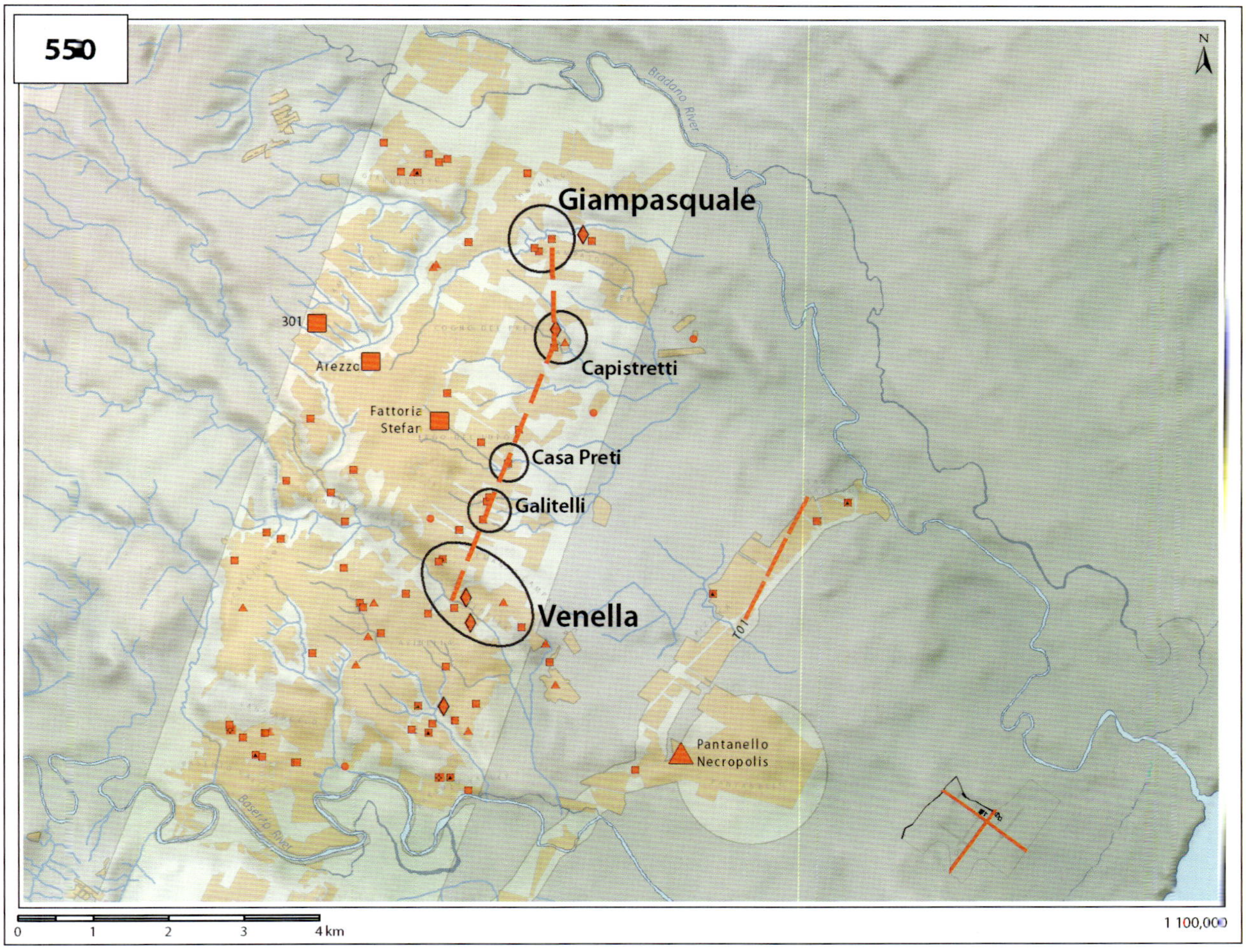

Figure 21.12 Villages in **550** along possible alignment. (JT)

possible that other farmhouses and another cluster were located beside the large drainage farther north, the last before the Fosso San Marco, along which lie the farmhouses of the Giampasquale cluster. Again, there is a gap—a much larger one—in the survey coverage. This area has been particularly badly compromised by recent quarrying operations.

The wide expanses of nearly flat land between drainages and clusters in the Central Plateau were punctuated with single isolated sites: site 170F (the first and only farmhouse in Lago del Lupo in **600**) lies 450 m to the north of site 182F/T in the Venella cluster and 500 m to the south of site 254F in the Galitelli cluster. Halving the distance between them is site 211F, just 300 m from site 254F and the drainage. Site 211F, close to the headwater of a parallel small drainage, could be considered the southernmost farmhouse in the Galitelli cluster. The 1.2 km between Casa Preti and Pozzo Solfegna was bisected by site 503F, a possible satellite of the cluster. The isolated second magnitude farmhouse 501F is 450 m to the north of site 503F on a smaller drainage parallel to the Fosso Solfegna. These intervals of open space between the clusters and isolated farmhouses in the southern half of the chain are never more than 0.5 km. It is conceivable, given the gaps in the survey coverage there, that such was also the case in the northern half.

The NE–SW spacing of the farmhouses was dictated by the topography and the positions of the drainages and sources of water. From east to west there were far fewer constraints (*Fig. 21.12*). They could have been located, in theory, at any point along the drainages. Why do they line up the way they do in a nearly straight line? A closer look at the farmhouses of the Galitelli, Casa Preti, and Capistretti clusters (*Map 6*, Vol. IV) reveals that they are all located on or very near the same contour line, running across the Central Plateau. This is confirmed by elevations for

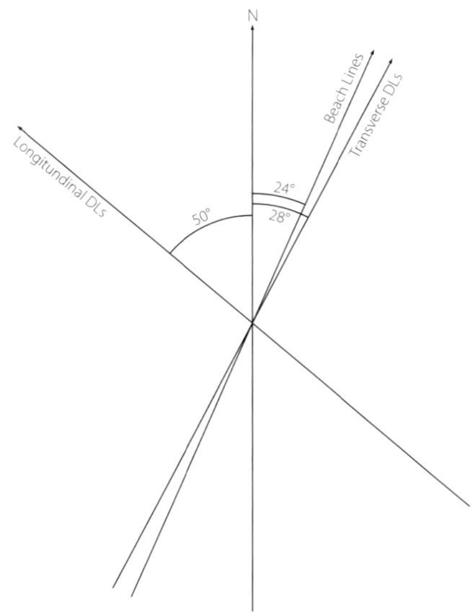

Figure 21.13 Angles between longitudinal and transverse division lines and beach line.

the individual sites, which range from 56 to 70 masl with a majority between 61 and 68 masl (*Appendix A*). All the contour lines across the expanse of the plateau follow essentially the same orientation, which is approximately 24° east of north (*Fig. 21.13*). This is in fact the orientation of the regularly spaced ancient beach lines. Springs occur naturally along these lines. So there is a practical reason—again, the presence of water—to explain the siting of the farmhouses. But why are almost all the sites in **550** located on the same beach line? Do they line up along a road? That would be a natural inference. Roads do not necessarily have to follow the most direct line between two points, but level courses are definitely preferred, and that is what the beach line offered. There may, however, be more behind this orientation than at first meets the eye, and this will be discussed in the synoptic view of **550** in the context of the division lines (see below).

Farmhouses. The spacing between farmhouses in the Galitelli cluster was closer than in the Early Village Clusters: 70 m between 254F and 274F, 170 m between 274F and 275F, and 200 m between 275F and 542F. Only in the case of 211F to the south of the center of the core was the space of its nearest neighbor in the Galitelli cluster (254F) greater than normal for members of a cluster. It may have been an outlier, like 222F and 503F near Casa Preti.

This impression of a tighter central core with more distant satellites is even stronger in the Casa Preti cluster: site 224F is 70 m from 244F, which in turn is 70 m from 245F, and 248F is an equal distance (70 m) from 245F. The outlying sites like 503F and 222F have the more traditional spacing. The significant farmhouses which include these two and 224F, in fact, form a triangle 500 m on each side which surrounds a compact core of second magnitude sites. This is a more complex situation than we have encountered before. Given the greater uncertainty in dating the second magnitude sites to this period, there is a danger in over-interpreting it. That said, it can still be suggested that Casa Preti is a small, highly nucleated settlement, centered around a good source of water. Given the narrow distance between sites in the core, it is unlikely that the area around the dwellings served as more than a garden (or a sheepfold) for the inhabitants of these four farmhouses. The areas to cultivate would have been found in the extensive surrounding plains.[19] The Capistretti cluster, the smallest of the three, by contrast conforms to the pattern of the more established clusters, such as that at Giampasquale or Venella.

Necropoleis. At Casa Preti, all five necropoleis (all undated) are located together in an area separated from the central core of houses by deep drainages at the headwaters of the Fosso Appio (*Fig. 21.9*). This cluster of separate necropoleis, which is unique in the study area, suggests a veritable village of the dead.

Significance, land, and continuity. The Central Plateau between the Bradano and Basento was probably previously uncultivated land—at least by Greek farmers—when the villages of Lago del Lupo grew up in the mid-6th century BC. It was not quite virgin land, however, since site 43F had been there since **600**. How did the farmhouses compare in significance with those of the older established settlements? A glance at the table of average FMCE values (*Appendix C7*) would suggest that on the average the farmhouses were slightly less significant (2.07) than the average for the study area as a whole (2.14). They were slightly less significant than those of the farmhouses scattered along the Conca in Avinella (2.11) or Venella (2.03), for example, but considerably less significant than those in the Early Village Clusters along the main river valley at Giardinetto (2.29) or Sant'Angelo

[19] A possible parallel is the tiny agricultural villages of Medieval Europe with their open fields, which were divided into strips and cultivated by individual owners dwelling in the village. See M. Johnson 1996, 73.

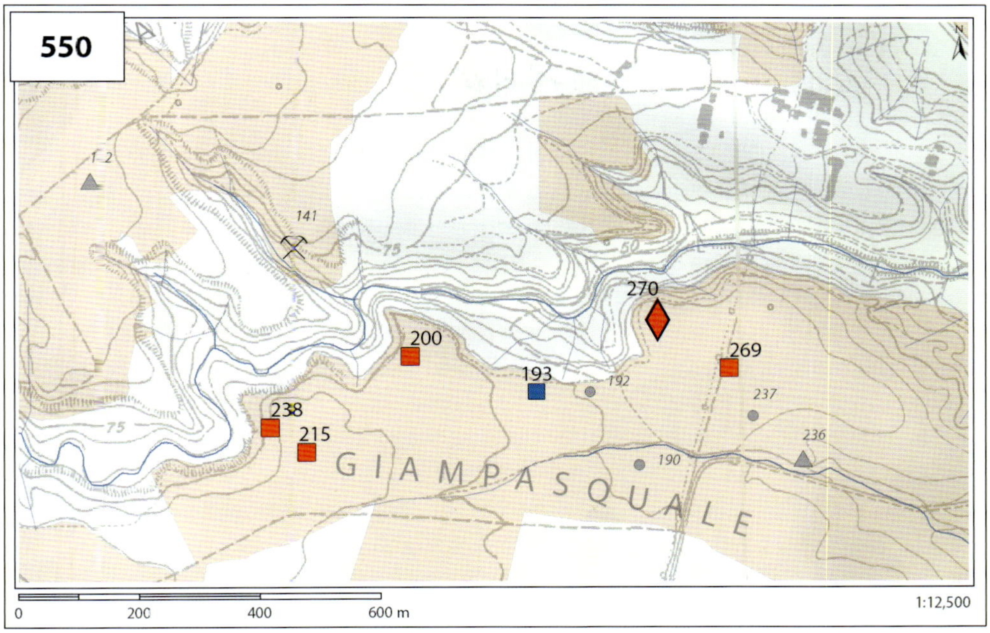

Figure 21.14 Giampasquale in **550**. See Fig. 21.1 for map key. (JT)

to be some social and economic stratification in the villages of the Central Plateau.

Giampasquale (*Fig. 21.14*). The chain of clustered and isolated farmhouses that stretched northward across the Central Plateau from the Venella terminated in the Giampasquale cluster, the least developed of the original five villages dating from **600**.

The areas along the north side of Fosso San Marco, facing the settlement along its southern edge were very incompletely covered by the survey as noted earlier, largely because of the presence of an extensive farming operation at the mouth of the drainage where sites of the Bronze Age were located.[20]

Farmhouses. In **600** there were only three second-magnitude farmhouses and a sanctuary at Giampasquale. Two of the three (269F and 200F) became significant in **550**, and to these were added two more significant sites (238F and 215F). This was the greatest percentage increase in any cluster in the mid-6th century BC. The pattern of development was similar to that of Avinella and the Venella: the new sites occupied points along the southern rim of a deep drainage, the Fosso San Marco, each on a point of land—though in one case, 238F and 215F, closely paired with another site set back from the valley rim. The arrangement finds its analogies in the Sant'Angelo cluster already in **600**.[21]

(2.42). Could the apparent difference in material success represented by FMCE score depend on the type of crops raised? The areas under cultivation by the inhabitants of Giardinetto and Sant'Angelo were likely to have been the alluvial terraces, as opposed to the marine terraces of Lago del Lupo with their relatively thin mantle of sandy gravely soil that covered thick deposits of clay. The soils of the Central Plateau, as noted earlier, are particularly adapted to the cultivation of cereals such as barley, whereas the heavier and more fertile soils of the valleys can support a more varied agricultural regimen including bread wheat (*Triticum aestivum compactum*). Despite the fact that the Metapontines represented themselves on their coinage with a head of barley, wheat was preferred for human consumption.

There are three significant farmhouses in the Galitelli cluster, and two each in the Casa Preti and Capistretti clusters. Sites 258F (2.21) and 266S (2.21) in the latter are equally significant, but in each of the other two clusters a single farmhouse, 254F (2.86) in Galitelli and 224F (2.47) in Casa Preti, is considerably more significant than the other components of the cluster, with FMCE values well above the averages for Lago del Lupo (2.14) and the study area as a whole (2.14). Both 254F and 224F are significant for two 50–year periods only. There would appear

[20] Of the Greek period is site 547N, which served as a burial site in **400** and **350**. This area has long been known to have been particularly attractive to settlers in the Bronze Age (see Gazetteer and Ch. 17, 575). Farther up the valley was a quarry of unknown date (site 141).

[21] On the ridge separating the Fosso San Marco from the Fosso dell' Acqua Fetente, nearly 2 km from the mouth from the Fosso San Marco, was another pair of farmhouses, 57F and 58F: one on the valley edge, the other (securely dated to **550**) 150 m away but on the Acqua Fetente side of the ridge.

Necropoleis. Giampasquale is unique among the clusters in having no dateable necropoleis. There is only a single undated example, site 236N.

Sanctuaries. The sanctuary at Giampasquale was among the earliest to be frequented and the first to be forgotten. Its active life was essentially limited to the 6th century BC. It suffered like the rest in the crisis of **550**, but shows only sporadic signs of life thereafter.

Significance, land, and continuity. The most significant site in this cluster and one of the most significant in the study area in **550** was the sanctuary 270S (FMCE value, 2.71). The sanctuary was probably the first Greek site in this area of the chora. The mid-6th century BC was its modest heyday. It was occupied in **500** when very few other sites were. Farmhouse 269F had a long history, longer than any other in the cluster. It was continuously occupied from the 6th to the early 2nd century BC. Its FMCE value (1.88) was in the lower range for farmhouses in **550**, but by **500** it was among the more significant farmhouse sites in the study area. It is not a general rule that one member of a pair is less significant than the other, but it is the case with the pair 238F and 215F, with FMCE values of 1.86 and 2.44, respectively (*Appendix B*). Located farthest from the mouth of the Fosso San Marco, these had a brief lifespan that began in the middle of the 6th century BC and ended together a century later.

Upper Venella (*Fig. 21.15*). While farmers sought new fields and established permanent homes in the Central Plateau, they also pushed deeper along the major drainage through the heart of the chora between the Bradano and Basento, the Venella and its tributaries, the Fosso Bufalara and the Fosso Gaudella, which join the main stream of the Venella below the steep bluff of Pietra San Giovanni (*Fig. 21.16*). This commanding position attracted settlers beginning in the Neolithic.[22] In **550**, the valley rims along the tributaries were the preferred locations for residences. The heights above the valleys, which offered water for livestock, good forage, and a holding area, would have been especially attractive for animal husbandry. The traditional name Bufalara (from *bufala*, a water buffalo whose milk is used to make a superior cheese), indicates that this area continued to appeal for these reasons down to the present.

This is the least compact of the clusters so far considered. A factor contributing to its greater dispersion is the width and depth of the valleys that divide the plateau. This settlement has sites in five different localities: Bufalara (304F, 487F/T, and 488F); Venella (486F, 491F, and 492F); Lago del Lupo (131F); Avinella (310F); and Scarciullo (365F, 366F, 375F, 376F, and 511F). These wide neighbors are considered a cluster because they all line the edges of the interconnected drainages and they are much nearer to each other than to the other clusters (Venella, upper Conca in Avinella, or Galitelli in Lago del Lupo; see *Map 6*, Vol. IV). Springs, indicated by thick stands of willows, line the edges of these valleys (*Fig. 21.17*).

Farmhouses. In **600**, there was a single securely dated farmhouse (488F) on the high plateau between the two tributaries of the Venella, along with three second magnitude sites (487F/T, 376F, and 365F). By **550**, there were nine significant farmhouses plus four of the second magnitude within 1 km of the junction of the streams (*Table 21.5*). Though individual farm-

Site no.	Period			
	600	550	500	450
488F	1.0	1.0	0.1	4.6
487F/T	0.5	0.8	0.7	
486F	0.4	3.0	0.7	
492F	0.3	1.7		
491F	0.1	0.9		
131F	0.6	1.4	0.1	2.8
304F	0.1	1.4	0.7	2.5
310F	0.3	2.2	0.5	3.3
365F	0.5	1.3	1.2	2.3
375F	0.3	1.7		1.6
511F	0.1	0.7	0.1	0.4
376F	0.5	0.5		
366F	0.3	1.7		

Table 21.5 Upper Venella **550,** with EAW values rounded to the nearest tenth through **450.** Dark green = significant farmhouses or sanctuaries (EAW ≥ 0.95) and active necropoleis (EAW ≥ 0.45). Light green = second magnitude farmhouses (EAW ≥ 0.45, but ≤ 0.95).

[22] A "Neolithic domestic structure" was situated here. It was also the site of the most important Medieval settlement village or *casale* from the 11th to the 15th c. AD (see Ch. 34, 1165). The standing remains of the Medieval period are impressive. Except as a burying ground this particular point, however, seems to have been ignored by the Greeks.

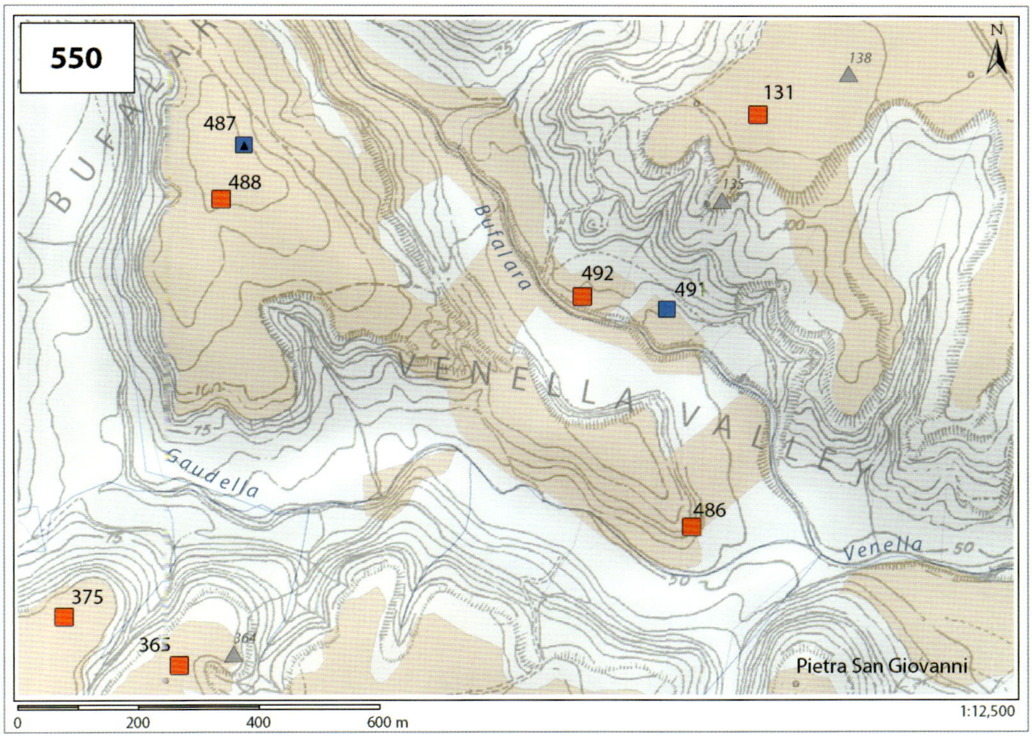

Figure 21.15 Upper Venella in **550**. See Fig. 21.1 for map key. (JT)

Figure 21.16 View from Pietra San Giovanni. (ML)

houses can be quite far from their nearest neighbors, they are rather concentrated along the south bank of the Gaudella. The individual sites (511F, 375F, 365F, 376F, and 366F) are separated by distances of 200 to 300 m, with the average just slightly under 250 m. Each occupies a small headland or stands at the end of a drainage. The peninsula between the Gaudella and Bufalara and the valley below were occupied by two pairs of sites and a loner: 488F and 487F/T, separated by 100 m; 491F and 492F, 150 m apart; and site 486F, which sits all by itself at the very tip of the peninsula above the stream junction. The pairs are 600 m apart, and the lone site 400 m from the middle pair. Along the north bank of the Bufalara are three farmhouses including the excavated farmhouse Cogno del Pero, which has been dated to **550** and has a well-preserved plan.[23] It lies between site 304F, 300 m to the northwest, and 131F, 550 m to the south (*Map 6*, Vol. IV). Between the latter and 304F lies a wide and deep ravine joining the Bufalara. To the southwest, on the south bank of the Gaudella, lies the most isolated site belonging to this settlement, 310F. It lies 400 m to the southwest of Pietra San Giovanni with its panoramic view up and down the three valleys (*Fig. 21.16*). The spacing of farmhouses along the Gaudella and in Scarciullo is re-

[23] Adamesteanu 1974a, 82–84

Figure 21.17 View of upper Venella, eastern bank of the Gaudella. The presence of numerous springs along this valley is indicated by thick clumps of reeds growing around them. (JCC)

markably close to that in Avinella and along the Venella.

At its southern end, the narrow ridge between Bufalara and Gaudella, where both of the earliest sites in the area are located (488F and 487F/T), becomes a wide, naturally defended plateau overlooking the Venella valley. At its north end, the ridge narrows and could easily be cut off with a short fence or hedge. The surface area of the plateau, which slopes very gradually to the precipice, is a little under 20 ha: an ideal location for a farmer or herder who desired privacy. This site was continuously occupied for three and a half centuries, until **300**. The southernmost projection of this ridge (at a lower elevation) is quite narrow but also has site 486F at its tip. A shepherd living here would not have had much land to cultivate, but it was an excellent vantage point to take in the three valleys at a glance. This strategically placed lookout, however, was abandoned in the next half-century and was never reoccupied—another sign, perhaps, of a shift from pastoralism to agriculture in this area of the chora.

Necropoleis. There are no necropoleis in this loose cluster that can be securely dated to **550**, but there are at least three undated examples that are very close to farmhouses: 510N is 150 m from 511F, 138N is ca. 150 m from 131F, and 364N is ca. 100 m from 365F. All of these farmhouses had relatively long lifetimes, so that while we can be reasonably sure that they were associated with the necropoleis, we must remain ignorant about the periods in which the burial sites were utilized. The burials beside the second magnitude farmhouse 487F/T could belong to its occupants in **550** or to those of the 4th century BC. In any case, this cluster was not like Giampasquale, without evidence for burial. In later periods, necropoleis increased in the upper Venella.

Significance, land, and continuity. Three sites in the upper Venella are among the five most significant farmhouses in the study area in **550**, with FMCE values of 3.00 or greater (see *Appendix B*).[24] This is surprising

[24] They are sites 492F (3.57), 366F (3.20) in the valley, and 486F (3.20) at the point. The other two belong to the older clusters—sites 459F in Giardinetto and 155F in Sant'Angelo.

considering that this was one of the more remote and rugged areas of the chora in the mid-6th century BC. All three of these sites had very brief lifespans of no more than a half-century.[25] The other sites in the upper Venella had a combined average FMCE value of 1.61 in **550**—much lower than the average of the study area as a whole at this time (2.14)—and were occupied in subsequent years.

Other Possible Clusters, 550

Besides those just described, there were other sites that may have belonged to clusters in the study area in **550** that, because of the incompleteness of the coverage, could not be recognized as such. Four areas with more than one farmhouse and necropolis are located very close to unsurveyed areas. They may well have formed clusters as extensive as the ones described above. One is located in the southeast corner of the Bradano-Basento transect, one in the Pantanello transect, and two in the narrow SNAM transect (*Fig. 21.1*).

The mouth of the Conca. Three farmhouses and an undated necropolis occupy one corner of a 25-ha area between the Conca and a secondary drainage to the west. Farmhouses 530F/K and 531F/T are separated by 150 m and by an undated necropolis (472N) that they may have shared. From 530F/T comes some of the earliest Greek pottery in the study area (see Ch. 20). It was possibly a center of ceramic production already operating in the 6th century BC. These farmhouses occupy the slopes (with elevations of 30 and 32 masl) leading down to the Basento, while the third, 485F (15 masl), is actually situated on the T_2 alluvial terrace in the valley. It is about 70 m from another undated necropolis (484N) which may have been associated with it. The ancient Basento Road, followed by the modern SS Basentana, passed near all four sites. The mouth of the Venella and the Sanctuary of Artemis at San Biagio are not far off.

Pantanello. In a narrow stretch of the SNAM transect in the Pantanello locality, two farmhouses (414F and 726F/T) were located along the Basento Road (see *Fig. 21.1*). This road led from the coast up the Basento valley and was lined by burials for much of its length. Here in Pantanello, the modern *tratturo*, a wagon track that connected the town of Bernalda with the coast, follows the ancient road.[26] Site 414F lay directly alongside the *tratturo*; site 726F/T, 100 m to the southeast. At a greater distance to the south and east of the *tratturo* were located two second magnitude farmhouses like 726F/T: 449F, approximately 450 m from 414F; and 417F, about 350 m east of 449F.

Pantanello has been intensively cultivated with modern equipment since the 1950s, so site preservation in this area is not as good as in the localities farther inland, where irrigation and deep plowing reached more recently. In the small area under consideration here, three sites (469UGF, 729UGF, and 412UGF) were so badly compromised by agricultural activity that their function could not be identified. Those farmhouses that could be identified as such and date to **550** (including the second magnitude sites) are regularly spaced, like those of the other clusters: 414F is approximately 300 m from 726F/T, and 449F is 350 m from 417F. The 300-to-350-m-spacing, further, continued to be maintained in succeeding periods. Despite having been more mangled than others, these sites probably comprised a cluster along the Basento Road that had its origins in **550**, with fuller development in the next century.

Pizzica. The SNAM transect cuts through Pizzica approximately midway between Pantanello on the Basento side of the study area and San Salvatore on the Bradano side. Pizzica consists mainly of marine terraces. There is one very large and deep drainage in Gravaglione, and next to it lies the only significant farmhouse in this period. The location of site 734F/T beside a major drainage conforms to the pattern observed in **550** in the early clusters and in Lago del Lupo. Two other farmhouses of the second magnitude (728F/T and 730F/T) were located approximately 400 and 600 m to the south of the Gravaglione. If these farmhouses belonged to a cluster, it was at least a very loose one.

San Salvatore. Four farmhouses, three of them dating to this period, were found at the extreme northern end of the SNAM transect. All but one occupied elevated

[25] The small amount of black-gloss pottery present happened to be precisely datable. Had the same quantities been spread out over a longer interval, a much lower FMCE value would have resulted. Nonetheless, the fact that the high values are concentrated in this area does indicate that a number of sites with pottery dated to the period occurred in one of the more challenging areas, and that they soon disappeared.

[26] A stretch of the ancient road was discovered right beside and, at one point, under the *tratturo* during the excavation of the Pantanello necropolis. See Ch. 28, 1045, and Carter 1998b, 34–39.

points at the edge of the marine terraces overlooking the valley of the Bradano. The current streambed is only 300 m away to the north. Sites 272F/T, 742F/T, and 762F/T, on the basis of experiences elsewhere in the study area, are located precisely where sites might be expected, that is, near the spring interface. 742F/T is separated by a little less than 300 m from 762F/T, and by about 350 m from the isolated farmhouse 272F/T. Each has associated burials beside the dwelling. Site 741F, by contrast, is 500 m from its nearest neighbor (742F/T) and not close to any easily recognizable drainage, spring, or necropolis. It is, however, very close to a modern drainage ditch (*Colatore D*, and, coincidentally, the office of the Consorzio Bonifica of Metaponto, which has been charged with the drainage of the Metapontino since the 1950s). All four sites, occupying as they did favored positions along the valley of the Bradano, continued to be occupied into the 3rd century BC and, in one case (762F/T) into the 2nd century BC. Was this a cluster? Probably so, but as is the case with the other sites in the SNAM transect, the maximum width of the surveyed area is less than twice the average distance between farmhouses, so no definitive judgment can be made.

Isolated Sites, 550

The single sites considered here are, with one exception, located in areas that were well and thoroughly surveyed. If they appear on the map as isolated, it is in all probability because they were. Still, when second magnitude sites are considered, they may not be as remote as they may appear (*Fig. 21.1*).

Site 228F, San Marco. Like the west side of the Fosso dell'Acqua Fetente, the eastern bank opposite Giardinetto was almost devoid of sites in the Greek period. Of the four farmhouses that occupied the steep slope of the promontory jutting out into the Bradano valley, three were not occupied until the late 5th century BC. The exceptional site 228F is situated at mid-slope (42 masl, about the same elevation as a number of sites in Giardinetto). It was one of the earliest significant sites in the chora, but was also a relatively short-lived one. It went out of existence at the end of the 6th century BC—a creature, it would seem, of the agricultural prosperity of that century and a victim of its decline in **500** (see Ch. 22). This is no nice pocket of fertile and gently sloping land like the Giardinetto. It is rather a high plateau, to the east of a small drainage, that overlooks the valley below. The San Marco promontory is naturally defended on three sides by steep cliffs with a narrow neck separating it from the Central Plateau behind.[27] The modern Masseria San Marco, 1 km still further east on the same plateau, undoubtedly lies over ancient sites and itself probably represents several centuries of history.

Site 301F, Campagnolo. This site occupies the center of a flat tongue of land between two drainages. It is very near the western boundary of the Bradano-Basento transect, so whether there were neighbors to the west cannot be known. Site 302F, 100 m distant at the bend of one of the drainages, was not occupied until a century later. Both sites were cut off from those in the extensive plateau of Lago del Lupo to the east by the headwaters of the deep ravine created by the Fosso dell'Acqua Fetente. Site 301F was significant in **550** and continued to be so, with a break in **500** until the early 3rd century BC. It is situated 50 m to the south of the central division line (L15), which runs through the above-mentioned drainage and beside site 302F (see discussion of division lines below).

Sites 58F, 70N, and 81N, Cogno del Prete. Site 58F sits on a point on the south side of the Fosso dell'Acqua Fetente, all by itself. It was a less significant site in **550** (FMCE value 1.89) than most, and its fortunes did not improve. Site 70N, on an adjacent point 600 m to the south of 58F, was flanked by necropoleis: 81N, also active in the mid-6th century BC, and the undated 78N. A contemporary second magnitude farmhouse (88F) 100 m away may well have made use of 81N, or possibly 78N, but there are clearly too many necropoleis here for the resident population. This situation suggests a possible parallel situation to that encountered at Casa Preti in Lago del Lupo. The concentration of burials in one area may have served a concentration of residences in another. The cluster of farmhouses still farther to the north, consisting of sites 54F, 65F, 66F, 67F, and 68F—all of which, with the exceptions of 54F (a second magnitude farmhouse in **400**) and 68F (significant for the first time in **450**)

[27] This was surely the reason for its appeal to the Bronze Age occupants and the settlers of a Medieval agricultural village (see Ch. 17, 575, and Ch. 34). The eight Medieval farmhouses are spaced less than 100 m apart (sites 85, 86, 89, 90, 91, 92, 93, and 94) in an area no more than 300 m in diameter (see discussion, Ch. 34, 1166). Curiously the fertile Giardinetto attracted no such settlement in the Middle Ages.

are unfortunately undated. The hypothesis of a Greek agricultural village with separate, communal area for burial, though attractive, is ultimately unverifiable. With the more probable arrangement at Casa Preti, there is a hint, at least, of a mid-6th–century settlement pattern that differs fundamentally from that observed elsewhere in the study area.

Site 43F, Cogno del Prete. Near the center of the Central Plateau, farmhouse 43F stood out in **600** by virtue of the remarkable distance—2 km—between it and its nearest significant neighbor, site 170F, and by the absence of any readily identifiable water source. By **550**, that distance had been considerably reduced. Site 222F is 800 m to the south, and three second magnitude sites, 283F, 772F, and 221F, are closer still. Site 43F does not fit into any category or easily defined niche. There must have been among the early colonists a few rugged individualists ready to endure greater hardships in return for greater independence.

Site 103F, Scarciullo. The Fosso Sant'Angelo attracted remarkably few settlers compared with the other major drainages—the nearby Conca, for example. Site 103F, significant in **600** and continuously so for four centuries down to **250**, stands out among the half-dozen other sites that the deep, three-forked stream impressively divides (see *Fig. 21.1*). There is one undated necropolis, site 81N, and an undated farmhouse, site 101F, at the headwaters of the middle fork, midway down the east fork on its east bank, as well as an undated farmhouse, site 106F, near the northern source of the main stem. Not far from the excavated farmhouse and ceramic production site of Sant'Angelo Vecchio (and the 18th-century masseria) were two undated necropoleis, 380N and 560N. The only farmhouse securely dated to this period is 377F, a second magnitude farmhouse in **550** located at the point where the middle fork joins the main stem. Like several new sites of this period, it disappears from the record entirely in the next period. Site 103F, isolated in **600** like 43F in Lago del Lupo, by contrast remained isolated for the rest of its history, yet another enduring example of obstinate individualism.

This area of the chora, in particular the flat plateau of Scarciullo, was very lightly settled for most of the Greek period, when other areas—notably the Central Plateau—were being in-filled. It was not until **300** that three significant farmhouses and one second magnitude one suddenly appeared in the middle of the Scarciullo plateau. The reason for the scarcity of sites on this plateau until this late date is almost certainly the lack of convenient sources of water. It, unlike the Central Plateau, is perfectly flat and there are no beach lines where springs might occur.[28]

Sites 622F and 635N, Demanio Campagnolo. These sites, which lie too close to the edges of the surveyed areas to be sure that they were alone, were considered earlier in the discussion of the Venella cluster.

Synoptic View, 550
Outstanding Characteristics of the Chora

The half-century between 575 and 525 BC was, as we have seen, one of dramatic population growth in the chora, during which its agricultural production expanded greatly. This wealth made possible ambitious building projects in the city, and its source was symbolized by its distinctive coinage bearing the six-rowed ear of barley that began to be minted in about 550 BC. The Greek poleis of Magna Grecia in the 6th century BC were expanding, none more aggressively than Sybaris, whose influence was felt from Poseidonia on the Tyrrhenian to Metaponto on the Ionian coast and beyond. Clashes resulted, in one of which Metaponto, allied with Croton and Sybaris, triumphed over Siris, its neighbor to the south. Whatever the consequences—a vast territorial gain or a political influence over a powerful rival—the victory probably contributed to the vibrancy of the Metapontine economy and the expansive mood of the citizens.

A sudden great increase of population, wealth, and probably of territory in **550** must also have resulted in social tensions. The impression of a degree of equality among the pioneer settlers in **600** is not as strong in **550** to judge from the growing disparity of FMCE values of farmhouses in the chora (*Appendix C8, 550*), suggesting a growing gap between the most prosperous and subsistence-level farming families. In the rich burials of the first half of the 6th century BC in the Crucinia necropolis, which was located immediately outside the walls of the city, we have the first tangible evidence of an aristocracy (see below, Asty, Monumental tombs).

[28] That problem was resolved, as will be suggested below, by the introduction of dug wells here and elsewhere in the flatter areas of the Central Plateau. See Ch. 25, Wells.

The possibility that Metaponto had a tyrant shortly after mid-century has been suggested and is supported by some documentary sources as well as the archaeological evidence of burials and inscriptions (see Asty, Monumental tombs, below). If so, Metaponto's tyranny would have been roughly contemporary with those of mainland Greece, though among the earliest examples in Magna Grecia or Sicily. The presence of an aristocracy among the inhabitants of the countryside, however, is not so easily detected. Besides the material differences in farmhouses, we may point to the existence in the 6th century BC of localized cults and sanctuaries of greater and lesser importance at sites fairly evenly distributed throughout the chora.

The Growth of Clusters

On the local level, this rapid expansion of the middle years of the 6th century BC can be documented in detail in the study area. The number of farmhouses increased from 19 to 64 (or 3.3 times).[29] The number of necropoleis between **600** and **550** grew at a lower rate—from 8 to 13, or 1.6 times as many as a half-century earlier (*Appendices C1a* and *C1b*). The growth of the clusters and isolated sites in the study area was not the same throughout. The clusters established in **600** grew, but not at uniform rates, and formerly remote regions where at most an isolated farmhouse or two could be found in **600** had become among the most active.

In the established clusters, such as Giardinetto, Avinella, and Venella, the significant sites increased in line with the chora as a whole; but in Lago del Lupo and upper Venella, the count of farmhouses rose from a few examples in **600** to eight in **550**. Sant'Angelo, which was the most precocious of the late-7th- to early-6th-century BC clusters, increased at the lowest rate, about 1.6 times. It would appear that by **550** Sant'Angelo was already a mature community.

The wave of new settlers in the Central Plateau and the upper Venella in the 6th century BC arrived in what had previously been relatively remote areas. The lower and middle course of the Venella valley had attracted settlers in **600** but only one, the family residing at 488F, had ventured farther inland. The broad expanses of the plateau of Lago del Lupo had been home for only two farm families (43F and 170F), but by **550** each of these areas had more occupants than the established village of Sant'Angelo.

The patterns of settlement in **550** mostly followed those of **600**, but there were new developments also. Some clusters experienced infilling (like Giardinetto, where the phenomenon is most obvious), increasing in number of dwellings but not in overall area. Infilling consistently halved the distance between sites. This is true also in the long narrow cluster of Sant'Angelo. In the clusters that grew up along the tributaries (the Conca, Venella, and the Fosso San Marco), new sites were added in a linear extension maintaining a spacing that was regularly on the order of 300 m, occasionally as much as 400 m, but rarely less than 200 m. The distances between farmhouses are greatest in the most remote area, the upper Venella.

In the new clusters of the Central Plateau (Galitelli, Casa Preti, and Capistretti) that spread across the high plateau in the Lago del Lupo and Cogno del Prete localities, the spacing between significant farmhouses remained the same, but the second magnitude farmhouses in Casa Preti were very close together. On the average, less than 100 m separated them. The necropoleis of Casa Preti, instead, are grouped all together by the drainage that flows between them and the dwellings, effectively segregating them. The separation of the living and the dead in clusters resembles the long-established practice of the asty, in contrast to the equally strong tradition in the chora of burying near the residence.

None of the new clusters in **550** included a site identified as a sanctuary. 266S in Capistretti gave no indication of being anything but a farmhouse before the 4th century BC, which is the date of the numerous votive plaques recovered there. Sanctuaries, not a part of the early Giardinetto and Sant'Angelo clusters along the major river valleys, appear to have been the most central and earliest element of the clusters that grew up along tributaries of the rivers in Avinella, Venella valley, and Giampasquale.

Pairs of Sites

The phenomenon of farmhouse pairs separated by 100 m or less appeared in **600** at Sant'Angelo, and was widespread by **550**. It occurs at Giardinetto, Sant'Angelo, upper Venella, the Venella, in Lago del Lupo (Galitelli), and Giampasquale.[30]

[29] If farmhouses of the second magnitude are also counted, the increase is still over twice as great. In general, adding these less securely dated sites moderates extreme fluctuations.

[30] In **600**, Sant'Angelo, 285F/286F and 155F/154F. In **550**, Sant'Angelo, 13F and 12F/K; Giardinetto, 459F and 519F; upper Venella, 349F and

It is reasonable to assume that the land around a farmhouse constituted an allotment, the original kleros to which every citizen was entitled and required to possess as a condition of citizenship.[31] This plot may not have been all the property that the occupants may have farmed or grazed, but it was probably an important part of it. The spacing between farmhouses gives at least an approximate idea of the size of the plot, which may have been different in different periods. In **550**, the interval of ca. 300 m occurs frequently, corresponding to a circular area of ca. 7 ha or a square of ca. 9 ha.

When two dwellings are separated by no more than 100 m (as is the case with these pairs) an alternative explanation is needed. The Greek custom of "partible inheritance" would require an equal division among heirs. An original allotment to a family or oikos at the death of its head would, in theory, have been divided in equal portions between his male heirs, who—as heads of families in their own right — would have required dwellings. If there were more than one son, or daughter if there were no sons, a division of the property would be required, and the fractioning of the original allotment would begin.[32] This is one rather obvious explanation of pairs, but others are also possible. When the two earliest pairs at Sant'Angelo (285F/T and 286F, and 155F and 154F) appeared in **600**, they were almost alone in this spacious and fertile niche beside the Basento. Building together may have been a deliberate choice of the heads of two different families for reasons of security in these early years, or to share more easily labor and resources. The histories of paired farmhouses seem quite individualistic and particular, but several general patterns are observable.

In the Giardinetto pair, 459F and 519F, one of the two farmhouses, 459F, appeared quite suddenly in **550** and did not survive long, whereas the other was more or less continuously occupied for four centuries. Site 459F had an exceptionally high FMCE value (3.36) during its brief existence, while 519F had among the lowest in Giardinetto for most of its long duration. The one location, it would appear, produced brief material success for its occupants in a period of rising general prosperity; the other, subsistence at a relatively low but stable level. (Because of the way FMCE is calculated,[33] they were perhaps more nearly comparable in **550** than their FMCE scores would indicate.) The pairs in Avinella and the Venella appear to follow this pattern. Other pairs seem to follow more nearly parallel chronological trajectories, such as the two early pairs in Sant'Angelo (285F/T and 286F; and 155F and 154F). In each, one farmhouse has a larger FMCE value than the other, though the apparently weaker of the two succeeded at a lower level over a longer period of time. (Again this may depend, in part at least, on the way the FMCE is calculated.) The later pair in Sant'Angelo (12F/K and 13F) dating to **550** behaved more like that in Giardinetto, as did a pair at the mouth of the Venella (530F/K and 531F/T) with one component (531F/T) having a higher FMCE value in **550**. A third possible variant is that represented by the pair in Lago del Lupo, 254F and 274F. One has a higher FMCE value in **550** than the other and both continue into the 5th century BC, but not beyond it, as did the older pairs in Sant'Angelo. The pairing, fission, or gemmation of sites in the mid-6th century BC is found in practically all areas of the study area where sites were concentrated in clusters.

The Changing Shape of Settlement

A fundamental alteration in the overall shape of settlement took place in **550**, as exemplified by the strip of farmhouses stretching across the high plateau from the Venella to Giampasquale. Whereas the clusters of the preceding period formed around and took their essential shape from the natural topography, the three small clusters in Lago del Lupo and Cogno del Prete, considered as a whole, obey another logic. This is despite the fact that in each case they are, like the earlier clusters, centered on drainages and as a whole their location across the chora follows an ancient beach line at a constant elevation. The aim of this settlement seems to differ from that of the previous period. The original clusters laid claim to lands defined locally by natural features. The strip across the chora appears to

348F; Venella, 181F and 182F/T; Lago del Lupo (Galitelli), 254F and 274F; and Giampascuale, 238F and 216F.

[31] Asheri 1966, 7–12.

[32] See the useful discussion in Lacey 1968, Ch. 6, on family and property with special reference to Athens.

[33] See Ch. 4 and Ch. 21, n. 24. Single period sites contain large numbers of associated artifacts dated within a single 50-year period. It is the concentration of the material that makes the difference and causes certain ratios used in the algorithm generating FMCE, DMCE, and LAW to be high. It is important to realize that FMCE values are not driven by these ratios in most cases. Of the 20 highest FMCE values for 450 BC, only seven have a FMCE value above the mean for all significant farmhouses in that period.

be a conscious attempt to claim a whole, previously unsettled territory. This could have been done by placing farmhouses or clusters of sites opportunistically where springs or drainages occur, but that is not what seems to have happened. The placement of sites follows a more or less straight line running northeast to southwest across the chora.

A vast area to the west of this line remained virtually untouched for at least a half-century. Why was it not settled at this time? The soil was extremely fertile as the large number of farmhouses that were located in Lago del Lupo testifies. Could it be that it was reserved as ἱερὰ χώρα for the gods?[34] Or instead was this land the property of powerful aristocrats who employed others, perhaps living in the surrounding villages, to work their land? The fact that this area, as will be argued in Chapters 23 and 25, was apparently redistributed (γῆς ἀναδασμός)—not once but twice—might weigh in favor of the latter interpretation. Both are plausible, but neither may be right. The area may not have been developed at this time for entirely different reasons.

Whatever the case, as a straight line of farmhouses clearly shows, a fundamental change has taken place. Geometry has entered the picture, as it will in the layout of the urban center on the coast. The outlines of a system of intersecting axes, which will become a dominating aspect of the chora in the next century, can already be detected.

Roads

Dirt tracks, paths, or primitive roads would have left little archaeological trace that could be detected at ground level or in aerial photographs, but the alignment of sites in Lago del Lupo, as noted above, strongly suggests that the clusters grew up along such an axis of communication. How else could the farmers who lived at these sites, remote from the major valleys, have reached their neighbors and eventually the city? Aerial photography was for the first time regularly employed by Dinu Adamesteanu to investigate the road systems in the Metapontino—some in use from the Neolithic period. Roads and tracks (*tratturi*) for humans and animals were identified along the coast and behind it, connecting Metaponto and the indigenous centers of the interior with the colonial urban centers along the coast and with each other.[35] More recently, excavations, notably at Pantanello (1984)[36] and Pizzica (1999), have uncovered sections of ancient roads that correspond to the division lines detected by other investigators in aerial photographs (see Ch. 28). Though only uncovered for a short stretch of a few meters, the road flanking the longitudinal division line L09 in Pizzica had ruts left by wagon wheels in the local conglomerate near or perhaps under the division line feature visible in the air photos. It appeared to have the precise orientation of the division line and hence that of the system of drainage ditches flanked by roads or paths that has been dated to the first half of the 5th century BC. The orientation of the drainage canal and the division line was confirmed by excavating extensive sections of two portions of it, separated by 150 m, and a third intermediate portion. This road, we have argued (Ch. 28), served a quarry for conglomerate stone of the sort frequently used in constructions of the late 7th and 6th centuries BC.

The earliest burials along the Basento Road in the area of excavation at the Pantanello necropolis date to the first quarter of the 6th century BC, though the road, in some form, surely existed earlier. The early farmhouses at the mouth of the Conca and at Sant'Angelo probably lined its extension up the Basento valley.[37]

The road that crossed the chora in Lago del Lupo, however, was of a different sort. While it did not ignore topographical constraints, it maintained a constant direction and was virtually level as it passed through the clusters with an average elevation of 64 masl tracing a direct course across the chora (*Appendix A*). A NE–SW thoroughfare across the chora, connecting the villages of the Central Plateau with the Venella and the Fosso San Marco would have responded to a necessity of the inhabitants of the Early Village Clusters—communication with each other and with the asty on the coast. The NE–SW road across the interior opened it to future development.

[34] As was famously the case on the Lacinian promontory where the whole peninsula was reserved as the temenos of Hera for her sacred cows and no farmhouses were constructed, as in fact was confirmed by archaeological field survey. Carter 1985b.

[35] Adamesteanu 1963, 50–51.

[36] Carter 1998b, 34–45.

[37] The Basento Road followed the course of the river valley, keeping close to the valley floor (ca. 20 masl) at Sant'Angelo and rising slightly near the mouth of the Venella (ca. 30 masl) before dipping again to the level of the valley (15 masl) at Pantanello. The road rose just before the Pantanello necropolis, where it climbed a steep little slope (20 masl), then traveled up and down across the rolling countryside in the direction of the coastal road and the Temple of Hera (Tavole Palatine) on the banks of the Bradano. Alberto Prieto's study of the aerial photos is the basis for this interpretation of the road pattern (see Ch. 31).

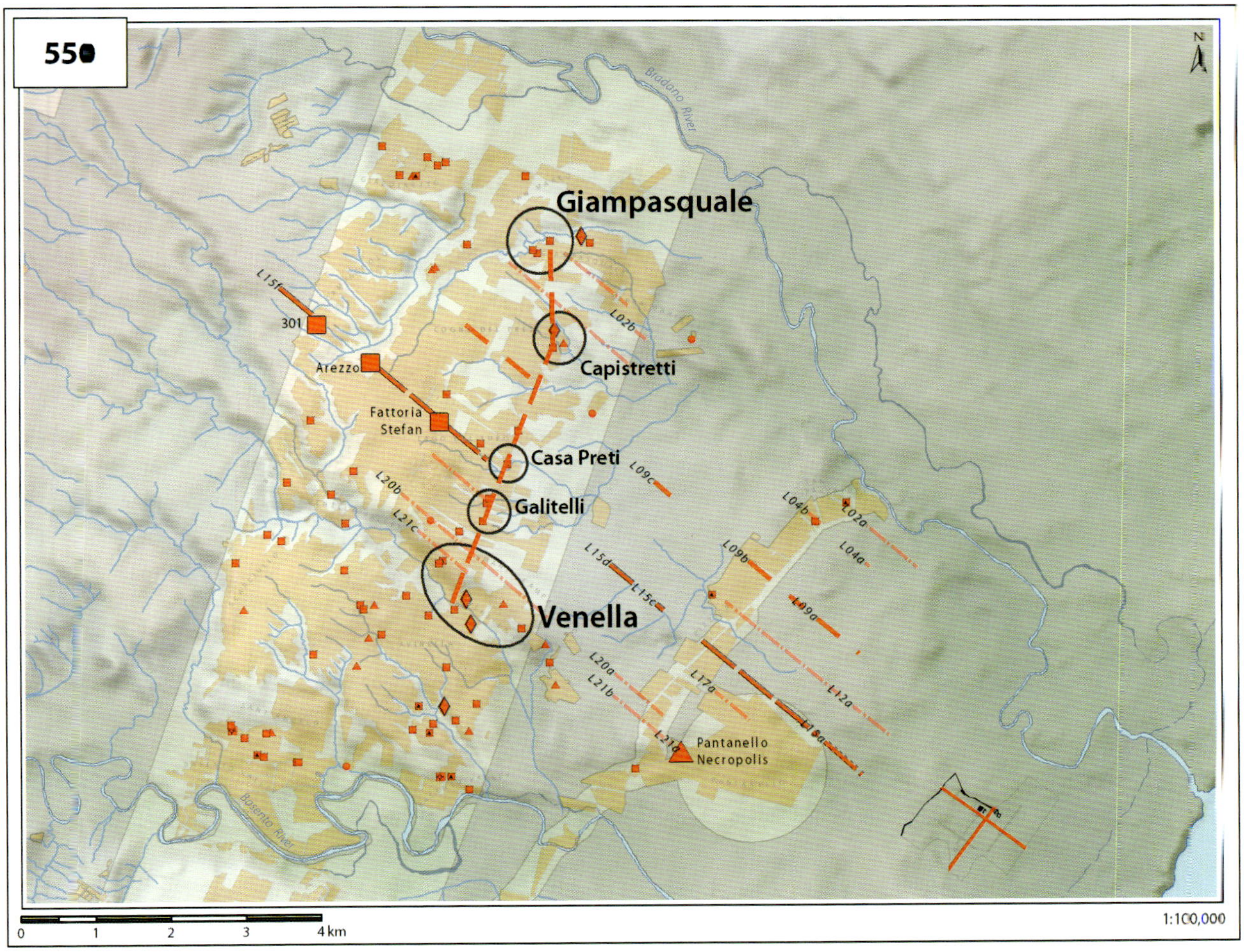

Figure 21.18 Villages in **550** and possible longitudinal axes, corresponding to later division lines. (JT)

If in fact it was intersected by one or more east-west roads, communication between other areas of the Central Plateau and the coast would have been greatly facilitated. In the hypothetical system proposed here (*Fig. 21.18*), the NE–SW road across the Central Plateau is plausible, and there are arguments to be made for the existence of east-west linear axes already in the mid-6th century BC. The section of road corresponding to division line L09, excavated beside the quarry in Pizzica, is of probable 6th-century date (Ch. 23).[38] The principal division line L15 that splits the chora between the Bradano and Basento into two equal halves was flanked by sites dating to the mid-6th century at its farthest extent: in Campagnolo (site 301F) and Close to the City in Pizzica (site 728F). There are other examples of east-west lines which may have existed in the 6th century. Though the evidence is slim, it indicates that the organization of the chora on geometric principles had begun in this period, contemporaneously with the introduction of an orthogonal plan in the asty (see below) in its mid-6th–century BC reorganization.

The Shift to Higher Ground

Water—its presence or the lack thereof—was perhaps the single most important factor determining the shape of settlement in all periods. It was as important in **550** as a half-century earlier. What distinguished the pattern in the two periods, more than any other, was the decisive shift of settlement to higher ground (see Ch. 19). The marine terraces, rather than the river valleys, were the preferred destinations of new settlers. Why should this be so? It does not appear that the fertile river valleys had lost their appeal. The number of new sites there more than doubled from **600** to **550**, and space for new settlers would have been limited. As a consequence, the rate of occupa-

[38] See Ch. 28, 1035, 1051.

tion of new lands on the terraces grew. Twenty-two percent more significant farmhouses could be found on the marine terraces in **550** than in **600**, and 7% fewer in the valleys. The attraction and popularity of the sloping valley walls suffered to an even greater degree in **550** than those of the valleys themselves. This sort of terrain was best adapted to viticulture or olive cultivation and to pasturing animals, especially sheep and goats.

Metapontine farmers were surely responding to an increased demand for cereals when they turned to the higher, drier ground. It was easier to plow with a team of oxen and it was normally better drained. The Central Plateau offered a great expanse of fertile fields that, if cultivated at all before by the Iron Age or earlier prehistoric populations, had had ample time to recover. The soil was more adapted to barley than to wheat, which does better in the valleys.

The magnitude of the shift from valleys to high plains can perhaps best be appreciated by considering the relative percentages of sites and densities in each of the four major geomorphological units in **600** and **550** in relation to the percentages of the total land surface that each of these units represents (*Appendices C11a* and *C11b*). Though there were more significant farmhouses in **600** on the Pleistocene marine terraces (which comprised 56% of the study area), the density of farmhouses was much lower there (0.3 sites per km^2) in the earliest period than that on the alluvial terraces which comprise 19% of the total surface (0.7 sites per km^2). The valley walls were the only unit to attract settlement in direct proportion to the amount of that land available. Five farmhouses, 26% of the total, occupied 24% of the total study area. The mass movement deposit, as noted earlier, was in a class by itself.

By **550**, the ratio between farmhouses and types of land forms had been radically altered. The number of farmhouses on the marine terraces had risen, so that nearly two-thirds of the farmhouses occupied this land form. The number of farmhouses in the valleys had also grown—it doubled in the intervening 50 years—but now constituted only a fifth of all farmhouses. The farmers or farmer/shepherds who chose to live on the valley walls increased relative to the earliest period but did less well than all the others. In **600**, their homes represented a quarter of the total in the chora; by **550**, this percentage had been halved.

Significance

High and low FMCE values in **550** were not evenly distributed over the study area, as the graph and table for this period show (*Appendices C7* and *C8*). Some clusters, like Giardinetto (with six significant sites in 550) and Sant'Angelo (with eight) continue to have average values higher than that for the chora as a whole. If FMCE scores are used with due caution as an indicator of material success in a particular period, then these two early compact and dense clusters collectively continue to be markedly more successful than the others that were expanding into new territory. The three new clusters in the Central Plateau combined (11 significant farmhouses) are below the chora average in **550**.[39] Avinella (combined lower and upper) with nine significant sites is slightly below average. The upper Avinella, despite its remoteness at the upper end of a drainage, expanded in **550** and is slightly more significant than any of the Early Village Clusters except Sant'Angelo. In this case, as in Lago del Lupo and upper Venella, the newness and short duration of several sites in each account in part for the high FMCE score. Upper Venella, deep in the interior at the headwaters of that tributary (with nine significant farmhouses) was, with Lago del Lupo, the area of the most rapid growth in the middle of the 6th century BC, but despite having five farmhouses with FMCE values over 3.00, its average is still less than that of the chora as a whole. The phenomenon of sites with great significance—in the upper Conca to some extent but to a much greater degree in upper Venella—may reflect a speculative fever.[40]

Necropoleis

The ratio of burials to farmhouses in **550** and throughout the history of the chora was not uniform across the study area (*Table 21.6*). In **550**, the locality with the lowest ratios (fewest necropoleis in proportion to farmhouses) was Lago del Lupo, which had rapid growth in this period; this is also the locality with the highest percentage of undated necropoleis.[41] Only

[39] The mean of 2.14 in **550** is substantially higher than that of 1.74 in **600**. In only two other periods was it higher: in **300** when the expansion of rural settlements was at its all-time greatest, and in **100** when the chora was under Roman control.

[40] Though the frontiers that existed in the Metapontino and in Magna Grecia were particular to their time and place, they have some characteristics of frontiers in general, and more particularly, of the moving frontiers of the New World. Cf. Webb 1951, esp. 1–28

[41] In the whole of the study area there are 87 undated necropoleis (54% of the total).

	550					All periods (600–50)					
Locality	**Total necs. ≥ 0.45**	**F/Ts ≥ 0.95**	**Total burials (Ns + F/Ts)**	**Total farms ≥ 0.95**	**Ratio Burials/F**	**Necs. (dated/ undated)**	**Total farms (all types)**	**Ratio N/F**	**Total F/Ts**	**Total burials (Ns + F/Ts)**	**Ratio Burials/F**
Villages											
Giardinetto	1	1	2	6	0.33	12 (7/5)	22	0.55	3	15	0.68
Sant'Angelo	1	1	2	8	0.25	8 (3/5)	24	0.33	1	9	0.38
Avinella	3	2	5	9	0.56	20 (10/10)	26	0.77	2	22	0.35
Lower	1	2	3	3	1.00	15 (7/8)	17	0.88	2	17	1.00
Upper	2	0	2	6	0.33	5 (3/2)	9	0.56	0	5	0.56
Venella	2	1	3	6	0.50	19 (10/9)	34	0.56	1	20	0.59
West	0	0	0	3	—	7 (3/4)	13	0.54	0	7	0.54
East	2	1	3	3	1.00	12 (7/5)	21	0.57	1	13	0.62
Giampasquale	0	0	0	4	—	1 (—/1)	8	0.13	1	2	0.25
Central Plateau											
Lago Lupo	0	1	1	8	0.13	24 (6/18)	59	0.41	6	30	0.51
Cogno Prete	3	0	3	3	1.00	10 (5/5)	29	0.34	0	10	0.34
Far from the City											
Campagnolo	0	0	0	1	—	3 (2/1)	11	0.27	0	3	0.27
Bufalara	0	0	0	2	—	1 (—/1)	12	0.08	1	2	0.17
Scarciullo	1	0	1	4	0.25	5 (2/3)	20	0.25	1	6	0.30
San Marco	0	0	—	1	—	6 (1/5)	8	0.75	1	7	0.88
AVS	1	0	1	4	0.25	14 (8/6)	18	0.78	2	16	0.89
Close to the City											
Pizzica	0	1	1	2	0.50	11 (10/1)	8	1.38	3	14	1.75
Pantanello	0	0	0	1	—	16 (8/8)	16	1.00	3	19	1.19
DBSBS	1	2	3	5	0.60	11 (2/9)	17	0.65	5	16	0.94
Totals	13	9	22	64	0.35 (avg.)	161 (74/87)	312	0.52	30	191	0.67 (avg.)

Table 21.6 Ratio of farmhouses to places of burial in **550** and averages for **600–50**, by locality.

six, or 25%, of the necropoleis there had any dateable pottery associated with them. Some early necropoleis without Archaic fine wares can be dated very approximately by the amphorae in their assemblages (*Table 21.7*). In most localities the dated and undated burials are about evenly divided. A possible explanation of the much higher percentage of undateable burials in Lago del Lupo is that this has always been one of the most intensely cultivated areas of the chora, both in ancient and modern times. Had Lago del Lupo been more like the other localities, we might have expected about half, or a dozen dateable necropoleis.

The localities with the lowest ratios of necropoleis to farmhouses in **550**, Scarciullo and Sant'Angelo, are among the more remote from the city. This may seem counterintuitive. One might expect these settlers to be more likely to live permanently on their farms and therefore to have their own burying grounds, since they presumably would not have wished to transport the deceased a long distance—for example to the asty—for interment. Alternatively, the occupants of these farmhouses may not have been permanent residents of the chora, though there is no indication from the pottery assemblages of these sites that they

547N	San Marco	Corinthian A 700–450 BC
48N	Campagnolo	(Corinthian B 500–450 BC) Corinthian A 700–500 BC
782N	Lago Lupo	Corinthian A 700–500 BC
439N	Pantanello	(Corinthian A 550–450 BC) Corinthian A 700–450 BC
457N	Giardinetto	Corinthian A 700–450 BC
467N	Giardinetto	Corinthian A 700–450 BC
471N	Giardinetto	Corinthian A 700–450 BC – Corinthian A 700–450 BC
522N	Giardinetto	2 Corinthian A 700–450 BC

Table 21.7 Some early necropoleis, dated by amphorae.

are in any way different from farmhouses closer to the city. Still, they may have used their farmhouses seasonally.

Equally puzzling is the fact that farmhouses in localities nearer the city appear to have a higher proportion of necropoleis to farmhouses in **550** and in all periods. For this there is no single explanation (see Ch. 23). When we come to consider the relation of dwelling and the place of burial here and elsewhere in the chora in **550**, the Casa Preti cluster stands out. In the Early Village Clusters this relationship is direct. The small necropoleis are located a short distance—100 m or even less—from a farmhouse to which they are clearly related. At Casa Preti all five necropoleis (all undated) are located together in an area separated from the central core of houses. This is the first evidence of segregation of the living and the dead to be encountered in the chora, and it will not be the last. The parallel with modern Italian agricultural villages with their separate necropoleis[42] is obvious.

As noted in the discussions of isolated sites at the far western edge of the study area there were larger necropoleis, like 70N in Cogno del Prete, that have no farmhouses close by. These may have been common burial grounds for widely scattered sites, closer to the frontier. The largest necropoleis without an obvious connection to a farmhouse is 401N along the Venella, which may have been closely associated with the Sanctuary of Artemis at San Biagio and the population that frequented it. Those with ratios closer to 3 to 1 (one burial place for every three farmhouses) that characterized the study area as a whole in this period are Avinella, Cogno del Prete, Giardinetto, Pizzica, and Venella.

[42] Colorfully referred to as *il paese della pansa in aria*, "the town whose inhabitants have their bellies up."

Sanctuaries

An unusually large number of rural sanctuaries have been identified in the chora of Metaponto,[43] The scarcity of precise information about them has given rise to much speculation about their functions, date, and nature. Various attempts have been made to classify them, and theories have arisen about their role in the birth of the polis.[44]

The Temple of Hera (Tavole Palatine), the Sanctuary of Artemis and Zeus Aglaos at San Biagio, and that of a female deity at Pantanello[45] were important sanctuaries that were strategically placed in the chora.[46] Their sites were among the first occupied in the late 7th and early 6th centuries BC; in the 5th and 4th centuries BC, they were dignified by monumental architecture in stone.[47] They were certainly not the only such sanctuaries, though the evidence for similar ones at other strategic points in the chora, such as Pantano[48] and Sant'Angelo Grieco, while very suggestive, is either incomplete, or has not been published.[49] San Salvatore on the Bradano and San Basilio on the Cavone are possible sites for monumental temples, but systematic excavations have never been carried out.

Besides their varying degrees of monumentality, the rural sanctuaries with the longest continuous periods of use have in common their placement at points along the roads that linked the interior with the coast. The Temple of Hera (Tavole Palatine) and Pantanello are each located at points where rivers reach the coastal plain. Their monumental temples in the 5th century BC housed cult statues in marble. It would be logical to expect that the entrance to the Cavone valley, which marked the southern extent of the chora, would have been similarly emphasized. The sanctuary at San Biagio marks the juncture of a major tributary, the Venella, and the Basento. It, too, had a monumental peristyle building and a marble cult statue.[50]

[43] Carter 1994a. For the chora of Croton see Genovese 1999; for Poseidonia, see Cipriani 1996; for Sybaris, see Novellis 2003; and in general, see R. Leone 1998.

[44] The most ambitious to date is *La naissance de la cité grecque*, updated and expanded as *La nascita della città greca* (De Polignac 1991).

[45] A volume in this series on the sanctuary at Pantanello is in progress.

[46] Mertens 2006, 47–49, 88–89.

[47] For the column drum and foundations at San Biagio, see Giardino and De Siena 1999b, 346–48; for the column drums and other architectural elements at Pantanello, see Carter 1994a, 177–78; 2006a, 147–48.

[48] De Siena 2002, 36; Schojer 2001; 2002.

[49] Carter 2006a, 147–48.

[50] See Nava 2000, 689–91. The farthest extent inland of the divided chora along the Basento was probably marked by a monument of con-

Sanctuaries along the lesser tributaries, with the exception of San Biagio, were more modest.[51] The surviving architectural elements (when present) associated with these sites consist exclusively of terracotta revetments and antefixes. These smaller sanctuaries, like all of the above-mentioned major rural shrines, are located in close proximity to springs. Places of the simplest sort with dedications of pottery—a cup, a figurine—are also located by springs.[52] The essential elements of rural sanctuaries like Pantanello must have been present here in very reduced form at a number of sites and they could hardly be more different from cults of the urban center.[53] Even the simple types were celebrated by poets.[54] There is a continuum, from the grand to the entirely unprepossessing.

About the major and better known examples a degree of generalization is possible. Sites like San Biagio, Pantanello, and the Temple of Hera (Tavole Palatine) were established near important sources of water, and accessible from the main roads (*Map 6*, Vol. IV). They were connected and could be easily reached from the asty. With the sanctuaries of the city, they formed a tangible bond between asty and chora.[55] In the case of the Tavole Palatine, this link was architectural as well. In some if not all cases, as De Siena has pointed out, the cults of the city had their counterparts in the chora: Hera, on the banks of the Bradano at the Tavole Palatine, Artemis at San Biagio, Apollo at a shrine just outside the walls at Pizzarieddo.[56] Artemis may have been worshipped in the asty at Oikos D in the 6th century BC[57] and its successor Temple D from the early 5th century BC, or, alternatively, in Oikos C[58]—but certainly at San Biagio[59] in a setting that recalled her cult place at Lousoi in Arcadia, on the border of Achaia.[60]

The terracotta frieze from a small cult building, as noted briefly in Chapter 20, dates to the fourth quarter of the 7th century BC. It depicts the departure of Achilles for the Trojan War in a chariot drawn by winged horses.[61] This was an unmistakable allusion to the place of origins and heroic ancestry of the Greek settlers, and so it would have been understood by the Metapontines and their non-Greek neighbors, the indigenous peoples of the interior. They could both appreciate to some degree and imitate rather well the mythological subject matter and form of Greek figural decoration,[62] as the mid-6th–century BC revetments representing heroic combat from Serra di Vaglio, 100 km from the coast, well illustrate.[63] In the early years of the chora especially, the sanctuaries were constant reminders of common beliefs and ancestral connections with the homeland. There the inhabitants of the surrounding countryside could gather on festival days, for sacred processions, such as that represented on the frieze of Oikos C in the asty. They would have made use of the roads, like the Basento Road leading to San Biagio, to transport images of the gods between her (or his) rural and urban shrines.[64]

In an early study of Metapontine rural sanctuaries, it was suggested that the fairly regular spacing of rural sanctuaries might reflect an administrative function for them and that their placement near the mouths of drainages was chosen because these were natural magnets for population as well as conduits to the interior.[65] What was lacking at the time of this analysis, however, was knowledge of how they related spatially to the contemporary farming population of the chora. Now that the farmhouses located by the survey have been accurately dated, it is possible at least to begin to describe that relationship in a limited number of instances. It is clear, for instance, that the

siderable pretension if the mid-5th–c. BC Doric capital found reused in the excavation of the farmhouse at Sant'Angelo Grieco belonged to it. Carter 2006a, fig. 4.17. See also below, Ch. 27, 909.

51 See discussion, Carter 1994a, 180.

52 A fine example of a simple cult place is that beside the spring at the foot of San Teodoro and Incoronata in the Basento valley, which was excavated by Antonio De Siena. It consists of a square basin, in which were deposited a bird sacrifice, a terracotta figurine, and a plaque representing a standing female with *polos*, who carries an oinochoe in her right hand and a lamb in her left, which dates the site to the 5th or 4th c. BC. A second, small sacred place of Archaic date was found nearby.

53 Osanna 2008, 926, Pl. 18a, b.

54 Cf. the epigram by Leonidas of Taras (see Ch. 18, 615).

55 Mertens 2006, 217; Osanna 1992, 48–54.

56 Giardino and De Siena 1999b, 347.

57 Giacometti 2005, 83.

58 Giacometti 2005, 27. She identified the divinity worshipped in Temple D as Aphrodite. Mertens-Horn (1992, 61) has argued that Oikos C was a sanctuary of Athena.

59 Olbrich 1979.

60 Giacometti 2005, 5.

61 Mertens-Horn 1992, 28–44.

62 Mertens 2006, 48.

63 Mertens-Horn 1992, 75–82.

64 Giacometti (2005, 68) sees the scene as symbolizing the rite of passage of ephebes to the new reality—that of citizens and knights, *politai-hippeis*. The procession is a tradition that persists even today in certain parts of southern Italy and Sicily, where the connection with the ancient past is particularly evident as at Croton. There a procession takes place annually between the Cathedral (Duomo) and the Chapel of the Madonna at Capo Colonna, the site of the extra urban Temple of Hera (see Carter 2006a, 158, with bibliography).

65 Carter 1994a, 181–82.

area around San Biagio and the mouth of the Venella had attracted settlers from the beginning. Much the same was the case with the Pantanello sanctuary and probably with the others, but these are the only two that lay in or beside the study area.

ἔνθεν καὶ ἀρηϊφίλοις
ἄνδρεσσιν <ἐς> ἱπποτρόφον πόλιν <τ'> Ἀχαιοῖς
ἕσπεο, σὺν δὲ τύχᾳ
ναίεις Μεταπόντιον, ὦ
χρυσέα δέσποινα λαῶν·
ἄλσος τέ τοι ἱμερόεν
Κάσαν παρ' εὔυδρον...

(Artemis) your point of departure.
Martial Achaeans escorted you out
To the town where mares nurse colts,
And now Metapontion makes you thrive
O gold Queen of the land;
And the rippling Casas[66]
Rims your inspired grove—[67]
Bacchylides, *Epinicia* 11.113–119

A further function of the sanctuaries, it has been suggested, was as a base for "powerful aristocratic families," as suggested by Bacchylides' poem.[68] The Sanctuary of Artemis and Zeus, for example, was located at the mouth of a fertile inland valley, perfectly adapted, as the poet says explicitly, for the raising of horses, the aristocratic farming pursuit *par excellence*. It was also an exceptionally rich sanctuary to judge by the quality of the votives, one that attracted dedications from beyond the confines of the territory as the coins from other Greek colonies found among the votive offerings testify. Its architectural importance may reflect not only the wealth of the families inhabiting this corner of the chora, but even more its role in the religion of the polis as a whole. Artemis was one of the principal gods of Metaponto. She was the ruler of boundaries between the cultivated chora and the wild (the eschatia) and the divinity who presides over the integration of young people, both male and female, into adult society and a large number of other functions reflected in her epithets and the iconography of votives from the deposits around the spring. Artemis and Hera were the protectors of the polis in the chora,[69] emphasizing the unity of chora and asty in the political system of the Achaian settlers.[70]

Four sites were identified as Archaic sanctuaries during the survey on the basis of both high counts of ritual vessels—miniature kraters, oinochoai, phialai, and others, and on a larger scale, louteria and thymiateria—and terracotta figurines representing goddesses, gods, and their coteries and offerants. The range of subjects is detailed in Chapter 13.[71] In the 6th century BC, the predominant subjects are standing or seated female divinities. These surveyed sanctuaries are similar to those prominent ones just described in several respects: they are located near sources of water; all are likewise among the earliest sites in their respective areas of the chora and all are located in clusters of farmhouses where they occupy central though not dominant positions. They differ in important ways: the spacing between sanctuaries and nearby farmhouses is uniform. There were, without exception, no remains other than building stones and roof tiles typical of farmhouses. The assemblages of pottery recovered from these sites and those from farmhouses, apart from the very high percentages of votive material, are quite comparable (see Ceramic assemblages, below). Making the distinction between farmhouse and sanctuary has often been difficult. On site 506F, for example, 35 fragments of terracotta figurines dating to the 4th century BC were recovered,[72] but only a single fragment of a miniature vessel. In the final analysis, site 506 was rejected as a sanctuary. Had there been more miniature vases, it might have been otherwise.

Similar doubts arose in the case of Metapontine farmhouses with large quantities of terracotta figurines, excavated in the 1960s. Barberis, who studied the terracotta figurines, was puzzled by the 200 fragments of figures, 180 of which were dateable to the late 7th and 6th centuries BC from the Arezzo farmhouse at the northern edge of Lago del Lupo.[73] She concluded that they had constituted a votive deposit,

[66] Casas and Casuentus are earlier names for the Basento.

[67] Text from R. Jebb 1905, *Bacchylides: the poems and fragments*, Cambridge University Press. Trans. from Fagles 1961, 34.

[68] De Siena 1999b, and Montepaone 1986.

[69] These are well summarized and discussed by Olbrich (1976 and 1979), Osanna (1992, 48–50), and Giacometti (2005, 50).

[70] Osanna 1992, 46–53.

[71] See also the study by Barberis (1995).

[72] A number of the fragments seemed to be from the same figurine.

[73] Barberis 1995, 16–18. Unfortunately, little is known of the excavation of this farmhouse. Cf. Amore 1992–1993, 77–78. Amore's study of the existing records and materials constitutes the fullest documentation of these early farmhouse excavations.

perhaps separate from and certainly antedating the 4th-century BC farmhouse where they were found. Of the eight farmhouses located in our study area whose materials Barberis examined, five had terracotta figurines, and of those five, three had figurines of the Archaic period.

The potential for confusing sites identified as sanctuaries with farmhouses or vice versa obviously owes much to the fact that over 20% of farmhouses had some terracotta figurines and miniature vases. These have generally been viewed as material remains of family cult practice within the confines of the household.[74] The actual family cult place has been very convincingly identified at Fattoria Fabrizio.[75] This is a universal feature of life in the countryside, not limited in time or place. It can be seen in the chorai of Croton and throughout Magna Grecia wherever the Greek countryside has been investigated.[76]

Very occasionally, architectural elements—above all antefixes—make up a part of excavated and surveyed farmhouse assemblages. A 6th-century gorgon head antefix, for example, was found in the excavation of the Pascucci-Torraca farmhouse in Saldone on the Bradano side of the chora outside the study area. Such antefixes are generally found at sanctuary sites like Pantanello, the Tavole Palatine, or San Biagio.[77] Its discovery on the site of a farmhouse was interpreted by Adamesteanu as a sign of the architectural pretension of some farmhouses of the 6th century.[78]

A major factor contributing to the uncertainty about the nature of these sites is the fact that both farmhouse and sanctuary sites also seem to have served as centers of votive terracotta production.[79] A large number of terracotta figurine fragments dating to the 4th century were recovered from site 266S. It was subsequently discovered that this site was a production center for Greco-Italic amphorae of this period (see Ch. 12). The two fragments of miniatures offer slight support for the designation of this site as primarily a place of worship. In all essential respects apart from the votive figurines and plaques (Ch. 13, Nos. 87–137), it resembles a farmhouse.

Ceramic Assemblages, 600/550

Many sites have multiple periods of occupation, so their assemblages represent the wide variety of forms existing in these periods. By limiting the analysis to sites which have the greatest possibility of representing a single period, we can hope to observe the major changes over a span of six centuries. To that end, a restricted group of significant farmhouses that date to a specific period was selected for comparison in terms of functional groups. The following criteria were used for inclusion in this group: first, the assemblages had to be sufficiently large that all classes were represented and most of the more common forms were present. Second, the dateable pottery had to be largely concentrated in this period. This has also been done for the periods **450**, **300**, and **100**, for which the functional groups analysis (Ch. 19) has been performed.

Survey farmhouses—*Functional groups 600, 550.* The results for the period of **600/550** are represented in Figure 21.19 (and in tabular form in *Appendix C19a*). The number of vessel fragments for each of the various functions is expressed as a percentage of total identified fragments. As is evident from the table, there is a degree of correlation among the farmhouse sites. The percentage of drinking vessels, for example, varies between 22% and 32% with outliers at 19% and 42%; short-term storage, between 17% and 30% with outliers of 13% and 36%. Correlation is what we ought to expect of similar units such as farmhouses, which are essentially engaged in the same basic activities. Statistical analysis of the same data using the chi-square test (see *Appendix D, Table D11,* chi-square 8) compares the twelve farmhouses and finds essentially a statistically significant difference between them. Cross tabular analysis reveals, however, great similarities among the sites in terms of functional groups, with only a few exceptional cases. The degree of correlation is higher, as also could be expected, between sites with large quantities of identifiable shapes, such as sites 285F/T, 286F, and 155F. Not all sites contain

[74] See Ch. 13, Conclusions, and also Barberis 1995, 36.

[75] I am grateful to Elisa Lanza Catti for this information. Pers. comm., Feb. 2010

[76] Good examples have been discovered in the chora of Chersonesos on the Black Sea. Carter et al. 2000, 727–29.

[77] Tavole Palatine: Adamesteanu 1974a, 56 (gorgon head antefix). San Biagio: Tempesta 2005, 86, and n. 16. Pantanello: Carter 1983a, 455–56, Nos. 26 & 27, Fig. 31.

[78] Amore 1992–1993, 64. If the antefix mold from farmhouse 489F is any indication (Ch. 13, Cat. No. 341), this was not limited to farmhouses of the Archaic period. The antefix with an image identified by Ammerman as Io is a refined product of the 4th c. and quite similar to ones discovered in the excavations of the larger rural sanctuaries.

[79] Adamesteanu early on pointed to the evidence that the figurines and other votive materials from San Biagio were produced on the site. Adamesteanu 1974a, 60–61.

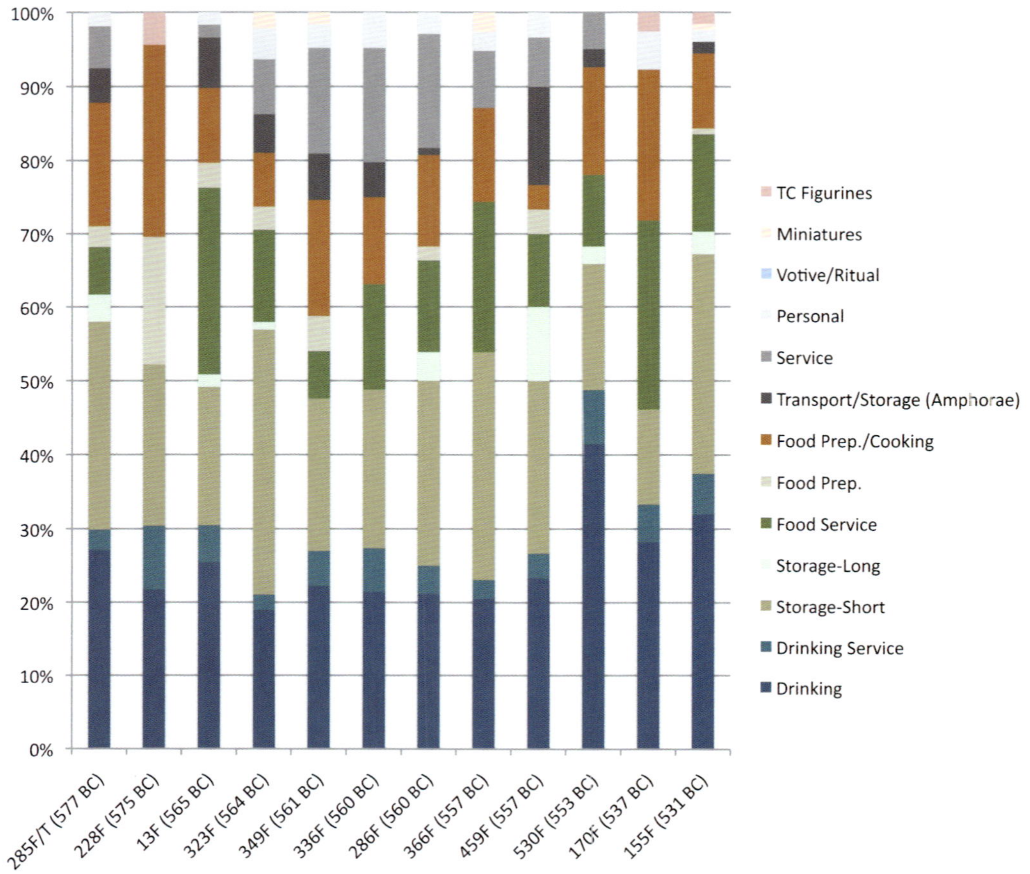

Figure 21.19 Comparison of functional groups between selected farmhouse assemblages, **550**. Percentages calculated from total identified fragments, with best date based on fine-ware pottery.

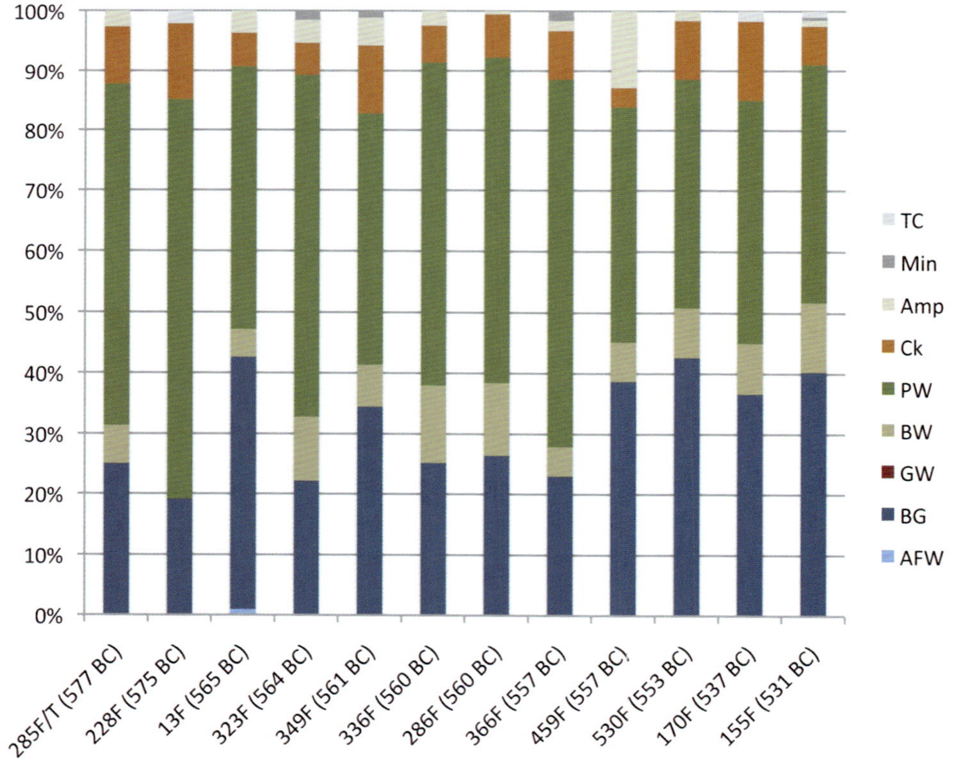

Figure 21.20 Comparison of ware classes between selected farmhouse assemblages, **550**, showing percentages based on total fragments.

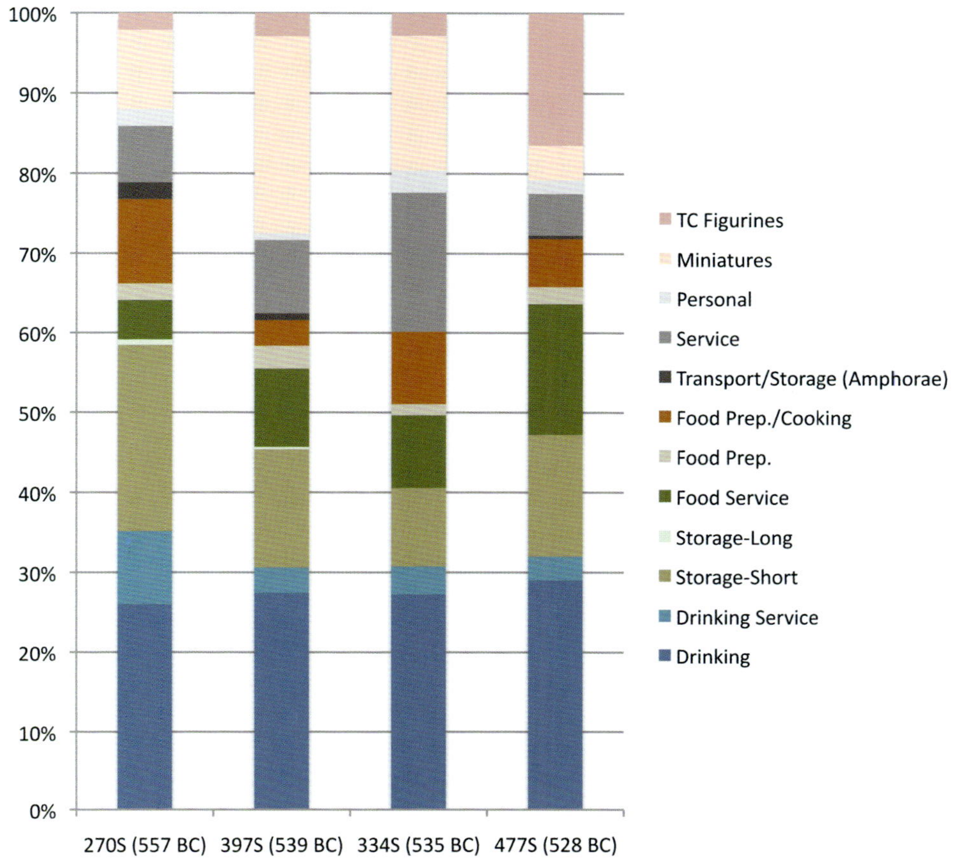

Figure 21.21 Comparison of functional groups between selected sanctuary assemblages, **550**. Percentages calculated from total identified fragments, with best date based on fine-ware pottery.

examples of all functional groups (e.g., 228F with only 23 identifiable forms). In some of the smaller assemblages, only five or six of the possible eleven groups are present, but no site is missing drinking cups, storage vessels, or cooking ware, and only one or two had no food service (platters), amphorae, or drinking service vessels (larger black-gloss vessels for holding liquids).[80]

In broad terms, approximately 30% of all material recovered from the surface of the site belongs to drinking (drinking cups and skyphoi, and drinking service seen as hydriai and oinochoai), 30% was destined for some sort of storage, 10% includes principally objects of personal hygiene and personal devotion. The final 30% relates to solid food (food service, such as ubiquitous lekanai; food preparation, often mortars; and cooking). This represents the typical household of the middle of the 6th century BC. The following pages will show the range of material goods that define the household and how the life it reflects changes in succeeding periods. And change it does, though only very gradually.

It is interesting to note here that certain types of jugs, lekanai, and mortars have a wide popularity in **550** and others apparently do not, at least in the farmhouses considered.

Ware classes. An alternate way of comparing assemblages is by considering the ware classes. This is the first information that appears in individual assemblage tables. There are relatively few classes: black-gloss (BG), fine wares (other than black gloss, e.g. red-figure ware), banded ware (BW), plain ware (PW), cooking ware (Ck), amphorae (Amp), miniature (Min), and terracotta objects (TC). Whereas the functional grouping takes into account only those fragments whose form has been identified, the ware classes include all the pottery from the sites, thus, the counts are higher. The two methods complement each other. The comparison of farmhouses of

[80] The numbers of vessels in each of the groups at selected sites are summarized in the individual assemblage tables in Volume IV.

600/550 by ware classes (*Fig. 21.20*; *Appendix C19b*) shows again a great similarity among the farmhouses in this respect.

Sanctuary assemblages. The individual types of vessels and groups of vessels from sanctuary assemblages (*Fig. 21.21; Appendix C20*) are remarkably similar in many respects to the farmhouse assemblages. The black-gloss drinking vessels from each compare closely in type, form, and quantitative proportions. The kotylai that are present in the sanctuary 477S assemblage are not, however, a type found at farmhouses. The black-gloss cothon (properly an exaleiptron) from 397S is a rather rare shape in any context. It had a ritual function,[81] but was found in burials and sanctuaries (e.g. in the Pantanello sanctuary). Other ritual vessels, such as louteria used in purificatory rites, have also been recovered from farmhouses (e.g. 285F/T in Sant'Angelo). Ritual objects are widely distributed throughout the chora, but it is the greater concentrations of them that make these four Archaic sanctuary sites stand out from the farmhouses and necropoleis of the period.

A comparison of the functional groups at sanctuaries with those typical of farmhouses displays fundamental similarities but also some differences (see *Table 21.8*). The relative proportions of the group concerned with drinking are slightly more elevated than those for farmhouses. Storage is limited essentially to short-term and constitutes 16% of the identifiable vessels from the sites. Pithoi and amphorae used for long-term storage were present, but in negligible quantities. Vessels concerned with food service and preparation, including cooking, account for another 19%. Of the remaining 32% of the vessels and objects, 20% consist of votive vessels and terracotta votive figurines with clearly ritual function. It is possible that the remaining 10% of vessels assigned to service (principally small bowls) and some of those classified as personal may have had ritual functions as well. The bowls, for example, are also frequent finds at excavated sites like the Pantanello sanctuary. The lekythoi, labeled personal, were used as containers for offerings of oil and were a component of a large percentage of burials of both men and women in the Archaic and Classical periods. Together, nearly a third of all material from these four sites could be related to religious activities.

	Farmhouse 600–550	Sanctuary 600–550
Drinking	25%	27%
Drinking Service	5%	5%
Domestic Storage–Short	24%	16%
Domestic Storage-Long	2%	< 1%
Food Service	13%	10%
Food Prep.	3%	2%
Food Prep./Cooking	14%	7%
Transport/Storage (Amphorae)	4%	1%
Service	7%	10%
Instr. Domesticum	0%	0%
Personal	3%	2%
Votive/Ritual Total	2%	20%
(Miniatures)	(1%)	(14%)
(TC figurines)	(1%)	(6%)
(Other vessels)	(< 1%)	(< 1%)

Table 21.8 Comparison of averages of functional groups from the 12 farmhouse and 4 sanctuary assemblages **600–550**. The parentheses enclose subtotals for the votive/ritual function.

While the composition of a sanctuary assemblage is, in the overview, not unlike that of a farmhouse, there are 10% fewer storage vessels and 11% fewer vessels used for food preparations and serving, and the difference of 20% is made up entirely by votive objects. Had the sanctuary sites been incorrectly identified in the field as farmhouses, serious doubts would have arisen during the analysis of the pottery. Are there other differences in the two types of assemblages that are marked by the broad grouping into functional groups? The evidence is at hand to begin to answer that question, thanks to the attention paid to small differences in the typological analysis. The coarse wares from sanctuaries differ from those of farmhouses in their numerical proportions and also in their specific forms. (The forms are detailed in the individual assemblage tables in Volume IV.) It is perhaps not surprising that there was little provision for long-term storage on a site where the principal function was the periodic performance of ritual. Communal ritual eating and drinking were a major aspect of the life of a Greek sanctuary,[82] and this is reflected in the ceramics.

[81] Cf. Ch. 29, grave goods from site 737N tomb 23.

[82] Rabinowitz 2004; 2009.

Figure 21.22 View of the Fattoria Fabrizio excavation, 1980. (JCC)

These ritual meals should also have been distinguished from the everyday nourishment of a farm family. Is this reflected in the types of vessels used? Many of those used for drinking and serving drinks in sanctuaries are the same as at farmhouses, but among those used for the preparation and serving of food there are some quantitative, as well as typological differences (see Ch. 9 on plain/banded ware). There are many more examples of *dinoi* found in a wider range of types in the sanctuary assemblages than in the farmhouse assemblages. This cannot be solely a result of the relatively larger sizes of the sanctuary assemblages, since other types of vessels—e.g., basins, jugs, and lekanai—are proportionately equal on farmhouse sites and sanctuary sites. Might the *dinos* have had a particular role in the rituals of the sanctuaries? Both types of sites make great use of jugs, and many are of exactly the same type (e.g. J1 and J14 types), but there are certain jug shapes that appear in the sanctuary assemblages (e.g., J5, J16, J20, and J21), and not on any of the twelve farmhouses considered here. This might be a reflection of the greater size of sanctuary samples, but some jugs that appear at farmhouses were not in use at these four sanctuaries (J2, J4, J12, J18, and J19). The same apparently occurs with the lekanai, though not with the carinated lekanai, plate/lekanai, or lekane/basins. Two lekane types (Lne 6 and Lne 11) occur in sanctuaries only, and three (Lne 5, Lne 8, and Lne 10) only at the farmhouses analyzed here. Does one group consist of lekanai or of jugs especially intended for use in rituals and made especially for that market, and the other consist of vessels for everyday use only? This is a reasonable working hypothesis. The evidence for this sort of analysis, as noted, is present in the individual assemblage tables, and it is hoped that it will stimulate further research.

Assemblage from an excavated farmhouse. Excavation in the Metapontino in the last half-century has included examples of the three principal types of sites identified in the survey. In Chapter 20, some of these excavated necropoleis and sanctuaries were described. Excavated farmhouses that date to the early years of the

Figure 21.23 Reconstruction of Fattoria Fabrizio. (ALP)

settlement of the chora are rare. The main reason so few Archaic farmhouses have been found is that the materials of many early farmhouses were appropriated to construct their successor(s) on the same site. This was the case with Fattoria Fabrizio, which was discovered in a preliminary survey season in 1980 (*Figs. 21.22–23*). The material appearing on the surface was purely Archaic. Subsequent excavation revealed that the farmhouse's main period of occupation was the 4th century BC.[83]

The farmhouse at Cogno del Pero is one of only two exclusively Archaic farmhouses in the study area. It was constructed of mudbrick on a socle of river cobbles. All of the internal spaces were covered by a tile roof; there was no indication of a courtyard. The excavator, Adamesteanu, only made brief mention of the date: "the numerous vases and fragments belong to the 6th century, most probably to the middle of that century. The majority of the diagnostic fine ware consisted of Ionic cups of local production."[84] The structure was not completely excavated but the basic arrangement is clear: rooms of more or less equal size in three parallel rows, the same basic plan as that of Fattoria Fabrizio.

The other known example of a purely Archaic farmhouse was found by ICA at Sant'Angelo Vecchio.[85] It was a very modest, one-room structure that had no subsequent occupation. The pottery from that site has been analyzed and it is possible to compare the percentages by ware class (*Table 21.9*) with the average of the twelve surveyed farmhouses. They resemble each other quite closely except that a great deal more cooking ware was excavated at this site, as a percentage of the whole, than was collected from the surface of the survey of farmhouses. A comparison by functional group is also possible and the percentages are presented in tabular form (*Appendix C21*).

Ware class	Farmhouses average, 600/550	SAV, House on Hill
FW	1%	<1%
BG	31%	25%
BW	8%	4%
PW	49%	46%
Ck	8%	22%
Amp	4%	2%
Min	2%	
TC	2%	

Table 21.9 Comparison of excavated farmhouse assemblages.

Villages

The early settlers of the chora for the most part chose to settle close to each other in what we have called Early Village Clusters. By **550** these had grown into centers of a half-dozen or more farmhouses separated by a considerable amount of space, as described above. It is a small village by modern standards, but villages come in many sizes. How would the Greeks have described such a settlement? The term village would be translated as κόμη, but are these really what the Greeks would have recognized as such? They seem more like hamlets, but there is no obvious Greek term for this.[86] Perhaps after all, κόμη would not be a complete misnomer for our villages. The brief investigation that follows finds quite a few points in common between the classic definition of Aristotle and our villages.[87]

Villages in Magna Grecia. Aristotle defines the village (*kome*) as an apoikia of the household, formed of sons and sons' sons.[88] The kinship of members is a defining characteristic of the home or village at its most basic level. He adds that it is not an association for the satisfaction merely of daily needs. The polis, the culmination of all partnerships, is composed of several villages and it alone attains self sufficiency. And since self-sufficiency is an end and a chief good, the polis is a natural growth and man is by nature a political animal.

This can be read as a biological evolutionary chart, beginning with the first partnership of Adam and Eve

[83] Carter 2006a, 138–41. Fattoria Fabrizio is currently being restudied for publication by Elisa Lanza Catti. In light of this recent work, the possibility that the later plan closely follows that of the 6th c. BC is being seriously considered.

[84] Adamesteanu 1974a, 82, 85, plan 85; 1983, 54; Osanna 1992, 54, Fig. 21 (using the same plan as Adamesteanu). See Amore 1992–1993, 80–81, where it is referred to as Fattoria Bufalara. It is located along the southern rim of the Acqua Fetente in the Cogno del Prete locality. See also discussions of two 6th- to 4th-c. BC farmhouses, Musillo and Pascucci-Torraca, excavated by G. Uggeri in the Saldone locality. See Amore 1992–1993, 57–64.

[85] Known as "the House on the Hill," the structure was excavated by Ingrid Edlund-Berry (see Edlund 1986). It will be included in the overall study of Sant'Angelo Vecchio in a future volume in this series by Francesca Silvestrelli.

[86] M.H. Hansen has suggested χωρίον, a very vague concept signifying "a spot, a place, a district" according to Liddell and Scott.

[87] The classic modern work is "Kome. A study in how the Greeks designated and classified settlements which were not Poleis" (M.H. Hansen 1995).

[88] *Pol.* 1252b, 16-1253a3.

and ending in the poleis of historical times. Athenians thought that their city was born of a union of villages, a *synoikismos*. The evolution of poleis in the Western Greek world was different. Every indication, based on the careful study of the archaeological evidence of several of the more important earlier apoikiai (e.g., Megara Hyblaia,[89] Syracuse), is that they came into being at a specific point in time. Metaponto, though one of the later examples, appears to have followed this pattern. The area of the city and the principal spaces within it were established early on (see Ch. 20, Asty).

The Greeks believed, according to Aristotle, that the population of early Greece lived in villages, *kata komas*, and that these were agricultural villages of a rural society. Until recently, examples of ancient agricultural villages were archaeologically almost unknown in Magna Grecia. Even now they are not common. Amastuola in the chora of Taras, a recent discovery, would be a good candidate.[90] It was in existence soon after the foundation of Taras and it occupied (like Cozzo Presepe and Incoronata in the Metapontino) the site of an earlier indigenous village. Another excavated example would be Pantano in the Bradano valley, which unfortunately was incompletely studied and remains unpublished (see Ch. 20).[91]

A central question is what purpose did these villages serve? What needs, beyond the daily requirements of a family, did they fulfill? It has been suggested in the case of the chora of Poseidonia that the villages far from the city center—which in this instance were not active until a century or more after the foundation—served the needs of farmers engaged in raising specialized crops like olives, adapted to the hilly terrain. These cultivators required heavy equipment that needed to be protected, as opposed to grain-growers, who, according to the theory, did not need to live near their fields.[92] A motive for living close to one another thus was protection, and perhaps sharing resources and labor. It has also been suggested that these outlying villages may have been occupied by successive waves of settlers (the epoikoi), who had to be accommodated on what may have seemed less than prime agricultural land.[93]

The village clusters in the chora of Metaponto.
Date. Previous views on the form and date of the settlement pattern in the chora of Metaponto have been well-summarized by Greco.[94] In short, the earliest settlements were sanctuaries. There were no villages in the chora, only isolated farmhouses, which were not occupied until a century after the asty. We now know in the case of Metaponto that farmhouses and related necropoleis were among the earliest structures in the chora, antedated perhaps only by a few sanctuaries, and that the early farmhouses in the study area were at first concentrated in five areas, at relatively great distances from the asty (6 to 11 km), in what we have called the Early Village Clusters.

The remains of what were identified by Michele Lacava as *villagi* were still visible in fields relatively closer to the ancient city.[95] Unfortunately there is no indication of the function or the dates of these villages. Lacava left fairly accurate positions for several of them, and they have been placed on the map of Early Village Clusters (*Fig. 21.24*, indicated by dotted circles). Permanent occupation of the chora nearer to the city did not begin in earnest until the 5th century BC (Ch. 23), and this is the probable earliest date for the villages noted by Lacava and represented on the map. We are ignorant of possible clusters in areas Close to the City not covered by the survey, such as Saldone where there were farmhouses, necropoleis, and perhaps an early sanctuary. Only three significant farmhouses dating to **550** were found by the survey in the transects (SNAM and Pantanello) nearest the city, and this is indicative of a lower rate of occupation there in the early years.

The dates of the fine pottery from the farmhouses and necropoleis prove beyond doubt that the villages came into being at approximately the same time as the earliest Greek settlement on the site of the asty. There was at least one sanctuary, that of Artemis at San Biagio, at this time, but probably others also. The four Archaic sanctuaries identified by the survey are among the earliest sites in the study area, but they are not appreciably earlier than the first farmhouses.

Distances—village to asty. The central place in the Greek polis was the asty.[96] The asty at Metaponto

[89] See discussion of the problem by E. Greco (2001, 173–75, with relevant bibliography). For an opposing view, see Osborne 1998.

[90] Marugg 1996a; 2002.

[91] Schojer 2001; 2002.

[92] E. Greco 1979.

[93] See Aristotle 5.2.10–11, on friction between newer and older settlers, with reference to Sybaris and its successor, Thourioi. Cf. Diodorus Siculus 7.11. See also Vallet 1968, 78.

[94] E. Greco 2001; cf. also Osanna 2001.

[95] Lacava 1891, 91–92.

[96] See Bintliff 2002a and 2002b.

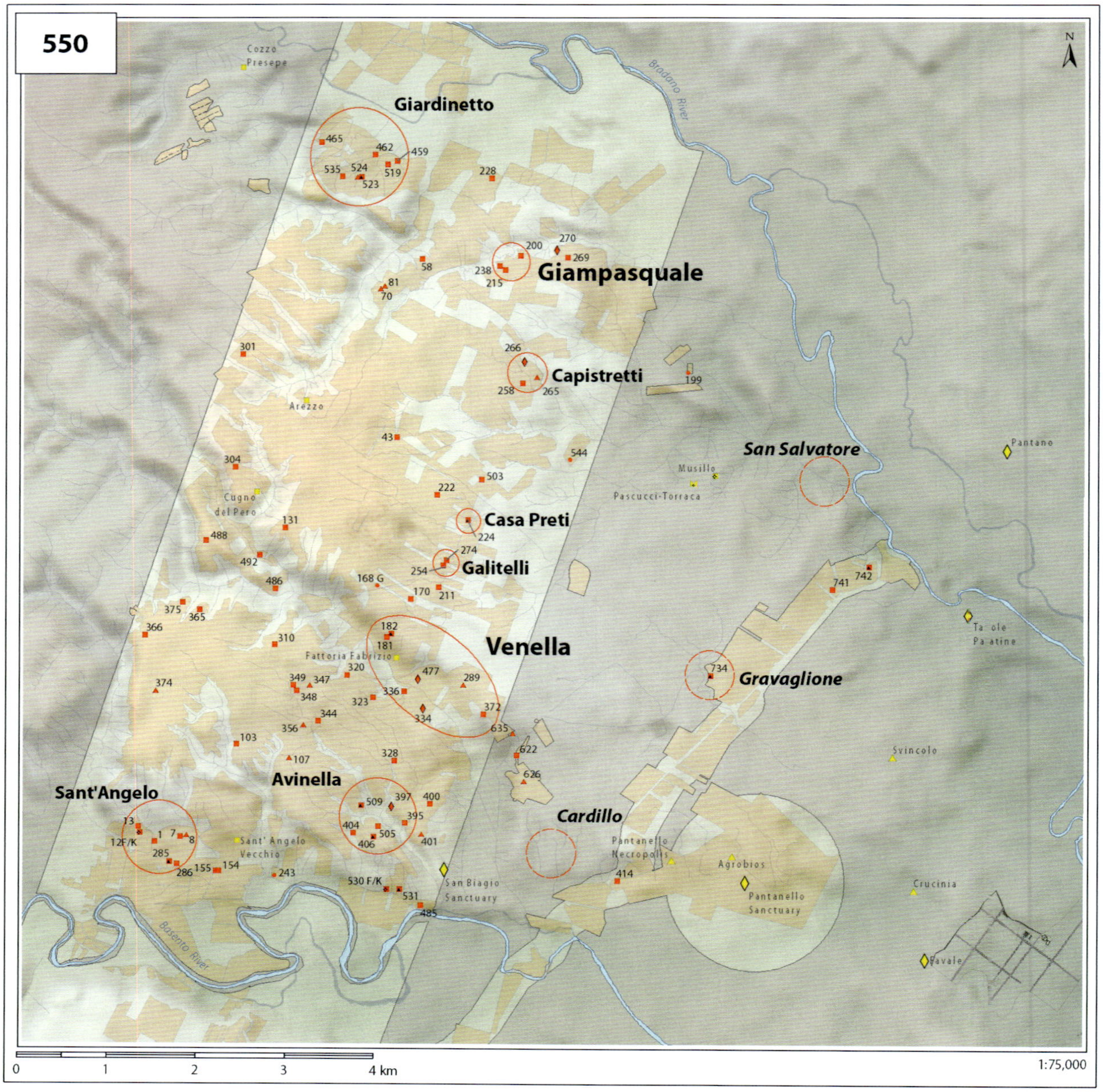

Figure 21.24 Villages in **550** with Early Village Clusters, clusters on the Central Plateau, and, in dotted circles, those identified by Lacava (Cardillo, Gravaglione, San Salvatore). (JT)

might have resembled a large village in 600. It was, however, already the home of the gods of the polis and the place where the community of early settlers met as a group on the ikria. All of the securely oldest concentrations of farmhouses, the Early Village Clusters, the complete ones and the possible ones, as noted, are distant from the asty: approximately 11 km (as the crow flies) to Giardinetto, 8.5 to Sant'Angelo, 8 to Giampasquale, and 7 km for Avinella and the Venella. There is a limit to the distance that a farmer would walk to cultivate his field. For example, for the high plains of Arcadia in the Peloponnesos, near the traditional homeland of those who settled the Metapontino, it has been empirically determined to be between 5 and 7 km.[97] The Early Village Clusters at Avinella and the Venella were thus at the limit and the others went beyond. Those of the early settlers living in the asty in the late 7th century BC could have cultivated the fields within this range and returned to the city for the night. They may have built tem-

[97] Hodkinson and Hodkinson 1981, 281. Cf. also Jost 1999, 224–25.

porary shelters, but would not have needed to reside permanently near their fields. The villages, in the 6th century at least, must have served the needs of a rural population that could not easily reach the asty.

Distances between villages. The distance between the individual villages is relatively short. Since the Avinella and Venella are located about 1 km up the valleys of the Conca and 2.5 km up the Venella, respectively, the easiest land route between them was across country, but the easiest land route to the city was somewhat longer—about 7 km along the Basento Road. Giardinetto, the most distant village, is separated by 2 km from Giampasquale, which is 8.5 km from the asty along the Bradano. We cannot be sure that there were not clusters farther inland beyond the study area to the west, for example at Cozzo Presepe, the site of a large partially excavated village. It was already a village in the Iron Age (see Ch. 17).[98]

On the Basento side, in the first phase of settlement, Sant'Angelo, the farthest from the city at 8.5 km, is about 2.5 km from Avinella, which in turn is 1.5 km from the village in the Venella. The three villages that appeared in the Central Plateau for the first time in 550, Galitelli, Casa Preti, and Capistretti are, on the southern side of the Plateau, more closely spaced. Only 1 km separates Galitelli from the Venella, and 0.8 km from Casa Preti. The longer intervals—1.5 km between Casa Preti and Capistretti and 1.2 km between Capistretti and Giampasquale on the northern side of the Plateau—are comparable to that between Avinella and the Venella. The farmhouses that compose these villages are generally more closely spaced, and, in the case of Casa Preti, appear to have had an area for burials separate from the farmhouses. The rather looser association of farmhouses lining the edges of the valley in the upper Venella in 550 was separated by greater distances from the chora and the asty. Its center is 8.5 km from the asty, 4.5 km from Giardinetto on the Bradano, 3 km from Sant'Angelo on the Basento, and 1.5 km from the Venella village on the middle stretch of that valley.

Distances between farmhouses in the village clusters. The space between individual components of the majority of the village clusters is greater than what might be expected in the sense in which "village" is generally understood. They are not like Medieval agricultural villages,[99] or the modern Mediterranean ones where the houses are huddled together, often with the church in the center, and with the cultivated land surrounding. In the chora of Metaponto, with some possible exceptions, the basic unit is not the house with distant fields, but the agricultural plot, the *kleros*. The buffer between farmhouses varies, but it averages between 300 and 400 m—though in the special case of pairs of farmhouses, for example, it is much lower. There would have been, in theory, enough cultivable land around most to provide a subsistence living or better. We do not suppose that this was the only land to which the occupants of the farmhouses had access. The villages are placed so as to be conveniently near more extensive areas of relatively level land in the Central Plateau—easily within the 5 to 7 km range suggested by the studies of Arcadia,[100] arguably the homeland of at least some of the settlers. The villages of the Metapontino in the first century of settlement behave like secondary central places, as satellites of the asty, but being distant from the asty, and in some cases well beyond the 7-km limit, they must have enjoyed a fair degree of independence.

Function

Agricultural functions. The Avinella, the Venella, and upper Venella clusters are located along valleys that lend themselves ideally, but not exclusively, to pasturing animals. They are also adjacent to fertile, wide, and relatively level plateaus. The more gentle slope and southern exposure of Sant'Angelo would have been suitable for raising both grapes and olives.

Ceramic production. The special functions that villages appear to have include the production of ceramics, indicated in the survey by the presence of ceramic slag and misfired vessels. A number of the sites identified as F/K and simply as kiln, with no indication of a residential structure, occur in villages from an early period. Of course, there is no requirement for a kiln to be in the middle of a population center.[101] Ceramic production in villages and on more isolated sites could have served not only "daily needs," to use Aristotle's phrase, but could also have provided a surplus for trade.

[98] It was occupied for much of the 6th c. only to be deserted and reoccupied in the 4th c. BC by Lucanian villagers. *Cozzo Presepe*, 207–43.

[99] M. Johnson 1996; see n. 18, above.

[100] Jost 1999, 225.

[101] Even today, traditional kilns in southern Italy can sometimes be found in the middle of fields, convenient to clay deposits.

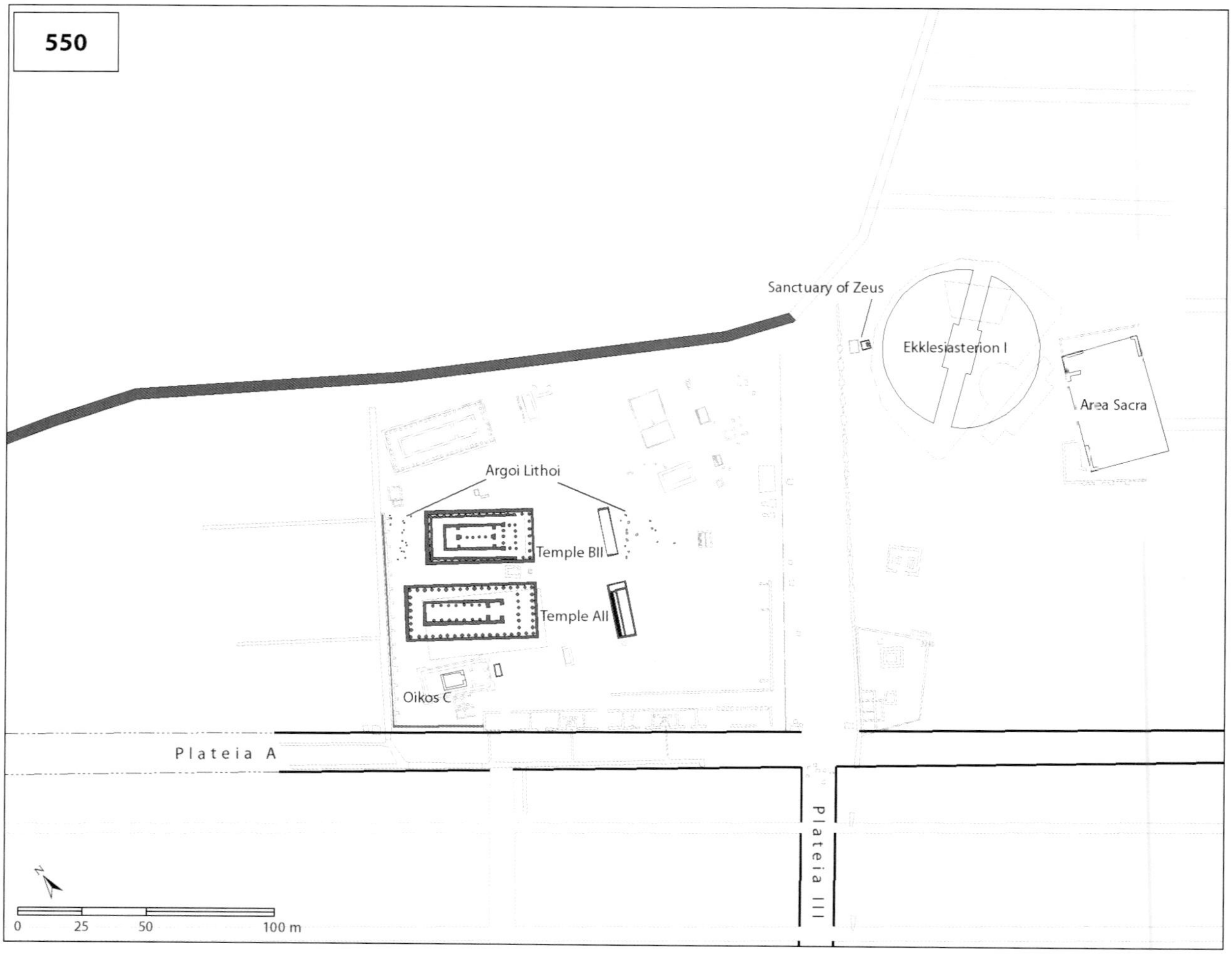

Figure 21.25 Asty in **550**, plan and detail of sanctuary and agora. (JT based on Mertens 2006, Fig. 270)

Religious functions. Sites 270S in Giampasquale, 297S in Avinella, and sites 334S and 477S in Venella clearly had a religious function that set them off from the typical farmhouses and would seem to have exceeded the daily needs of a single family. At the same time they were much less important than the large rural sanctuaries, like Pantanello and San Biagio, that functioned as extensions of the cults of the asty in the chora. They may be expanded cults of the original family or families that first settled these villages.[102] If so, why do we not see similar cult places at Giardinetto and Sant'Angelo? Why would some villages have had a religious function and others not?

A sanctuary may well have existed at some point at or near Sant'Angelo.[103] The failure to find any traces of a cult place at Giardinetto, I believe, was not the result of chance or because they were overlooked. Giardinetto was one of the best-preserved and most thoroughly surveyed villages in the study area. We cannot, however, exclude the possibility that there may have been a sanctuary downslope, buried under the alluvium of the Bradano valley floor.

Villagers. Who, after all, lived in the villages? Up to now we have assumed that they were Greeks. This question is rendered more complex in the context of Western Greece and Magna Grecia by the well-documented presence of indigenous people in the same areas settled by the Greeks, and sometimes the very same sites. Without the evidence of excavated buri-

102 There is some evidence of family-based religions observances of a more formal sort in the form of disks with the symbols of family groups, the so-called *dischi fittili* for phratry festivals at Metaponto. Loprete and Bini (1989) discuss the evidence for Herakleia and Taras as well as Metaponto.

103 See Sanctuaries, above. See also Carter 2006a, 147–48; 1994a, 180–81.

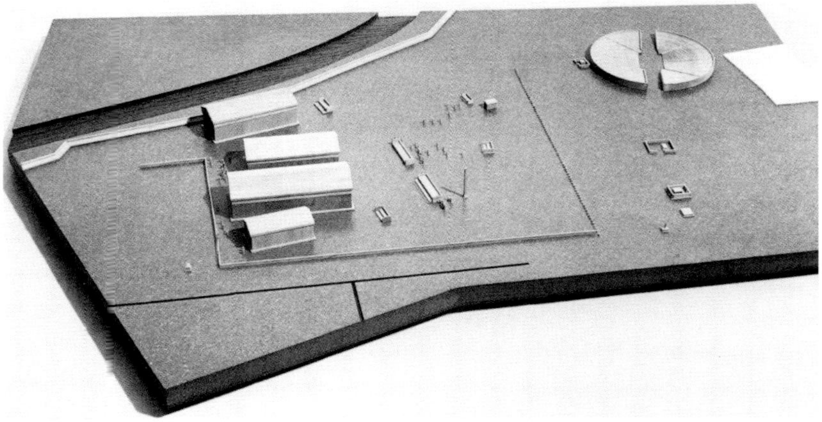

Figure 21.26 Model of sanctuary and agora, **550**. (After Mertens 2006, Fig. 274; courtesy Dieter Mertens)

als, the question of whether the early occupants were Greeks or indigenous peoples using mainly Greek-style pottery cannot be finally resolved.

Though some doubts may linger about the Greekness of the villages, the fact remains that apart from Incoronata and Cozzo Presepe (Ch. 17), there is no evidence of a consistent indigenous presence in the area of the future chora before the arrival of the Greeks in the second half of the 7th century BC. No Greek village rose on or near the site of an earlier indigenous settlement, as was clearly the case at Amastuola in the chora of Taras (see above) and in the chora of Sybaris at Francavilla,[104] nor is there at Metaponto any sign of a settlement after the mid-7th century[105] of a subaltern or slave population as in the chora of Poseidonia.[106] We can conclude, in the absence of any evidence to the contrary, that the occupants of the villages were Metapontines, who worked their own fields (*autourgoi*) as the Hodkinsons have concluded for the Greeks of the area around Mantineia.[107]

The Asty in 550

While the chora was expanding in **550**, the asty was acquiring a monumental appearance (*Figs. 21.25* and *21.26*). Two imposing peripteral temples were erected (*Figs. 21.27–29*), as was an equally, if not more impressive, place of public assembly that was unique for this early period and unrivaled in any period in the Greek world for its sheer size (*Fig. 21.31*). Less immediately obvious, but fundamental, was the plan of the city, which was organized on geometric principles (see Street Plan and *Fig. 21.30,* below). The plan's impact was felt throughout the city at all levels of society.

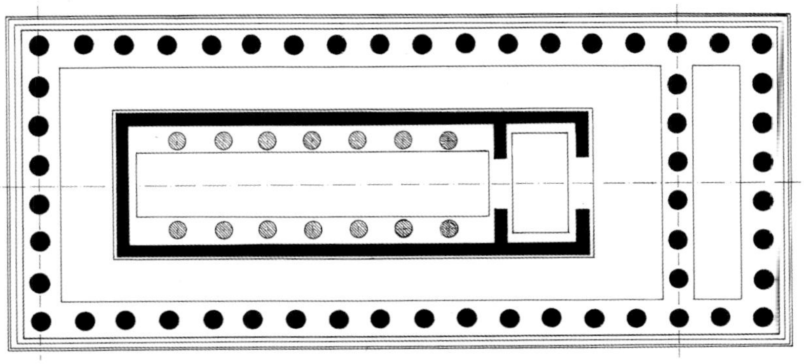

Figure 21.27 Plan of Temple of Hera II (A2). (After Mertens 2006, Fig. 260; courtesy Dieter Mertens)

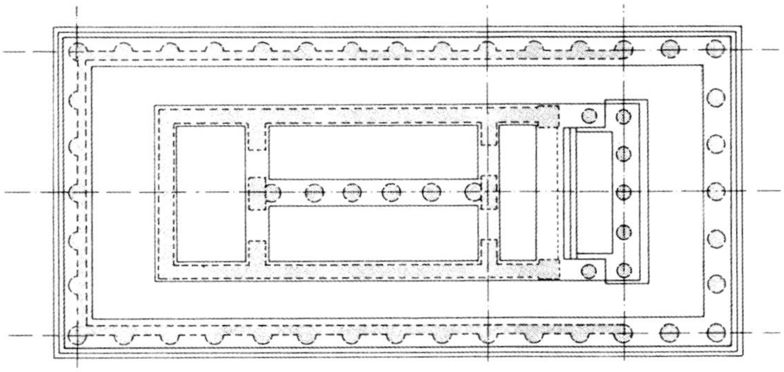

Figure 21.28 Plan of Temple of Apollo II (B2). (After Mertens 2006, Fig. 261; courtesy Dieter Mertens)

The challenges faced by the inhabitants of the ancient city and chora were very much the same: how to organize the land in clearly demarcated portions, how to provide adequate thoroughfares for the transportation of goods, animals, and people, and, especially pressing at Metaponto, though not a unique problem in this region, how to drain the excess water from the surface of the asty and the fields of the chora.[108]

[104] Maaskant-Kleibrink 2000, 165–84. Carter 1994a, 175; Carter 2006a, 74–78.

[105] Cf. Ch. 17, the enchytrismos burials at Incoronata Azienda Agricola.

[106] Avagliano 1988, 329–30; E. Greco 2001, 186–88.

[107] Hodkinson and Hodkinson 1981, 279.

[108] Mertens 1998, 132.

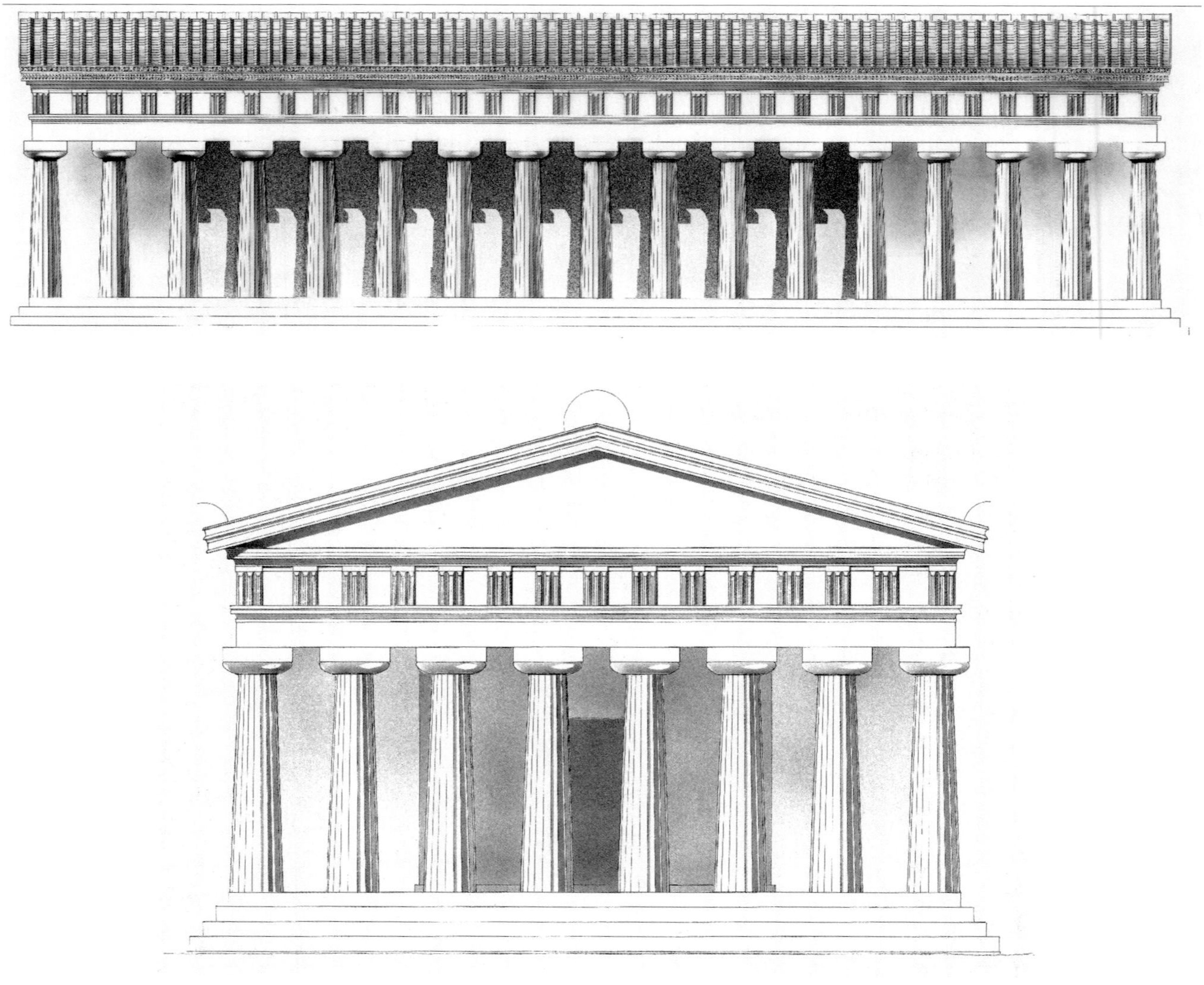

Figure 21.29 Temple of Hera II (A2) reconstruction. (After Mertens 2006, Figs. 262 and 263; courtesy Dieter Mertens)

The early buildings with important functions for the community were located on higher ground, in the north side of the area that was enclosed by the Archaic city wall. The area of the urban sanctuary was approximately 2.3 ha, and that of the agora, 7 ha (*Fig. 21.25*). The rest of the urban space (roughly 150 ha) was available for the dwellings of the inhabitants and for their workplaces (e.g. the kerameikos). It has been suggested that this large area was rather sparsely occupied, with settlement limited to the higher, drier sites, as efforts were made to drain the pools of stagnant water via ditches that emptied into the flanking rivers. The long-anticipated excavation of the residential quarters (on land expropriated already in the 1990s) will, it is hoped, resolve this uncertainty and much else about the development of Metaponto.

The Temples

The sanctuary area began to fill up in the first half of the 6th century BC (*Fig. 21.25*), following the construction of Oikos C—the first structure in the asty not primarily of wood—and its altar in about 600 BC. Other cult buildings were added, including Oikos D on the site of the future peripteral Temple D. Construction had begun on the first of a pair of temples in the Doric order, Temple of Hera I (A1),[109] which was envisioned on a large scale already by 570–560 BC. It was oriented to the east, with its east-west axis at an angle of approximately 57° west of north—close but not quite parallel to Oikos C at approximately 61° west of north. Work soon stopped on Hera I after

[109] Temple A1 is thought to have been dedicated, like its successor, Temple A2, to Hera, the principal divinity of Achaian Greek foundations in southern Italy (including, certainly, Croton and Poseidonia).

a few courses of the peristyle had been laid; it is not known why construction was interrupted. According to Mertens, work was begun "slightly later" on a second monumental temple, Apollo I (B1), Apollo Lykeios,[110] beside it, which had an appreciably different, more northerly orientation.[111] Work on Apollo I progressed as far as foundations for the inner colonnade before work was stopped, again for unknown reasons. The altars of both unfinished temples maintained the orientation of the altar of the already venerable Oikos C.

As Mertens has demonstrated, the orientation of Apollo I is the first and most visible evidence of a new era in the asty. The temple was aligned with the major east-west axis through the city, as designated by Plateia A, which runs from the West Gate (Porta Settembrini) to the agora.[112] This, as both De Siena (who excavated soundings along it) and Mertens have urged, was a major thoroughfare from the beginning, despite the fact that the earliest road surface to survive does not date before the 5th century BC.[113] De Siena argues that the earlier surfaces were removed in successive operations of leveling and filling, which were part of the ongoing effort to channel water out of the city and into the rivers (see Ch. 23).[114]

In about 540 BC, some twenty or thirty years after ground had been broken for Hera I, work resumed on the site and the building was quickly completed (*Figs. 21.26* and *21.27*), making use of monolithic column drums prepared for its neighbor to the north, Apollo I. The new building, Hera II, was indeed impressive, measuring 20.55 x 19.82 m, with eight columns in two rows across its eastern façade, and an interior colonnade on each side of the cella.

As Mertens has shown,[115] this type of plan was characteristic of Sicilian temples, especially that of the Temple of Apollo at Syracuse. Hera II also shared another feature peculiar to the temple at Syracuse—a dedicatory inscription on the principal façade (*Fig. 21.33*). In contrast to its predecessor on the site, Hera II was oriented so as to be almost perfectly parallel with Apollo I and Plateia A; the altar, however, maintained the orientation of its predecessor, as did all the altars of the peripteral temples, past or future in the sanctuary (in what has been termed the "sacred" orientation).[116] The altar was remodeled with a crowning triglyph and metope frieze that ran its whole length of just over 20 m. On the narrow ends were pediments with a Doric entablature. An effective architectural conceit, it accurately replicates the design of the building behind it.[117]

Work on the monumental Temple to Apollo recommenced only slightly later (ca. 530 BC) than that on the Temple of Hera II. The plan was modified by extending the building to the east to bring it more nearly into line with Hera II, and to make room for an additional row of columns across the front (*Fig. 21.28*). The central row of columns and the adyton at the rear were retained from the original plan.

The truly distinctive feature of Apollo II was its complete enclosure, with an exterior wall punctuated by engaged half columns on three sides—their first use in the history of Greek architecture, according to Mertens.[118] This created in effect a series of four naves, approached through the open colonnade of the eastern facade. What rites and functions this was designed to accommodate must remain a matter of speculation, but oracular performance may well have been among them. This was surely a major feature of the later 5th-century BC Manteion of Apollo in the agora (see Ch. 23).

The Street Plan

The temples, grand though they were, were merely one element of an all-embracing reorganization and redesign of the city,[119] along with a new plan that was not merely functional, but also highly symbolic of a new order. This was reflected in the orthogonal street grid that came into being at this time (*Fig. 21.30*), which characterized the city for the rest of its existence as a Greek asty and even into the period of Roman domination (2nd and 1st centuries BC).[120]

At first, the investigations of the plan in the 1950s were, like those of the division lines (and sites) of the

110 Temple B1 was dedicated, as De Siena has shown, to Apollo Lykeios (De Siena 1998, 162). The proof is the large number of rough stones (*argoi lithoi*) and stelai found at both ends of the temple: some were inscribed to Apollo or with the name of the dedicant (Doepner 2002). After nearly two centuries, the erroneous identification of Apollo with Temple A has at last been corrected.

111 Mertens 2006, 157.

112 Adamesteanu 1987.

113 Mertens 1998, 127.

114 De Siena 1998, 160.

115 Mertens 2006, 151.

116 Mertens 1999, 275.

117 Mertens 2006, 155.

118 Mertens 2006, 153.

119 Mertens 2006, 158.

120 De Siena 1998, 159–61.

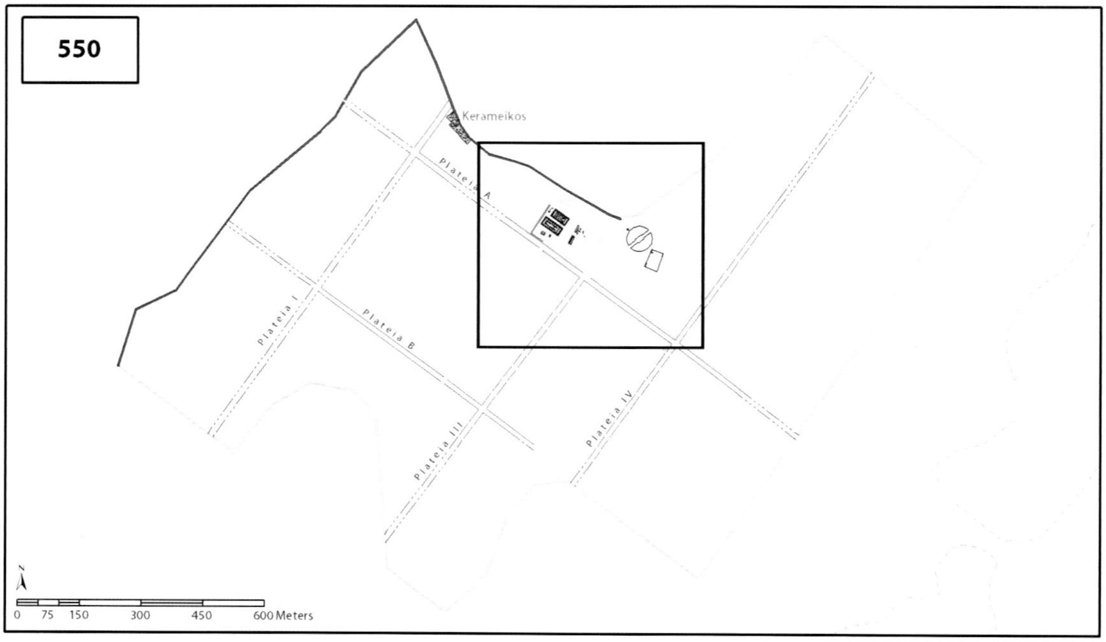

Figure 21.30 Street plan of the asty. (JT based on Mertens 2006, Fig. 276)

chora, almost totally reliant on the evidence of aerial photography. The features of the city plan, visible as anomalies in the black and white photographs, included the walls, wide streets (plateiai), narrow streets (stenopoi), and the outlines of the castrum (see Ch. 26). There were gaps and nothing could be determined from the photos about the dates of these features without excavation, which up to this point had been limited to the "Temple of Apollo" (Temple of Hera II). Early reconstructions of the plan show a grid of long narrow strips, interpreted as insulae 35 m wide, which were separated by the stenopoi, oriented SE–NW on an axis approximately 50° west of north. According to the interpretations of the photographs, these were intersected at right angles by as many (in some accounts) as five plateiai, the broad streets running NE–SW.[121] SE–NW-running plateiai were also recognized, and in the case of Plateia A, actually revealed through excavation. Investigations in the intervening years have clarified aspects of the plan but left many points in doubt. The two westernmost NE–SW plateiai (I and II), for example, have been shown to be no earlier than the 4th century BC, when the asty underwent a major reorganization (see Ch. 25).[122] A recent extensive campaign of geophysical prospection has added much detail to the plan,[123] and well-targeted soundings in the unexcavated residential quarter have revealed Plateia B, located 386 m to the southwest and parallel to Plateia A.

How much of this plan originated in the mid-6th century? Two main elements, the principal axes and the size of the blocks, were established at this time. The major NE–SW Plateia III measures 22 m across (compared to the 15 m width of all the others). This is generally agreed to have been the principal artery from the beginning of the city connecting the residential quarter with the agora and separating the agora and sanctuary areas, thus more important even than Plateia A. The two of them together define the grid insofar as it is presently known. The possibility remains, however, that Plateia A was not as important as it appears, and that the other principal generator of the grid was located at some remove in the still unexcavated residential quarter.

The dimensions of the agora were defined, and so at least was one dimension of a city block. The agora stretched from east to west 263 m,[124] between Plateiai III and IV, and that was the length of a city block in the north central section of the asty (*Fig.*

121 A limited geophysical investigation in the 1960s helped to confirm the results of the aerial photographic studies.

122 Mertens 2006, 159–60; Nava 1999, 698–703.

123 Mertens 2006, Fig. 277; Nava 2002.

124 De Siena 1998, 159–65.

21.30).[125] What were the NE–SW dimensions of a city block? Was the system of narrow strips separated by stenopoi 5.5–6 m wide (with blocks 35 m in width and various lengths, including the now firmly established 263 m for the blocks south of the agora) an invention of the 6th century BC? Or is it rather the implementation in a later period of a concept that would certainly have appeared anachronistic in the 5th or 4th centuries BC?

The arrangement of parallel narrow strips (*per strigas*) has good early parallels in the Sicilian colonies, foremost among them that of Selinos, the object of a recent and extraordinarily impressive investigation directed by Mertens.[126] The plan, dated to the early 6th century BC, has strips 32.8 m or 100 Sicilian Doric feet wide measured from the axis (that is, including the width of the stenopoi). The "100 footer" width of the block, as Mertens points out, reflects not a street plan, but the underlying division of the land into lots of equal width, in essence a land division.[127]

The similarity of the Metapontine plan *per strigas* and that of its sister colony, Poseidonia (founded by Achaians from Sybaris at about the same time, in the late 7th or early 6th century BC) has long been recognized.[128] The date of this plan, too, has been in doubt, again because of limited confirmation through excavation of the details of the grid, likewise known primarily as a result of aerial photographic investigations. The grid of long, narrow blocks at Poseidonia follows the major NE–SW axis that parallels the agora and sanctuary to the east, as Plateia A at Metaponto, but rotated 90% with respect to the cardinal points. The east-west running plateiai (measuring 12, 18.2 and 10 m wide, respectively, from north to south) divide the plan into three large strips about 300 m wide. The central plateia, like Plateia III at Metaponto, separates the sanctuary with its two great temples of Hera from the agora. This division results in blocks whose length is 273 m, divided one from the other by stenopoi 5 m wide.[129]

The results of soundings in recent years have strengthened the conviction that the plan of Poseidonia is Archaic, dating probably to the last decade of the 6th century BC. The similarity between the two plans is striking. It is perhaps worth noting that none of the peripteral temples at Poseidonia, however, is aligned with the grid of plateiai and stenopoi. Their orientation, instead, seems to result from an earlier road system that linked the north gate of the city, the Porta Aurea, with the south gate, the Porta di Giustizia. An axis perpendicular to this led from the west gate, the Porta Marina, to the Sanctuary of Hera.[130] Neither the plan of Metaponto nor that of Poseidonia—even if realized in part at a later date—show the influence of later developments in urban planning.

The width of city blocks in the Archaic period of Selinos, Poseidonia, and Metaponto (allowing that those strips had their origins in the 6th century BC, and to pick only the most immediately relevant examples) was about 35 m (slightly smaller at Selinos). These may all be related to the division of their respective chora, but the dimensional relations within the current limits of our knowledge, both of divisions in the chorai and the asty, must remain a matter of speculation.

Ekklesiasterion

The enormous building that was erected in the northwest corner of the agora about 550 BC made as strong a statement about the ideals, beliefs, and ambitions of the Metapontines as the two monumental Doric temples. A rectangular orchestra, measuring approximately 12–14 x 18–20 m,[131] was enclosed by a circular wall measuring 62 m in diameter and 2 m high. The space between the two was filled with earth brought in from elsewhere and graded to slope down to the central space that was reached by two dromoi, each 8 m wide, on the short sides of the rectangular space (*Fig. 21.31*). The long axis marked by the dromoi was

[125] Mertens 1999, 284. The easternmost of the four, Plateia IV, has 6th-c. BC levels and together with Plateia III defines the eastern and western limits of the agora. On the south, the agora was limited by the city wall. Its northern limit, at least in the 4th c. BC, would have been the northern extent of the Hellenistic stoai that subsequently marked its southeastern edge, though that is still unknown (Ch. 25).

[126] Mertens 2006, 173–92, esp. 173–75.

[127] Foxhall (2002) has noted that the 210 m squares could be subdivided into 14 strips of 0.315 ha. which she identifies as the amount of land that a pair of oxen could plow in a day, a number that she supports with many examples. It may be a very conservative estimate. A normal day's work load for oxen in the Metapontino in the first half of the 20th c. was about three times this, or about a hectare a day. The same point concerning land divisions was made by Boyd and Jameson (1981) with regard to Halieis in mainland Greece. There, the basic city block was a square of 210 m on a side, a basic measure of agrarian dimensions, not only at Metaponto (see Ch. 28), but also at Chersonesos (Carter et al. 2000, 713). These are the only poleis for which this measure has actually been identified in the chorai.

[128] De Siena 1998, 165.

[129] Mertens 2006, 164–65.

[130] Theodorescu 1988.

[131] Mertens 2006, 161–63.

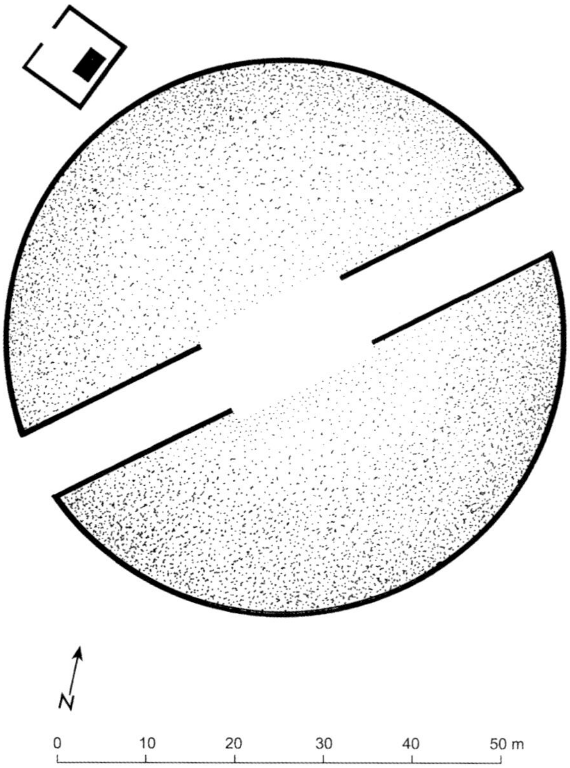

Figure 21.31 Plan of Ekklesiasterion I and Temenos of Zeus.

not aligned with any other obvious feature of the mid-6th–century BC city plan.

Mertens estimates that the structure was designed to seat more or less 7,500.[132] It was the largest place of public assembly ever constructed by the Greeks before the 5th century BC, surpassing the estimated 5,000 person capacity of the Pnyx (ca. 500 BC) in Athens.

The structure's function has been a source of divergent opinions. It was at the time of its discovery baptized as the "Ekklesiasterion," an appellation that has survived despite serious afterthoughts. It was nonetheless understandable since the other circular buildings known at Poseidonia (5th century BC) and Akragas (probably 4th century BC) had, without any question, served as meeting places for the assembly. All are later in date, quite a bit smaller, and located in Magna Grecia or Sicily, perhaps not coincidentally.[133] The principal reasons for second thoughts about its label were the implications that this would have for the constitution of Metaponto in the 6th century BC—and even earlier, if Ekklesiasterion I is considered a monumental version of the ikria on the site in the late 7th century BC. The existence of an assembly of the demos did not fit into the preconceived notion of historians about when and where democratic government by the assembled citizens began.[134]

If not a meeting place for an assembly, then what? It has been suggested that it served for non-political public events such as choral contests or agones,[135] which were widely held throughout the Greek world in the 6th century BC and before.[136] Such events, it has been argued, would have acted as focus for the settlers, binding them more closely together and giving them a sense of community and common culture and background, especially important in the early years as roots were being established in far flung areas of the chora. The facts that one of the very earliest structures, the ikria, probably served this purpose, and that a very great deal of resources were allotted to it in the mid-6th century BC, are proof enough of the importance of the activities that took place there.

Hansen has documented the cases where the assembly was known to have met in the theater,[137] and the possibility that this structure may also have served a political function even at a very early date should not be excluded. Both functions could have been accommodated in this very ample space. The possibility that the impetus for the monumental building of the mid-6th century BC may have come from a tyrant would have been perfectly in keeping with such a person's political ambitions and cultural-religious agenda.

Just outside the circle or the retaining wall of the building, a temenos with an altar and a stele bearing the inscription Διὸς ἀγορά was erected at about the same time as the Ekklesiasterion. This precise meaning of the appellation— literally, "Agora of Zeus"— has been much debated.[138] Is it a horos, a boundary stone referring to the agora as a whole? Does it place the Ekklesiasterion only under the protection of Zeus? Whatever its precise significance, the invoca-

[132] Mertens and De Siena 1982, 30–34.

[133] Hansen and Fischer-Hansen (1994) provide a useful list and discussion. See also Ampolo 1987, 92.

[134] The Athenocentric view has been challenged (E.W. Robinson 1997).

[135] Mertens 2006, 162–63.

[136] See Moretti 2001, 27–35. The fragment of a vase by Sophilos illustrating ikria is cited as iconographical evidence for seating for the agones. Cf. Hansen and Fischer-Hansen 1994, 47, for other vase representations of 6th-c. BC meeting places.

[137] Hansen and Fischer-Hansen 1994, 51–53.

[138] De Siena 1998, 151–56.

tion of Zeus makes whatever went on here of the highest possible importance for the community as a whole. It has been suggested that the Ekklesiasterion was conceived as an agora, that it was a symbol of the community,[139] including naturally those citizens resident in the chora. The wealth generated in its fields, after all, made possible the construction of the temples, the organization of the chora, and the very building itself.

Monumental tombs

A nucleus of some 10 chamber tombs was discovered by De Siena in the heart of the urban necropolis of Crucinia in the early 1990s.[140] These tombs were built of stone and contained burials with sumptuous grave goods dating from the late 7th century BC to about 530 BC. There is nothing like this display of wealth and finery at Metaponto in any other context, urban or rural. Three generations of one or more families were interred here, all adults who lived sometime between the end of the 7th century and the middle of the 6th century BC. The funeral bed (*kline*), sacrificial instruments, the *machaira*, a sword, and oinochoe and phiale in bronze suggest activities connected with sacrifice and purification. The female buried in tomb 238 wore a headdress of gilded silver with relief decorations that included representations of horses and goats. De Siena is certainly correct in suggesting that these were personages of high status who carried out functions that were important to the community, perhaps priests and priestesses.[141] Among the burials was a gilded bronze helmet (*Fig. 21.32*)[142]—again without parallel at Metaponto. Burials with armor of any sort in the Metapontino are unknown at this or any time before the Lucanian burials of the 5th and 4th centuries BC (see Ch. 25).

A connection between these discoveries and documentary evidence—late but almost certainly genuine—for the existence of a tyrant at Metaponto, was soon recognized.[143] He was named in the ancient sources as Archelaos. Though the story of his assassination at the hands of the tyrant slayers

Figure 21.32 Bronze helmet from Crucinia (h 49.2 cm). (Image courtesy of the St. Louis Art Museum)

Antileon and Hipparinos resembles closely the well-documented tradition of the Athenian tyrannicides, Harmodios and Aristogeiton, that is by itself no reason to discount it. De Siena has argued convincingly that two burials, in a chamber with four places, and a superimposed monument base, were the burial places of Metaponto's tyrant slayers.[144]

The tyrant Archelaos has been associated with the self-aggrandizing inscription on the architrave of the Temple of Hera II—αὑτοῖ καὶ γενεί "to himself and his family" (*Fig. 21.33*),[145] and the impetus for the resumed construction of the great temples as well as the monumental place of public assembly has been attributed to him. His panoply, including the helmet, underscored his role as a victorious military leader and the occasion for this would have been present, as the conquest of the territory of Siris by Metaponto and her allies certainly occurred near the mid-6th century BC. The victory would have added to the wealth of the polis and presumably to the popularity of the leader.

The tyrant may have risen to power in consequence of his military success. He may have been the

[139] Mertens 2006, 161–62.

[140] De Siena 1999b, 233–36.

[141] De Siena 1999b, 234. See Guzzo 2008 (2009).

[142] It is now in the St. Louis Art Museum. De Siena 1999b, 235, with bibliography.

[143] Lombardo 1982.

[144] De Siena 1999b, 234.

[145] De Siena 1999b, 236.

Figure 21.33 Architrave, Temple of Hera (A2), with capitals and inscribed architrave (far right). (JCC; courtesy of the Soprintendenza)

guiding force behind the architectural transformation of the city and even the reorganization of the chora at this time, using the same geometric principles that were employed in the asty. We should not, however, as Parker has urged, emphasize tyrants' importance too much "except as agents of a broader tendency of the age."[146] The age of tyrants in Magna Grecia, of Telys of Sybaris, of Kleinias of Croton, and of Aristodemos of Kyme was, moreover, still to come, at the end rather than the middle of the 6th century BC. They were also prominently involved in staseis[147] in the poleis of this time, not simply as demagogic leaders backed by the disenfranchised against the aristocrats, but actually siding in some cases, such as that of Aristodemos of Kyme (Ch. 23), with the aristocrats.[148] With few exceptions (Aristodemos?), they were not noted for their achievements in building or the arts. Archelaos, if he existed, would have been a tyrant in the older tradition of Kleisthenes of Sicyon, or Peisistratos of Athens.

[146] Parker 1996, 67–68.

[147] Berger 1992, 64.

[148] Luraghi 1994.

22
Crisis in the Chora, 500

Joseph Coleman Carter

The Chora, 500

After a half-century or more of phenomenal growth, there were serious problems facing the Metapontines in the period 525–475 BC. This has been confirmed by archaeological evidence from the asty as well as the chora. For one, there was a sharp falling off, if not a total cessation, of building in the city in the areas that are known through excavation (mostly limited to the sanctuary and agora). As noted earlier, the residential quarter, covering all but 5% of the surface of the asty, remains unexcavated. When monumental construction resumed around the end of the first quarter of the 5th century BC, it took a new direction. This shift has been interpreted by the excavators of the city as involving, to some degree, a repudiation of the political and social establishment that had prevailed in the mid-6th century BC. The meager ancient sources (see Stasis, below) speak of upheaval at this time throughout the poleis of Magna Grecia, especially at Sybaris and Croton. These are classic examples of a political crisis (στάσις) that pitted the popular and more democratic elements of these cities against the entrenched aristocracies whose numbers included members of the powerful communities, the συνεδρείαι, of the followers of Pythagoras.[1] Sybaris was destroyed in 510 BC by Croton and its allies, perhaps including Metaponto. This event was followed by an uprising against the Pythagoreans at Croton,[2] which may have resulted in a redistribution of land, among other changes. Though Metaponto is not actually mentioned as an active participant in the conflict but only as the refuge for a prominent fugitive, Pythagoras himself, it would not be surprising at all for Metaponto to have suffered similar shocks to the existing political and social order.

There were also growing problems of another sort. Swampy conditions in the asty seriously affected the quality of life for its inhabitants. This problem was present continuously from the beginning, but became critical in the chora for the first time in this period. The level of the water table rose sometime in the late 6th or early 5th century BC, and this may have been exacerbated by excessive rainfall. A massive project of drainage was successfully completed by the middle of the 5th century. A natural disaster involving the very fields occupied by the rural population would have noticeably affected the settlement pattern. It is enough to compare the map of the chora in **550** with that in **500** (*Maps 6* and *7*, Vol. IV; *Figs. 21.1* and *22.1*) to realize the magnitude of the change. While it is not clear what impact stasis might have had on the population, a serious environmental crisis would surely have affected families residing in the chora and determined the areas in which they could live until it was resolved.

In the reconstruction of the settlement pattern in this period, as in all others, the original chronology has been followed (see Ch. 20).[3] In **500**, the number of significant farmhouse sites dropped precipitously from 64 to 14 (*Appendix C1a* or *Fig 16.5*). It would appear that a major exodus had taken place. The number of farmhouses is actually smaller than in the first years of the expansion in **600**. The necropolis count, however, remained essentially unchanged from the preceding period. This could mean that the farmers continued to frequent family burying grounds, and perhaps even to cultivate some fields, while living elsewhere. The absence of occupied farmhouses does not necessarily indicate that the land was totally abandoned, but perhaps only that it may have been exploited differently. This is one of a number of such questions that will be examined in the present chapter.

Whole areas of the chora previously occupied by farmhouses in **550** now appear to have been abandoned. The most dramatically affected were precisely those where the growth from the early to the mid-6th

[1] Berger 1992, 19–21.

[2] Von Fritz 1940; Minar 1942, Ch. 4.

[3] This period is one of two where the number of farmhouses occupied in the period would differ considerably depending on whether the original chronology of fine wares or the revised chronology is correct. The revised chronology would have resulted in a slightly higher number of farmhouses in **500** than in **550**, and a slight dip in **450**. The resulting narrative would be quite different in **500**, but requires relatively slight modification in **450**.

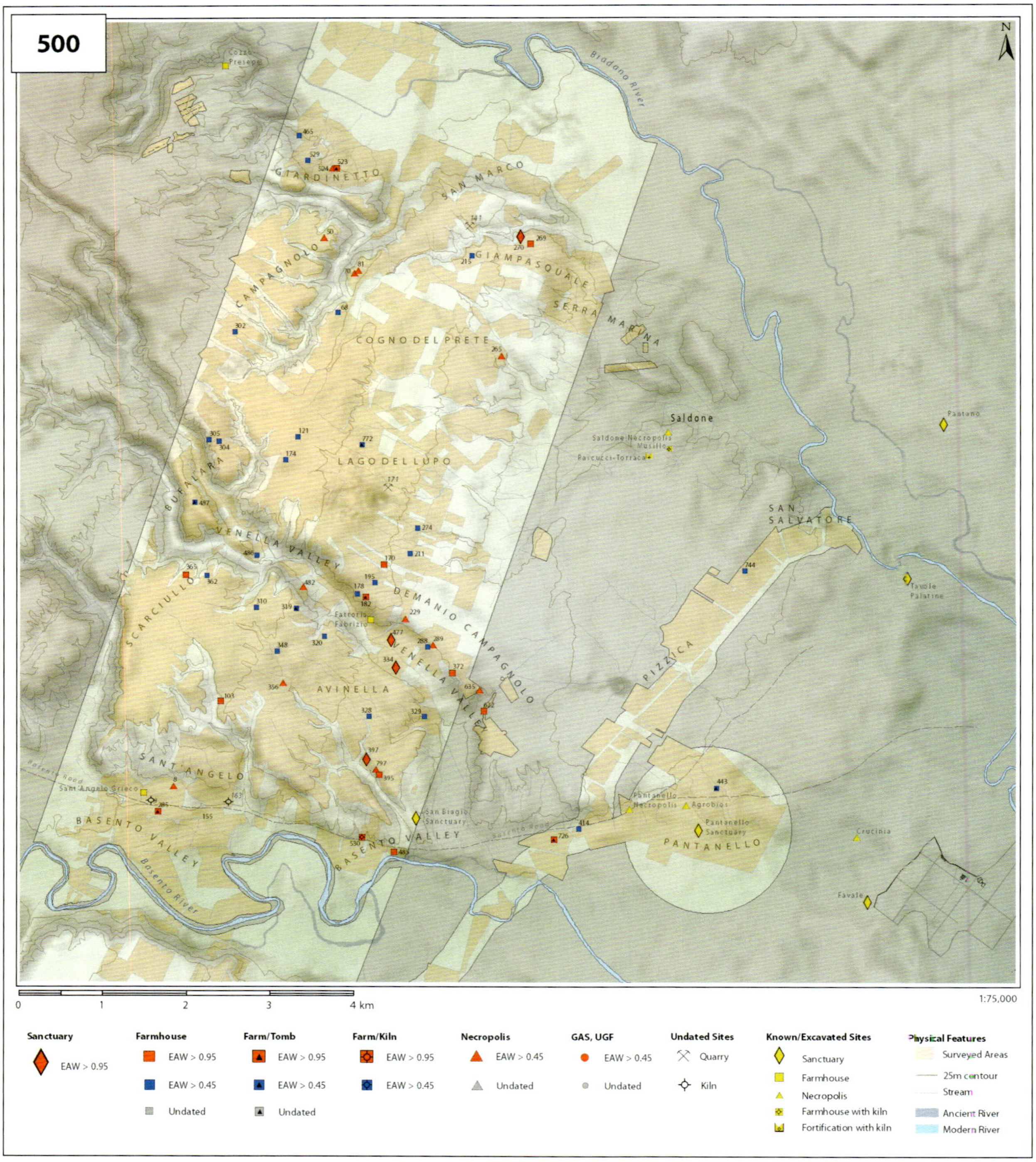

Figure 22.1 The study area in **500**. (JT)

century BC was most apparent—in the area of Lago del Lupo, the upper Venella valley, and the partial clusters at Pantanello, Pizzica, and San Salvatore (cf. *Figs. 21.2* and *22.2*). All the Early Village Clusters were seriously affected, but they did not disappear entirely. Of the 14 farmhouses that were significant in **500**, five were located in the Early Village Clusters of Giardinetto, Sant'Angelo, Avinella, and Giampasquale.[4] An equally large number, which included those in Sant'Angelo, could be found along the Basento valley and the Basento Road.[5] This included the only newly

[4] At Giardinetto, 523F/T; at Sant'Angelo 155F, 285F/T; at Avinella 395F; at Giampasquale 269F.

[5] The sites are 485F, 530F/K, 726F/T, 285F/T, and 155F.

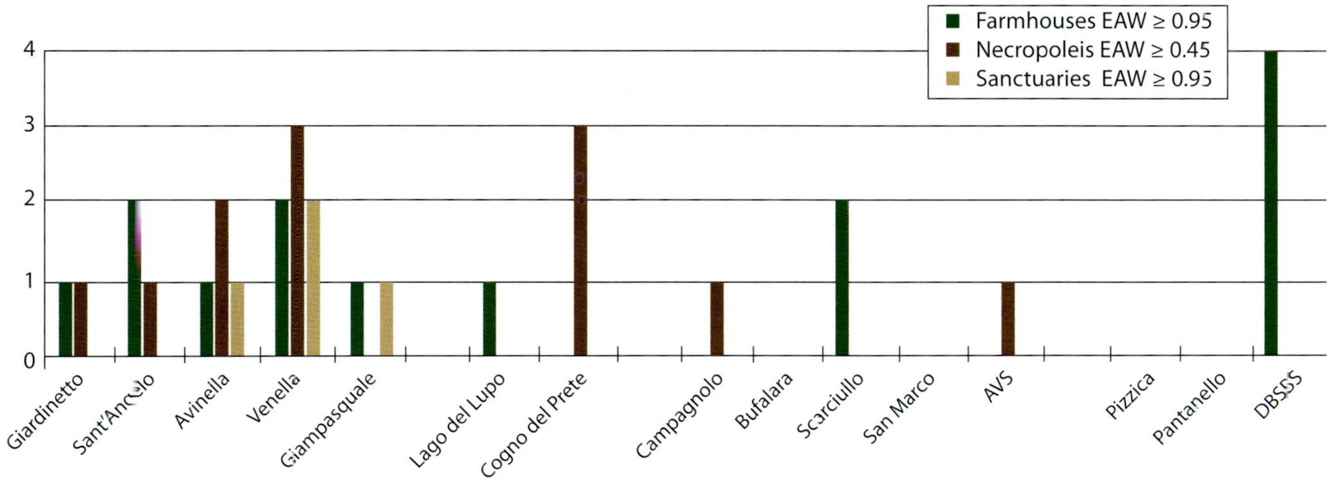

Figure 22.2 Distribution of farmhouse, necropolis, and sanctuary sites, **500**. Totals: significant farmhouses, 14; necropoleis, 12; sanctuaries, 4.

significant site in **500**, site 726F/T. Three were scattered along the length of the east bank of the Venella,[6] and two in the upper Venella, in remote Scarciullo.[7]

Pottery and Dating

The number of significant farmhouses declined from 64 in **550** to 14 in **500**, a decrease of nearly 80% (*Appendix C1a*). This reflects the sudden decrease in the quantity of dateable pottery assigned to the 50-year period between 525 and 475 BC. The total EAW for all farmhouses in the chora fell abruptly from 192.02 in **550** to 46.76 in **500**, a decline of 76% (see *Table 19.6*). This is not due to the way individual fragments of pottery have been dated, since the dating method described in Chapter 4 was designed to smooth gaps that could be produced by the dating of one or even a number of pottery forms.[8] The decline instead is due to much lower rates of production and/or consumption of fine pottery by the occupants of the chora.

Of the 14 significant sites in this period, only two (155F and 523F/T) have an EAW value greater than 2.0 (*Appendix A, 500*).[9] Most of the other 12 are barely at the threshold value of 0.95. The average EAW value in **500** is 1.58, compared with an average of 2.42 in **550** (based on 64 sites).[10] The number of second magnitude farmhouses (EAW between 0.45 and 0.95) also decreased, but less dramatically—from 44 in **550** to 27 in **500** (39%). More telling is the fact that the ratio of second magnitude farmhouses to significant ones was nearly two to one (see *Fig. 19.3*). Was the pottery on these 27 farmhouses indicative of some human activity on the site? Here we face a problem posed by the method of dividing all the dateable pottery into 50-year date bins, and of setting and maintaining an arbitrary threshold (EAW ≥ 0.95) for considering a site active. Frequently, date ranges fall into two (or more) date bins and this is very much the case in **500**. Not enough pottery dated to **550** or to **450** had a portion of its range in **500** large enough to bring the total EAW for more than 14 sites above 0.95.

The second magnitude farmhouses, according to our definition, are not active, nor can we be sure that they were even occupied since almost all were either previously significant or became so subsequently. The pottery that contributes to their EAW values may more accurately reflect activity at these sites in the earlier or later period. The possible exceptions, a handful of second magnitude sites,[11] had these values only in **500**. Only in this case may we be reasonably sure that they were occupied, but using fine pottery at a lower level, and therefore not significant.

[6] 622F, 372F, and 182F.

[7] 365F and 103F.

[8] Buffers of an additional 20 years on either side of the date range were assigned specifically to avoid the effect that too close a conventional dating might create—namely, a sharp break in what might otherwise have been a continuous evolution of black-gloss forms.

[9] Of the pottery fragments with date ranges overlapping the period **500**, only about a quarter have date ranges that are mainly in that period. Approximately 38% have date ranges that cover mainly the half-century **550**, and another 30%, the period to **450**. Another 7% has its weight evenly divided between the half-centuries around **500**.

[10] Without the two high values of sites 155F (6.31) and 523F/T (2.31), the average drops to 1.12.

[11] 329F, 362F, 443F/T, 487F/T, and 529F. See Appendix A.

Areas Affected

A detailed consideration of the areas and the sites that continued to exist, albeit much more modestly, in **500** may yield some clues as to the reasons for the apparent abandonment of the chora at this time.

Giardinetto. The whole eastern half of the cluster, which was densely populated in **550**, was abandoned by **500**. The sole remaining significant farmhouse is 523F/T, on high ground (52 masl) in the center of the bowl-shaped cluster. There are two second magnitude farmhouses: 529F, a new one, and 465F, one of the two significant farmhouses in **600** now reduced to this status in **500**. The eastern half of the cluster on the lower terraces with a spring had previously attracted more significant farmhouses. Not one was left in **500**: the survivors are in the western half on higher ground (*Fig. 22.3*).

Figure 22.3 Site 189, an unidentified and undated Greek feature (UGF) in Giardinetto, looking north across the Bradano valley. (CDA)

Sant'Angelo. Sant'Angelo was the most densely inhabited of the original settlements of **600** with the largest concentration of significant farmhouses. There were seven in **550**, but only two (285F/T and 155F) were significant in **500**. No other farmhouse could even meet the minimum threshold of a second magnitude farmhouse. Necropolis 8N, however, was active, as it had been in **550** and would continue to be in **450**. Whereas at Giardinetto the lower-lying sites were abandoned, the reverse occurred at Sant'Angelo. The two surviving sites were along the valley's edge at 19 to 23 masl. They lay beside the Basento Road, which continued to attract sites in this period as in the preceding half-century. The Basento valley stands out in **500** as one of two parts of the study area—the Venella being the other—that show unusual persistence in the face of adversity. Perhaps it is not coincidental that these areas were among the earliest to be settled (e.g., 285F/T and 530F/K near the mouth of the Venella and earliest of all, the Sanctuary of Artemis on the Venella: see Ch. 20, The Chora in **600**, An Alternate View).

Avinella. The desertion of the slopes overlooking the Conca both at the three forks and in the upper stretches was all but complete by **500**. These sites in **550** were all at high elevations (58–60 masl) in the lower Conca; and at 91–97 masl in the upper. Of the 13 significant farmhouses in Avinella active in **550**, only 395F remained in **500**. The sanctuary 397S and the nearby necropolis 797N, which surely was associated with this farmhouse and first came into use in **500**, are the only other sites with an EAW value greater than 0.95. The necropolis 356N in the upper Conca was in use in **550** and again in **500**, after which it was abandoned forever. Three necropoleis served the combined Avinella settlement in **550**. In **500** that number was reduced by one; still, their survival rate was much better than that of the farmhouses. As at Giardinetto and Sant'Angelo, so at Avinella: the survivors were the earliest significant sites.

Venella (*Fig. 22.4*). The low-lying, lush valley that proved so attractive to the first settlers seems to have lost its appeal a century later, at least temporarily. In this we may see reflected a decline in pastoralism as it yielded to cereal farming, which itself was facing problems.

The settlements on the west and east sides of the Venella had very different histories in **500**, again

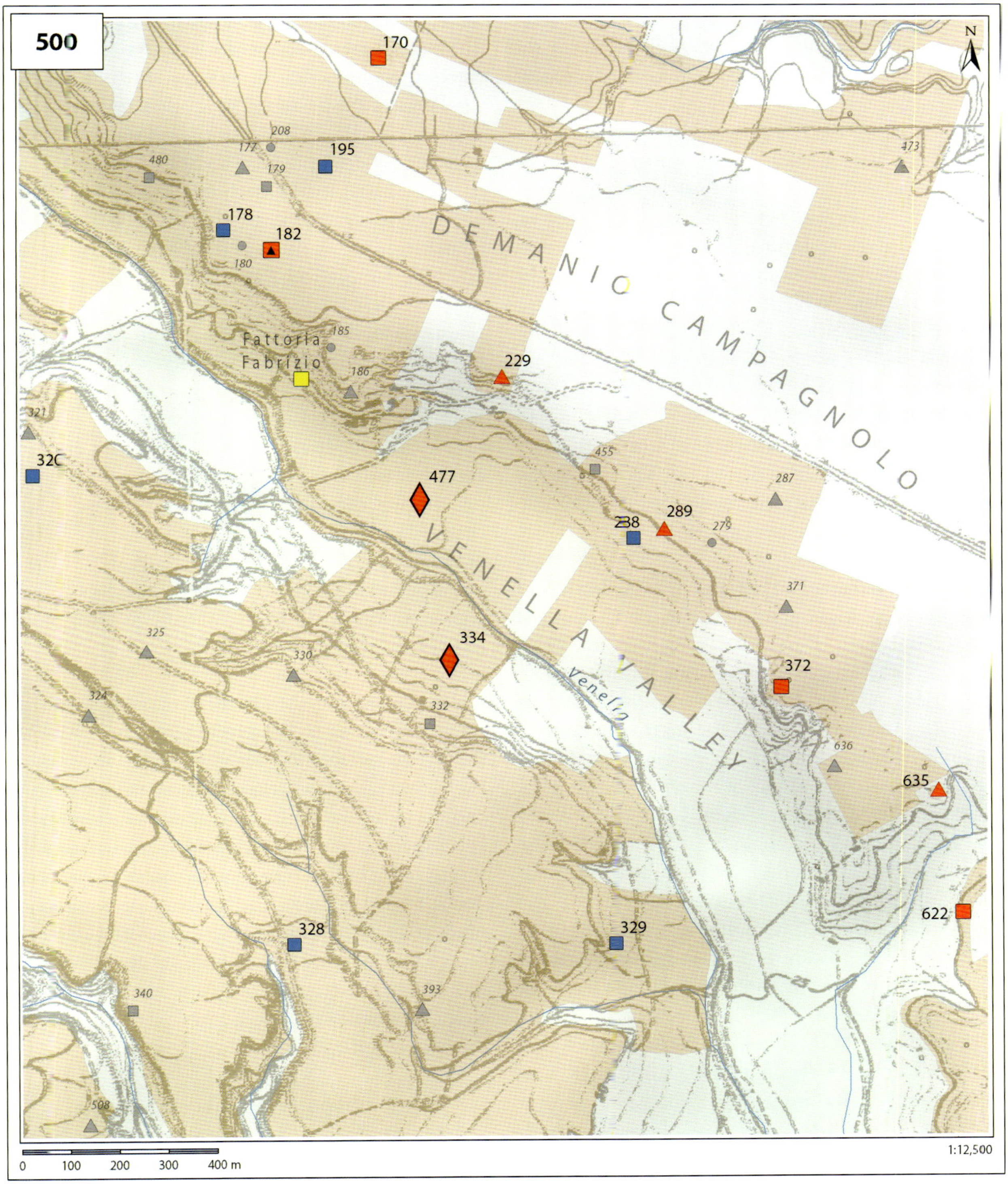

Figure 22.4 Venella in **500**. See Fig. 22.1 for map key. (JT)

demonstrating a certain independence. The west bank settlement could count three significant farmhouses in **550**. One of these, 320F, the least significant in **550**, dropped to second magnitude status in **500**; the others no longer appeared even as second magnitude sites.

On the east, where farmhouses stretched out along the valley's edge, two of the least significant farmhouses in **550**, 372F and 182F/T, survived as significant in **500**. It would appear that having a low value in **550** in east Venella was not a disadvantage in

500. The contrast with the farmhouses on the west is instructive. The previously well-to-do, except for 320F, were no longer even second magnitude. Location also mattered.

The reason for the relative well-being of the farmhouses of east Venella even in difficult times could be their position at the edges of a well-drained plateau with a favorable exposure to the sun.[12] The sites on the western edges of the valley were less favored in this respect. The selective abandonment of sites on the west banks of the Avinella and the Venella perhaps had a similar motivation. The two sanctuary sites remaining in the valley surely owed their continued existence to their roles as places of worship for the farmers who continued to hold land in the surrounding countryside. Two other nearby significant farmhouses in **500** (170F and 622F) outside the Venella cluster proper demonstrate a tenacity evidenced elsewhere in the chora only along the Basento Road.

The Central Plateau. Not a single significant site of the eight occupying Lago del Lupo or Cogno del Prete in **550** remained in **500**. Three of them, 211F, 274F, and 266S, existed as second magnitude sites to return in **450** and later. In fact, with the sole exception of site 170F, the entire area of the marine terraces in the Central Plateau and farther east in Pizzica and San Salvatore had no significant farmhouses. Even the pioneer farmhouse 43F in Cogno Prete disappeared below the 0.45 EAW threshold in **500**, as did all of the second magnitude farmhouses of the previous period.

In Galitelli, the southernmost cluster of the Central Plateau (see Ch. 21 and *Fig. 21.12*), the two sites that were significant in **550** were either reduced to second magnitude or disappeared completely in **500**. The situation was much more drastic for all the sites in the Casa Preti cluster. At the three sites that were significant in **550**, 224F, 503F, and 222F, and the three second magnitude sites, almost no fine ware with a date range falling in this period was recovered.

In the Capistretti cluster it was much the same: the significant site 258F was not documented by a single sherd of fine ware after **550** for another century. Only the necropolis 265N may have continued in use in this period. The pattern here also appears to be related to the geographical location of the sites. The sites on the south side of the Central Plateau, like significant site 170F, managed to survive slightly better than the rest. The central area, where the Casa Preti cluster was located, was and is rich in springs, as noted in Chapter 21. Perhaps it was an excess of the very element that attracted settlers in **550** that caused them to retreat in **500** (see Environmental Conditions, below).

The almost total desertion of the most agriculturally productive area of the chora is brought home by a comparison of the maps of the Central Plateau in **550** and **500** (see *Figs. 21.1* and *22.1*).

Giampasquale. Of the four farmhouses along the valley of the Fosso San Marco that were significant in **550**, only the exceptional farmhouse 269F remained so in **500**.[13] Several sites were completely without fine pottery in **500**. These sites shared the fate of others in the lower and upper Conca, which lined a deep valley penetrating the area of the upper plateau. Many of these likewise were virtually without any fine ware dateable to **500**. The early sanctuary 270S, the only one documented by the survey of this side of the chora, continued to be significant in **500**, but soon lost that distinction.

Isolated sites (*Fig. 22.1*). The remaining two significant farmhouses in Scarciullo (365F and 103F), and the active necropoleis 70N and 81N in Cogno del Prete, were far from their nearest neighbors in **500** and for most of their history. Their isolation may have been their salvation, with space and inaccessibility acting as a buffer during the probable conflicts in Metapontine society of this time. Scarciullo, surrounded on four sides by deep ravines and steep bluffs, is perhaps the most remote part of the study area. Although these were previously prosperous areas of the chora, Scarciullo and the Central Plateau rank among the least significant in **500**.

Farmhouses and Geomorphology

The 14 farmhouses that were significant in **500** were located mainly in two of the four principal geomorphological units (*Appendix C11a*), the Pleistocene marine terraces (seven sites or 50% of the total) and the alluvial terraces (four sites or 29%). The proportions do not differ significantly from the previous two 50-year periods, but in the next period, **450**, there are major shifts to higher ground, with 72% of significant

[12] Farmers who use this area today have commented that crops ripen two weeks before those in the Venella valley itself.

[13] Site 269F was continuously occupied and significant from **550–200**.

farmhouses on the marine terrace and only 12% in the river valleys. The permeable marine terraces, as discussed earlier, are particularly suited to the cultivation of winter-sown cereals; the thicker soils of the alluvial valley floors, to a wide range of crops including cereals; and the valley walls, to pastoralism or olive production.[14] The fertility and abundant springs of the limited area of the mass movement deposit, included entirely within the Giardinetto locality, were a magnet for Greek settlers.[15]

Given the small number of sites involved, the percentages should be viewed with circumspection, but, viewed in context, three longer-term trends emerge: (1) the declining significance of valley walls that begins in **500** and continues until the Roman period, which we may interpret as reflecting a decline in pastoralism; (2) the relatively high significance of the river valleys until **500**—but not afterwards; (3) the increasing importance of the marine terraces. Their average FMCE values in **500** is below that for the chora as a whole. From **450** onward it increased until by the Roman period it was well above average. The mass movement deposit in Giardinetto is the only geomorphic unit that never saw a noteworthy decline in its FMCE value over the whole history of the Greek occupation of the chora and beyond (*Appendix C7*).

Necropoleis

The farm population declined by at least 80% in **500**. The number of active necropoleis, however, remained almost the same as in the preceding period. There were 12 necropoleis and 14 farmhouses (including four with burials). The ratio of burials to farmhouses of roughly 1:1 is without parallel in the history of the chora (see Ch. 21, Necropoleis).

These active necropoleis are located, for the most part, in areas where there were populous settlements in **550**, and in some instances where settlement survived to some degree in **500** (*Fig. 22.1*). Necropoleis 229N and 289N are located in Venella not far from the sanctuary 477S and 182F/T. This area, as we have seen, showed greater resilience in **500** than most. The necropolis 797N in Avinella lies about 70 m distant from the exceptionally long-lived farmhouse 395F, which was active in **500**, and it is less than 200 m from the sanctuary 397S.

The early established clusters in Giardinetto and Sant'Angelo persisted in **500**. Each had an active necropolis, 524N and 8N, and each had a farmhouse with associated burials, 523F/T and 285F/T. Burials were associated with the only farmhouse to become significant in **500**, 726F/T, along the Basento Road.

There are few isolated necropoleis in **500**. The group of three necropoleis in Cogno del Prete and Campagnolo (50N, 70N, and 81N), on the opposite bank of the deep gorge of the Acqua Fetente, as noted earlier, may always have been truly isolated. They may have served families living elsewhere. Nothing may have changed in **500**. That is less clearly the case with 265N, which lies within the Capistretti cluster.

The lone necropolis, 482N, in the valley floor of the upper Venella seems truly isolated. The nearest Greek farmhouses are above, on the edge of the valley and at least 300 m distant.[16] The persistent use of necropoleis in areas where the resident population had been decimated would indicate that the families that had originally settled these areas, particularly the Early Village Clusters, continued to frequent their lands.

In terms of the sheer quantity of dateable artifacts associated with necropoleis, the trend of increasing numbers is decisively interrupted in **500** (*Tables 19.5* and *19.6*). From the equivalent of 23.31 fine-ware artifacts for all 13 necropoleis in **550**, the number falls 61% to 11.75 for 12 necropoleis in **500**.

Excavated necropoleis in the countryside. Examples of excavated necropoleis and burials of this period in the chora indicate that only adults were buried, often with expensive grave goods and containers (*Figs. 22.5* and *22.6*), and that burials were not numerous compared to the following periods. They give the impression of belonging to an elite few. This result is based on a large sample, the Pantanello necropolis, which is closer to the asty than the necropoleis of the survey.[17] It will be interesting to see how this sample compares with the much larger one from areas immediately around the city: the Crucinia necropolis excavated in the 1990s and the burials along the coastal road, SS106, excavated in the last few years (2006–present).[18] The families who continued to use the necropoleis actively in **500** may have resided in the city, as has been suggested. A migration to the city in this period seems

[14] See Delano Smith 1979, 166–72, on land use.

[15] It was in the Late Roman period (see *Fig. 32.4*) but not in the Medieval period (see *Fig. 34.2*).

[16] The farmhouses in the valley are Medieval; see Ch. 34, 1165.

[17] Carter 1998f, 175–91.

[18] See Ch. 25, Urban Necropoleis.

Figure 22.5 Sarcophagus burial T 292 along the Pantanello Road in the Pantanello necropolis, at the moment of its discovery in 1986. (JCC; Carter and Hall 1998, 307)

to have occurred far to the east, in the chora of Taras, after a political upheaval resulted in a defeat for the aristocracy and the rise of a more democratic form of government.[19] If there was a similar move at Metaponto it may also have been related to broad political and social changes.

Sanctuaries

Rural sanctuaries in the survey. The four early sanctuaries—397S in Avinella, 334S and 477S in the Venella valley, and 270S in Giampasquale—were not immune from the forces that caused the widespread decline in the resident farming population. To judge by the amounts of fine ware deposited at these sanctuaries dateable to **550** (the equivalent of 83.67 fine-ware vessels) and to **500** (18.79; a decline of 78%) (*Tables 19.5* and *19.6*), these sanctuaries were much less frequented in this period. The drop in EAW values at all four was between 75% and 85% (see *Appendix A*). None of the sanctuaries ceased, but in contrast to the farmhouses, none subsequently regained the status enjoyed in the middle of the 6th century BC. The fluctuation of FMCE values confirms this (*Appendix B*).

[19] Osanna 1992, 14–15.

The small sanctuary 270S in Giampasquale was hardest hit, and dropped from sight after **500**. With the exception of a single fragment dated to the second half of the 4th century BC, no later pottery was found on the site. Of the original four, only one (477S) returned to something of its former importance in **450**. Its FMCE value fell below the average until **400**, as it joined the others in decline. It occupied a favorable position in the Venella valley, beside a spring, but so did its less significant neighbor, the sanctuary 334S. Both of these were located at low elevations and would have been subject to periodic flooding when the Venella rose over its banks, as it often did and does. The sanctuary 397S, high on a bluff overlooking the Conca in Avinella, behaved much like 477S in terms of FMCE values through **450**, after which it resembled 334S. Essentially all were places of very little significance after **450**.

Figure 22.6 The grave goods from T 292 included a bronze mirror supported on an Ionic column (h 30.2 cm), and two black-figure vases, a pelike by the Nikoxenos Painter ca. 510 BC (~ 29.5 cm, diam. 23.3 cm), and a lekythos by the Leagros Group (h 31.8 cm). (CW)

Why did these sanctuaries lose importance in the 5th century BC? The reasons may have to do with their location and size. All were situated in close proximity to the earliest farmhouses, which were established in the chora and all at a considerable distance from the city—270S at roughly 8.5 km; 334S, 477S, and 397S at roughly 6.5 km—unlike the sanctuaries located along major arteries.[20] Equally important, perhaps, is their small size and their lack of architectural pretension. The surface finds do not include a single piece of architectural terracotta. They were an integral part of their respective communities with religious, social, and perhaps political functions, but were probably more modest and limited to smaller social groups.

Although the communities they served did not disappear, they suffered severely. These begin to expand immediately after **500** with large numbers of ex novo farmhouses (see Ch. 23), but can we be sure that the newest farmhouses were occupied by the same families? The evidence of the necropoleis would indicate that many were not. New farmhouses, and above all new necropoleis, characterize **450**. The reason for the sanctuaries' disappearance may indeed be that they were closely tied to the families of the earliest settlers and the groups that formed around them. As those lost importance and began to die out, the religious focus shifted to cults of the renewed and more democratic city, and to the larger rural sanctuaries like those at San Biagio and Pantanello.

Excavated sanctuaries in the countryside. The sanctuary at Pantanello and a handful of others, including one recently discovered at Pantano (see Ch. 20), differ from those identified by the survey in two important respects: first, their sites continued to be a focus of activity and exist throughout the history of the chora, and in several cases well into the period of Roman domination of the area. This was thanks to the excellent springs at their hearts.[21] Second, they were enhanced by monumental buildings and embellished with terracotta revetments. Monumental stone structures were added in the 6th, 5th, and 4th centuries BC. Besides Pantanello and Pantano, the major rural sanctuaries include the Temple of Hera on the Bradano (Tavole Palatine) and the Sanctuary of Artemis and Zeus at San Biagio on the Venella. There may have been others at Sant'Angelo Grieco and Cozzo Presepe, and possibly at Saldone and San Salvatore, though, as noted, all that remains in these cases are tantalizing fragments of monumental architecture reutilized in structures of later date. Virtually all traces of others that were noted in investigations from the 19th century down to the 1970s have disappeared.[22] The larger sanctuaries survived the downturn at **500** and soon regained their former renown. They may have started under the protection of a well-established family in the region, like San Biagio, discussed earlier, but became in the 5th century extensions in a sense of the cults of the city (Ch. 20, Sanctuaries, and Ch. 21, Sanctuaries). They were also located strategically by important all-weather springs with, no doubt, perceived therapeutic properties. As such, they were an important part of a wide community and could survive the possible eclipse of the families whose patronage they enjoyed in the early years of the chora.

Worsening Environmental Conditions in the City and Chora

The problem of excess water in the asty, already an issue from the beginning, grew more serious with time. By the mid-6th century BC there was a system of drains, following the at-first orthogonal street grid (see Ch. 21, Asty). Canals were dug along the streets, and the great NW-SE drain to the west of the sanctuary and the Temples of Hera and Apollo were laid out.[23] A second phase, with stone-lined sewers (*cloacae*), was motivated by a rise in the water table in the 5th century BC. Soundings have shown that this improved system made it possible to maintain the structures of the 6th century BC.[24]

The effects and the timing of the rise in the water table in the chora are well documented. The flooding of sites in the river valleys and their permanent disappearance under water and layers of alluvium have been

[20] There may have been other sanctuaries close to the city where the areas surveyed were very much more restricted; it is clear, however, that these were not situated along major arteries, but must have been far up smaller valleys. The two Early Village Clusters, Giardinetto and Sant'Angelo, which do not have examples of these small sanctuaries, are located along the major river valleys, where they would have had easier access to the larger sanctuaries.

[21] In the case of Pantanello, the site continued after the mid-3rd c. not as a sanctuary but rather first as a farmhouse and, later in the 2nd c., as a ceramic production center in use until the 1st c. AD. There is evidence of a Late Roman presence, and of visitors in the 11th c. AD. See Chs. 26 and 27. This site will be the subject of a future study in this series.

[22] Carter 1994a.

[23] Sconfienza 1996, 47–48.

[24] Mertens 1985, 649, 654. A much more elaborate system of *cloacae* was constructed in the 4th c. BC, replacing and expanding the earlier ones.

Figure 22.7 Hypothetical reconstruction of the rural sanctuary at Pantanello in its late-6th-c. phase. (ALP)

documented by the recent excavations (2000–2002) in the path of the pipelines that cross the entire chora from south to north. The low-lying site at Pantanello is particularly informative, where the Archaic sanctuary was located beside an all-weather spring set off by temenos walls. The first cult building, a stoa surrounded by an extensive cobbled terrace, rose to the west of it in the 6th century BC (*Fig. 22.7*).[25] Excavation revealed that a thick layer of grey clay containing no archaeological material lay immediately above the cobble pavement and extended across the entire site. This stratum was covered with a thin layer of gravel, and an occupation level well documented with pottery of the 4th century BC. The grey clay is alluvial, similar in texture and composition to those layers periodically deposited over the whole of the lower river valleys and coastal plain, right down to present (*Fig. 22.8*).[26]

The Archaic stoa was simply abandoned and a structure of the same length and orientation was built on higher ground just meters to the north. This new building can be dated by pottery to ca. 400 BC.[27] The earlier sanctuary was destroyed by a flood and the replacement was built after a century or more; a site at a higher elevation farther up the same slope was chosen. This is one of a number of indications from excavations that the water table rose suddenly and permanently at the end of the 6th or beginning of the 5th century BC. The event that buried the sanctuary was a serious one with lasting consequences.

Abbott's study of the geomorphology (Ch. 2) shows why this could happen in this period. The landforms through which the Bradano, Basento, and Cavone, as well as the Agri and Sinni flowed, are highly susceptible to erosion. By studying carefully the alluvial deposits in the valley, and accurately dating them with C14 analyses and the cultural materials included in them, Abbott was able to establish that the periods of maximum erosional activity were the Neolithic and Bronze Ages. These were the first periods of intense agriculture and removal of forest cover, after which there was enough of a pause for a stable soil to have been created in the process known as pedogen-

[25] See Carter 1994a for more description and plans.

[26] *Metaponto* I, Fig. 111, a photograph of the alluvial deposit in the area of the sanctuary illustrates this well.

[27] The dating is that of Keith Swift (pers. comm., 2009), who is preparing the pottery of the sanctuary volume for publication (see n. 21, above).

Figure 22.3 The urban sanctuary after the flooding in early spring 2011. (Courtesy of the Soprintendenza)

esis. Excavations on the slope above the Pantanello necropolis produced evidence of massive erosion and gullying in the Classical period. Quantities of pottery dating to the 5th and 4th centuries BC in the gully fill provide an unusually precise date for such an event.[28] The next major episode of pedogenesis did not occur until near the end of the period of Roman dominance of southern Italy.

Abbott's conclusions differ from those of previous investigators in that he places responsibility for the erosion and the consequent creation of swamps along the lower courses of the rivers and clogging of the deltas on man's direct intervention.[29] The major conclusion of immediate concern here is that such activity coupled with tectonic shift or bradyseism of the coast would naturally have resulted in a rise in the water table. A rapid rise in **500**, at the end of a century of agriculture on a scale never before seen in this area, presented immediate problems, as we have seen, for the sites located at low elevations on the margins of the river valleys, like Pantanello, and in the river valleys themselves.

What is not clear, however, is how the fields at higher elevations—particularly those in Lago del Lupo that flourished in **550** but were totally abandoned in **500**—could have been affected to such a degree as to require a massive project of drainage with parallel, deep canals (Ch. 28) on a chora-wide scale, but this is in fact what clearly happened.[30] We can only speculate about the reasons. The most probable would seem to be the existence of a period of exaggerated meteorological extremes, such as torrential rains that would have, year after year, flooded the fields and drowned the crops. In areas of thin soils, such as the Central Plateau, as Delano Smith observes, "even quite small variations in the levels of the water table could stunt plants' growth and hence yields."[31] There is a remote possibility that parching drought required irrigation. This would have been extremely unusual in Greek agriculture, but then so would drainage of fields. The project might conceivably have served to irrigate the fields as well as drain them, although there was no obvious upstream source to supply the canals. The canals may conceivably have trapped the run-off from rare rains and distributed it throughout the fields.[32] We do not at this point have the necessary evidence to decide, but future studies, most promisingly of ancient pollen, will hopefully provide some well-dated evidence for climate change.[33]

Metaponto could not have been a unique case in Magna Grecia, though its ambitious solution to its water problems may well have been. Something analogous may have occurred in the chora of Siris, which is geologically similar to the Metapontino. There was excessive water in the fields of its successor, Herakleia, as we learn from the Herakleia Tablets, which date to the late 4th–early 3rd century BC and refer to sacred lands.[34] This is one of the very few ancient texts that leaves us an evocative description in non-poetic terms of actively cultivated fields. Tablet I, which describes the new rules for renting the lands of Dionysos, describes swampy conditions and boundary markers that must be kept from disappearing into the muck, as they did before (ll. 57–59), as well as the management of water running between plots in canals (τράφοι = τάρροι) that must be kept clear and not be blocked by dams (ll. 131–134).[35]

[28] Abbott 1998, 490–501.

[29] See Brückner 1980 and Neboit 1977.

[30] Cf. Ch. 2, and Delano Smith (1979, 337–40), who concludes that a small change in the water table can produce big changes in the environment in coastal areas. She notes that in the thin soiled areas, which would include the Central Plateau, when the water table is close to the surface, even a small variation in its level could stunt plants and the yield (185–91, especially 187).

[31] Delano Smith 1979, 187.

[32] Prieto 2005, 180–85.

[33] Mercuri 2009 and Ch. 19.

[34] Uguzzoni and Ghinatti 1968.

[35] Arangio-Ruiz and Olivieri 1965, XIV.645. Texts in Uguzzoni and Ghinatti 1968, 15 and 18, with translations into Italian and commentary. For an English translation, see Prieto 2005, 246–52.

The chora of Siris/Herakleia was likewise located in the Bradano Trough, framed by two rivers—the Cavone (ancient Akalandros) and the Sinni (ancient Siris/Semnus)—and traversed by a third, the Agri (ancient Akiris), which likely all behaved very much like the rivers of the Metapontino. About the Cavone we are well-informed, as it was part of Abbott's study. The wide plain of Sybaris traversed by the Crati (ancient Crater) would probably have had similar if not more serious problems.[36] The geologies of Taras to the north and of Croton to the south were substantially different. Both were located on higher ground, the first on the limestone shelf of the Murge region, and the second on the southern extension of the Apennines.[37]

Staseis in Magna Grecia

The desire for change and new leadership felt throughout the Greek world in the 6th century BC grew in intensity in its final decades and in the first half of the 5th century BC. One common result was stasis, or internal war between fellow citizens, which occurs when a number of citizens form themselves into a faction for the attainment of a political end by legal or illegal means.[38] This spirit manifested itself early on in Western Greece and in distinctive ways. The conflicts naturally centered on questions of land, the basic measure of wealth in the western Greek poleis as elsewhere. There was tension between the aristocracy of the first settlers, those without land (ἄποροι), and those who came later (ἐποίκοι). Stasis is often framed in terms of the opposition of aristocrats and the people (δῆμος), but much depends on what was meant by those terms. Are these the landless or the whole mass of non-elite citizens? Aristotle seems to mean the latter. Thus the use of the term stasis in this period would not be anachronistic, as is sometimes claimed.[39]

The presence of Greeks of different extraction and non-Greeks and their relations with the polis in question was an important variable, particular to the Greeks who left the mainland, whether in Western Greece or elsewhere.[40] Tyrants, the agents of change who arose in such situations of conflict, were quite common in the West, and often acted to liberate the demos from obligations to the aristocracy.[41] If the presence of numerous Pythagoreans, whose role in the political life of the poleis of Magna Grecia was considerable in the last two decades of the 6th century and beyond, did not cause upheaval, it certainly contributed to the unrest. Some Pythagoreans were non-Greeks, which further complicated the situation.[42] Though its members were aristocratic, the Pythagorean groups were often in opposition to the aristocracy of the first settlers.

By **500**, many of the apoikiai of Western Greece had become fully evolved poleis. In the case of Metaponto, a hundred years, roughly three or four generations, separated the current inhabitants from the early settlers. So it is perhaps no surprise that staseis and the careers of most of the tyrants date from the last decades of the 6th century or early 5th century BC. The appearance of a Metapontine tyrant (see Ch. 21) already in the mid-6th century is something of a surprise. Even if at the beginning the distribution of land and wealth had been equal, or nearly so,[43] still, discontent may have already been in the air by the mid-6th century. There would also have been the question of distribution of the conquered land of Siris ca. 550 BC.

For southern Italy, documentary evidence of stasis is very limited, and in the case of Metaponto, which was certainly not immune to these wider currents, it is practically nonexistent. Fortunately, we are slightly better informed about her neighbors, Sybaris, Taras, and Croton. Limited though they might be, Pythagorean sources—always to be handled with caution—describe uprisings against the Pythagoreans in the poleis founded by the Achaians along the southern coast of Italy, including Metaponto and Croton.[44] For this passage, Iamblichus describes the popular revolt led by Kylon against the politically dominant Pythagoreans.

> καὶ λεγόντων ἐξ αὐτῶν τῶν χιλίων Ἱππάσου
> καὶ Διοδώρου καὶ Θεάγους ὑπὲρ τοῦ πάντας
> κοινωνεῖν τῶν ἀρχῶν καὶ τῆς ἐκκλησίας
> καὶ διδόναι τὰς εὐθύνας τοὺς ἄρχοντας ἐν
> τοῖς ἐκ πάντων λαχοῦσιν, ἐναντιουμένων

[36] Mertens 2006, 436–37, with references.

[37] On the geomorphology of Croton, see Abbott 1997; see also Chs. 1–2.

[38] Berger 1992, 12, quoting Newman.

[39] E.W. Robinson 1997, 37–38.

[40] Berger 1992, 65; cf. Aristotle, *Pol.* 1303a 25–35.

[41] Berger 1992, 64.

[42] Mele 1981.

[43] The FMCE analyses suggest gross inequalities did not arise in **600**, or even in **550**.

[44] See Iamblichus *VP* 257–58; E.W. Robinson 1997, 76.

δὲ τῶν Πυθαγορείων Ἀλκιμάχου καὶ Δεινάρχου καὶ Μέτωνος καὶ Δημοκήδους καὶ διακωλυόντων τὴν [258] πάτριον πολιτείαν μὴ καταλύειν, ἐκράτησαν οἱ τῷ πλήθει συνηγοροῦντες. μετὰ δὲ ταῦτα συνιόντων τῶν πολλῶν διελόμενοι τὰς δημηγορίας κατηγόρουν τῶν αὐτῶν ἐκ τῶν ῥητόρων Κύλων καὶ Νίνων. ἦν δ' ὁ μὲν ἐκ τῶν εὐπόρων, ὁ δὲ ἐκ τῶν δημοτικῶν.

And when, from the council of the Thousand, Hippasus, Diodorus, and Theages spoke in behalf of all citizens having a share in the political offices and in the assembly, and of having public officials give accounts of their conduct to those who had been elected by lot from all citizens, the Pythagoreans Alcimachus, Deinarchus, Meton and Democedes opposed this proposal and sought to prevent the inherited constitution from being abolished. Those who were champions of the common people prevailed. Thereupon, when the people assembled, the politicians Cylon and Ninon, apportioning between themselves the thrust of their speeches, launched an attack on the Pythagoreans. The former of these was from the wealthy class, the latter from the ordinary folk.[45]

Iamblichus, *VP* 257–258
Stasis at Croton, ca. 509 BC[46]

Iamblichus' statement is thought to be based on the relatively reliable Timaios.[47] In all likelihood, the democratic movement started in the wake of the Kylonian revolt shortly after 510 BC and spread throughout many of the cities with Pythagorean clubs, including Metaponto.[48] The somewhat fuller accounts of staseis involving Pythagoreans at Sybaris and Croton may shed some light on the possible reasons and outcome of that at Metaponto. According to Diodorus, sometime before 511 BC the tyrant Telys persuaded the demos at Sybaris—in this case only the *aporoi*, it would seem[49]—to confiscate the property of the rich and send them into exile,[50] the usual fate of the losing side. They sought refuge in Croton, and returned with the victorious army of Croton when it destroyed Sybaris in 510 BC. According to Pythagorean sources, Pythagoras had accused Telys of persecuting his followers at Sybaris, so it is possible that stasis there may have involved pro- and anti-Pythagorean aristocrats,[51] and not the demos of Diodorus' account.

πολλῶν δὲ κακῶν κατὰ τὴν πόλιν καὶ τὴν χώραν ὄντων, εἰς κρίσιν προβληθέντων τῶν φυγάδων καὶ τρισὶ πόλεσι τῆς ἐπιτροπῆς παραδοθείσης, Ταραντίνοις, Μεταποντίνοις, Καυλωνιάταις, ἔδοξε τοῖς πεμφθεῖσιν ἐπὶ τὴν γνώμην ἀργύριον λαβοῦσιν, ὡς ἐν τοῖς τῶν Κροτωνιατῶν ὑπομνήμασιν ἀναγέγραπται, φεύγειν τοὺς αἰτίους. προσεξέβαλον δὲ τῇ κρίσει κρατήσαντες ἅπαντας τοὺς τοῖς καθεστῶσι δυσχεραίνοντας καὶ συνεφυγάδευσαν τὴν γενεάν, οὐ φάσκοντες δεῖν ἀσεβεῖν οὐδὲ τοὺς παῖδας ἀπὸ τῶν γονέων διασπᾶν.

Since many evils existed throughout the city and land, the fugitives were brought to trial, and the power to decide the case granted to the three cities: Tarentum, Metapontium, and Caulonia. And it was decided by those who had been summoned for a verdict, after they were bribed, as is registered in the records of the Crotoniates, that those who were guilty should go into exile. Having thus won the trial, they banished all who were discontented with the existing state of affairs, and banished, at the same time, their families, claiming that one ought not be impious and separate children from their parents. And they abolished debts, and redistributed the land.[52]

Iamblichus, *VP* 262
Redistribution of land, ca. 450 BC

[45] Text and trans. Dillon and Hershbell 1991.

[46] For date, see Minar 1942, 61.

[47] E.W. Robinson 1997, 76 n. 46.

[48] Minar 1942, 72–73.

[49] 12.9.2. Demos was the word chosen by Diodorus. We do not know whether it was that of his sources.

[50] Berger 1992, 32.

[51] *Ibid.*

[52] Text and trans. Dillon and Hershbell 1991. See Minar 1942, 57–60, for date, which is disputed. The passage may refer to events in the mid-5th c. BC.

In the aftermath of the defeat of Sybaris, stasis arose in victorious Croton, and, predictably, it was about land. The Pythagoreans refused to accept the demos—this time the word choice is that of Iamblichus and his source[53]—under the leadership of Kylon, an aristocrat who had a personal grudge against the Pythagoreans because he was not admitted to the club. The opposition to the Pythagoreans may have included, besides those of the aristocracy of the "first settlers," some discontented popular elements who felt they were owed a greater share of the spoils. That this was to an important degree a democratic movement is indicated by its outcome, as reported by Iamblichus, using Timaios (Iamblichus, VP 262). First and most important was the redistribution of land, and also the cancellation of debts, which often went together in the resolution of stasis and are characteristically democratic. These were followed by the opening of the magistracies to all citizens and holding the magistrates responsible through the process known as *euthyna*, a rendering of accounts, by representatives drawn from the people by lot.[54] The latter two measures were also adopted after the stasis in Taras of 473 BC.[55] A further outcome was the violent overthrow of the Pythagoreans and their loss of influence.[56] Pythagoras himself had perspicaciously sought refuge in various poleis and was accepted in Metaponto.

While the local aristocracies of many poleis of southern Italy contained adherents to Pythagorean beliefs, it is unlikely that the Pythagoreans had authority elsewhere comparable to what that they had in Croton before 510 BC, and regained for a while in the first half of the 5th century BC. This was probably true also of Metaponto, where Pythagoras is reported to have died about 494 BC in the Temple of the Muses, where he sought sanctuary and where, according to one Pythagorean source, he was starved to death.[57]

In the confusion resulting from the expulsion of the Pythagoreans at Croton and the dissolution of their clubs, the tyrant Kleinias seized power. Though his dates are disputed, his brief career may coincide with the mid-5th–century BC expulsions of the Pythagoreans, also followed by such chaos and instability that the citizen body appealed to the motherland, Achaia, in order to reestablish its own version of a democratic constitution.[58] Our source this time, Polybios, is probably recounting the events of these years when he states that the Kaulonians, Crotonians, and Sybarites soon adopted the laws and customs of the Achaians.[59] Metaponto is not mentioned among them. Robinson has interpreted this silence to suggest that Metaponto was founded with the same democratic constitution as the Achaians, and because it was not altered subsequently, there was no necessity for Achaian intervention in these circumstances either in 510 BC or later in the 440s.[60]

Two other examples of stasis in this period illustrate how widespread and varied this phenomenon, its causes, and its results could be. The tyrant Aristodemos of Kyme (Cumae) was a successful military commander who won the favor of a democratic faction, murdered some of his aristocratic opponents, and in turn met the same end. He had promised his followers a redistribution of land, the cancellation of debts, freedom of speech (ἰσηγορία), and liberty (ἐλευθερία).[61] During his period in power, 504-491 BC, he carried out a variety of public works projects, including what must have been a huge, propagandistic drainage project.[62]

Taras remained an aristocracy in the manner of its motherland, Sparta, until 473 BC, when defeat at the hands of the local, indigenous Iapygians eliminated much of the aristocracy. According to Herodotos, the casualty figures were enormous, "the greatest loss of life in battle the Greeks ever suffered."[63] Tarentine victories over the same foes were recorded on two dedications at Delphi that Pausanias described.[64] From this catastrophe a democracy emerged, almost certainly as a result of stasis. The aristocracy, which did not give up power easily, had to share some public offices with persons chosen by lot, though they retained the rest.[65]

Each of these examples of stasis has its own particular characteristics, but there are some common factors that may be useful in attempting to understand

[53] Dunbabin 1948, 366.

[54] E.W. Robinson 1997, 76.

[55] Berger 1993, 53.

[56] E.W. Robinson 1997, 76, and n. 46.

[57] On Pythagoras at Metaponto, see below, n. 76.

[58] E.W. Robinson 1997.

[59] Polybios 2.38.6–39.8; E.W. Robinson 1997, 76–78.

[60] E.W. Robinson 1997, 76, n. 45.

[61] Berger 1992, 22, with bibliography.

[62] Cerchiai 2000.

[63] Herodotos 7.170. See also Diodorus 11.52.

[64] Pausanias 10.10.6. See also Cordano 1995.

[65] Berger 1992, 53.

what may have happened in Metaponto in **500**. Most staseis produced exiles. The losing side could not be permitted to stay around while a new order was attempting to establish itself. This could have had its impact on demography. If this were to manifest itself in the archaeological record, it would very likely be in a fall in the number of burials, especially of adult males.[66] It may, in part at least, have been responsible for the abandonment of the farmhouses in the chora in the massive desertion of **500**.

In all of the staseis in poleis founded by Achaians, and also in the case of Kyme, a redistribution of land was reported. In the case of Croton it involved not only the conquered territory of Sybaris, but also that of the exiled aristocrats. At Kyme, the ascendancy of a tyrant resulted in public works projects, specifically in a massive drainage ditch surrounding the city, of which, however, no archaeological traces have been definitely identified.[67] Finally, in cases of stasis in the Achaian poleis, Pythagoreans were sometimes centrally involved. Their presence in the population of Metaponto is generally admitted, but on the whole minimized. The archaeological evidence of a growing number of burials in the chora suggests that they may have been more numerous than previously thought. Although there is no way of knowing for sure whether the burials belonged to Pythagoreans or to initiates in one or more of the mystery cults that were sweeping southern Italy beginning in **500**, they are set off by distinguishing traits from the majority: they are predominantly of women and are among the most costly burials in the chora in terms of containers and grave goods.[68]

A popular demand to redistribute property at Metaponto would, in analogy with Croton, have been opposed by the Pythagoreans and their natural allies, the aristocrats who were the descendants of the first settlers. This assumes that the Metapontine aristocracy was composed of old landowning families who accepted the Pythagoreans. The presence of the Pythagoreans in Croton created tension as the old aristocracy, the Thousand, felt that it was losing control.[69] This also assumes that the Pythagoreans always behaved in the same way in all circumstances. The situation in Metaponto in the first half of the 5th century BC was particular and unlike that of the other poleis of southern Italy. The great project of land reclamation was community-wide and its success depended on the participation of the entire population, which in the end could not be denied its aspirations (see Ch. 23). Redistribution of the land may even have been accepted with minimal opposition from the old families. When the established democracy had to fight against tyranny a century later in the period of Dionysios I, the Pythagoreans came out on the side of democracy.[70] The presence of the Pythagoreans in Metaponto is well documented for the 5th century BC and later, even if very little is known about them.[71] Equality, in theory, was important to the Pythagoreans.[72] They considered the limited, good; the unlimited, bad, and they speculated on equality, both geometric and arithmetic.[73]

The Asty, 500
The Residential Quarter

The apparent desertion of large areas of the chora in **500** raises the obvious question, where did the farmers and their families go? In part, at least, they must have found refuge in the asty. As long as we remain ignorant of the development of the residential quarter of the city, however, there can be no definitive answers. The huge area (ca. 140 ha) reserved for the residential quarter was, as Mertens and De Siena have argued, probably very selectively settled during the 6th century BC. There was space and there were certainly some amenities, such as drainage at this early date and, quite possibly, a system for the distribution of water to individual dwellings.[74] The pattern of isolated settlements in villages in the distant chora was not the only one in early Metaponto. As the survey has shown, the chora closer to the city had relatively few inhabitants in the middle of the 6th century BC, and few places of burial. It seems likely that many of these close-by fields within a 5-km radius of the city were cultivated by farmers who resided in the city. The rush to occupy these lands with permanent dwell-

[66] Carter 1998e, 154.

[67] See Ch. 2[illegible], Caputo 1997, 332–33.

[68] A. Bottini 1992; Carter 1998f, 184; 2006a, 177; 2006b.

[69] Berger 1992, 20.

[70] Von Fritz 1940, 97.

[71] Cf. Iamblichus' list of Pythagoreans; Lo Porto 1980; Lombardo 1995. Ancient sources provide some names and occupations, including mathematicians (e.g., Hippasos), philosophers (e.g., Philolaos), and doctors.

[72] Minar 1942, 118.

[73] Delatte 1921, 59ff.

[74] Sconfienza 1996.

Figure 22.9 Temple of Hera (Tavole Palatine). (After Mertens 2006, Fig. 366; courtesy Dieter Mertens)

ings did not begin until the 5th century BC, when land was redistributed (as we shall argue; see Ch. 23, Bonifica), and families previously without land had an opportunity to own it.

Excavation has produced some evidence that the long blocks divided by the narrow streets, the stenopoi, were already in existence in the Archaic period.[75] In **500** the street grid may have been modified, perhaps to accommodate an immigration from the chora. A tantalizingly enigmatic and very rare late literary source for Metaponto mentions the "stenopos of the Muses" on which was located a temple of the Muses—the building where, according to Pythagorean sources, Pythagoras was starved to death. This event, if true, would have taken place sometime in the first decade of the 5th century BC.[76]

Temple of Hera in the Chora

There was little building activity in the urban center in **500** that has left traces. The projects of the previous half-century—the two great Doric temples, Ekklesiasterion I, and above all the street plan—defined the basic outline and appearance of the city for all time. In the chora, a fine Doric temple, an advanced and trend-setting design, according to Mertens,[77] was begun in the first decades of this period and completed before 500 BC. At a distance of 3 km from the asty, the Temple of Hera (Tavole Palatine) rose on the right bank of the Bradano River on high ground overlooking the coastal plain and the northern boundary of the chora (see *Figs. 16.1* and *16.7*). The plan was peripteral, originally with 12 columns along the flank and 6 at the ends (*Fig. 22.9*). Built of the same limestone used for the temples of Hera and Apollo in the city, it differs from them with its generally shorter proportions, slender columns, and wider spaces between columns—"modern" elements, according to Mertens, that it shares with the last important Archaic temple of Poseidonia, the Temple of Athena.[78] Despite the temple's relatively late date, the capitals clearly resemble those of Hera II in the asty, and its footprint is almost exactly half the size (16.13 x 33.24 m compared to 20.5 x 49.82). Like the temples of the city, the structure may have fallen into disrepair in this period, and, like some other sanctuaries of the chora, it may have been less frequented in this period.

[75] Nava 1999, 698–703.

[76] Minar 1942, 73; Vallet 1974, with references.

[77] Mertens 2006, 162.

[78] Mertens 2006, 217, Figs. 359, 362, 364–66. See also the most recent studies by Mertens (2010) and De Siena (2010).

Besides a dearth of new building in the asty, basic maintenance on earlier structures was apparently not being performed. Not only the temples, but also the ekklesiasterion suffered.[79] Undoubtedly, the work of draining the excessive groundwater, a continual problem, went on as the insulae in the residential quarter continued to fill out. Potential evidence for this will lie buried until the large-scale excavation of the residential quarter is undertaken.[80]

Kerameikos

The kerameikos, the potters' quarter, was a center of production for fine pottery, including Ionic cups of the B2 type.[81] It occupied a space inside the asty at its extreme northern limit, about midway between the sanctuary and the western gate (Porta Settembrini). It appropriated two insulae north of the Plateia A.

The kerameikos was in continuous use from **550** to **300** and became an important center for the production of figured pottery in **450** and especially in **400**. The potter's craft had already played an important role in the economy of the city by **500**. Its products reached deep into the interior by the mid-6th century BC; so did the artisans themselves, as witnessed by a dedication to Herakles by the potter Nikomachos, who was active in the last quarter of the century. A pyramidal column in terracotta, found in the territory San Mauro Forte on the right bank of the Cavone, bore the inscription,

Χαῖρε, ϝάναξ *h*‹έ›ρακλες
ὅ τοι κεραμεύς μ᾽ ἀνέθε̄κε
δὸς δέ ϝ᾽ὶν ἀνθρόποις
δόξαν ἔχε̄ν ἀγαθ‹ά›ν
Νικόμαχός μ᾽ ἐπόε̄.

«Salve, Herakles signore,
a te il *kerameus* mi ha dedicato;
concedi che abbia buona fama fra gli uomini;
Nicomaco mi fece».[82]

[79] Mertens 2006, 162.

[80] Nava 1999, 698–703.

[81] The area of the kilns discovered in the 1970s and excavated by D'Andria (1980b) has been the subject of several excellent studies: Cracolici 2004a; Silvestrelli forthcoming a. The publication of Silvestrelli's comprehensive study is eagerly awaited.

[82] Inscription on a terracotta obelisk in the Achaian alphabet, dated 525–500 BC. Giacometti 2005, 131–33, with text and full bibliography. "Hail, lord Herakles. To you the potter has dedicated me. Grant him a good reputation among men. Nikomachos made me." It is rare to hear a voice from the past and a craftsman's at that. The decorators of vases in Metaponto's kerameikos did leave unintended and unself-conscious proof of their individuality, their fingerprints, which offer an unparalleled record—for now at least—of the procedures in pottery workshop (Cracolici 2004a).

The Mint

The coinage of Metaponto is particularly attractive and plentiful. It documents all periods of Metapontine history, from its first appearance in the mid-6th century BC,[83] down to the late 3rd century BC. The silver coins of the second half of the 6th and first half of the 5th centuries BC were struck using the incuse technique, whereby the relief design on the obverse is repeated intaglio on the reverse. This method was adopted as well by the other Achaian poleis: Sybaris, Poseidonia, and Croton. It persisted at Metaponto and Croton long after it had disappeared elsewhere.[84] For approximately a century, a single subject was represented: the six-rowed ear of barley, accompanied by grasshoppers or other symbols, and the inscription META (*Fig. 22.10*). The slow stylistic development and unbroken progression of the type, created by numismatists,[85] and based on die links, suggests a degree of stability and continuity in the governance of the city, despite difficult conditions, and the fifty years centered on 500 BC is no exception. Some of the finest examples of the die-cutters' art have been assigned to **500**. It has been suggested that the choice of barley as a symbol reflects the commercial position of Metaponto as a major exporter of grain. There may be reason to doubt that this was its original function, or at least its principal one in the 6th century. Coinage in its early years, according to Finley,[86] performed primarily non-commercial functions of a political

Figure 22.10 Metapontine silver stater (obverse), ca. 520–510 BC. (Courtesy of the American Numismatic Society, 1941.153.68)

[83] Noe and Johnston 1984, 49; Stazio (1999, 457) makes it decades later, ca. 530 BC.

[84] Noe and Johnston 1984, 36.

[85] Stazio (1999, 459, 461) notes a brief interruption ca. 430 and ca. 300 BC, after Cleomenes and a possible parallel series with Pyrrhos (see Ch. 25).

[86] Finley 1979, 36.

or political-psychological character. The question of who minted these coins is rarely asked.[87] Who had the authority to represent the city in this way? Daverio Rocchi has argued that coins played a role in the rise of Peisistratos, tyrants, and other leaders, during whose careers coins began to be common in the Greek world.[88] Could the still-shadowy figure of the Metapontine tyrant Archelaos be behind the early issues? Or might it be one or more heads of the aristocracy of the first settlers?

[87] An important clue might be the source of silver, but that has not as yet been determined for the coinage of Metaponto.

[88] Daverio Rocchi 1973.

The fractional coinage of Metaponto, which dates primarily to the 5th century and later, in contrast to that of Sybaris and Croton was extremely varied and rich from the beginning, and it had a practical function.[89] According to Stazio, it played a key role in small exchanges between farmers.[90]

Despite the natural and man-made crises of this half-century, by the end of **500** a degree of stability in the social structure and economy had been regained, and the stage for a second great flowering in **450** was set.

[89] Noe and Johnston 1984, 46.

[90] Stazio 1974a, 73–74.

23
Reform and Rebirth: The Mid-5th Century BC

Joseph Coleman Carter

The middle years of the 5th century BC were the most creative and happiest, depending on one's circumstances, in the history of the polis. This could rightly be called the democratic century. There was much intellectual and religious ferment. The asty was reinvented architecturally and embellished with art under a popular government, operating in a much-improved meeting place. The drainage of the chora, a great project of public works—one of the most ambitious in the Greek world at that time—was undertaken and successfully completed by mid-century. This achievement, it is argued, was part of a radical reorganization of the chora including a redistribution of land.

With the timely resolution of the pressing problems of the previous 50-year period, the fortunes of the chora reversed as rapidly as they had declined. This is immediately clear from a comparison of the distribution maps of **500** and **450** (*Maps 7–8*, Vol. IV; see *Figs. 22.1* and *23.1*). The number of significant farmhouses surpassed the level of the mid-6th century BC, the first rapid expansion of chora and city, but was accompanied by a slower increase in active necropoleis from 12 in **550** to 21 in **450** (compare *Fig. 23.2* with *21.2*). This is not surprising, as the ratio of rural burial places to farmhouses is lower in periods of rapid expansion than in more settled times.

In this period of the repopulation of the chora, of revitalized agriculture—particularly of cereal production—and of the increase of Metaponto's wealth, reflected in the growing significance of individual farmhouses and more obviously by the city's revived building activities, some areas of the chora were developing more rapidly than others. Following the pattern of previous chapters, these areas will be visited again and analyzed in detail beginning with the Early Village Clusters, then the areas of the Central Plateau (Lago del Lupo, Cogno del Prete), the localities of the study area farthest from the city (Campagnolo and Bufalara), and finally, the locations closer to the city (Pantanello and Pizzica).

The Chora in 450
Early Village Clusters

Giardinetto (*Fig. 23.3*).

Farmhouses. There are six significant farmhouses in this village cluster in **450**—the same number as in **550**—with a configuration very similar to that of the earlier period. Four of the farmhouses that were significant in **550** continue to be so in **450** (465F, 523F/T, 462F, and 519F). Sites 465F and 523F/T, together with 524N, were the two earliest significant farmhouses in Giardinetto in **600**. Site 465F continued to grow in significance after **450**, while 523F/T never regained its earlier prominence, and 462F and 519F quickly fade in importance (see *Appendices A–B*). The farmhouses with associated burials (515F/T and 521F/T/K) were the two newest additions to the cluster, ex novo on their sites. The surface assemblage at 521F/T/K was particularly varied. Besides human bone, there was slag, evidence for ceramic production (see Gazetteer). This activity may be a source of some of its success.

Necropoleis. The necropolis 524N was in use continuously from **600** to **450**, but not thereafter. 471N was added in this period and grew in importance down to **300**. The rates of burials, necropoleis, and farmhouses with associated burials rose from 2:5 in **550** to 5:6 in **450** (*Appendix C3: Giardinetto*).

Significance, land, and continuity. Even with the newly significant farmhouses in **450** (515F/T, a single period site, and 521F/T/K), the average FMCE value for Giardinetto in **450** fell from 2.29 in **550** to 2.05 in **450**. After Sant'Angelo this is the highest score among the Early Village Clusters (*Appendix C7*).

Essentially the life of the village picked up where it left off during the general recession of **500**, the only notable change being a slight shift away from the west to the lower, more gentle slopes on the east in this natural basin. The farmhouses at Giardinetto remained throughout its history among the most significant in the chora.

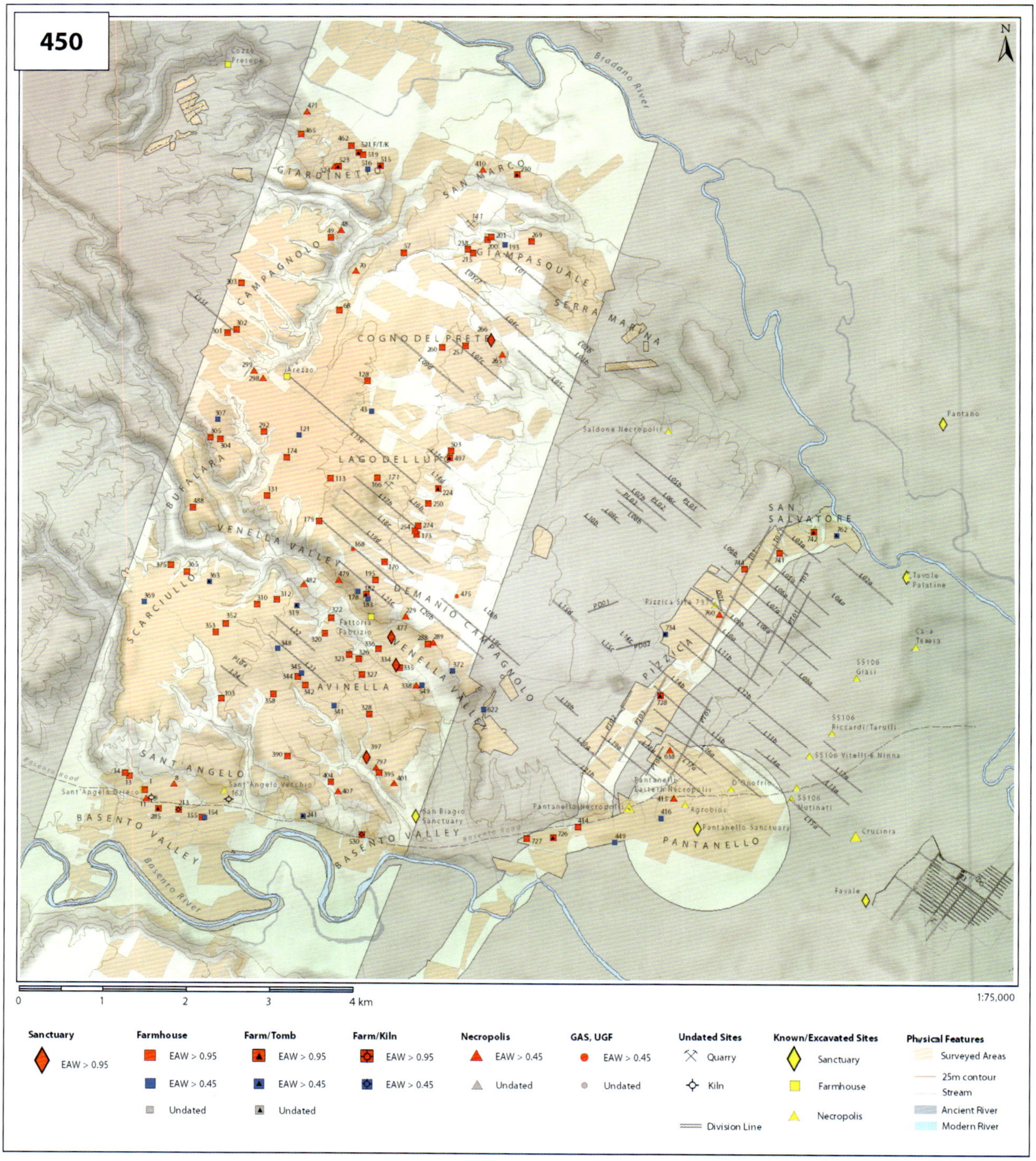

Figure 23.1 The study area in **450**. (JT)

Sant'Angelo (*Fig. 23.4*).

Farmhouses. Three of the eight significant farmhouses of **550**—1F, 13F, and 155F—returned to significance in **450**. One, 285F/T, seems to have been unaffected by the crisis of **500**. Another four were either definitely abandoned (7F and 12F/K) or had a long hiatus in their occupational histories (154F and 286F).

One of the two sites that became significant for the first time in **450** (14F) remained so for only half a century—only to disappear permanently, like a number of single-period sites that came into being in **550**; the other, 213F/K, had a relatively long existence and was probably engaged in ceramic production at some point to judge by the presence of slag on the surface

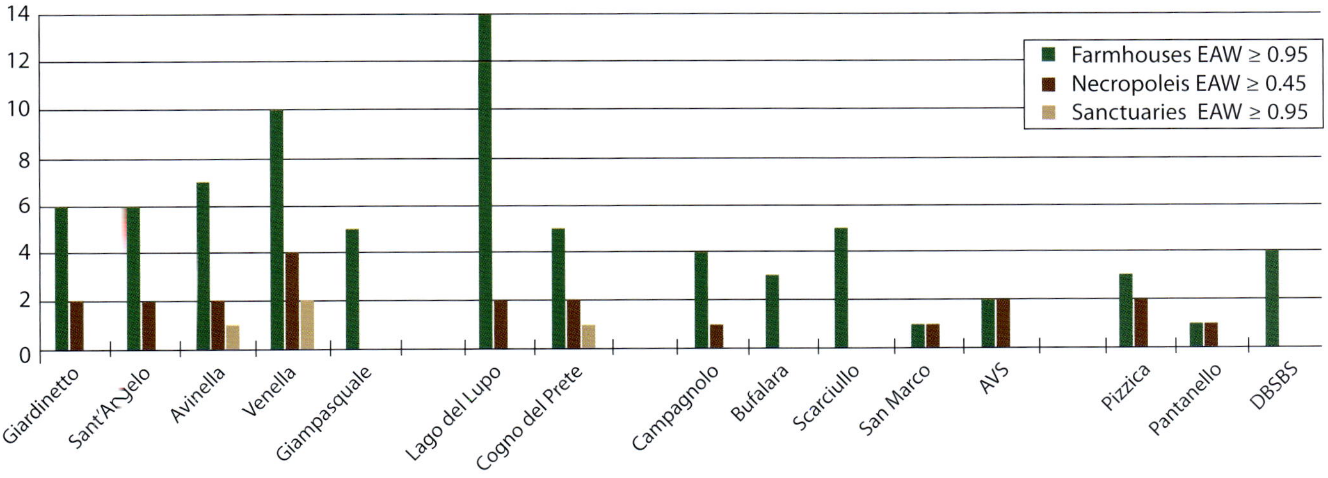

Figure 23.2 Distribution of farmhouse, necropolis, and sanctuary sites among localities in **450**. Totals: significant farmhouses, 76; necropoleis, 21; sanctuaries, 3.

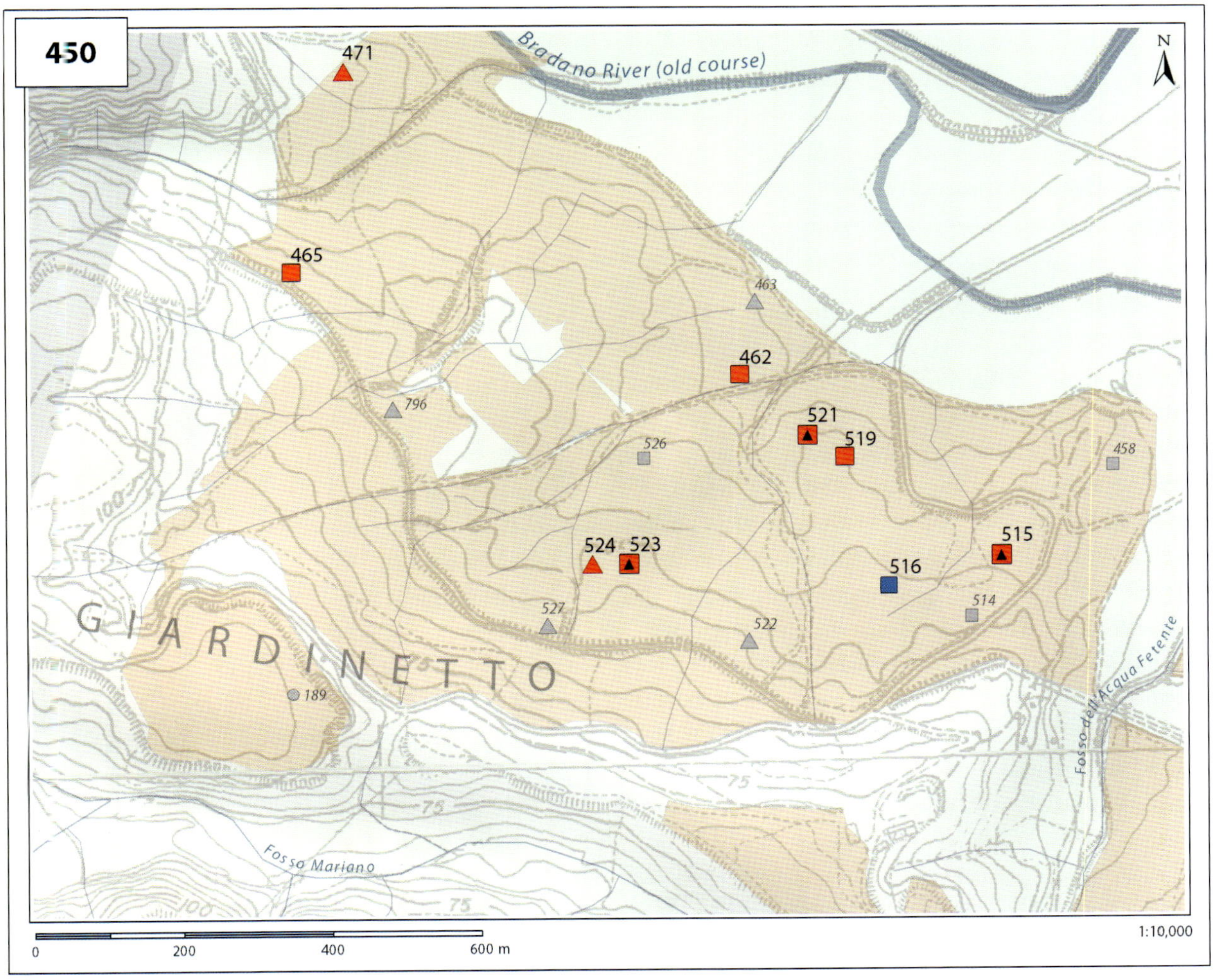

Figure 23.3 Giardinetto in **450**. See Fig. 23.1 for map key. (JT)

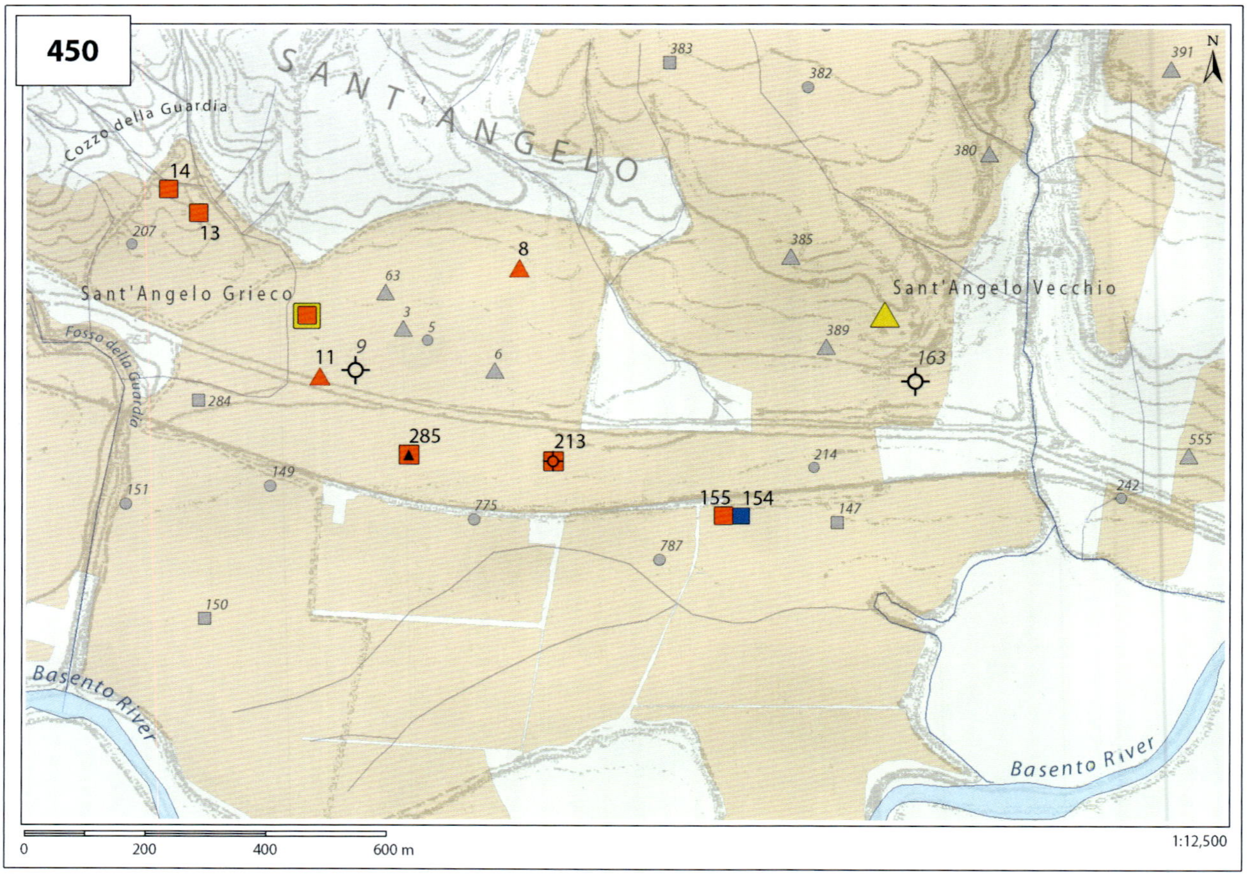

Figure 23.4 Sant'Angelo in **450**. See Fig. 23.1 for map key. (JT)

(see Gazetteer, Vol. IV). Centers of ceramic production of all periods, as noted earlier, lined the Basento valley and a respectable number have been excavated.

Necropoleis. Necropolis 8N, which was active in **550** and **500**, continued to be so in **450**, but not after. 11N was added in **450**, but was in use in this period only. Among the farmhouses with closely associated burials, 285F/T was the most significant site in the chora in **600**, and continued to be in **450**, but not afterward. The ratio of burials to places of residence was 1:2 (*Appendix C3: Sant'Angelo*). The pattern of burials growing in relation to farmhouses is similar to that at Giardinetto, but that at Giardinetto continues in the next period, while Sant'Angelo burials decline.

Significance, land, and continuity. Sant'Angelo has the highest average FMCE value among the Early Village Clusters in **450**, and is surpassed in the study area only by the remote and smaller Campagnolo (Far from the City). In the revival of the chora in **450**, it suffered a net loss of sites with respect to the previous century.

Avinella (*Fig. 23.5*).

Farmhouses. Significant farms along the Conca valley and its tributaries decreased from eight to seven between **550** and **450**. This represents a conspicuous recovery from the immediately preceding period when only one farmhouse (395F), a necropolis (356N), and a sanctuary (397S) remained from the prosperous and rapidly expanding mid-6th–century BC settlement. What is noteworthy about the recovery is the shift of the settlement towards the interior. The greatest number of sites in **550** was concentrated around the confluence of the three forks of the Conca below the sanctuary. All of the farmhouses that were no longer significant in **450** (400F, 505F, 406F/T, and 509F/T) had been located here. Two were never to be reoccupied, and the other two, both farmhouses with related burials, were reoccupied only much later. Of the four earlier farmhouses that continued to be significant in **450**, two were located at the three forks (404F and 395F) and these also had unusually long lives. They are part of a remarkable group in Avinella that was continuously in use from the 5th century BC and

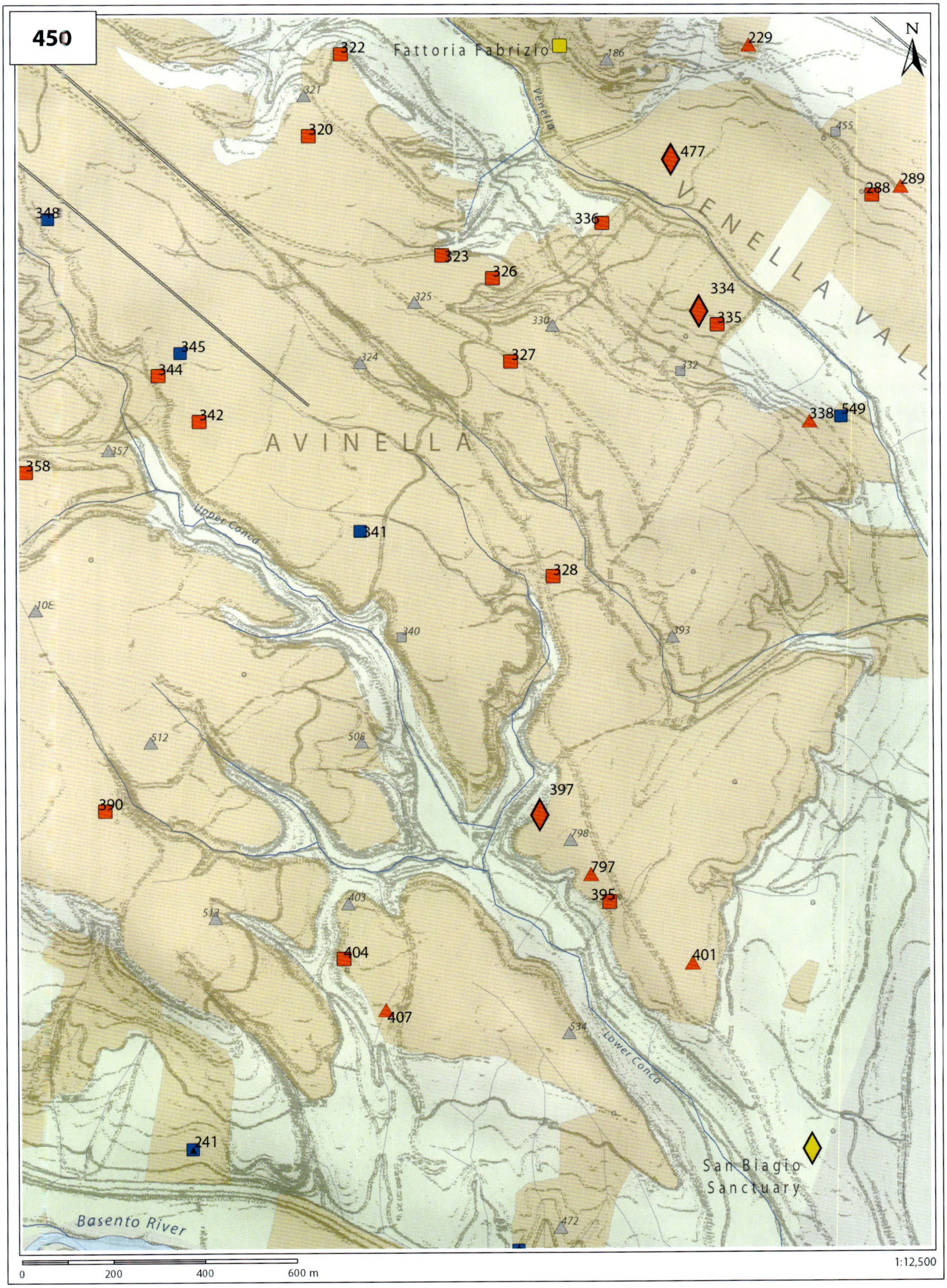

Figure 23.5 Avinella in **450**. See Fig. 23.1 for map key. (JT)

earlier, well into the period of Roman domination of the chora in the 2nd and 1st centuries BC. In this respect, Avinella, with its numerous interconnected drainages, is quite unique and its continued vitality may indicate a renewed interest in animal husbandry in the later period. Two other survivors of the 6th-century BC expansion up the Conca were 328F and 344F, located in the upper reaches of the drainage. They, by contrast, had very short lifespans and disappear from view after **450**.

The new sites, those without any previous history of occupation, are all located in the upper Conca and along its tributaries. Their positions on the edges of these drainages put them close to the flat, fertile fields of the Central Plateau with a constant supply of water from springs and a natural passage to the main river valleys and the Basento Road. Their main activity probably included the cultivation of the fields on the Central Plateau in addition to herding. Two, 390F and 358F, were significant from **450** to the 3rd century BC, and a third, 342F, as late as the 2nd century BC.

Necropoleis. There was a net decline in the number of burial places in Avinella from **550** to **450**. Both farmhouses with associated burials (406F/T and 509F/T) ceased to be significant in **450**. Both were located in the lower Conca. Necropolis 797N, added in **500**, continued in use in **450**, but ceased abruptly thereafter, as did 407N, which was added in this period. Both were located, like the farmhouses with burials, in the lower Conca. There were two active necropoleis in the upper Conca (347N and 356N) in **550**, but none in **450**, though this changes in the next period. The ratio of places of burial to farmhouses is 2:7 in **450** (*Appendix C3: Avinella*). With the exception of Giampasquale, this is the lowest among the Early Village Clusters in this period.

Sanctuaries. The significance of the sanctuary at the three forks, 397S, continues to decline in **450**. It had been less affected by the crisis in **500** than the other three Archaic sanctuaries (*Appendix A*).

Significance, land, and continuity. Three new farmhouses (342F, 358F, and 390F) were added in **450**—all in the upper Conca. They were among the least significant farmhouses in Avinella and the study area as a whole. The average FMCE value for the upper Conca (*Appendix C7*) was slightly higher than that for the lower Conca. Combined, they rank with Venella and Giampasquale among the least significant areas of the chora, surpassing only Scarciullo.

Venella (*Fig. 23.6*).

Farmhouses. When the rather loose clusters of sites, which grew up on the west and east banks of the Venella in the 6th century BC above the two sanctuaries (334S and 477S) located on its floor, reformed after **500**, their spatial distribution was noticeably altered (*Fig. 23.6*). The number of significant farmhouses increased sharply from the combined total of six in **550** to nine in **450**. This is in contrast to Giardinetto and Avinella where the number of farmhouses showed no increase over **550** and, in the case of Sant'Angelo, a slight decline. By **450**, four new farmhouses ranged along the west bank to the north and west of the sanctuary sites 334S and 477S. The three farmhouses 323F, 326F, and 327F form an unusually tight group separated by 150 to 200 m. The new farmhouse 335F is only 50 m from the west bank sanctuary 334S. The normal spacing in previous periods between farmhouses and between farmhouses and sanctuaries is closer to 300 m, though there were always a few exceptions (including, of course, the paired sites). Proximity to constant all-weather springs was an absolute necessity before the introduction of wells (probably in the 4th century BC) and was certainly a factor in all periods. These sites are, not surprisingly, consistently located around prominent secondary drainages into the valley. On the east bank, by contrast, the wide spacing of sites encountered in **550** persists in **450**.

Necropoleis. There were four active necropoleis in the preceding period and four (229N, 289N, 338N, and 479N) in **450**. Two were located at opposite ends of the east side of the valley's edge, and the third on the west bank at its southern extreme. The farmhouse with burials 182F/T (*Fig. 23.7*) continued in use in this period, so that the ratio of burials to dwellings was 1:2 (*Appendix C3: Venella*).

Sanctuaries. Both sanctuaries (334S and 477S) had steeply declined in significance in **500**; 477S recovered better in **450**, and was more frequented than the other Archaic sanctuaries.

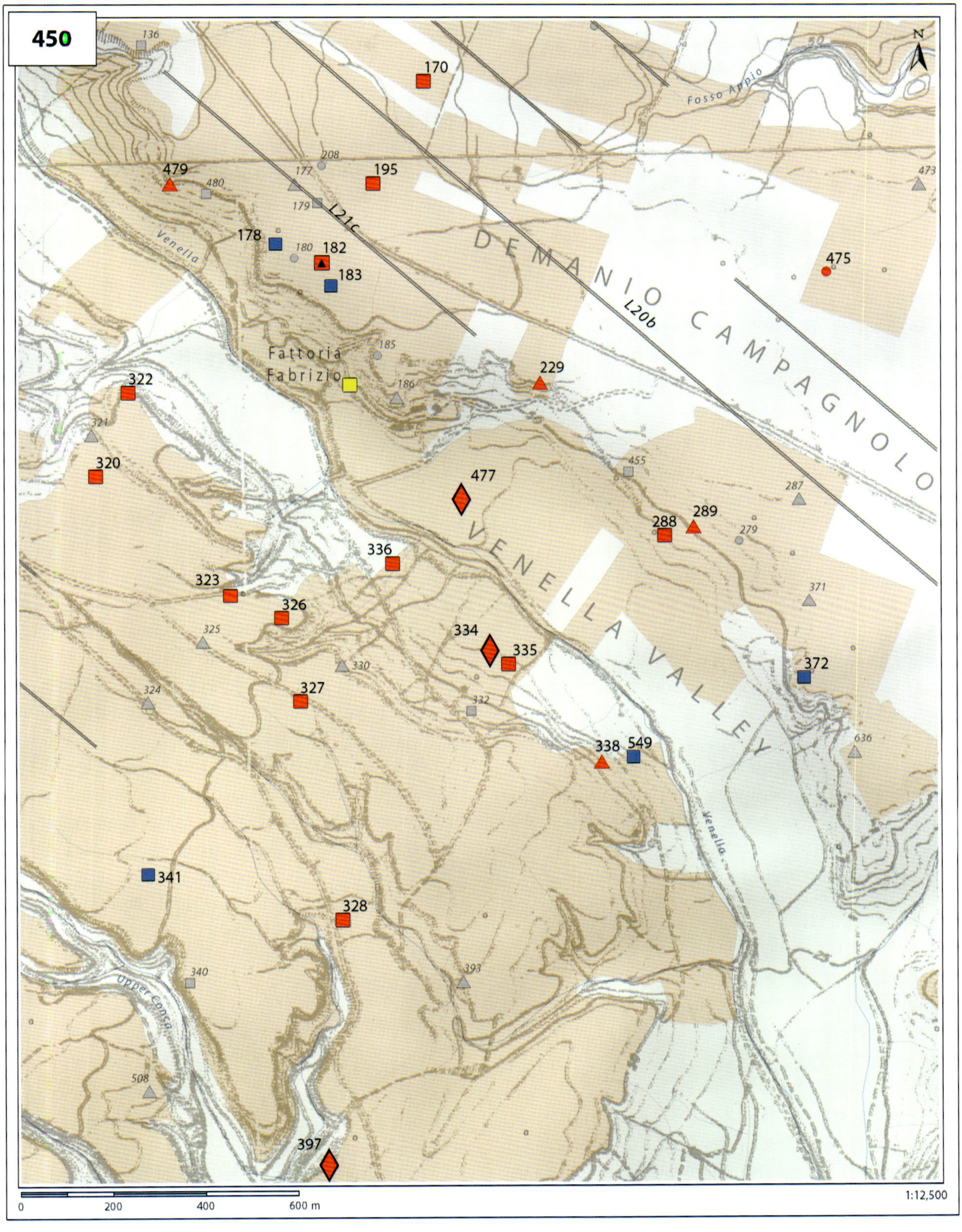

Figure 23.6 Venella in **450**. See Fig. 23.1 for map key. (JT)

Figure 23.7 Part of the assemblage from farmhouse site 182F/T in Venella, which was continuously significant from **550** to **400**. (MLG)

Significance, land, and continuity. The new and the older sites on both banks of the Venella have generally lower FMCE values than the average for the chora as a whole in this period, a tendency noted already in the 6th century BC (see Ch. 21). This is true of the new sites (e.g., 327F, 326F, 335F on the west and 288F on the east) and also of some of the sites previously occupied in the 6th century BC. Site 323F, whose FMCE score in **550** was among the highest in the chora, was in this period reduced to the level of its near neighbors, 326F and 327F, which are among the least significant sites in the Venella. The reason may be that the land provided quite sufficient returns for the farm family occupying 323F in **550**, but when divided among three (326F, 327F, and 323F) in **450**, the result was a more nearly equal, but lower level of material well-being—reflected to some degree, if imperfectly, by their respective FMCE scores (*Appendix B*).

Giampasquale (*Fig. 23.8*). At first glance, little change took place in this village cluster between **550** and **450**. All four of the farmhouses significant in **550** were reoccupied after the hiatus in **500** and a fifth was added, 201F. Although a smaller cluster than others, with 269F and the sanctuary 270S still significant in **500**, it weathered the crisis relatively better than the others. By **450**, however, the sanctuary appears to have been definitively abandoned. The appearance of stability at Giampasquale suggested by the numerical parity of the mid-6th– and mid-5th–century BC farmhouses is, however, illusory. Change, in fact, was taking place, but of a particular sort, seen here but not elsewhere in the study area at this time.

Farmhouses. In the previous century, the proximity of 238F and 215F (separated by only 70 m or so) suggested a pair similar to those found at Sant'Angelo as early as **600**. As in those cases (285F/T-286F and 155F-154F), both sites were of closely comparable significance.

Necropoleis. There was only one undated necropolis, 236N, in Giampasquale.

Sanctuaries. The sanctuary 270S, the smallest of the four Archaic sanctuaries, seems not to have been visited after **500**. A single dateable fragment of black gloss pottery was deposited there in the middle of the 4th century BC.

Significance, land, and continuity. At Giampasquale, both components of the pair 238F and 215F continued to be significant but the difference in the FMCE values is notable—215F is well above the average for the period, 238F very much below. Some disparity in their FMCE values existed in **550**, but this was magnified a century later (see *Appendix B* for individual values). The new farmhouse in **450**, 201F, was only 50 m from 200F. The earlier site was as significant in **450** as it had been in **550**, but 201F was even less significant than 238F and over the two centuries of its existence had consistently lower FMCE values than its very near neighbor. The appearance there of unequal pairs in **450** may reflect a social reality. We can only speculate about its nature—land owner and dependent laborer or tenant family? The possibilities are many, but this phenomenon seems to have been limited, among the Early Village Clusters, to Giampasquale.

The Central Plateau

The Central Plateau (*Fig. 23.9*), unlike the Early Village Clusters, developed entirely ex novo in the mid-6th century BC. There were only two previously occupied significant sites in this vast area: 170F in the southern part of Lago del Lupo, within easy reach of the older settlement along the Venella; and the isolated 43F in Cogno del Prete. The three villages, Galitelli, Casa Preti, and Capistretti, that grew up in **550** were located near springs and drain-

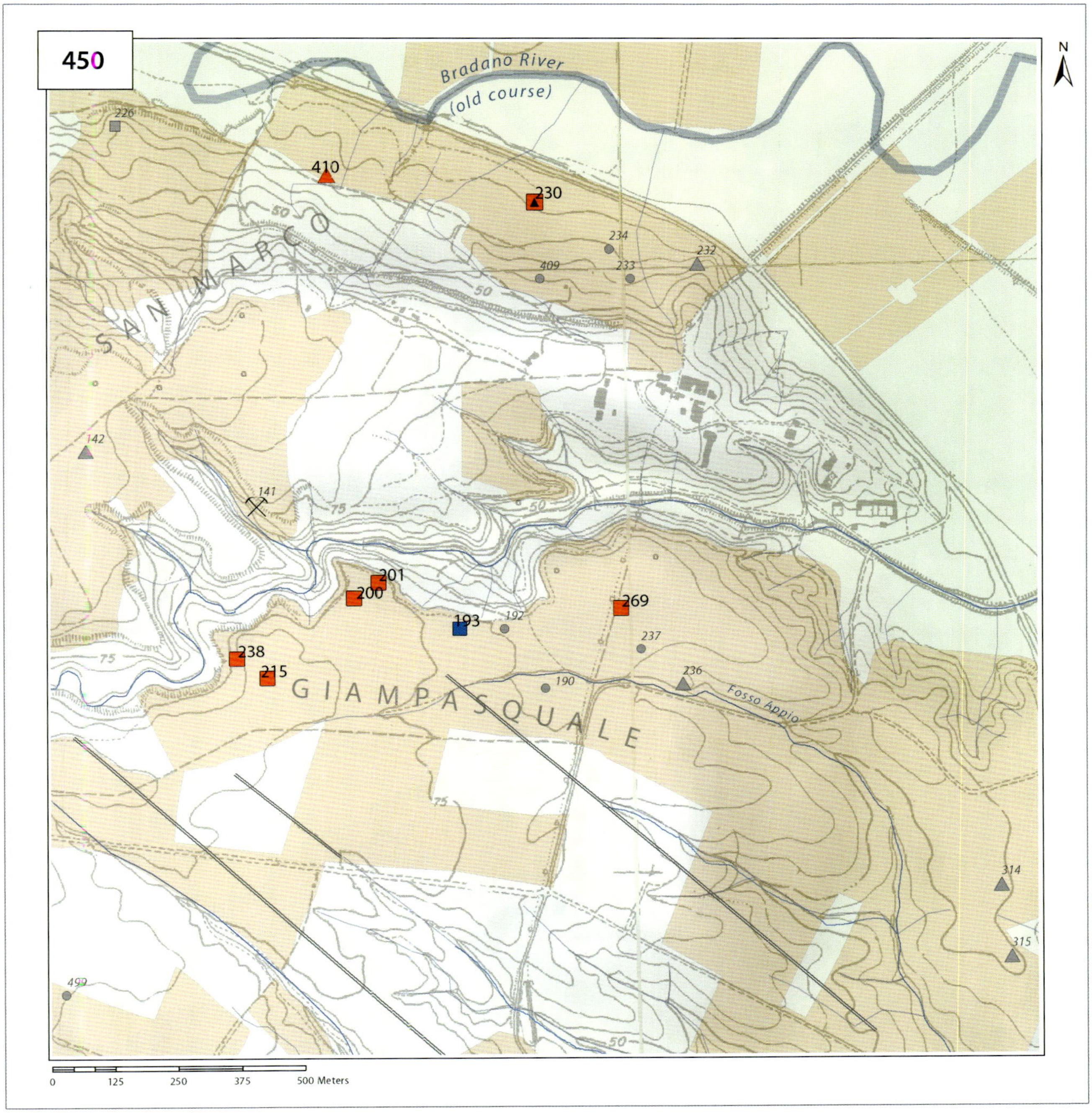

Figure 23.8 Giampasquale in **450**. See Fig. 23.1 for map key. (JT)

ages. All were extinguished in **500** with the single exception of 170F. In the mid-5th century BC, five of the ten farmhouses in the area of these villages were reoccupied, and another five appeared for the first time, completely ex novo (*Appendix C6*). This is a much higher percentage of ex novo farmhouses than in the five Early Village Clusters, but lower than the average for the Central Plateau as a whole, which had the highest percentage of completely new farmhouses for all areas of the chora in **450** (68%), and in every subsequent period as well. The 50-year period **450** was pivotal for the whole chora, and the Central Plateau played a key role. The discussion of the villages follows the format on the Early Village Clusters. The widely spaced farmhouses and their importance will become apparent in the synoptic view of the chora, below.

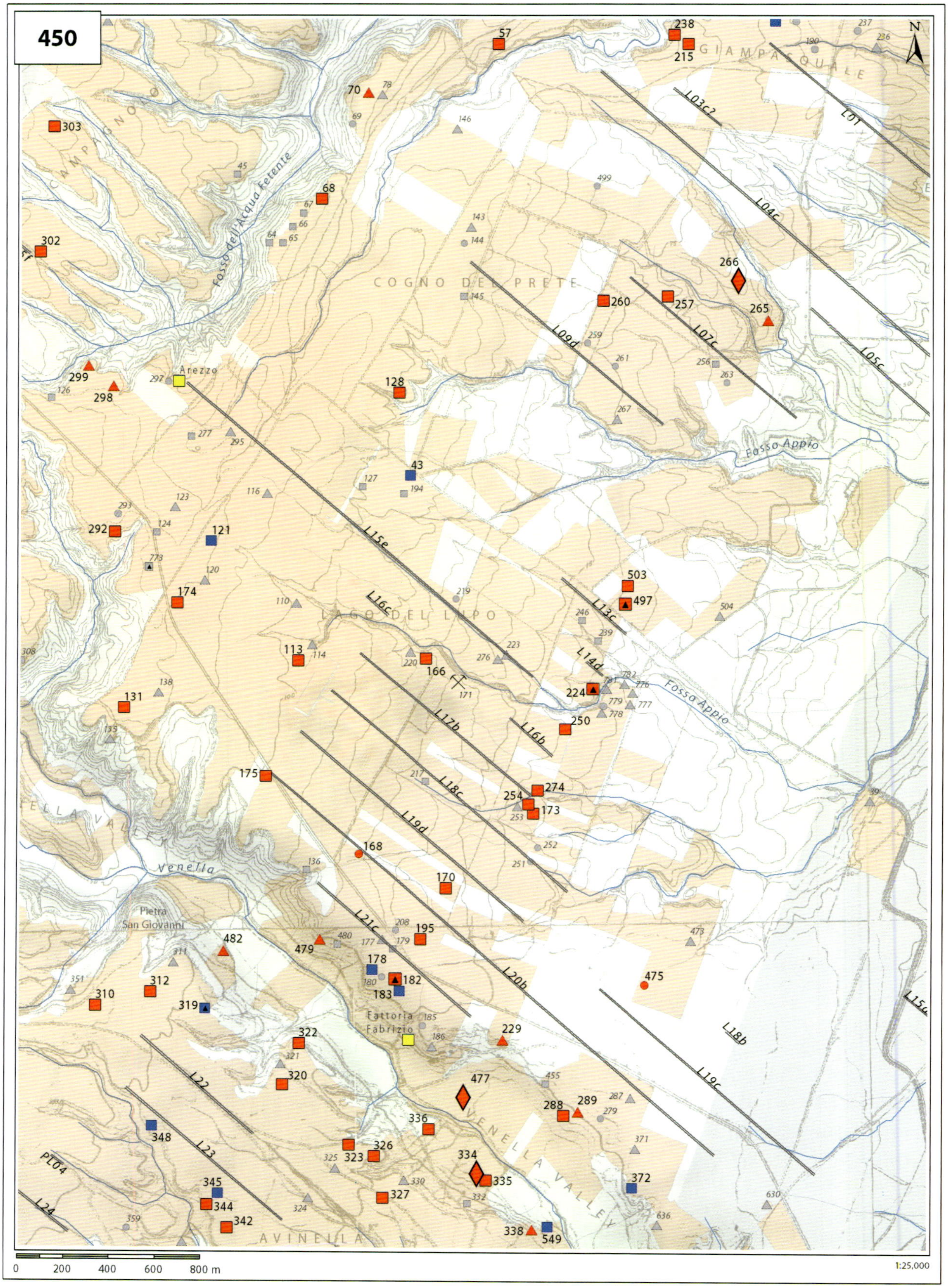

Figure 23.9 The Central Plateau in **450**. See Fig. 23.1 for map key. (JT)

Figure 23.10 Site of farmhouse 173F in the Central Plateau, significant from **450** to **200**.

Two villages of the Central Plateau (*Fig. 23.9*). The villages of Galitelli and Casa Preti in Lago del Lupo had a similar development in the mid-5th century BC (see *Table 21.4*). In Galitelli two of the three farmhouses that were significant in **550** returned to significance in **450** and they were joined by the ex novo 173F (*Fig. 23.10*), so that the population in both periods remained the same. In Casa Preti the population actually grew. To the two farmhouses that were again significant (224F/T and 503F) a third and fourth (250F and 497F/T) were added. There were no dateable necropoleis in either village but Casa Preti had two farmhouses with associated burials, and so the ratio of burials to farmhouses was 2:3. These were two of the three active places of burial in all of Lago del Lupo that could be dated to **450**, which had a considerable number of undated necropoleis.

Significance, land, and continuity. The average FMCE value in **450** for the three farmhouses in Galitelli was 1.83; that of the four in Casa Preti, 2.13. The impression that of the two, Casa Preti was the more thriving, is supported by these numbers. The two new farmhouses there are among the most significant in the study area. Galitelli instead seems about as prosperous as Avinella, Venella, and Giampasquale in this period.

Galitelli and Casa Preti had as many, or more, second magnitude sites in **550** as significant farmhouses—two in Galitelli and three in Casa Preti—which suggests that they may have been larger centers in the 6th century BC than they appeared when considering only significant farmhouses. As a result of the abandonment of the second magnitude and outlying sites in **450**, the villages become more compact, though less populous than they had been a century earlier.

Outside the villages of the Central Plateau. The remarkable change of the mid-5th century BC in the Central Plateau was the proliferation of sites outside the Village Clusters (*Fig. 23.1*). Seven out of eight significant farmhouses were new in **450**. Along the Venella there were four new farmhouses in **450**: 195F, 175F, 174F, and 292F. Newly significant farmhouses were located on the three large drainages that penetrate deep into the Central Plateau and flow into the Fosso Appio: just to the west of the Casa Preti village (113F, 166F), at the head of the largest stream that flowed between Casa Preti and Capistretti where the Fosso Solfegna arises (128F), and on the southern branch of the Capistretti tributary (257F, 260F), whose northern branch flowed below 266S. Two sites (113F and 128F), remote and isolated in the heart of the Central Plateau, continued to be significant down to the period of Roman domination in the 2nd century BC, as did the newly significant 260F, 173F, and 497F/T, in the villages. The central area of the chora was very active in **450**, but many sites displayed great stability even well beyond the heyday of the chora in the 4th century BC.

The Other Areas

Far from the City (*Fig. 23.1*). The northwestern corner of the Bradano-Basento transect is one of the most rugged parts of the study area. It is deeply dissected by the northern branch of the Acqua Fetente, which separates Giardinetto from the main landmass, and by the main stem of the Acqua Fetente and the Fosso San Marco. Between these latter two, on a narrow spit of land overlooking them, appeared two farmhouses newly significant in **450**: 68F and 57F. Site 49F on the western bank of the Acqua Fetente in Campagnolo forms an equilateral triangle across the valley with sides of approximately 900 m long, with 68F and 57F. The spit broadens toward the north, on a low slope facing the Bradano sits 230F/T, separated from the others by 1.5 km. It was occupied for a single half-century. In what may be considered an extension of the Central Plateau, because they are separated from it only by the headwaters of the Acqua Fetente, are three closely grouped significant farmhouses, 301F,

302F, and 303F, the latter two of which remained significant for this period only. Site 301F (significant already in **550**) and 302F, which came into being at the same time that the system of drainage of the chora was being completed, lie on opposite sides of the westernmost extension of the central division line.

The upper reaches of the Venella valley, at the point where two tributaries, the Bufalara and the Gaudella, come together, was densely occupied for the first time in **550**, like the area of the villages on the Central Plateau, and completely abandoned in **500**. When reoccupied in **450**, the settlement had shrunk relative to its mid-6th–century extent. All the farmhouses (488F, 365F, 375F, and 304F) but one had been previously occupied in the earlier period. The new farmhouse 305F was significant for a single period. Farther east along the Venella, 310F, first settled in the mid-6th century BC, overlooks the vale from the heights beside Pietra San Giovanni. It was joined in **450** by 312F, located back from the valley's edge in an open, level area, one well suited to cereal cultivation.

The area south of the upper Venella, outside the Sant'Angelo and Avinella clusters, was sparsely populated in general before this period. Of the seven significant sites here in **450**, four are new. Site 322F on the bank of the Venella and to the north of the western Venella cluster shares an extent of flat land with 320F. Like the other sites in Avinella, a good percentage of the farmhouses is long-lived compared to the rest of the chora. There are no single-period sites among these new sites. Site 352F, like 320F and 103F in Scarciullo, endured for centuries. Sites 352F and 353F, new in **450**, occupied an ample flat plain between the upper Conca and the Fosso Sant'Angelo drainage. It is strange that the flat open fields on either side of its steep and rather forbidding gorge were apparently not attractive to farmers until much later. Despite the numerous drainages in Avinella, there are extensive areas of relatively flat land and the new sites in particular were located to take full advantage of these terraces, well-suited to the plow.

Close to the City. Four farmhouses in **450** are all that were identified in the very partially surveyed localities closer to the city, Pantanello, Pizzica, and San Salvatore. Of these, 728F/T and 744F were new in **450**. The two older sites (414F and 741F) were continuously occupied (or nearly so) down to **300**, but not beyond—a trajectory replicated by only one of the two new sites, 728F/T. Both new sites, in contrast to the older ones, were located not far from the division line canals L06a/L06b and L15b. The outstanding development in these areas near the city in this period was the sudden appearance of necropoleis where none were active earlier.

A few significant farmhouses in the Basento valley in **450** lie between the mouth of the Venella and Pantanello, at a distance that is intermediate between the distant farmhouses just discussed, and those Close to the City, treated below. 530F/K, one of the earliest farmhouses in the chora (Ch. 20), was significant from **600** to **450**, but not thereafter. A new site in **450**, 727F, is significant then and not again until much later. The somewhat unusual histories of these sites, located along a principal thoroughfare linking the ancient city and the interior, may well have been affected by their low elevations in a valley subject to frequent flooding.

Synoptic View of the Chora, 450

The overall distribution of 76 significant farmhouses followed, in the main, the pattern established a century earlier. The clustering that appeared early on in **600** resumed and became a constant feature of the organization in the chora. By **550** restricted areas of open, relatively flat plateau were occupied by clusters and more or less isolated farmhouses. In **450** there was a move into areas previously settled lightly, if at all. All this should appear obvious from a comparison of the period maps. A change not so easily recognized—one with important consequences for Metapontine society—was taking place in the second and third quarters of the 5th century BC.

The new settlers. When the chora filled again with farmhouses and necropoleis in the mid-5th century BC, it was not a case of the former occupants plus some newcomers relocating to previously occupied sites. Half of the farmhouses in **450** were completely new settlements on land that had no trace of a former structure (*Appendix C6*). That is not to say that the land had not been formerly cultivated, but simply that there were no remains of a structure on it or any pottery that could be dated to an earlier period. The other 38 farmhouses were located on sites that had been occupied in the mid-6th century BC or before. These were the residences of early settlers. We have no way of knowing, however, whether their occupants in **450** were descendants of the first settlers or newcomers unrelated to the former owners.

Many of the long-lived sites are located in remote areas like Scarciullo and Avinella, where they would have been somewhat more self-sufficient, out of the public view and therefore probably better insulated than most from the economic and political currents of the day. Of the thirteen farmhouses of **600**, nearly a quarter of the total were abandoned before **500** and never reoccupied (see *Appendix B*, from which the continuity of significant farmhouses can easily be read). Another ten of these early farmhouse sites were abandoned after **450**, and fifteen more by **400**, at which point 60% of these early sites were no longer in use. A very small number (320F and 395F in Avinella, 488F in the upper Venella, 200F and 269F in Giampasquale, 465F in Giardinetto, 414F in Pantanello, 103F, 365F, and 375F in Scarciullo) continued in use, interrupted only in **500**, from the 6th century through the 4th century BC. Excavation has moreover uncovered much evidence of the rebuilding of farmhouses on previously occupied sites.

We cannot be sure—in fact it is highly unlikely—that the same families and their descendants lived in the same house or on the same site for as long as two centuries, much less five (e.g. 395F in Avinella, which was significant from **600** to **150**). For one, the necropoleis corresponding to family groups indicate uninterrupted use for 50 or 100 years (*Appendix C14*) in most cases.[1]

In **450** half of the resident farm population consisted of families making a new beginning. For some the attempt was apparently unsuccessful. Six of these new farmhouses were abandoned after **450** and never again occupied. These are similar to the single-period sites of the mid-6th century BC (see Ch. 21, Giardinetto, upper Venella). In the mid-5th century BC, as in the mid-4th, there were proportionately more single-period sites in remote areas, such as Campagnolo, than in the other regions. They seem to come into being in response to a perceived opportunity, such as an expanding market for Metapontine grain in **550**, or, as will be argued, a sudden availability of land as a result of a new organization of the chora in **450**.

[1] Two to five generations' duration of family burial plots was to be the case in the excavation of the Pantanello necropolis (Carter 1998e, 153).

Prime farmland. The spatial distribution of farmhouses in the study area in successive periods indicates clearly that some areas of the chora were favored over others. From one period to the next the pattern constantly changes. One way of tracing both large-scale and meaningful shifts is by comparing the numbers of significant farmhouses in each locality and region with the total number in the whole study area in successive periods (*Appendix C5*). Between **550** and **450** there is a relative percentage decrease in the population of the Early Village Clusters, from 52% in **550** to 45% in **450**. The largest decreases took place, as might be suspected, in Sant'Angelo (from 13% to 8%), but also in Avinella (from 14% to 9%). The only village that really grew was Venella, which increased from 9% to 13%, matching the percentage decrease in both Sant'Angelo and Avinella.

The other region that registered a loss with respect to the other three was that of the localities Close to the City, Pizzica, Pantanello, and the combination of scattered sites in five localities, DBSBS (Demanio Campagnolo, Basento valley, San Salvatore, Bradano valley, and Serra Marina). The region consisting of four localities Far from the City and parts of three others, AVS (Avinella, Venella, and Sant'Angelo), remained very much as it had been in **550**, with about 20% of significant farmhouses.

The Central Plateau grew strongly from **550** to **450**, with the prime agricultural zone Lago del Lupo accounting for much of the gain. Considering the percentage increases within individual localities between **550** and **450**, the largest percentage gain (300%) was made in the Campagnolo locality, but this is also the result of the very small numbers involved: a net gain of three farmhouses for a total of four. Among the more populous localities, Lago del Lupo (14 sites in **450**) was up 75%, Cogno del Prete (5 sites) up 66%, and the Venella (10 sites) up 66%. This is indicative of which areas of the chora were the most sought-after in **450**—the Central Plateau and the valley of the Venella along its southern border.

Successful farmers. In the expansion of the mid-5th century BC it is reasonable to expect that some farmers, as always, did better than others. Using FMCE values, our gauge of the relative material success of the farmhouses, we shall consider how these values were distributed among the population of the chora in

450. The average FMCE value for all 76 farmhouses in **450** is 1.97, considerably lower (by about 8%) than the average value for the 64 farmhouses of **550**. In fact it is the lowest average value recorded until **200** (*Appendix C7*). How should we reconcile this with the great revival of the countryside in **450**, which saw a net increase of sites in most localities and regions? More farmhouses, but more modest, might suggest that changes had taken place in the way that material wealth was distributed, and that the new population was more egalitarian.

The distribution of FMCE values in **450** is illustrated, along with those for all twelve periods, in Appendix C8. Like the distribution of farmhouses over the landscape, its value as an interpretive tool is fully realized only when it is compared with distribution in other periods. A chi-square test indicates that there is a difference between distributions in all periods that is statistically significant (*Appendix D,* chi-square 1 and cross-tabulation, *D4*). The cross tabulation shows that there are many more than expected low FMCE values in **550** and **450**, but many fewer than expected high values in **450**.

If there are relatively few very successful farmers in **450**, who were they and where were they located? Were farmhouses with higher and lower FMCE values in **450** evenly distributed over the regions and localities? The answer is no: the localities Close to the City have the higher FMCE average in **450** (*Appendix C7*). In this table weighted averages have been computed for each locality and region. From this point on the primacy of this region is unchallenged (except in **300**, and in the periods of Roman administration, **100** and **50**). The second place in **450** is taken by the localities Far from the City.[2] Campagnolo was a previously little-occupied extension of the Central Plateau. There was a single significant farmhouse (301F) there in **550**. Despite its isolation it must have been appealing to farmers desiring what may have been comparatively virgin land. Apparently, the flat terraces of Campagnolo rapidly lost their appeal, as there is only one site there in the next 50-year period.

The area ideal for plow-based agriculture, the Central Plateau, is just average in terms of FMCE value. The mean, based on 19 values, is equal to that for the study area as a whole. If any area were egalitarian, this is it. There are few very high or very low values. This is an area where the division lines and the canals of the bonifica are concentrated, and it is also where the largest numbers of ex novo sites spring up in **450**.

The Early Village Clusters, whose material success appeared the greatest in **550**, rank last in **450**, with an average, based on 34 values, well below that for the whole study area, despite high averages for Giardinetto and Sant'Angelo. The generally low values for farmhouses in Avinella and Venella may reflect a relative downturn in the mixed agricultural and pastoral economy that we suspect characterized that area of the chora. There are areas suitable for cereal cultivation, but they are dissected and far less extensive than those of the Central Plateau, Lago del Lupo in particular. We may conclude that in **450**, as in every period, location mattered. As in most periods, it paid to be close to the city.

Farmhouses and burials. The increase of necropoleis in **450** reflects the change in the overall settlement pattern, which sees more families residing permanently in the chora. It also reflects a change of attitude about burial.[3] As the study of the Pantanello necropolis, with its 320 burials covering three centuries from **550** to **300**, showed, beginning in the second quarter of the 5th century BC, not only were mature adults buried, but also increasing numbers of younger adults and children in the same burial plots as the seniors. In contrast to the 6th century BC, the burials of the 5th century exhibited a less uniform use of expensive containers; instead, humbler inhumations in tile tombs or in the bare ground became increasingly frequent. There was an evident democratization of burial, in concert with similar trends throughout the Greek world.

Necropoleis

The emphasis up to now in describing the development of the chora has been on the farmhouses, rightly so. While the importance of the necropoleis, which constitute 28% of the sites in the study area, has been acknowledged, they have not so far received commensurate attention. In **450**, there is a 61% increase in the number of burials compared to an 18% increase in farmhouses. From **450** their numbers grow relatively

[2] A closer look at the components of this average reveals that three of the eight single-period sites in this period (which often have quite high FMCE values) are concentrated in Campagnolo. Were the single-period sites to be removed from Campagnolo, the average for the localities Far from the City (including eight sites from Bufalara and Scarciullo) would be below average for the study area as a whole.

[3] A similar conclusion was reached independently in the examination of burials in a very large necropolis in the chora at Pantanello (Carter 1998c, 90).

faster than that of the farmhouses (*Appendix C1a*). The increased space devoted to necropoleis in this and the following chapter reflects their fundamental place in the settlement of the chora.

Necropolis statistics. The fresh wave of settlers in **450** established permanent homes and continued the practice of burying the dead in close proximity to the living. Half of the significant farmhouses in the study area at this time were ex novo, but the percentage of new necropoleis was even higher at 57% (12 of 21) (*Appendix C26*). This percentage is only surpassed in **400** (68%), after which there is a steady decline in ex novo necropoleis.

These new burials were not evenly distributed throughout the chora (*Appendix C15*). In **450**, half (5 out of 10) of all new burials occur in the Early Village Clusters. In the Central Plateau two of the four necropoleis dateable to **450** are new, but in Lago del Lupo 75% of the 24 necropoleis are not dateable, compared with 54% for the study area as a whole, making the statistics for this area of somewhat questionable value. The burials Far from the City and Close to the City, though they are few relative to the other areas, are all new. None of the necropoleis in these areas that were active in the 6th century BC were reactivated in the 5th. For example, necropolis 374N in Scarciullo was abandoned after **550**, and not used again for burial for two centuries. This is not the case in Early Village Clusters where there is continuity in burials, at 524N in Giardinetto and 8N in Sant'Angelo. In the Central Plateau, both of the necropoleis (70N and 265N) in Cogno del Prete that were active in **550** continued to be frequented in **450**.

Some necropoleis remain in continuous use for as long as two centuries; others for only a half-century (*Appendix C14*). If continuity of burial in a single necropolis indicates that the same family made uninterrupted use of it for the whole span, it would imply that the rural society was quite conservative. Two centuries means at least six generations buried their dead at the site of 524N in Giardinetto, or that one or at most two generations used 299N in Lago del Lupo. Whereas the latter seems likely, especially in an area like Lago del Lupo that was developing rapidly with an influx of new farmhouses, six generations in a village seems extreme. Still, it is hard to imagine that two different families might have used the same plot without an interruption. The example of the Pantanello necropolis indicates that half a century, at least, passed between the abandonment of a family nucleus and its appropriation by a different group.[4]

Duration of necropoleis. The diagram in Appendix C14 illustrates the history of burial in the 74 separate dateable necropoleis in the chora. We cannot say this is the history of all burial, because there were 28 farmhouses with closely associated tombs (F/Ts) in which the burial component could not be isolated and dated separately. These farmhouses with tombs comprise over a quarter of the total burials in the chora and could alter the picture in any period between **600** and **250**, in terms of numbers of burials in any given locality, and of their distribution among the various localities. For example, there are no necropoleis active in **600** in Sant'Angelo, but there is a farmhouse (285F/T) with associated burials that may already have begun in that period.

The most important information about necropoleis in **450** is that, of the eight necropoleis that began to be used in **600**, only half continued in **450**, and only one in **400**. Similarly, of the five necropoleis that started in **550**, only two continued in **450**. If their continuous use over long periods indicates that the same family or families were using them, then we may see the fortunes of the original pioneer families reflected—their decrease and eventual extinction. Sites 401N in Avinella, 81N in Cogno Prete, and possibly 635N in Venella continued to receive burials with interruptions for four centuries, but these cases are special and will be discussed below. This interpretation of the diagram, in terms of family history, also implies that the early families continued to bury in these necropoleis in **500** when the chora was home to very few.

The essential facts that can be read from the diagram (*Appendix C14*) are that many of the families who buried in **600**, **550**, and **500** ceased to do so after **550** and **500**. Their number trailed off in **450** to almost none, though some burying grounds revived after gaps of a half-century or more, as in the excavated Pantanello necropolis.[5]

By far the largest influx of new burials occurred in **450**. Of the first wave of farm families in the mid-5th century BC, however, less than 50% continued to use these new necropoleis for more than half a century. This result appears to confirm an early impression

[4] Carter 1998f, 218–20.

[5] Carter 1998f, 218–21, with reference to the mid-4th c. BC.

based on the farmhouse analysis that the influx in **450**, especially in the Central Plateau, was not long-lived. Looking ahead, the survival rate from the second wave of settlers in **400** was much higher: 15 out of 20 necropoleis continued in use without interruption for a century, and nine—nearly 50%—lasted for at least a century and a half.

Five exceptional necropoleis. A group of five necropoleis stands apart from the rest because of the relatively large amounts of pottery and other remains found on the surface: 81N[6] and 70N in Cogno del Prete, 471N in Giardinetto, 401N in AVS, and 415N in Pantanello.[7] From three of them (471N, 401N, and 415N), exceptional quantities of figured pottery were recovered. Three (401N, 81N, and 70N) began to be used in the early occupation of the chora in **600** or **550**, and the other two (471N and 415N) are ex novo in **450**. None of the other 69 dateable necropoleis has an EAW greater than 3.0, but all of these at one point or another in their histories surpass 7.0, and 401N in **350** had an EAW value of 21. The possibility that this group of necropoleis served a larger community, unlike the necropoleis often associated with individual farmhouses, is reinforced by their topographical positions.

Site 415N in Pantanello is not near any documented farmhouse, and 401N, the most impressive of all in terms of both the quantity and the quality of the pottery, lies midway between the sanctuary 397S in Avinella and the Sanctuary of Artemis in San Biagio. Its long continuous history parallels that of the more important of the two sanctuaries at San Biagio and suggests a connection. Among the five, 471N seems part of the fabric of the community in Giardinetto, close to a contemporary farmhouse (465F), though not as close as most of the others in this Early Village Cluster. The large concentration of limestone tomb slabs at 401N suggested to Cesare D'Annibale, who led the survey in 1982, that this was a large necropolis at one time. He conservatively estimated ten tombs of various types. Some of these slabs were still evident in recent revisits to the site, but much material had been destroyed or removed.

The somewhat isolated positions of these five necropoleis and the concentrations of impressive tomb markers—far above the average for all other necropolis sites—define a special class that was probably utilized by a large well-to-do family or by a larger community of various family groups, as was the Eastern necropolis at Pantanello (see below).

Excavated necropoleis.

Survey necropoleis. Four necropoleis discovered by the survey team were subsequently excavated, and thus provide a rare opportunity to compare results and judge the accuracy of the survey's prediction about the size and periods of activity at necropolis sites.

Necropolis 415N was fully excavated in 1983. It was identified as an eastern extension of the Pantanello necropolis, and subsequently referred to as the Eastern necropolis at Pantanello.[8] It was, as the survey indicated, unusually large, and its period of usage was confirmed as beginning in **450**.[9] The other necropoleis include the much larger Crossroads necropolis at Pantanello,[10] located at the intersection of Division Line Road (L21) and the Basento Road (west), and the two small necropoleis in Pizzica, 736N and 737N, along division line L06, discussed at length in Chapter 28.

The pottery on the surface of both Pizzica sites included fragments of large, dateable red-figure vessels. The survey results indicated that 736N began to be used in **400** and was abandoned by **250**, while 737N was occupied only in **400**. In this last case, the study of the pottery instead proved that it was in use for two generations in **450**, in the middle rather than the late 5th century BC, though red-figure pottery from at least one burial destroyed by the earth-moving operation that brought the material to the surface dated to **400**, like the earliest material from 736N.

Pantanello. A comparison of the chronological development of a large sample of excavated burials can serve as a useful confirmation of the trends that the surveyed necropoleis suggest. This line of research becomes more valuable when there are multiple sources of information—other necropoleis in various areas of the chora.

[6] **450** is the only period from **550** to **250** when 81N was not active.

[7] This necropolis was excavated soon after its discovery by the survey team (see n. 9, below).

[8] Carter 1998b, 27.

[9] The excavated Eastern necropolis at Pantanello (Carter and Hall 1998, 397–428) was in use from **450** continuously down to **300**. There was a single deeply interred enchytrismos burial, T 151, of the mid-6th c. Pottery of this period does not appear in the survey assemblages. See also Ch. 20, Enchytrismos burials.

[10] Excavation began immediately after its discovery in 1982 by the survey team. It was not given a survey site number.

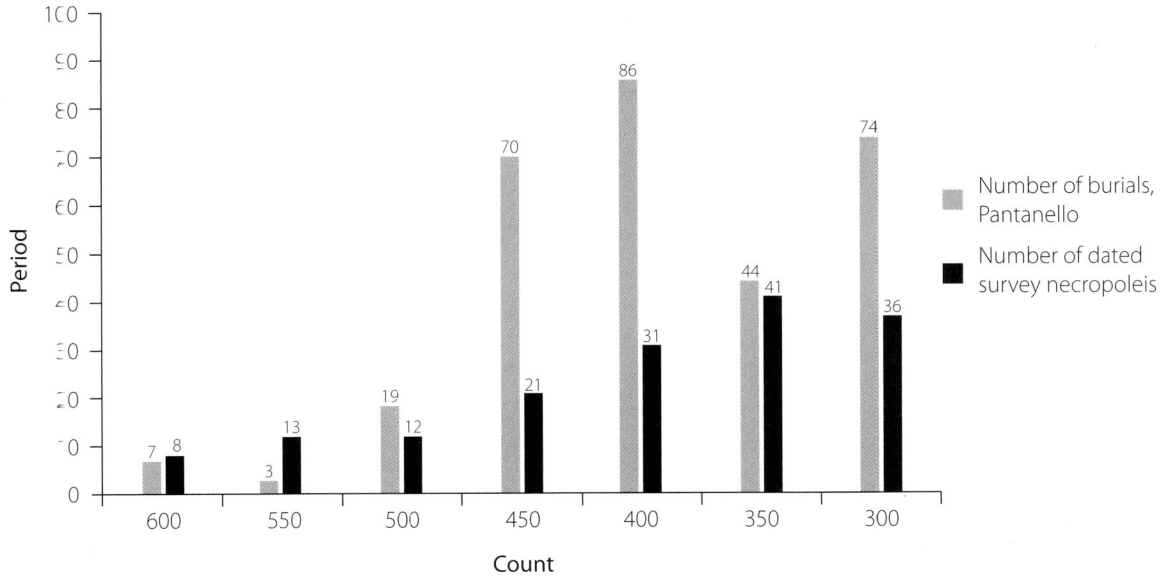

Figure 23.11 Pantanello necropolis burials (303*) and dated survey necropoleis (68**) (EAW > 0.45) by 50-year periods, compared. *The midpoint of the weighted date-range at Pantanello is used and each burial is counted in one period. **Each necropolis is counted in all periods in which its EAW ≥ 0.45.

303 dated burials from the Eastern and Crossroads necropoleis at Pantanello are the basis for this analysis. The burials of the Pantanello necropolis constitute the largest sample[11] so far excavated in the chora.[12] Can the chronological development of this necropolis be considered representative of the chora as a whole?

Apart from the middle of the 4th century BC, the growth in the number of necropoleis and the increase in the number of burials in the Pantanello necropolis match each other closely: there is difference in the size of the increases, but the trend is consistently upward (*Fig. 23.11*). The middle of the 5th century BC was a decisive period throughout the chora. The dramatic rise in the number of families burying at Pantanello in **450** reflected a trend that has been revealed by the survey. Now we know that between **500** and **450** necropoleis in the Early Village Clusters nearly doubled: half of the ten necropoleis there in **450** were new in this period, all necropoleis in the localities Close to the City became active for the first time in **450**, and in the Central Plateau half of the four dateable necropoleis were new in **450**. The mid-5th century BC marked a new era in the chora.[13]

Developments at the Pantanello and Pizzica necropoleis in **450** have broader implications for the chora as a whole. The first burials lining the sides of Division Line Road (L21) at Pantanello occur in **450**, a strong indication that this axis was established at least as early as the mid-5th century BC. This road was a part of the system of canals and roads that were created both to drain and to divide the chora (see Ch. 28). These burials, none of which date before **450**, are essential evidence for dating this system or at least the parts of the system that have been excavated. The burials discovered in Pizzica along division line L09b begin in **450** (see discussion in Chs. 28 and 29) and the majority date to that half-century, though, as at Pantanello, burial here also continued.

Ceramic Assemblages

Farmhouses. Analysis of the farmhouse assemblages in **450** follows the pattern seen in assemblages of **600/550**. The percentages within each functional group of the 13 assemblages (*Figs. 23.12–23.13; Appendix C23a* and *b*) are comparable, though perhaps not as closely as they were for the farmhouse assemblages of the 6th century BC.[14]

The percentages of functional groups in **450**, though close to those for farmhouses of **600/550**,

[11] For a plan of the Pantanello necropolis, see Fig. 20.29. See also Ch. 25, Necropoleis.

[12] Smaller excavated and published necropoleis of this period include Saldone (30 burials) and Sant'Angelo (8 burials). The number of burial sites excavated in the chora has been greatly augmented by the excavations of the Soprintendenza since 2000. It continues to grow.

[13] The burials in the three nuclei at Saldone and in the single nucleus of burials at Sant'Angelo Vecchio tell the same story. For Saldone and Sant'Angelo, see Carter and Hall 1998, 429–47.

[14] See *Appendix D*, chi-square 9, and Ch. 25, Assemblages.

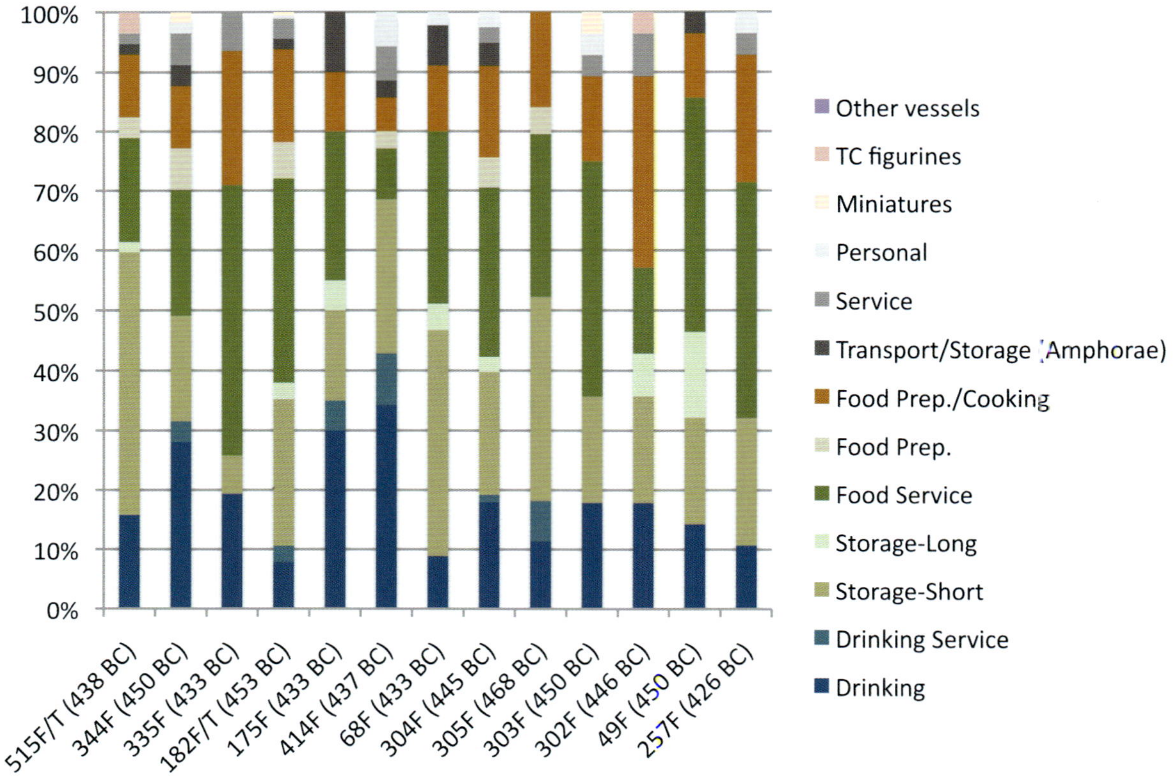

Figure 23.12 Farmhouse assemblages in **450**: functional groups—counts and percentages.

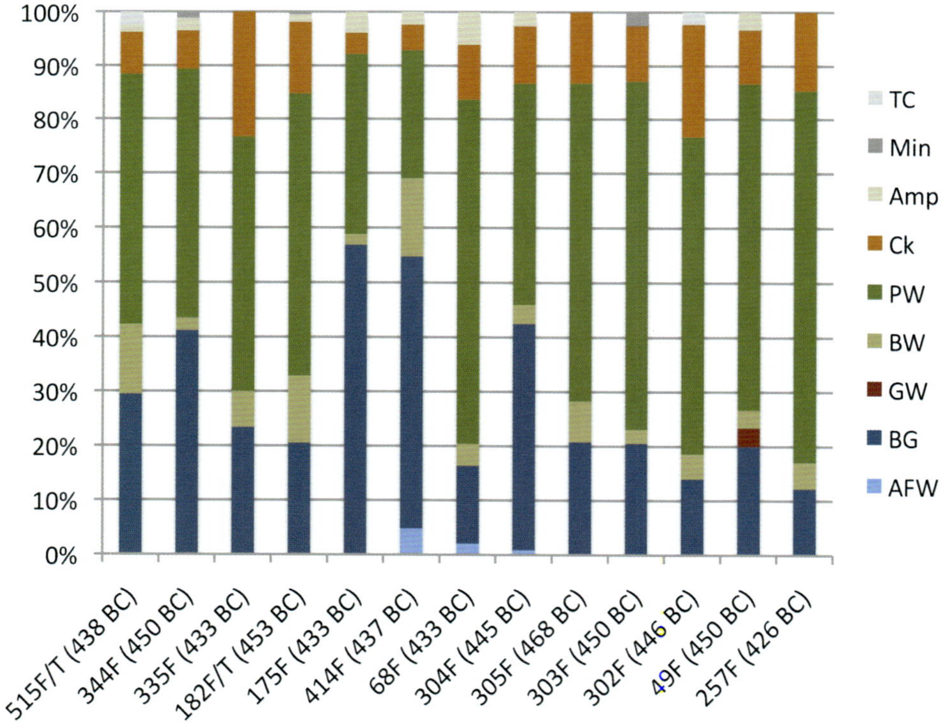

Figure 23.13 Farmhouse assemblages in **450**: ware classes—counts and percentages.

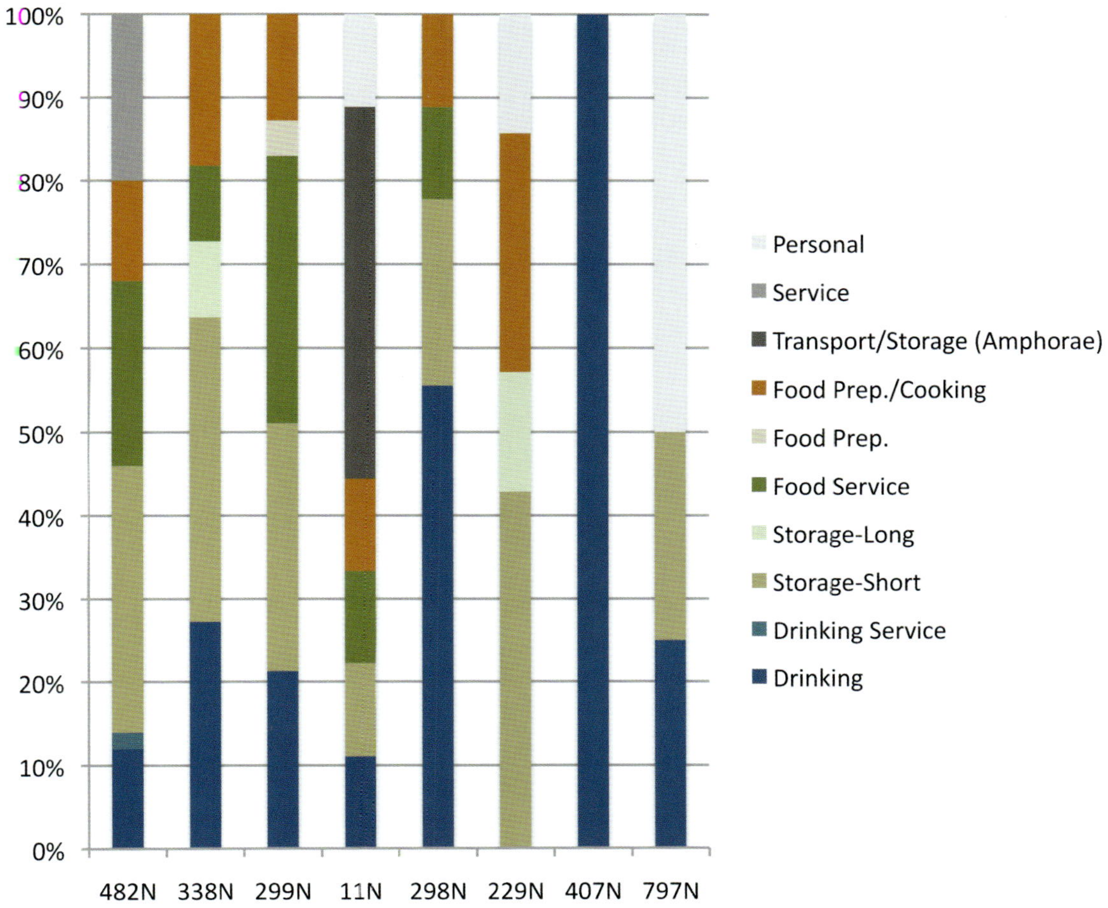

Figure 23.14 Necropolis assemblages in **450**: functional groups—counts and percentages.

are different in several respects. This is evident in the graph (*Appendix C17*) that compares the averages of functional groups from four periods (**600/550**, **450**, **300**, and **100**). There is a lower percentage of drinking vessels in **450**, and a higher one of food service. Both of these changes carry over into the two later periods. There is also a decrease in **450** in the percentage of service vessels including the numerous one-handlers that might be used for dipping, and hence are related to drinking. Relatively minor differences between periods are also observable in a graph of the averages of ware classes (*Appendix C18*).

Necropoleis. An analysis identical to that for the farmhouses was carried out for eight necropoleis assigned to **450** (*Fig. 23.14*). Except for 482N and 299N, these sites have less than a dozen identifiable forms, and so statistics for these need to be treated with considerable caution. The two most fully documented necropoleis 482N and 299N, (*Fig. 23.15*) have similar profiles, but there are sizeable differences in the percentage of drinking vessels present in each. At 482N it is 9% lower than at 299N, but the difference may be due to the fact that some vessels served multiple functions. If

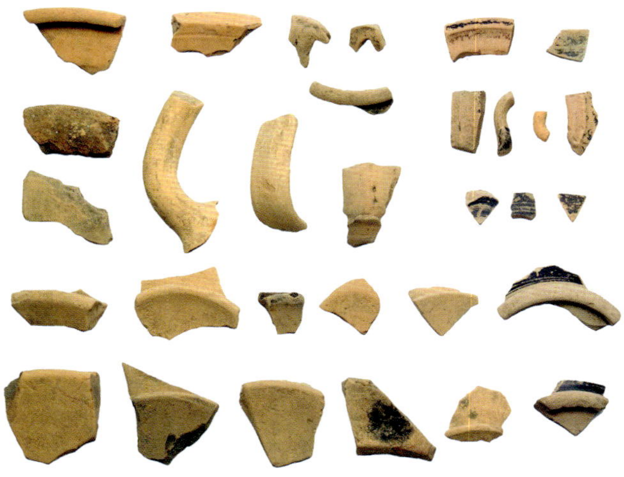

Figure 23.15 Part of the assemblage from 299N, a necropolis in Lago del Lupo, active only in **450**. (MLG)

the one-handlers at 482N were used as cups instead of general service vessels, then there would be no difference in the percentage of drinking vessels between the two necropoleis. The percentage of cooking ware in these two necropoleis (12–13%) corresponds well with that of 70N a century earlier (*Appendix C28*).

A comparison of the average percentage for these two periods (*Table 23.1*) shows that drinking and drinking service may have been better represented in burials of **550** than in those of **450**. Though the average for food service—lekanai and the like—is lower in the early period, the percentage for the best-documented site, 70N, is practically the same as the average for **450**. Likewise, the high average for cooking ware in **550** was elevated by a single site, and the percentage of cooking ware at 70N is close to that for cooking ware in **450**. In sum, the percentages in the various functional groups in necropoleis for **550** and **450** parallel rather closely those for farmhouses in the same periods.

Period	600/550	450	300
Drinking	32%	21%	18%
Drinking Service*	8%	< 1%	12%
Storage Short-term	26%	29%	21%
Storage Long-term		< 1%	7%
Food Service	6%	14%	17%
Food Preparation	1%	< 1%	0%
Food Prep./Cooking	22%	16%	16%
Transport/Storage Amphorae	1%	1%	4%
Service	0%	1%	2%
Instr. Domesticum	0%	< 1%	< 1%
Personal	3%		2%
Ritual/Votive	< 1%		1%

Table 23.1 Necropolis assemblages **600–300**: average functional groups. Necropoleis with fewer than 5 identifiable forms are excluded. *Elevated levels of this functional group may be due in part to the custom of using kraters and other large containers of liquids as grave markers.

A different and fuller picture of a necropolis assemblage emerges from the study of the large ones (401N, 471N, and 635N), which can only be generally dated to the late 5th and 4th centuries BC (*Table 23.2*). The most obvious difference between these three and the less fully documented necropoleis of **450** is not so much the diminished percentage of drinking vessels, but the high percentage of drinking service—in the cases of 401N and 471N, the highest of any functional group. The forms typical of drinking service, often decorated in the red-figure technique—pelikai, amphorae, and hydriai—were numerous in these assemblages. These vessels were often too large to fit comfortably in tombs and were probably not part of the grave goods. They were probably used for drinking service during the ceremonies and libations at the tomb, and were often erected over the tomb afterward as grave markers.[15] Figural vessels in these shapes are most often found at the larger necropoleis, where more ostentatious rites may have been common, and not so often in small family burial grounds. In other respects the assemblage from these necropoleis is very similar to those of the more modest necropoleis.

Site	401N	471N	635N
Best Date	353	341	319
Total Fragments	231	113	129
Drinking	(18) 17%	(5) 11%	(9) 21%
Drinking Service	(32) 30%	(17) 37%	(5) 12%
Storage Short-term	(12) 11%	(4) 9%	(7) 17%
Storage Long-term	(1) 1%		
Food Service	(20) 19%	(9) 19%	(6) 14%
Food Preparation			(5) 12%
Food Prep./Cooking	(7) 7%	(3) 7%	(2) 5%
Transport/Storage Amphorae	(8) 7%	(6) 13%	(2) 5%
Service	(5) 5%		(1) 2%
Instr. Domesticum	(0) 0%	(1) 2%	
Personal	(3) 3%	(1) 2%	(4) 10%
Ritual/Votive	(0) 0%		(1) 2%
Unknown	(125) 54%	(67) 59%	(87) 67%
Known	106	46	42

Table 23.2 Necropolis assemblages: functional groups averages—large necropoleis of the 5th and 4th c. BC.

Figured Vessels from Necropoleis and Farmhouses

Worthy of special attention are the figured vessels. They are found in survey contexts—not only at necropolis sites, but also at farmhouses—in **450**, though they are even more numerous in **400**, and continue to appear regularly in **300**.[16] The 230 fragments of figured pottery collected during the survey from 96

[15] Carter 1998d, 115–17.

[16] Figured vessels begin to appear in survey assemblages in **500**, and in excavated necropolis assemblages already in the early 6th c. (see Ch. 20).

sites include mainly red figure (RF) and Gnathian (Gna), but also a few fragments of black figure (BF), the less well-known Xenon Group (XG), and black-on-buff ware (BoB), a local product contemporaneous with Gnathian (Ch. 8) not to be confused with the mostly imported black figure of the 6th and early 5th centuries BC. In a number of cases the fragments have been large enough not only to identify the type of decorations, but often the shape of the vessel and the stylistic milieu (e.g., Early Lucanian, Late Apulian) to which it belongs—and in some rare cases the painter himself: e.g., the Pisticci Painter and the Darius Painter. Though small in number as well as size, these fragments are disproportionately important for gauging the social and economic level of the rural population. The excavation of larger and smaller rural necropoleis at a variety of locations in the chora has demonstrated that figured vessels, often of the highest quality, were placed in the graves of Metapontine farmers. Figured pottery was a prestige item. As a result of the survey, we know this practice was widely diffused, and occurred in all four of the principal regions of the study area, in comparable quantities. Even those localities at considerable distance from the city, where few necropoleis were to be found, are represented (see *Appendix C29*).

The most surprising result of the ceramics research is that fragments of figured pottery turn up in practically equal numbers where the families resided. In fact there is an equal percentage (20%) of farmhouses with figured pottery (56 out of a total of 284 farmhouses without burials) and of necropoleis (32 out of a total of 161 necropoleis). That of farmhouses with burials (7 out of a total of 28 F/Ts) is only slightly higher (25%). It is clear that the more costly and prestigious ceramics, far from being restricted to the special occasion of burial, were household objects for a sizeable portion of the population. All the excavated Metapontine farmhouses contain examples of them, and in the case of Fattoria Stefan, in considerable quantities.[17] The widespread consumption of figured pottery in the countryside leaves no doubt that the economic and social level of the rural community in Metaponto was in no way inferior to that of those who dwelt in the asty. This section is a brief survey of the types of figured pottery that occur commonly in the chora of Metaponto from **500** to **300**.

Black-figure pottery. Only three identifiable fragments of this fine pottery all dating to **500** were recovered by the survey team. All were found on necropolis sites: Giardinetto (524N), Cogno del Prete (265N), and Campagnolo (50N). The fragment from 50N (Ch. 8, Cat. no. 1) is dated to the late 6th century BC, while the other two fragments date to the late 6th or early 5th century BC and come from necropoleis active when the chora was in crisis. The scarcity of figured pottery in surveyed necropoleis dated before **450** contrasts sharply with the relative abundance from **400** onwards. Excavation results indicate that the use of figured pottery in burials of the early period (see Ch. 20, Necropoleis and Sanctuaries) was more common than the survey would indicate.

Red-figure pottery. Pottery decorated in the red-figure technique is found on all but 14 of the 96 sites with figured pottery. In contrast to the black-figure pottery, none of the fragments recovered by the survey can be identified as Attic. The appearance of red figure in survey assemblages in the periods **450/400** corresponds with the initial period of locally produced red-figure pottery by the early masters of the Proto-Italiote. Several of them, it appears, were actually working at Metaponto together with itinerant Attic vase painters, who established temporary bases there in the 5th century BC.[18]

The great mass of the red-figure fragments belong to the end of the 5th century and the 4th century BC. Among the vase painters represented are the Pisticci Painter (*Fig. 23.16*), the Sisyphos Painter, and the Dolon Painter, who decorated large vessels that were used by residents of the chora as tomb markers. Examples of the works of these painters and others were first identified in excavations of necropoleis in the indigenous interior. With the great increase of excavated burials in the chora it is now clear that the chora was a major market for this fine pottery. Numerous examples were recovered from the Pantanello, Saldone, and Sant'Angelo necropoleis, and excavations since 2006 have dramatically increased the numbers.[19]

In the 4th century BC, red-figure vessels in the Apulian style make their appearance, and by the mid- and late 4th century BC, Middle and Late Apulian style vessels are the only red-figure found in the chora.

[17] This farmhouse will be the object of a future study.

[18] Denoyelle 2008, 342–43; De Siena 2008, 590–91.

[19] De Siena 2007, 435–45; Osanna 2008, 925–31.

Figure 23.16 Red-figure lebes gamikos from tomb 354 by the Pisticci Painter, ca. 430 BC (h 17.8 cm). (CW)

Whether they were all imports or some actually locally produced in the kerameikos of Metaponto is now the subject of ongoing research and debate.[20]

Gnathia, Xenon Group, and black-figure ware. Gnathia style pottery was produced at various centers in southern Italy in the second half of the 4th century BC, and it is virtually certain that Metaponto was a production site (*Fig. 23.17*). Gnathia vessels were covered with black gloss, with designs, often floral motifs, overpainted in white, red, and yellow. This material is present in the chora in much smaller numbers. A total of 26 fragments was distributed among 23 sites, 18 farmhouses, 4 necropoleis, and one farmhouse with burials. Only one site—a large necropolis (401N)—yielded more than one fragment. Whereas red-figure pottery was found in slightly greater quantities at necropoleis and farmhouses with burials (F/Ts), and was as likely to turn up at a necropolis as at a farmhouse, Gnathia pottery was primarily found in a residential context.

Just two fragments of pottery were identified as belonging to the Xenon Group, a type of pottery with a leaf pattern in red painted on reserved buff-colored clay that appears to have been created for the indigenous market. One, a fragment of a kantharoid vase (Ch. 8, Cat. no. 77), was found at a farmhouse with burials, which was in all likelihood also a pottery production site on the Bradano side of the chora (521F/T/K) in Giardinetto. The shape, not one commonly employed at Metaponto, had a wide distribution in interior Lucania and Apulia. The other (Ch. 8, Cat. no. 76) was the only piece of figured pottery from a site identified as a sanctuary in Cogno del Prete (266S), also located on the Bradano side. The Xenon Group pottery dates to the 4th century BC generally; the kantharoid vase in particular to the second quarter of that century.

Two skyphos fragments decorated with leafy tendrils are in a black-figure technique that is particular to Metaponto, referred to as black on buff (BoB). Both come from a single farmhouse site (312F) in Avinella (Ch. 8, Cat. nos. 98 and 99). The standard black-gloss pottery from the site indicates it was occupied first in the mid-5th century BC, again in the 4th, with the highest concentration of securely dateable material belonging to **300**. This is the probable period of the black-figure ware.

Figure 23.17 Gnathia-style hemispherical bowl from tomb 71, Pantanello necropolis, ca. 300 BC (diam. 15.7 cm). (CW)

Shapes. The shapes of large vessels that can be identified from their fragments—all decorated in the red-figure style—are principally kraters, hydriai, pelikai, and amphorae. The recognizable krater forms and their counts are: volute (9), calyx (4), bell (2), and column (1). Seven other fragments have been generally classified as kraters. All six fragments of amphorae are of the sort known as pseudo-Panathenaic. There

[20] Silvestrelli, pers. comm., July 2008.

were also six examples each of fragments from hydriai and pelikai.

The majority of the fragments attributed to large decorated vessels comes from sites identified as necropoleis. A more limited number of large vessel fragments was recovered from farmhouses with tombs (F/Ts), where their function could well have been the same. At 728F/T there were at least four kraters. Slabs of the sort that were used for cist tombs—the monumental tomb type, in use in the 5th century, but particularly popular in the 4th century BC in the chora—were also present on the site. The occurrence of large, decorated vessels in a funerary context should be no cause for surprise, but when they turn up consistently in domestic assemblages, other possible functions have to be considered. Three red-figure krater fragments were found at the venerable farmhouse 395F in Avinella, one of the longest-lived farmhouses in the chora. Gnathia bell-krater fragments occur at a necropolis (464N) in Giardinetto, and at three farmhouses (250F, 326F, and 727F). In two cases these were the only large vessels from the farmhouses. At farmhouse 250F, in the Casa Preti village in Lago del Lupo, besides the Gnathia bell krater, there was a red-figure pseudo-Panathenaic amphora fragment. We have already noted that Gnathia-style pottery is found mainly at farmhouse sites, which suggests that the bell kraters were a normal part of a service used for symposia, on special or ceremonial occasions in the lives of the farm families. The other Gnathia vessels found at farmhouse sites consist of skyphoi (395F, 342F in Avinella, and 521F/T/K in Giardinetto). The existence of fine pottery in the home was evidently not widespread. The evidence is concentrated somewhat in Avinella, but is also documented in Lago del Lupo (250F), where a red-figure pseudo-Panathenaic amphora may have occupied a place of honor in the household beside the Gnathian bell krater. The family that occupied farmhouse 727F in the Basento valley possessed a Gnathia bell krater, and those of 413F in Pantanello, a red-figure pelike.

By far the most common small vessel with figural decoration is the skyphos, which accounts for more than half of the total of small figured vessels. Some were in the red-figure technique (*Fig. 23.18*). The majority of these drinking vessels, however, were decorated in the Gnathian style, and all but one were found on sites identified as farmhouses. All of the C-type skyphoi were in the red-figure technique; only a single example of the A type, in Gnathia, could be securely identified. Seven examples assigned to the general category of skyphoi were all in the Gnathian style.

Figure 23.18 Red-figure skyphos from tomb 330-7.

Lekythoi, represented by a single black-figure example dateable to **500** and three squat lekythoi in red-figure, are in distant second place numerically among the smaller decorated vessels, ahead of two lekanis (lid) examples, and single examples of a bombylios, patera, lekanides, and lebes gamikos. All of these forms would not be out of place in a necropolis assemblage. Three of the functional groups that typify burials in the Pantanello necropolis[21] are present among the fragments, but only three—the single type-C skyphos fragment, one of three squat lekythoi, and the bombylios (attested also in the Pantanello necropolis)—were found in sites identified as necropoleis. It would seem that the smaller decorated vessels were actively used in the homes of these farming families. Why were the smaller vessel shapes, which are often encountered in excavated burial contexts, rarely found at burial sites identified by the survey, while the larger, decorated vessel shapes were often recovered? Possibly because the vessels used as grave goods, predominantly smaller, were more deeply interred than the large markers above them. They had a much better chance of staying underground, especially in areas where traditional plowing was practiced, at a depth of 20–40 cm below the surface, as was the case in much of the study area until the 1980s. The study of the assemblages from **450** has served, as it did in that of **600/550**, to demonstrate the similarity of the individual farmhouses and necropoleis and, in most respects, the continuity with the 6th century. The excursus on figured pottery presents an ample element of the assemblages illustrating the economic and cultural level of many of the occupants of the chora in the mid-5th century—an unexpected and much welcome result of the analysis of the essential data of the survey.

[21] Carter 1998f, 186–87 and passim.

Bonifica

By far the most historically important result of the survey is the discovery of the mid-5th century BC relationship between the division lines and the dwellings and burial places of the many new occupants of the chora. The conclusion that the chora underwent a major transformation of the sort best described by the term bonifica is the subject of this special and major section of Chapter 23. The primary meanings of the Italian word *bonifica* are drainage and land reclamation. It describes a type of undertaking that made its first appearance in Magna Grecia and Sicily at the end of the 6th and first half of the 5th centuries BC, with a subsequent long history in Italy from the Roman and Medieval periods down to the present.

Metaponto

The Metapontine bonifica, first identified as such by Giovanni Uggeri forty years ago,[22] is one of the earlier known examples and has certainly left the most extensive traces on the landscape of any from the ancient world. The Late Archaic and Early Classical *bonifiche* in Sicily and Magna Grecia that are known mainly from literary sources or numismatic evidence have a number of features that they share with the Metapontine examples. Basic land reclamation is usually water management such as the taming or re-routing of rivers, but in the case of Akragas, the bonifica was exceptional, as it involved the taming of destructive winds. Bonifiche were important events that naturally involved the gods—in particular river gods—who were celebrated and represented on coins commemorating the event. They involved the leading personalities in the poleis, and were often accompanied by a distribution or redistribution of land. This last feature is a defining characteristic of the bonifica at Kamarina, and, as we shall argue, also at Metaponto.[23]

The extensive system of canals with flanking roads discovered by aerial photography and excavation shaped the chora in the first half of the 5th century and for the next two centuries. They are the subject of a separate detailed study in Chapter 28.[24] The division lines, ca. 210 m apart, cover much of the

Figure 23.19 Metaponto stater with river god Acheloos, ca. 450 BC (Hirmer Fotoarchiv München; after Kraay 1996, Pl. 82, Fig. 230).

nearer chora (*Maps 18–22*, Vol. IV). In the study area between the Bradano and Basento, they are concentrated in the Central Plateau and they are particularly dense in the southeastern part of Lago del Lupo. Apart from the villages of **550**, there were surprisingly few farmhouses or burials in this area before **450**.

The bonifica, with its kilometers of canals, would have taken a number of years to realize, a fact supported by the dating evidence for individual stretches of the system. There are early-5th-century BC burials along the lines in Pantanello, and votive deposits of later date. The striking of the Acheloos coin (*Fig. 23.19*) almost certainly commemorated the successful completion of the whole project, also celebrated with games in honor of the god—the AΨEΛ◇Ϟ◇ AE⊕Λ◇N of the coin legend. He is not the personification of a local river, as on the Sicilian coins, but the father of all rivers.

The Acheloos coin is part of a group of three Metapontine silver staters, dated by Stazio to the mid-5th century BC.[25] Herakles and Apollo beside his laurel (see *Fig. 23.29*) appear on the other two.[26] Herakles might have been considered an artificer of the bonifica as on the coins of Selinous, and the image of Apollo, which may be a representation of a statue in his sanctuary in the agora (see below, Asty), is connected with the shadowy figure of Aristeas, and perhaps through him with Pythagoras. All this is extremely speculative but, if true, would make the possible parallel with the bonifica of Akragas, reputedly orchestrated by Empedokles, more compelling.

Besides drainage, the canals served another function as a visible divider of the land, a function that

[22] Uggeri 1969.

[23] Later examples in Central Italy, such as the system of underground channels (*cuniculi*) along the Via Appia, in the Pontine marshland, or at Cales, certainly seem to be connected, at least with the arrival of new settlers in these areas. See Quilici Gigli 1997.

[24] The most comprehensive treatment of the subject is in Prieto 2005, 176–85.

[25] Stazio 1974, 79–85.

[26] Noe and Johnston 1984, 55, 70–71, Nos. 311–20, Pls. 23–24.

550				450			
Locality	Earlier farmhouse site no.	Nearest division line	Distance (m)	Locality	New farmhouse site no.	Nearest division line	Distances (m)
Cogno del Prete	266S	PL01	11	Cogno del Prete	68F	L10b ext	3
Giampasquale	200F	L01	14	Cogno del Prete	257F	PL02	6
Lago del Lupo	274F	L17b ext	33	Lago del Lupo	113F	L18c ext	6
Pizzica	741F	L04b	34	Cogno del Prete	260F	PL03	10
Giampasquale	215F	L02b ext	36	Lago del Lupo	503F*	L12b ext	13
Lago del Lupo	254F	L17b ext	38	Cogno del Prete	128F	L12b ext	18
Lago del Lupo	131F	L21c ext	41	Campagnolo	49F	L07c ext	19
Giampasquale	238F	L02b ext	43	Cogno del Prete	57F	L05c ext	20
Bufalara	304F	L02b ext	54	Lago del Lupo	175F	L20b	28
San Salvatore	742F/T	L02a ext	56	Campagnolo	302F	L15f	34
Campagnolo	301F	L15f	66	Lago del Lupo	166F	L16c ext	35
Avg. distance between farmhouses and division lines			**39**	Pizzica	728F/T	L15b	38
				Pizzica	744F	L06b ext	43
				Lago del Lupo	173F	L17b ext	57
				Venella	288F	L21c ext	64
				Avg. distance between ex novo farmhouses and division lines			**26**

Table 23.3 Distances of farmhouses to division lines, their extensions (ext), and potential division lines (PL) comparison of **550** and **450**. *All farmhouses listed in **450** are ex novo except 503F in Lago del Lupo.

the first students of the aerial photography, Schmiedt and Chevallier, recognized immediately.[27] The name division lines, though only a partial description of their function, has, nonetheless, stuck, and for reasons of convenience been retained here.

A second very visible phenomenon in the chora in **450** was the sudden appearance of many farmhouses and necropoleis, ex novo, in sites never before occupied. This is particularly pronounced in the part of the study area, southeastern Lago del Lupo, where there were remarkably few farmhouses in **550** (and none in **500**). The arrival of numerous settlers, divided from each other by physically impressive boundaries in an area previously slightly settled and apparently undivided, strongly suggests a distribution (δασμὸς) of previously unoccupied land, or more likely, a redistribution (ἀναδασμὸς) of land belonging to owners who did not reside on it.

The essential elements—drainage, a land division, and a new population—of a complete bonifica with a redistribution of land, like the historically attested example at Kamarina,[28] were present at Metaponto. All that was missing was an historian to describe it and call it by its proper name. In lieu of an historian's text, we shall attempt to show that these phenomena were indeed related, by demonstrating the close spatial relationship between the new settlers and the new property lines: the canals.

Division lines and ex novo sites. The excavated examples of necropoleis at Pantanello and Pizzica lie along division lines L21 and L09. The earliest burials along L21 in Pantanello date to the first half of the 5th century BC, as do the earliest burials from Pizzica (all of the burials at site 737N are dated between 480 and 440 BC). The excavations carried out by the Soprintendenza confirm that this period saw the first use of the canal system,[29] and it corresponds to that of the sudden influx of ex novo farmhouses in the very area where the division lines are most densely concentrated.

In **450** there is an observable tendency for new farmhouses to be located closer to division lines than those built earlier (*Table 23.3*). This is true of all localities the division lines traversed, but is especially

[27] Schmiedt and Chevallier 1959.

[28] Diodorus Siculus 11.76.6.

[29] Nava 2002, 735–43; 2002, 740–42; 2003a, 664–67.

Locality	Site	Reference	Distance (m)
Lago del Lupo	299	L16	23
	298	L16	23
Cogno del Prete	48	mdpt L06/L07	45
	70	PL03	15
	81	mdpt L07/L08	36
	265	PL01	61
Venella	229	L21	7
	289	L21	18
DBSBS	41	mdpt L12/L13	47
Pizzica	760	L09	50
	638	L17	55
	415	L19	31
	732	L14	3
Pantanello	446	mdpt L15/L16	32
	439	L16	9
	429	L18	19
Pizzica	736 (exc.)	L09	4
	737 (exc.)	L09	17

Table 23.4 Distances from necropoleis **450–400** to nearest division line, potential division line, or midpoint between division lines Red for necropoleis of **450**; blue for those of **450** and earlier; and black for those of **400**.

true of the area of the Central Plateau where the lines are most concentrated.[1]

There is an even closer spatial relationship, as will be seen, between the lines and necropoleis identified in the survey (*Table 23.4*). Establishing a secure chronological connection between the canals and the burials of **450**, however, is rendered problematic by two facts: (1) the number of dated necropoleis is relatively small and the undated ones are more numerous in the crucial area, the Central Plateau, than anywhere else in the chora; and (2) the lines are often not continuous. The interruptions do not necessarily correspond to points where they would have crossed over a valley created by a drainage canal, where discontinuity would be expected.

What appear to be segmented division lines may correspond precisely to a segmented system of canals. The proximity of necropoleis to division lines has been observed in a few instances where excavation was undertaken, it can be demonstrated for the numerous necropoleis identified by the survey. The spatial analysis by Dana is reproduced here:

> Two adjustments will be made in order to explore the spatial relationships between necropoleis and the network of division lines, separated from each other by ca. 210 m. Necropoleis of all periods, dated and undated, that lie within 120 meters of any division line will be included in the analysis, and the division lines that show up in the aerial photos will be extended. The procedure is as follows.
>
> We attempted to answer the question of whether or not necropoleis were near to division lines, beginning with a subset of division lines, those within the study area aligned along the primary division line axis. We constructed a set of *extended division lines*, connecting line segments that were aligned. With a layer containing all sites identified as necropoleis, we performed a *spatial join*, resulting in a table of perpendicular distances between each necropolis and the nearest division line. We then selected those 75 necropoleis within 120 meters (about half the distance between parallel division lines) of a division line. The resulting distribution is bi-nodal, with a peak at about 5 meters and another at 60 meters [*Fig. 23.20*].
>
> Fifty necropoleis are within 60 meters of a division line and 25 between 60 and 120 meters from the nearest line. A chi-square test for independence was performed that suggested that for 20-meter histogram bins the observed distribution had a 1.3% probability of occurring by chance [*Fig. 23.21*].[2]

This analysis proves that the relationship between necropoleis and division lines was not random. For the period in which drainage canals were dug (**450**), and for the two centuries they were active, the relationship was close. We may reasonably conclude that a new group of people are burying in the Central Plateau in **450**, and probably also later in Avinella, where division lines were also discovered (see Ch. 25, Avinella).

[1] The distance between new farmhouses in **550** and the position of the division lines, which were not realized as canals until **450**, is somewhat wider, but still close (cf. the distance between ex novo farmhouses in **350** and **300** and the division lines). This relationship in **550** should support the case made in Ch. 21, 702–03, for the existence of NW to SE axes in the Central Plateau already in the 6th c. BC.

[2] P. Dana, unpublished report, 1 June 2009.

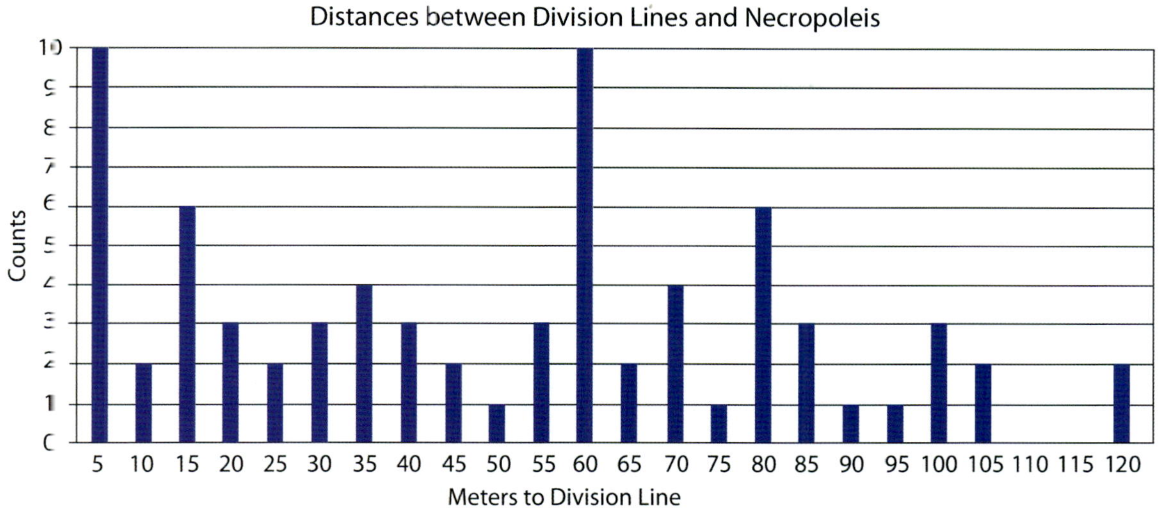

Figure 23.20 Distances from necropoleis to division lines.

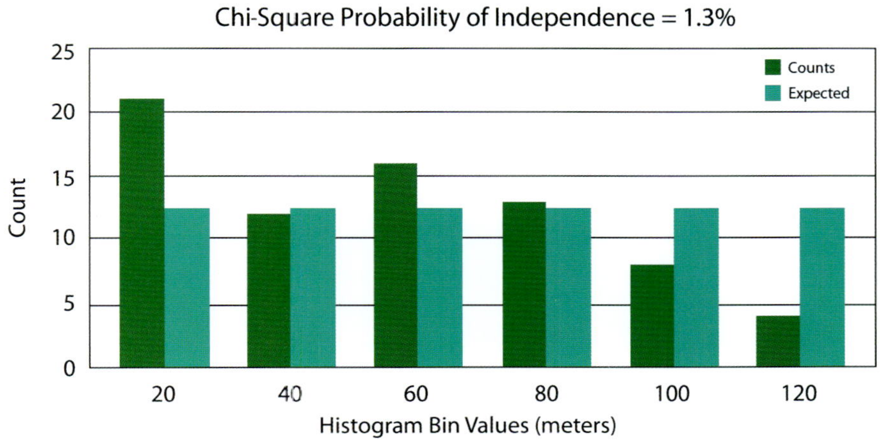

Figure 23.21 Importance of distance distribution.

Possible Parallels

Kyme (Cumae). The earliest example of a large-scale water project for which we have any evidence is that of the tyrant Aristodemos of Kyme (Cumae) in the late 6th century BC. This project followed, consciously or not, the lead of other tyrants, such as Polykrates, who had water channeled to Samos through the Eupalinos tunnel, or Peisistratos, who had aqueducts built to bring water to the Enneakrounos, and perhaps had the agora drained. Aristodemos' efforts differed from these in an important respect. If Plutarch's account can be believed, the tyrant had the chora of Kyme surrounded with a circular ditch, making the citizens of all classes move some dirt.[32] Although deemed useless and unnecessary by Aristodemos's enemies, and Plutarch's hostile source, such a project may have been intended to perform the function of draining, as flooding rather than lack of water was a perpetual problem in this area. The circular form of the ditch—τάφρος was the general term used—has suggested to Cerchiai that it was something more than just a useful bonifica,[33] that it was also the central element of a propaganda campaign to make the tyrant a re-founder of the city. He compares it to the model of geometric space in the Athens of Kleisthenes, with the idea of a circle—a center of a homogenous space—equated with *isomoiria*, an equal division of resources.[34] Unfortunately, no clear trace of this ditch has been found in the area, though there

[32] Plutarch, *Moralia*, 262 A-B.

[33] Cerchiai 2000.

[34] Lévêque and Vidal-Naquet 1996, 81.

is a possible channel in the plain of Licola. It has been claimed that the redistribution of land by Aristodemos to enfranchise former slaves resulted in a substitution of new rich for old rich.[35]

Akragas. It is difficult to say whether or not the verses of Empedokles, whose career spanned the middle quarters of the 5th century BC, and accounts of his concern for the crops and fields of Akragas and Selinous[36] reflect real experiences like that in the chora of Metaponto. They are heavily colored by the legends and fantasy that grew up around his name and those of the other famous civic men of this time: Asparis, Epimenides, Aristeas, and Pythagoras. The accounts do, however, indicate that philosophers thought about how to resolve practical problems of grave concern to their poleis, and provided resources to that end. Empedokles' intervention at Akragas seems to have involved a rather impractical-sounding solution for destructive winds,[37] as well as a promise to teach how to stop winds that ruin crops and how to moderate the effects of too much rain in some periods and not enough in others. After the death of the tyrant Theron and expulsion of his son, the regime in Akragas was democratic, and Empedokles was said to have supported it.

Selinous. Pairs of rivers flanked three of the most prominent colonies of the southern coast of Sicily—Akragas, Selinous, and Kamarina—a circumstance that is closely paralleled by Metaponto and Siris-Herakleia on the Ionian coast of Italy. Empedokles brought about the diversion of two streams at Selinous, removing a cause of pestilence that had affected the health of the population. There was a general celebration on the rivers' banks, and Empedokles was honored as a god.[38] The rivers were represented on the coins of Selinous, which takes its name from one of them. The first coins of Selinous bear the image of a celery stalk, and a golden stalk of celery was dedicated at Delphi, like the golden harvest of Metaponto. Personifications of the rivers sacrificing at the altar of Asklepios appear on a tetradrachm dated 467–445 BC. The river god is young, naked, and horned. There is a cock on the altar and a bull—the traditional symbol for rivers—on the base. The god holds a sprig of laurel and a phiale.[39] Not only were the rivers commemorated on the coinage of Selinous, but also the swamp (Εὐρυμέδουσα).

Kamarina. Pindar's fifth Olympian Ode is the second of two honoring Psaumis of Kamarina.[40] After praising the victory, he passes immediately to a description of the charioteer's newly populated hometown: its sacred grove of Athena, the river Oanos, the swamp between the rivers (ἐγχωρίαν τε λίμναν), and the holy channels (σεμνοὺς ὀχετούς) through which the Hipparis River waters the people and quickly puts together a grove of sturdy houses, drawing the community (δᾶμον ἀστῶν) from weakness to light. Brunet had interpreted Pindar's poem in the light of changing climatological conditions that had resulted in floods following periods of extreme drought.[41] Climatic and environmental conditions justify the bonifica of the river and the modification of the swamp, which was documented in the early exploration of Kamarina in the 19th century by the great pioneer archaeologist of Magna Grecia and Sicily, Paolo Orsi.

Representations of the river Hipparis as a handsome young male appear on the coins of Kamarina, along with that of Kamarina herself, the nymph of the namesake swamp, as a beautiful young female on the back of a swan. The positive image of the swamp reflects the swamp's beauty and utility as a wetland teeming with life.[42] This ancient attitude has not been shared by other ancient and most modern proponents of bonifiche, who have viewed swamps either as sources of pestilence, or, when drained, as potential farmland. The personification and veneration of the swamp at Kamarina and at Selinous, however, was not a local peculiarity.

After the death of the tyrant Hieron in 466 BC, a number of cities of Sicily, like Akragas, reasserted their independence. Kamarina was one of these, and in 461/460 BC the descendants of the original Geloan

[35] P. Caputo 1997, 332. See also Mele 2009, 134–35, and G. Greco 2009, 421.

[36] Diogenes Laertius 8.60.

[37] Diogenes Laertius 7.59–61. He had the population make goatskin bags and stretch them across a valley to stop the wind. The wind mills that now rise in the plains of Croton to generate electricity may not have seemed as out of place to the ancient as they do to some of the modern inhabitants of the Crotonese.

[38] Diogenes Laertius 7.70.

[39] Kraay 1966, 298, Pl. 66, No. 186. See also Kraay 1966, 298, Pls. 67 and 68, Nos. 187–90.

[40] Race 1997, 94–99.

[41] J. Brunet 1971; Cordano and De Stefano 1997, 299.

[42] Borca 1996.

settlers returned and divided up the land into plots (κατεκληρούχησαν) as did practically all the cities, according to Diodorus Siculus.[43] This was probably a redistribution of all the chora, ex novo, because of all the damages inflicted during the war that restored the democracy.[44] Pindar's ode was composed in 452 BC.[45]

Both the ancient city and the chora[46] of Kamarina have been the subject of archaeological investigation since the days of Orsi. Since the 1980s much has been revealed about the orthogonal grid plan of the city, which was divided into long strips like that of Metaponto. The discovery that the division of the chora is an extension of that of the city, and that drainage canals flank the divisions of the chora, has close parallels in the chora of Metaponto. The Kamarina bonifica, which was also contemporaneous with the Metapontine, is directly relevant to this study. The excavators relate all elements of this unfinished project to the "democratic," post-tyrannical "reforming" of the polis promoted by the Geloans in 461/460 BC.[47]

What are the common features of these four possible parallels? The importance of rivers is clear. This is understandable in a society that is based on agriculture and where there are climatic extremes like heavy rains in fall and winter, and drought in summer. The concern with rivers in the early 5th century BC in these areas, as well as in mainland Greece, likely reflects problems caused by these extremes, and more fundamentally, a desire to do something about them. Rivers were sought out by the founders of these apoikiai. They placed the asty by the mouth, and in a number of cases between the mouths of flanking rivers. Such a location had its advantages, but could become a problem, as it did in the case of Selinous, and of course of Metaponto.

The rivers were always personified and often the objects of cult. Their centrality to the life of the polis was further emphasized by the large number of representations of rivers on the coins, beginning already in the 6th century BC. In many of these cases the coins are the most tangible and easily dated evidence for bonifiche. Good archaeological evidence is, unfortunately, much rarer and often much harder to identify. The re-routings of rivers, as suggested by ancient texts, have left few recognizable traces. There are indications of older, abandoned river courses in aerial photographs, but dating these is very difficult. The fullest evidence for the drainage projects are the ditches (τάφροι), only rarely mentioned in sources.[48]

There were certainly valid reasons for bonifiche—disease and crop failure—but one wonders in how many cases there were real emergencies, and not simply a natural desire to improve conditions as part of rational planning for the economy, health, and well-being of the population. What emerges from these case studies is that the wise and efficient management of water was a concern and that there were members of the community in these poleis who were willing to take the lead. They might be tyrants, leaders of the demos and, in some unusual cases, philosophers (Empedokles will not have been the lone example if our intuition that Pythagoras was in some way involved at Metaponto is correct). Water problems, in fact, may have been secondary in some cases. The bonifica of Aristodemos, if such ever existed, seems to have had primarily a political motive. Instances where a bonifica and a distribution or redistribution of land seem to be part of a single project are the possible examples of Kyme and Kamarina. What makes Kamarina an important comparison for Metaponto is the new division of the land, fossilized by permanent boundary markers consisting of, as the excavator insists, roads flanked by water channels. By the time of the Roman colonization of Campania, this combination became very widespread.[49] The settlements of veterans in the Late Republic and Empire provide a plethora of examples where centuriation and drainage of the land were combined.[50] To the early example of Kamarina we may now confidently add Metaponto, the most ambitious and fully realized, perhaps, of all.

The Asty, 450

The second quarter of the 5th century BC witnessed the second great period of building in the urban cen-

[43] Diodorus Siculus 11.76.4–6.

[44] Asheri 1966, 52.

[45] J. Brunet (1971) discusses also the date, the question of authenticity and the interpretation of terms like the ὀχετοί, in the light of local conditions.

[46] Di Stefano 2001.

[47] Cordano and Di Stefano 1997, 300.

[48] See the Herakleia Tablets: Uguzzoni and Ghinatti 1968, 18; Arangio-Ruiz and Olivieri 1965, 4, 645, 1.130. Prieto 2005, 236–58, has the text and the English translation.

[49] Quilici Gigli 1997, 198.

[50] Leveau 1997 is a model of interdisciplinary research.

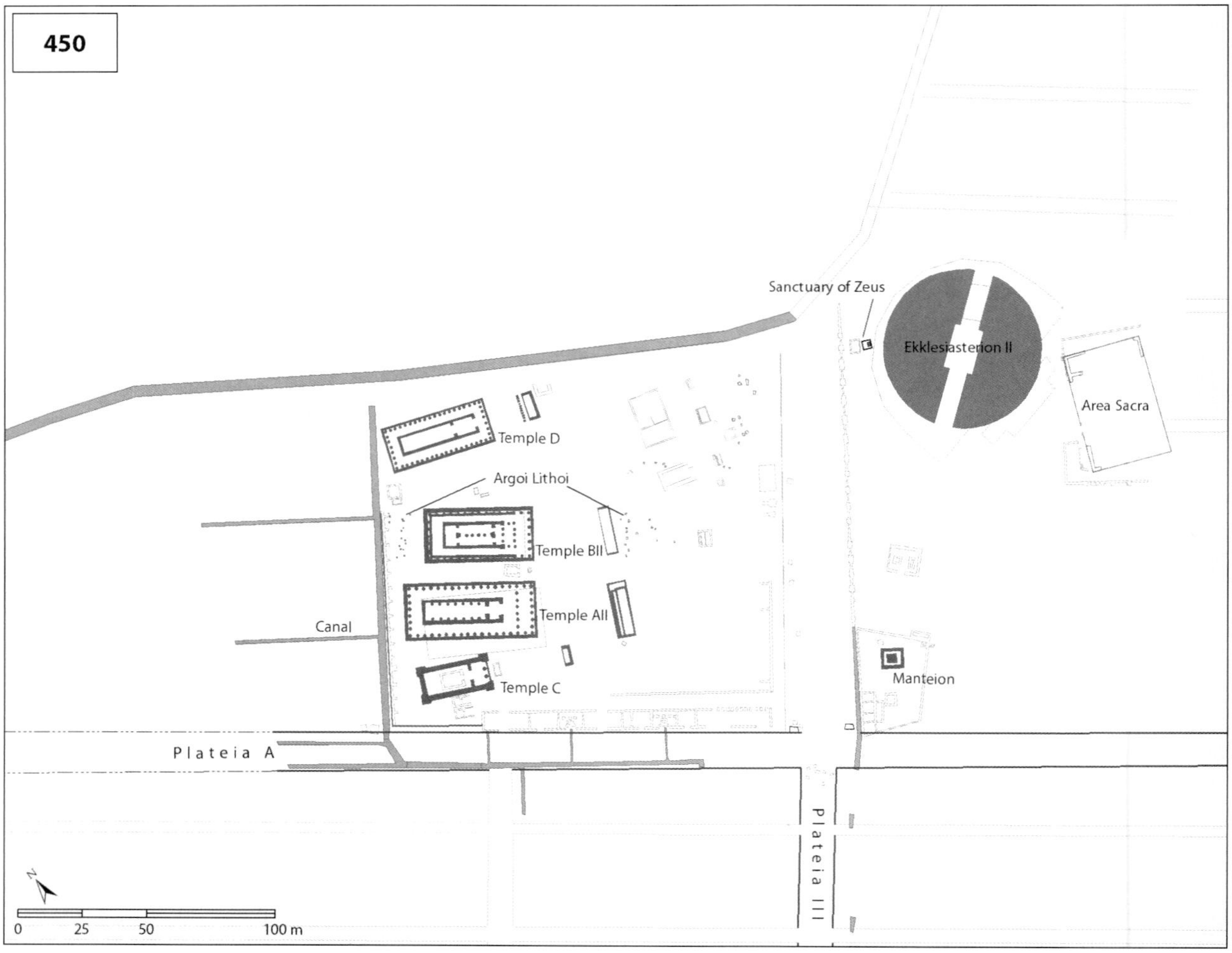

Figure 23.22 Plan of the asty in **450**. (JT, based on Mertens 2006, Fig. 270)

ter after a half-century of virtually no activity. This half-century saw a major transformation in the appearance and infrastructure of the city. Much of this activity was directed at remodeling, modernizing, and repairing earlier structures, although two completely new cult places were created ex novo (*Fig. 23.22*).

Ionian Influence in the Sanctuary

Remodeling the temples. The great Doric temples of the mid-6th century BC, which were dedicated to Hera and Apollo Lykeios, were given new, brightly colored terracotta revetments for the geison, and a sima with a lotus and palmette frieze punctuated with expressive lion-head spouts to carry off the rainwater (*Fig. 23.23*).[51] Whereas the use of painted terracotta decorative elements in the superstructure declined in other parts of the Greek world as wooden elements were replaced with stone, in Greek southern Italy they took on new life in the first half of the 5th century BC, especially in Metaponto, which had always been a center of innovation. This new set of revetments, astonishing for its vivacity and the high quality of workmanship, replaced its simpler 6th-century predecessor, which was impressive even without lion-head spouts.[52] Constant exposure to the elements made periodic repairs or total replacement a necessity.[53] According to Mertens, however, these updated versions are part of a larger transformation, that is, a widespread infusion of Ionian forms[54] and spirit in the architectural projects that began in the second quarter of the 5th century BC at Metaponto and about this time in other centers of Magna Graecia. He believes

[51] Illustrated in color: Mertens 2006, Fig. 548 (Temple B2), Figs. 550 and 551 (Temple A2).

[52] Mertens 2006, Figs. 258 and 259b.

[53] Tempesta (2005, 84) suggests that the lifetimes of terracotta antefixes was about 30 years.

[54] The Ionic cymation and Lesbian leaf moldings in the new revetment of Temple B2 is a good example: Mertens 2006, 251, Fig. 454.

Figure 23.23 Color reconstruction of revetments of Temple of Hera (A2). (After Mertens 2006, Fig. 551; courtesy Dieter Mertens)

this may reflect the arrival of immigrants from Asia Minor,[55] presumably as a result of the failure of the Ionian revolt there in the 490s. It is certain from the survey results that this was the beginning of a period of major demographic increase in the chora. This suggestion, if it is indeed the case, would indicate one possible source of the new settlers.

In the chora, the identical revetments made from the very same molds used for the Temple of Hera in the city were employed to decorate the considerably smaller Temple of Hera on the Bradano (Tavole Palatine). That the latter was about half the size did not, as Mertens observes, seem to matter to its restorers.[56] It may have been incongruous but in this way the cults of the city and chora were visually associated in a synchronous act of embellishment and renewal.

Figure 23.24 Terracotta sima and lion's-head spout dating to the mid-5th c., from the Sanctuary of Artemis and Zeus at San Biagio. (VM, ICA Ancient Crossroads; courtesy of the Soprintendenza)

A still smaller cult building at San Biagio had its gaudy Late Archaic revetment, featuring a sima of alternating red and white palmettes and tendrils on a black background interrupted by blue maned lion's-head spouts,[57] replaced by one of appropriate scale with naturalistic features in the finest Classical style (*Fig. 23.24*).[58] Framed by a frieze of lotus and palmettes in blue and red, this sima contrasts sharply in its crisp and elegant simplicity with its Archaic predecessor.[59] The remodeling at San Biagio took place in the middle of the 5th century BC, when the cult of Zeus was linked to the very heart of the city in the agora and precinct of Zeus with the erection of a stele to one of its titular deities, Zeus Aglaos.

While the older temples in the city were being refurbished and embellished, one entirely new building, Temple D, was erected on the site of an earlier Archaic building of which few traces remain (*Figs. 23.25–26*). It contrasted sharply with the two older monumental temples to Hera and Apollo, and the monumentalized outer frame of Temple C, in that it was in a local version of the Ionic order, but it followed the orientation of Temple C. This return to the earlier orientation, which the great building program of the mid-6th century had ignored, has been seen as a deliberate renunciation in architectural terms of the form of government that had created the earlier monuments—presumably the rule of the tyrant, or the oligarchy. It was under their guidance that the visually dominating structures had been created, and, thus, this may have represented a return to architecture of a more egalitarian past.[60] The Ionic order of Temple D, a curious melange of heterogeneous elements, has been interpreted by Mertens as a bow to the tastes

[55] Mertens 2006, 305.

[56] The stylobate on the Temple of Hera [A2] is 20.55 x 49.82 m compared to 16.13 x 33.24 m for the Tavole Palatine. Mertens 2006, 151, 217, 303.

[57] Mertens 2006, Fig. 456.

[58] Mertens 2006, Fig. 552.

[59] Adamesteanu 1974a, 62 and 63, for color photographs.

[60] Mertens 1998, 136.

Figure 23.25 Reconstruction of Ionic Temple D: a) front, b) long side. (After Mertens 2006, Figs. 543 & 544; courtesy Dieter Mertens)

Figure 23.26 Restored capitals and entablature of Temple D in the Ionic order, built early in the first half of the 5th c. BC. (JCC; courtesy of the Soprintendenza)

of a new population of immigrants from the east executed by craftsmen with no first-hand experience of the Ionic order.[61] This was not an isolated monument. Ionic influence can be seen elsewhere along the southern and western shores of Magna Grecia in the arts and architecture of the Late Archaic and Early Classical periods—for example in the capitals on the pronaos of the Temple of Athena at Poseidonia[62] and the Ionic temple in Contrade Marasa at Locri.[63] The three votive columns that were erected in the sanctuary of Metaponto at this time—the largest 1.4 m in diameter and 14 m high—were also Ionic. These freestanding monuments would have recalled the columns of the Naxians at Delphi.[64]

Marble cult statues and other sculptures. The renewed wealth of the polis and its close contacts with mainland Greece, especially Athens, is evident in its unusually lavish use of imported Greek marble, and in the extremely high quality of craftsmanship evident in those rare pieces of sculpture and architecture that escaped the lime kilns that proliferated around the site in relatively recent times. The similarly rare sculptures of earlier periods were first in terracotta and then in limestone, which also had to be imported but from neighboring Apulia.

Around and over the venerable but tiny Oikos C was erected an elegant structure in marble, of which survives a very large number of marble roof-tile fragments, and a sima with lion's-head spouts and palmette acroteria.[65] The building, which engulfed and enshrined a monument of Metaponto's first years, is referred to as Temple C after the transformation. It may have housed the cult statue, fragments of which were recovered from excavation on the site.[66] The shoulders of a draped female divinity, an acrolithic statue[67] of monumental scale, are preserved, but not enough of the drapery to suggest a firm date for the sculpture.[68]

Marble was not reserved for the urban center, but appears also at important cult places in the chora. In this period, the Sanctuary of Hera on the Bradano was not only given a new terracotta revetment that echoed that of the Temple of Hera (in the city), but also received a marble cult statue, of which the life-sized, draped midsection found near it has been preserved.[69] This could, as has been suggested, have been the occasion of the creation of a cult statue in marble of Hera within the walls, though there is no proof that this was so, since so little of that image has been preserved. This Hera overlooking the Bradano is not the only cult image in marble to be found in the chora. A fragmentary acrolithic cult statue, apparently a female, was found in the kiln deposit (perhaps awaiting conversion to lime) of the Roman period at Pantanello,[70] which occupies a position analogous to the Temple of Hera (Tavole Palatine) at the beginning of the Bradano valley. Again, there is not enough of the sculpture remaining to support

[61] Mertens 2006, 296–302, 304–05, 334.

[62] Mertens 2006, 244, Fig. 432, 222–26.

[63] Mertens 2006, 305–09.

[64] Mertens 2006, 304.

[65] *Metaponto* I, 72.

[66] *Metaponto* I, 74, Fig. 69.

[67] A good parallel for the technique is the well preserved marble head of Apollo set into a body of limestone or wood coming from Crimisa near Croton. Langlotz and Hirmer (1965, 284–85) date it to the late 5th c. BC. A better idea of what a cult statue of this period might have looked like is provided by the very small scale (21.8 cm high) statuette of a seated female divinity in marble from Garaguso in the *proschoros*, or hinterland, of Metaponto. Some years ago it was stolen from Potenza but returned and can now be seen in the Museo Provinciale di Potenza. Langlotz and Hirmer (1965, 266, pls. 52–53) saw Ionian characteristics in it. They date it to the period 480–460 BC. It was found together with the little marble model of a temple in which it stood. Langlotz and Hirmer thought it more likely that both were imported from Greece than carved from a block of imported stone. There is, however, another possibility—that the stone was imported, and carved in Garaguso, or more likely Metaponto. M. Mertens-Horn (2001, 71–88), in her excellent preliminary study of the stone sculpture from the excavations at Metaponto, concludes that an expert sculptor from an artistic center in Magna Grecia or the Greek mainland was active in Metaponto in the Late Archaic period. An inscribed base with the name of a famous Aiginetan bronze sculptor of animals, Theoupropos (whether it was he or another with his name), was found in the sanctuary area. While it does not resolve the question of where his work was carried out it does point to a well established mainland school of sculptors associated with Metaponto.

[68] This sculpture, one of a handful of cult statues (together with the Apollo of Crimisa) found in or near the temple where the acrolith stood or sat, is, unfortunately, still unpublished and not included by Mertens-Horn in her study. No dimensions were given in the excavation report. The head and most of the neck were destroyed. Of the flesh surfaces only the right side of the neck and shoulder remain. A thick himation indicated by a few broad folds covered the back and back of the neck and was originally drawn up over the head. Opaque and viscous, this drapery resembles no other in style so much as that of the Lapiths on the west pediment of the Temple of Zeus at Olympia; cf. Ashmole 1972, Figs. 61, 68, etc. This would point to a mid-5th–c. BC date for the cult statue. This was a veiled goddess, more likely a Hera (or possibly a Demeter) than an Artemis or Athena, the two divinities whom scholars have suggested for Temple (and Oikos) C. Cf. Giacometti (Artemis, 1999) and Mertens-Horn (Athena, 2001). It is certainly possible that the statue was originally associated with the Temple of Hera, just to the north of Temple C, and had been left temporarily in its findspot in preparation for its total destruction in the kiln.

[69] Mertens-Horn 2001, 76.

[70] Carter 1983a, 478, Fig. 54. A capital in limestone of a type very similar to those crowning votive stelai of the urban sanctuary was found in a Late Archaic sanctuary at Pantanello (Carter 1977, 849; cf. Doepner 2002, Met A65).

Figure 23.27 Marble head of a young woman in the Severe Style of the first half of the 5th c. BC, identified by Mertens-Horn as Io, priestess of Hera. (Courtesy of the Soprintendenza)

a stylistic dating, but the choice of the material and the obvious attention being paid to important rural cult places at this time could be cited in favor of a mid-5th–century BC date.[71]

Civic pride is reflected in other sculptures executed for the urban center: the torso of an over-life-sized (estimated 2.2 m high) marble statue of a god, the smaller scale head of an image of Io (*Fig. 23.27*), and a head of a young male identified by Mertens-Horn as the wrestler Alexidamos, whose victory at Delphi was immortalized by Bacchylides (Ch. 21). These are among the precious few images to survive the scavengers.[72] Io, the priestess of Hera and beloved of Zeus, was the progenetrix of the Danaoi, the ancestors of the Metapontines as Bacchylides reminded his audience. She was also a subject of the *Prometheus* of Aeschylus produced in 460 BC, the period of the severe style of which this lovely head is an outstanding example. Mertens-Horn is right to make these associations, and her suggestion that the play could have been produced in the amphitheater known today as Ekklesiasterion II is very attractive.[73] Io was again associated with the sanctuary in the 3rd century BC, when she was represented in the series of terracotta antefixes that probably decorated the stoa along the southern edge of the sanctuary.

Dedications at Olympia and Delphi.

> Προελθόντι δὲ ὀλίγον Ζεύς ἐστι πρὸς ἀνίσχοντα τετραμμένος τὸν ἥλιον, ἀετὸν ἔχων τὸν ὄρνιθα καὶ τῇ ἑτέρᾳ τῶν χειρῶν κεραυνόν· ἐπίκειται δὲ αὐτῷ, καὶ ἐπὶ τῇ κεφαλῇ στέφανος, ἄνθη τὰ ἠρινά. Μεταποντίνων δέ ἐστιν ἀνάθημα, Αἰγινήτου δὲ ἔργον Ἀριστόνου· τοῦ δὲ Ἀριστόνου τούτου διδάσκαλον, ἤ καθ’ ὅντινα χρόνον ἐγένετο, οὐκ ἴσμεν.
>
> A little farther on is a Zeus turned towards the rising sun; he holds an eagle in one hand and in the other a thunderbolt. On him are set spring flowers, with a crown of them on his head. It is an offering of the people of Metapontum. The artist was Aristonus of Aegina, but we do not know when he lived or who his teacher was.[74]
>
> Pausanias, *Elis* 1.22.5–6

We learn from Pausanias that an Aiginetan sculptor, Aristonos, was chosen by the Metapontines to execute a statue of Zeus for the Sanctuary of Zeus at Olympia. It showed him crowned with spring flowers, holding an eagle in one hand and hurling the thunderbolt with the other. Pausanias did not know the date of the sculptor's career, but the first half of the 5th century BC would be a fair guess.[75] That is the probable date of another Metapontine commission, a votive of a "golden harvest" at Delphi, remarked on by Strabo.[76] The base of this monument, with its cuttings for individual golden sheaves and dedications by the Metapontines, has recently been recognized by C. Vatin. The work was signed, in Vatin's reading of

[71] There are certainly other possibilities, such as the late 4th and 3rd c. BC, when there are signs of monumental (temple?) building on the site. The evidence for this building will be discussed in the study on Pantanello, forthcoming in this series.

[72] Mertens-Horn 2001, 75–80.

[73] Mertens-Horn 2001, 77–79.

[74] Text and trans. Jones and Ormerod 1954 (Loeb).

[75] Cf. the Zeus of Ugento, a bronze from about 500 BC in just such a pose; *Greci in Occidente*, Cat. No. 77.

[76] Strabo 6.1.15. See Ch. 17, where the full passage appears.

the stone, by an Athenian goldsmith, Theophilos.[77] This is a plausible suggestion, as wealthy Metapontines clearly sought the best talent available to make displays at the major pan-Hellenic sanctuaries.

Athens and vase painting. The influence of Athens on Metaponto, in politics as well as culture in general, begins to be felt already in this period. A cult of Athena Hygeia, evidenced by an inscription of the 5th century BC, is another indication of Ionian—specifically Athenian—influence at this time. The goddess's cult is primarily attested in Athens and almost nowhere else. The inscription was found in the Sanctuary of Metaponto to the south of Temple C, which may have been Athena's.[78]

The new cultural orientation of the Achaian colony in the second and third quarters of the 5th century BC is reflected also in the kerameikos of Metaponto, where vases decorated in styles with close affinities to Attic vase painters of the circle of the Polygnotos Painter and others began to be produced about 440 BC.[79] Attic vessels of the period 470–460 BC were found among the fragments from the kiln deposits, and it has been suggested that they served as models for the local craftsmen.[80] Denoyelle has proposed that the early vase painters of Metaponto and Taras, whose style is distinctively Magna Grecian, trained with Attic craftsmen before coming to southern Italy. This was not as part of an historical migration, such as the Athenian colonization of Thourioi, as has long been believed, but rather as a response to market demand and part of a process that began earlier and extended over a long period.[81] The first of these Athens-trained craftsmen producing Early Lucanian pots was the figure known as the Pisticci Painter (see *Fig. 23.16*).[82] Some of the vessels of the period 470–460 BC are very puzzling: one wonders whether some lekythoi found in Metapontine burials are Attic imports or made by an Attic craftsmen using local clay.[83] Not enough has been done yet to analyze these clays, but there is intense interest by students of the kerameikos in doing just this.[84]

Agora

Ekklesiasterion II. Ekklesiasterion I, which had been completed in the mid-6th century BC during the building program that gave the urban center its first monumental form, suffered partial destruction in the period 525–475 BC.[85] Its exterior face was held in place by a low ring of stones that had partially collapsed.[86] We can only speculate about the reason for this—whether it was the result of natural causes, or of temporary abandonment of the structure and cessation of the functions for which it had originally been created. Ekklesiasterion II was not a repaired version, but a new building erected on the footprint of its predecessor (*Fig. 23.28*).[87] The exterior retaining wall was raised to a height of 3 m, and the slope of the auditorium increased from 4° to 7° to improve visibility of the central space. This long narrow space, measuring 12.80 x 19.05 m, was entered from its narrow ends, as in Ekklesiasterion I, by two dromoi, 8 m wide. The orchestra was approached by three steps of stone, and stone seats replaced the simple earthen slope of the earlier buildings. The architects ingeniously made the transition between the circular shape of the auditorium and the rectangle of the orchestra by arranging the seats nearest the orchestra in ellipses, which gradually became rounder until they formed circles in the higher rows of seats, conforming perfectly to the plan of the exterior wall. This impressive feat was accomplished 150 years before Euclid derived his formula for the ellipse. According to Mertens, the solution was not mathematical but empirical and practical. String compasses were used to draw segments of circles, which formed the ellipses in sections; the joints were disguised within the aisles.[88]

There may be legitimate doubts about the functions of Ekklesiasterion I in the mid-6th century:

77 Vatin 1991, 53–58.

78 Giacometti 2005, 80–83; De Siena 2008, 584–85. Artemis has been suggested by Giacometti, who believes that it was paired with the Sanctuary of Artemis and Zeus at San Biagio (2005, 70–71). The idea that each of the major cults—Hera, Apollo, Artemis, and Athena—had corresponding cult places in the chora is attractive even if the details are in question (cf. also, Giardino and De Siena 1999b, 347).

79 Trendall 1970, 218–19.

80 Mannino 2006, 431–35.

81 Denoyelle 2008, 342–43.

82 Trendall 1990, 217.

83 *Necropoleis*, 664–66; e.g. the lekythos from T.218, drawing (121) 666, photos 218–20, 345. Cf. also Mannino 2006, 434; 1996, 364; D'Andria 1980b, 125.

84 Silvestrelli forthcoming a, 1, n. 1.

85 Described by the excavators as "thrusts from the fill of earth" (*spinte del terreno di riporto*) used to create the artificial slopes. Mertens 1998, 136.

86 Mertens 2006, 162.

87 Mertens 2006, 334–37, for a full description with his latest thoughts.

88 Mertens 1991.

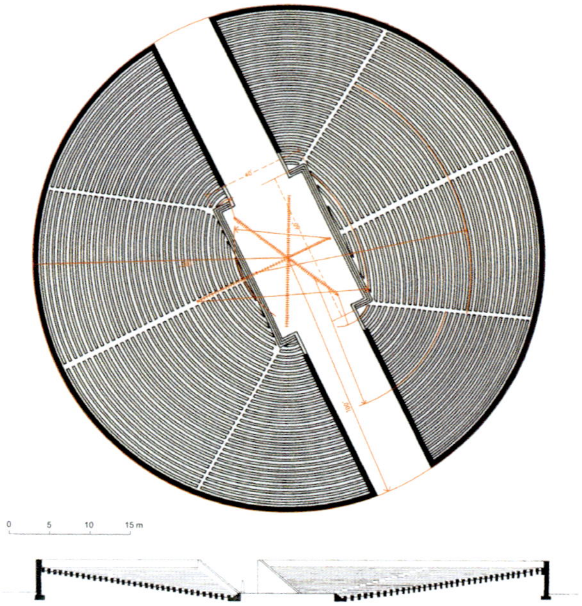

Figure 23.28 Plan of the Ekklesiasterion in its Late Archaic phase (ca. 500 BC), with seating plan. The front rows are concentric ellipses surrounding the central rectangular orchestra. (After Mertens 2006, Fig. 598; courtesy Dieter Mertens)

whether it served as a gathering place just to celebrate festivals with musical competitions (*agones mousikoi*) in honor of the gods, or if it was used to discuss practical issues relating to the community as a whole (see Ch. 21). It would have been unusual indeed if the new Ekklesiasterion II had not been intended and used for both. Excavations, remarkably, have documented the chalk lines (*grammoi*) used by the chorus, on the orchestra floor.[89] Hansen has remarked that large buildings of this sort can be assumed to be for theatrical performances, but could be used for other purposes as well.[90] About the political function of the ekklesiasterion of Poseidonia-Paestum, dated by the excavators to 480–470 BC, no doubts have been expressed. The plan and development of that polis so clearly mirrors that of Metaponto that it would be unreasonable not to assume a political function for Ekklesiasterion II in the 5th century BC, despite the scarcity of documentary evidence for an ekklesia or for an assembly, in any other poleis of Magna Grecia or Sicily.[91]

The temenos of Zeus. Along the northwestern edge of the agora, bordering the very wide north-south plateia that is the principal axis (*cardo*) of the city plan, stretches a line of pilasters dating to the 4th century BC. It visibly demarcated the area of the sanctuary, with its temples and shrines, from the "agora of the gods,"[92] which contained the Ekklesiasterion and other large and small cult places. The sacred space (*temenos*) between the plateia and pilasters and the western retaining wall (*analema*) of the Ekklesiasterion was tiny in comparison with the great building with which it was already securely associated in the middle of the 6th century BC.[93] The temenos contained an altar and a stele of porous stone inscribed with two words in the Achaian alphabet, ΔΙΟΣ ΑΓΟΡΑ ("Agora of Zeus" literally, or alternatively "of Zeus Agoraios"). Depending on the reading, it may mean that the whole of the agora, including the Ekklesiasterion, was placed under the protection of Zeus. More likely, though, the reference is just to the *temenos* of Zeus Agoraios (see Ch. 21, Ekklesiasterion).[94]

Between the second and third quarters of the 5th century BC, the temenos was enlarged.[95] A second stele, inscribed with ΔΙΟΣ ΑΓΛΑΟΣ ("of Zeus Aglaos"), also in the Achaian alphabet, was erected on the opposite side of the altar.[96] A new wall on the western side of the Archaic temenos enclosed it and other small monuments, giving the area a more unified and pleasing appearance. The wall, which enclosed the new stele, incorporated a wooden screen or transenna and was pierced by a wide door opening to the south. The altar may have been remodeled at this time.[97]

Zeus Aglaos was worshipped in the Sanctuary of Artemis at San Biagio, as evidenced by an identical inscription found at that site, which was reused in a Roman tomb lid.[98] Artemis and Zeus shared cult

[89] Mertens 2006, 336–37, Fig. 603.

[90] Hansen and Fischer-Hansen 1994, 65–67 (Metaponto), 52–53 (in general).

[91] There is evidence for the existence of a magistracy charged with evaluating the opinions expressed in the ekklesia of Metaponto. These were the so-called ἀκοαστηρες (Dubois 2002, 98). Hansen and Fischer-Hansen list all poleis for which there is either documentary or architectural evidence of an ekklesiasterion. Of the 14 examples, three are in Magna Grecia (Metaponto, Poseidonia, and Rhegion) and one in Sicily (Akragas). Only three, Metaponto, Poseidonia, and Akragas, have a circular plan. The ekklesia met in theaters in Syracuse and other sites in Sicily and nearby Taras. Hansen and Fischer-Hansen conclude that special buildings for meetings of the assembly were a feature, principally, of Western Greek cities (Hansen and Fischer-Hansen 1994, 48–50; 55–75).

[92] Giacometti 2005, 120.

[93] Giacometti 2005, 117–19, with preceding bibliography.

[94] For the controversy, see De Siena 1998, 151 n. 21.

[95] De Siena 1998, 150–56. Plans, 150; date, 155 n. 35.

[96] Giacometti 2005, 65.

[97] De Siena 1998, 150–56, 168–70, plan on 150; see 155 n. 35 for the date based on pottery.

[98] Adamesteanu 1974a, 64–65; De Siena 1998, 168.

places in the Peloponnesos, and it has been suggested that the epithet Aglaos refers to the chthonic character of the god.[99] The very visible association of Zeus Aglaos with Zeus Agoraios in the political center of the polis around 450 BC may reflect a strengthening of relations between asty and chora in this period.[100]

Of all the cults of the countryside, that of Artemis and Zeus was perhaps the best known. Much was made of its association with an aristocratic family in Bacchylides' ode for Metaponto's athletic hero, Alexidamos (see Ch. 21). If we are correct in interpreting the sanctuaries as the centers of local administration of the chora and bases for the aristocratic families of the early settlers, then the transfer to the asty of the name and epithets of Zeus Aglaos has a definite purpose in religious and political terms.[101]

The inclusion of the inscription of the rural Zeus Aglaos in the Sanctuary of Zeus in the agora could be interpreted as a confrontational attempt to concentrate religious authority in the administrative center of the polis, presumably under a democratic regime. There is, however, an attractive alternate explanation. The inscription could, instead, be understood as a symbolic and subtle affirmation of the cohesion of polis society not its division along class lines. The cults of the countryside are significant as indicators of the self-definition of rural groups. They do not assert their independence with cults particular to their locale, but "recognize the dependence of the deme on the polis and assert a claim to be constituent of the polis."[102]

In this way, we may see the appearance of Zeus Aglaos of San Biagio beside Zeus Agoraios as a reaffirmation of the unity of the polis after the political upheaval of the preceding half-century (see Ch. 22). It would recognize the role of aristocratic families resident in the chora—perhaps in reduced economic circumstances—in the community of the polis. Certainly the prestige of the family of Alexidamos was still high after his victory in wrestling at Delphi in 458 BC.[103]

The Manteion. A much more extensive temenos in the southwest corner of the agora was created ex novo in the mid-5th century BC (see *Fig. 23.22*). There was no evidence of an Archaic building on the site. The quadrilateral temenos, which was subsequently surrounded in the 4th century BC by a 2-m high wall, included important structures: a cult building open to the sky (later roofed), a well on one side, and a base for a large statue. The excavator, De Siena, discovered a large number of gilded bronze laurel leaves at the level of the well head,[104] near the entrance to the diminutive cult building. With that remarkable discovery as crucial evidence, he has very convincingly argued that this was the shrine of Apollo, which Herodotos described in detail in a digression on Aristeas of Proconnesus' mid-5th century visit to Metaponto.

> Ταῦτα μὲν αἱ πόλιες αὗται λέγουσι, τάδε δὲ οἶδα Μεταποντίνοισι τοῖσι ἐν Ἰταλίῃ συγκυρήσαντα μετὰ τὴν ἀφάνισιν τὴν δευτέρην Ἀριστέω ἔτεσι τεσσεράκοντα καὶ διηκοσίοισι, ὡς ἐγὼ συμβαλλόμενος ἐν Προκοννήσῳ τε καὶ Μεταποντίῳ εὕρισκον. Μεταποντῖνοι φασὶ αὐτὸν Ἀριστέην φανέντα σφι ἐς τὴν χώρην κελεῦσαι βωμὸν Ἀπόλλωνος ἱδρύσασθαι καὶ Ἀριστέω τοῦ Προκοννησίου ἐπωνυμίην ἔχοντα ἀνδριάντα πὰρ' αὐτὸν ἱστάναι· φάναι γὰρ σφι τὸν Ἀπόλλωνα Ἰταλιωτέων μούνοισι δὴ ἀπικέσθαι ἐς τὴν χώρην, καὶ αὐτὸς οἱ ἕπεσθαι ὁ νῦν ἐὼν Ἀριστέης· τότε δὲ, ὅτε εἵπετο τῷ θεῷ, εἶναι κόραξ. καὶ τὸν μὲν εἰπόντα ταῦτα ἀφανισθῆναι, σφέας δὲ Μεταποντῖνοι λέγουσι ἐς Δελφοὺς πέμψαντας τὸν θεὸν ἐπειρωτᾶν ὅ τι τὸ φάσμα τοῦ ἀνθρώπου εἴη. τὴν δὲ Πυθίην σφέας κελεύειν πείθεσθαι τῷ φάσματι, πειθομένοισι δὲ ἄμεινον συνοίσεσθαι. καὶ σφέας δεξαμένους ταῦτα ποιῆσαι ἐπιτελέα. καὶ νῦν ἕστηκε ἀνδριὰς ἐπωνυμίην ἔχων Ἀριστέω παρ' αὐτῷ τῷ ἀγάλματι τοῦ Ἀπόλλωνος, πέριξ δὲ αὐτὸν δάφναι ἑστᾶσι· τὸ δὲ ἄγαλμα ἐν τῇ ἀγορῇ ἵδρυται. Ἀριστέω μέν νυν πέρι τοσαῦτα εἰρήσθω.
>
> Such is the tale told in these two towns.
> But this, I know, befell the Metapontines in Italy, two hundred and forty years after the second disappearance of Aristeas, as reckoning made at Proconnesus and Metapontium shows me: Aristeas, so the Metapontines

[99] Giacometti 2005, 123.

[100] De Siena 1998, 169–70.

[101] De Siena 1998, 169.

[102] Osborne 1985b, 181–82.

[103] Besides the ode of Bacchylides, a statue group of two wrestlers in the sanctuary may also refer to the event. Mertens-Horn 2001, 79–80.

[104] De Siena 1998, 156–59; 167–68, cf. also Carter 2006a, 216–17.

say, appeared in their country and bade them set up an altar to Apollo, and set beside it a statue bearing the name of Aristeas the Proconnesian; for, he said, Apollo had come to their country alone of all Italiot lands, and he himself—who was now Aristeas, but then when he followed the god had been a crow—had come with him. Having said this, he vanished away. The Metapontines, so they say, sent to Delphi and inquired of the god what the vision of the man might be; and the Pythian priestess bade them obey the vision, saying that their fortune would be the better; having received which answer they did as commanded. And now there stands beside the very image of Apollo a statue bearing the name of Aristeas; a grove of bay trees surrounds it; the image is set in the market-place. Suffice it then that I have said thus much of Aristeas.[105]

Herodotos, 4.15

The vividness of the description and Herodotos' statement that he had compared the records in Proconnesus and Metapontium (ὡς ἐγὼ συμβαλλόμενος ἐν Προκοννήσῳ τε καὶ Μεταποντίῳ εὕρισκον) supports the suggestion made by Prieto and Polleichtner that Herodotos may have visited Metaponto between 445/444 and 433/432 BC.[106] The historian was one of the new colonists that Perikles sent to Thourioi, and the visit would have been in part a diplomatic mission on behalf of that city and Athens to make an alliance with Metaponto against the Tarentines, in the struggle to add the territory of Siris to their own and keep the Tarentines out of this grain-producing land. It is difficult to overestimate the importance of this contemporary testimony for the political and religious history of Metaponto in the mid-5th century BC.

One of the three subjects chosen for the first series of double relief staters—the others were Acheloos (see *Fig. 23.19*) and Herakles as noted above—was Apollo. He stands in a frontal pose (*Fig. 23.29*)—a more graceful *contrapposto* than that of Acheloos. His head is in profile to the right, with weight on the right leg. His left arm is lowered and in his hand is a bow, which touches his left thigh. His right elbow appears to rest on his hip, the forearm extends forward, and in that hand he holds a branch with very large laurel leaves. It has been suggested that this coin may represent—with some liberties—the divine image or agalma of Apollo,[107] which, according to Herodotos, stood in the agora beside the andrias of Aristeas and the laurel tree. The coin dates to the 440s, according to Stazio, not the 470s as earlier thought.[108] The style of the representation is mid-5th century BC at the earliest.[109] A marble torso of an over-life-sized nude male, previously identified as a *kouros* and discovered near the Temple of Hera, has recently been identified by Mertens-Horn as that of a divinity. She calculates that it stood 2.16 m high and suggests that it belonged to the statue of Apollo in the Manteion.[110]

Figure 23.29 Silver stater of Metaponto, ca. 450 BC, possibly representing the agalma of Apollo holding a laurel tree that Herodotos described as standing in the Sanctuary of Apollo in the agora of Metaponto. Approximate diam. 2.3 cm. (Hirmer Fotoarchiv München; after Kraay 1996, Pl. 82, Fig. 232)

Either the coin image is an original representation, or, if it does reproduce even approximately an agalma of Apollo, there must have been two. I would suggest that the fragment belonged in the Temple of Apollo, closer to its findspot than the Manteion. The material is marble, like that of the fragmentary acrolithic cult statue from the area of Temple C, while the god's image in the Manteion may have been bronze, the material preferred by sculptors of Aigina

[105] Text and trans. Godley 1957 (Loeb).

[106] Prieto and Polleichtner 2008, 190–91.

[107] De Siena 1998, 157; pers. comm., 2009.

[108] Stazio 1974, 83–84, dates it so, but denies that this could be even a freely interpreted version of the cult statue; see Carter 2006a, 235 n. 61.

[109] Close to the Doryphoros, ca. 440 BC (Stewart 1990, Figs. 378–79), with the exaggerated attention to the muscular divisions of the thorax that is typical of the Niobid Painter's vase representations of male anatomy about a decade earlier. Arias/Hirmer/Shefton 1961, Fig. 175 (Apollo slaying the children of Niobe).

[110] Mertens-Horn 2001, 75, Fig. 81. Mertens-Horn is undoubtedly correct in dating the sculpture to the period 480–470 BC. If it had been the image in the Manteion then it is unlikely that the coin image reproduces it. The position of the missing arms, as reconstructed by Mertens-Horn, would correspond to those in the coin but the pose and anatomical style do not. The torso is much closer to that of the Kritios Boy than that of the Doryphoros.

whose commissions from the Metapontines in this period are well documented. In any case, the sculpture would have been very new when Herodotos, in all likelihood, saw it.

This new cult place of Apollo represented a new direction in the religion of the Metapontines. The two references to it in ancient literature, Herodotos and a fragment of Theopompos,[111] leave little doubt that it was an oracular cult with strong Delphic connections. The latter recounts that a dancing girl, Pharsalia, was torn apart (διεσπάσθη) by the seers (μάντεις) in response to a voice emanating from the bronze laurel tree that the Metapontines had set up at the time that Aristeas lived among them, after his return from the land of the Hyperboreans.[112] The connections of the Manteion with Delos, but also with Delphi, were strong from the beginning. Aristeas was associated with Trophonios and his oracular cult, along with the figures Zalmoxis and Pythagoras.[113] Very little, as Bonnechere has argued, separated oracles and mystery cults.[114] Beliefs about death and rebirth that have been associated in scholarship with Dionysus, Orpheus, and Pythagoras were widespread in southern Italy at this time. With the creation of the Manteion in the mid-5th century BC, these cults were given a public place in the very heart of the asty. Evidence for their diffusion throughout the chora from the late 6th century BC has been documented by the burials of probable initiates and it continues to grow.[115]

The central location of the Manteion in the agora, like the tombs of oikists and the altar of Apollo Archegetes—traditionally located in the agora[116]—has a more overt political sense. It places the andrias of Aristeas, and by association if not explicitly, Pythagoras,[117] on a spot where honors were usually accorded the founder of an apoikia. The opposition of the new cult of Apollo to that of the Apollo Lykeios, with its connections with the older aristocracy and the tyrant, is implied by the choice of a site apart from the monumental temples and chora, and closer to the assembly place of the demos. It is underscored by the unpretentious design of the temenos and its distinctively different features.[118]

Street Plan and Drainage

The changes in the street plan (see *Fig. 23.22*) between the 6th century BC and the second half of the 4th century, when its plan was reorganized, are largely unknown because excavations beyond the areas of the sanctuary and agora have been limited thus far to soundings.[119] It is clear from these soundings, however, that a great effort was exerted by the Metapontines in the asty during the mid-5th century BC to deal with the problems of drainage, which were as severe, if not more so, as those in the chora. Soundings across the main east-west Plateia A indicated that at this time the previously undulating ground surface was leveled, but it still preserved faithfully the orientation and dimensions of the earlier streets. According to the excavator, the slight hillocks and dunes of the original topography were leveled and the shallow valleys filled in with enormous quantities of sand and gravel. The effect of this work was to make the streets nearly 0.5 m lower than the sanctuary. The same operation no doubt was performed on other major streets, both SE-NW and NE-SW arteries, with the goal of improving the drainage of surface water into branches of the rivers that drained the city.[120] This cannot have been entirely successful, and it was not until the late 4th century BC that large masonry drains were introduced (Ch. 25).

The Asty: Major Trends

The city center of Metaponto inherited much from the period 575 to 525 BC. Its organization and most of the large structures in the main public areas of the sanctuary and agora are those of an Archaic city. The atmosphere and feeling of the center, however, changed markedly and much of what was added at this time seems either aimed to recall and celebrate a distant past, or to challenge deliberately the more immediate past and announce a "new founding." The Ionian influence, particularly in architecture, contrasts with the Doric spirit manifested in the grand temples

[111] Theopompos *apud* Athenaeus 13.83.

[112] The stated reason for her murder was that she had accepted a crown of golden laurel leaves stolen from Delphi during the third Sacred War, which dates that episode probably to the mid-4th c. BC.

[113] Bonnechere 2003, 171.

[114] Bonnechere 2003, 175.

[115] Carter 1998c, 67–69; Carter 1998f, 184, 209–10; Carter 2006a, 175–85; A. Bottini 1992.

[116] Graham 1964, 29.

[117] Giacometti 2005, 45.

[118] It cannot be an accident that the temenos has no right angles, and all the walls were of different lengths. The walls were raised in the late 4th c. BC, shielding the whole from view. Mertens 2006, 212.

[119] It is this 4th-c. phase that has been mapped by aerial photography and magnetometry.

[120] De Siena 1998, 159–61.

of the bygone era. Metaponto was participating in a stylistic movement, and perhaps also a population shift that was affecting all of Magna Grecia. The artistic influence of Ionia is evident in the remodeling of earlier buildings, notably the Temples of Hera and Apollo, as well as in new constructions. The influence of Athens—and probably Attic artisans—is particularly apparent and it had an economic impact, as in the second half of the 5th century BC. Metaponto became a center for the diffusion of red-figure vases through its chora, the indigenous settlements in the interior, and farther afield.

The elaborate embellishment of Metaponto's earliest shrine that survived into the 5th century BC is clearly an attempt not only to honor Artemis, but also to glorify her venerable shrine in a much larger and very expensive marble frame. Considering the amount invested in the material, the older building could simply have been replaced with a new one that harmonized better with the new designs in the city. But that would surely have defeated the purpose.

The origins of the settlers, their descent from Achaian heroes, was brought home by the sculptures: Io was chosen for the sanctuary, and a contemporary Metapontine of recent fame, Alexidamos. With such images, the polis made evident its pride in the past and confidence about the future. The display of wealth, produced by its fields, more limited and more equally distributed than before, rivaled that of the city a hundred years earlier. It takes on different forms at this time, such as the patronage of foreign artists, and self-advertisement with important and memorable dedications at Olympia and Delphi. Some of the wealth was used for vitally important infrastructure projects, not for show. The maintenance of the earlier monumental structures, especially the Temples of Hera and Apollo, as well as those new constructions in the 5th century BC like Temple D, must have been a great burden on the public finance.

By bringing the cult of Zeus from the Sanctuary of Artemis in the chora into the heart of the asty, and by creating marble cult statues for at least two rural shrines, the message is the unity of the asty and the chora under divine protection. The disappearance of the smaller rural shrines may signal the decline in power of those families that established them in the very early years of the colony. A new direction in religion, which has no roots in the previous century, is the oracular cult of Apollo in the agora. Here, the importance of the mystery cults that were sweeping Magna Grecia—especially popular in the chora of Metaponto at the end of the 6th century BC[121]—and the new belief about the afterlife associated with them were given public recognition. This was emphasized, not only by the architectural form, but also by the location of the cult at the very crossroads between the oldest and most central streets of the asty, the NE-SW Plateia III, and the oldest SE-NW street, Plateia A, stretching out into the chora.

[121] Recent excavations in the suburban necropoleis (De Siena 2008) are revealing more plaster lined graves, which have been argued to be those of initiates in the mysteries of Dionysos; see n. 115 above.

24
The Chora in 400

Joseph Coleman Carter

The period 425–375 BC has generally been viewed as the beginning of a long decline for Metaponto that was reversed, though only temporarily, in the second half of the 4th century BC. Politically, it is clear that Metaponto was under pressure from Dorian Taras, her powerful neighbor to the north. The establishment of the Tarantine-sponsored colony of Herakleia on the site of Siris in 432 BC appears to have been an attempt to encircle Metaponto and block her expansion to the south, echoing the Achaian strategy towards Taras several centuries earlier. Metaponto's alliance with Athens some time in the latter half of the 5th century BC has been interpreted as a response to the Tarantine threat. A passing remark by Thucydides about renewed stasis raises the specter of internal strife at Metaponto, which again would certainly have involved the chora.[1] The extensive building activity in the first half of the 5th century BC had no sequel in the second half. In fact, there is firm evidence from the first half of the 4th century BC that the Ekklesiasterion, entirely remodeled and embellished in **450**, had fallen into disrepair—itself perhaps a sign of stasis and crisis for the democratic government or even of foreign dominance for a period. Both possibilities have been suggested.

This was, nonetheless, a period of intense activity and creativity, as evidenced by the outstanding production of the kerameikos of Metaponto during the heyday of Early Lucanian vase painting. The survey evidence from the chora reveals it as one of steady but unequal expansion in terms of settlement, with a setback in the Central Plateau, an area of vital concern for the economy of the Metapontino (*Map 5*, Vol. IV; *Figs. 24.1–2*).

The Chora in 400

The relation of sites to the topography and to each other are always important aspects of the changes that will be examined, as in preceding chapters, in some detail with reference to the local maps, as well as the larger-scale map for this period.

[1] Thucydides 7.57.11.

Early Village Clusters

Giardinetto (*Fig. 24.3*). Giardinetto in **400** is not surprisingly the most flourishing of the Early Village Clusters.

Farmhouses. The farmhouses grew from six to eight, including two with burials in **400**, an increase of 33%. Two of the new additions (518F and 460F)—entirely ex novo farmhouses—join 519F and 521F/T/K on the more gently sloping eastern side of the village. Only 465F, among the earliest farmhouses at Giardinetto, and 529F, a third ex novo farmhouse, occupy the extreme western slope.

Necropoleis. Between **450** and **400**, the number of necropoleis in Giardinetto doubled from two to four. On the west below the farmhouses appear three new necropoleis, in addition to 471N that was new in **450**. The arrangement with the necropoleis largely clustered on the lower slopes to the west suggests that this part of the village may have functioned as an area especially reserved for burial. Since two of the farmhouses have burials closely associated, the ratio of places of burial to residences is 3:4. This is perhaps because Giardinetto, the Early Village Cluster farthest from the city, may have had few options but to bury its dead nearby. It remained through its history as the cluster with the highest ratio of burials to residents (*Appendix C3: Giardinetto*).

Significance, land, and continuity. The average FMCE value for Giardinetto in **400** (*Appendix C7*) is the second highest among the Early Village Clusters in this period after west Venella. The highest individual values are those of 465F on the extreme west, the ex novo 460F on the extreme east, and 521F/T/K, which by this period probably engaged in ceramic production, which may in part account for the fact that its FMCE value was the highest of all.

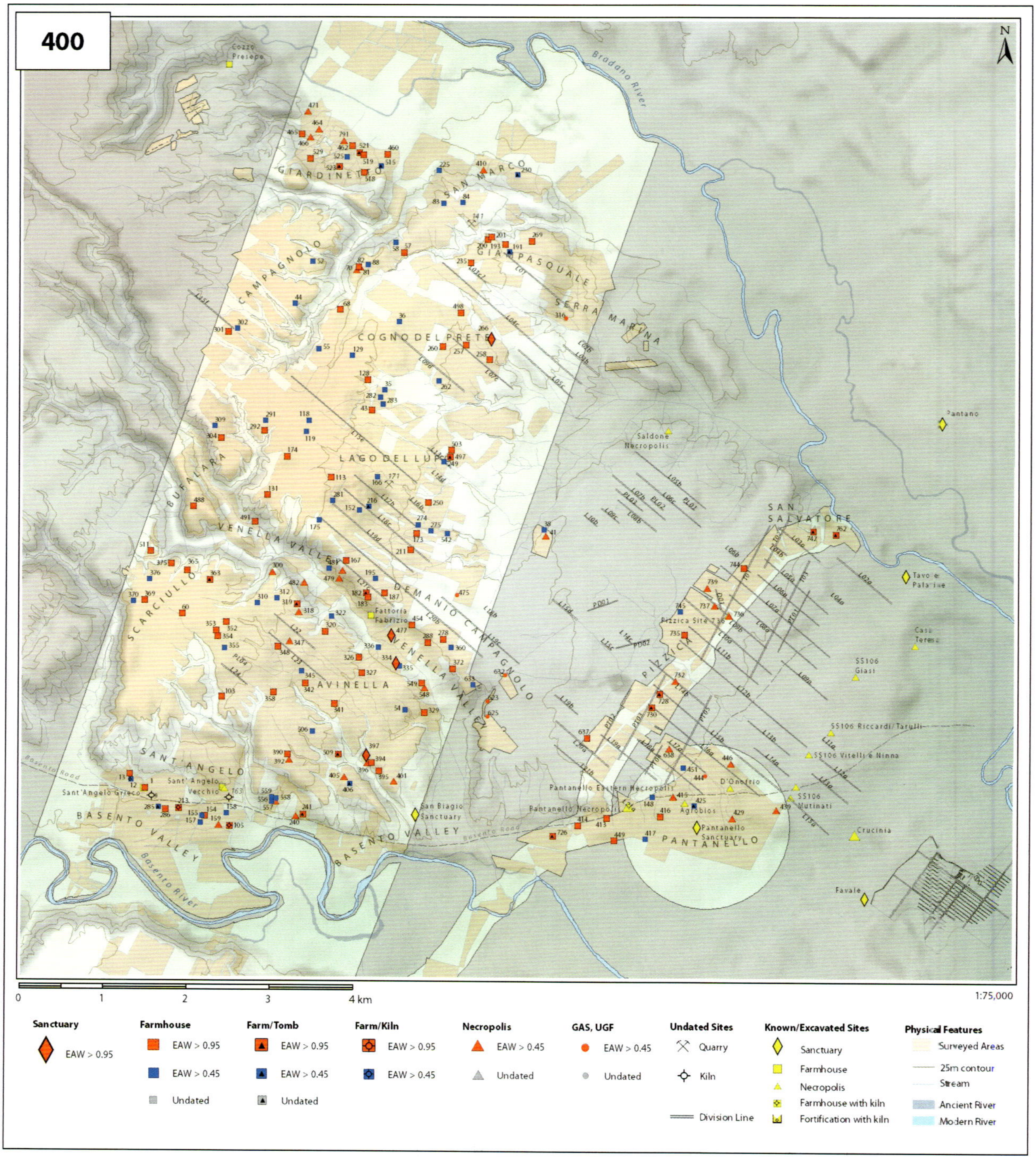

Figure 24.1 The study area in **400**. (JT)

Sant'Angelo (*Fig. 24.4*).

Farmhouses. The number of significant farmhouses in this cluster did not increase in **400**. In fact there were fewer farmhouses in both **400** and **450** than there were in **550** (see *Appendix C3*). In **450** and earlier, the farmhouses were concentrated, for the most part, on the higher round in the northwest corner of the cluster. In **400**, there is a noticeable shift to the southeast, closer to the old course of the Basento River. Site 105F/K, a farmhouse that incorporated a kiln, perhaps more of a center for the production of ceramics than an agricultural establishment, became significant for the

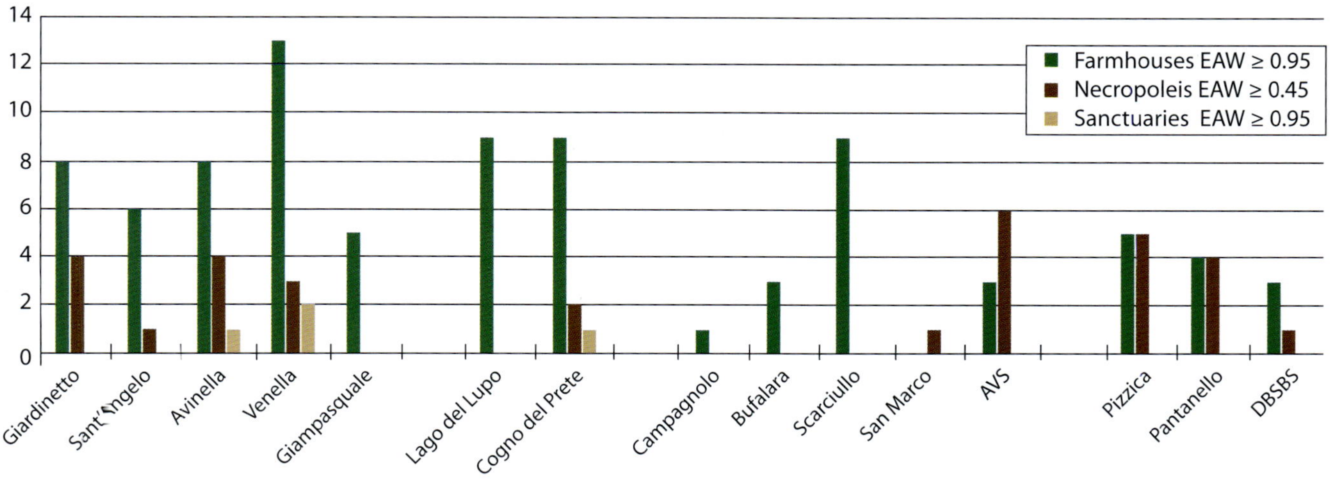

Figure 24.2 Distribution of farmhouse, necropolis, and sanctuary sites among localities in **400**. Totals: significant farmhouses, 81; sanctuaries, 3; necropoleis, 31.

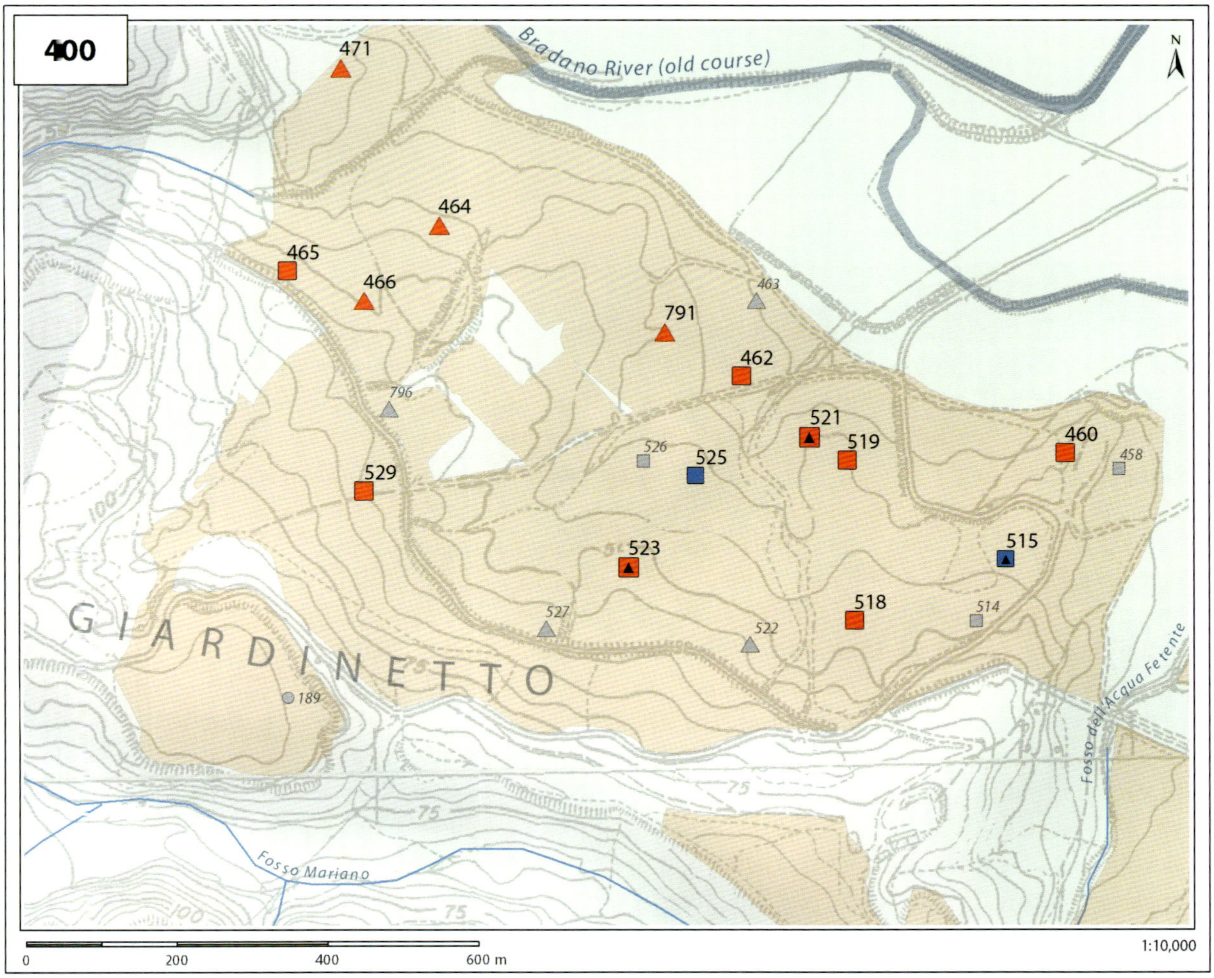

Figure 24.3 Giardinetto in **400**. See Fig. 24.1 for map key. (JT)

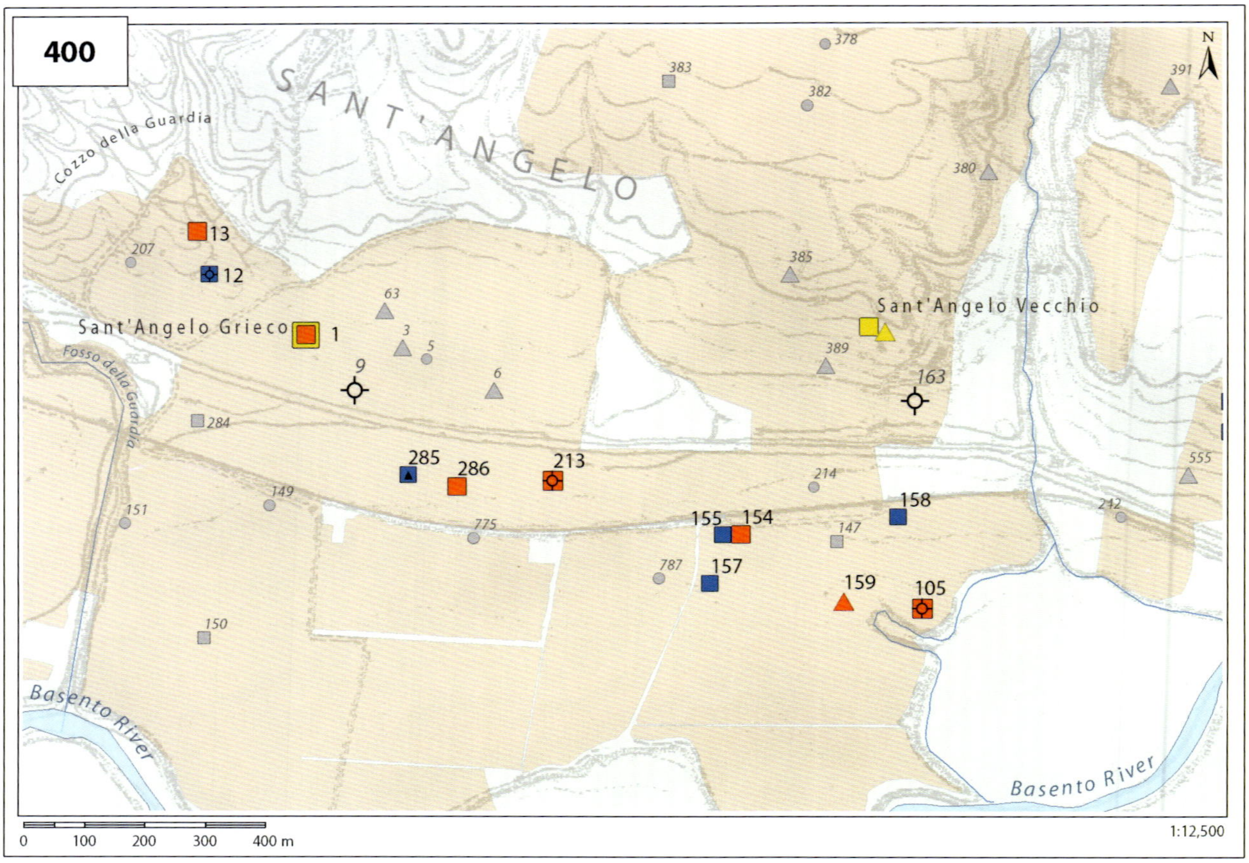

Figure 24.4 Sant'Angelo in **400**. See Fig. 24.1 for map key. (JT)

first time in **400**, and it continued to be an important place through the remaining history of the chora. It was located at the end of what was an inlet on the north bank of the old stream, which may have served as a small port or landing.

In this period two pairs of sites—285F/T and 286F, and 155F and 154F, prominent in the 6th century BC—begin to act like binary opposites. Site 285F/T is significant in **450** but not in **400**, and the reverse is true for 286F. Sites 155F and 154F follow the same precise pattern. It may be simply a coincidence but since both pairs were significant in **600** and **550**, and 285F/T and 155F were in **500** and **450**, the alternation in **400** from the traditionally more significant site to the less may in some way reflect the declining fortunes of the families involved, together perhaps with that of traditional agriculture in this area. Neither 285F/T nor 286F is ever significant again after **400**.

Necropoleis. There was only one active necropolis in **400**, down from two in the previous period: 159N, which was about 150 m to the west of 105F/K, with which it was probably paired. It was active only in this period. The farmhouse with burials 285F/T, significant since **600**, ceased to be in **400** with the result that the ratio of burials to farmhouses was only 1:6, the lowest at any point in the earlier history of the village.

Significance. The average FMCE value for Sant'Angelo fell from 2.20 in **450**, the highest among the Early Village Clusters, to 1.88 in **400**, very nearly the lowest among the villages. Only upper Avinella is lower in this period. The reason is the decline of FMCE values at individual sites 1F and 13F and the disappearance of significant sites 155F and 14F, which was ex novo in **450**. The average was held up by 213F/K and brought up by the new 105F/K, both involved in ceramic production. The decline of Sant'Angelo as a community begins in this period.

Avinella (*Fig. 24.5*).

Farmhouses. The number of significant farmhouses in Avinella increased very slightly in **400**, from seven to eight. Along the lower Conca, the density at the three

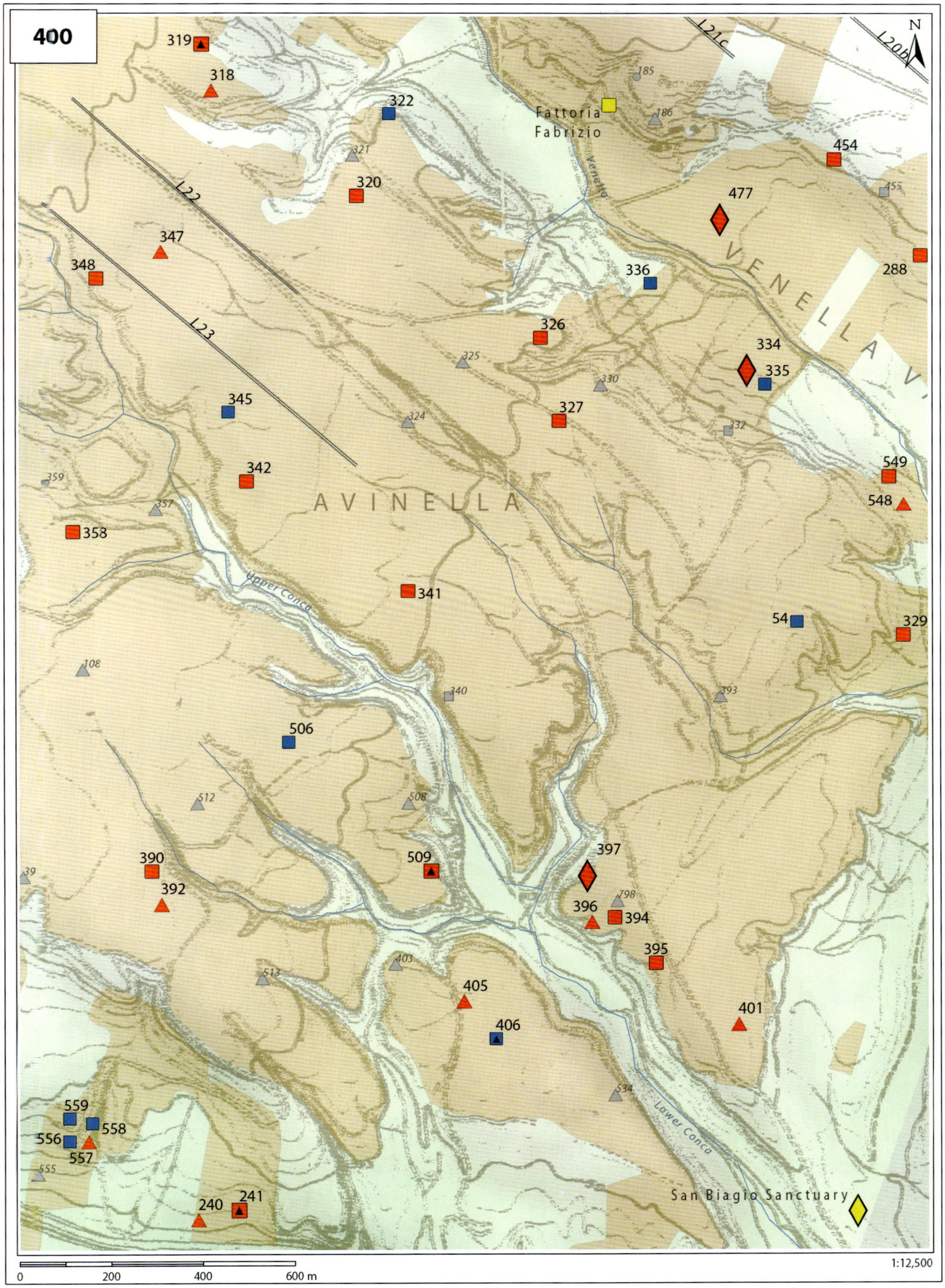

Figure 24.5 Avinella in **400**. See Fig. 24.1 for map key. (JT)

forks is greater in **400** than in **450**. Sites 509F/T and 394F were added, but the previously significant 404F at the end of the west fork ceased temporarily to be so in **400**. Site 328F at the end of the east fork disappeared forever. The broad area to the north, which the occupant of 328F had most likely exploited, may have passed under the control of the farm family that had just installed itself on 341F. Among the sites that were added around the stream junction in **400** was 394F. Two necropoleis were added as well.

In upper Avinella, the more dispersed settlement ca. 1 km farther upstream, an equal number of farmhouses was added and lost. Site 344F, one of the pioneering sites of the early 6th century BC, was part of a pair in **450**, but no longer significant in **400** or thereafter. Its place was taken, presumably, by the surviving member of the pair, 342F. The number of significant farmhouses in upper Avinella remained the same from **450** to **400** but the settlement shifted farther upstream, reaching its earlier limits with 348F and 347N.

Necropoleis. Two necropoleis were added in **400**, doubling the previous number. Necropolis 396N was perhaps paired with 394F, while 405N was situated on the west side of the Conca. Two other necropoleis—797N (perhaps belonging to the family in 395F) and 407N—seem to have gone out of use in this period. A new necropolis, 392N at the farther tip of the west branch of the Conca, was probably created by the occupants of 390F, a new farmhouse in **450**. In the upper Conca, necropolis 347N and the farmhouse with which it was probably associated, 348F, were reactivated in **400**. The ratio of burials to residences in this period is 1:2 (*Appendix C3: Avinella*).

Significance. The average FMCE value of the farmhouses at Avinella, 1.90, is consistent among the Early Village Clusters (*Appendix C7*). In **400** only Sant'Angelo is lower. Still, it is not far out of line for the study area as a whole, 2.01. There are few values much higher than this average (except Close to the City). The outstandingly high value among the individual farmhouses in Avinella in **400** is site 348F, at the very head of the Conca drainage in upper Venella.

Venella (*Fig. 24.6*).

Farmhouses. The Venella cluster in **400** stretched along both banks of the Venella, extending farther down to the west bank than it had ever done before. The total number of farmhouses on both sides of the stream rose from 10 in **450** to 13 in **400** (*Appendix C3: Venella*), an increase of 30% compared to that at Giardinetto. More than half of them are ex novo foundations. They include five on the east bank (167F, 182F/T, 183F, 454F, and 278F) and two on the west (549F and 329F). This is the highest percentage of new farmhouses among the Early Village Clusters, by far (*Appendix C6*). Only two localities Close to the City and the nearby AVS Far from the City surpass it.

New groupings of sites along the west bank, including 326F and 327F, and to the north, 549F and 329F, form subclusters. This same development can be perceived on the east. To the north, 187F and 183F (both ex novo), join 182F/T opposite the sanctuary 477S. On the east bank, a looser group of three centered around 288F is formed by two ex novo farmhouses, 454F and 278F. Site 372F probably belonged as well. No more than 350 m separates any one of these farmhouses. A third possible cluster may once have existed around 167F to the far north, as beside the farmhouse there were two active necropoleis, 176N and 479N, and a second magnitude farmhouse, 481F.

The rapid growth of east Venella in **400** was surely not unrelated to the vicissitudes of the neighboring Central Plateau in **450** and **400**. Lago del Lupo contracted sharply in **400**. It is possible that the drop in the number of farmhouses in Lago del Lupo may in fact be responsible in some way for the sharply increased number of farmhouses along its southern border in Venella. It cannot be pure coincidence that while six farmhouses were lost in Lago del Lupo in **400**,[2] six were added in east Venella. All but one of these (372F) were ex novo in **400**. It would seem that in some way they were compensating for or absorbing the losses of their neighbors to the north.

Necropoleis. Both of the new necropoleis in Venella in **400** are in close proximity to new farmhouses: 548N is approximately 70 m from the new farmhouse 549F on the west bank, and 176N is 120 m from the new farmhouse 167F on the east. Other connections with residential units may have existed: for example, new farmhouse 278F may have been paired with undated necropolis 287N only 110 m away.

[2] 170F, 175F, 254F, 274F, 166F, and 224F/T.

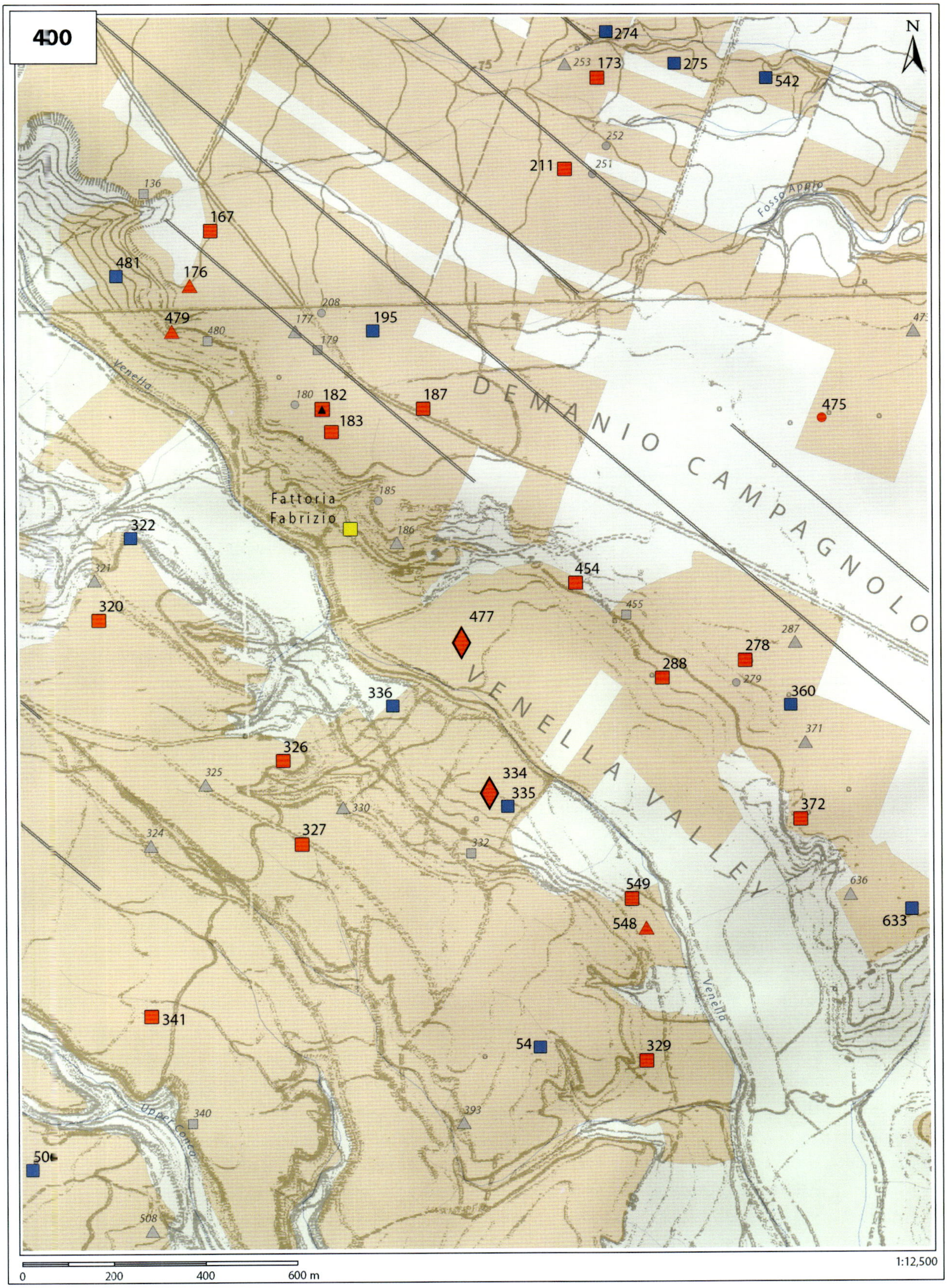

Figure 24.6 Venella in **400**. See Fig. 24.1 for map key. (JT)

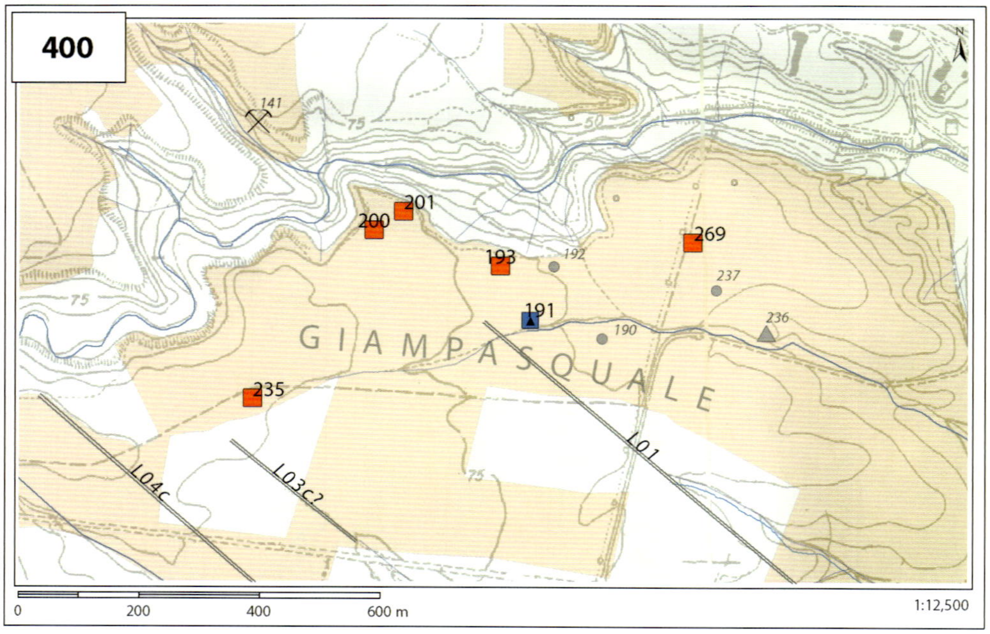

Figure 24.7 Giampasquale in **400**. See Fig. 24.1 for map key. (JT)

There are many more farmhouses with necropoleis in the Early Village Clusters in **450** and **400** than previously, but the fact remains that many of the new farmhouses in this period seem to have had no necropolis nearby (*Appendix C3: Venella*). There is a good chance that the families that occupied these farmhouses buried elsewhere, perhaps in one of the larger necropoleis, like 401N, not far from Venella above San Biagio (see Ch. 23).

Sanctuaries. The sites of the two sanctuaries 334S and 477S that flourished in the 6th century were barely significant in **400**.

Significance, land, and continuity. The average FMCE value for Venella in **400**, 2.08, is like that of Giardinetto: comfortably above the average for the Early Village Clusters and the study area as a whole (2.01) (see *Appendix C7*). The most significant sites individually are 187F on the east bank and 549F and 329F in the new cluster in the southwest. All are ex novo foundations in **400**.

Expansion might have been expected on the eastern side, where the Venella borders the Central Plateau, especially after the bonifica of **450** and influx of new settlers there. Instead, both east and west grew in this period and prospered relative to much of the rest of the study area. Though more dissected, there are nonetheless ample, broad, and level fields adapted to cereal cultivation also in west Venella.

This period in the Venella is marked by the reemergence of regular, incremental distances between farmhouses. This is evident in the sub-cluster in the west. The distance separating 326F and 327F is approximately 200 m; between 329F and 549F, 350 m. On the east, about 200 m separate 182F/T from 187F; 300 m lie between 454F and 288F. 200 m separate 288F and 278F; 400 m lie between 288F and 372F. Limits of 200 or 300 m (and multiples of these) repeat themselves in the newly found clusters. The appearance of a studied division of the land would be in keeping with the spirit of the bonifica of the previous period. These measures between farmhouses in fact compare closely with those evident already in the 6th century. A possible implication: the original dimensions of a kleros were revived there during the reorganization of the chora in **450**.

Giampasquale (*Fig. 24.7*). The stability of the cluster at Giampasquale is unique.

Farmhouses. There were constantly five significant farmhouses on the south bank of the Fosso San Marco in **450** and for two centuries. Of those that remained from the previous period (200F, 201F, 269F), 269F had been continuously occupied since **600**. Two, 215F and 238F, disappeared definitively from view after **450**, and there were two completely new sites in Giampasquale in **400**: 193F, which occupied a central position between the pair 200F and 201F, and the easternmost and earliest site 269F. The other ex novo site, 235F, was located at the opposite extreme, to the west and set back from the valley's edge. This was the period of the maximum expansion of the cluster to the west and it maintained this outline for the next century and a half.

Necropoleis. No necropoleis were active at Giampasquale in this period or any other.

Significance. The average FMCE value for the cluster in **400** was 1.99 (*Appendix C7*). Among the Early Village Clusters, this is midway between Giardinetto and the Venella on the high end and Sant'Angelo and Avinella on the low. Among the individual farmhouses, the three farmhouses with the highest values are 193F, 200F, and 269F, all with values well above the average for the cluster and the study area. Those of 201F and 235F are well below it.

The Central Plateau

The Central Plateau was the only one of the four regions to have fewer significant farmhouses in **400** than in **450** (*Appendix C4*). Why did Lago del Lupo, whose fertile fields had attracted so many farmers in **450**, suddenly lose its appeal? This is all the more mysterious given the continued rapid growth in the adjacent, comparable fields of Cogno del Prete. Some indication of the reason may be found by examining the distribution of the farmhouses that disappeared in **400** from Lago del Lupo, and of those that remained.

Lago del Lupo, Cogno del Prete (*Fig. 24.8*).
Farmhouses. In **400**, there were eight farmhouses in Lago del Lupo compared to 14 in the previous half-century, a decrease of 43%. In Cogno del Prete, on the other hand, nine farmhouses were present where in **450** there had been five, an increase of 80%. In Cogno del Prete, the rate of occupation of significant farmhouses increased dramatically in **400** and remained high. The five ex novo farmhouses in **450**—the only ones there in that period—continued in **400**.

Necropoleis. There were no necropoleis in Lago del Lupo dateable to **400**; in Cogno del Prete there were two, as in the preceding period. In the two localities combined, there were only two places of burial and 17 farmhouses. The very low ratio—approximately 1:8—is due in large part to the fact (noted earlier) that three quarters of the necropoleis in Lago del Lupo could not be dated.

Significance, land, and continuity. In **400**, Lago del Lupo saw a sharp decrease in farmhouses that had been significant in **450**. By **400** they were not generally scattered throughout this large area, as were the ex novo farmhouses of the new settlers in **450**. They clustered instead in an area centered on a concentration of division lines (*Fig. 24.8*). Seven out of the twelve farmhouses that had been significant in this area in **450** had been abandoned by **400**. In fact, all the significant farmhouses abandoned in Lago del Lupo in **400** are located in this area. What made this fairly restricted area so much less attractive, say, than the adjacent land to the north and west? Were the dense concentration of division lines and the canals especially prone to being flooded or waterlogged? Or did the citizens who had established new homes there when land was distributed or redistributed in **450** simply fail or decide to abandon their homes? As with the crisis of **500**, the issues could have been environmental, economic, or political, and in the present state of our knowledge it is impossible to decide which were ultimately the determining reasons. This area continued to be one of relatively dramatic change down to the period of Roman domination in the 2nd century BC.

This situation might have existed elsewhere, but unfortunately the survey results cannot provide further examples that would illuminate such phenomena in other areas where division lines are concentrated, such as Pizzica and Saldone, because survey coverage there is slight or nonexistent.

There were other areas where farmhouses occupied in **450** had been abandoned in **400**. One example is along the western bank of the Venella, extending to Pietra San Giovanni (see *Fig. 24.1* and *Map 9*, Vol. IV). There, six of nine farmhouses significant in **450** were not so in **400**. Five of nine significant farmhouses along the Basento Road stretching from Sant'Angelo to Pantanello had likewise been abandoned. The low elevations and proximity to a flood-prone river suggest the possibility that seasonal and violent flooding may have caused their occupants to relocate to higher ground or leave the countryside altogether. In this instance, the case for natural causes appears particularly strong. In fact, a number of farmhouses on the high plains of Avinella and Scarciullo to the north were either occupied for the first time in **400** or reoccupied in this period. New farmhouses are in the majority on the east bank of the Venella. Five out of seven farmhouses located there were ex novo. New farmhouses, as noted, were built along the Venella in **400**, at the same time as the new farmhouses of the previous period in Lago del Lupo were abandoned. Perhaps families in one area had reason to resettle in the other.

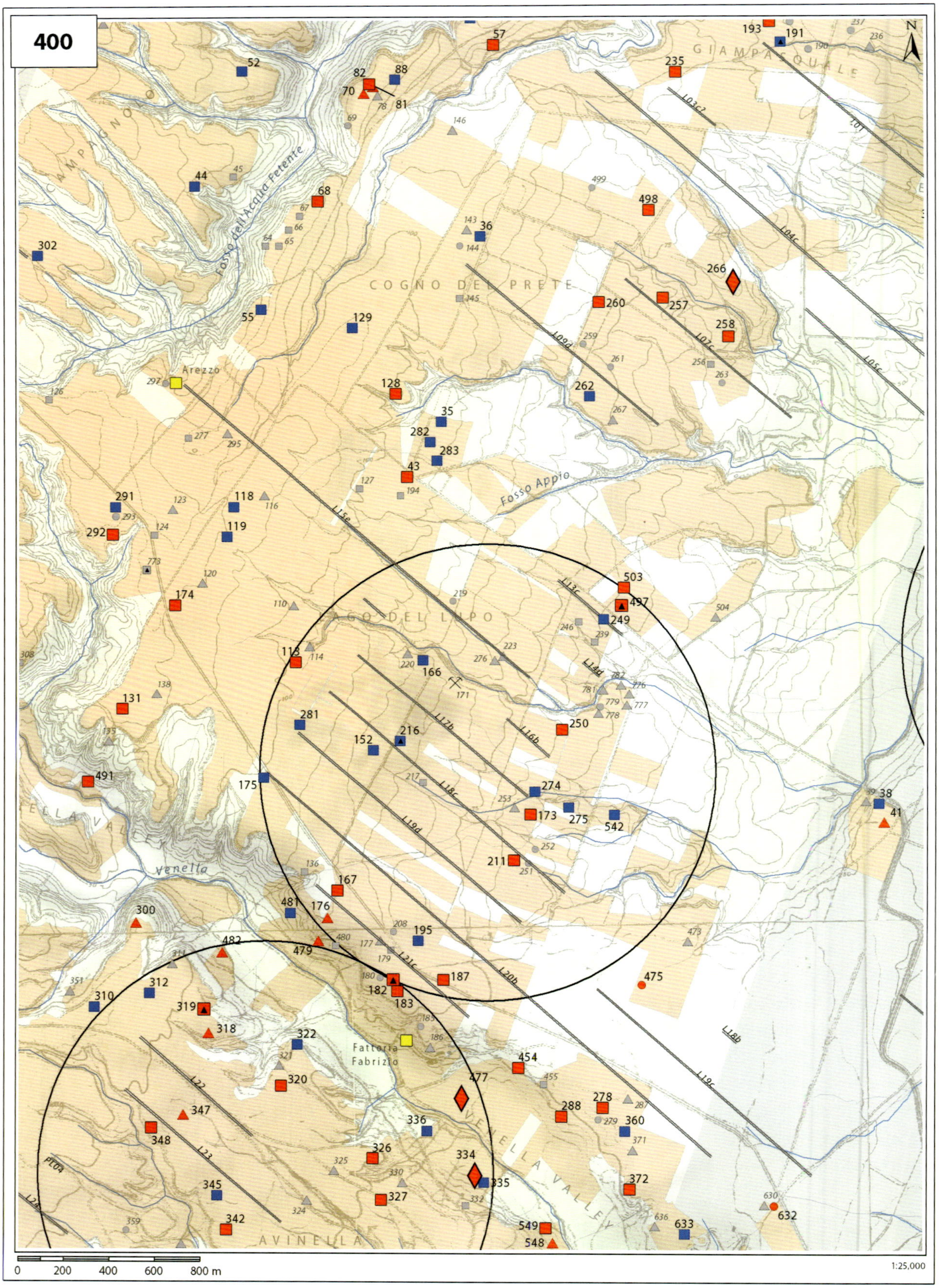

Figure 24.8 The Central Plateau in **400**. Lago del Lupo is within the area where division lines are concentrated. See Fig. 24.1 for map key. (JT)

The Villages of the Central Plateau. Half the farmhouses in the villages of Galitelli and Casa Preti that were significant in **450** were no longer so in **400** (224F/T, 254F/T, 274F). The Capistretti cluster in Cogno del Prete, more dispersed than either of the two in Lago del Lupo, doubled its population in **400**, from two farmhouses to four. One of the additions was 258F (previously occupied in **550**); the other, 498F, was completely ex novo. As we have seen, this side of the Central Plateau was enjoying growth while the other languished. Both of the villages became progressively more active in the boom years of the second half of the 4th century BC, far surpassing their earlier levels of occupation. The contrasts between these two areas of the Central Plateau will be fully explored in the next chapter.

Far from the City

The experiences of these lightly settled localities in **400** were quite diverse. The rate of occupation of significant farmhouses in Campagnolo declined precipitously from four in **450** to one in **400**.[3] Site 302F, newly significant in **450** and paired with the venerable 301F, declined in **400** and was never again significant. These two sites were notable for their very close relation to the northern extension of the principal division line L15, the only line of the larger system of the Central Plateau that is visible to the northwest of the formidable divide created by the Acqua Fetente.

San Marco was without a significant farmhouse in this period; Bufalara (with a total of eight significant farmhouses, three of which were occupied in the periods **450** and **400**) remained largely unchanged. Though it lost one site (305F), it gained another, an ex novo farmhouse (491F), both located on the northern branch of the Bufalara valley. There was a slight increase in AVS, and a notable increase in the previously lightly settled Scarciullo—as striking as the decrease in Campagnolo. There was a very high concentration in **400** of ex novo farmhouses in Scarciullo and adjacent areas (e.g., 511F in Bufalara) on the southern edge of the Gaudella valley, an area that became increasingly exploited in the following two 50-year periods (see Ch. 25).

[3] Since we are dealing with relatively very small numbers, comparisons and percentages can be misleading.

Close to the City

This region saw by far the greatest increases in resident population. In **400**, there were 13% more significant farmhouses occupied than in **450** (cf. *Figs. 23.2* and *24.2*). In Pizzica, as noted above, and in the Bradano valley, there was a high concentration of ex novo sites and only one site (741F) among the four that were significant in **450**, that ceased, temporarily, to be so in **400**. Had it been possible to survey a wider area in Pizzica, it would have proven to be one of the most densely occupied in the chora in **400**. In Pantanello, all of the farmhouses that appeared in **400** (413F, 449F, and 416F) were ex novo. The new site 413F and the well-established 414F lie on the Basento Road, about 350 m apart. The new site 449F is about 300 m to the south of 413F. Site 416F arose in **400**, about 300 m to the south of the Basento Road, and about 600 m distant from sites 413F and 449F. All of the new farmhouses in Pizzica and Pantanello are uniformly significant down to **300**. There is every reason to see the settlement of these areas in **400** as a concerted effort with long term consequences, as decisive for this period and region as that of the Central Plateau in **550**.

There was a concomitant increase in the number of active necropoleis in the localities Close to the City in **400**. In Pizzica, the count rose from two in **450** (including the excavated necropolis 737N) to four in **400** (including also the excavated necropolis 736N; see Ch. 28). Besides the large necropolis of the Pantanello Crossroads, but including the Eastern Necropolis (previously survey site 415N), the count in that locality jumped from one to four necropoleis, three of which were concentrated to the east of the Pantanello sanctuary in an area where there were only two farmhouses, neither of which was significant. The ratio of places of burial to places of residence was 1:1, the highest to date in the study area.

A Synoptic View of the Chora

Population trends. The number of significant farmhouses in the study area increased from 76 to 86 between **450** and **400**. This is an increase of 13%, compared to the 18% increase between **550** and **450** and the 27% that will occur between **400** and **350**. Between **450** and **400** the number of dated necropoleis rose from 21 to 31, an increase of 48%. The percentage increase in necropoleis in all periods is generally far higher than that of the farmhouses. Besides a growing rural population, this trend may in part reflect the further "democratization" of burial that began in the mid-5th century BC.

These basic numbers seem positive enough, but the distribution of the 86 farmhouses among the four principal regions reveals local trends that are not uniformly so (*Appendix C5*). The proportion of significant farmhouses in the Early Village Clusters in relation to the chora as a whole rose very gently in three out of the five clusters between **450** and **400**.[4] Viewed in relation to the other regions over the course of the history of the chora, these are very stable areas. Sant'Angelo did not lose any farmhouses in the century between 425 and 375 BC—it simply did not gain any new ones. The situation was precisely the same in Giampasquale. The more dynamic clusters are Giardinetto and Venella.

The changes in the two localities of the Central Plateau, by contrast, are both more dramatic and, at the same time, contradictory. The prime cereal growing territory of Lago del Lupo lost five significant farmhouses, while Cogno del Prete gained four. Together they represent 20% of all significant farmhouses in the study area in **400**, down from 25% in **450**. The growth of Cogno del Prete in **400** is noteworthy. Though its surface area is only two-thirds that of Lago del Lupo, it had an equal number of active farmhouses in the period, but this unusual growth had no sequel. The farm population of Lago del Lupo, however, more than doubled in the next period. Both areas belong to the cereal growing heartland of the chora. The reason for a decline of one of these neighbors, and a sudden rise in the farm population of the others could well be related to the outcome of the bonifica of **450** when this area was redivided.

The farm population of the localities Far from the City, two of which (Campagnolo and Bufalara) can be considered extensions of the Central Plateau toward the interior, also generally decline as a percentage of the total farm population. The most affected was Campagnolo, which lost three of the four farmhouses that existed in **450**. An unexpected increase took place in Scarciullo, an extensive plateau on the Basento side of the chora, where the nearly perfectly level surface would have been ideal for the plow. Still, its full potential was for some reason realized only later.

[4] The rate of occupation is the number of significant farmhouses occupied as a percentage of the total number in a locality. Cf. Ch. 19, 637.

Of the four regions, that comprising the localities Close to the City stands out as showing the greatest consistent increase in farming population. The trend begun in **450** continues in **400**, and it is most noticeable in Pantanello, the closest to the city.

A slightly different and complementary perspective on the growth and decline of the farming population can be had by considering the relative rates of occupation of the significant farmhouse sites in each of the localities of the four major regions (*Appendix C4*).

The highest rate of occupation of significant farmhouses identified in the Early Village Clusters occurs in Giampasquale, which was constant at 71% for two centuries from **450**. Among the more dynamic and less predictable clusters, Venella enjoyed a 57% occupation rate, increasing 14% from **450**. The growth occurred principally along the east bank of the Venella. This area was one of the few that flourished in **500**, but it lagged behind the rest of Venella in the recovery of **450**. The increase in the rate of occupation of significant farmhouses in **400** among the Early Village Clusters is considerable at 7%, but it is much less than the 13% increase in the same period in the localities Close to the City.

The rate of occupancy of significant farmhouse sites in Cogno del Prete in the Central Plateau had reached its maximum already in **400**. The same is true of Scarciullo, and also at Pizzica and Pantanello. In these selected localities the population saw a significant and sustained upward move in **400**.

Ex novo: farmhouses and necropoleis. The mid-5th century BC saw many new farm residents in the study area (*Appendix C6*). Half of the farmhouses extant in **450** had been constructed on new sites. The highest percentage of these was in the Central Plateau, but three quarters of those in Campagnolo were new and there were larger than average increases in fertile Pizzica and in the Venella as well. An even higher percentage of the necropoleis in this period were on entirely new, not simply reactivated sites: exactly half of the burial sites in the four principal regions were new. This is a clear indication that new people were moving into the chora.

In **400** the picture was much changed and the changes were significant. The percentages of new farmhouses remained high in both the Early Village Clusters and the localities Far from the City, but in the Central Plateau, which had led in the previous half-century in new farmhouse construction, it had fallen to an all time low by **400**, with only two new farmhouses in Cogno del Prete. Lago del Lupo suffered a net decline in its resident population and was the only locality besides Campagnolo to lose farmhouses This was the area in which many families settled as part of a new distribution of the land, and their withdrawal may signify a change in the political climate. In the same period, the new farmhouses in the region Close to the City grew, as did the number of new necropoleis. The period of **400** saw a new direction in the settlement of the chora, as the localities Close to the City, which had been virtually ignored in **450**, became suddenly attractive. They also became the most successful, as measured by FMCE values. The ratio of new to old necropoleis in the region Close to the City surpassed that of all other regions, though new necropoleis proliferated also in the Early Village Clusters, especially Giardinetto, Avinella, the Venella, and in AVS—a decided contrast with the situation in Lago del Lupo.

Significance.

The Early Village Clusters. Between **450** and **400**, the average FMCE values for the Early Village Clusters rose from 1.92, slightly below the average for the study area, to 2.01 (*Appendix C7*). Not all the villages shared in this equally, and some actually declined. Giardinetto and Venella, especially west Venella, modestly surpassed the average for the study area as a whole, but Sant'Angelo suffered a sharp decline of 15%. The gain of Venella was offset by the loss at Sant'Angelo, as one of the previously most successful villages began a decline that was only interrupted a century later in **300**. The average for the Early Village Clusters, which has almost half of the active farmhouses in the study area in **400**, is exactly equal to that of the study areas as a whole. The Central Plateau and the localities Far from the City especially were below that average. Only the localities Close to the City exceeded it (by 31%).

The Central Plateau. The number of farmhouses in Lago del Lupo was greatly reduced from the preceding period (see *Appendix C3*). Among the nine survivors, four were of above average significance (250F, 292F, 497F/T, 503F) and three of these were located in the Casa Preti village cluster; the fourth (292F) was in the extreme northwest, beside a minor tributary of the Venella. Two farmhouses with FMCE values close to the average (2.01 in **400**) were located in the Galitelli village. The farmhouses of lesser significance (113F, 131F, 174F) were located in the upper northwest corner of the Central Plateau near the more significant 292F, and like it, not associated with a village.

At Cogno del Prete in the northern half of the Central Plateau, the average of nine values (*Appendix C7*) is 1.89, decidedly below the average for the study area in **400**. Four are well below average, and three somewhat above it. Three (257F, 258F, and 260F) of the four more significant farmhouses were located, like those in Lago del Lupo, in a village cluster, Capistretti. Site 266S, identified initially as a sanctuary, but probably a farmhouse and a ceramic production center, has an FMCE value (2.17) considerably above the average.

Far from the City. The farmhouses in localities Far from the City in **400** are among the least significant. This represents a sharp decline from the previous period, when in spite of their small numbers their average FMCE value was second only to the localities Close to the City. The only exception to this general picture of decline in **400** is the Scarciullo locality. Though the average for this locality is well below that for the study area, there are two farmhouses with relatively high values (353F and 60F) located on an ample terrace at the headwaters of the Conca and the Fosso Sant'Angelo.

Close to the City. The area of the chora within relatively easy reach of the city emerged already in **450**, with the rapid rise of new farmhouses there, as the most significant of the four regions of the study area. The trend continued in **400**. The average value for the five farmhouses of Pizzica (2.50) is higher than that of any other locality before **300** (*Appendix C7*). The exceptionally significant sites in DBSBS (326F/T, 742F/T, and 762F/T) are all farmhouses with associated burials located in the Basento valley along the Basento Road as it turns onto the coastal plain.

At Pantanello, the closest locality to the city and the locus of a disproportionately high number of larger and smaller necropoleis, the average FMCE value of the four farmhouses in **400** is as low as the lowest in the other three regions of the study area. This area, despite or perhaps because of its proximity to the asty, never became an important residential settlement.

Necropoleis. In **400** there were 21 necropoleis in the study area, an increase of 48% with respect to the preceding period.

Necropolis density. The highest densities of both dated and undated necropoleis in the study area are found in four of the Early Village Clusters (all but Giampasquale). The highest density occurs on the east bank of the Venella; the lowest, in localities Far from the City—Campagnolo and Bufalara, where farmhouses also are more scattered (*Appendix C15*). The discrepancies in the densities of necropoleis and farmhouses seem to show that those farmers who lived farther from the city were not so inclined to bury near their dwellings as those in the Early Village Clusters. This is not what one might intuitively expect. The opposite situation, though not so exaggerated, is found in localities Close to the City (Pizzica and Pantanello). The density of farmhouses is less than that of necropoleis in Pizzica and only just equal in Pantanello. In this respect, this region is unique. The disparity between necropoleis and farmhouses—almost one to two—all over the chora would have been expected also there.

The undated necropoleis, more than half of the total, are widely distributed (*Appendix C15*). There is parity between dated and undated in the Early Village Clusters, slightly more dated than undated Close to the City, and slightly less Far from the City. In the Central Plateau, however, the undated necropoleis make up the majority, with only eleven out of thirty-four dated necropoleis. Those localities and areas with the highest ratio of undated to dated necropoleis are Lago del Lupo, San Marco, and DBSBS. We may only speculate about the reasons.

Ratio of places of burial to farmhouses. In the study area in **400** the ratio between the 31 necropoleis and the 86 farmhouses is slightly more than 1:3. When the farmhouses with burials (F/Ts), which number 12 in this period, are added to the places of burial, the ratio becomes precisely 1:2 (see *Appendix C13*).

Those localities that come closest to a 1:1 farmhouse to necropolis ratio are the lower Avinella, the east bank of Venella, and Pantanello. San Marco and AVS have a somewhat lower ratio, but still well above 1:2. Avinella, east Venella, and the localities Close to the City best approximate the ideal of one place of burial per family. Where did families whose residences do not have closely associated necropoleis or burials dispose of their dead? This also remains a matter of speculation.

The Asty in 400 *(Fig. 24.9)*

Growth in the chora continued in **400** with a steady overall increase in the number of occupied farmhouses and necropoleis. The city, however, witnessed a slowdown in building between **450** and **350**, during which no new temples or public monuments were begun.[5] This hiatus has been interpreted as a sign that Metaponto was experiencing hard times.[6] This could be exaggerated, especially as regards the second half of the 5th century, though it would appear that in the early 4th century BC Metaponto had serious economic and political problems. There were nonetheless numerous signs of vitality, and not just in the chora. In this period the kerameikos in the asty was particularly active, the mint continued to produce coinage of good quality, and there was some building, which, though limited to infrastructure, the drainage system, and fortifications, must have absorbed considerable resources that might otherwise have been directed to showier but less essential projects.

Drainage

The construction projects were a sustained response to long-term problems. Drainage of the city was a concern from the start, and the first phase of the elaborate system was already in place by the mid-6th century. The second, intermediate phase has been dated to the end of the 5th and beginning of the 4th century BC. In this period, the first cloacae were created in stone, along the plateiai.[7] The impetus for this effort was the sudden dramatic rise in the water table in the 5th century BC throughout the chora (see Ch. 22), which affected all low-lying settlements and undermined the health of the population in general. This bonifica

[5] De Siena 2001b, 35.

[6] Mertens 2006, 334, speaking on the "Drastic transformation of the condition of life around the middle of the fifth century BC."

[7] E.g., the cloaca in the area known as proprietà di Taranto, *Metaponto* I, 250–51 and 306–11. This is the system that was substantially preserved, to the degree that it could be, and was reactivated during the excavation of the city in the 1970s. It had continued to be used for the maintenance of the archaeological park. In its current state it dates to the city's last great building period in **300**. See also Sconfienza 1996, 47–48, and Adamesteanu 1973b, 33.

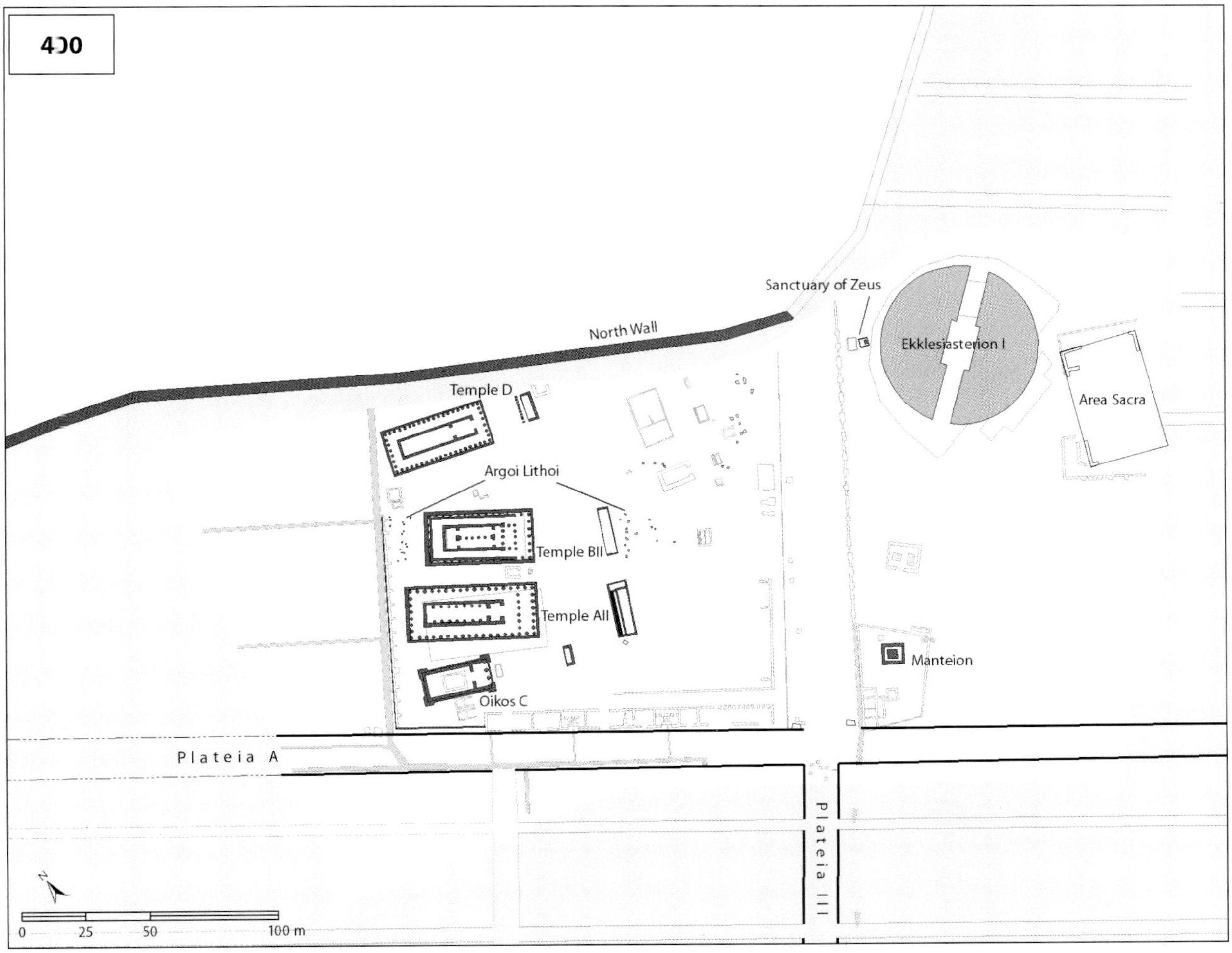

Figure 24.9 Plan of the asty in **400**. (JT, based on Mertens 2006, Fig. 270)

within the city culminated in the 4th century BC with the construction of canals in masonry.

Fortification

Metaponto's first defensive wall ran along the western limit of the city, beyond which lay the open chora (Ch. 20). It was a simple construction of local stone. Apparently no need for anything more extensive or substantial was felt before the 4th century BC. Beginning already in the late 5th, the potential threat posed by the rivalries between neighboring poleis was largely replaced by immediate threats: the energized Lucanians of the interior who were expanding toward the fertile coastal plain, and the ambitious tyrants of Syracuse who were intent on creating an empire in southern Italy (see Metaponto and Magna Grecia, below). These plus changing military tactics, including siege warfare, required extraordinary and sometimes hurried preparations.

Like the drainage system, the fortifications of Metaponto grew in stages. By the late 4th century BC the walls completely encircled the city, and sections of them have been excavated on all four sides.[8] In the 4th century BC constructions, parallel lines of squared ashlar blocks 3 m apart, were regularly joined by cross walls. The space between, in this typical *emplecton* construction, was filled with rubble. On the south side where five courses were identified, the height has been calculated at 1.6–1.8 m, but it is probable that there were more courses.[9] The surviving walls on the north, west, and south,[10] which show evidence of several phases and repeated repairs, have been dated to the second half of the 4th century BC.[11] The signs of hasty

[8] Sconfienza 2005.

[9] *Metaponto* I, 289–94.

[10] We use the conventional terms north and south walls, although their true orientation is northwest and southeast.

[11] Good building stone is in short supply in the immediate surroundings of Metaponto, so that the excavators were not surprised to find reused stones, even in the simple Archaic wall. De Siena 1998, 145.

construction on both the north and south sides were obvious. On the east between the asty and the sea, as on the west, the work was apparently more careful. It has been dated to the end of the 5th or early 4th century BC. The lower courses of the external faces were huge blocks of conglomerate; in the interior, for reasons of economy no doubt, the stones were of various types and dimensions. The whole was given a handsome exterior upper storey in fine limestone.[12]

The main SE–NW avenue, Plateia A, passed through the western gate toward the chora. The passage at this point had existed as early as the early 6th century BC.[13] In the 4th century, the gate presented an opening almost 5 m wide that was flanked on either side by adjacent pairs of square towers.[14] The north or Tarantine gate by the kerameikos has also been excavated, and at least one other gate identified in aerial photographs. Though now sadly reduced, the fortifications of Metaponto in their final form in the late 4th century BC must have been both impressive and solid. This was noted by Pausanias some 500 years later.[15]

The Kerameikos

From the 6th century onward, pottery was produced at a number of sites within the city's walls, as well as in the villages or at isolated sites throughout the chora.[16] In certain periods, a significant portion of this production was concentrated in a triangular area situated between the city wall on the north, the first SE–NW stenopos on the south, and first wide NE–SW thoroughfare, Plateia I. This area, excavated in the 1970s,[17] has been the object of extensive studies. Ongoing research is creating a uniquely detailed picture of workshop organization and revealing technical aspects of production.[18] The kerameikos was located south and east of the sanctuary, set apart from the religious center and the main residential quarter. It lay in the shadow of the north city wall, and was conveniently placed with access to major roads leading out into the chora through both the north and west gates. The kilns, deposits of waste and misfired pottery (*scarichi*), and related structures of four workshops are distributed chronologically from the second half of the 6th to the early 3rd centuries BC. They occupied an area of nearly 1600 m^2, the equivalent of 1% of the total surface of the city.

Figure 24.10 Banded ware squat lekythos from tomb 350 (h 18 cm). (CW)

Figure 24.11 Unglazed table amphora, cooking ware fabric from tomb 78, Pantanello necropolis, dated 300 to 275 BC (h 27 cm). (CW)

The extensive, though far from complete, excavations in the area of the kerameikos revealed a workshop of the 6th century BC, the oldest yet documented in the city.[19] There were coeval production centers in the chora.[20] The material from the Archaic deposit included B2-type Ionic cups and a range of shapes including lekanai, one-handlers, olpai, and hydriai, many of which are also present in the assemblages of farmhouses in the chora in 550 and 500. In addition,

[12] A. Bottini 1988a, 309, reporting on the excavations by De Siena.

[13] Adamesteanu 1987.

[14] De Siena 2001, Fig. 32b.

[15] Pausanias 6.19.11.

[16] Silvestrelli forthcoming a.

[17] D'Andria 1975.

[18] Silvestrelli 2004; Cracolici 2004a.

[19] Silvestrelli 2004, 111; Silvestrelli forthcoming a, 6–8.

[20] A. Bottini 1994, 705–06, Pl. 50.

black-figure lekythoi,[21] lebetes gamikoi, and phialai were produced here. Similar forms turn up in the survey assemblages in the chora: the lekythoi occur in very limited quantities, exclusively in necropoleis. This varied production implies an extensive knowledge of clays and firing temperatures, as well as a highly complex system of organization.

The operation of the kerameikos in the second quarter of the 5th century BC is documented by banded ware (*Fig. 24.10*), unglazed table ware (*Fig. 24.11*) and cooking ware. In the 460s, the potters of Metaponto, directly inspired by Athenian models, began to produce pottery in the red-figure technique. The Early Lucanian (or "Protoitaliote") style begins with the work of the Pisticci and Amykos Painters early in the second half of the 5th century BC. Though the location of their workshops has not as yet been discovered at Metaponto, their presence in the city, though it cannot be absolutely proven, is next to certain. Vases by the Pisticci Painter (e.g., *Fig. 23.16*), including his earliest, according to Trendall,[22] have been found in great numbers in burials in the chora[23] and in necropoleis of surrounding indigenous settlements such as Pisticci. Vases by the Cyclops Painter, an associate of the Pisticci Painter, whose surviving work is much rarer and sometimes mistaken for Attic, are also found in the Metaponto area (see Ch. 29; site 736, T 19-01 and T 19-03). The Amykos Painter is represented by misfired fragments from an early, uncontrolled excavation in the city.[24]

Although black-gloss production has been well documented for this period in the kerameikos, the work of the first generation of red-figure vase painters is not well represented. The next generation, however, is very much in evidence. Deposit One (scarico 1) has yielded misfired and discarded fragments attributed above all to the Dolon Painter and, to a lesser extent, to the Creusa Painter,[25] whose careers cover the first two decades of the 4th century BC. They are tied to the earlier tradition of Metapontine pottery by the choices of decorative schemes and shapes, among which figure Campana kraters and nestorides. The latter, a shape of indigenous origin, was introduced by the Amykos Painter.[26] The kerameikos finds demonstrate that some production was clearly aimed at indigenous markets.[27] This is important evidence for the relationship between the city and surrounding indigenous populations.

A single burial in Pizzica (site 736, T 19) yielded a pelike (Ch. 29, T 19-02), lebes gamikos (Ch. 29, T 19-05), and skyphos from the Dolon and Creusa Painters workshop (Ch. 29, T 19-09), and a *dinos* (Ch. 29, T 19-06).[28] The latter is extraordinary for its shape and ivy-leaf decoration. The pelike may be by the Dolon Painter himself, and the lebes and skyphos by the Creusa Painter. The works of both painters have been found also in numerous burials in the Pantanello necropolis.[29] It is clear that the Metapontine potters found eager markets close at hand.

The influence of Attic potters and painters on those of Metaponto was very strong throughout the 5th century BC. A number of figured vases are so like the Attic ones that they were thought at first to be imports. The two squat lekythoi from site 736, T 19, are almost indistinguishable from the work of an Athenian, the Washing Painter (Ch. 29, site 736, T 19-01 and T 19-03). Burn and Denoyelle agree, however, that they are probably the work of the Cyclops Painter, a Lucanian who has been judged by Burn to be deliberately copying Athenian styles and motifs. The black-gloss pottery of this period produced in some kerameikos workshops imitated typical shapes—the kylix (*Fig. 24.12*) and olpe—and the stamped decorations of Attic models, but according to Silvestrelli,[30] there is more than just passive copying here. She sees this as an adaptation

Figure 24.12 Stamped black-gloss kylix from tomb 115, Pantanello necropolis, dated 430–390 (diam 14 cm). (CW)

[21] E.g. palmette lekythos from T 218. Pantanello necropoleis, no. 20 Carter and Hall 1998, 245.

[22] Trendall 1990, 217; cf. Silvestrelli 2005, 115–16.

[23] Burn 1998, 613–24, on the vases by the Pisticci, Amykos, Dolon, and Creusa Painters, which represent the single largest group of figured vases in the necropolis.

[24] D'Andria 1976, 447, figs. 78–81.

[25] Silvestrelli 2005, 117.

[26] Silvestrelli 2005, 118–19.

[27] Silvestrelli 2005, 120.

[28] See detailed description and discussion in Ch. 29.

[29] Burn 1998, 614–15, Tomb 51.

[30] Silvestrelli forthcoming a, 10.

Figure 24.13 Black-gloss "corded" beaker from tomb 330, Pantanello necropolis, dated 414-336 BC (h 12 cm). (CW)

within a consolidated tradition that is expressed in the maintenance of traditional Metapontine forms such as the lebetes gamikoi and ribbed oinochoai. Continuity in local traditions is maintained through the plain/banded ware forms such as the lekanai and jugs, and the cooking ware lopades that are such a basic part of farmhouse assemblages. There are also several quite original forms, such as the "corded" beakers (*Fig. 24.13*).[31]

The production of black-gloss as well as red-figure vases in the late 5th and early 4th centuries BC was the largest by far in the area of the kerameikos. With interruptions, manufacture of fine wares and all the other classes continued through the second half of the 4th and into the 3rd century. It included vessels in a local version of the so-called Gnathia style.

There are also indicators of metallurgy, sometimes on the same site (for example on the south side of the agora), a pattern also seen in the chora at Pantanello in the 2nd century BC Already in the 6th century BC it seems, this area proved attractive to artisans in clay and metal. Their proximity to the grand public space of the agora is reminiscent of the kerameikos in Athens, with its nearby metalworking quarter and cult place of Hephaistos. At Metaponto, this area would appear to have been a street of shops in the heart of the city, not a circumscribed and isolated locale. On the north side of the sanctuary existed another type of ceramic production: the workshop attached to the residence.[32]

Ceramic production, at least in the kerameikos, reached its peak in terms of quantity and creativity in 400.[33] The history of pottery production reflects fluctuations in one specialist area of the economy, and does so over a long period of time.

The Mint

Another potential gauge of economic activity is the productivity of the mint. With a few interruptions, it covered a span of approximately three centuries, and offers a nearly continuous record of the polis in many spheres. It is an indication of the commercial activity and the economic health of the polis, and also a record of important events in its life, as we have already seen in the notable case of the image of Acheloos and the representations of the statues of Apollo. The range in types of Metapontine coins, their varying aesthetic quality, and the inscriptions are a potentially rich source of information about the cults and public currents and the artistic taste of the community. Not least, the study of buried hoards provides evidence for major historical events.

The coins form a more or less continuous sequence from the mid-6th century to the 3rd century BC. A gap, however, between the very first double relief issues—with the full figures of Acheloos, Apollo, and Herakles—and those with the heads of gods at a slightly lower weight standard, could be significant.[34] The new series which broke sharply with this past started up after a pause of some years, probably about 430 BC, and is both on a lighter standard and of less careful workmanship.[35] There were only half as many dies. The quantity grew probably as a result of the increasing defense spending as the Lucanian pressure escalated and the generals arrived with their mercenary armies from Greece and Epiros beginning in the 340s BC. Should this decline be interpreted as a sign of relatively harder times in this roughly eighty-year period between 430 and 350 BC? The sharp cutback in all but essential public building together with the relatively modest growth in the rural population in this period would fit well with such an interpretation.

The Metapontine spirit of independence in its coinage is evident in the variety of gods and heroes chosen for the obverse once the changeover had been made. Other cities favored one divinity. The ear of barley almost never absent from Metaponto's coins was moved to the reverse. The obverses of the new series included busts of Apollo and Herakles, which were already present on early double relief series. In addition, Dionysos, Zeus, and Athena were used, as

31 Silvestrelli forthcoming a, Fig. 9, No. 16. Cf. a similar beaker from a Pantanello burial: Elliott 1998, 667–68, Fig. M15 (T.330).

32 Silvestrelli 2004, 110, Fig. 2 B.

33 Mertens 1999, 276–77.

34 The hiatus in the coinage of Metaponto and a similar one at Croton has been connected with the foundation of Thourioi and the events surrounding it (Stazio 1974a, 80).

35 Stazio 1974a, 85, and Stazio 1999, 460, noting that the new series was far less abundant between 430 and 350 BC than was the incuse coinage over an equal span of time; cf. Johnston in Noe and Johnston 1984, 91.

Figure 24.14 Silver stater (obverse) representing Apollo Karneios or possibly Dionysos Eryphios, 420–400 BC. (Courtesy of the American Numismatic Society, 1944.100.4937.obv)

Figure 24.15 Silver stater (obverse) representing a goddess (Demeter?) by the die-engraver Aristoxenos, 400–380 BC. (Courtesy of the American Numismatic Society, 1944.100.4960.obv)

well as a male figure with ram's horns and goat's ears generally identified as Apollo Karneios—but who just might be Dionysos Eryphios (*Fig. 24.14*).[36] More frequent than the rest was a figure identified by inscription as Demeter, with grain woven into her hair, but who, in some cases, might just as easily have been Kore-Persephone (*Fig. 24.15*).[37]

The divinity notable for her absence is Hera, to whom the second oldest and the largest of the temples standing in the sanctuary had been dedicated. Was this a deliberate omission? Demeter may have resonated better with the new settlers of the land redistributed in the mid-5th century BC than the austere goddess of an earlier era. Though her cult place has not been positively identified, the worship of Demeter at Metaponto is documented by a series of terracotta images found in the sanctuary area towards its western and northern edges.[38]

Pythagorean tradition has it that after Pythagoras' death, his dwelling on a stenopos in the city was converted to a temple to Demeter, though it is more likely that her cult place, as in most poleis of Magna Grecia and Sicily, was outside the city walls in the chora.[39] The Metapontines clearly had a special veneration for this goddess and for her daughter, one already expressed in the 5th century BC in the dedication made at Eleusis by Aristodamos of Metaponto.[40] The coins remain, however, the most visible and attractive evidence of this devotion.

The willingness of Metaponto's mint to innovate can be seen in the production of fractional coinage from the beginning,[41] and in the introduction of bronze coinage, already late in the 5th century BC. Bronze coinage greatly facilitated the production of fractions. Its adoption at this time reflects current social and economic reality. In a broad perspective, a coinage with a wide range of denominations ought to correspond to a society where wealth is more widely and evenly distributed.[42] A variety of denominations lower than the stater, at first in silver, were an outstanding characteristic of Metapontine coinage. The smaller denominations indicate that the authorities involved in minting wanted to encourage transactions even of very modest import. Metapontines, one might conclude, were unusually eager to monetize transactions, and were more intensely involved in small scale trade than was the case among the other poleis of Magna Grecia. The introduction of bronze coinage at the end of the 5th century BC represents an escalation of this tendency. Metapontine bronze coinage, as Johnston observes, was more diverse than that of most of her South Italian neighbors.[43]

It has been suggested that fractional coinage was a reflection of a fundamentally agricultural economy.[44] If so, why does it not occur elsewhere, as most of the poleis of Magna Grecia lived by agriculture? It has also been suggested that the fractional coinages would have benefitted the activities and the savings of artisans, such as the potters of the kerameikos and others.[45] It may not be a coincidence that the bronze fractions, though they became extremely numerous only in the late 4th century BC, had already been introduced in the period when the production of the kerameikos peaked at the end of the 5th and beginning of the 4th century BC. The impression of Metaponto, from its coinage from the mid-5th century onward, as

[36] Kraay 1966, 307. Carter 2006a, 172, and n. 121.

[37] Kraay 1966, 307, identifies the figure in 24.15 simply as "a goddess."

[38] Calabria 2005.

[39] Giacometti 2005, 91–94.

[40] Giacometti 2005, 95.

[41] Noe and Johnston 1984; Johnston 1983; 1989; 1990; 1998.

[42] Stazio 1974a, 106.

[43] Johnston 1989, 121.

[44] Stazio 1974a, 101.

[45] Lepore 1974b, 336.

Johnston has put it, is that "of a city that is minding its own business, joining in Greek cooperative efforts when necessary."[46] A number of poleis of Magna Grecia ceased to mint coins in the period 450 to 350 BC. The output of Metaponto, as well as that of Taras, Thourioi, and Velia, however, remained heavy.[47]

Ekklesiasterion II

About 375 BC, the multifunctional meeting place of the assembly went out of use. Parts of it—the seats and other architectural elements—were recovered from a necropolis that lies on the road to Taras.[48] This is significant for the political history of the polis. The assembly would probably have been forced to meet elsewhere if it continued to meet at all. Some scholars have imagined a barbarian incursion, perhaps even the capture of the city at this time, supporting their arguments with slender evidence from disputed readings of Strabo,[49] and an inscription in the Oscan language with a doubtful provenience.[50] Such a dramatic turn of events is not supported by any ancient text or by the evidence of the continuously expanding occupation of the chora, the smooth and apparently uninterrupted flow of coin issues, or the creativity of the kerameikos.

Commercial relations with the populations of the interior would have benefited both the Greeks of Metaponto and the indigenous peoples. In the early decades of the 4th century BC some Metapontine coins are similar to those of Terina, whose coins were widely diffused in the Lucanian-occupied interior, and several coins of Terina appear to be imitations of Metapontine ones. This has been taken as a sign that Metaponto and Terina were in competition for the markets of interior Lucania.[51] The kerameikos at this time was producing vessels for those markets, the traditional indigenous forms such as the kantharoi and nestorides, in a fabric known as the "Xenon Group," which show up primarily at sites in the interior.[52] The damage to the Ekklesiasterion is unlikely to reflect an historic event that did not have long-term consequences. A sudden attack and withdrawal would hardly explain the reuse of the stone, which instead implies hard times. This question should be reexamined in the light of all the relevant evidence and in the broad historical context. The aim of the next section is to open the perspective on the expanding world of the second half of the 5th and 4th centuries BC.

[46] Johnston 1990, 61.

[47] Kraay 1966, 302.

[48] Mertens 2006, 436; cf. A. Bottini 1988a. Until the evidence is fully published it would be risky to give it excessive weight.

[49] Strabo 6.15. Guzzo 1983; cf. Musti 1988, 143; Maddoli 1974.

[50] Lombardo (1987b, 254) did not believe that the Poldi Pezzoli helmet proved that a Lucanian garrison occupied Metaponto.

[51] Noe and Johnston 1984, 61, followed by Stazio 1974a, 87–88.

[52] Silvestrelli 2005, 120. See Ch. 8.

Metaponto and Magna Grecia, 450 to 350 BC

> ...ἀφικνοῦνται ἐς Μεταπόντιον τῆς Ἰταλίας.
> καὶ τοὺς Μεταποντίους πείσαντες κατὰ
> τὸ ξυμμαχικὸν ἀκοντιστάς τε ξυμπέμπειν
> τριακοσίους καὶ τριήρεις δύο καὶ
> ἀναλαβόντες ταῦτα παρέπλευσαν ἐς Θουρίαν.
> καὶ καταλαμβάνουσι νεωστὶ στάσει τοὺς τῶν
> Ἀθηναίων ἐναντίους ἐκπεπτωκότας·
>
> ... they arrived at Metapontum in Italy. There
> they persuaded the Metapontines to send
> with them, in accordance with the terms of
> their alliance, three hundred javelin-men
> and two triremes, and taking up these they
> sailed along the coast to Thuria. At Thuria
> they found that the faction opposed to the
> Athenians had recently been expelled in a
> revolution;[53]
>
> Thucydides, 7.33.5

> ...Ἰταλιωτῶν δὲ Θούριοι καὶ Μεταπόντιοι,
> ἐν τοιαύταις ἀνάγκαις τότε στασιωτικῶν
> καιρῶν κατειλημμένοι, ξυνεστράτευον...
>
> ... but of the Italiots the Thurians and the
> Metapontians took part in the expedition, being reduced at this time to such
> straits by party crises that they could not do
> otherwise...[54]
>
> Thucydides, 7.57.11

Were it not for the brief mention by Thucydides of the reception of the Athenians in Metaponto at the outset of the second Sicilian Expedition in 413 BC, we would have no explicit word of the city at all before the arrival of the foreign *condottieri* in the second half of the 4th century BC.

In the last half of the 5th century BC, Taras, in the power vacuum left by the decline of Croton,

[53] Text and trans. C.F. Smith 1958 (Loeb).

[54] Text and trans. C.F. Smith 1958 (Loeb).

emerged as the preeminent military and economic power among the poleis of southern Italy. To imagine because we have practically no news about Metaponto in this period, that it was merely "minding its own business," or that it was simply a silent and passive partner of its powerful neighbor to the north would in both cases be erroneous. Lepore traced the influence of Taras on Metaponto back to the 470s.[55] It has more recently been reevaluated by historians who see evidence of Metaponto's independence, and by archaeologists, who can now distinguish the artisan production of the two poleis more clearly than was possible a generation ago.[56]

The foundation of Herakleia in 433/32, essentially an extension of Taras into the former territory of Siris, would undoubtedly have put Metaponto in a disadvantageous position. The strategy of the Achaians in Antiochos' description of the founding of the apoikia (Ch. 17)—of sending settlers to the site of Metaponto to block Tarantine advance into the rich chora of Siris—was turned on Metaponto, and at this point, two centuries later, left her isolated and vulnerable. The failure of the sources to mention any attempt by the Metapontines to thwart this action has been interpreted by historians arguing from silence as a demonstration of Metaponto's impotence at this time, the beginning of the period, as they see it, of virtual subjugation of the polis to the will of the Tarantines. Metaponto's response may have been to continue to support Thourioi.

The presence of regimes with democratic, pro-Athenian leanings in Metaponto and Thourioi in the last three decades of the 5th century, however, effectively encircled Herakleia and a balance of a sort was achieved without warfare. By this time the common interest of the Italiote poleis was self-preservation and a united front against the Lucanians.[57] Metaponto could maintain a degree of independence, as will be seen in its relations with Alexander the Molossian, without embracing a destructive war with its powerful neighbor to demonstrate it. Understanding of the political history of the region has become more nuanced, especially as knowledge of indigenous cultures of southern Italy has increased dramatically in the last four decades, and with greater attention paid to underlying social and economic forces.

[55] Lepore 1974a, 317.

[56] Adamesteanu 1999, 346; Musti 1988, 141–42.

[57] Lombardo 1987, 55.

The struggle between oligarchic and democratic factions throughout the poleis of Magna Grecia came violently to the fore with the refoundation of Sybaris, and the foundation of Thourioi several years later in 444/443 BC.[58] As would be expected, the question of fair distribution of land was central.[59] In the territory of Thourioi, the principle of *isomoiria* was strictly applied, at least in theory.[60] In the chora of Metaponto in **450** and **400**, we may see a more or less successful application of that principle. Metaponto and Thourioi were closely allied in their opposition to Tarantine expansion and remained so into the 4th century BC. Their closeness even a century later is revealed by their comparable relations with Alexander the Molossian, and by the similarities of their coinages.[61] It would appear that they were first drawn together by the alliance with Athens and thereafter by shared, though not universally embraced, democratic values that may explain the remarkable loyalty between the two cities.

We learn from Thucydides that Metaponto and Thourioi alone among the poleis of Magna Grecia opened their gates to the Athenians in 413 BC. This event reveals not only their isolated position in Greek Italy, but also the deep fissures that existed within the societies of both poleis. The Metapontines, it appears from Thucydides' brief description, had to be persuaded under the terms of the alliance to hand over ships and soldiers for the Sicilian adventure.[62] The pro-Athenian party had come to power in Thourioi and from them the Athenians asked for even more. Thucydides comments later that the Metapontines and Thourians took part because they had no other option due to the climate of stasis in each.[63] Shortly thereafter an anti-Athenian reaction occurred in Thourioi.[64]

Within a decade or so,[65] the legend "Homonoia" appears on Metaponto's coins with a female head that is indistinguishable from those identified by inscription as Demeter.[66] This and the similar inscriptions

[58] Berger 1992, 32–33.

[59] Aristotle *Pol.* 1303a 32–35.

[60] Asheri 1966, 14.

[61] See Ch. 25, 863–66.

[62] Thucydides 7.33.5-9.

[63] Thucydides 7.57.11-14.

[64] Berger 1992, 33.

[65] The coin is among the first that has the signature of Aristoxenos, the premier die engraver of southern Italy, active at the end of the 5th or beginning of the 4th c. BC. Noe and Johnston 1984, 62–63, Pls. 32–33, No. 420.

[66] E.g., Noe and Johnston 1984, 78, Pls. 32–33, No. 421, from same die as No. 420 with "Homonoia" altered by the introduction of a torch symbol and recutting of letters.

with "Hygieia," "Soteria," and "Nika" were at first taken as epithets of Demeter but are now recognized as personifications of these abstract concepts. As Giacometti observes, this is perhaps the first appearance in the Greek language of the term that stands for civic harmony or concord.[67]

Why did Metaponto commission a special coin with Homonoia, the first work for them by an outstanding artist, Aristoxenos?[68] Ghinatti proposed that the concord intended was that between Metaponto and its neighbors.[69] What was commemorated then was its entry into the Italiote League, a continuation of the Achaian League under the leadership of Taras that was formed at the end of the 5th or beginning of the 4th century BC.[70] There is no documentary evidence to support Metaponto's membership, and Musti has thought it extremely unlikely, but had also no reason to exclude the possibility.[71]

In Stazio's view,[72] the concord intended was internal, and he proposed that the inscription and the first appearance of personified Concord marked the resolution of the factional strife evident at the time of the Sicilian Expedition, but that certainly existed before and after the episode noted by Thucydides. As Lepore suggested, there need be no contradiction, if the foundation of the Achaian League, as described by Polybios decades earlier, was retrospectively commemorated.[73] The league brought to an end, at least temporarily, the internal discord in the member cities and, at the same time, united them in a common defensive alliance for the first time. Such a novel event needed to be recognized and celebrated.

The fall of Poseidonia, Metaponto's sister city, to the Lucanians in about 400 BC must have made a deep impression on all the Greek poleis of Magna Grecia. The site had great strategic value because it controlled the best passage from the Tyrrhenian coast to the Ionian shore. Though the conflict was long and fierce, recent scholarship has tended to view the event more as a takeover than an abrupt break with the past, in which the conqueror already seems to have absorbed the organizational and military skills of the conquered.[74] A degree of continuity has been observed in the continued functioning of Poseidonia's ekklesiasterion as a seat of government.[75] There is cultural continuity as well, as evidenced by 4th century BC Paestan pottery, which continued in the red-figure tradition of the Greek poleis of Magna Grecia. Lucanian society was, as Lombardo puts it, "highly structured" with the elites firmly in control,[76] but as Strabo noted, democracy prevailed in times of peace. In times of war the assembly elected a *basileus*, or king, to lead them.[77]

This example may help us to understand the changes that took place at Metaponto and Herakleia at the end of the 5th century BC. Lucanian pressure on the Ionian coast—part of the general expansion of Italic people towards both the southern and western coasts of Italy—began as early as the late 440s BC and with it the Greek opposition, as exemplified by the campaigns of the Spartan strategos or general Kleandridas on behalf of Thourioi.[78] These hard battles, where superior tactics gave the Greeks an advantage, foreshadowed those of the *condottieri* a century later. By that time, the Lucanians had absorbed much and could turn these borrowed tactics against the Greeks. A Lucanian presence, on a very limited scale and not at all aggressive, can be seen as early as the end of the 6th century BC. There was a small but growing number of burials with distinctly Lucanian weapons and armor, and these continued into the 5th and early 4th centuries BC.[79] Scholars have perhaps exaggerated their significance.[80] It is surprising to find a Lucanian *cinturone*, or dress belt, in a grave of a young man with typical Greek strigils (*Figs. 24.16–17*). The belt is not there casually, and the man may be a foreign husband buried in the family plot with his Metapontine wife.[81]

[67] Giacometti 2005, 96–97. It has been thought to have originated with Pythagoras himself, who, according to Iamblichus, exhorted the council (*synhedrion*) of Croton to build a temple to the Muses to preserve the existing Homonoia. Iamblichus *VP* 45; Giacometti 2005, 97. Can it be purely coincidental that the first appearance of this personified virtuous state of concord should have the features of Demeter, whose place of worship at Metaponto, if the sources, Iamblichus and Favorinus, are to be believed, was the former temple of the muses and home of Pythagoras himself? Cf. Giacometti 2005, 90–91, with sources.

[68] Noe and Johnston 1984, 61–65, 78, Pl. 32, No. 420.

[69] Ghinatti, cited by Stazio 1974a, 86 n. 43.

[70] The date 393 BC is cited by Lombardo 1987b, 249.

[71] Musti 1988, 142.

[72] Stazio 1974b, 334.

[73] Lepore 1974b, 335. Lepore accepted the late 5th to early 4th c. BC dating of the coins.

[74] Lombardo 1987, 55.

[75] Mertens 1987, 564.

[76] Lombardo 1987b, 249.

[77] Strabo 6.1.

[78] *Storia della Magna Grecia* II, 402–03.

[79] A. Bottini 1993b, 23, 181.

[80] A. Bottini 1988a, 308.

[81] Carter 1998e, 157; 1998f, 212; Carter and Hall 1998, 277, T.106 in Family Group 5.2 in Nucleus 5.

Figure 24.16 The burial of a male, 25–30 years of age, wearing a Lucanian belt, in an earthen, plaster-lined grave. T. 106, Pantanello necropolis. (JC)

Figure 24.17 Bronze ceremonial belt, typical of Lucanian or Samnite warriors in the 5th–4th c. BC (w 75 cm). T 106, Pantanello necropolis. (CW)

The Italiote League was formed, it would seem, as an immediate response to the fate of Poseidonia.[82] As a defensive alliance it obligated all the member cities to come to the aid of any member in case of Lucanian aggression. The military forces—civic militias, as Lombardo observes—were still made up of citizens drawn from the landowners, rather than mercenaries. Leadership was provided by elected *strategoi* of the individual member poleis, who were compelled to serve on pain of death. There was no integrated command, and the results were disastrous at first[83]—until Taras took over the direction of the League. What began as an alliance against the Lucanians soon expanded to offer a united response to the threat posed by the designs of Dionysios I on southern Italy. The great man of Syracuse had begun to move against the Greek cities already in the early years of the 4th century BC. Rhegion and Locri fell in 379 BC, at which point the Tarentines moved the headquarters from the Sanctuary of Hera on the Lacinian promontory (Capo Colonna) and later to Herakleia.

With Dionysios, the poleis of Magna Grecia experienced for the first time a new kind of warfare involving mercenary soldiers, alliances with indigenous peoples, and siege tactics, all of which had lasting effects. Dionysios' aggression, and his goal of empire, was also new, and foreshadowed the later expansion of Macedon in the east and Rome in Italy. The only polis at this time with sufficient resources to oppose Dionysios was Taras under Archytas. A Pythagorean philosopher and friend of Plato, he skillfully guided the internal and foreign policy of Taras for much of the first half of the 4th century BC. Archytas preferred diplomacy to confrontation, and early on supported Dionysios' colonizing efforts in the Adriatic. Despite past tensions between Metaponto and Taras, relations with Metaponto, we may suppose, were good. Archytas spent time in Metaponto in his early years of formation as a Pythagorean, and the Metapontine Pythagorean Euphrones, at first hostile to Dionysios, became his friend.[84] It is likely that Metaponto aligned with Archytas' policy toward Dionysios. Under Taras' influence, Metaponto, as Lepore suggests,[85] may have become more democratic. Archytas' policy of seeking civic concord by trying to diminish the tensions between rich and poor would have resonated well with the democratic elements of Metaponto's population.

Politics in the Chora

The main fact of Metapontine history in the last quarter of the 5th century and the first half of the 4th to emerge from this brief survey is the existence of stasis that would have pitted the democratic pro-Athenians against the remnants of the aristocracy as well, perhaps, as the Pythagoreans, still very active at Metaponto. The search for homonoia, or civil concord, more an ideal than a reality, was actively pursued and its commemoration on the coinage of the polis suggests that

[82] Ghinatti 1961–1962.

[83] Lombardo 1987b, 250.

[84] Lepore 1974a, 318; Valerius Maximus 4.1.1 and Polyaenus 5.2.22.

[85] Lepore 1974a, 318–21.

it may actually have been achieved, if only briefly. The events of the final years of the 4th century BC (Ch. 25) show that stasis was a permanent condition at Metaponto as in Magna Grecia and the Greek world. It was particularly widespread and intense in this period and throughout the 4th century. At the center were always questions of land: possession, equality of distribution, inalienability, confiscation, and redistribution.[86]

The desertion in **400** of Lago del Lupo and perhaps of other areas like it elsewhere in the chora, which were divided and distributed to new settlers by a democratic government in **450** may well reflect the stasis in the citizen body in the last quarter of the 5th century BC that Thucydides thought important enough to mention. This area of Lago del Lupo, as we shall see, was again divided and redistributed in **350** and **300**.[87] It was a prized possession, and if this interpretation is correct, a football in the struggle between factions. Other considerations may have played a role, but one thing is clear: no other region of the study area had to endure the same dramatic ups and downs. The destruction of the meeting place of the democratic assembly in the early 4th century BC may be another manifestation of stasis, and not of external interference, as it has usually been interpreted.

Metaponto up until this time, apart from participating in the conquest of Siris in the mid-6th century, had generally avoided the serious conflicts that embroiled other poleis of Magna Grecia. Metaponto's relations with Taras after the foundation of Herakleia in the 430s were not hostile and she seems to have followed a cautious and independent course of action that continued into the next century and beyond. "Minding her own business" became more difficult in this period that it had been in the past.

Metaponto was now part of a much more complex and interconnected world. Was this half-century a period of decline? It has been argued that Taras dominated Metaponto artistically. The production of the kerameikos and the mint at this time argue for creativity and independence, and the chora itself makes the case that there was steady growth throughout this half-century. The abandonment of a part of the Central Plateau is only part of the story. Growth in other areas Close to the City compensated while the population in the Early Village Clusters and localities Far from the City was steady. Prosperity (measured by FMCE values) did not grow in this period, but neither did it decline.

[86] Asheri 1966, 60–108, in particular.

[87] See Ch. 25, 828–33.

25
The Refounding of Metaponto

Joseph Coleman Carter

The history of the chora as well as the asty in the century from 375 to 275 BC will be treated as a single continuous development. Up to this point the 50-year periods have corresponded to major changes in both. As will be clear in the following discussion, the transforming moments occur halfway through the third and in the fourth quarters of the 4th century BC. The last half of the period **350** (350–325 BC) saw the revival of independent Metaponto that coincided with a significant intervention from outside, the arrival of Alexander the Molossian in Magna Grecia. In the first half of the period **300** (325–300 BC) the Metapontines created a new and splendid urban center with a state-of-the-art theater and added a monumental frame of stately stoai to the vast agora. The renewed urban center and increased productivity of the chora together amounted to a re-founding of the polis. They indicated a new sense of purpose and pride with an expanded citizen body that included newly enfranchised citizens. Like the stoai and the theater, this vision was never to be completely realized. The closing years of the 4th century BC led up to another intervention, this time disastrous: that of Kleonymos of Sparta (303 BC). In all of the period **350** there is exceptional growth with respect to the preceding period, especially in the Central Plateau. In **300**, although the population of the chora reached its all time high, this was in a sense the dénouement of the preceding half-century, with signs of greater material success but of an ever more limited autonomy and of economic decline.

Following the pattern established in previous chapters, the analysis of the settlement in **350** and **300** begins with an overview of the numbers.

The Chora in 350 and 300

Between **400** and **350** the total number of farmhouses in the chora increased from 86 to 109 (26%), compared with only a 13% increase between **450** and **400**. Between **350** and **300** the number grew from 109 to 142 (30%) (*Maps 10* and *11*, Vol. IV; *Fig. 25.1*; *Appendix C1*).

This growth, as usual, was not evenly distributed among the four regions of the chora. Between **400** and **300** the percentage of significant farmhouses in the Early Village Clusters fell from 47% in **400** to 41% in **350**, and further still, to 38% in **300** (*Appendix C5*). The most marked decreases were in Venella, which lost just one site, but 4% with respect to the other localities. Sant'Angelo also declined. Giardinetto did not grow at all but steadily maintained more or less 9% of occupied sites. Avinella, which gained five significant farmhouses, alone among the Early Village Clusters increased its share of them.

Much of the growth in **350**, as will be clearly seen, was concentrated in the Central Plateau and in the most volatile part of it, Lago del Lupo. The number of significant farmhouses increased by 140% from 9 to 22 from **400** to **350**. It had witnessed a mass exodus between **450** and **400**, which was not only recouped in **350**, but the population of significant farmhouses had grown by nearly 60% with respect to **450**. Lago del Lupo, with 15% of the surface of the study area, in **400** accounted for 10% of the population; in **350**, an astounding 20% of the total (*Appendix C5*).

Settlement is sparser, numbers are smaller, and the gains less dramatic, but the region Far from the City, particularly Campagnolo and Bufalara, in **350** gave a foretaste of the great expansion of this region in **300**. By contrast, the region Close to the City, where the numbers of significant farmhouses reached an acme in **400**, remained static and unchanged from then until **300**. In **300**, the percent of significant farmhouses located in the Early Village Clusters was smaller than in **350**. Although the total number increased from 45 to 54, there were greater increases in the two regions Far from the City and Close to the City. In **300**, the period that saw the greatest expansion in the history of the chora, the Central Plateau suffered nearly as large a percentage loss as the Early Village Clusters, despite the fact that the total number of farmhouses located in both Lago del Lupo and Cogno del Prete reached their all

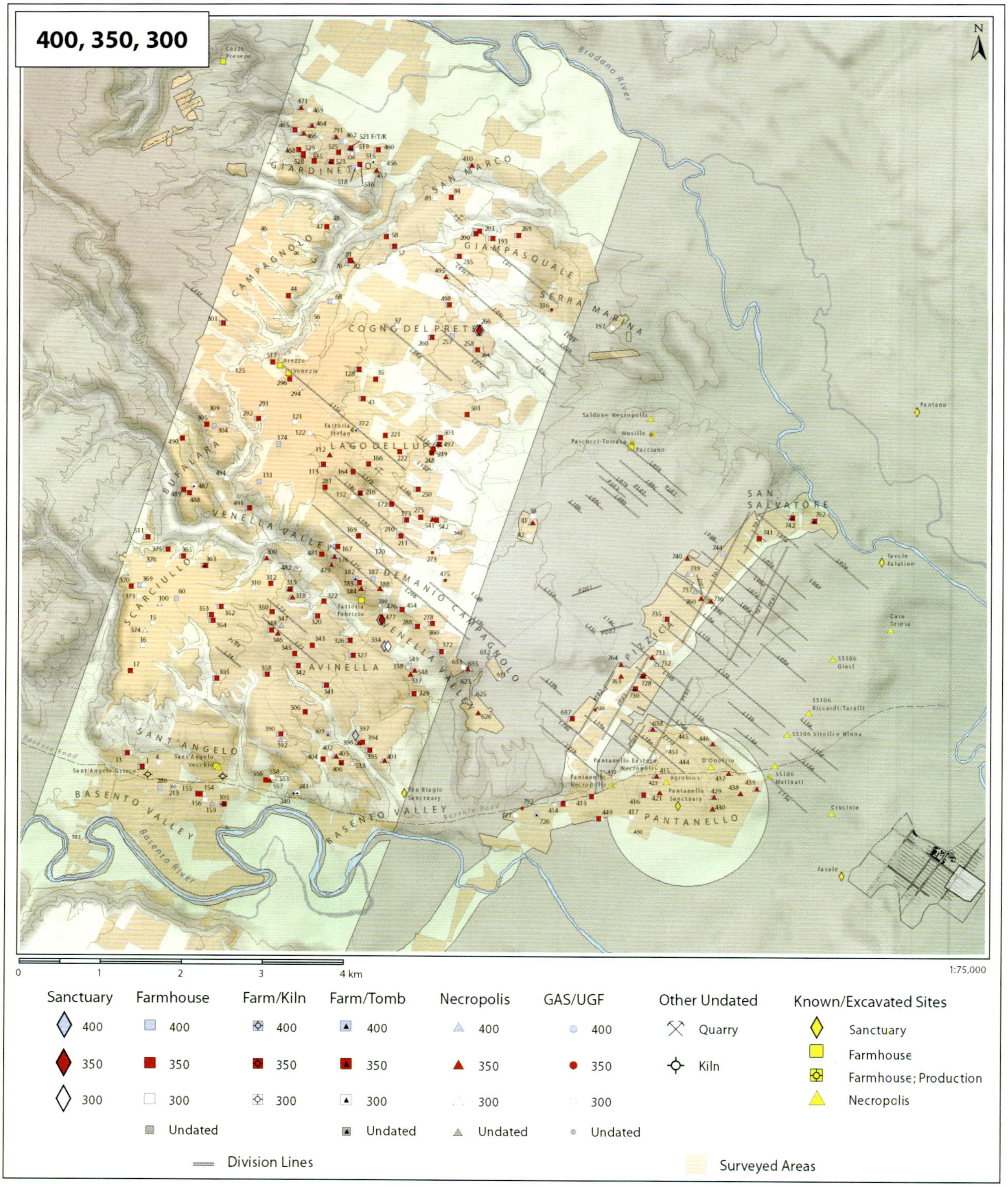

Figure 25.1 Study area transition map, **400, 350, 300**. (JT)

time highs. The growth was occurring elsewhere, at the extremes, Far from and Close to the City.

To define the large changes taking place in this century, a slightly different approach was taken here. The maps, graph (*Fig. 25.2*) in this chapter, and the Transition Tables (*Appendices C30–31*) attempt to highlight the transitions between 50-year periods by illustrating both periods side by side and comparing them with the preceding period, **400**. In this way it is, we hope, easier to envision the changes taking place.

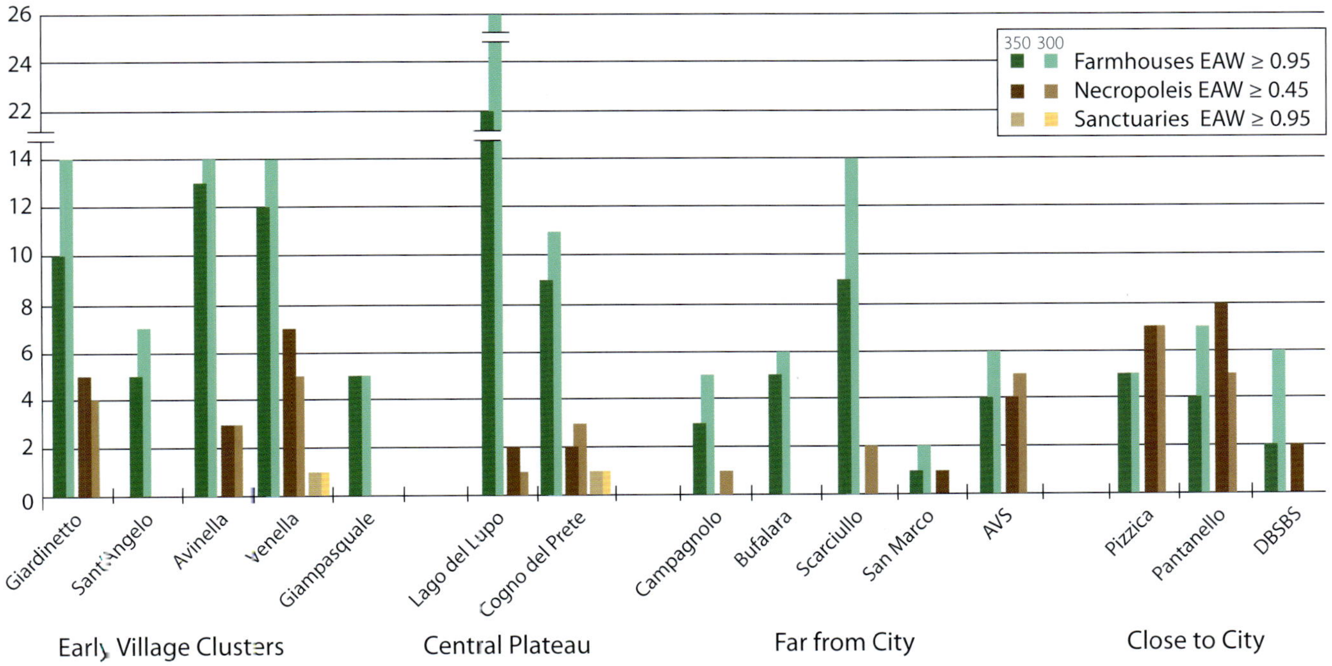

Figure 25.2 Distribution of farmhouse, necropolis, and sanctuary sites among localities in **350** and **300**.

Early Village Clusters

Giardinetto (*Fig. 25.3*). Between **400** and **350**, three farmhouses and no necropoleis were lost, but five newly activated farmhouses took their place, of which four were ex novo. Thus there was a net gain of two farmhouses and one necropolis. The farmhouses newly active in **350**, sites 468F, 516F, 525F, 528F, and 535F (see *Appendices C3, C30–31*) were concentrated in the center and the western half of the natural basin that encompassed the settlement, an area not very much occupied in **450**, **400**, or even in **550**. This is a steep and certainly less attractive area than the gentle sloping terraces on the east. Sites 468F and 528F were ex novo and they, like the older sites, are at a high elevation around the upper edge of the natural basin. They cluster quite closely around the head of the western of the two drainages, leaving 465F isolated. 529F and 528F are a pair, separated from 468F by ca. 75 m and from 535F by ca. 150 m. Only one necropolis, 464N, was added in **350**. It was ex novo and on the west below site 465F. The three farmhouses abandoned in **350** were all on the east side. The farmhouses that appear there in **350**, sites 525F and 516F, are ex novo, and were located conveniently near the eastern drainage. Here the arrangement is more spacious, with distances of ca. 150 to 200 m between sites, except for the isolated site of 460F, which is about 300 m from its nearest neighbors.

Over the course of about a half-century, Giardinetto's farm population increased by four families, to 14, but with one necropolis fewer among them. Now the abandoned sites (468F and 535F, which were new in **350**) are all on the west, and five of the six new farmhouses are located in the east. The two ex novo farmhouses (469F and 456F) occupy the opposite extreme west and east ends of the villages and the ex novo necropolis, 520N, is on the east.

Before **350**, "pairs" of farmhouses, separated by ca. 100 m or less, occurred in Giardinetto only in **450** (sites 521F/T/K and 519F). In **300**, two more pairs (518F-516F and 519F-521F/T/K) join the pairs of **350**, 529F-528F, and another possible pair 515F/T and 456F. In earlier periods, pairs consisted of sites with a considerable disparity of FMCE values (e.g., 519F-521F/T/K in **450**). This is not the case here, with only 519F-521F/T/K repeating the earlier pattern. In the present instance the differences are much less pronounced, and in one case, 518F-516F, there is perfect parity, each with a FMCE value of 2.5, slightly below the average for the period. What is the reason for the sudden increase of farmhouses in Giardinetto in **300**?

In the case of the pair 516F-518F, the site of 518F was occupied in **400**, but not in **350**; 516F was occupied ex novo in **350**. It is possible that one of the

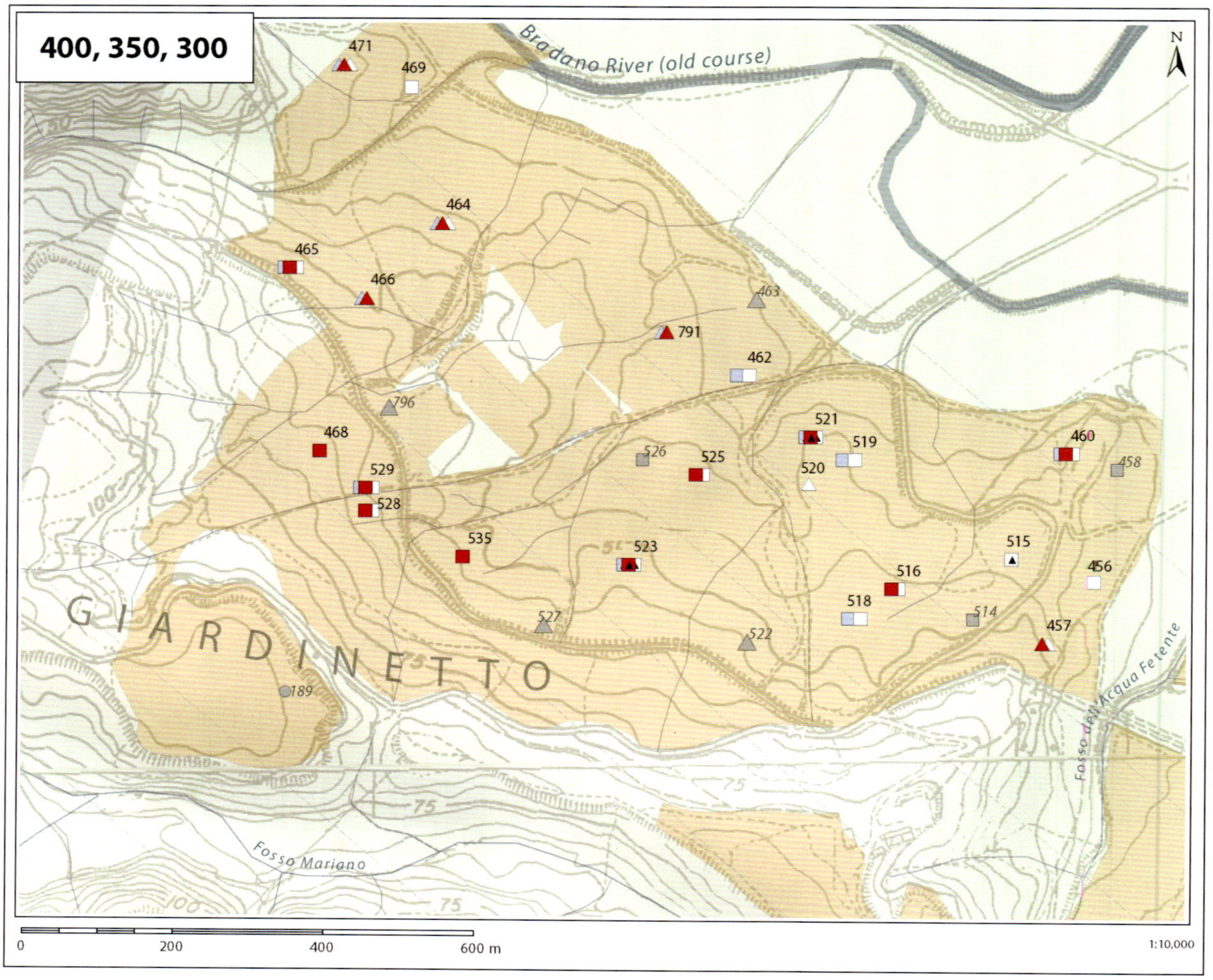

Figure 25.3 Giardinetto transition map, **400, 350, 300**. See Fig. 25.1 for map key. (JT)

pair was constructed while the closely neighboring site was abandoned, but this is not the case with the other pairs. The ex novo site 456F arose in **300** as a single-period farmhouse next to an older site, 515F, that had, apparently, been vacant for a century. The pairs of similar structures were, it seems, erected or refurbished in the same period and thus there were two families living in an area which had in the past supported only one. Could they have been related?

On the west side the spacing between farmhouses is very close to 300 m, a distance that is documented throughout the chora in earlier periods (see Ch. 21). On the east this same distance separates sites 523F/T and 518F, one of the pair 518F-516F. There are ca. 300 m between 518F and the ex novo 456F, one component of the pair 456F-515F/T. Approximately 150 m, or half this ubiquitous measurement, separates 456F from 460F, which is ca. 300 m to the east of 519F, a component of the pair 519F-521F/T/K. Around the eastern drainage, the sites are more closely bunched. Site 521F/T/K is ca. 150 m from 525F and ca. 120 m from 462F. The impression is strong that on the eastern side of the village, where the topography was suitably level, areas that used to accommodate a single farmhouse were shared by two farm families of comparable means, to judge by the nearly equal FMCE values of their farmhouses.

Except for 529F, the farmhouses of the west side of the village had above average FMCE values. Four of those sites—525F, 521F/T/K, 460F, and 456F, a one-period site—had FMCE values over 3.0, the top quarter of the FMCE range for the whole study area in **300**. These sites, in two cases components of pairs, are separated from their nearest neighbor, besides the other members of the pair, either by approximately 150 m or 320 m.

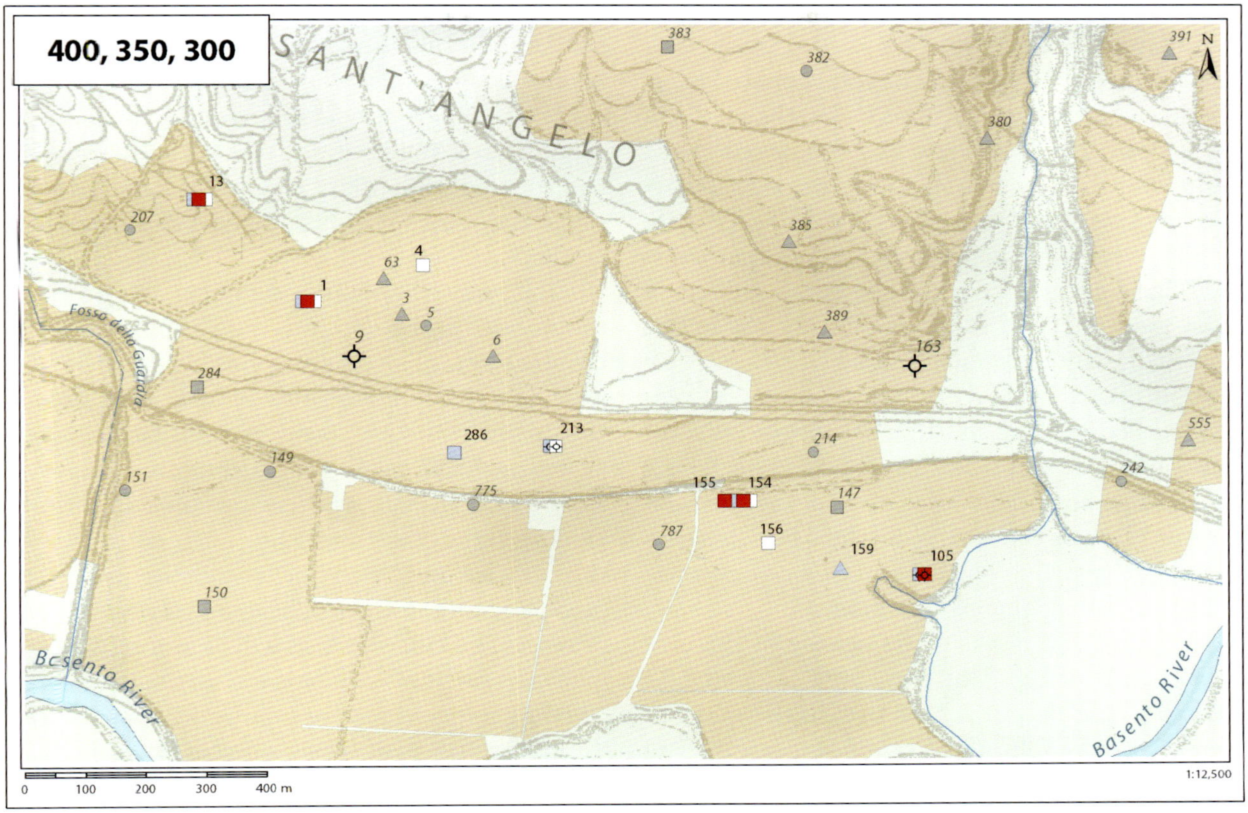

Figure 25.4 Sant'Angelo transition map, **400, 350, 300**. See Fig. 25.1 for map key. (JT)

The population of the chora increased in **350** and again in **300** by 20% or more. In the case of the Giardinetto village no more space was occupied; the rates of occupation in already established space simply increased from 47% in **400** to 82% in **300** (see *Appendix C4*). This, to judge by the FMCE values, did not result in impoverishment of the already existing families or the newcomers. It would be reasonable to suppose that the space that had previously been sufficient for one family, but now was occupied by two, with no apparent disadvantage to either, was not all the land available to either. The thought that the settlers of Giardinetto in the mid-6th century BC may have had land to cultivate either in the valley below or in the Central Plateau to the south, would seem even more reasonable for the 4th century BC.

Sant'Angelo (*Fig. 25.4, Appendices C3, C30–31*). The farm population at this Early Village, after having regained some but not all of its mid-6th–century BC numbers after **500**, lost two farmhouses between **400** and **350**, and gained one. It presents a clear contrast to the exuberant growth of Giardinetto throughout its later history. In **300**, while many other areas of the chora thrived, Sant'Angelo gained several sites but remained the only one of the original villages to have a lower population than it had in **550**.

The settlement in **350** and **300** remained, as it had been in **600**, divided between the rise at the foot of the Cozzo della Guardia and lower ground close to the Basento. Three of the four structures there, 105F/K, 213F/K, and 156F, were of above average significance in **300**, and it may well be, as suggested earlier (Ch. 23, Sant'Angelo), that ceramic production and distribution from a small port on the river was responsible for the site's continuing modest prosperity. Curiously, though, necropoleis were never numerous at Sant'Angelo, and they disappear entirely after **400**. That did not happen elsewhere. Dated examples are, as noted, few in Lago del Lupo, though not undated ones, and they are generally quite scarce in the localities Far from the City, but every village and most localities had active necropoleis in all periods. It would appear that the population of Sant'Angelo after **400** was either not permanently resident or was disposing of their dead in some as yet undiscovered spot.

Avinella *(Fig. 25.5; Appendices C3, C30–31,* **compare 400, 350, and 300***).* In no part of the study area, except for Lago del Lupo, did the rural population record a greater percentage increase in **350** than it did in Avinella. Only one farmhouse was abandoned, while six were added, of which four were ex novo. All of these completely new settlements (506F, 343F, 345F, and 350F) were in the upper stretch of the Conca valley. The cause of the sudden interest in this area could well be the same as that for the emergence at this time of many new farmhouses in Lago del Lupo. One indication is that 343F, 345F, and 350F are all very close (between ca. 70 m and 120 m) to two sections of division lines L22 and L23. This area of Avinella is a relatively large and arable piece of land. Its extension to the southeast, by contrast, is dissected by tributaries of the Conca and the Venella, and as a consequence the division lines do not extend there.

The period **350** marks a real change in the very restricted area of approximately 80 ha[1] around L22 and L23. To put this in perspective, in the mid-6th century BC, there were three significant farmhouses in the upper Avinella cluster, 349F, 348F, and 344F, and all were located on the very edge of the upper Conca. The one necropolis was about 150 m from 348F and was probably associated with it. In the revival of the chora in **450** and **400**, 344F was reoccupied for a single 50-year period before disappearing forever; 348F, on the bank of the Conca and only ca. 50 m from L23, remained significant from **400** to **300**. In **400**, there were four significant farmhouses (348F, 342F, 341F, and 327F) and two necropoleis (347N and 318N) within 200 m of the division lines. By **350**, there were eight: 342F, 343F, 345F, 348F, 350F, and at the limits of this area, 312F (in AVS, not a part of the Avinella cluster), 327F (actually in the Venella cluster), and 341F. Site 327F is about 150 m to the north, not of L22, but of the very straight drainage that clearly was its continuation to the southeast, where, after 700 m, it flows into the Venella.[2] L15e and L15f in Lago del Lupo, and L03c in Cogno del Prete, are comparable examples.

It is not a coincidence, I believe, that there were eight significant farmhouses in this approximately 80-ha area at this time, when maximum use was being made, as we have seen at Giardinetto, of all available sites. Each farmhouse, on the average, would have had approximately 10 ha available at hand, a measure suggesting close to the hypothetical ideal size of a plot or kleros.[3]

The only farmhouse surviving from the 6th century BC in the area was 348F. Four of the eight farmhouses near the division lines in Avinella in **350** were ex novo in **450** or **400**—at the same time that the division lines and many ex novo farmhouses appeared in Lago del Lupo. Three more ex novo sites were added in **350**, at the same time that thirteen ex novo farmhouses appear in Lago del Lupo. Both this smaller area in Avinella and the prime cereal producing area in Lago del Lupo experienced sudden expansion in **350**, which, as will be argued, was not so much the result of economic forces, though they were certainly important, as of political actions, including the enfranchisement of new citizens.

The distribution of farmhouses in lower Avinella in **350**, by contrast, was little changed from the 6th century BC. Sites still cluster around the three forks of the chora. Three of the six significant farmhouses of **350** had been occupied in the 6th century BC.[4] Two others were ex novo in **450** and **400**. None were ex novo in **350**. All of the start up sites in this period, including 506F, occupying a generous plateau on the right bank of the upper Conca, were in the upper Avinella (*Figs. 25.6–7*).

No farmhouses in either upper or lower Avinella were abandoned in **300** and only one, 402F, an ex novo site, was added in the lower course of the Conca. The mid-4th century BC saw the important changes in Avinella, especially the upper part, and in **300** the material conditions of life, as measured by FMCE, had decidedly improved.

The farmhouses of Avinella, both upper and lower, are well below the average FMCE value (2.04) for the chora as a whole in **350**. The ex novo farmhouses of upper Avinella were the lowest of the low.[5] Only one

[1] The equivalent of 10 or 12 hypothetical plots. See Ch. 21, 701, where using the distance 300 m, the approximate distance between many farmhouses, sizes of square hypothetical plots of 9 ha or circular ones of 7 ha were calculated. They would not have been all the land a farmer may have cultivated or controlled.

[2] This is, by the way, one of the clearest illustrations of the phenomenon (see Ch. 31, 1120) of a man-made ditch creating a drainage that continues along the axis defined by the division line. See Schmiedt and Chevallier 1959.

[3] 7–9 ha, calculated from the common distance of 300 m separating many sites. See Chs. 21, 701, and 23, 750.

[4] Sites 404F, 406F/T, 390F, 506F, 354F, and 395F.

[5] This is surprising, considering the generally higher FMCE values of start-up sites in other areas and periods, many of which like 345F and 350F had relatively short lifetimes. The ex novo farmhouses in Lago del Lupo follow this pattern.

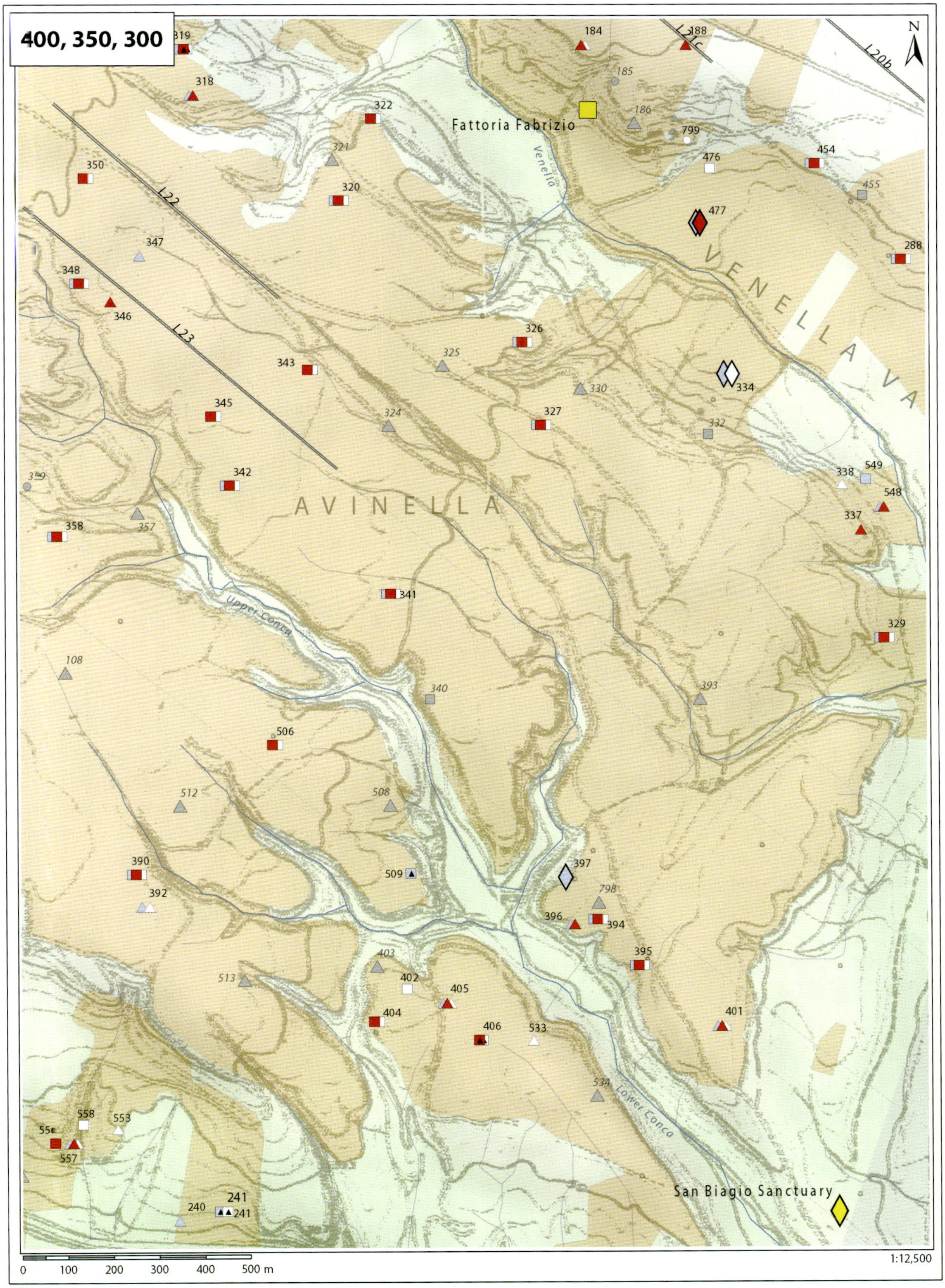

Figure 25.5 Avinella transition map, **400, 350, 300**. See Fig. 25.1 for map key. (JT)

Figure 25.6 Site 532F, a farmhouse in lower Avinella of second magnitude in **300**, and its associated necropolis.

Figure 25.7 Site 533N, also active in **300**. 532F and 533N are separated by ca. 120 m and overlook the Basento valley

farmhouse, 348F (the only one occupied in **550**), of the seven in upper Avinella in **350**, had an FMCE value greater than the average for the study area as a whole. The situation in the lower Avinella was much the same, with a single above average site, 395F. By **300**, Avinella, both upper and lower, was relatively prosperous.[6] Does such a dramatic turnaround have parallels anywhere else in the study area? Yes, practically everywhere, but the case of Cogno del Prete in the Central Plateau stands out. One might instead expect the older, more established farms of the lower course to be more materially successful than the new ones. I would suggest that the reason is location: topography and proximity to wider areas of arable plateau land.[7] The ex novo farmhouses required some time to develop, but by **300** they were often the equals of, if not superior to, the older farmhouses.

Venella (*Fig. 25.8; Appendices C3, C30–31*). After reaching a high in **400**, the farm population of Venella, west and east, declined slightly from 13 to 12 significant farmhouses. The only other area to experience a population loss at this time was Sant'Angelo. Cogno del Prete in the Central Plateau was unchanged, as were Pizzica and Pantanello.[8] All localities Far from the City increased, and the largest increases of all were just to the north of Venella in Lago del Lupo. Why should Venella, along with Sant'Angelo, have lost population, while other areas, including adjacent ones, remained constant or gained?

The loss of a single site, after all, is not very serious. Venella, as a whole, might be considered static, like the localities Close to the City. It did not decline much in **350**; it simply did not participate in the growth that is evident elsewhere. The great population boom in the chora in **300** seems to have passed Venella by.

Between **400** and **350**, Venella lost three farmhouses, one in the west (549F) and two in the east (183F and 187F). It gained only two, 322F on the west bank and 481F on the east. The really unusual change, however, was the addition of four necropoleis, three of which (337N, 184N, and 188N) were ex novo in **350**. The total number of necropoleis in Venella increased from three to seven (or 133%). Venella in this period has a higher ratio of necropoleis to farmhouses (7:12) than any of the other localities of the study area including Giardinetto (1:2).

The settlement of Venella extended from an original core formed by two early sanctuaries, sites 334S and 477S, along both sides of the valley. In **350** the east bank was lined by nine farmhouses, one (182F/T) with associated burials. This number includes Fattoria Fabrizio.[9] They are spaced somewhat regularly, though not equidistantly. At the northern limit is 481F, a new farmhouse in **350**, and two older necropoleis (176N and 479N). The valley's edge beyond is deeply incised by several drainages. Though most of the farmhouses

[6] Only one site, 406F/T, in the whole extended village was way below the average. Three farmhouses, two in lower Avinella (394F and 404F) and one in upper (506F) are in the upper 25% of values for the study area as a whole. Half were above average and four were only slightly below.

[7] In the discussion of FMCE values for upper and lower Avinella in **550** (Ch. 21), it was suggested that perhaps the newness of a number of farmhouses in upper Avinella could explain why its average value was greater than that of the lower. If newness were the reason, then it did not affect the average in **350**. The two halves of Avinella are perfectly equal, but in **300**, the average value of upper Avinella is 2.79, comfortably above the average for the study area, while that of lower Avinella, 2.53, is slightly below. (A similar imbalance between upper and lower, largely thanks to one site, 344F, occurred in **450**.)

[8] Site 726F/T, like Sant'Angelo along the Basento valley but closer to the coast, was no longer significant in **350**, resulting in a loss of one farmhouse in DBSBS.

[9] Ch. 21, 687.

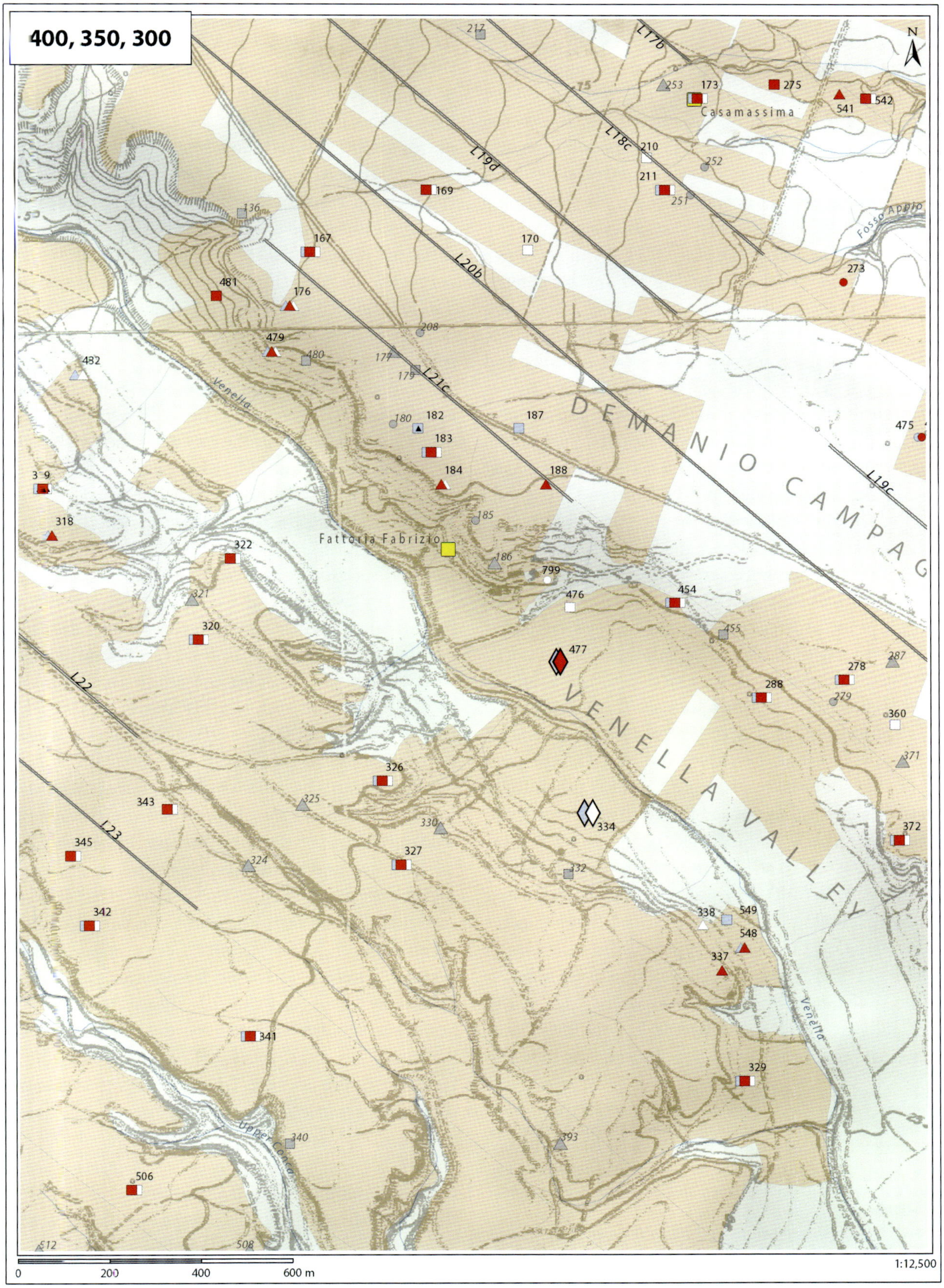

Figure 25.8 Venella transition map, **400, 350, 300**. See Fig. 25.1 for map key. (JT)

of this cluster hug the valley's edge, or like 481F are actually on its slope, some sites extend some distance into the Central Plateau. Site 167F, a new site in **400**, is one of these. It is 220 m to the north of 481F. The distance to site 169F is 300 m. It is tempting to see here a possible expansion of this village well beyond the valley's edge. Besides these two farmhouses and the necropolis 188N, however, there are no others more than ca. 50–100 m from the valley's edge.

Some 600 m separates site 481F from its nearest neighbor to the south, site 183F, and its related necropolis 184N.[10] The farmhouse 183F dates from **400**; the necropolis from **350**. Fattoria Fabrizio, which was occupied already in the 6th century and again now in the middle of the 4th century BC, is ca. 130 m to the southeast of 184N, which may have served it. The valley's walls are particularly steep here. Fabrizio occupies a narrow terrace about equidistant between the valley's edge and its bottom.[11] The valley slope remains steep till it reaches site 454F to the southeast, after which it broadens out and is more gentle. The distance between 454F, new in **400**, and 288F, which was new in **450**, is ca. 300 m. Site 288F sits on a point projecting from the valley's edge and has a good view of the other farmhouses to the north and south. Distant, ca. 400 m farther to the south, is site 372F, occupied discontinuously from the 6th to the 1st century BC.[12] The settlement likely continued down the valley, but the survey did not. For the purpose of this analysis, the southern limit is taken to be a deeply incised valley caused by a drainage similar to the one that marks its assumed northern limit near 481F. The last site to the south is the necropolis 635N (active already in **600**). It is quite possible, as suggested earlier (see Ch. 21), that another Early Village Cluster was located along the lower course of the valley, with the Sanctuary of Artemis at San Biagio at its heart.

The spacing of sites along the eastern edge of the valley suggests that the settlement here may have reached its maximum in **400**.[13] Expansion into the Central Plateau and Lago del Lupo does not seem to have been attempted. I would suggest that the reason there was limited growth after **400** was not because no more space was available. The valley's edge could have been subdivided, like the eastern side of Giardinetto, in **300**. By **350**, however, Venella was a mature residential community. Because of its proximity to the Central Plateau, and its participation in the many changes going on there, Venella east seems quite a different place in this period from Venella west, just across the stream. The settlement on the west bank by **350** was not what it had been a century earlier.

Virtually nothing on the west bank changed between **350** and **300** (*Appendix C31*), except that the western of the two sanctuaries, 334S, became again briefly significant—a reflection, perhaps, of the new direction of religion in the city and chora.[14] Necropolis 337N, ex novo in **350**, disappears from the record together with 548N, and in their place a necropolis from **450**, 338N, becomes active again. There is a net loss of a single necropolis.

A discrepancy in FMCE values between farmhouses on the east and west side of the Venella appeared already in the 6th century BC. It did not become a pattern. In **450**, the averages on both sides were well below that for the study area as a whole; by **400** they were both slightly above; in **350**, slightly below. Only in **300** does the difference between the two sides in terms of FMCE emerge again, and, as in the 6th century BC, the west bank farmhouses appear to have been more significant than the average of the study area, and of those in the east (*Appendix C7*).

The average values, however, mask an interesting discrepancy within these two groups of sites. Individually the farmhouses of the west Venella are all either average for the study area or slightly above average. There is only one farmhouse, 322F, with a really low value. On the east, however, three farmhouses have very high values, over 3.0, while three have very low ones (average 1.67); only a third of the farmhouses are average. In no other restricted area of the chora are the differences in material remains of neighboring farmhouses so obvious.

Giampasquale. Apart from a flurry of activity in the mid-5th century, there were no changes at Giampasquale between the revival of the chora at that

[10] 481F is only about 200 m, however, from 167F, which is not as close to the valley's rim as the other sites in the cluster. 481F would seem to be more related to development in the Central Plateau than in the Venella.

[11] The site and the others along this slope enjoy a refreshing breeze from the early afternoon until evening in summer, and a good exposure to the sun in winter.

[12] It is one of only 13 farmhouses recorded by the survey in the entire study area for the 1st c. BC (see Ch. 27, 895)

[13] Seven ex novo farmhouses are here, nearly as many as in Lago del Lupo in **450**, and more than any other part of the study area in **400**.

[14] See discussion of Oikos E in the asty, below.

Figure 25.9 Lago del Lupo, looking to the northeast from the air. (SMT)

time and its decline in the 3rd century BC. That is a remarkable fact in itself. Also remarkable are its high average FMCE values from **350** to **300**. In **350**, only Pizzica (2.59) had a much higher average. Giampasquale (with 2.20) was practically the equal of Giardinetto (2.18) and Lago del Lupo (2.22). In **300** only Pizzica (3.04) and Lago del Lupo (2.79) surpassed Giampasquale (2.69)—all were well over the very high average for the chora (2.57).[15] Although its FMCE scores in **450** and **400** were unremarkable, the general prosperity of the descendants of this original village in the 4th century BC may owe something to its stability.

The Central Plateau

It is enough to look from the eastern edge of the Venella north across Demanio to Lago del Lupo and Cogno del Prete to understand why it was, and is, attractive to farmers (*Fig. 25.9*). The flat surface extends as far as the eye can see, interrupted by drainages that are, however, invisible at eye-level. It is, as we have noted more than once, ideally suited to plow-based agriculture and well furnished with springs, especially in the southeast (*Fig. 25.10*). Lago del Lupo is by far the largest locality in the study area, one and a half times greater than next largest, Cogno del Prete, the other major component of the Central Plateau.

Lago del Lupo (*Fig. 25.10*). Between **450** and **400** the number of farmhouses in Lago del Lupo fell from 14 to 9, a decrease of 36%. That decline was very abruptly reversed in **350** when the farmhouses numbered 22, an increase of 144%. This was the largest percentage increase in the study area in the history of the chora, far surpassing the 75% increase here between **550** and **450**. The necropoleis here—never a large number because so many in this area are undated—increased from zero in **400** to two, the maximum number and equal to that in **450** and **300** (*Appendices C3, C30–31*).

In **350** there were a total of 15 newly active farmhouses of which 13 were completely ex novo. One of the necropoleis, 541N, was ex novo. Only two farmhouses and one necropolis were abandoned in this period. Thus, there was a net gain in **350** of thirteen farmhouses and one necropolis. A large percentage of the ex novo necropoleis are clustered in the southeast area of Lago del Lupo, where the preserved division lines are also concentrated, and which witnessed a similar growth in **450**. This convergence of division lines and ex novo farmhouses is extremely significant and will be examined in detail later in this chapter in the discussion of the redistribution of land in **350**.

[15] This material prosperity carried over into the period of decline that began in **250**. Ch. 26, Giampasquale.

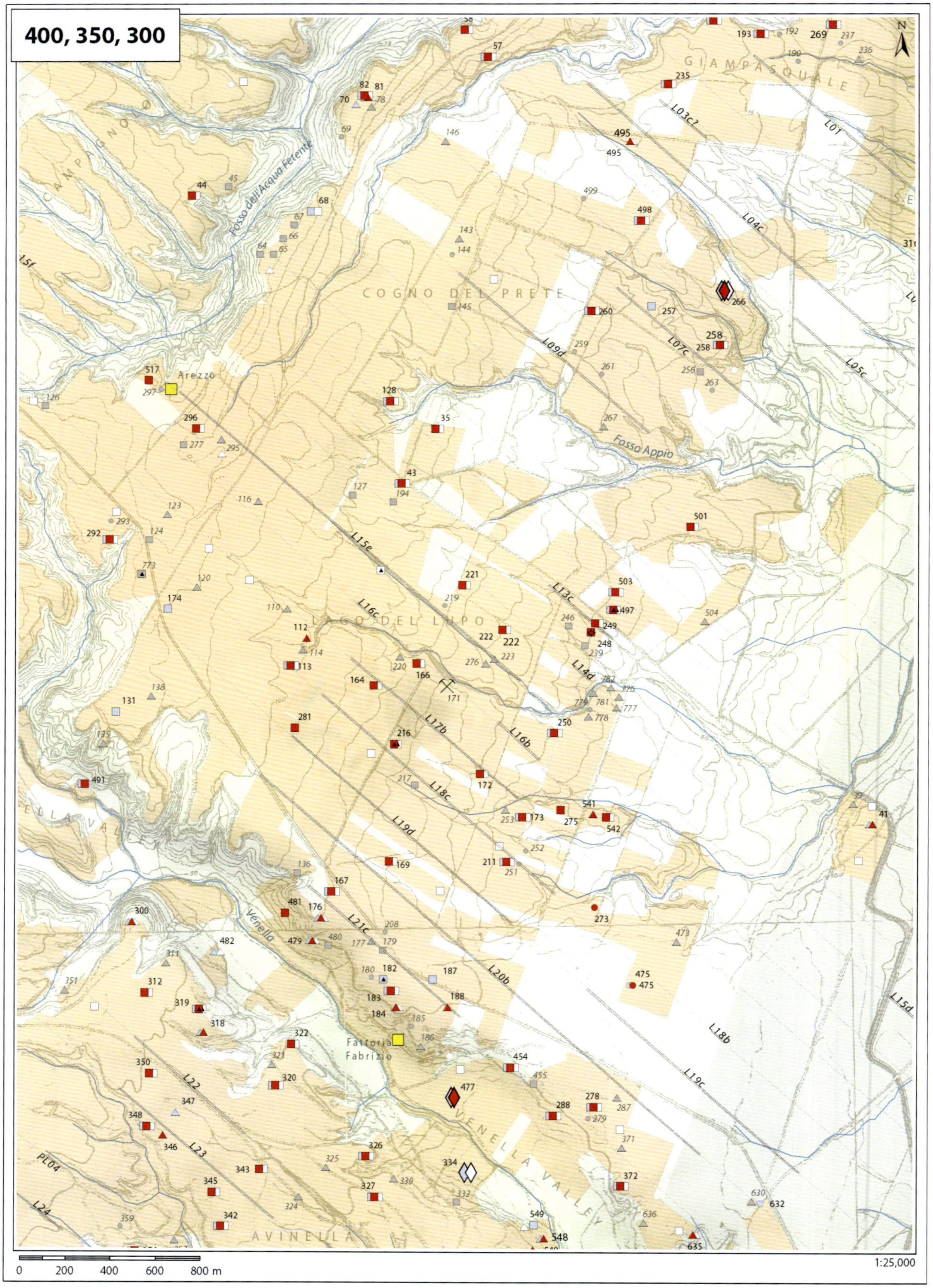

Figure 25.10 Central Plateau transition map, **400, 350, 300**. See Fig. 25.1 for map key. (JT)

In **300** the total number of farmhouses in Lago del Lupo climbed to 26. Nine of these were newly active and included 7 ex novo farmhouses (see *Appendix C31*). Only three farmhouses (275F, 281F, and 517F) were abandoned in this period. There was thus a net gain of four farmhouses. Two necropoleis were abandoned, but one, 294N, was added ex novo, for a net loss of one necropolis in **300**. The ex novo farmhouses, in contrast to the previous period, are about evenly divided between the southeast and northwest halves of Lago del Lupo.

Cogno del Prete (*Fig. 25.10, Appendices C3, C30–31*). The smaller area of the Central Plateau, Cogno del Prete, had nine farmhouses and two necropoleis in both **400** and **350**. The situation was not quite static as two farmhouses, 58F and 35F, became newly active; the latter, ex novo. The one newly active necropolis 495N was also ex novo. In this same period two farmhouses, 68F and 257F, and one necropolis, 70N (which had been continuously in use from **600** to **400**) were abandoned. The new farmhouse, 35F, was located on the southern side of Cogno del Prete near the venerable farmhouses 43F (active since **600**) and 128F, an ex novo farmhouse in **450**. The new necropolis, 495N, is in the very northwest corner of the Central Plateau in a somewhat isolated position not far (ca. 70 m) from division line L04c.

In **300**, when the farmhouses increase from 109 to 142, or 30%, and the necropoleis decline from 41 to 36, or 12% throughout the study area, the number of farmhouses in Cogno del Prete rose from nine to eleven, or 22%, while the necropoleis rose from two to three, 50%.

The newly active farmhouses in this period were two, 37F, an ex novo foundation, and 68F; the necropoleis 56N was ex novo, and remained active only during this 50-year period. The new farmhouse 37F was located ca. 50 m to the north of division line L09d, and it enjoyed a brief existence in this period.

Far from the City

The localities Far from the City (Campagnolo, Bufalara, Scarciullo, San Marco, and AVS) contain fewer, more widely scattered farmhouses (*Maps 10* and *11*, Vol. IV). Each has a distinct topography that obviously influenced the settlement pattern. Campagnolo is dissected by many smaller and larger drainages into Fosso dell'Acqua Fetente, the northernmost of which separates Campagnolo from Giardinetto. They are truly impressive barriers. Bufalara consists of two narrow strips of high plateau, almost islands, between three drainages that come together to form the Venella just above Pietra San Giovanni (site 300N). The Bufalara and the Guadella to the south divide the locality from Scarciullo.

The area of Scarciullo (*Figs. 25.13* and *25.14*) is nearly as large as that of Avinella. It is divided by the east branch of the Fosso Sant'Angelo and by the upper courses of the Conca, which meet near site 358F (actually in Avinella), while site 352F at the northern tip of the Conca is in Scarciullo. The extensive plain that lies south of the Venella between Avinella and Scarciullo is divided by the main, middle fork of the Fosso Sant'Angelo into eastern and western halves—the eastern gently sloping; the western unusually flat. Previously we have discussed only the northern portion of Scarciullo along the edge of the Gaudella. With the exception of isolated sites like 103F (among the first settlements in the study area), the rest was virtually ignored by the settlers of the 6th and 5th centuries BC despite its size and obvious attractions. That began to change in the 4th century BC, and did so dramatically in **300**.

The combined farm population of Campagnolo, Bufalara, and Scarciullo increased between **400** and **350**. That of Campagnolo and Bufalara jumped up, while Scarciullo remained unchanged from **400**. The movements up and down of the population in Campagnolo in **400** and **350** mirror fairly closely those of Lago del Lupo. In **400**, when Lago del Lupo suffered a sharp decline, Campagnolo did not gain, while all of the others did. In **350**, Campagnolo was second in growth behind Lago del Lupo in the study area.

Campagnolo (*Fig. 25.11*). Two sites, 44F and 47F, sprang up ex novo along the north edge of the Acqua Fetente. Beside the main division line L15 stood 301F. This increased the farm population of Campagnolo, but not to the level of **450**, much less of **300**. The relation of the farmhouses to the topography is so close that it is possible to deduce the approximate size of the plot of land associated with the farmhouses. The new farmhouses are at the very edge of the plateau overlooking the deep gorge. Sites 47F and 44F are the sole occupants of small plateaus, with very narrow necks

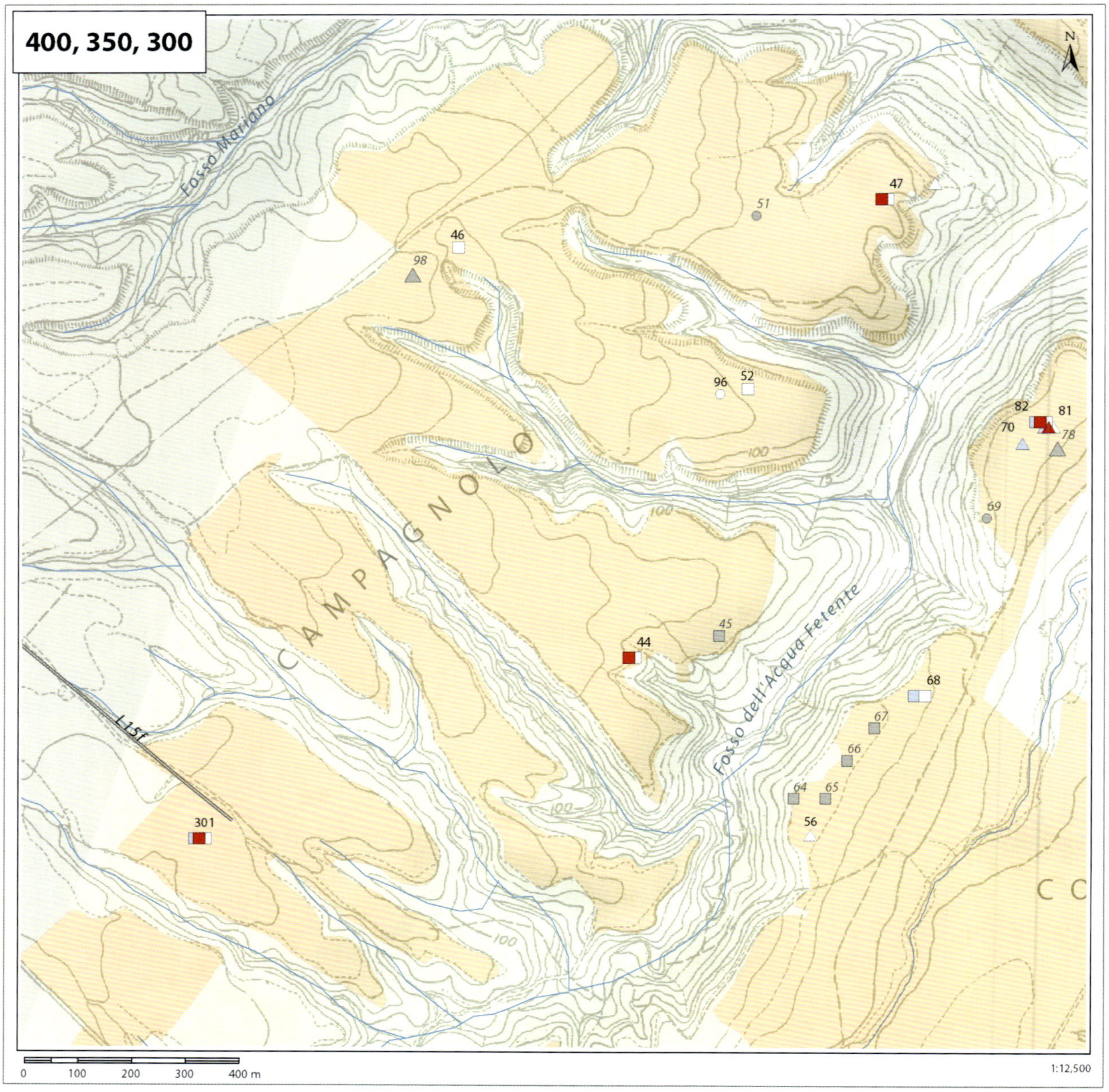

Figure 25.11 Campagnolo transition map, **400, 350, 300**. See Fig. 25.1 for map key. (JT)

connecting each site to the broader plateau behind. Fortunately, the whole surface of these promontories has been surveyed so we may be sure that the occupants of 47F, at least, were alone on land that was very probably all their own property. In the case of 44F, we cannot be sure if this was one family's holding, because the undated farmhouse 45F, just ca. 150 m to the east, also occupies the point. The land on the small plateau behind 47F measures 25 ha; that behind 44F, 24 ha.

Two more farmhouses, 52F[16] and 46F, appeared ex novo in Campagnolo in **300**, on the previously untouched plateau between 47F and 44F. Site 52F was located at the edge of the plateau overlooking the Acqua Fetente. A distance of ca. 600 m, the length of the plateau, separated 52F from 46F, which commanded its narrow neck. The necropolis 48N, ca. 80 m from 47F, was first active in **450**, and it continued to be active until **250**. The undated necropolis 98N, some 35 m away, likely served the

[16] 52F was 2nd magnitude in **400** and **350**.

family occupying 46F. This was evidently a very good choice of site, as 46F became the second most significant farmhouse (FMCE 4.0) in the study area in **300**, after 296F in Lago del Lupo (4.32). The average FMCE values for Campagnolo and for the entire region are only slightly below those for the study area as a whole. Averaging masks the fact that there is a single farmhouse in each period with a high FMCE value—44F in **350** (2.52) and 46F in **300** (4.0); the majority are well below the average.

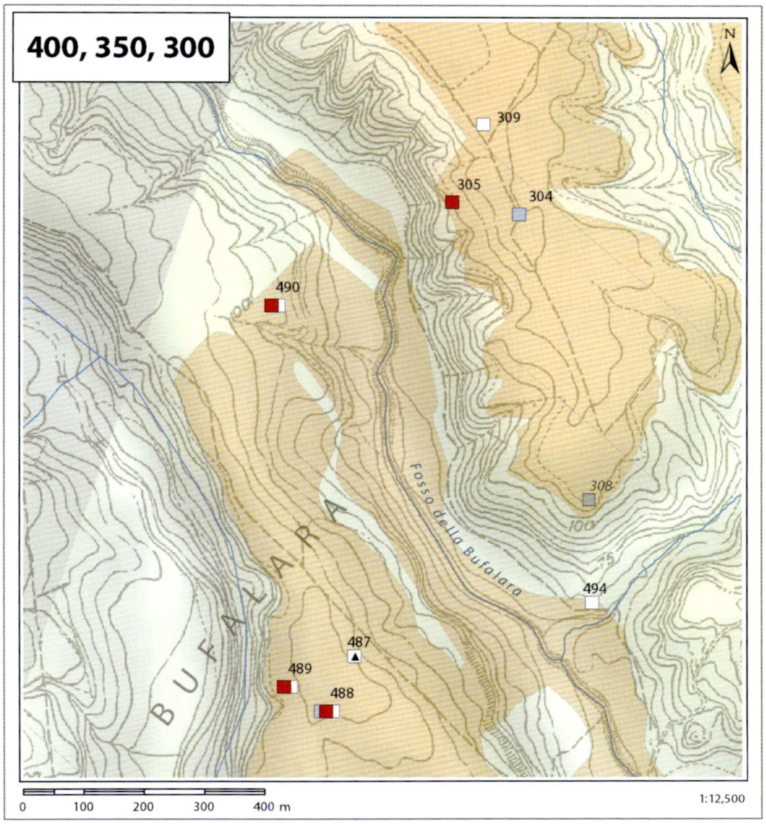

Figure 25.12 Bufalara transition map, **400, 350, 300**. See Fig. 25.1 for map key. (JT)

Bufalara (*Fig. 25.12*). The farmhouses of Bufalara in **350** are located on similar small plateaus, virtually peninsulae of relatively level land surrounded on three sides by steep ravines. At the center and narrowest part of the more northern of these peninsulae, site 304F, occupied already in **550**, continued to be significant from **450** to **400**. In **450**, another, 305F, was erected just 120 m to the west and the two coexisted only in that period. Subsequently, they were alternately significant in successive 50-year periods until **350**. By **300**, both had ceased to be significant, but at this point, 120 m to the north, a third, 309F, of well above average significance, came to occupy the narrow neck of land. It was clearly a desirable site that was ideally situated to control both halves of the point. Only in **450** would the occupants of 304F and 305F have had to share it.

A very similar situation occurred on the other larger point between the Bufalara and the Gaudella. Site 488F occupied its lower end in **550** and again in **450**, and was then continuously significant until **250**. It was alone until **350**, when 489F, around 120 m to the southeast, and at the same time 490F at the western extreme of the survey coverage, were constructed ex novo. In **300**, a third farmhouse, 487F/T, was built ca. 120 m to the northwest of 489F. The three (488F, 489F, and 487F/T) formed an equilateral triangle at the narrow point. The newcomers presumably would have shared the broad southern end of the plateau with 488F and the northern end with 490F, which had occupied it since **350**. This close grouping of farmhouses seems to imply a high degree of cooperation among farm families on the limited arable surface in this distinctive, rather inaccessible naturally defended landscape.[17]

One of the farmhouses, 489F, in the cluster on the larger peninsula in **350** had a higher FMCE value than the other two (488F and 490F), which were well below average for the study area at this time. A half-century later, the three at the top of the peninsula were again all significant, but their FMCE values were all slightly below average, while the isolated site 490F was well above average. Site 309F across the Bufalara to the north was the only farmhouse in Bufalara ever to have an FMCE value greater than 3.00.

Scarciullo (*Figs. 25.13–14*). The northern edge of the Scarciullo plateau, as noted, had been occupied already in the 6th century BC. The farmhouses 375F and 365F were situated on points projecting into the valley of the Gaudella to the north and the Fosso

[17] A third "peninsula," on the south side of the Gaudella, was only partially surveyed. On its southern tip, in a position analogous to sites 52F and 47F, is a farmhouse, 511F, and an undated necropolis 510N that probably served it.

della Guardia to the west. For some reason, the rim of the plateau and the edge of the drainage known as the Fosso Sant'Angelo did not attract settlers, and occupants of 103F continued to enjoy complete solitude as they had since **600**. With the revival of the chora in **450** the areas occupied in the 6th century BC were so again, and there were two ex novo farmhouses, 352F and 353F. Placed back from the edge of two valleys (the upper Conca and the east fork of the Fosso Sant'Angelo), 353F and 354F command an open plain. By **350** a cluster of three farmhouses (352F, 353F, and 354F) was situated across a bottleneck between two areas of fertile plain. The broad plain that runs southward between the Fosso della Guardia and the Fosso Sant'Angelo as far as the Cozzo della Guardia was occupied by farmhouse 370F at its northern end in a spot that had been occupied since the 6th century BC, and by 17F near its southern end. The broad tabletop of flat land had not seen a farmhouse before, though there was a necropolis, 374N, that was among the earliest in the study area. By **300**, the settlement was positioned to exploit the plateau and surrounding areas in a very efficient way. A cluster of three sites (369F, 370F, and 373F) commands the northern approach to the flat plateau. The earlier group of three (352F, 353F, and 354F) continued to control the plateaus to the east at the bottleneck. The most strikingly original feature is the very straight line of farmhouses running the length of the western plateau (15F, 16F, 17F, and second magnitude 18F).

Figure 25.13 Aerial view of Scarciullo, looking to the southwest. (SMT)

The transformation that took place between **350** and **300** is very revealing. It shows that the phenomenon of many ex novo sites in a limited area was not unique to Lago del Lupo or Avinella. The method of organizing the chora emerged very clearly here because it takes place in what was essentially virgin land. Though the settlement trails that of Lago del Lupo by a half-century, it is remarkably similar and equally successful. One of the four ex novo farmhouses (373F) form a triangular-shaped village with 370F, like that 1 km to the east (352F, 353F, and 354F). One, 376F, is located along the northern edge of the plateau ca. 250 m from the new village; the remaining two, 15F and 16F, are aligned along perfectly straight axes with 17F, that almost perfectly bisects the western plateau. Though there were no division lines visible in the aerial photographs, the effect is as if there were a set here. The Scarciullo plateau is like the Central Plateau in miniature, with its own equivalent of the central division line. The spaces between the three farmhouses starting from 17F on the south are: 350 m from 17F to 16F; 300 m from 16F to 15F. This progression continues, as 15F is separated from 100N, a new site in **300**, by 250 m. It was identified as a necropolis by the survey team, but the site was small and badly damaged, and some doubt remains about its true nature. The identifying feature was a curved tile like those used as burial containers in **300**, but which might equally well have been used to line a well (see Wells, below), in which case it could well have been a farmhouse site.[18] The distance between the sites decreases as the plateau broadens toward its northern end, which may have been intended to equalize the size of the virtually identical, flat land occupied by the farmhouses.

Large areas of the Scarciullo plateau in the north and east remained devoid of farmhouses. It might well be that the farmers who cultivated them lived, like those in Bufalara, in the triangular-shaped villages, and commuted to their respective plots on foot. This transformation, and full occupation of the territory, resulted in a dramatic rise in the FMCE values for

[18] C. D'Annibale, pers. comm., 2007.

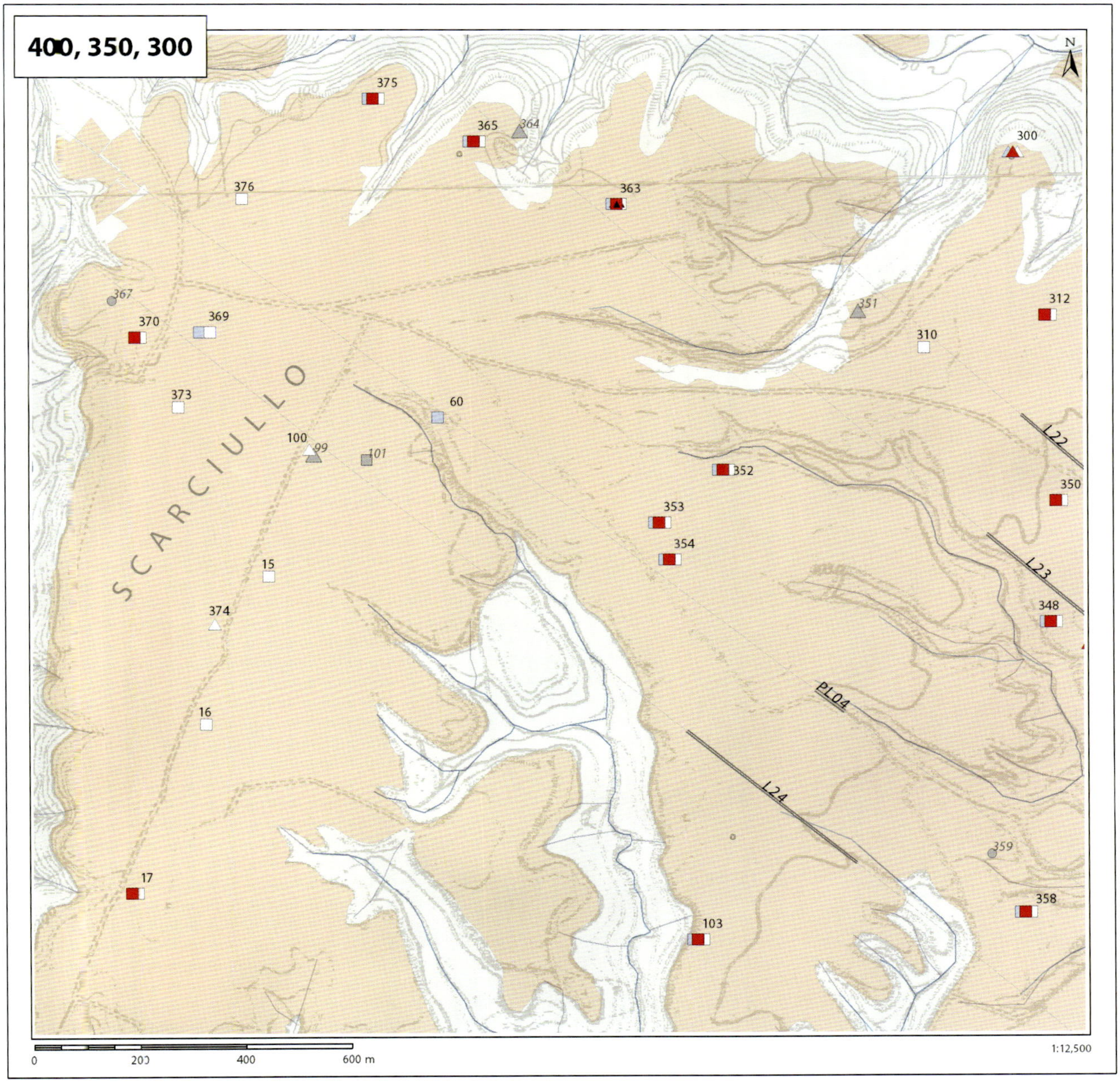

Figure 25.14 Scarciullo transition map, **400, 350, 300**. See Fig. 25.1 for map key. (JT)

the sites that comprised the settlement. Interestingly, the highest values occur in the older village to the east, and at 15F in the center of the western plateau. In this period, measured in terms of FMCE values, Scarciullo went from being one of the least successful localities in the study area to become, with Lago del Lupo, one of the most prosperous.

Close to the City

Pizzica (*Fig. 25.15*). Between **400** and **300** the number of farmhouses remained the same, three, but the number of necropoleis increased to seven. Together with the two farmhouses with associated burials, the ratio between places of burial and of residence grew from 7:5 in **400** to 9:5 in **350** and **300** (*Appendix C3*). The sudden appearance of necropoleis Close to the City and the rapid growth in their numbers, has been noted (Ch. 23), but that is an extreme situation, exceeding even the 2:1 ratio at Pantanello (see below).

While the number of farmhouses in Pizzica remained the same between **400** and **300**, five, their average FMCE value increased substantially from

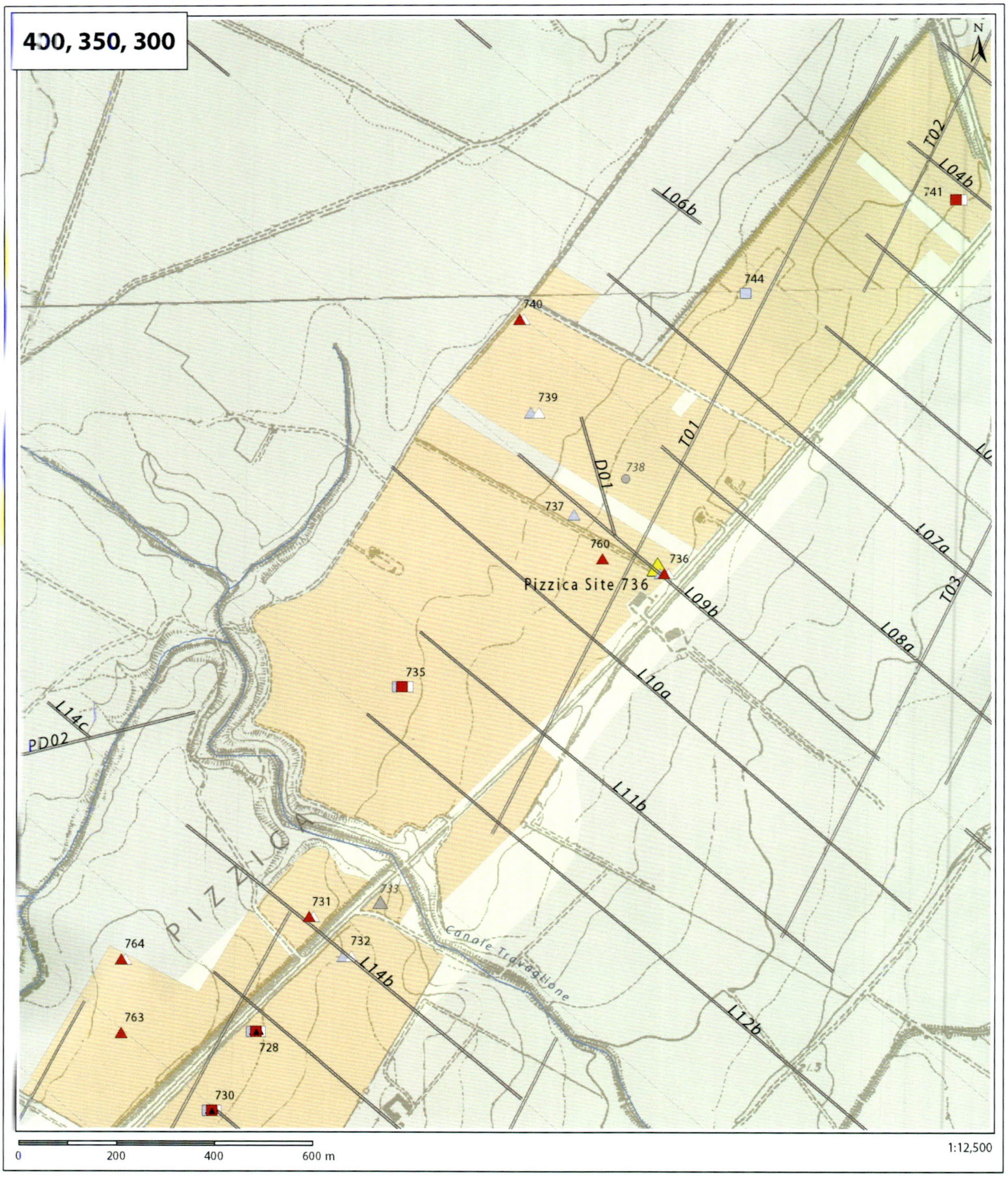

Figure 25.15 Pizzica transition map **400, 350, 300**. See Fig. 25.1 for map key. (JT)

2.59 in **350** to 3.04 in **300**. The average was the highest in the study area by a wide margin in **350**, and continued to be so in **300**, but, while no farmhouses in **350** had been below average and several were exceptionally high (728F/T and 730F/T),[19] in **300**, one farmhouse, 741F, had a value that was below average. Both 728F/T, first occupied in **450**, and 730F/T in

[19] What is reflected in these cases is a sharp increase in the date-dependent variables (see Ch. 4), and specifically in the dateable pottery. These were farmhouses with associated burials. Not surprisingly, fragments of figured pottery, which can be precisely dated to a specific period, were recorded at both, along with exceptionally high counts of BG pottery fragments.

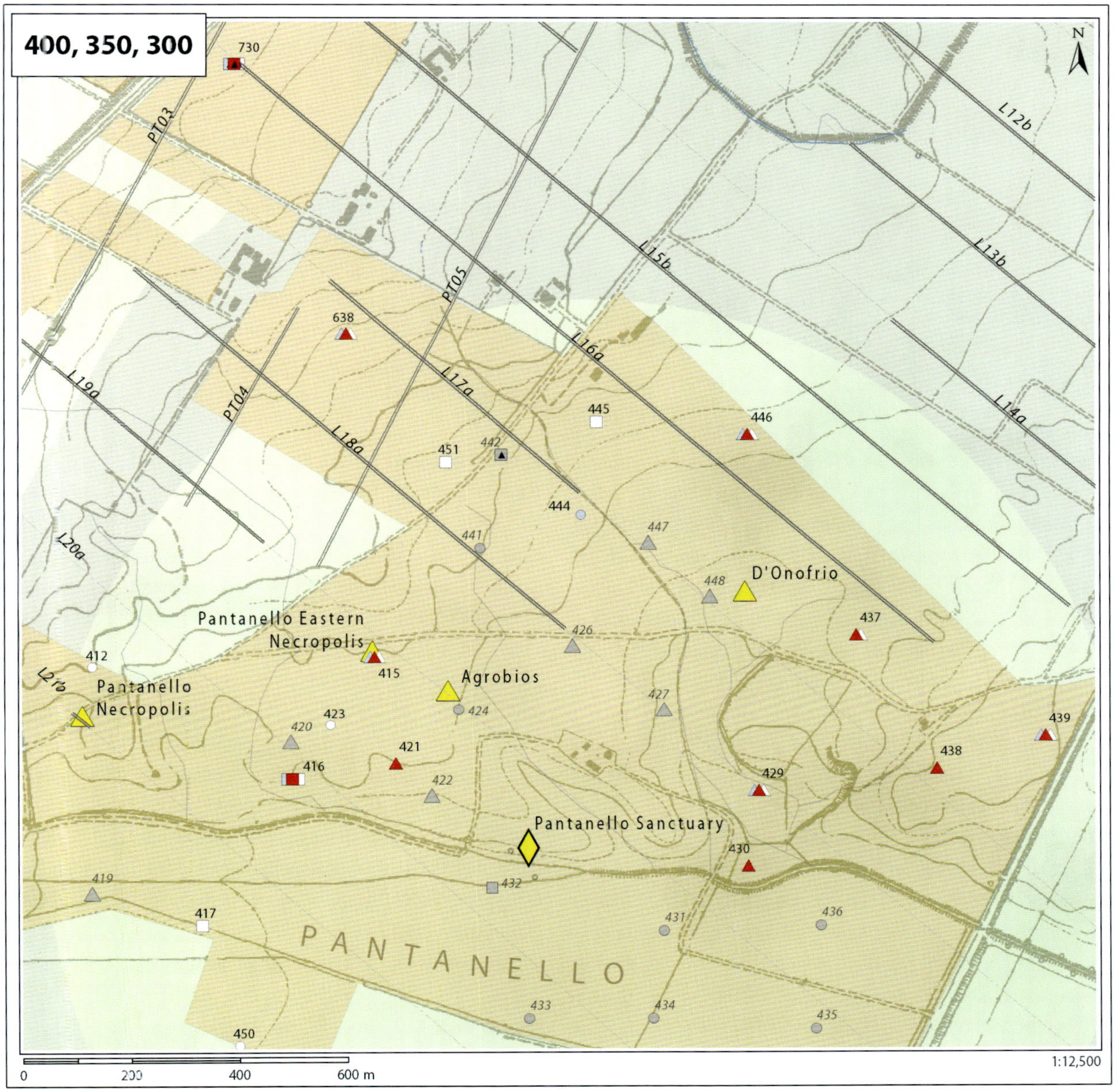

Figure 25.16 Pantanello transition map, **400, 350, 300**. See Fig. 25.1 for map key. (JT)

400 lie on preserved sections of L15 and L16, respectively, very near their intersections with the possible transverse line PT03.

Pantanello (*Fig. 25.16*). In **350** there were four farmhouses and eight active necropoleis in Pantanello, a ratio of burials to residences of 2:1, an unusual, exaggerated imbalance in favor of burials in this period encountered elsewhere only at Pizzica and AVS. All three of the farmhouses added in Pantanello were ex novo constructions. There had been no dateable farmhouses anywhere near the sites of 451F, 417F, and 445F in Pantanello.

The FMCE values of the four farmhouses in Pantanello in **350** are, in contrast to those of its near neighbor Pizzica, among the lowest in the study area. The decline in the significance and size of the farmhouses in this area may reflect a change in the agricultural economy—perhaps a shift away from the grape and olive that paleobotanical research in the Pantanello sanctuary indicates were in decline in the 4th century BC in comparison with cereals.[20]

[20] Carter 2006a, 23–29; for the shift to monoculture of cereals in the late

Figure 25.17 Assemblage from site 446N. (MLG)

Pantanello stands out from the other localities of the study area not only for the high number of necropoleis in relation to farmhouses,[21] but also for their durability. No necropoleis went out of use there in **350**. Remaining continuously in use from **400**, in some cases through **250**, were burial sites 446N (*Fig. 25.17*), 429N, 439N, and the subsequently excavated 415N.[22] The four (437N, 438N, 430N, and 421N) that were added were all ex novo. The sites (437N, 438N, 430N, and 421N) were all discovered in an area between division lines L15b and L20a, and east of a line parallel to and 375 m to the east of the possible transverse line PT05 that was devoid of farmhouses, except for undated 432F. The survey recorded in this area eight necropoleis with sufficient pottery to provide dates and nine undated necropoleis. Not included in this count are the excavated necropoleis at the Pantanello Crossroads, the Agrobios necropolis, the Eastern necropolis 415N,[23] and the D'Onofrio necropolis.[24]

Redividing the Land

A profound transformation took place in the Central Plateau in the middle decades of the 4th century that the statistics presented earlier in this chapter have only suggested. This section presents the arguments for a second sweeping reorganization of the chora.

No locality in the study area had a more eventful history than Lago del Lupo. The expansion across the Central Plateau in the mid-6th century BC, its complete desertion at the end of the 6th and beginning of the 5th century BC, the bonifica of the mid-5th century, of which the southeastern portion of Lago del Lupo was a particular target and beneficiary, and the distribution and probable redistribution of land to a large number of new settlers, make this locality the most complex and rewarding to study. The developments of the periods **350** and **300** were as innovative—not to say as radical—as before and decidedly more comprehensive, affecting areas all across the chora.

Ex novo farmhouses and the division lines. Fourteen ex novo farmhouses appeared in the Central Plateau in **350**, many of them (64%) in the southeastern corner of Lago del Lupo where the division lines are concentrated. This is half of all the new farmhouses in the study area in this period.[25] Another seven appear there in **300** (*Table 25.1*).

We can be certain, after the excavations of 1999–2001 (Ch. 28), that the actual sections of division lines correspond to drainage canals, and I believe that the hypothetical extensions probably corresponded to roads or paths along these axes.

The average distance between actual lines and ex novo farmhouses in **350** and **300** is 44 m. If we consider only the sites in the area where the lines are concentrated (249F, 172F, 164F, 210F, 248F/K, and 152F), with ca. 1-km-radius circle in the southeastern corner, the average distance is 37 m but the majority (5 out of 6) are much closer,[26] averaging 24 m. This is a little more than a tenth of the distance between the division lines in this area. The sites that are within this area are indicated with a circle in the table.

The average distance between ex novo farmhouses and extensions of division lines is 40 m. This distance compares closely to those between excavated farmhouses of this period and the division lines. Fattoria Stefan (*Figs. 25.18* and *25.19*) directly overlooked line L15e (20–25 m away), and Fattorie Arezzo and Venezia were very close to the same segment of L15, near its northern end (*Fig. 25.20*). In **450**, the aver-

4th c. BC, see especially p. 27.

21 The Pantanello survey was dealt with in summary fashion, without the precise information about the dating of the sites, and without the benefit of the study of aerial photos by Prieto (Ch. 31 in Vol. III) in the publication of the Pantanello necropolis (Carter 1998b, 50–55).

22 The Pantanello Eastern necropolis began to be used already in the 6th c. BC, though this period was not documented by the survey.

23 Excavated by the Soprintendenza in 1986.

24 Carter 1998b, 45–46 (map).

25 In **450**, there were 8 ex novo farmhouses in Lago del Lupo and 5 in Cogno del Prete, for a total of 13 for the Central Plateau, making up a third of all ex novo farmhouses in the study area in **450**.

26 Site 152F, an ex novo site of **300**, is 73 m from the nearest section of division lines. It is the outlier. Without it, the average would be 24 m.

	Farmhouse	Distance to nearest (m)	Division line or extension
		Ex novo **350**	
○	249F	1	13c
○	248F/K	44	13c
	296F	80	15e
○	172F	29	17b
○	164F	32	17b
	501F	1 <	10b ext
○	221F	56	14d ext
	517F	46	15f ext
	542F	8	16a ext
○	275F	76	17b ext
○	216F/T	35	18c ext
○	281F	30	19d ext
○	169F	74	20b ext
	35F (Cogno Prete)	43	12b ext
		Ex novo **300**	
○	210F	16	18c
○	152F	73	18c
	122F	36	17b ext
	121F	44	17b ext
	125F	45	17b ext
	291F	9	18c ext
	37F (Cogno Prete)	62	09d ext

Table 25.1 Distances between farmhouses and division lines.
○ Within a 1-km radius circle where division lines are concentrated.

age distance between ex novo farmhouses and division lines—actual survey sections and extensions—is comparable.[27]

We have considered ex novo farmhouses of **350** and **300** as a group in this discussion, but it should be clear that the period that saw the arrival of the greatest number of new farmhouses, as well as the closest relationship between these and the division lines (both sections and extensions) in the area of Lago del Lupo where the division lines are densest, was **350** (*Table 25.1*).

In **300**, the farming population in Lago del Lupo and the chora generally rose to its all time high. The number of ex novo farmhouses in Lago del Lupo, however, actually fell from 13 to 6. Of these completely new sites, only two (152F and 210F) were located in the southeast corner, the scene of greatest new growth in **350** (*Fig. 25.19*). Four farmhouses were constructed in the northwest corner of Lago del Lupo, up to now only sparsely settled. They line up along the northern extensions of the same two lines, L17b and L18c, beside which the new sites in the southeast arose in **350**. The spatial relationship between the new farmhouses and the lines, was, unsurprisingly, very similar to what it had been for farmhouses begun in the preceding periods. Though there are only half as many ex novo sites in the Central Plateau in **300** as there had been in **350**, their placement in relation to the division lines appears to follow very closely that of the preceding periods.

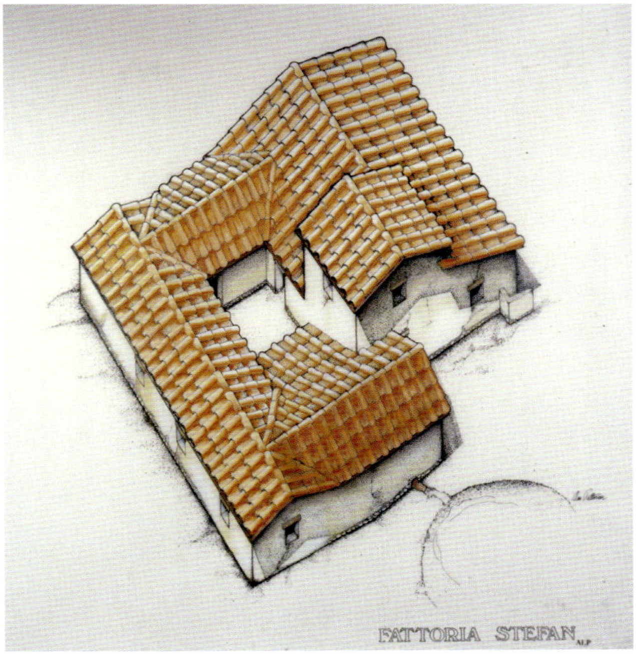

Figure 25.18 Reconstruction of Fattoria Stefan, the farmhouse under the second magnitude site 772F dated to **300**. It was located right beside main division line L15e (Fig. 25.10). Its nearest neighbor was 43F in the center cluster. (ALP/EH)

Spatial organization in 350. The proximity of so many farmhouses newly constructed in **350** to sections of division lines is an important aspect of the mid-4th–century BC settlement of the Central Plateau, but not an entirely unexpected one. A century earlier in **450**, drainage canals, highly visible indications of property lines, were dug and new farmhouses constructed. The case that these two phenomena were linked was made stronger by a statistically significant relationship be-

[27] The average in **450** is 26 m. Cf. Table 23.3.

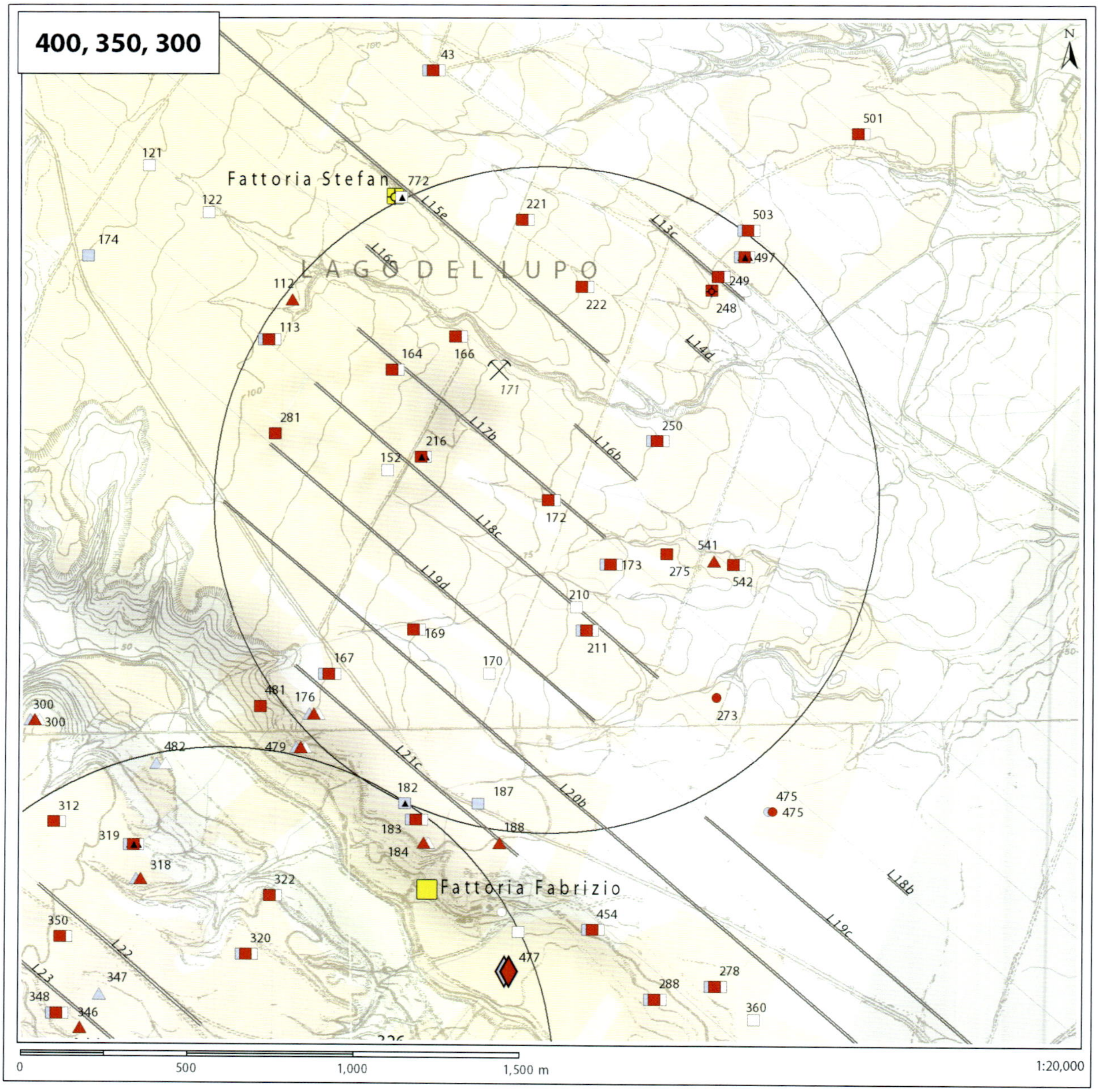

Figure 25.19 Lago del Lupo (and Avinella) in **400, 350, 300**. One-km radius circle around concentration of division lines. See Fig. 25.1 for map key. (JT)

tween the lines and the large number of necropoleis in the Central Plateau. In **350**, this interpretation is all the more believable because there are many more new farmhouses and new necropoleis securely dateable to the mid-4th century BC.[28] The conclusion to be drawn, that a second distribution of land took place a century later in the same area—in this case clearly a redistribution—is supported by regular patterns in the distribution of the new farmhouses of **350**.

In the renewal of the Central Plateau in **450**, two-thirds of the active farmhouses there were on new sites, and of these, a third were located in the original villages of the 6th century BC, Galitelli and Casa Preti. Those outside the village clusters were, with few exceptions, widely separated from one another (see *Fig. 23.9*). The space separating six of the farmhouses[29] is very regular, ca. 600 m. In **450**, the place-

[28] 112N (350), 541N (350), and 294N (300). 112N and 541N are in the area of concentrated division lines (*Fig. 25.19*). 112N is ca. 30 m from the extension of 17b and 541N is ca. 20 m from the extension of 16b. 294N is in the western region of Lago del Lupo and not close to division line segment 17e.

[29] Site 174F, 113F, 175F, 166F, 173F, and 170F.

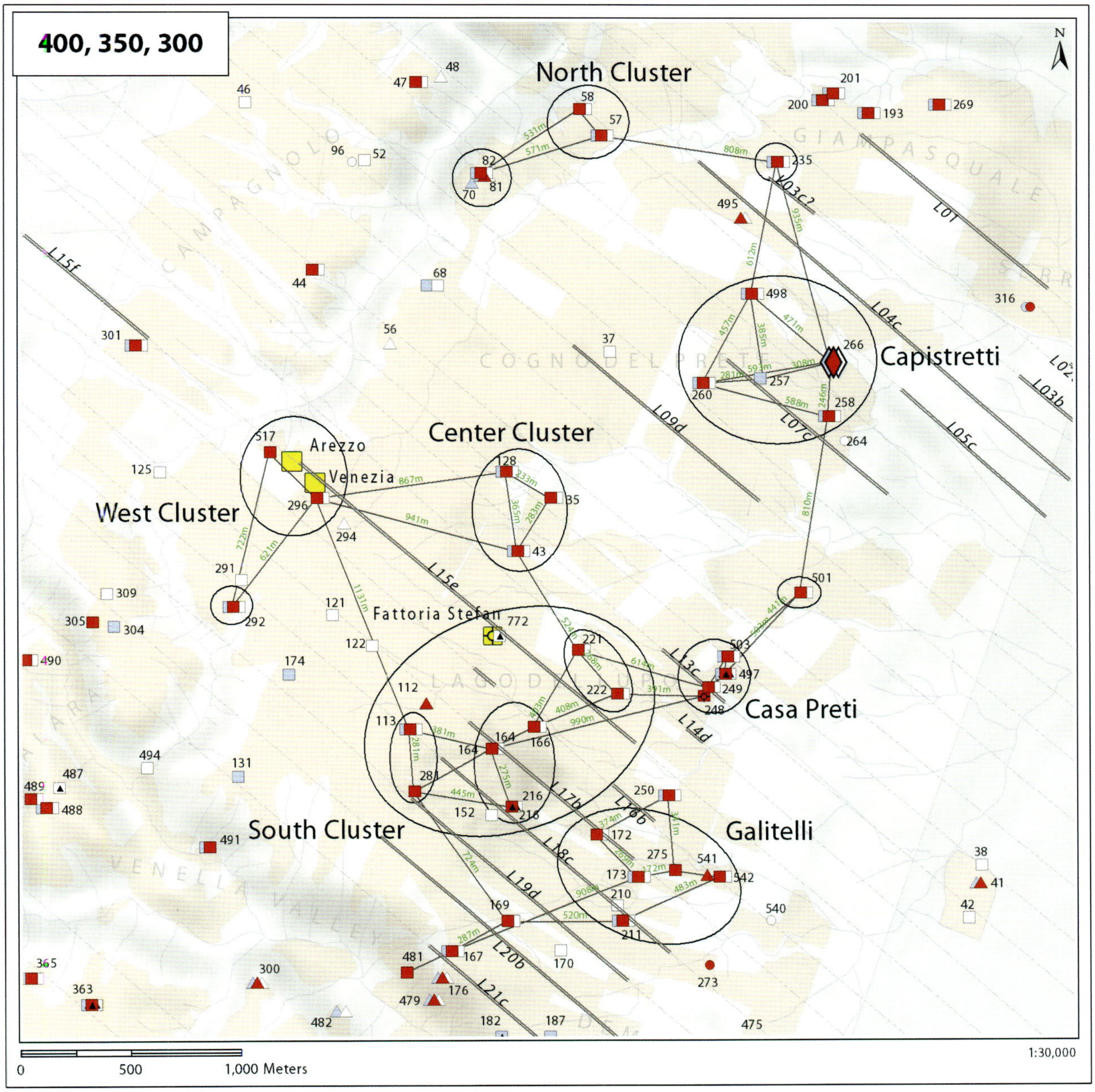

Figure 25.20 Transition diagram **400–350**. See Fig. 25.1 for map key. (JT)

ment of farmhouses not belonging to villages was very clearly determined by the presence of water, whether along the edge of the Venella (292F, 131F, 175F) or beside drainages (113F, 166F). The villages, of course, were originally sited beside abundant resources. In **350**, the distribution of new and active farmhouses in the Galitelli and Casa Preti village clusters and the open spaces of the Central Plateau is different from that in **450** in some fundamental ways. For one, the area where the division lines are thickest, just to the west of the villages, became the focus of new building in **350**, although this area was avoided in **450**. The 22 active farmhouses in Lago del Lupo in **350** are dispersed throughout the countryside in a way that may at first glance appear random, but with detailed analysis—beginning with a consideration of the distance of each farmhouse to its nearest neighbor—was clearly very carefully planned (*Table 25.2*).

The new clusters, 350. A study of the distances separating neighbors reveals hypothetical new groupings of significant farmhouses with characteristics that

	Farmhouse	Nearest Neighbor	Approximate Distance (m)
Central Plateau			
Lago del Lupo	167F	*169F	300
	169F	211F	520
Galitelli Cluster	211F	173F	210
	173F	*275F	180
	*275F	*542F	200
	173F	*172F	270
	250F	*275F	340
250F, isolated	250F	*248F	480
	*172F	*216F	400
South Cluster	*216F	*164F	280
	*164F	*166F	210
	*164F	113F	380
	113F	281F	280
	*166F	222F	400
	222F	*221F	280
Casa Preti Cluster	*248F	*249F	50
	*249F	497F	100
	497F	503F	80
501F, isolated	*501F	503F	440
	*501F	258F	810
Capistretti Cluster	258F	266S	230
	266S	498F	480
	498F	260F	460
	235F	498F	610
	*221F	43F	530
Center Cluster	43F	*35F	280
	*35F	128F	230
296F, isolated	296F	128F	870
West Cluster	*296F	*517F	300
296F, isolated	*296F	292F	620
235F, isolated	235F	57F	810
North Cluster	57F	58F	150
Isolated farmhouse	82F	57F	570

Table 25.2 Distances to nearest neighbor farmhouses in the Central Plateau in **350** (*ex novo). Grey rows indicate subclusters.

are similar to those of the older village clusters in the Central Plateau (*Fig. 25.10*). The new clusters in 350 (*Table 25.2* and *Fig. 25.20*) are concentrated in the southern half of the Central Plateau, but not exclusively. The South Cluster, which actually consists of three small subclusters of three, two, and two farmhouses, is located at a distance of approximately 400 m from both the Galitelli and Casa Preti clusters. The subcluster to the south consists of 281F and 113F, separated by 280 m. The three components of the subcluster in the center, 216F, 164F, and 166F, are 275 and 215 m apart; it is 380 m distant from the south subcluster, and ca. 400 m from the subcluster to the north, which is composed of 221F and 222F, 270 m apart. Patterns very similar to those in the South Cluster are encountered in the Center Cluster, and in the smaller West and North Clusters. The distance between component farmhouses in each case is ca. 200–280 m.

The single most recurrent measurement between sites is approximately 280 m (three out of four times in the South Cluster). Broadly summarized the measurements are gathered in Table 25.3.

Site/cluster	Range (m)	Average (m)
Galitelli (3)	180–270	197
South Cluster (4)	210–400	252
Casa Preti (3)	50–100	77
Center Cluster (2)	230–280	255
West Cluster (1)	300	—
North Cluster (1)	150	—
Capistretti (3)	230–480	390

Table 25.3 Distances between component sites in the clusters of the Central Plateau in **350**.

The distance separating farmhouses varies from cluster to cluster. The peculiar compactness of Casa Preti and its separate area of necropoleis, and the dispersiveness of Capistretti have been noted before.[30] Similar measurements found in the South Cluster and the rest are quite often encountered in the more established clusters such as the Galitelli and in Venella east, and others. Compare especially the area of the division lines, where the ex novo farmhouses are about 220 m apart and approximately 400 m separate the pairs (*Fig. 25.20*).

There is greater variation in the distance between the seven clusters, and between them and three isolated sites (292F, 501F, 250F), plus 235F in Giampasquale. They fall generally into three groups that range between approximately 340 and 520 m, between 810 and 900 m, and between 580 and 620 m apart. Beginning at the southeastern corner of the Central Plateau with

[30] Ch. 21, 689–90.

From	To	Distance (m)
Galitelli (*172F)	South Cluster (*216F)	400
Galitelli (275F)	Isolated site (250F)	340
Isolated site (250F)	Casa Preti (*248F)	480
Casa Preti (248F)	South Cluster (222F)	400
Casa Preti (503F)	Isolated site (501F)	810
Isolated site (501F)	Capistretti (258F)	810
Capistretti (498F)	Giampasquale (235F)	610
Giampasquale (235F)	North Cluster (57F)	810
North Cluster (58F)	Isolated site (82F)	530
Central Cluster (43F)	South Cluster (*221F)	530
Central Cluster (128F)	West Cluster (296F)	870
West Cluster (296F)	Isolated site (292F)	620

Table 25.4 Distances between villages and clusters in the Central Plateau in **350**.

the Galitelli Cluster and, moving northward, to the south and west, the distances separating the clusters are fairly regular. They are listed in Table 25.4.

In summary, an overall pattern of settlement emerged in **350** that was consistently followed in large areas of the Central Plateau. The topography of the areas where it developed lacked the characteristic focus of the Early Village Clusters—the deep drainage, and the natural hollow that the early settlers found attractive. These new settlements are situated in the heart of the territory they exploited. Communication between them and the urban center was almost certainly facilitated by the creation, or at least amplification, of the system of roads or paths flanking the drainage canals of the bonifica of **450**, which surely continued to function much as before, as drainage and boundary divisions. There is no reason to believe that this pattern was not widespread in similar terrains not included in the survey, for example, in the area of the south of Casa Preti and Galitelli.

A study in contrasts. In Cogno del Prete to the north, apart from the Capistretti village cluster, the space between farmhouses is much greater (*Fig. 25.20* and *Table 25.3*). Between the ex novo farmhouses (128F in the center of the Central Plateau and those on the edge of the Acqua Fetente, 68F and 57F), there are distances of about 900 and 1020 m. Approximately 770 m divides 57F from site 235F and the Early Village Cluster at Giampasquale.

In the Capistretti cluster, the quadrilateral polygon with sides of roughly 600, 480, 450, and 230 m formed by sites 266S, 498F, 260F, and 258F took shape in **400** with the revival of the 6th-century BC farmhouses 258F and 498F. It is considerably larger than contemporary clusters in Lago del Lupo. The space between its individual components is about twice as great as that of the clusters to the south, and is more nearly equal to the space that separates clusters, not farmhouses. Only two of these four sites, 266S and 258F, are near drainages. In **350**, it appears, man-made drainages may have attracted people to the Central Plateau as possible sources of water.[31] Two farmhouses in the Capistretti village cluster of **350** (260F and 498F) are not near division line segments or natural drainages, as are most of the other active farmhouses in Cogno del Prete. What convenient sources of water were available to them? The system of canals could, in any case, only have supplied part of the water needs for the farmhouses—for irrigation of gardens or for animals and only at certain periods of the year. It is unlikely that they could ever have provided drinking water for the human inhabitants.

The same question must be asked of other farmhouses in Cogno del Prete (43F), Giampasquale (235F), and Lago del Lupo (113F, 174F) that are neither near natural drainages nor division line segments. Where did they obtain water for drinking and other needs? Most of these farmhouses date at the earliest to **450** (174F, 113F, 260F) or **400** (43F, 498F, and 235F). It is tempting to believe that in these cases the vital water was subsurface and was supplied entirely by wells. They may not have been so intensively involved in agriculture as their neighbors to the south. Well water may have sufficed for livestock when the animals were held closely to the farmer's dwelling.

There are no natural physical barriers dividing the northern from the southern half of the Central Plateau. It is one continuous surface with an apparently similar history of settlement, until the 5th century BC. Cogno del Prete is about half the size of Lago del Lupo.[32] Both have been thoroughly surveyed and have comparably large areas in the northwest where settlement from the beginning and throughout their history was sparse to non-existent. The history of the settlement of Cogno del Prete and Lago del Lupo af-

[31] Prieto 2005, 170–76, 180–85.

[32] 3.55 km^2, compared to 7.08 km^2, but a higher proportion of it has been surveyed: 2.79 km^2 (79%), compared to 4.15 km^2 (59%).

ter the 5th century BC, however, is a study in contrasts with its larger neighbor to the south.

In **450** both localities seem to have led the rush to occupy villages and more isolated sites abandoned in **500**. In Lago del Lupo, the settlement was expanded especially to include areas unoccupied before, in the open spaces to the northwest of the villages of **550**. In Cogno del Prete, all of the five farmhouses in existence in **450** were on new sites; in Lago del Lupo, by contrast, only 57% were ex novo farmhouses. The combined number of ex novo farmhouses in the Central Plateau showed that Cogno del Prete played a major role in the fundamental changes taking place in **450**. In **400**, as we have observed, a complete reversal took place in the Central Plateau. There were only two ex novo farmhouses in Cogno del Prete, and none in Lago del Lupo. Lago del Lupo had 14 occupied farmhouses in **450**, nearly three times as many as Cogno del Prete. In **400** that number had dropped to nine, the same as its smaller neighbor to the north. The number of farmhouses in Cogno del Prete did not rise again until **300**, and then only by two, a change of 22%. In **350** the number of farmhouses in Lago del Lupo, as we have noted, jumped two and a half times to 22, a rise of 144%. The contrast between the two shows Cogno del Prete to be a very stable place that reached its maximum level of occupation in **400**, and for the next century saw little change. It had one of the lowest percentages of ex novo farmhouses among sizeable localities in the study area after **450**, when it had the highest. Cogno del Prete, we may surmise, reached a sort of equilibrium, a maximum capacity for growth, limited, it would appear, not by available space or natural factors.[33] It is tempting to conclude, given the low number of new farmhouses, that growth was limited by a certain innate resistance to change among the inhabitants of this region. This is a very subjective impression, but consider Giampasquale, further to the north. With its population, constant at five farmhouses for two centuries, this small locality would appear to be a model of conservatism. Behind the appearance of immobility, however, even at Giampasquale modest changes were taking place, at least in **400**, that had some resonance in Cogno del Prete.

Spatial organization in 300. Though by most measures the population of the Central Plateau reached its peak in **300**, the real period of growth, as we have attempted to show, was **350**. The configuration of farmhouses that appeared for the first time in **350** changed little in the last part of the 4th century and early 3rd century BC. The Central Plateau grew least of all the four major geographical regions in **300** (*Appendix C4*), and the clusters along the southern edge of Lago del Lupo, south of division line L17b, remained virtually unchanged. Five of the seven ex novo farmhouses in the Central Plateau in **300**, however, are located there.

A north-south line of farmhouses clustered in the villages had been a feature of the Central Plateau from **550**. The connection was reinforced in **300** and the distance between the cluster on the east bank of the Venella and the Galitelli Cluster was diminished with the addition of site 169F in **350** and the reactivation of site 170F in **300**. Sites 170F and 169F are only 250 m apart and both are located between lines L19d and L20b, so that they would almost seem to be northern extensions of the Venella cluster. In any case, site 170F is also only 340 m from 211F, the southernmost member of the Galitelli Cluster. There is virtually no separation between these two clusters by **300**.

One farmhouse, 275F, ex novo in **350**, was abandoned in **300** and its loss compensated by 210F, ex novo in **300**. This kept the farm population of the Galitelli Cluster stable at five families in **300**. About 80 m separated the new 210F from 211F (a site already in use in the 6th century BC and from **400** down to **300**). This spacing within a village was previously encountered at nearby Casa Preti and in the Early Village Clusters Giardinetto and Sant'Angelo. Such pairings of farmhouses separated by 100 m and often less, as we have seen, regularly occur in the villages from the 6th century BC onward.

As we move northward across the flat expanse of the Central Plateau in **300**, there are fewer signs of change since the previous period. At the village of Casa Preti there was none at all. The two isolated sites (250F, 501F) to the south of it remained as before. A new farmhouse, 37F, was added 480 m to the west of site 260F in the Capistretti Cluster. This is typical of the distances separating the farmhouses making up this cluster, so we may say that Capistretti grew by one farm family. It was there for a half-century around **300**, but then was abandoned forever. Some 830 m away across the plain to the northwest of site 37F, in a space unpunctuated by evidence of human

[33] A similar situation existed in Venella east on the opposite side of Lago del Lupo in this period.

occupation of any sort, site 68F (*Fig. 25.10*) came to life again after a pause of half a century. Site 68F, at the northern end of a village whose other four constituent elements are undated, was 550 m to the south of site 82F. This seems to be a standard distance between sites in the northern half of the Central Plateau where sites are almost always isolated. It is about three times the normal distance between pairs of sites. The pair 58F and 57F, 150 m apart, with 68F and its related village, are the exceptions. This is the sum total of activity north of division line L16b.

The largest group of ex novo farmhouses in the Central Plateau in **300** occurs in Lago del Lupo to the west of the line of villages (*Fig. 25.20*). The South Cluster, formed ex novo in **350**, lost its southernmost component, 281F, whereby the most southern of the subclusters became a single, somewhat isolated site, 113F, 380 m to the south of the central subcluster (composed of sites 164F, 216F/T, and 166F). That subcluster instead grew with the appearance of ex novo site 152F. This center subcluster now consisted of four sites, two of which (the new 152F and 216F/T) were a pair separated by only 130 m. The surface pattern of the South Cluster of **350** changed slightly but the count remained the same.

In the previously empty space to the west of the South Cluster, two new farmhouses (122F and 121F, spaced 220 m apart) form a new cluster. The distance between 122F and 113F in the South Cluster is 420 m, the typical distance between subclusters, and hence this might be considered a satellite of the South Cluster.

The clusters of farmhouses continued to grow. Farmhouse 292F had perhaps not been entirely isolated for all of its continuous history from **450** to **350**. The dates of the neighboring farmhouses 124F and 773F/T cannot be known. It certainly ceased, however, to be alone on its point of land overlooking the Venella valley in **300**, when the ex novo farmhouse 291F was erected barely 120 m away—another pair like 216F and 152F, consisting of an older and new site, as seems always to be the case. This pair is separated from the Southwest Cluster by 450 m. Thus a continuous chain of pairs of farmhouses, separated from each other by 350 to ca. 450 m was created in **300** that stretched 2.5 km from Galitelli to within ca. 600 m of the western edge of the Central Plateau. At the very edge, a narrow space between the headwaters of the Acqua Fetente and a tributary of the Venella, a single farmhouse in this period, as far as can be known for sure (since 126F, ca. 80 m away, is undated), was erected in **300**. Site 125F, separated from its nearest neighbors (291F, 301F and 296F) by approximately 600, 580, and 550 m. These distances are on the scale of the northern Central Plateau and the banks of the Acqua Fetente. The regular repetition of similar intervals throughout this area of the Central Plateau shows that the organization of the chora here, as elsewhere, was deliberate and little was left to chance.

FMCE in the Central Plateau. The region Close to the City, throughout most of its history, is the region with the highest FMCE values (*Appendix C7*). In the course of the 4th century BC, however, the Central Plateau took the lead by a wide margin with an average of 2.72 compared to 2.58. The distribution of these generally higher values over the landscape is not random, and detailed examination again reveals a complex situation from which some tentative conclusions may be drawn.

In **350** the farmhouses of Lago del Lupo had the third highest average FMCE value in the study area. The high values are concentrated in the three original village clusters and in the West Cluster. The exception is the Galitelli Cluster. Four of its five farmhouses have values well below the average. In the mid-6th century BC, the Galitelli Cluster had the highest FMCE average of the three original clusters in the Central Plateau. It did not recover as well as the others from the crisis of **500**, and from **450** onwards had by far the lowest values of the three. Did this reversal of fortune reflect the changing pattern of occupation in this area of the chora, from larger land holdings to more numerous smaller ones? The farmhouses of the Casa Preti and Capistretti Clusters farther north have values well above the average for the study area, in **350**, with one exception in each case. The isolated farmhouses 250F and 501F on either side of Casa Preti have comparably high values.

Among the neatly formed clusters of **350**, there is quite a range—from the extreme high average FMCE of the West Cluster to the abnormally low average of the North Cluster (*Table 25.5*). More typical and at the low end is the South Cluster. Its average (1.97) is nearly identical to that of Galitelli (1.94). Can it be a coincidence that the low values occur in the area of

	Site	350	300
	Venella		
range	167F, *169F	2.13-2.41	1.52-3.49
average	167F, 169F	2.27	2.65
	Galitelli		
range	*542F, 211F	1.73-2.24	2.24-2.92
average	172F, 173F	1.94	2.61
Site 250F		2.50	2.19
	Casa Preti		
range	*248F, *249F	1.63-3.11	2.48-3.70
average	503F, 497F	2.39	3.06
Site 501F		2.38	3.48
	Capistretti		
range	498F, 258F	1.84-2.53	2.87-3.65
average	258F, 266S	2.23	3.21
Giampasquale	235F	2.03	3.00
	South Cluster		
range	113F, *221F	1.38-2.63	2.10-3.21
average	222F, 221F	1.97	2.56
	Center Cluster		
range	35F, 128F	1.88-2.51	2.48-3.38
average	35F, 128F	2.15	2.84
	West Cluster		
range	*296F, *517F	2.52-3.41	4.32
average	296F	2.97	
	North Cluster		
range	*58F, 57F	0.84-1.70	0.90-2.66
average	58F, 57F	1.27	1.78
	Southwest Cluster		
range	292F	2.64	2.75
average	291F, 122F		
	(Isolated)		
	*Site 122F		3.50
Site 82		1.85	
Average for study area		**2.04**	**2.57**

Table 25.5 Average FMCE values for clusters in the Central Plateau. *Ex novo sites.

the greatest concentration of division lines, and of ex novo sites in 350? I believe not. Here the spaces between houses is smaller than in any area of the Central Plateau, with on the average 280–300 m between components of two and three farmhouse subclusters, and 400 m between the subclusters.

It is tempting to think of farmhouses with a high FMCE as somehow more prosperous than most, but this may not necessarily be the case. The land around them was more extensive and this may have been dictated by the particular needs of the farmer. In the case of the farmhouses in the Galitelli and the South Cluster there seems to be a correlation between the smaller space between farmhouses and the generally smaller FMCE values of the farmhouses. Does a correlation exist between the size of the area around a farmhouse and the size of FMCE value? The larger they are the higher the value? That would appear to be case.[34] The West Cluster, consisting of two farmhouses with exceptionally high FMCE values, is in fact one of the most isolated settlements, separated from its nearest neighbors by a distance of 600 m or more. Let us see how this works when several sites are close to each other.

In the South and West Clusters, in particular, one value in a subcluster consisting of two or three farmhouses is consistently higher than the other member or members. This was frequently the case in the pairs that have been identified elsewhere and in different periods. It was suggested (Ch. 21) that the pairs might have belonged to related families or to family members—father and son, or younger brother. If this were the case, then it might be expected that the farmhouse belonging to the older and more established member of a family would have the higher FMCE value. This could be the case in the Center Cluster, with the pair 128F and 35F, but it is not so in the South Cluster with 113F and the ex novo 281F, so the explanation must be sought elsewhere.

The average FMCE values of the farmhouses of Capistretti and Case Preti are intermediate (*Table 25.5*). The farmhouses of Capistretti are separated by larger distances, ca. 400-600 m, than those of the clusters in the southeastern part of Lago del Lupo around Galitelli. In Casa Preti four farmhouses are located within a radius of ca. 130 m, the densest grouping of more than two farmhouses anywhere in the chora. As was suggested earlier, the houses of Casa Preti, in contrast to those of other villages, were clearly not surrounded by all the property that their

[34] It should be recalled that scatter size is used to calculate FMCE (Ch. 4), but this is independent of the size of the land controlled by a farm family, based on the distance between farmhouses.

owners probably cultivated. Their fields could possibly have been located to the north in the relatively unoccupied space between Casa Preti and Capistretti.

The suggestion that the FMCE value is correlated with the size of the associated territory to be cultivated is borne out by the isolated sites, like 501F between Casa Preti and Capistretti and 250F between Galitelli and Casa Prete, and especially 292F in the far west of Lago del Lupo. The undeveloped areas surrounding these sites in **350** are on the order of 20 to 30 ha.

In the fifty years centered on **300**, there were no serious changes in the way the Central Plateau was organized. Settlements were simply more intensive than that of the preceding period. The farm population of the Central Plateau increased in **300**, but at a much lower rate than that of the localities Far from the City, where eleven farmhouses appeared ex novo. In terms of increased significance, no region compared with the Central Plateau, and only Pizzica continued to surpass Lago del Lupo. The two original village clusters, Casa Preti and Capistretti, were again in the vanguard in **300** with average FMCE values that were, respectively, 25% and 19% higher than average (2 57) for the study area as a whole. The average of all of the clusters of the Central Plateau, except for the North Cluster, and all of the isolated farmhouses, also were higher. Site 296F in the West Cluster in **350** had, in the intervening years, been left isolated in a large area in the northwest corner of the plateau. It had the highest FMCE value of any farmhouse in the chora. The only other farmhouse equal to or greater than 4.0 was 46F in Campagnolo to the north of the Acqua Fetente (see Campagnolo, above).

Settlement pattern and social change in 450 and 300. The Central Plateau, more obviously than anywhere else in the study area, was the theater in two periods of change, **450** and **300**, that fundamentally affected the society of the polis. The large-scale shifts in the settlement pattern have been interpreted in both periods as a distribution or redistribution of land, and consequently of wealth. In **450**, land changed hands and many new farmhouses were created; in **350** this was again the case, as either landless citizens, or new citizens recently enfranchised to fill the ranks for the defense of the polis, took possession of the land. Similarities and differences in the patterns in the two periods, noted briefly earlier, will receive a more thorough and comprehensive discussion in this section.

There were many more sites—and they are more widely spaced—in the Central Plateau in **350** than in **450**. It is enough to compare the maps of the Central Plateau in these periods (see *Figs. 23.9* and *25.10*). It seems that the amount of land surrounding the active farmhouses in **350** was very much smaller.[35] Was the greatly increased population in the Central Plateau and the entire chora in **350** a result of the fragmentation of larger properties or more intensive cultivation, reflecting a growing market and demand for Metapontine crops? Could it have been due to improvements in agricultural crops, such as the introduction of alfalfa,[36] or technology, such as the introduction of wells? All or some of these factors may be behind the population boom. Whatever the explanation, the fact remains that the amount of land associated with the distribution of land in **450** was considerably greater than in **350**.

In **450**, the expansion beyond the original villages of **550** was largely limited to the western corner of Lago del Lupo, not far from the Venella. In **350**, much of Lago del Lupo was occupied by new farmhouses. There was expansion of the settlement also in Cogno del Prete, though vast tracts had no active farmhouses.

The effect of the reorganization of **450** on the FMCE values for Lago del Lupo was not exceptional. The average for the whole Central Plateau equaled precisely that for the study area as a whole. In **350**, the average FMCE value for Lago del Lupo was among the highest in the study area, significantly greater than that of Cogno del Prete in this period. The reorganization of the land in this always rich agricultural area would appear to have benefitted the many there and brought them a material prosperity that surpassed that of the few in the northern Central Plateau.

Redistribution in other areas—Bradano to Basento. The nexus of division lines and newly constructed farmhouses in **450** and **350** is the main argument for a large-scale redistribution of land in the Central Plateau. The most densely concentrated area of division lines where survey coverage was most complete is in Lago del Lupo. The other areas of concentrated

[35] For example, the four farmhouses in **450** that define a polygon (131F, 175F, 113F, and 174F) are separated by approximately 500, 580, 540, and 690 m. Only one of these farmhouses (131F) was not ex novo in **450**. In **350**, as we have seen, the pairs of sites are separated often by as little as 200–300 m. In the South Cluster, there are four such groupings.

[36] See Ch. 19, Agricultural Potential of the Land.

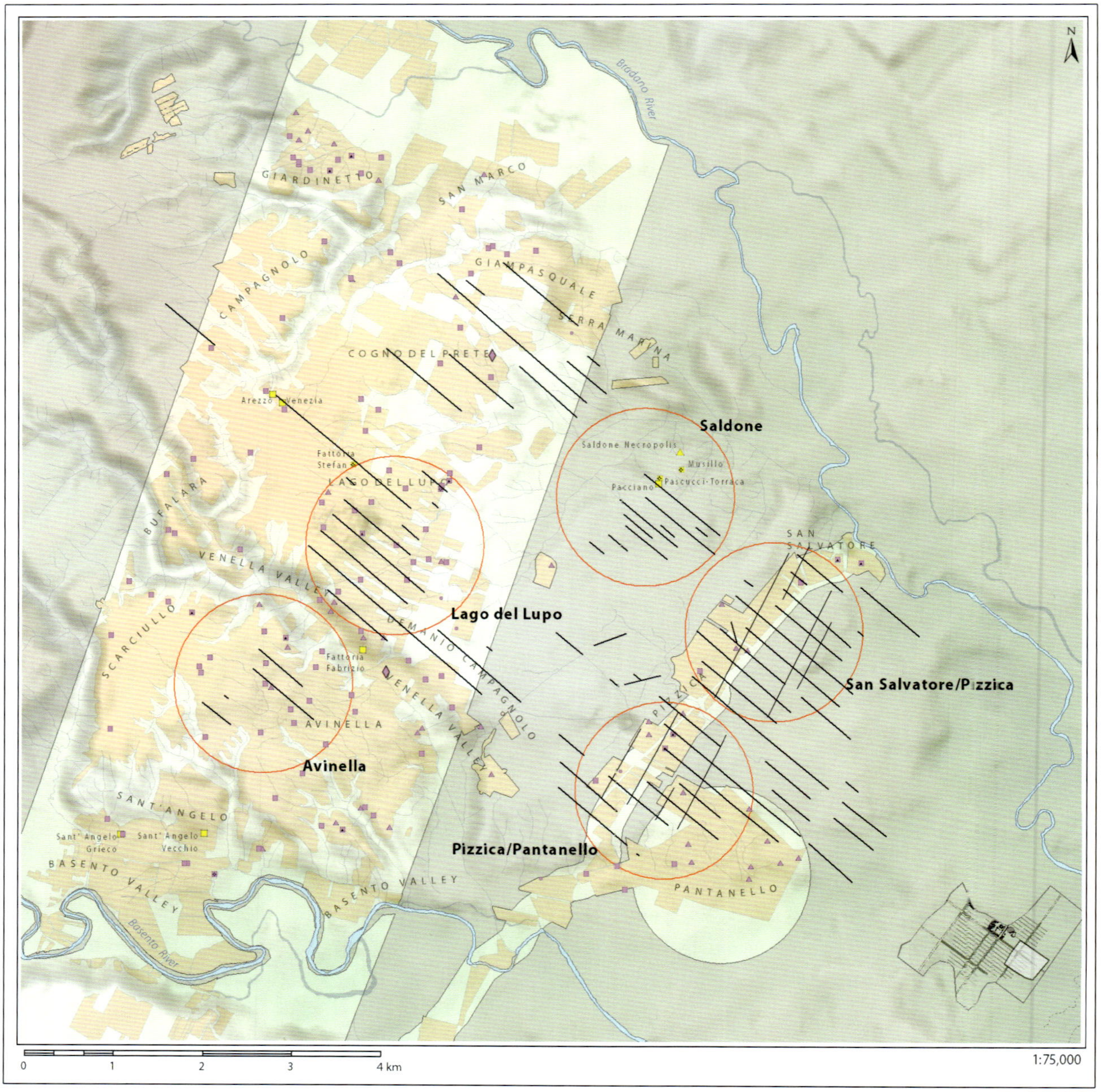

Figure 25.21 Areas of concentrations of division lines. The sites included are all farmhouses and necropoleis active in **400**, **350**, and **300**. (JT)

lines (*Fig. 25.21*) were only incompletely surveyed—if at all—so the question of whether the land redistribution there was more widespread than the Central Plateau cannot be finally resolved.

The other areas of maximum concentration of lines are:

San Salvatore/Pizzica. More division lines, both longitudinal and transverse, are located here than in any other part of the chora between the Bradano and Basento. Though only 17% of this area was covered by the survey, the percentage of farmhouses was comparable to Lago del Lupo. None can be dated to **350**. Here necropoleis dating to **450** and later have been excavated right beside the line (see Ch. 28).

Pizzica/Pantanello. This locality is second only to San Salvatore/Pizzica in the total length of preserved division lines. Proportionately, there are far fewer farmhouses there than in either Lago del Lupo or San Salvatore/Pizzica, and though there are many more necropoleis in very close proximity to the lines or their extensions, the majority are undated. The 1-km radius areas in Pizzica/Pantanello and the locality Cogno del Prete are practically equal in terms of surface area covered by the survey, and they are remarkably alike in terms of the number of farmhouses active in **350**.

Saldone In this extension of Pizzica to the north, the aerial photographs indicated segments of six division lines and three possible division lines. This area was not touched by the survey, but three farmhouses, Musillo, 6th and 4th to 3rd centuries BC, Pascucci-Torraca 6th and 4th to 3rd centuries BC, and Pacciano, 4th to 3rd centuries BC,[37] were excavated here in the 1960s, and the Saldone necropolis in 1976.[38] The burials, dated mainly to the second half of the 5th century BC (like those at Pizzica, mentioned above) are aligned with the axes of the division lines.

Avinella The division lines in Avinella are relatively few. There are many sites within a 1-km radius of site 343F, an ex novo site in **350**, but only four farmhouses and two necropoleis in **350** are within 50 to 100 m of a division line.

What emerges from comparisons of these areas is that there is a correlation, across the chora, between division lines and relatively high counts of farmhouses. In **350** in Lago del Lupo that relationship is intense. It is also clear, when considering the Central Plateau as a whole, that where there are not lines—in the two large areas in the western halves of Lago del Lupo and Cogno del Prete—neither are there farmhouses or necropoleis, or at least very few and widely scattered. The exception to this, the area of the Center Cluster, may provide the clue as to why this is the case. Two of the three active farmhouses there in **350** are located not far from the major drainages in the center of the plateau, a western headwater of the Fosso Solfegna and ultimately of the Fosso Appio and the Bradano.

The imperative for the early settlement in the chora in the 6th century BC was proximity to a source of water: a spring, a natural drainage, or stream. In the 5th century BC, a whole system of drainage was created to remove excess water from the surface. An early suggestion that this network of canals might indeed have been an irrigation system was at first ruled out because there was no obvious source of the water to supply the canals. They could, however, as noted above, also have functioned as reservoirs, trapping the runoff from winter and spring rains.[39] The canals could have carried off water in the fall rainy season and collected it during the spring and summer, when rains were infrequent. That hypothesis would be a likely explanation for the high density of sites around the lines in **350**. If this were indeed the case, then the function of the system between **450** and **350** changed or at least expanded. The drainage ditches would have become, to some extent, reservoirs, a source of water for the farmers who placed their dwellings in **350** conveniently close to them. This function would not have been as useful or necessary in **450**, when sites, even those outside the villages, were conveniently close to natural water sources.

Wells in the Central Plateau and elsewhere. An alternate source of water in **350** is likely to have been provided by dug wells. The use of wells cannot be verified from survey data, but there is one type of artifact found in excavation that might indicate a well. The wells of the late 4th century BC were lined with a terracotta cylinder, with a diameter ranging from 55–95 cm, made as two halves. Numerous examples have been excavated in the Pantanello necropoleis[40] (see *Fig. 25.25,* below) where they served as the containers for burials, and all are dated by the grave goods to 320–280 BC. An example was found in situ functioning as a well in the Pantanello sanctuary (*Fig. 25.22*).[41] Apart from one doubtful case (99N),[42] none has been documented by the survey and none in the excavation of a farmhouse, but it should be noted that most excavated farmhouses have been conveniently near natural sources of water. No cisterns have been found in the chora either, perhaps because with abundant water from other sources, including wells, they were not necessary. Water was quite near the surface throughout the chora of Metaponto. It was particularly close in the Central Plateau, since it occurs at the interface of the underlying Pliocene clay and the thin

[37] Amore 1992–1993, 58–64.

[38] Carter and Hall 1998, 429–42.

[39] Prieto 2005, see n. 31 above.

[40] Carter 1998c, 100–01, Table 3.16, Figs. 3.57–60.

[41] The well was part of the revival of the sanctuary in the second half of the 4th c. BC: Carter 2006a, 168; 1994, ill. 7.15. The well was installed over the 6th-c. BC spring. The excavation of the Pantanello sanctuary and its materials are being prepared for publication in this series.

[42] This site, mainly on the basis of the fragments of cylindrical well liners found on the surface, was interpreted as a burial site. See n. 18 above.

Figure 25.22 Well over the sacred spring in the Pantanello sanctuary. The terracotta well liner consists of two matching half cylinders, often used as burial containers. (see Fig. 25.25) (JCC)

mantel of sandy gravely sand that forms the beaches of the marine terraces that constitute the Central Plateau. The wells would not have had to be very deep.

Pantanello and Plots

This archaeologically very rich area of the chora provides a unique, detailed picture of the interrelation of division lines and sites, particularly necropoleis. The concentration of burials, and the virtual absence of dwellings of any period in this restricted area of the chora, is without parallel. It is as if there had been a prohibition against settlement here in effect from the very earliest days in the 6th century BC. The Pantanello sanctuary lies on the southern side of this area. Apart from the undated farmhouse 432F, there were no farmhouses within a 450 m radius of the sanctuary. It is possible that this area of the chora was the property sacred to the divinity or divinities worshipped there.[43] The apparent absence of evidence for sites of any type to the south of the sanctuary is probably the result of the flooding of the Basento valley and continual deposition of alluvial soil on the valley floor, still problematic there today.[44]

The close relationship of division lines and necropoleis in Lago del Lupo was the subject of detailed analysis earlier.[45] A higher percentage of necropoleis in Pantanello were located either in close proximity to, or directly on longitudinal division lines or their extensions (*Fig. 25.23*; *Map 11*, Vol. IV): the undated necropoleis 447N and 448N on the southern extension of L17a; 426N, 427N, and 429N (**400–300**) on the southern extension of L18a; and the undated 420N on the extension of L19a. The Eastern necropolis (site 415N) is close to the extension of L20a. The extensions of division lines, as noted earlier, are hypothetical and they may never have signaled the presence of a canal; it is more likely that they correspond to roads or *antomoi*.[46] Positive proof of this is the section of road flanked by burials in the Pantanello necropolis, excavated for a length of 150 m. There was no indication of it in aerial photographs, and it was added to the plan of division lines as segment L21a solely on the basis of the excavation results, and because it aligns perfectly with L21b.

The only surviving examples of transverse lines in the Bradano-Basento portion of the chora are to be found in Pizzica and Pantanello. Four appear in aerial photographs of Pizzica, T01–T03, and possible transverse line PT01. The 1999 excavation there confirmed the aerial photographs.[47] There are four more possible transverse lines in Pantanello (PT02–05). The rhombi formed by the transverse lines have long sides of ca. 630 m and short sides of ca. 210 m, the distance between longitudinal lines. Two of these together would have measured 420 x 630 m, a measurement identical to the largest divisions of the chora of Chersonesos.[48] A similar result was reached by Vatin in 1976,[49] but with far less evidence and less precise positions for the transverse lines.[50] A hypothetical reconstruction of the similar system in Pantanello has been attempted (*Fig. 25.23*). There is a single necropolis or farmhouse in several rhombus-shaped plots, and all dimensions

[43] The question of their identities is not an easy one to answer, and will be taken up in a forthcoming publication on the Pantanello sanctuary in this series. The Herakleia Tablets mention agricultural structures but no residential ones on the sacred lands of Athena Polias and Dionysos at Herakleia. See Uguzzoni and Ghinatti 1968, 116.

[44] The Pantanello sanctuary, which lay under 1 m to 1.5 m of alluvial deposit, in fact, almost escaped detection. Ch. 20, 668–69.

[45] Ch. 23, 769–70.

[46] Ch. 28, 1031.

[47] Ch. 28, 1037.

[48] Chtcheglov 1992, 230–35, and Carter et al. 2000, 713–14. The total area of these plots is ca. 26 ha, subdivided into six plots 210 m on a side.

[49] Adamesteanu and Vatin 1976.

[50] See Prieto 2005 and Ch. 31, 1120.

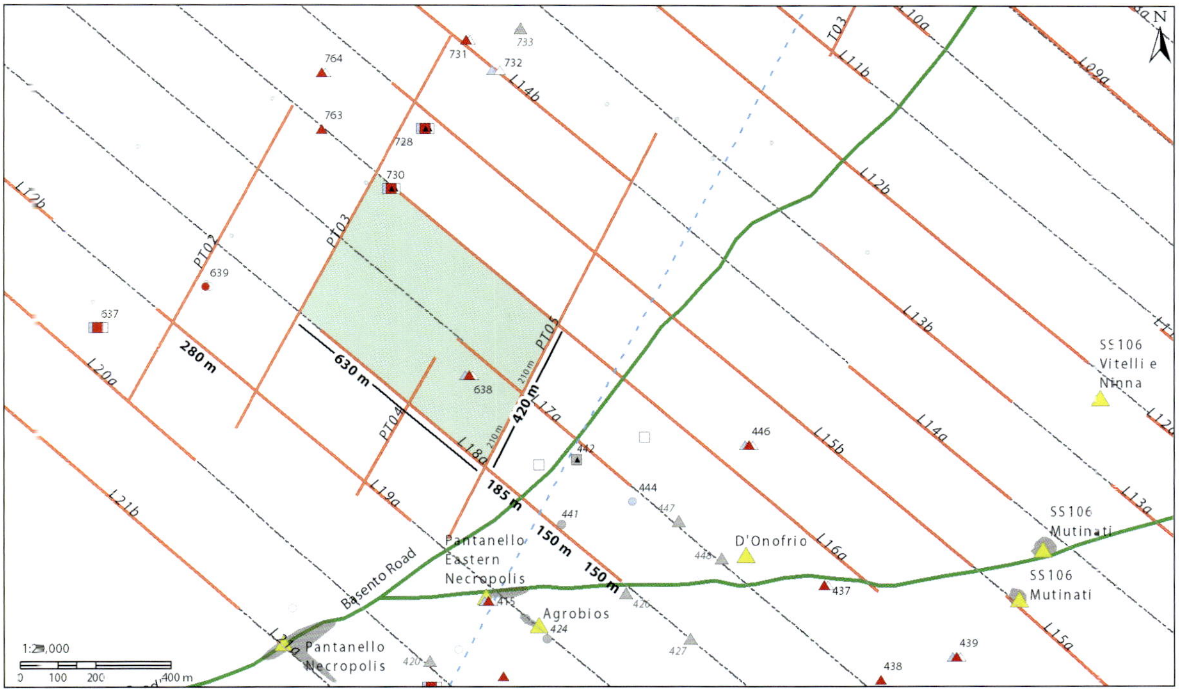

Figure 25.23 Diagram of Pantanello in **300**. (JT)

were established on the basis of lines visible in aerial photographs.

Necropoleis

The survey results reflect a great increase in the population of the chora in **350** and **300**, but do not indicate much about the nature and identity of that new population. Was the growth an internal expansion or the result of the addition of a new group or groups to the farming and citizen population? The large numbers of ex novo farmhouse sites in certain areas of the chora, especially in Lago del Lupo in **350** and in Scarciullo in **300**, strongly suggest a fresh beginning. Were these families newly resident in the chora, citizens formerly resident elsewhere in the chora or asty, or is it possible that some of them were newly enfranchised peoples formerly living in the indigenous settlements of the interior? The historical sources for southern Italy are silent about the extension of citizenship to neighboring native peoples,[51] but there are good parallels for it in this period[52] in Metaponto and Magna Grecia in the 4th and early 3rd centuries BC, and elsewhere in the Greek world.

The evidence from necropoleis, though not without problems, has proved to be valuable. In the case of Metaponto, the Pantanello necropolis (see *Fig. 20.29*) furnished detailed information about changing burial customs that led to the conclusion that the population using this necropolis was fundamentally different than the one that had used it previously in the 4th century BC.[53] In this section we shall again compare the necropoleis identified by the survey with the excavated burials at Pantanello.[54] Here the focus will be on the 4th century BC, and on comparing the duration of survey necropoleis and the frequencies and distribution of burials in the family plots at Pantanello. The aim will be to verify whether the changes taking place at Pantanello, where we have relatively full information, are encountered elsewhere in the chora where we have only the limited data of the survey.

Survey necropoleis. The survey identified 41 necropoleis active in **350** and 36 in **300**. As the diagram of necropoleis (*Appendix C14*) clearly shows, a number of burials were made in **350** and **300** in necropoleis that had been previously in use in earlier periods: three as early as **600** or **550**, and four that began in **450** and were continuously active till **250**. Three more that began in **450** had gaps of a half-century or more but revived in **350** or **300**. A total of 15 of the 20

[51] See Ch. 24, 806–07.

[52] Carter 1998e, 154–55.

[53] This theory was tentatively proposed in the study of the burials at Pantanello, *Necropoleis* 16–17, and further developed in the Jerome Lectures, 2000–2001; Carter 2006a, 230–32.

[54] See Ch. 23, 760–61.

necropoleis that started up ex novo in **400** continued, six to **350** and nine to **300**, and three more, after remaining idle in **350**, were again in use in **300**.

Of the 22 necropoleis dating entirely to **350** and to **300**, 16 were frequented for the first time in **350**. Only six belong exclusively to the last period of expansion of the settlement in the chora. This is a further sign that the revival of the 4th century BC had its real impetus in **350**, rather than **300**, despite the higher numbers of farmhouses in the later period.

The duration of necropoleis, their first use, and the gaps and cessations of burial reflect a similar pattern among the farmhouses, but the picture is more intimate because these necropoleis document directly the history of families in the chora. When activity at a necropolis ceases it may well signify that the family (or families) that buried there has either died out or moved and was burying elsewhere. A farmhouse, by contrast, may be continuously in use, but under different ownership.

Figure 25.24 Pantanello necropolis, a family burial plot in Nucleus 11, tomb 251, a sarcophagus with a female over 60 years, and tomb 252, the earthen grave of a female 25–30 years, both belonging to the second half of the 5th c. BC. (Carter and Hall 1998, 314–15)

Excavated necropoleis—Pantanello. The 18 separate nuclei of the Pantanello necropolis containing over three hundred excavated burials ranging in date from **550** to **300** show similar periods of intense activity, gaps, and abandonment (*Fig. 25.24*).[55] The record illustrates well the ebb and flow of population across these successive periods. At Pantanello, a large group of 5th-century BC families consisting of three to five generations ceased to bury at various times between 430 and 380 BC. Many areas of the necropolis were

Figure 25.25 Pantanello necropolis, two burials in containers made from terracotta well liners. Giuseppe Di Taranto, excavation foreman, stands above tomb 72. Tomb 71 (left) contains a female 20–25 years of age and especially fine and numerous grave goods, including golden earrings in the form of Erotes and Gnathia pottery (Fig. 23.17). Tomb 72 contains a female over 70 years and a single lebes gamikos. Both burials date to the late 4th or early 3rd c. BC. (Carter and Hall 1998, 264–65)

entirely unused or went out of use in **350**.[56] As the older groups gradually faded away, new ones came to take their places.

The great mass of fresh burials at Pantanello in **300** was concentrated in Nucleus 6 where only a few traces of earlier burials had survived, in a large assortment consisting of seven separate family groups. This was the greatest innovation of this period (*Appendix C27*).

Elsewhere in the Pantanello necropolis, after a gap of two or three generations, activity resumed in **300** in four of the nuclei where it had been interrupted for at least two generations (*Fig. 25.25*). The changes in the spatial organization of the Pantanello necropolis, the long gaps in the use of areas, the dying out of families, and the sudden massive incursion of burials in **300** have a pattern and timing that can be recognized in the

[55] Ch. 23, Fig. 23.11.

[56] Carter 1998f, 218–20.

shifting picture of burial in the chora outlined by the appearance of necropoleis in the survey record, the gaps in their use, and their disappearance (*Appendix C14*).

The well-documented burial customs and grave goods at Pantanello[57] and the change in spatial organization are a clear indication that there has been a real change in the population using it. The older groups, with some exceptions, have died out or gone away, and in their place is a new population, to some extent aware of the earlier one and to a degree respectful of it, but with distinctive traits and customs of its own.[58] This new population has been identified[59] as Lucanian, recently enfranchised to fill the depleted ranks of citizens.

Assemblages, surveyed and excavated necropoleis. The assemblages from the surveyed necropoleis compare in a number of important respects to those from Pantanello in general in **350**. They have in common the cups, with or without stems,[60] the table amphora, and the unguentarium, represented by numerous examples, and also some very specific forms that occur less frequently, such as the saltcellar, the convex-concave small bowl, and the lamp. An example, the assemblages from surveyed necropolis 81N, has been organized by the four functional groups recognized in the Pantanello necropolis:[61] oil vessels, covered containers, and large containers for liquids; tomb markers; vessels used in rites at the tomb; and other vessels and objects (*Appendix C37*). Though this necropolis had most of its black-gloss pottery concentrated in **300**—in fact its best date is 306 BC—there was material documenting use in earlier periods; it was active between **550** and **500**, and again from **400**, continuously to **250** (*Appendix C14*). The forms and types that are clearly of the period **300** are the unguentarium (U3b),[62] lamp, saltcellar (SA3a),[63] and dishes. The lekanai (Lne 12 and Pl/Lne 4b2)[64] and the hydriai (H1b)[65] would be at home in an assemblage from the middle of the 5th century BC. These were among the forms and types that had relatively long lifetimes (see *Appendix C34*).

Taken together, the evidence from the Pantanello necropolis strongly indicates—even if it does not prove absolutely—that another group had arrived in the chora in **300**. This is the conclusion that was reached early in the necropolis study. The chronological relationship between the survey necropoleis and those of the Pantanello family groups in **350** and **300**, the similarities in tomb types and grave goods, would extend this conclusion to the study area and the chora as a whole.

The urban necropoleis. Current views of burial at Metaponto and the relation between burials in the chora and those closer to the asty will have to be radically revised in the light of very recent discoveries. The purpose of this very section is simply to call attention to these discoveries.

The area around the asty on all sides, including the coast, was crowded with burials. Until recently only slightly more than 100 excavated over the years from the 1940s had been published. The very useful study by Rocchietti[66] synthesizes these results and compares them with those from the Pantanello necropolis, still the largest published sample of Metapontine burials. Not yet published are the hundreds of burials from the Crucinia necropolis excavated in the early 1990s by De Siena, which include the monumental necropolis of the late 7th and first half of the 6th century BC,[67] and others in the vicinity of the asty from Torre di Mare and Marinella, to Casa Teresa and Lo Svincolo. Early in the new millennium, public works projects such as the SNAM pipeline, those of the Ente Irrigazione (see Ch. 19), and the now since 2006 long overdue widening of the coastal road between Taranto and Reggio Calabria, the SS106, have resulted in the salvage excavation of many hundreds more burials. A total corpus of close to two thousand documented Metapontine burials, as of this writing, now exists. Apart from those few illustrated in the valuable report of the Soprintendenza at the annual *Convegni di Studi sulla Magna Grecia*,[68] this material

[57] Carter 1998f, 218–36.

[58] Carter 1998f, 233–35.

[59] Carter 2006, 219–223.

[60] Ch. 6, Cat. no. 98, and an example from tomb 80-86, Pantanello necropolis. Elliott 1998, 651.

[61] Carter 1998f, 186–87.

[62] See Ch. 9, Cat. nos. 267–69.

[63] Compare SA3 from the survey with Ch. 6, Cat. nos. 202 and 205, and an example from T111-7, Pantanello necropolis. Elliott 1998, 690.

[64] See Ch. 9, Cat. no. 61.

[65] See Ch. 9, Cat. no. 103.

[66] Rocchietti 2002. Cf. also Lo Porto 1966; 1981b.

[67] Ch. 21, 725.

[68] See the preliminary publications of Metapontine burials in *Atti Taranto*: Osanna 2008; De Siena 2006; M. Tagliente 2006 Nava 2004, 2003, 2002, 2001, and 2000.

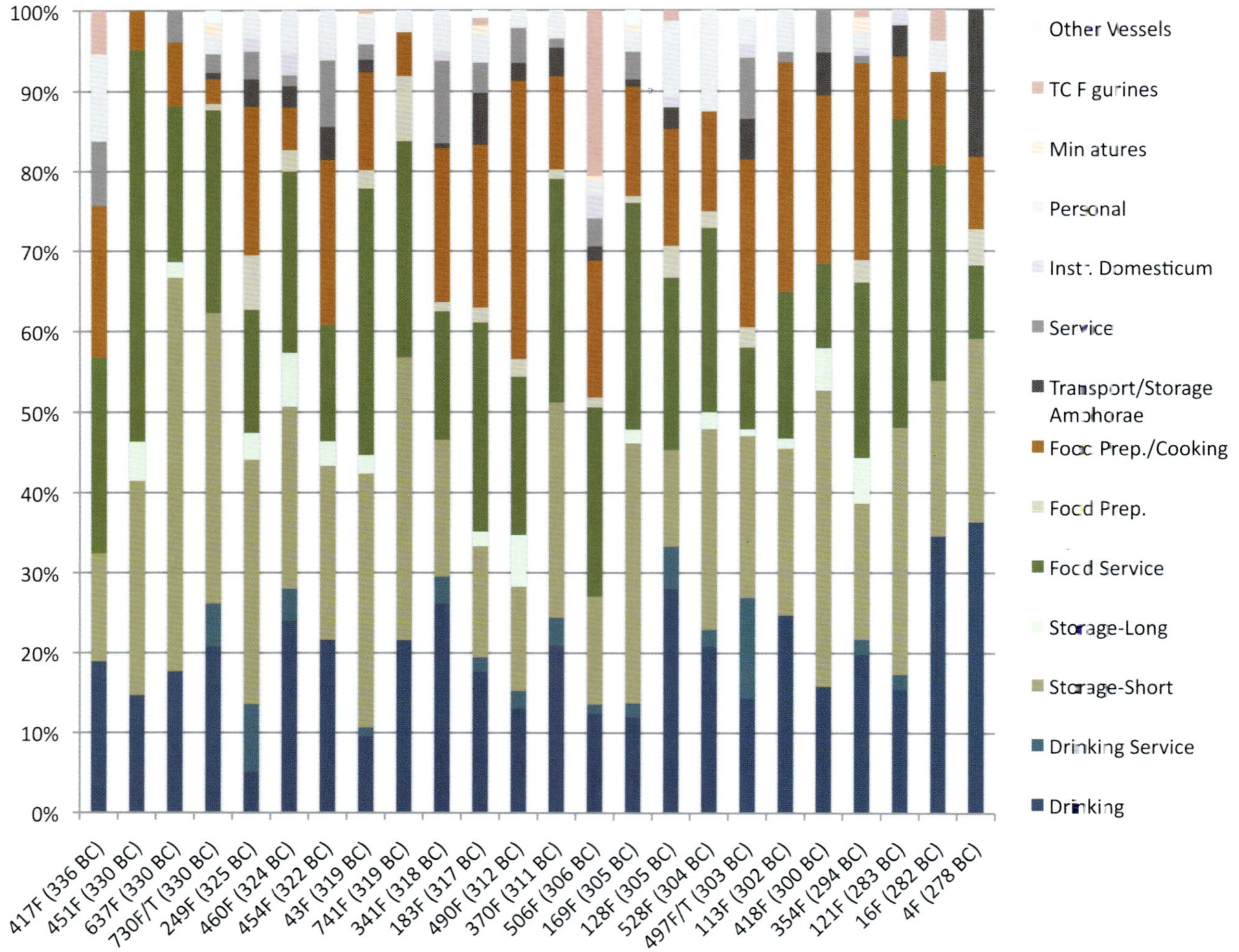

Figure 25.26 Farmhouse assemblages in **300**: functional groups—counts and percentages.

has not yet been published, making further attempts at synthesis premature.

What the latest discoveries show is that the whole length of the SS106, which must have corresponded fairly closely with the ancient coastal road,[69] was lined with burial areas from north of the Bradano,[70] to the Cavone and south of the Cavone. The distinction between "urban" and "rural" necropoleis, an arbitrary one at best, is valid but only in the geographical sense.

Ceramic Assemblages

The great expansion of rural settlement in **350** and **300** is documented not only by the greatly increased number of farmhouse sites, but also by a much enlarged corpus of ceramics—of black-gloss pottery, plain and banded wares, and cooking wares. These last categories of pottery can be dated usually only when associated with other dateable materials, in most cases black-gloss pottery. Some of the otherwise undateable pottery appears on sites that are new in this period (*Appendix C6*)—there are 31 ex novo farmhouses in **350** and 32 in **300**—and these provide a *terminus ante quem non* for this material. The composition of the mass of pottery from survey sites, both farmhouses and necropoleis, was analyzed in terms of principal functional groups and ware class percentages for the 6th century (Ch. 21), and the mid-5th century (Ch. 23). In the course of two centuries, or even a century, it would be reasonable to expect changes, though not necessarily major ones, in the daily activities of farm families and their burial practices. These have been observed in the two earlier periods. The functional groups of 24 farmhouses (*Fig. 25.26, Appendix C24a*) and the ware class proportions of 13 farmhouses (*Fig. 25.27, Appendix C24b*) are compared here, following the

[69] See above, Pantanello.

[70] Nuzzi 2008.

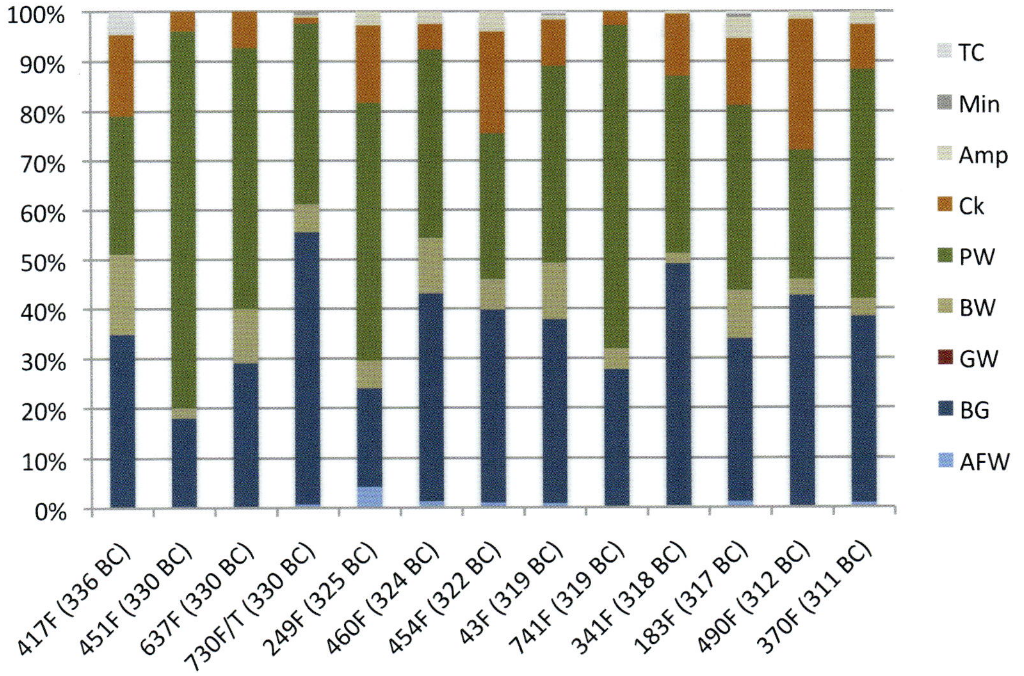

Figure 25.27 Farmhouse assemblages in **300**: ware classes—counts and percentages.

model of **600/550** and **450**. In addition, seven necropoleis of **300** were analyzed (*Fig. 25.28, Appendix C32*). The individual ceramic assemblage tables for these sites are reproduced in Volume IV. The sites have been selected because the date ranges of their black-gloss pottery components are concentrated in this period. Although certainty is never possible when dealing with survey assemblages, we may be reasonably confident that a high percentage of other classes of material is also of this period. The other criteria employed in selecting these sites are: that the assemblage is large enough to be reasonably representative, and that the percentage of identifiable forms is high.

Surveyed farmhouse assemblages, 300 (*Appendix C24a* and *b*). The average total count of pottery fragments with identifiable forms is over 80, and most of the 24 farmhouse sites in **300** have at least 40. Some sites have over a hundred, and one, site 43F, has 262. In the case of site 43F and several other large assemblages, there are relatively smaller quantities of black-gloss pottery from several periods. The assemblage from site 43F has black gloss of almost every period from **600** to **50**. More typical is site 460F with black gloss from **400**, **350**, and **250**, besides the much larger concentration in **300**. A group of a half dozen farmhouses (sites 169F, 354F, 121F, 16F, 49F, 249F) has no black-gloss pottery that does not belong to the period **300**. The banded and plain wares, the cooking wares, and other materials in these assemblages, therefore, have a higher probability of dating to the period.

A comparison of the percentages of pottery representing the principal functional groups shows a remarkable degree of consistency, as did the assemblages of the mid-6th and mid-5th centuries BC. Most of the ceramic vessels used by farm families, not surprisingly, have to do with the consumption and serving of food, water, and wine, and with the storage, preparation (including cooking), and service of liquid and solid comestibles destined for the table or the market. The vessel of general usefulness ("service") around the house, or those employed in personal hygiene or cosmetics, are numerically of minor importance in comparison. Evidence of a family cult, in the form of miniature votive vases, larger votive vessels (louteria, thymiateria), and terracotta figurines have been recovered at many, but by no means all, farmhouse sites.

No farmhouse of any period was not well furnished with drinking vessels. Nearly a quarter of all the pottery fragments from farmhouses of the 6th century BC were of cups or skyphoi. In **300**, the percentage is lower, and functional groups, principally food service, have come to play a correspondingly greater role. In **300**, the average percentage of drink-

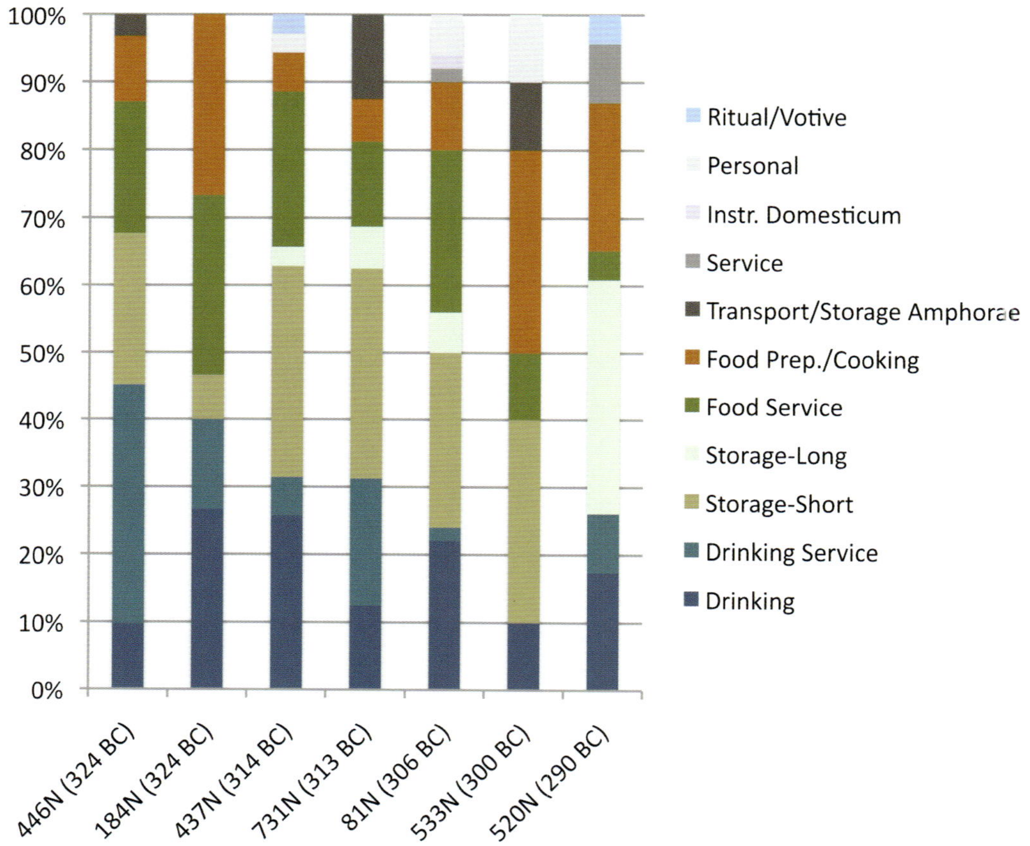

Figure 25.28 Necropolis assemblage, **300**: functional groups.

ing vessels in the assemblages is 19%, excluding the highest (36%) and lowest values (5%). The two highest values appear in quite small assemblages (those of 16F and 4F), which are less likely to be representative than larger ones. The lower percentage of drinking vessels in the otherwise apparently normal assemblages from site 249F must have another explanation that is not immediately apparent, but is not entirely alone. There is a low percentage (10%) also from the very large and long-lived site 43F. Some variation from the mean is to be expected. Such differences occur in the assemblages of **450**, but are very minor in **600/550**, where only two sites are more than four percentage points from the average of 23% for drinking vessels.

Large containers for liquids, such as kraters, which have been assigned the function of drinking service, are not common and fragments of them were identified on only 15 of the 24 sites studied. The category of short-term storage is represented by such shapes as the dinos, but principally by the jug and related forms with similar shapes—the hydria, table amphora, or amphoriskos—which in fragmentary form are practically indistinguishable. Short-term storage is the single largest functional group encountered on farmhouse sites. There is generally a lot of minor variation in the shapes of the vessels in this group. Many of the types of jug in use in **300** were also found in earlier farmhouses (see *Appendices C33a* and *b*, *C34*).

Long-term domestic storage is represented by the pithos and basin. The fragments of these large, heavy vessels were recovered from just 14 sites, or slightly more than half of the farmhouse assemblages considered here.

There are as many fragments of pottery representing the food service function on the farmhouses sites as the short-term storage, and the principal form is the deep platter known as the lekane. Lekanai occur in major variant types, the flanged lekane (more common in **550**), the carinated lekane, the lekane/basin, and the plate/lekane. There are even more variant types of the simple lekane than there are of the jugs. The jug and the lekane, which occur in both plain- and banded-ware fabrics, are the two most frequent forms in any farmhouse assemblage. Two other vessel types that may have served in food service, the plate and the bowl, are relatively infrequent.

The raw materials of meals were processed in basins, heavy duty lekanai, and in mortars. Pots, pans, jugs, casseroles, and other vessel shapes in cooking ware fabric make up, on the average, 16% of every farmhouse assemblage, surpassed only by food service, short-term domestic storage, and drinking in numerical importance.

Fragments of the major types of transport amphora—the Corinthian A, A', B, and the Greco-Italic—in use in the chora, either for the transport and storage of agricultural produce, or quite often as grave markers or grave goods,[71] appear in a high percentage of farmhouse assemblages, and always in excavated farmhouses,[72] though not necessarily in great numbers. In any case, they are better represented on excavated sites than on survey sites.[73] In this sample of 24 farmhouses dating to **300**, amphorae are represented in 16 of them.

Small bowls and one-handlers, lumped together under the catchall rubric "service," would have been useful for a variety of tasks, such as dipping (drinking service) or informal drinking, serving of condiments on the table, or storage of small objects in the living areas.

The lamp (*instrumentum domesticum*) would have been an innovation in the households of **300**—but not only in the house, as they appear also in the tomb (where it is very rare), and in the Pantanello sanctuary in contexts of the 4th century BC.[74]

The objects comprising the functional group designated "personal," the lekythos, unguentarium, and lebes gamikos—represented also among the functional groups of contemporary burials—the oil container and the covered container, are more frequent in the assemblages of **300** than in any previous period.

Unglazed miniature votive vessels, the krateriskoi and other shapes that appear in excavated Metapontine farmhouses, burials, and sanctuaries,[75] occur in six of the 24 assemblages; fragments of terracotta figurines in seven; and in four instances both together (43F, 183F, 506F, and 354F). Miniature votive vessels, a large scale terracotta plaque with a representation of a divinity, and black-gloss kantharos were found near the entrance to Fattoria Fabrizio that was occupied in **350** and **300**.[76] Four other survey farmhouse assemblages (130F/T, 183F, 169F, and 497F/T) have, in addition to the miniatures, large-scale ritual vessels—louteria and thymiateria. Altogether ten of the 24 sites in this analysis have some evidence for cult activity. It is probably a reflection of family worship within the home, for which there is evidence in every period. Multiple terracotta figurines and numerous miniatures at single sites designated as sanctuaries is a phenomenon of the 6th century BC.[77] There were four examples, but only two in the 4th century. At site 266S and possibly 506F we may be dealing with a place of community worship, or possibly a production center for votive plaques, or both combined. In view of the uncertainty about identifying these sites as sanctuaries, no separate analysis of sanctuary assemblages, like that for the Archaic sanctuaries, was attempted.

Comparison of vessel forms of 450 and 300. The calculated averages are useful as a rough guide to general trends that should be kept in mind as the ceramic assemblages for individual sites are examined. The analyses of the assemblages by functional groups reveal some changes in the kinds of activities taking place in the farmhouses, but there is overall a remarkable continuity across the centuries from **600/550** to **300**—at least as regards the quantities of objects and the general forms of objects documenting the functional groups. The detailed study of the pottery, however, makes it possible to go a step further in tracing change, or the lack of it, within each of the functional groups. Do the specific types of jugs, say, or cooking ware vessels, change over time?[78] Though a thorough treatment of this subject is beyond the scope of this book, this attempt was made here to suggest an approach and to offer some partial and provisional answers.

71 Carter 1998d, 126.

72 E.g., Carter 2006a, 139, Fattoria Fabrizio.

73 See below, Comparison of assemblages from survey and excavated farmhouses.

74 See the future study of this site in this series.

75 See the future volumes on individual farmhouses in this series. For miniatures in necropoleis, see Carter 1998f, 189. For sanctuaries, see Carter 2006a, 164, Fig. 4.36(A).

76 Elisa Lanza Catti, who is preparing this farmhouse for publication, believes that this entrance was sacred, placed under the protection of the divinity.

77 Chs. 20, 662–63, and 21, 706–09.

78 This line of investigation was suggested by the experience of studying the functional groups of burials in the Pantanello necropolis (e.g., Carter 1998f, 201–03), and observing that, though the functional groups in burials of the 5th and 4th centuries BC remained the same, very different shapes and types were employed not only in pottery but in every class of object in **300** compared to **450**.

In the analyses of the functional groups of the assemblages considered we were limited to those from twenty-four farmhouses, whose black-gloss pottery was mostly concentrated in the second half of the 4th century BC. For this analysis a sub-group of those farmhouses with an even more limited date range will be assessed (*Appendices C33* and *C34*). None of the half dozen farmhouse assemblages (sites 169F, 354F, 121F, 16F, 249F, and 40F) has any black-gloss pottery that falls outside **300**, and for the comparison group the same limitations apply. The black-gloss pottery in the six assemblages of **450**, sites 182F/T, 344F, 302F, 175F, 14F, and 304F, all lies between 475 and 425 BC.

Clearly the types of black-gloss pottery within the various forms of drinking vessels—cups, skyphoi, etc.—that are found in both periods are different, at least in details, and that applies as well to most, if not all, of the other black-gloss forms. This is to be expected since the typology and chronology of black-gloss pottery based on the study of forms is, after all, practically the sole basis for assigning a farmhouse to one period or another. Most forms of plain and banded ware are found in both **300** and earlier in **450**. Within the forms a good number of specific types also remain unchanged. The results of the comparison of these dozen assemblages appear in tables (*Appendices C33* and *C34*) which list all of the more common forms and types that appear both in **300** and **450**. The jug was in use in both periods and some types appear more often in one than the other. The same can be said of lekanai, and here the preferences are a little clearer. Based on this limited sample, the carinated lekane is more frequent in assemblages of **450**. The type 12 lekane seems more popular in **300**. The analysis could be extended but only up to a certain point, as there are a limited number of sites whose black-gloss components are so narrowly concentrated (see Ch. 9).

The comparison also makes clear which forms and types seem to be exclusive to each period. Forms and types that appear in **300**, but not **450**, include, for example, the dish,[79] an important black-gloss form that appears in four of the narrowly dated 4th-century BC farmhouses, but not before. The late introduction of the dish is confirmed by the results from the Pantanello necropolis,[80] as well as from the study of the excavated farmhouses. The distinctive minor black-gloss forms, the salt cellar and the lamp[81] (sites 354F, 121F, 249F), are innovations of **300**. Among the plain wares, the jar, which occurs both in plain ware and cooking ware,[82] apparently made its debut in **300**.

A number of components of the farmhouse assemblages from **450** do not appear in those of **300**. The graceful kylix, so frequently present at formal occasions for drinking in the Archaic and Classical Greek world, occurs in the assemblages of the 5th century BC but not afterward. The types of carinated lekane preferred in **450** were apparently no longer employed in **300**.[83]

*Necropolis assemblages, **300***. The assemblages of survey necropoleis with much of their black-gloss pottery concentrated in **300** (*Fig. 25.28*; *Appendix C32*) provide an instructive contrast with those of the farmhouses and an opportunity to trace the evolution of necropolis assemblages from the early settlement of the chora (*Table 25.6*). The functional group averages of necropoleis in **300** compare closely with those of farmhouses in the same period in four important categories: drinking, storage, cooking, and transport/storage (amphorae). Much of the pottery at burial sites, as noted earlier,[84] will have had much to do with rites at the tomb as well as grave goods.[85] The

Period	600/550	450	300
Drinking	32%	2[illegible]%	18%
Drinking service*	8%	< [illegible]%	12%
Storage short-term	26%	29%	21%
Storage long-term		< [illegible]%	7%
Food service	6%	14%	17%
Food preparation	1%	< 1%	0%
Food prep./cooking	22%	16%	16%
Transport/storage amphorae	1%	1%	4%
Service	0%	1%	2%
Instr. domesticum	0%	< 1%	< 1%
Personal	3%	< 1%	2%
Ritual/votive	< 1%	< 1%	1%

Table 25.6 Necropolis assemblages **600–300**, average functional groups. Necropoleis with fewer than 5 identifiable forms are excluded. *Elevated levels of this class may reflect the custom of using kraters and other large vessels as grave markers.

[79] D22, D23d etc, from sites 169F, 354F, 16F, 249F.

[80] Carter 1998f, 217. The dish appears for the first time in burials of the period 425–381 BC; cf. also Elliott 1998, 652–53.

[81] SA 3a, site 169F, 354F.

[82] Ja1a and Ja1b, sites 169F, 354F.

[83] CLne2b, CLne 3a, CLne 4.

[84] See Ch. 23, 763–64.

[85] For an arrangement and interpretation of the functional group according to categories established in the study of assemblages from the Pantanello necropolis, see Carter 1998f, 183–90.

rites included ritual meals on the site or offering of foods prepared elsewhere, hence the high percentage of cooking wares.[86] Amphorae were frequently used as grave markers from the 5th century onward,[87] and that is probably reflected in the higher level of transport and storage wares relative to farmhouses.

The most obvious deviation from a typical farmhouse assemblage is to be found in the category of drinking service—12% average at necropolis sites versus 2% at farmhouses. Larger vessels used for mixing, serving, or storing wine were regularly used as grave markers by the population of the chora.[88] The greater percentages of long-term storage, which may have included fragments of large coarse ware vessels, may overlap with the amphorae and be related to external markers of the burial, if not to the burial container itself, in the case of the enchytrismos burials. Food service vessels—lekanai are the most common form—are scarcer on burial sites than residential ones, and this could be an indication that more offering than feasting took place beside the grave than in the home. Cooking ware is particularly numerous in the fills over burials at the Pantanello necropolis.[89] More surprising is the lower level at necropolis sites of objects relating to personal hygiene, such as lekythoi, unguentaria, and lebetes gamikoi, which are typically prominent among the grave goods.[90] When the plow disturbs a burial site, the fill over the grave is the first material to be touched, then the burial below, which is often 1 m or more below the surface. Deep-plowing in recent years has brought the more deeply buried material to the surface more frequently (as is well-illustrated by the survey and subsequent excavation of the small necropoleis[91]). It is surprising nonetheless to see relatively little evidence of this class of grave goods in the necropolis assemblages, since the categories of drinking and drinking service—together comprising 30% of all the material collected—would also have included a high percentage of cups and containers for wine and water used as grave goods.[92]

The percentages among the functional groups of necropoleis from **600/550** to **300** show a high degree of continuity from one period to the next (*Table 25.6*), but still some trends are clearly evident. Drinking, toasts, and libations beside the tomb, as well as the deposition of cups and containers of wine and water in the grave, was relatively more important in the 6th century BC, and gradually declined thereafter. This is paralleled by a decline in **300** in short-term storage—a category that would have included table amphorae and jugs, also useful as drinking service.

The reduced percentage of drinking vessels in **450** and **300** is accompanied by a corresponding increase in food service, as if in compensation. It would be tempting to interpret this as a change in funereal rites, were it not for the fact that the increase is paralleled in farmhouse assemblages from **600/550** to **300** (*Appendix C17*). The percentage of cooking ware on necropoleis and on farmhouse sites was identical in **450** and practically so in **300**. If the identification of these site types from the 5th and 4th centuries depended alone on the average percentage of cooking ware recovered from them, they would be indistinguishable. In the 6th century, however, cooking ware on necropolis sites is only outnumbered by short-term storage and drinking vessels. In making such a statement, it is well to keep in mind that it is based on only three necropoleis, 356N, 70N, and 107N.[93] Also important is that the counts of pottery from necropoleis in all periods are low, with rarely more than 20 identifiable forms, except in the larger necropoleis of the 5th and 4th centuries. Even with the assemblages of smaller necropoleis, the high levels of drinking, drinking service (and probably drink-related storage), and of cooking wares clearly sets off necropoleis of the 6th century from those of later periods and from contemporary farmhouses.

Cooking ware. The evidence for one of the basic activities of the population of the chora, cooking, is ubiquitous at survey sites of all types, remarkably well-preserved for a surface collection, and has been carefully studied in Chapter 11. An opportunity was thus presented to carry out an analysis not only of the evolution of forms of this distinct fabric but also of the changing methods of preparing food over fire.

Cooking wares, made in a distinctive gritty fabric that withstands high temperatures, are found in virtu-

[86] Carter 1998d, 118–23.
[87] Carter 1998d, 127–29.
[88] Carter 1998d, 115–17.
[89] Carter 1998d, 118–19.
[90] Carter 1998f, 227–30.
[91] E.g., 736N and 737N in Pizzica. See Ch. 28, 1040–44.
[92] Carter 1998f, 228–29.
[93] See Ch. 21, 705. Cf. Assemblage Tables, Vol IV.

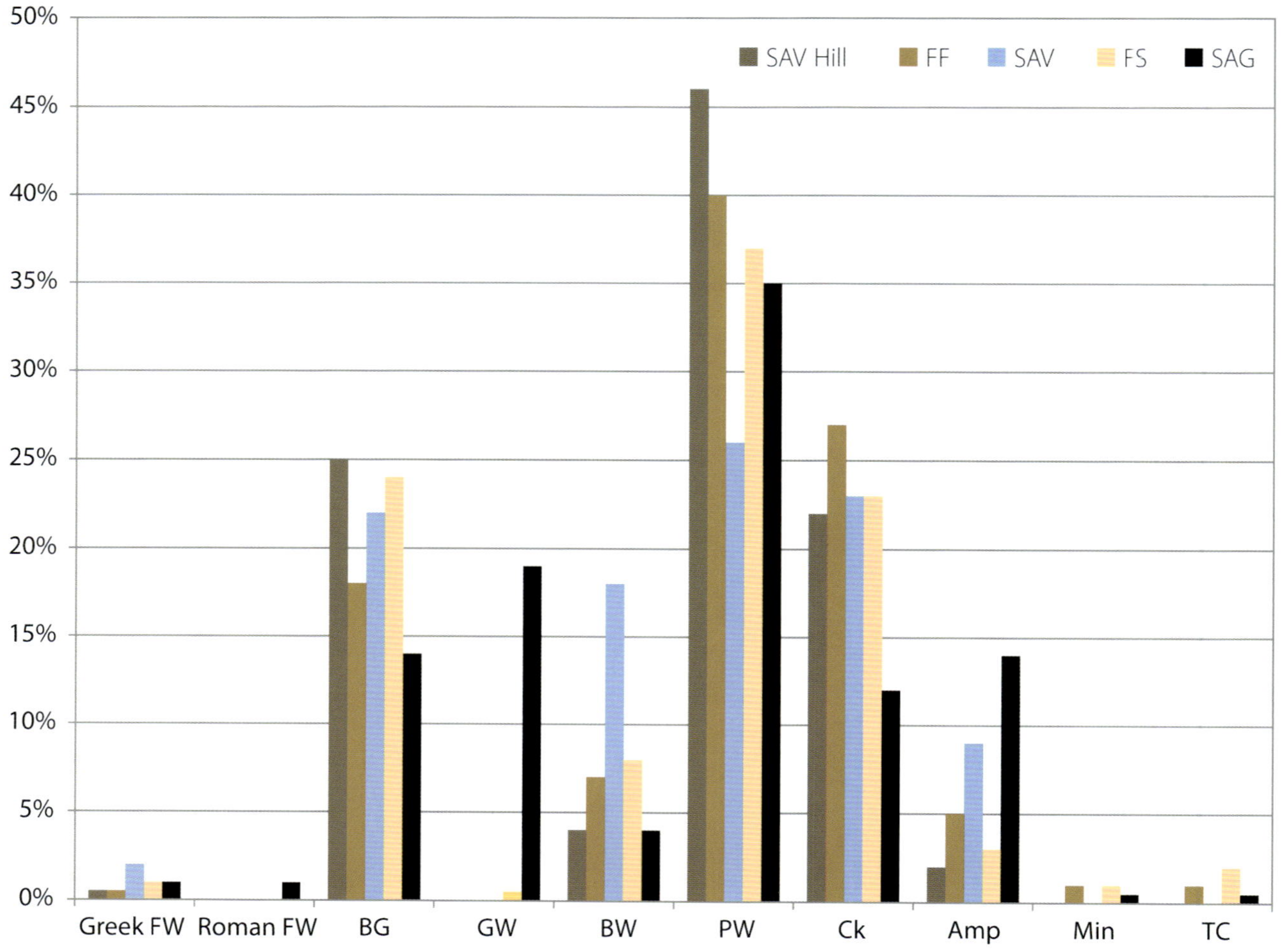

Figure 25.29 Excavated farmhouses: assemblages by ware-class proportions.

ally all farmhouse assemblages (see Ch. 11). There are only six out of 207 significant farmhouses from which not a single fragment of cooking ware was recovered. The assemblages of five of these six[94] are very small and contain only a handful of fragments of other wares in addition to enough identifiable black-gloss pottery to provide an EAW value ≥ 0.95 in at least one 50-year period. Cooking ware is present in 50 necropolis assemblages out of 74 sites with sufficient black-gloss pottery to be dateable. Cooking ware, in recognizable forms and types, is abundant at all eight sites identified as sanctuaries or possible sanctuaries.

A high proportion (26%) of the 2726 fragments of cooking ware collected from sites of all types could be identified as to form and type: sanctuaries (38% identifiable forms), farmhouses (27%), and necropoleis (11%; *Appendix C35*).

With the information about forms as a basis, an admittedly speculative attempt will be made to illustrate the methods of cooking employed in various periods, in the home, during the ritual drinking and feasting at rural sanctuaries, and in the rites performed at the tomb. The assemblages with the highest quantities of cooking ware are, of course, those of large sites that were in use over the course of centuries. In order to be able to trace a chronological development, however, this analysis must make use of sites with only one principal occupation that can be rather tightly defined chronologically on the basis of the black-gloss pottery component.

Thirty-five farmhouse assemblages were chosen for this purpose, ranging in date from the 6th to 1st centuries BC. The forms typically found on farmhouse sites include:

(a) The globular cooking pot—the Greek chytra—with either a collared rim (P4, **Nos. 26–27**), necked with a projecting rim (PJ1, **No. 28**), flared rim (PJ2,

[94] 125F, 131F, 230F, 28F, 469F, and 558F.

No. 29), or lid-seated rim (P1 and P3, **Nos. 32–37**) used to boil. A variant globular form with a broad mouth and projecting rim (P2, **Nos. 30–31**) is also represented in the assemblages, though it varies from the typical Greek chytrai.

(b) The casserole (C)—the Greek lopas—bearing lid seats, can be divided into four main variant forms: "shouldered" casseroles, with inclined rims and lid seats, which form a convex shoulder on the outer profile (CL, **Nos. 8–10**); "globular" casseroles, with sagging bases and lid-seated rims varying from inclined to vertical (CG or CC, **Nos. 11–17**); casseroles with concave walls, in which the concave outline of the walls may be out-turned or vertical, usually with a carinated lower wall (CC1, **Nos. 18–24**); and casseroles with carinated shoulders, represented by a single example with a collared rim and lid seat (CR, **No. 25**).

(c) Pans (PD), employed either to fry or bake, can be divided into four variants: flanged pans (PD1), characterized by projecting rims which vary from upturned (**No. 1**) to horizontal (**No. 2**) to down-turned (**No. 3**); pans with plain rims (PD2, **Nos. 4–5**); pans with thickened rims (PD4, **No. 6**); and pans with flared rims (PD5, **No. 7**).

Lids (Li), of which there are a great number, may be associated both with casseroles and the deeper globular forms—the plain-rimmed forms may have carried lids (mounted over the rims) like their lid-seated counterparts. In the absence of extant rim diameters it is not possible to associate the handles, the most recognizable part, with either casseroles or generally narrower globular forms.

It would be unwise to base firm conclusions about chronology on a data set that is as limited and lacunose as that contained in Appendix C35, but it does suggest some possible directions for further investigation. On the basis of the sites presented here, it would appear that pots of type PJ3 (**No. 40**) are the most common vessels in cooking ware fabrics on sites of the 6th century BC. Although this form, in a coarse fabric like those of cooking wares, may have been used for cooking liquids, it may simply have served as a storage vessel, rather than for cooking, and is equivalent to the plain and banded ware jug (J1a–e, Plain/Banded Ware Catalog, **Nos. 110–116**). The real cooking of solid foods at farmhouses in this period seems to be limited to the casseroles (lopades), and all three of the principal forms are present: the concave, the globular, and the shouldered. Boiling of food in chytrai or pots does not appear to have been done, based on the six 6th-century sites considered here and only comes in tentatively in the 5th century BC. This conclusion, however, could be questionable, as chytrai are documented at sanctuary sites of the Archaic period and the 4th century BC and they are found in at least one Late Archaic necropolis, site 524N. This site, however, like sanctuary 334S where chytrai are also found, continued to be frequented in the 5th century BC. The chytrai in sanctuaries might have had a special significance in certain ritual meals. In necropoleis they may have been the containers of enchytrismos burials, grave goods, or the remains of a funereal repast. Chytrai in miniature appear in a mid-5th–century BC burial at Pantanello.[95] A normal sized chytra was included as grave goods,[96] as was a casserole.[97] Examples of these and other cooking ware vessels were commonly found outside of the burial, where, as suggested above, they probably represent the remnants of a ritual meal at the tomb.

We may conclude safely, I believe, that the cooking options, even if not restricted to stewing in a casserole, were rather limited compared to later periods. There is no indication that pans for frying or baking were in use in the selected farmhouses before the 5th century BC, or that they were used at all at sanctuaries or necropoleis in any period. The pan, like the chytra, though present in **400** and **350**, becomes quantitatively important in **300**. In this period, boiling, stewing, and to a somewhat lesser extent, frying were generally practiced in the kitchens of Metapontine farmhouses.

The commonplace that cooking ware is very conservative is well-illustrated by this material. The same forms of pots that were in use in **600** are still in use in **300**. Variants may have been introduced late; the PJ1 for example seems not to have appeared in farmhouses in the chora before the 5th century BC. That is what the farmhouse data indicate but, as was the case with chytrai, this form appears in the same sanctuary context, 334S, which, however—in contrast to the other four Archaic sanctuaries—continued in use well into the 5th century BC. The older variants, it appears, did not become obsolete. The same is true of the casseroles, though the CG3 design arrives on the scene in the 4th century BC in the farmhouses analyzed here;

[95] Tomb 218.125. Carter and Hall 1998, 345.

[96] Inside tomb 12. Carter and Hall 1998, 256.

[97] In tomb 112. Carter and Hall 1998, 303.

it is found in some other 5th-century BC farmhouse sites,[98] all of which, however, also had a 4th-century phase. The CG2a shape that served the settlers of the 6th century BC is, nonetheless, still around two centuries or more later.

Entirely new forms, such as the cooking ware jar, are not documented in our selected farmhouse assemblages before the 4th century BC, but jars, too, turn up in some earlier contexts, in the sanctuary that was still frequented in the 5th century, 334S, and in a farmhouse that can be dated precisely to **450**. The exceptions illustrate the difficulties and uncertainties of reaching firm conclusions on questions of chronology, but also some of the exciting possibilities of using survey assemblages to trace evolving life in the ancient countryside.

Surveyed and excavated farmhouse assemblages: a comparison. The analyses by functional groups and ware class percentages of excavated farmhouses have made it possible to compare these with the assemblages of surveyed farmhouses within chronological periods—those in which the majority of dateable black gloss pottery is concentrated.

Excavated farmhouse assemblages. Five farmhouses in the study area have been excavated and at this writing are being prepared for publication. The assemblages have been studied by a team of specialists and it is possible, thanks to the coordinators of these projects, to make use of their provisional data concerning ware class percentages (*Fig. 25.29, Appendix C22*). The precise numbers will no doubt change as these studies are brought to completion, but the overall totals will not vary much.

The basic statistics for three excavated farmhouses, Fattoria Fabrizio, Sant'Angelo Vecchio, and Fattoria Stefan, that were occupied in **300** are considered in this chapter.[99] Though relatively small portions of the assemblages of Fabrizio and Sant'Angelo belong to earlier periods, they will be treated as a whole, as if of one period, **300**.

The percentage of black-gloss pottery in these assemblages is highest at the single 6th century House on Hill at Sant'Angelo Vecchio. Among the sites dateable to the late 5th to the early 3rd centuries BC, it increased gradually from the earliest, Fattoria Fabrizio at 18%, Sant'Angelo Vecchio, the 4th-century farmhouse, at 22%, to 25% at Fattoria Stefan. Sant'Angelo Grieco has 18% black gloss and another 18% grey ware.[100] Among the three with closely comparable dating, Fattoria Fabrizio, Sant'Angelo Vecchio, and Fattoria Stefan, the quantities of coarse ware (BW and PW combined) are almost equal: 47% (Fabrizio), 44% (Sant'Angelo), and 37% (Stefan). Likewise, the cooking ware quantities are very close, if not identical: 27% (Fabrizio), 23% (Sant'Angelo), and 23% (Stefan). There is more variation in the amphorae: 5% (Fabrizio), 9% (Sant'Angelo), and 3% (Stefan). There is a greater appearance of homogeneity among the excavated farmhouses than among the surveyed examples, perhaps because there is in most cases at least ten times the amount of pottery involved in the analysis. The profile of the excavated farmhouses shows a remarkable degree of sameness compared with the surveyed farmhouses, and as will be apparent, some very strong divergences from surveyed ones.

Comparison (Fig 25.30; Appendices C25 and C36). The greatest disparity between assemblages from surveyed farmhouses and the excavated examples occurs with the black-gloss pottery and the cooking ware. The black gloss from survey sites of **300** averages 39% in the total assemblage, as opposed to 22% from these excavated sites—nearly twice as frequent. Cooking ware from the survey in this period is only 11% of the total as compared to 25% for the excavated sites—less than half of the amount. The coarse wares (BW and PW combined) are, somewhat surprisingly, present in almost precisely the same proportions in both surveyed and excavated farmhouse assemblages. Amphorae are not numerous in either context, but three times more so in excavated than in survey assemblages.

There is a fairly obvious rationalization for this. Black gloss is the most highly visible and also the most prized material from the surface, as it offers the best indication of the date of a site. A special effort is usually made not to miss even the tiniest fragment. Cooking ware, on the other hand, is often very fragmentary and is the color of the earth so it is more likely to be overlooked. The disparity in the results from survey

[98] Sites 416F, 193F, and 468F.

[99] The studies of these farmhouses are being led by Elisa Lanza Catti (Fabrizio), Francesca Silvestrelli (Sant'Angelo Vecchio and Stefan), and Erminia Lapadula (Sant'Angelo Grieco). Thanks to all for their generosity in sharing preliminary data.

[100] The atypical assemblage from Sant'Angelo Grieco is discussed separately. See Ch. 27, 905–08.

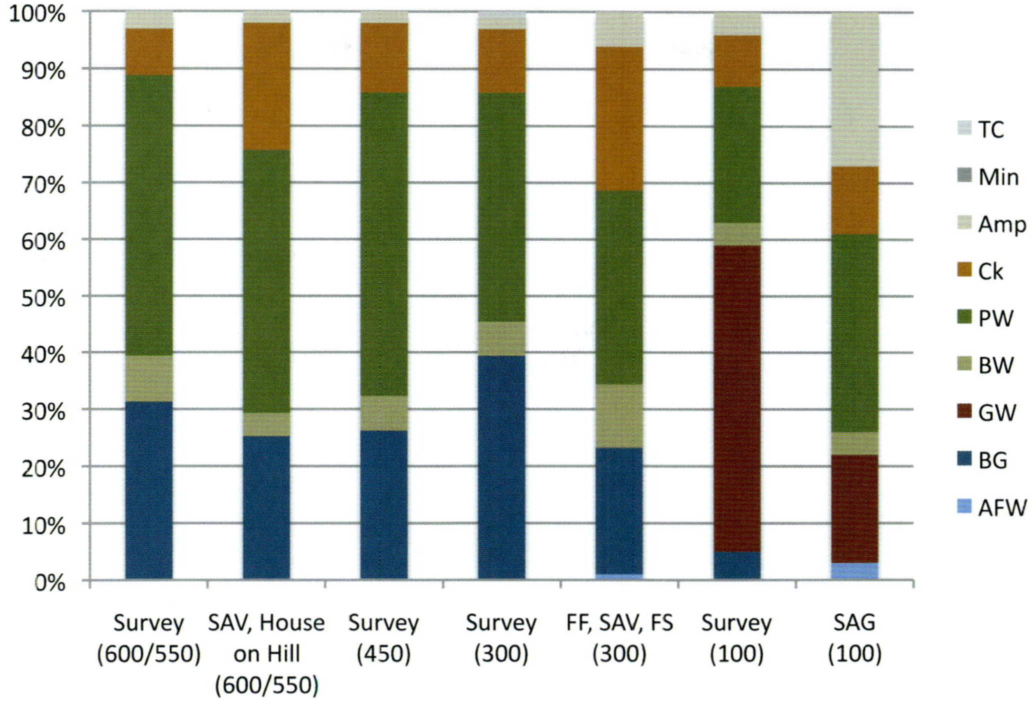

Figure 25.30 Comparison of surveyed and excavated farmhouse assemblages, ware class percentage averages **600–100**.

and excavation would certainly seem in part, at least, to be a by-product of the survey practices. The best that can be said is that the survey practices employed by the ICA team were consistent, and thus the results of many survey seasons can be readily compared. At the same time, there can be no doubt that excavation gives a fuller picture of the actual activities taking place at the site. This comparison provides a useful corrective when weighing the relative importance of materials and activities at survey sites.

The Asty in 350 and 300

The Urban Center: Agora and Sanctuaries

The three-century-old urban center of Metaponto underwent an extraordinary renaissance in the second half of the 4th century BC. As in the chora, so in the city—the population reached its zenith. Most areas of the residential districts were completely occupied, to judge from the remarkably successful and revealing results of the recent geomagnetic prospections and earlier soundings in several areas.[101] This is the period to which a majority of the city's network of narrow blocks, defined by narrow streets, the stenopoi, has been dated (see *Fig. 25.31* and *Map 11 (inset)*, Vol. IV). It is also when some earlier stenopoi were widened and became plateiai and when new ones, like Plateia II, came into being.[102] This was the period in which the city walls were in their most imposing state, and the system of drainage was functioning at maximum efficiency. The great cloacae at either end of the sanctuary, the capstones of this capillary system, were solidly constructed of masonry.[103] Elements of earlier monumental structures, including the temples, were reused. This was motivated by economic considerations, but it may also reflect a shift in religious beliefs and practices, and perhaps in the ethnic composition of the population. The kerameikos provided a wide range of ceramics. Among the fine wares, besides continuing to turn out mostly smaller scale red-figure vessels, Metapontine potters introduced a distinctive overpainted polychrome pottery, comparable to the Gnathia-style ware produced at Taras and other centers. By the time of Pyrrhos, late in the first quarter of the 3rd century BC, pottery production in the city had all but ceased.[104] The mint in the second half of the 4th century BC was busy minting silver coins to pay for mercenary soldiers and the generals

[101] Mertens 2006, 159, and Fig. 177. Cf. Osanna 2008, 929–30, Pl. 24, Figs. 46–7.

[102] Nava 1999, 701.

[103] Ch. 24, 798–99.

[104] Silvestrelli forthcoming a.

Figure 25.31 Model of the main sanctuary and western end of the agora, showing Metaponto as it would have appeared in the late second half of the 4th c. BC. (After Mertens 1985, Fig. 6b; courtesy Dieter Mertens)

called from Greece to command them. This is the major period for the bronze coinage of Metaponto,[105] with its wide range of small denominations, a feature that distinguishes the mint of Metaponto from that of all the other poleis of Magna Grecia.

The outstanding demonstration of this renewed energy and creativity—after a period of lassitude if not of decadence—was in the very heart of the city itself. In the Archaic and Classical Metaponto of the 6th and first half of the 5th centuries BC there had been two foci of civic activity and pride: the sanctuary, with the temples of the principal divinities, which was enlarged, embellished, and repaired in the 5th century; and the agora, with its smaller cult places and its great center for public meetings and communal celebrations, the Ekklesiasterion II.[106] In the second half of the 4th century BC, repairs to the temples of the sanctuary had become, in contrast to those of **450** which modernized their decoration,[107] perfunctory and very basic.[108] The western temenos wall was restructured, but nothing significant was added. Almost all of the attention of the citizens was focused on the area of the agora.

Theater. The building that arose on the site of the abandoned and partially despoiled Ekklesiasterion II sent a clear message of renewal and change, and at the same time of continuity and change—of continuity because of the type of building and its location, of change, because of its extraordinarily innovative form and the central position that it enjoyed in the wider building program. That program included a great stoa partially framing the vast space of the agora of the east, that if not in itself innovative, was, at the very least, extremely precocious and on a grand scale. As Dieter Mertens, who with Antonio De Siena excavated it and the previous structures on the site, has written, everything about the theater building set it off from previously known Greek theaters (*Fig. 25.32*).[109] Like its predecessor on this flat site, it was built on an

[105] Ch. 24, 803.

[106] Ch. 23, Ekklesiasterion II.

[107] Ch. 23, 774–75.

[108] A. Bottini 1988a, 307–12.

[109] Mertens and De Siena 1982, 34–40.

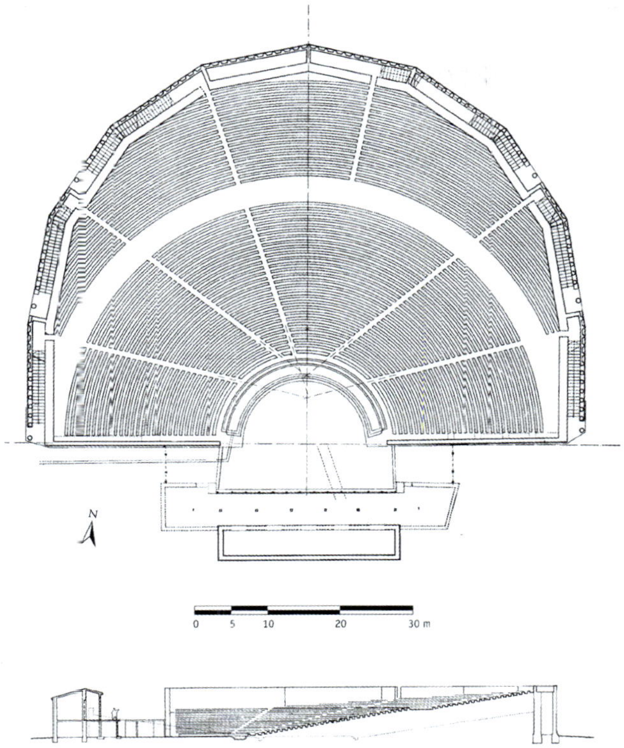

Figure 25.32 Plan of theater of Metaponto. (After De Siena and Mertens 1982, Fig. 16)

artificial earthen mound, the remnants of the earlier two amphitheaters. This unique solution, as Mertens observes, pointed forward to the permanent Roman stone theaters of the 1st century BC, which may have been directly inspired by these Greek precedents. The orchestra and koilon were a perfect semicircle, unlike other Greek theaters, and the stage building, orchestra, and auditorium were conceived as a unified whole as in the Roman theater. But the most daring invention of all was the carefully planned access to the auditorium (*Fig. 25.33*).

The earlier two buildings on that site had been surrounded by a blank circular retaining wall,[110] high enough to block any view of what went on within. It effectively cut the building off from its surroundings. It could only be entered along the narrow dromoi that pierced it along a single axis bisecting the rectangular orchestra. The whole arrangement conveyed a sense of exclusivity. The exterior appearance of the theater was deliberately planned as a contrast to the smoothly continuous and monotonous exterior of the two Ekklesiasteria.[111] The retaining wall for the earthen mound that supported the seats was shielded from view by an exterior curtain wall 5.5 m high (hence almost twice as high as the wall of Ekklesiasterion II) and separated from the koilon by 1.6 m. Its face was a polyhedron composed of ten facets, each measuring 15 m in length (*Fig. 25.34*).[112] Behind six of them, shielded from view from outside, were stairways that led from the exterior ground level to the top tier of seats in the koilon.[113] This was an entirely new and more efficient solution to the problem of entering and exiting a Greek theater, and one that was ultimately followed by the Romans.

The exterior wall was not left plain and flat, but articulated vertically in four zones.[114] A flat wall with crowning molding 2.7 m high (or half the total

Figure 25.33 Reconstructed internal stairway of the theater of Metaponto, leading to the modern observation deck. (JCC; courtesy of the Soprintendenza)

[110] Chs. 21 and 23, Figs. 21.33 and 23.28.

[111] Mertens 1999, 282–83.

[112] Mertens and De Siena 1982, Pl. 2 (actual state plan).

[113] Mertens and De Siena 1982, 16 (reconstructed plan and section).

[114] Mertens and De Siena 1982, Fig. 15.

Figure 25.34 Reconstructed rear wall of the *koilon* of the theater at Metaponto, with a second-story screen of engaged half-columns against the recessed wall and against pillars at the internal stairwells. The style of the facade was meant, perhaps, to harmonize with the stoas found on the east and south sides of the agora. (JCC; courtesy of the Soprintendenza)

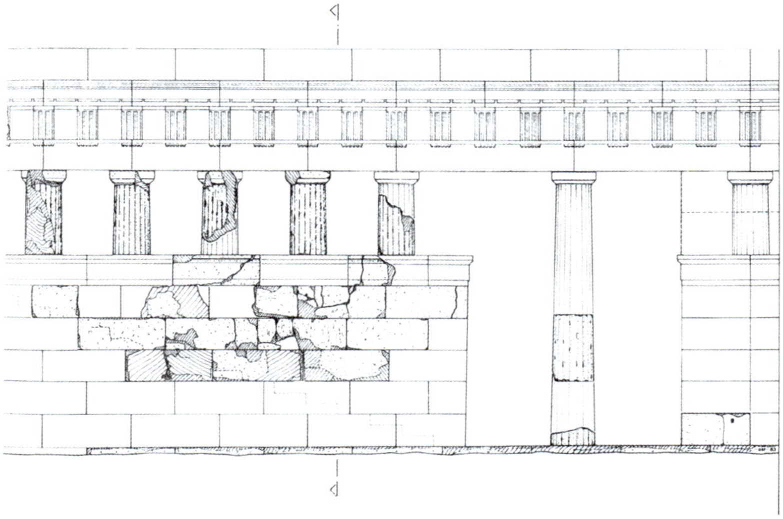

Figure 25.35 Theater exterior detail. (After De Siena and Mertens 1982, Fig. 15)

height) was surmounted by the top third of engaged Doric columns whose centers are 1.25 m apart. These in turn appear to support a Doric triglyph and metope frieze and cornice (*Fig. 25.35*). The fourth and uppermost element is a single course of blocks 40 cm high that forms a sort of attic. The wall is pierced by a 3-m-wide entrance, in the middle of which was a complete, freestanding Doric column with the same dimensions and proportions as the engaged column of which only the upper third was exposed to view. It is 3.8 m high and the spacing between it and the nearest engaged column on either side is twice that between engaged columns. The arrangement is both attractive and inviting. As Mertens has interpreted it, the design is a deliberate attempt to connect the building with its surroundings, to be inclusive rather than forbidding and remote, like its predecessors. The theater was part of a larger plan for the monumentalization and embellishment of the city, and this building would clearly have resonated with the other elements such as the stoai.[115]

A Tarantine model? How much of the innovation and inventiveness demonstrated by the construction of the theater was local and how much owed to the inspiration, and possibly to the design, of the theater of Dionysos in Taras, the foremost center for the performing arts in Magna Grecia? We shall never know. All that remains of Taras' two theaters are brief mentions in ancient texts.[116] We learn that the name for the orchestra in the larger of the two, the theater of Dionysos, was the "dromos." The entrances to the orchestra of the Ekklesiasterion at Metaponto have been described as dromoi. The theater of Dionysos, during the celebration of the Dionysia, was the scene where the Tarentine ekklesia received the Roman ambassadors in 282 BC, after the violation by the Roman fleet of the treaty of 302 BC, prohibiting it from sailing past the Lacinian Cape at Croton and the subsequent Tarentine attack on it. This historically important episode, which resulted ultimately in a Roman declaration of war and the Tarentine appeal to Pyrrhos illustrates well the other function of the theater, in Magna Grecia as in Athens and mainland Greece, as a place of public assembly.[117] The legacy to the Romans of the theater of Taras *in sensu lato* was great. Less than forty years after their victory over Pyrrhos and the Tarentines in 275 BC, Livius Andronicus, a freedman from Taras, was composing Latin versions of comedy and tragedy of the Greek type.

115 Mertens 1999, 282–83.

116 Wuilleumier 1939, 248–49.

117 Wuilleumier 1939, 103–04.

Figure 25.36 Terracotta plaque of Zeus from Sant'Angelo Vecchio, h 22.3 cm. (CW)

A statue of Zeus? The discovery of an extremely large statue base (measuring 5.5 x 7.25 m) in the temenos of Zeus, beside the theater at Metaponto, has suggested to its excavator, Antonio De Siena, the possibility that a colossal statue may have been intended for this spot.[118] De Siena adds that the statue may have resembled the Zeus brandishing a thunderbolt that the Metapontines erected in Olympia in the early 5th century BC.[119] The subject appears on a vase decorated by the Pisticci Painter, in the late 5th century BC.[120] Another possibility, a standing Zeus with scepter and with a pillar beside him on which perches an eagle, is suggested by a small terracotta plaque from the votive deposit at Sant'Angelo Vecchio (*Fig. 25.36*).[121]

The reference to the colossal statue of Zeus brandishing the thunderbolt that Lysippos created for the agora of Taras would have been obvious. Both the date and the appearance of the Tarentine colossus—the second largest in the Greek world after the Rhodian—have been much discussed. The most probable date of Lysippos' work, according to Wuilleumier, would be the period of the campaigns of Alexander the Molossian in Magna Grecia in the late 330s BC.[122] These are discussed in the following section on Metaponto and Magna Grecia in the 4th and early 3rd centuries BC. In these years the head of Zeus and the fulmen symbol appear both on the coins of Taras[123] and Metaponto.[124] The success of Alexander's campaigns against the barbarians must have restored, if not for long, the confidence of the Greek poleis. The Metapontine building program, coming after a period of prolonged inactivity if not decline, is, I believe, best situated in the period of Alexander and the years immediately following. It would have been financed by profits from the greatly increased productivity of the chora in **350**. The pressure from the Lucanians and their Samnite allies would have lessened in this period, as the Samnites were fully occupied with the Romans in the Second Samnite War (327–321, 316–304 BC). The coinage bearing the image of Alexander's protector, Zeus, and the statue in the agora, if one actually existed, would have expressed the Metapontine gratitude for his leadership.

Stoa and agora. The theater was the jewel, but the program envisioned a reshaping of the civic center of Metaponto on a large scale. Along the eastern edge of the agora arose a two-storey stoa of impressive proportions (*Fig. 25.37*).[125] The building was divided in at least two sections, between which the great NW-SE Plateia A passed on its way eastward to the port. The south section extended to the residential block that formed the south limit of the agora. This section measured 56.40 m. The continuation of the stoa to the north of the passageway for the Plateia A, a building of similar design, has only been excavated for a short distance. The passageway was given architectural emphasis by placing a column between the antae on either side of it. The projecting antae constricted its width to little more than 2 m. A similar solution was adopted for the southern end of the stoa, which

118 De Siena 2001b, 37–38; 1998, 153–54. Unfortunately, the base cannot be accurately dated.

119 Ch. 23, 778.

120 Carter and Hall 1998, 434, Tomb 12, 1.

121 Edlund 1979, Cat. no. 33 SA-79-509T.

122 Wuilleumier 1939, 281–82.

123 Wuilleumier 1930, 383–84, Pl. 26, Figs. 30–31.

124 Johnston 1990, 1, Pl. 2, 6.2–6.5, and 52, on the plausibility of the association of the fulmen symbol on the coins with Alexander the Molossian.

125 Mertens 1999, 283–84; 2001b, Fig. 24 (plan of the agora), Fig. 41 (plan of excavations).

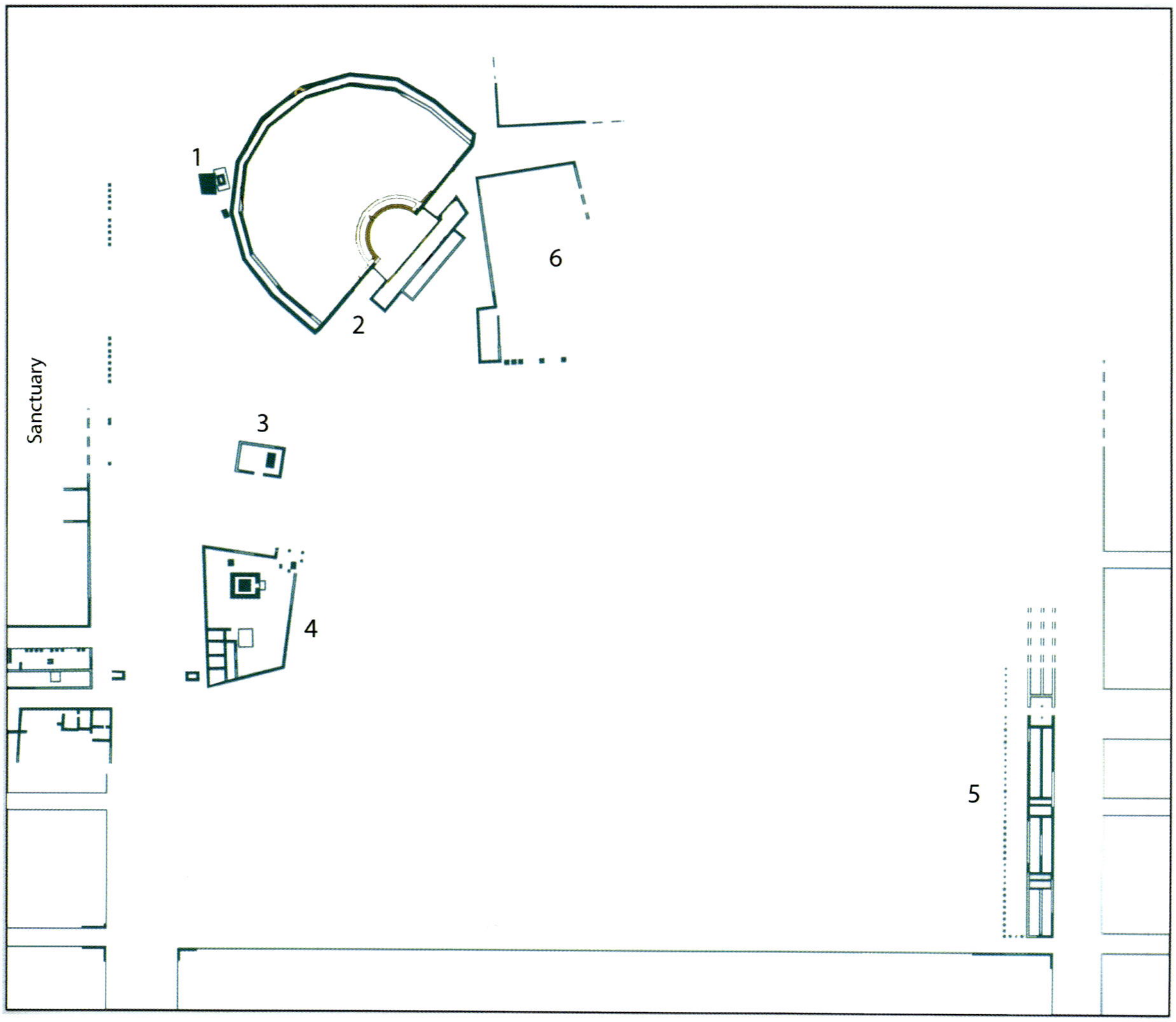

Figure 25.37 The agora in **300**: (1) Temenos of Zeus, (2) theater, (3) sacellum?, (4) manteion, (5) stoa, and (6) area sacra. (After De Siena 2001b, Fig. 24)

reached the stenopos that defines the agora's southern extent. The passageway here was nearly 5 m wide, which accommodated practically the whole width of the stenopos. The column was located near the center of the portico that fronted on the agora. The back of the building was occupied by shops separated by stairs that led to the upper floor. Mertens supposes that the stoa had Doric columns on the lower floor (partially preserved in situ), and Ionic columns along the upper storey. This arrangement of course became a standard feature of Hellenistic stoai (e.g., the Stoa of Attalos in the Athenian Agora), but this is certainly a very early manifestation. The columns of the upper storey were inserted into a parapet (as, for example, in the upper storey of the stoai surrounding the Temple of Athena at Pergamon).

The use of engaged columns, which becomes much more common in the Hellenistic period, already had a precedent in Metaponto two centuries earlier in the Temple of Apollo.[126] The engaged columns of the upper storey of the stoa are echoed in the screen around the theater, but those in the theater are quite unusual, especially in this period and in this part of the world. The closest contemporary parallels are to be found in the Macedonian tombs—in that of Philip II, but even closer is the Tomb of Leucadia.[127] There the wall of the lowest section of the exterior wall is divided at two-thirds the height of the Doric columns by a

[126] Ch. 21, 721, Fig. 21.28.

[127] Petsas 1966, Pl. A.

horizontal element. The engaged columns, however, are shown in full against a representation of an ashlar wall. In fact, the tomb façade is just that: it bears little if any architectural relation to the space behind it. In the theater screen, while the half-columns and triglyph-metope frieze play no structural role, the one complete column in front of every entrance stairway emphatically does. The columns between antae in the stoa are analogous. The sophistication of both these buildings puts them in the forefront of architectural developments in the Early Hellenistic period. Contacts with the Greek East, especially Macedon and Epiros, in this period are obvious in architecture and art. They extended perhaps to other areas more difficult to trace, such as livestock and forage crops.[128]

It is conceivable that stoai had at least been planned to extend along the whole of the eastern side of the agora. The agora measured 263 m from west to east,[129] and nearly as much from north to south. The total area was approximately 7 ha, which dwarfed the agora of Priene, planned in the mid-4th century BC, and was considerably larger than the south wing of the agora of Miletos, whose architectural growth had hardly begun at this time. The stoa was a regular feature of sanctuaries in much of the Greek world already in the Archaic period. In the civic sphere, in the agora, it becomes a dominant form only in the Hellenistic age. Although there are no Archaic or Early Classical stoai at Metaponto, the agora of Metaponto in this regard, also, appears to have been in the forefront of this development. There were good early precedents in the western Greek world. The agora of Megara Hyblaia[130] appears with framing stoai already in the 6th century BC—three centuries before the agora of Priene.

The agora of the Metapontines was framed on the east by a stoa, on the north by the city wall, on the south by a stenopos and block of the residential quarter, and on the west by the sanctuary with its monumental temples, which, though in need of repairs, at least formed a stately backdrop for the new civic center. We have spoken of the two principal features of this great creative effort. While in active use from the second half of the 4th century BC onwards, neither building was ever finally finished.[131]

At this time, too, other structures in the agora were refurbished. One was the Manteion and shrine of Aristeas, a spot no doubt venerated by the followers of Pythagoras, who, as we are reminded by a very interesting inscription, were active in the city in the late 4th and early 3rd centuries BC. The lead *defixio* that curses the workshop (*ergasterion*) of a Pythagorean doctor and his assistants was found in a monumental Hellenistic chamber tomb in the Crucina necropolis.[132]

Recent excavations have revealed better the area sacra to the east of the theater—essentially a sand dune surrounded by conglomerate blocks of the 6th century BC. This unusual holy spot, among the earliest shrines of the polis, was given a new architectural frame (measuring 58 x 140 m), access ramps constructed of ashlar blocks, and a new orientation that brought it more into line with the street grid.[133] The divinity worshipped there, De Siena suggests,[134] was Athena, for whom no securely identifiable cult place elsewhere in the asty had previously been identified. What went on in or around the dune is not known.

Between the area sacra and the theater a number of terracotta antefixes representing Dionysos and his throng (Pan, satyrs, and silens) was discovered, and it has been suggested that they decorated a "light" portico intended to emphasize the access to the scene building of the theater.[135]

The sanctuary and Temple E. The large space between the newly revitalized agora and the venerable, underutilized temples of the sanctuary was partially filled at this time by a number of new places of worship, including Temple E. Modest in appearance, these new cult places overlay even more modest monuments of earlier eras. The new ones reflect religious devotion that focused on the concerns of the Metapontine population of the second half of the 4th century and the early decades of the 3rd century BC. A votive deposit (*stipe votiva*) near Temple E contained a very large number of terracotta votive plaques of two types.[136] In one, a silen has one arm around an amphora and the other around a nymph who holds a cornucopia. Such plaques are extremely

[128] E.g., Carter 1998h; 2006a, 24 and n. 30.

[129] De Siena 1998, 164.

[130] Mertens 2006, 67–69, Fig. 80.

[131] A. Bottini 1988a, 305–12.

[132] Lo Porto 1980; Lombardo 1995.

[133] Osanna 2008, 930–31.

[134] De Siena 2008, 593–96.

[135] Tempesta 2005, 89.

[136] Postrioti 1996, 19–25.

common in the chora, at sanctuaries like Pantanello, where hundreds of fragmentary examples have been recovered,[137] in the votive deposits at Sant'Angelo Vecchio, at Cozzo Presepe,[138] and farmhouses and small rural sanctuaries throughout the chora (*Fig. 25.38*).[139] In this cult place in the very heart of the city Postrioti sees a new community of interest between city and countryside, and a new democratic spirit—shared by the inhabitants of the city and by the inhabitants of the countryside, some of whom, if our interpretation of the settlement pattern is correct, had just recently arrived there.[140]

Figure 25.38 *Left*, a fragment of a terracotta plaque from the dispersed votive deposit by the spring at Pantanello, representing a maenad or a nymph surrounded by grape clusters; h 10 cm (After Carter 1980b, 14, Fig. 11). *Right*, a more complete version of a Dionysiac subject similar to that represented on the left with a silenos carrying a krater in his left hand, and embracing a nymph with his right. This restored plaque comes from the excavation of the votive deposit at Sant'Angelo Vecchio. H 30 cm. (CW)

The diminutive Temple E (13.45 x 7.90 m), really an oikos like the early Oikos C, was perfectly on axis with its altar and both were aligned with the early-5th-century BC Temple D and its altar. De Siena has dated Temple E later than the altar and the plaques, on stratigraphic grounds. Both the temple and altar were part of a larger complex. The buildings to the north, also younger than the altar, were aligned with Temple E and may have served the cult of the divinity whose statue was presumably housed there.[141]

A pause in building activity has been postulated by De Siena,[142] after the major activity in the agora and the lesser activity in the sanctuary (repairs to temples and temenos wall, the altars and votive deposit associated with Temple E). To the later period, besides the construction of Temple E, may be assigned two further monuments: the tiny Oikos H between the Temples of Hera and Apollo, and the stoa along the south side of the sanctuary. Oikos H relates directly to the worship of Hera that in the earlier periods would have taken place in her Temple A (see *Fig. 27.16*). The rise in the water table and instability of the old temple may have made it unusable, and the observance for Hera would have been carried out at little Oikos H, which was elevated 0.5 m above the level of the Temple of Hera. The stoa faced Plateia A rather than the sanctuary and appears to have consisted of a series of rooms with wells to provide drinking water to the public. Of the stoa, little remains besides the foundation and the terracotta antefixes that decorated it.[143] They represent Io, the priestess of Hera, and an ancestress of the Achaians.[144] The structures in the southwest area of the sanctuary reveal a continuity with the past that serves as a reminder that perhaps not all elements of the population had converted, at least not entirely, to the relatively recently introduced and hugely popular cult of Dionysos.

Metaponto and Magna Grecia in the 4th and Early 3rd Centuries BC

The Major Trends

This was a period of momentous change for all of Magna Grecia. Metaponto, as all Greek poleis, found itself part of a far greater game with Macedon's rise

137 See the forthcoming volume on this sanctuary in this series.

138 Carter 1994a, 196–98; 2006a, 168–69, Fig. 4.41.

139 The other represents an enthroned female divinity flanked by sphinxes who has been identified plausibly with Persephone-Kore (Postrioti 1996, 145–47, and n. 149). The spheres of the two divinities—the fertility of the fields and the underworld—overlaps, and in fact so does their worship. This is reflected, Postrioti claims, in the plaques that represent a bearded reclining Dionysos-Hades and Persephone-Kore together with their offspring. This type, too, is very commonly found throughout the chora and in the asty (Postrioti 1996, 151–52. See Signore 1996 for an exhaustive treatment of the type).

140 If Postrioti's interpretation of the leggings worn by the silenoi is correct (Postrioti 1996, 151–52), the plaques allude to Marsyas, an historic and potent symbol of popular resistance to oppression.

141 Postrioti 1996, 14, 148–49, n. 170; see Inset, *Map 11*, Vol. IV.

142 Postrioti 1996, 148 n. 170.

143 De Siena 1993, 119.

144 Tempesta 2005, 84. See also Ch. 23.

to world power,[145] and with the rapid expansion of Rome toward the south of Italy that began in this period. In no period more than this one is the lack of documentary evidence about Metaponto more acutely felt. To attempt to understand the history of Metaponto and its chora, we must consider what is known about the political, social, and economic situation in general as reflected in more fully documented cases.

External threats, internal discord. The conflicts between individual poleis had receded in importance in the face of a common enemy, Dionysios of Syracuse, as he advanced along the southern coast of Italy subjugating the poleis in his path. After his death in 367 BC, the Greeks faced a rapidly growing threat from the Lucanians, Bruttians, Samnites, and, in the end, the Romans. How did these events impinge on Metaponto? The city was never a player in power politics, never had hegemonic aspirations, like Taras and the so-called condottieri that she called to the common defense. Caught between these and the Lucanians, Metaponto sought, as we shall see, to maintain a degree of independence till the end.

The struggle between parties within the Greek poleis of Magna Grecia and throughout the Greek world—already clearly evident in the uprising against the Pythagoreans of a century and a half earlier at Croton, Sybaris, and probably Metaponto[146]—came to dominate every aspect of their internal affairs. At Metaponto, as we have argued, the crisis provoked a redistribution of property in the first half of the 5th century. The conflict between the wealthy landowners and the demos, the rich and the poor, as Lombardo has argued, became more vitally important in the 4th century than the conflict between ethnic groups.[147] The second redistribution of land in the Central Plateau and probably elsewhere in the Metapontine chora can be seen as a response to stasis. The weakness of the agriculturally rich poleis was both military and political. The solution adopted for defense from external threat, a greatly increased reliance on foreign generals and mercenary soldiers, brought its own problems, since their aims only partially coincided with those of the poleis. Why did the poleis fail to fill the ranks of their previously all-citizen armies? One reason was an apparent shortage of citizen farmers, a phenomenon sometimes referred to as *oligandria*. This problem was not restricted to Magna Grecia. It was true of Sicily in this period and much of mainland Greece. The solution adopted by the Sicilian poleis under Timoleon and the Greeks of the mainland, in some cases of dire necessity, was to co-opt new citizens even from the non-Greek population,[148] as is documented by texts and supported by the results of archaeological research, especially the work of Adamesteanu at Gela and in southeastern Sicily.[149] Metaponto was a case in point in southern Italy, as I have argued before.[150]

Lucanians and Greeks. A major fact of social history in southern Italy in the late 5th and 4th centuries BC was the acculturation of indigenous peoples—a process not just of transmission but also of creation.[151] For the Ionian coast this process began early, with the arrival of the first Greeks, and is particularly evident in Archaic indigenous sanctuaries like Garaguso.[152] This is a trend about which archaeological evidence has been particularly informative and well-focused. The physical evidence in the 4th century BC is impressive. It includes the fortified village settlements like Pomarico Vecchio[153] inland from Metaponto and Serra di Vaglio near Potenza. They reveal an understanding of the more recent advance in siege warfare and building technologies. In this they probably received a stimulus from the presence of Alexander the Molossian (see below). An inscription in the Greek alphabet on one of the ashlar blocks at Serra di Vaglio identifies the magistrate, Nummelos, a Lucanian who authorized the work at Serra di Vaglio, and possibly, as Adamesteanu suggested,[154] also at Torre di Satriano. These and others are now securely dated to the second half of the 4th century BC. They are the centers on which depended the sanctuaries and farmhouses of the surrounding countryside.

Lucanian sanctuaries were located at strategic points in the open countryside, such as that at Chiaromonte and the more monumental sanctuary at Armento Serra Lustrante.[155] They were developed

145 Purcell 1994, 392.
146 Ch. 22, 738–41.
147 Lombardo 1987, 70.
148 Asheri 1966, 30ff, 90ff, e.g., the case cited by Aristotle (*Pol.* 1303a6), where Argos made citizens of perioikoi.
149 Adamesteanu 1958a, 35ff.
150 Carter 2006a, 223–24.
151 Purcell 1994, 396.
152 Morel 1970b.
153 *Pomarico Vecchio* I.
154 Adamesteanu 1974a, 190–97.
155 Barra Bagnasco and Russo Tagliente 1996, 186–93.

on terraces and the central feature was invariably a spring, the crucial point of contact between the upper and lower worlds. The divinity and the worshipper were predominantly female, as indicated by the very numerous terracotta votive figurines. Beside the terraces,[156] the architectural feature was a small cult building, an oikos, as at Armento. These sanctuaries were the scenes of ritual meals, as evidenced by the faunal remains and the abundance of cooking wares on the site. In this respect and in others, the sanctuaries were similar to those Greek sanctuaries[157] also found in the chorai of the Greek poleis. The Pantanello sanctuary at Metaponto,[158] and the rural sanctuary at San Nicola di Alba in the chora of Paestum on the western side,[159] are possible parallels.

The terracotta figurines representing females, made at kilns beside the sanctuaries, and the pottery have close connections with the production of Greek poleis. Small cups of shapes familiar in the kerameikos of Metaponto, the concave-convex cups and the one-handlers, predominate, but there are also large shapes, Gnathia kraters and louteria, that were employed in rituals.[160] Greek pottery, whether imported or imitated locally, is ubiquitous throughout the sanctuaries, fortified centers, and isolated farmhouses and their related necropoleis. The forms of fine wares, utilitarian wares and cooking wares, are in use everywhere, and everywhere they are same. The well-published sites of Roccagloriosa,[161] Forentum,[162] and Pomarico Vecchio[163] provide abundant comparisons.

Among the more precious offerings at these sanctuaries of the interior are the numerous coins. Many were minted in the Greek poleis on the coast, but also from as far away as Epiros, Carthage, and Rome, attesting to the importance of these cult places beyond the region and over several centuries.[164] The sanctuary of Mefitis at Rossano di Vaglio was the foremost example. The progressive acceptance and diffusion of the coins of the Greek poleis intensifies from the middle of the 4th century BC into the 3rd, and ends at the end of the 3rd century BC.[165] In the later 4th century BC it becomes more and more difficult to distinguish indigenous and Greek artifacts and the rituals in which they played a role.

After a century, more or less, the votives in sanctuaries cease. There are a few exceptions, like the central Lucanian or "federal" sanctuary at Rossano,[166] which continued to attract worshippers and expensive votives into the 1st century BC. After the mid-3rd century BC they lose their economic and political functions and cease to be a place of congregation for the population.[167] This is equally true of the sanctuaries in the chora of Metaponto, which lose their sacred character altogether and are replaced in the 3rd century BC by purely utilitarian structures, as at San Biagio and Pantanello (see Ch. 26).[168]

The face of the countryside in the interior of that part of southern Italy that began to be known as Leukania or Lucania was transformed by the appearance of numerous isolated farmhouses and their related small necropoleis from the middle of the 4th century BC and well into the 3rd. They are known mainly through the results of survey, in particular that along the valley of the Sinni,[169] for example, site 849 on the slopes of Monte Alpi at about 1,000 m above sea level near Latronico. This was a farmhouse, with its necropolis, site 850, located 300 m away. The recorded scatter of 700 m^2 at site 850 compares with those of Metapontine rural necropoleis (see Gazetteer in Vol. IV). The farmhouse at Tampone il Cupone, with a surface scatter of 1,000 m^2, is comparable to that of a small- to medium-sized Metapontine farmhouse. Apart from the greater altitude and mountainous backdrop, a visitor from the chora of Metaponto would have felt quite at home in such a setting, though a Lucanian farmer leaving the fertile terraces along the river valleys of the interior might more easily have adjusted to the relatively level marine terraces and the proximity of the sea. Indeed, Lucanians began to visit and settle in the chora of Metaponto as early as the 5th

156 Barra Bagnasco and Russo Tagliente 1996, 194–95, for plan and reconstruction.

157 Pontrandolfo (1996, 179) sees Samnite and Campanian parallels, but ultimately Greek models for terraced sanctuaries.

158 Carter 1994a.

159 Cipriani 1989.

160 Barra Bagnasco and Russo Tagliente 1996, 187.

161 Fracchia 1993; *Roccagloriosa* I and II.

162 *Forentum* I.

163 *Pomarico Vecchio* I.

164 Parente 2009, 46–48.

165 Parente 2009, 45.

166 Adamesteanu and Dilthey 1992.

167 Barra Bagnasco and Russo Tagliente 1996, 185.

168 De Siena 2005, 442–46. Carter 1994a, 196–98. The farmhouse belongs to the mid-3rd c.. The votive plaques were probably not associated. See the future volume on the Pantanello sanctuary.

169 Quilici and Quilici Gigli 2003, 121–25 and passim. Other surveys include those by the Universities of Turin (Barberis 1999) and Milan (Castoldi 2003).

century BC.[170] As Barra Bagnasco has remarked, the farmhouses represent a completely new organization of space in the interior,[171] and one that projects a vision of work and life shared by Greeks and Lucanians alike.

Besides the surveys, a growing number of Lucanian farmhouses are now known through excavation.[172] The farmhouses at Moltone near Tolve,[173] at Mancamasone near Banzi,[174] and at San Giovanni near Viggiano[175] had modest first phases in the second half of the 4th century BC. The original Tolve farmhouse covering an area of 390 m^2—very large in comparison to Metapontine farmhouses of this period[176]—consisted of three rooms opening onto a courtyard to the south. On the east a long room with a hearth was the center of domestic activities. Essentially the same plan was adopted for the first phase at Mancamasone. All three were greatly enlarged in the course of the 3rd century BC. By the middle of that century all three farmhouses had expanded significantly. The dwelling at Tolve, with the addition of a mosaic panel and central courtyard, changes character, and becomes more like the luxury residences familiar at Roccagloriosa, Pompeii, and the Greek east in this period.

As the sanctuaries of the interior went out of use, the farmhouses associated with them in the late 4th and early 3rd centuries BC underwent a transformation that reflected the enormous economic changes that came in the next centuries with the Romans—from small-scale farming to an agriculture based on intensive exploitation of large land holdings by a servile labor force.

The Historical Sources

Metaponto's disposition, as Johnston has remarked,[177] to mind its own business, has no doubt been partly responsible for the silence of ancient historical sources on Metaponto, even at the height of her prosperity in the Archaic period. Apart from a role in the conquest of Siris, the polis' avoidance of involvements in the power struggles that interested ancient historians was a constant feature, no less in the 4th and early 3rd centuries BC than in her earlier history. In fact, there are exactly two mentions of Metaponto in our documentary sources in the century under consideration. Both, not surprisingly, are in connection with the foreign generals, grouped together under the misleading term condottieri, who arrived in southern Italy on invitation from the Tarentines ostensibly to defend the Greek poleis from the barbarians—at first principally the Lucanians, and eventually the Romans.

The principal foreign protagonists came in three related pairs. The first pair, the Sicilian—Dionysios I and Agathocles—had minimal impact on Metaponto. The Spartans—Archidamos, the first and perhaps the least effective or damaging, and Kleonymos—were a disaster. Finally the Epirotes—Alexander the Molossian, the uncle of Alexander the Great, and his nephew Pyrrhos—were the last of the foreigners. Pyrrhos' defeat by the Romans spelled the end of independence for the Greek poleis, including Metaponto. The city is mentioned in connection with the Molossian and Kleonymos. This section will consider the possible interactions they had and their consequences. It does not aim to be exhaustive, but to focus on certain aspects that seem most relevant to developments at Metaponto, in the city, and the chora.

Alexander the Molossian. Alexander, as is well known and documented, had bigger dreams than simply defending Taras and her allied poleis, the remaining members of the Italiote League, from the aggressions of the Lucanians and Bruttians. He dreamed of an empire in the West that would rival that of Alexander of Macedon in the East. He arrived in 334 BC, several years after the ineffectual campaigns of Archidamos, and at first was successful in Apulia, then with the Lucanians to the west. He advanced as far as Paestum (the former Poseidonia, which had fallen ca. 400 BC, see Ch. 24), which he liberated from the Lucanians at least temporarily. This seems to have had a great effect on the Greek poleis, instilling new hope and optimism. At some point, probably to secure his supply line as he moved west, Alexander made a treaty of friendship with the Metapontines, Poediculi (Peucetians), and the Romans:

[170] See discussion above, Ch. 24, 806–07.

[171] Barra Bagnasco 1999a, 56.

[172] Russo Tagliente 1992.

[173] Tocco 1991; A. Russo 1993b.

[174] A. Russo 1993a.

[175] Nava 2003a, 686–89.

[176] A. Russo (1993b) connects the appearance of these farmhouses with a more specialized agriculture based on the grape and olive. An exception to the size issue may include the farmhouse at site 829 near the Cavone. A magnetometer survey was conducted on the site by Dan Davis of ICA in 2003 and he estimates that it occupied an area of 977 m^2; see *Ancient Territories 2003*, 77–80, esp. 79.

[177] Johnston 1990, 61.

Igitur cum in Italiam venisset, primum illi bellum cum Apulis fuit, quorum cognito urbis fato brevi post tempore pacem et amicitiam cum rege eorum fecit ... Gessit et cum Bruttiis Lucanisque bellum multasque urbes cepit; cum Metapontinis et Poediculis et Romanis foedus amicitiamque fecit.

On his arrival in Italy, his first contest was with the Apulians; but when he learned the destiny appointed to their city, he soon concluded a peace and alliance with their king ... He engaged also in war with the Bruttians and Lucanians, and captured several cities; and he formed treaties and alliances with the Metapontines, Pediculans, and Romans.[178]

Justin 12.2.5–6, 12

He established a base at Thourioi, which a century earlier, at the time of the Athenian adventure,[179] had found common ground with the Metapontines, in opposition to Taras and her colony of Herakleia. Alexander with these moves distanced himself from Taras, and she from him as she perceived and surely resented his greater ambitions. This, as several historians have argued,[180] never resulted in open conflict. The special nature of the relationship between Metaponto, Thourioi, and Alexander was manifested by the fact that Metaponto, according to one account, and Thourioi according another, made an extraordinary effort to recover the body of Alexander after his death at the hands of treacherous Lucanian exiles at the river Acheron.[181]

The effect that Alexander's presence and successes had on Metaponto was, I would suggest, far greater than simple loyalty to a fallen ally would indicate. The meager sources suggest that the Metapontines may have viewed him as a sort of second founder. The transformation of the civic center of the asty with the creation of the hugely impressive and innovative theater and stoa and the revival of the chora give the impression of a true renaissance and coincide with the period of Alexander's appearance in Italy and the years immediately afterward. This is not to say that he can take direct credit for these achievements. His main contribution would have been to have instilled some confidence in the Metapontines.[182]

Several modern historians have noted a sudden revival of interest in the mid-4th century BC in the early stories of the founding of the city.[183] The account of Strabo[184] presents multiple versions of the founding—one by the indigenous Metabos and records that there was an heroon to Metabos in the city.[185] Leukippos, another reputed founder, as Strabo reports, was sent for the purpose of creating a *synoikismos*, and succeeded because he was able to fool the Tarentines. He appears on the coins, the large distaters, that Metaponto struck to support Alexander's campaigns.[186] This high denomination was unique at Metaponto, and it may not be a coincidence that Thourioi also minted distaters at this time.

Who the founder (*oikistes*) was and which was the founding city became crucial questions in an apoikia when the citizen lists were modified, as they invariably were when there was a redistribution of land, as Asheri and Malkin have pointed out.[187] The question of who founded Metaponto could have arisen again in the mid-4th century because the citizen lists were probably being revised. I would suggest that the second major redistribution of land in the chora in the second half of the 4th century BC was the occasion that gave rise to this sudden interest in founders, and that the diverse founders mentioned by Strabo may reflect, in some way, the heterogeneity of the new settlers in this period. The new occupants of farmhouses in Lago del Lupo in **350** might have been citizens who had previously dwelt in the city; they may have been the urban poor and dispossessed for whom a more democratized government at Metaponto may have made provision, or they may have come from elsewhere, including the interior of Lucania.

We know almost nothing about the constitution of the Metapontines in any period, so the field for speculation is wide open. If they followed the example of the Tarentines, who in the 4th century BC modified their constitution under the influence of Archytas to become more democratic, then the first scenario, that

[178] Text and trans. J.S. Watson 1853 (Loeb).

[179] See Ch. 24, 805.

[180] For example, Giannelli 1969, 11–12.

[181] This moving story is recounted at length in Livy 8.24.

[182] The connection would be stronger and explicit if indeed a statue of Zeus, Alexander's patron, actually stood beside the theater (see above).

[183] Lepore 1974a, 322; Musti 1988, 147.

[184] Strabo 6.1.15.

[185] Mele 1998, 70.

[186] Johnston 1990, 57.

[187] Asheri 1966, 42–43; Malkin 1987.

the poor occupied the redivided land, is likely. This they could have done while maintaining a cautious independence from their powerful neighbor to the north. According to Aristotle, the Tarentines reached a delicate equilibrium between the wealthy (εὐπόροι) and the poor (πλῆθος) by offering land and an opportunity to govern to those with little property:

> καλῶς δ' ἔχει μιμεῖσθαι καὶ τὰ Ταραντίνων· ἐκεῖνοι γὰρ κοινὰ ποιοῦντες τὰ κτήματα τοῖς ἀπόροις ἐπὶ τὴν χρῆσιν εὔνουν παρασκευάζουσι τὸ πλῆθος· ἔτι δὲ τὰς ἀρχὰς πάσας ἐποίησαν διττάς, τὰς μὲν αἱρετὰς τὰς δὲ κληρωτάς, τὰς μὲν κληρωτὰς ὅπως ὁ δῆμος αὐτῶν μετέχῃ, τὰς δ' αἱρετὰς ἵνα πολιτεύωνται βέλτιον. ἔστι δὲ τοῦτο ποιῆσαι καὶ τῆς αὐτῆς ἀρχῆς μερίζοντας τοὺς μὲν κληρωτοὺς τοὺς δ' αἱρετούς.
>
> It is also a good plan to imitate the policy of the Tarentines. They get the goodwill of the multitude by making property communal for the purpose of use by the needy; also they have divided the whole number of their magistracies into two classes, one elected by vote and the other filled by lot,—the latter to ensure that the people may have a share in them, and the former to improve the conduct of public affairs. And it is also possible to effect this by dividing the holders of the same magistracy into two groups, one appointed by lot and the other by vote.[188]
>
> Aristotle, *Pol.* 1320b 9–16

The second scenario, that they were non-Metapontine immigrants, is perhaps the more likely in light of the general conditions of Metaponto in the first half of the 4th century BC. The economic decline of the polis in this period has been generally assumed (Ch. 24), and the concentration of land in the hands of fewer is a possibility. It is a reasonable deduction if the amount of available farmland is always the same but the number of actual occupants is smaller. Both scenarios should be reassessed in the light of the evidence from the chora. The construction of the walls in the 4th century BC was surely a response to the need for defense against real or perceived threats from outside.

Defenders, however, were more essential than walls, and to increase their numbers there were several possible options: hire mercenaries, who could be expensive and troublesome, or co-opt new citizens, who by receiving land would be obliged to defend it. A convenient source for new citizens would have been by now highly acculturated Lucanian populations in the hinterlands of the polis. The shock of transition would have been minimal, though probably not entirely negligible,[189] for both new and old citizens. The organization and economic systems of the polis were by now well known to the Lucanians, and the occupants of the chora and the asty had over the last century or so numerous encounters and probable intermarriages with the indigenous populations, or at least with the Lucanian elites,[190] a number of whom, including women, were known to be Pythagoreans.[191] Could the rediscovery of the indigenous founder Metabos have been an attempt to legitimize the new citizens?

Support for the case that some at least of the new occupants of the chora were Hellenized Lucanians comes from the interpretation proposed for the evolution of burial customs in the Pantanello necropolis in the 4th century BC.[192] The abandonment of many family burial plots in the first half of the 4th century and the reuse of only some of these by apparently unrelated people and the creation ex novo of a very large multifamily nucleus 6 indicates, as clearly as archaeological evidence alone can, the arrival of at least one new group of people in the chora at this time.[193]

The co-option into the citizen body of Greeks and probably non-Greeks was not unprecedented for the Greeks in the West. The rapid expansion of Sybaris in the 7th century was attributed to the fertility of the land and because they "kept granting citizenship to many aliens" according to Diodorus Siculus.[194] There is no reason to believe that the Greeks of later times did not follow suit when the need arose. The successful revival of Sicily in the second half of the 4th century BC, after the destruction wrought during the long

[188] Text and trans. Rackham 1932 (Loeb).

[189] Asheri (1966, 25–27) discussed the difficult and protracted process of accepting new citizens.

[190] For indigenous elements in the population of Herakleia, A. Bottini (1987b, 280) cites the non-Greek names in the late-4th-c. BC Herakleia bronze tablets.

[191] Mele 1981.

[192] See above, Excavated Necropoleis—Pantanello, 842–43.

[193] See above, Necropoleis, 841, and Carter 1998f, 218–36.

[194] Diodorus 12.9.2.

drawn out conflict between Greeks and Carthaginians, was well-known throughout the Greek world before Alexander the Molossian arrived in southern Italy. The leadership in the first case was provided by a foreigner, Timoleon, whose career in Sicily began in 344 and ended with his blindness and death in the mid- to late 330s BC. He enlarged the citizen body, brought in new settlers, and divided the land. Some of the new citizens came from Greece, and others could very well have been Hellenized Sicels. This was not explicitly stated by Plutarch, Diodorus, or their sources, perhaps because enfranchising non-citizens had become the most simple and natural way in mainland Greece and presumably in the Greek West as well.[195] The renaissance of Sicily under Timoleon was not limited to the cities of the coast. The archaeological investigations of Adamesteanu[196] and Orlandini have documented the farmhouses and tombs of the interior and of small centers like Agyron, Diodorus' home town, whose renewed sense of civic pride and solidarity was expressed by no less than the construction of a new theater. Plutarch records that the revived cities of the coast, Gela, Akragas, and Kamarina, referred to Timoleon as the new oikistes. Though perhaps motivated by different goals, both Alexander and Timoleon adopted the slogan "eleutheria." The example of Timoleon in any case cannot have been lost on Alexander. The co-option of the Hellenized Lucanians into the citizen body of Metaponto, however, was probably not announced policy of Alexander's or something that he would have publicly supported. After all, he had been called by Taras ostensibly to fight Lucanians.

Kleonymos. The period between the death of Alexander the Molossian and the arrival of Kleonymos of Sparta was one of relative peace and prosperity for the Metapontines, because both the Lucanians and the Romans, embroiled in the Second Samnite War, were otherwise occupied. This is the most probable period for the construction of the theater of Metaponto, the monumentalizing of its agora with the stoa, and general revival. It was a period of changing alliances. The Lucanians and the Tarentines, and the Lucanians and Rome, may have been allied with the Metapontines.[197] One thing is clear: Rome becomes a constant presence in southern Italy at this time, having established a stable base at Luceria in 315 BC, to be followed by colonies at Venusia in 291 BC and Paestum in 273 BC (see map, *Fig. 26.18*). If the Tarentines had been wary of the Romans before, with Luceria they must have concluded that Rome would become the next big threat. With the Roman victory in the Samnite War in 304 BC, Taras, afraid of having to face the Romans in addition to the Lucanians, called the Spartan general and royal, Kleonymos. Like his fellow countryman and predecessor, Archidamos, Kleonymos was, as Giannelli remarked, a general without an army and a prince without a throne. He was also anxious like all of the condottieri to build his glory and political base, as much as possible with other peoples' money.[198]

Kleonymos arrived punctually in 303 BC with a force sizeable enough to impress the Tarentines and the Lucanians. The Lucanians made peace and friendship with the Tarentines;[199] Metaponto did not, perhaps because she felt betrayed by this alliance. The punishment of the Metapontines by Kleonymos is recalled at length by Diodorus.[200] This is the second mention of Metaponto in the 4th century BC and the last until the arrival of Hannibal. The episode can only be understood, as Giannelli has argued, in the light of the tensions between the property owners and the demos.

> ...τῶν δὲ Μεταποντίνων οὐ προσεχόντων αὐτῷ τοὺς Λευκανοὺς ἔπεισεν ἐμβαλεῖν εἰς τὴν χώραν καὶ τῷ καιρῷ συνεπιθέμενος κατεπλήξατο τοὺς Μεταποντίνους. παρελθὼν δ' εἰς τὴν πόλιν ὡς φίλος ἐπράξατο μὲν ἀργυρίου τάλαντα πλείω τῶν ἑξακοσίων, διακοσίας δὲ παρθένους τὰς ἐπιφανεστάτας

[195] In the case of Larissa in northern Greece in 220 BC, it is stated that Philip V conferred citizenship on metics, foreigners, and freed slaves, to promote an increased cultivation of the land; a distribution or redistribution of land in this case is implied. The rights of previously established citizens, as Asheri illustrates, were preserved by assigning to the new citizens the previously undistributed or less desirable lands. Asheri 1966, 29–31, with citations of this and other examples of *politographia*, or the enrolling of new citizens.

[196] Especially Adamesteanu 1958a. Cf. Vallet 1968, 107.

[197] The very confused record of alliance is intelligently discussed by Giannelli (1974, 354–55), but the true situation remains obscure. Cf. Musti 1988, 146.

[198] Giannelli 1974, 369.

[199] Giannelli 1974, 361–62, thinks that the Lucanians had a treaty with the Romans in 303 BC, which they abandoned when Kleonymos arrived. The defection of the Lucanians further caused Rome to make the treaty with Taras, the one that was broken when the Roman fleet sailed beyond the Lacinian cape into Tarentine waters in 282 BC.

[200] Diodorus 20.104.3–4.

ἔλαβεν εἰς ὁμηρίαν, οὐχ οὕτω τῆς περὶ τὴν πίστιν ἀσφαλείας χάριν, ὡς τῆς ἰδίας ἕνεκεν λαγνείας. ἀποθέμενος γὰρ τὴν Λακωνικὴν ἐσθῆτα διετέλει τρυφῶν καὶ τοὺς πιστεύσαντας αὐτῷ καταδουλούμενος·

...and when the people of Metapontum did not come over to him, he persuaded the Lucanians to invade the territory of the Metapontines and, by making a simultaneous attack himself, intimidated them. Then, entering their city as a friend, he exacted more than six hundred talents of silver; and he took two hundred maidens of the best families as hostages, not so much as a guarantee of the city's faith as to satisfy his own lust. Indeed, having discarded the Spartan garb, he lived in continued luxury and made slaves of those who had trusted in him;[201]

Diodorus Siculus, 20.104.3–4

Kleonymos' enemy, as it turned out, was not the Lucanians whom he was called to fight, but an independent minded Greek polis at least nominally an ally of Taras, which he was supposed to defend. The passage is open to a variety of interpretations.[202] Giannelli doubts that Kleonymos would have had to persuade the Lucanians to "fall upon" the chora. What if the Lucanians had been on good or at least non-hostile terms at the time? The manuscript tradition of the text of Diodorus includes the reading συνεμβαλεῖν, to fall upon together, in a late version, though most codices have simply εμβαλλεῖν. In the generally accepted reading, Kleonymos could have stood back and pretended to be neutral, while the Lucanians attacked the chora. The text says that he only presented himself as a friend when he came to the city. This sounds redundant because he turned on the Metapontines together (συνεπιθεμένος) with the Lucanians and terrified them. A combined attack on the chora would presumably have caused the greatest harm to the largest property owners, a prosperous Metapontine elite, who may have actually been on better terms with the Lucanian elite than with the democrats within the city. We can imagine that within the city, which would have had a higher percentage of artisans and the propertyless than the chora, he would have been more likely to be received as a friend than in the chora. The burden of the tribute that he demanded would have fallen on the property owners who also lost their daughters.

This unorthodox tactic of threatening of an ally convinced the Tarentines that they had erred, even more than in the case of Alexander, in their choice of Kleonymos as a defender. He, convinced that he had accomplished his mission when he had brought the Metapontines into line, turned his attention elsewhere (to Sicily, Corcyra and the Po valley, where he lost an army), much, no doubt, to the relief of the Tarentines and the other poleis of southern Italy.[203]

The spirit of independence manifested by the refusal to join either Taras or Kleonymos was evident also in Metaponto's coinage. In the period of the condottieri Metaponto continued to use her own system in contrast to Taras, which adopted the Attic system for her gold coinage to be on the same standard as Philip II, to facilitate payment to the generals and their mercenaries.[204] That spirit, if not quelled, was perhaps redirected after the experience of Kleonymos. There are some indications, again in the numismatic record, of a possible rapprochement with Taras.[205] The wealth of Metaponto before Kleonymos must have been very great, as the sum he demanded, **600** talents—15 tons!—of silver was exorbitant by any reckoning.[206] The sum she owed Kleonymos would certainly have limited Metaponto's ability to go it alone.

Pyrrhos. There is no word about how Metaponto passed the two decades between the withdrawal of Kleonymos and the arrival of Pyrrhos in southern Italy—the last two decades of Metaponto's independent existence, except for short-lived episodes, including the Second Punic War and the Social War. In 298 BC the Lucanians called on the Romans for an alliance, excusing themselves for a previous breach of faith. This was presumably their withdrawal from the treaty with Rome, just before the arrival of Kleonymos in 304 or 303 BC. Rome went along, perhaps to prevent the Lucanians from joining the Samnites in the third Samnite War, 298–290 BC. Thanks to the war, at least temporarily, the pressure from the Samnites and Ro-

[201] Diodorus Siculus 20.104.3-4. Geer 1954.

[202] Giannotta 1980, 13.

[203] Meloni 1950b, 115–21; Giannelli 1974, 367–69.

[204] Stazio 1974a, 90–91. See also Meloni 1950b, 114.

[205] Johnston 1990, 45.

[206] Johnston 1990, 3.

mans, if not the Lucanians, was off the Greek poleis. This would have been another opportunity for building, had the Metapontines enough resources to continue the work on the theater and stoa. Both projects had been suspended before completion (see above). It seems likely, however, that Temple E and the stoa on the west side of the sanctuary date from this period.

When in 282 BC the Lucanians were again menacing Thourioi and the poleis of the Ionian coast, Thourioi, perhaps in view of past unfriendly relations, preferred Rome to Taras. Other Greek poleis, among which Lokri, Croton, and Hipponion followed suit. Certainly their internal political situation, as always, played a critical role. Rome's sympathies for the propertied classes were by now well known. The stage was set for the final struggle. In 281 Taras organized a coalition of all Greek cities that did not already have Roman garrisons. Among her allies were a certain number of barbarians, including, probably, the Samnites, Lucanians and Bruttians. Thourioi and Herakleia are mentioned by name; Metaponto was absent from the list, a fact which has led to further speculation.[207] Was she being feistily independent again, or just too exhausted and depleted to be able to muster a force? A Tarentine expedition, with the support of the demos, ousted the Roman garrison from Thourioi.[208]

Provoked by the Tarentines' attack on its fleet and disrespectful treatment of its embassy, Rome declared war on Taras, and Taras after some understandable hesitations, called on Pyrrhos as the general less occupied and most able to lead the allies, according to Plutarch.[209] Like his uncle, Alexander the Molossian, Pyrrhos had dreams of empire, in Sicily, perhaps North Africa, when he answered the call. The story of Pyrrhos need not be retold here, except where it is relevant to Metaponto. Pyrrhos must have passed Metaponto on his way to the battle of Herakleia. The Roman commander, Aemilius Barbula, who previously ravaged the chora of Taras, almost certainly did transit the chora. A brief passage of Frontinus[210] has him leaving the coast of Lucania to avoid the missiles of the Tarentines, launched from arrow throwers (*scorpiones*), mounted on ships. His route necessarily was up the Bradano through the chora of Metaponto, to winter camp in 281/280 BC at the newly founded Venusia on the border between Apulia and Lucania (see *Fig. 26.18*).

The few precious pieces of information for Metaponto's stance toward Pyrrhos come from its coinage. Metaponto, alone of the Greek poleis, did not issue coins on the reduced standard used by Taras and Pyrrhos to pay the troops. Those reductions took place earlier in Pyrrhos' Italian campaign. So it would appear that Metaponto, in contrast to its treatment of Alexander, did not contribute monetarily to the Pyrrhos-led effort. If Metaponto, after Kleonymos, were governed by the democratic faction, it should have favored Pyrrhos. One is left to surmise either that Metaponto's finances were in a very bad way, or that the sitting government was sympathetic to Roman interests, or at least opposed to intervention. After Pyrrhos' narrow but nonetheless huge victory at Herakleia in 280 BC, Tarentine civilization underwent a very short-lived revival. If Metaponto had sided with Pyrrhos, the effect on the city might have been similar. Tarentine engravers reused Epiriote emblems, the laurel-crowned head of Zeus and the fulmen on the city's coinage, and added Athena Promachos and the elephants of Pyrrhos. At Metaponto, Pyrrhos' ephemeral victories did not produce a similar, observable effect, nothing like the one the city felt with its participation in the success of the Molossian a half-century earlier. In any case, there was hardly time, if there had been the will, to celebrate, as the Roman domination of southern Italy was secure by 275 BC, when Pyrrhos returned to Greece. It is likely, as we shall see, that a fortress was constructed in the eastern part of the agora (Ch. 26) and a Roman garrison imposed almost immediately.[211]

[207] Lévêque 1957, 304. Wuilleumier 1939, 107. In fact, the text, Florus 1.13.1–2, cited by both Lévêque and Wuilleumier, mentions no Greek city at all.

[208] Wuilleumier 1939, 103.

[209] Plutarch *Pyrrhos*, 13.

[210] *Strategemata* 1.4.1.

[211] De Siena 1993. See Ch. 26, 886.

26
A Century of Decline

Joseph Coleman Carter

The Chora in 250 and 200

The year 275 BC marked a turning point in the history of Magna Grecia. After inconclusive encounters with Roman armies in that year, Pyrrhos withdrew from Italy. Following his death in 272 BC, Taras, the only force in southern Italy capable of offering resistance while retaining some independence, was compelled to give up Herakleia and its hegemony over the Greeks and indigenous peoples, and later to furnish ships to the Romans for the First Punic War (262–241 BC). Taras' allies, presumably including Metaponto, though she is not mentioned by name, were treated in a similar fashion.[1] Metaponto and Thourioi both received Roman garrisons during the Second Punic War (218–201 BC), according to Livy,[2] but we cannot be sure that one was not already in place at an earlier date in the 3rd century BC. That is what the excavations of the "castrum," an intrusive and hastily constructed earthen fortress with moat in the heart of the city, would indicate.[3]

275 BC is also the beginning of the eighth 50-year period into which the 600-year history of the chora has been divided. The strong downward inflection in the population curve of the chora that occurred in the first half of the 3rd century BC surely reflects a political and military situation in southern Italy, though the decline had begun earlier and accelerated after this crucial historical date.

Early Village Clusters, 250

Of the four regions, the Early Village Clusters, the most geographically diverse, were the least negatively affected by the crisis of the mid-3rd century BC (*Figs. 26.1–26.2*).[4] Measured in terms of loss of significant farmhouses, the Early Village Clusters, which lost 55%, managed better than the other three, though not by much. The Central Plateau was down by 62%, localities Far from the City by 61%, and those Close to the City by 78%. The average loss for the study area as a whole was 61%. The average FMCE value for the Early Village Clusters, which in **250** was slightly below the average for the study area as a whole, suffered the second largest drop—after the region with the highest in **300**, the Central Plateau (*Appendix C7*).

Giardinetto (*Fig. 26.3, Appendix C3*)**.** The number of significant farmhouses dropped from 14 to 5 (-64%), the necropoleis from 4 to 1 (-75%). The numbers of farmhouses in **250** was just one less than the total in **550** and **450**.

The eastern half of the cluster, which was the more densely occupied in the 4th century BC, lost 70% of its population, while the western section declined by only 50%. As a result of the thinning, the average distance between sites in this, the densest of the Early Village Clusters, increased from 156 to 325 m. In the cases of two sites that were much closer, 525F and 521F/T/K, the presence of water drew them together. Ceramic production centers like 521F/T/K seem to have a higher survival rate than farmhouses without this component, as is evident also at Sant'Angelo.

All of the farmhouses that survived to **250** were very significant in **300**; none that were not of above average significance in **300** made it to **250**. By **200**, all that remained of the cluster were two second magnitude sites, 465F and 521F/T/K, and no necropoleis. This, perhaps the most fertile and favored location in the study area, was almost totally abandoned in **200** until Post-Medieval times.

Sant'Angelo (*Fig. 26.4, Appendix C3*)**.** Of seven significant farmhouses in **300**, two remained in **250**. There were no active necropoleis in Sant'Angelo after **400**. Both surviving sites, 105F/K and 213F/K, were primarily involved in the production of ceramics, like 521F/T/K in Giardinetto.[5] Ceramic production,

[1] Wuilleumier 1939, 133–41; Lévêque 1957, 546.

[2] Livy 22.15.

[3] The question of its date has been much debated. See De Siena 1990b, with relevant bibliography. See also De Siena 1993.

[4] For the period **250**, see *Map 12* in Vol. IV.

[5] The venerable tradition of this activity along the Basento valley, and in this area in particular, has frequently been noted, most recently in Ch. 25, 813.

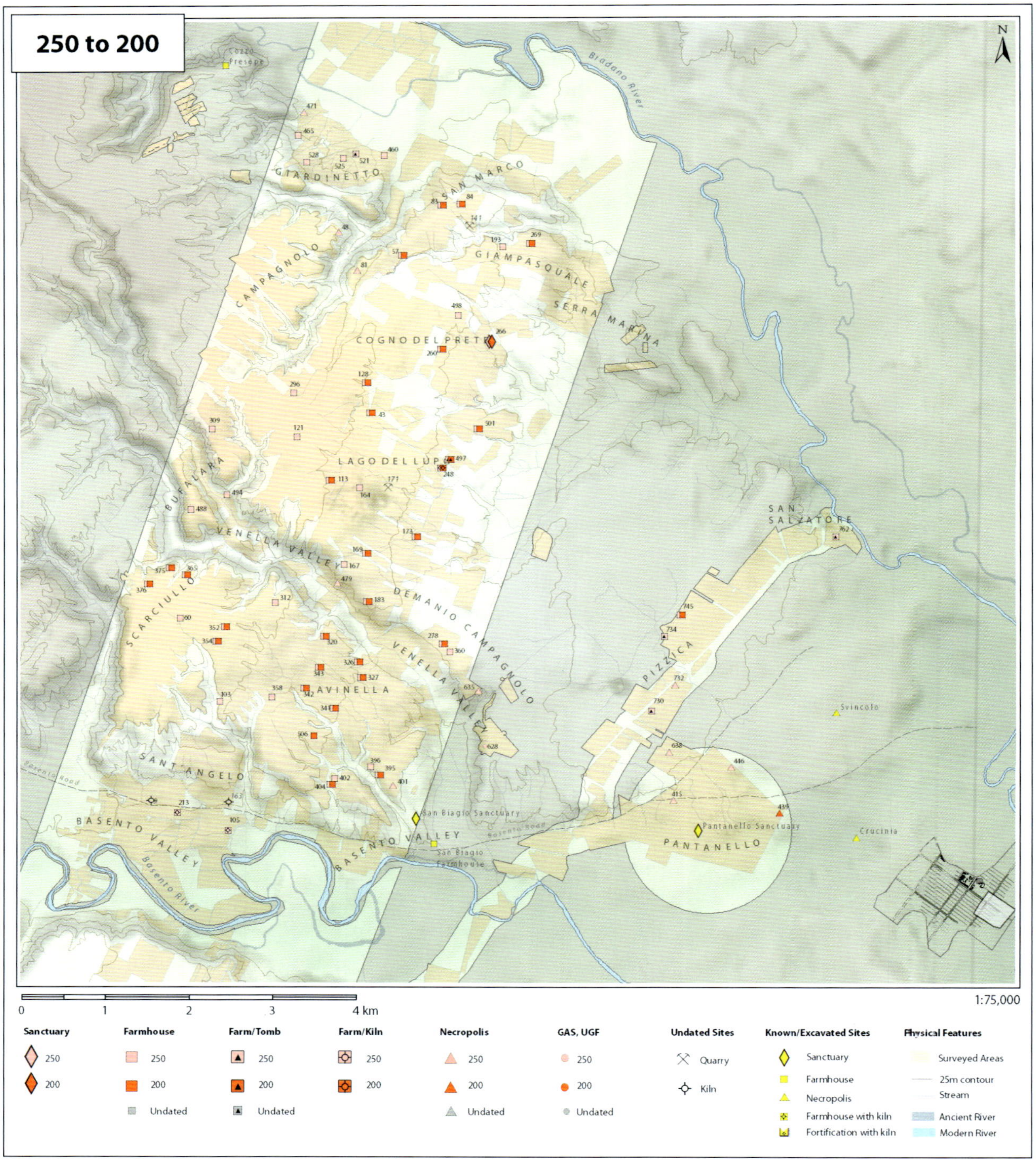

Figure 26.1 The study area in **250** and **200**. (JT)

probably including amphora and possibly roof tiles, was valued in this period. If intended for local consumption, this would indicate that agriculture, albeit on a reduced scale, probably was still a major activity in this area of the chora.

The surviving structures in Sant'Angelo continued to be very significant, more so than all but a handful of sites in the study area in this period. 213F/K was one of four sites to have an FMCE value greater than 2.60, when the average for the chora was 1.99. By **200**, both sites had become second magnitude, and the cluster at Sant'Angelo effectively no longer existed.[6]

[6] Sites 1F and 105K/F were reoccupied and flourished in the 2nd c.. See Ch. 27, 898.

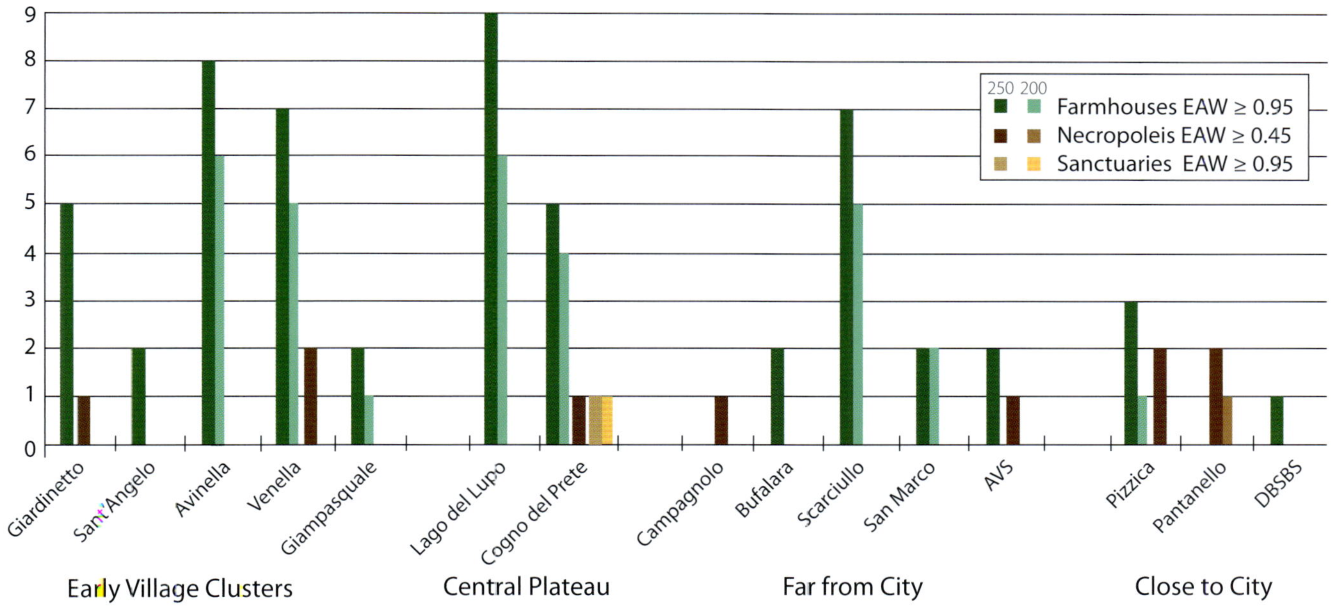

Figure 24.2 Distribution of farmhouse, necropolis, and sanctuary sites among localities in **250** and **200**.

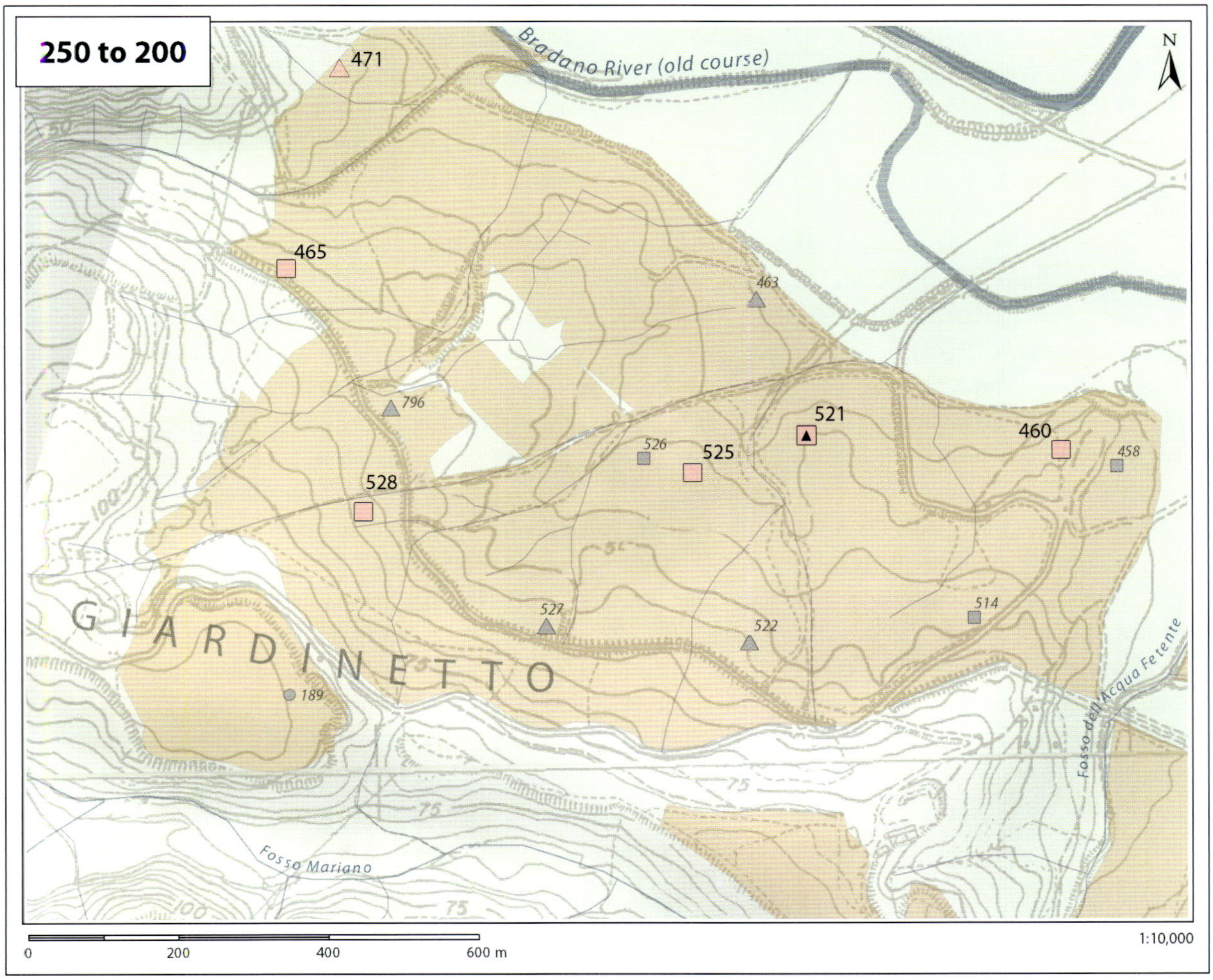

Figure 26.3 Giardinetto in **200** and **250**. See Fig. 26.1 for map key. (JT)

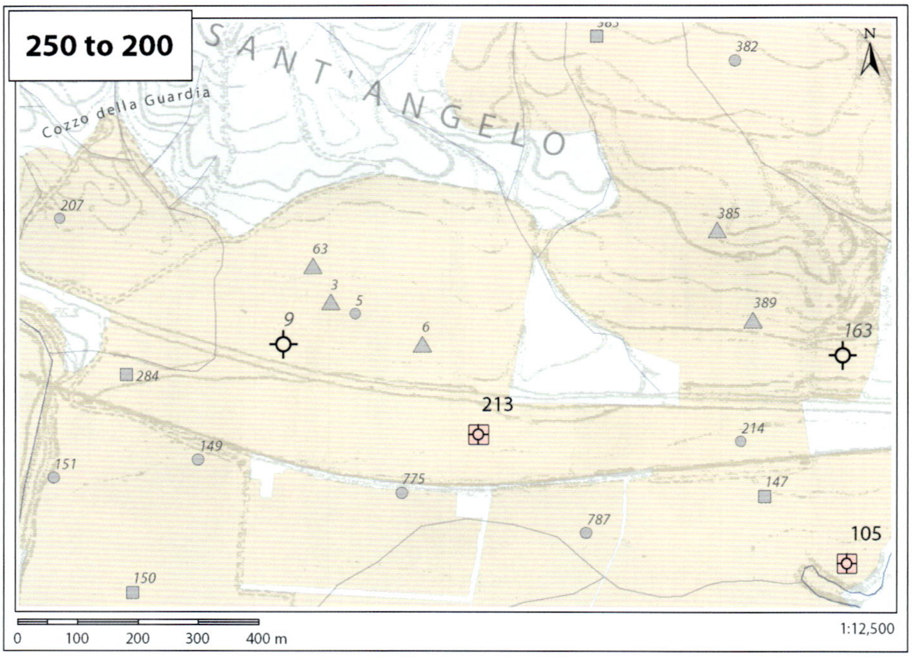

Figure 26.4 Sant'Angelo in **250**. See Fig. 26.1 for map key. (JT)

Avinella (*Fig. 26.5, Appendix C3*). This cluster and Venella, both located in the interior, lost relatively fewer farm families in **250** than the other Early Village Clusters. The level of the population was just one farm family less than it had been in the mid-6th century BC.

The cluster around the three forks on the lower Conca was only slightly diminished. The upper Conca did not fare so well. Particularly striking is the disappearance of three of the four sites that sprang up in **350** around the extant division line segments (L22 and L23) in the plateau between the upper Conca and the Venella.[7] Three of four farmhouses along the upper Conca drainage, however, remained, and 506F (*Fig. 26.6*) became a second magnitude site that re-emerged as significant in **200**. It would appear that between **300** and **250**, the division lines lost their importance in Avinella.[8]

The three small necropoleis (392N, 405N, and 533N) that were in use in the lower Conca had been abandoned by **250**. The average FMCE values were well below the average for the study area, in contrast to Giardinetto and Sant'Angelo where they were comfortably above the average. The survival rate of farmhouses is better in Avinella than in Giardinetto and Sant'Angelo in **250**, but their FMCE values have declined sharply, especially in the upper Conca, where the average, 1.65 (see *Appendix C7*), is among the lowest in the study area. Only two of the eight significant farmhouses remaining in Avinella in **250** are above average. Both the number of farmhouses in Avinella and their FMCE values remained remarkably level from **250** to **200**. This apparent ability to hold on, albeit at a relatively lower level than the more significant sites along the major river valleys, may well reflect the diversity of the agricultural economy that seems to have characterized these inland valleys from the 6th century BC. It was probably not based exclusively or primarily on cereal production, but incorporated animal husbandry and specialized crops such as the olive and the grape. The occupants of these valleys were better prepared to subsist in hard times than the larger scale farmers of the Central Plateau and those in the more exposed villages of the river valleys.

Venella (*Fig. 26.7; Appendix C3*). The number of farmhouses in this area dropped by half from fourteen to seven between **300** and **250** (just one farmhouse more than the total in **550**); the five necropoleis were reduced to two, both in east Venella, 479N at the far northwest of this long, narrow cluster, and 635N at the opposite extreme. No necropoleis remained in **200**, though all but two[9] of the seven farmhouses survived to **200**. The decline affected the east Venella slightly more: four of nine farmhouses remained—a survival rate of 55%—while in west Venella it was three of five. The average distance between farmhouses nearly doubled in the east between **300** and **250**, as a result of this attrition: from 300 to 550 m. In west Venella the distance between the three surviving farmhouses remained the same in **250**.

The average FMCE value in west Venella (2.61) in **300** was considerably higher than in the east (2.44), but by **250** that situation was just reversed with the average in the east (1.97) greater than that in the west

[7] See Ch. 25, 814.

[8] See below, Lago del Lupo, 874.

[9] The two located at the extremes of east Venella, 167F and 360F.

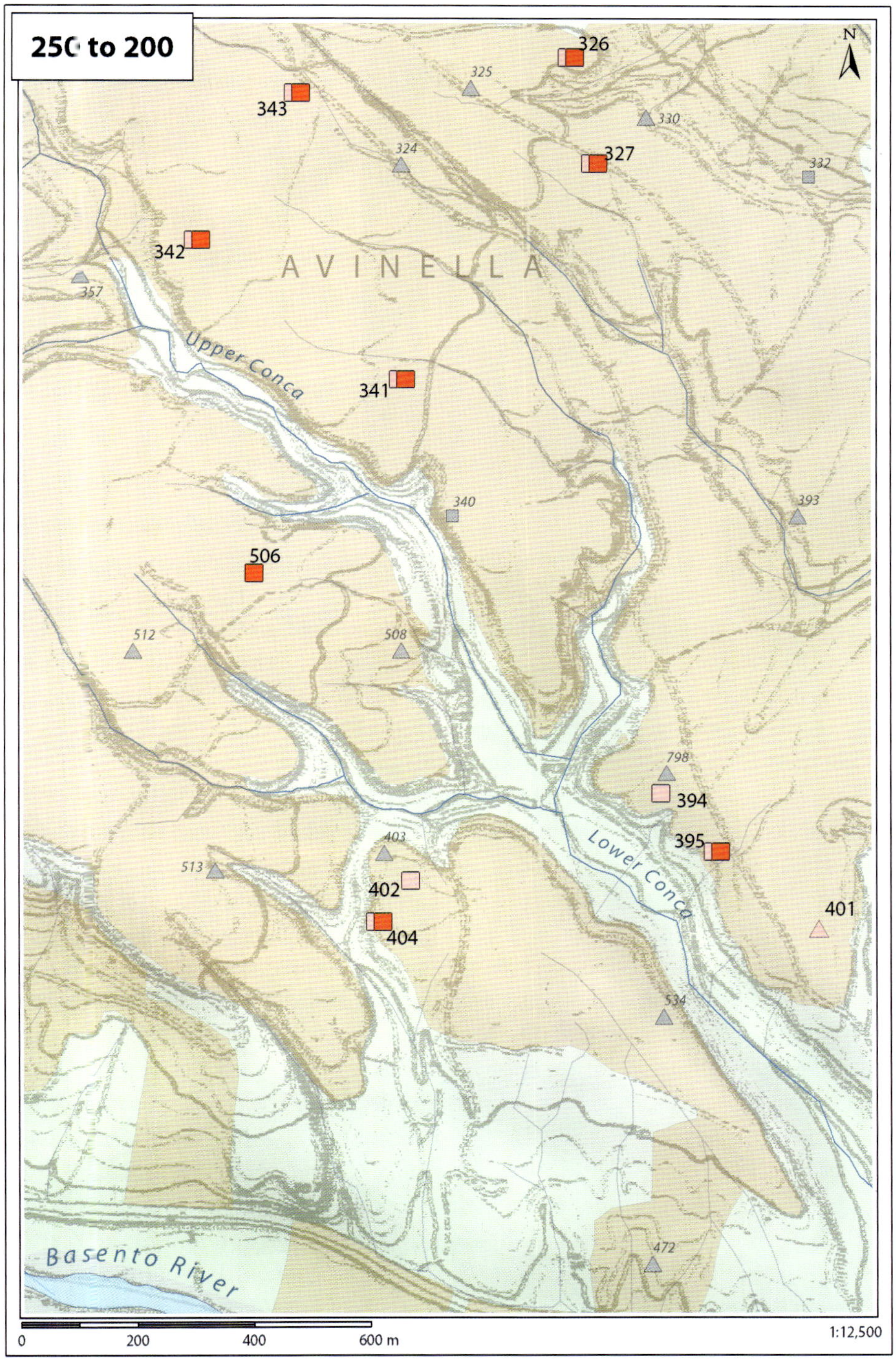

Figure 26.5 Avinella in **250** and **200**. See Fig. 26.1 for map key. (JT)

(1.79). Nonetheless, the two farmhouses with the highest FMCE values in Venella, 167F and 183F, are both on the eastern bank, though actually located on the flat terraces of the Central Plateau.

The configuration of farmhouses on both sides of the Venella remained essentially the same from **250** to **200**. The significance of individual farmhouses was basically unchanged from one period to the next, and these values were very close to those in Avinella. The two clusters were in fact remarkably similar not just in these two periods. Their shared characteristics, in particular the diversity of their agricultural economies and their remoteness and relative inaccessibility, probably explains, in part at least, their longevity.

Giampasquale (*Fig. 26.8, Appendix C3*). Like Avinella and Venella, Giampasquale is above the edge of a secondary stream, but in contrast to them, it is close to the point where the tributary joins the big valley, and hence is more exposed. The population had remained unchanged at five farmhouses for two centuries, after which it fell to two in **250** and one in **200**. Unlike Avinella and Venella, it was abandoned for good after **200**.

The two farmhouses of **250**, 269F and 193F, were located approximately 300 m apart—193F on the bank of the deep drainage, and 269F, as ever, commanding its own plateau of approximately 18 ha. Among the most significant farmhouses of **300**, site 269F had one of the longest lifetimes of any site in the study area.

The Central Plateau

Lago del Lupo (*Fig. 26.9; Appendix C3*). Only nine of the 26 farmhouses in the most populous locality in

Figure 26.6 Site 506F in Avinella, upper Conca. (CDA)

the chora's heyday, **300**, were still active in **250**, a loss of 65%. The dateable necropoleis of which there was just one example, 294N, in **300**, completely vanished from the scene. The rate of attrition had slowed by **200** and seven of the nine farmhouses of **250**, or 77%, were still active, including 248F/K, another possible example of a surviving ceramic production site.

Cogno del Prete (*Fig. 26.9; Appendix C3*). The northern half of the Central Plateau was affected in much the same way, or at least to the same degree, as the southern in this crisis, which is a little surprising considering the rather independent development paths it followed in the previous century. Of the 11 farmhouses in **300** only five remained, a loss of 55%. This was somewhat less severe, granted, than that of Lago del Lupo. Three necropoleis were active in Cogno del Prete in **300**, but only 81N appears to have survived for a while (*Fig. 26.10*; *Appendix C37*). In **200** four of the five farmhouses of **250** continued to be frequented. This is one of the highest survival rates anywhere in the chora. Taken together there was a loss of only 29% of the farmhouses in the Central Plateau between **250** and **200**. This would indicate that the cereal crops of the Central Plateau continued to play a role in what remained of the agricultural economy of the chora.

The average distance between a farmhouse and its nearest neighbor in Lago del Lupo nearly doubled between **300** and **250** from 331 to 657 m. The distance between neighbors in Cogno del Prete, always rather great, increased by a smaller amount (only by about 27%) compared to the 38% for the study area as a whole. The concentration of sites in Lago del Lupo and in the villages of Galitelli and Casa Preti was much reduced in **250**. Only one, 173F, of three farmhouses in Galitelli, and two (248F/K and 497F/T) of four in existence in **300** in Casa Preti remained in **250**. In the more dispersed village of Capistretti in Cogno del Prete, the production center and possible sanctuary 266S was the only structure still in use.

The most important change in the spatial distribution of sites is the very great decrease in the number of farmhouses in close proximity to the preserved segments of division lines in Lago del Lupo. Nineteen sites were within 110 m of a division line in **300**; by **250** that number had been reduced to five. A sharp decline in the use of drainage canals, and a failure to maintain visible property lines[10] in **250** would have had some serious implications both for the division of the property as well as for the type of agriculture that was being practiced in the areas where the division lines were concentrated.

Lago del Lupo had the highest average FMCE score, after Pizzica, among the separate localities in the study area in **300**, and it had the farmhouse with the highest value of all time, 296F. In **250**, only three of the nine farmhouses in Lago del Lupo have FMCE values below the average for the study area as a whole, while three are among the most significant in the study area (*Appendix C7*).[11] The farmhouses of Cogno del Prete had, as usual, lower scores than their neighbors to the south. The 25% decline in average FMCE values as a whole for the Central Plateau was right in line with that of the Early Village Clusters (23%) and the localities Far from the City (27%). Only the farmhouses in localities Close to the City retained their high level of significance, losing just 3%. This result, however, is based on a total of just four farmhouses.

The FMCE values in **250** were not, on the whole, much lower than those of **350**. In some cases they

[10] The situation may have resembled that in the chora of Herakleia in the late 4th or early 3rd c.. Uguzzoni and Ghinatti 1968, 91–96.

[11] 296F (the most significant in the study area in **300**), 497F/T, and 501F.

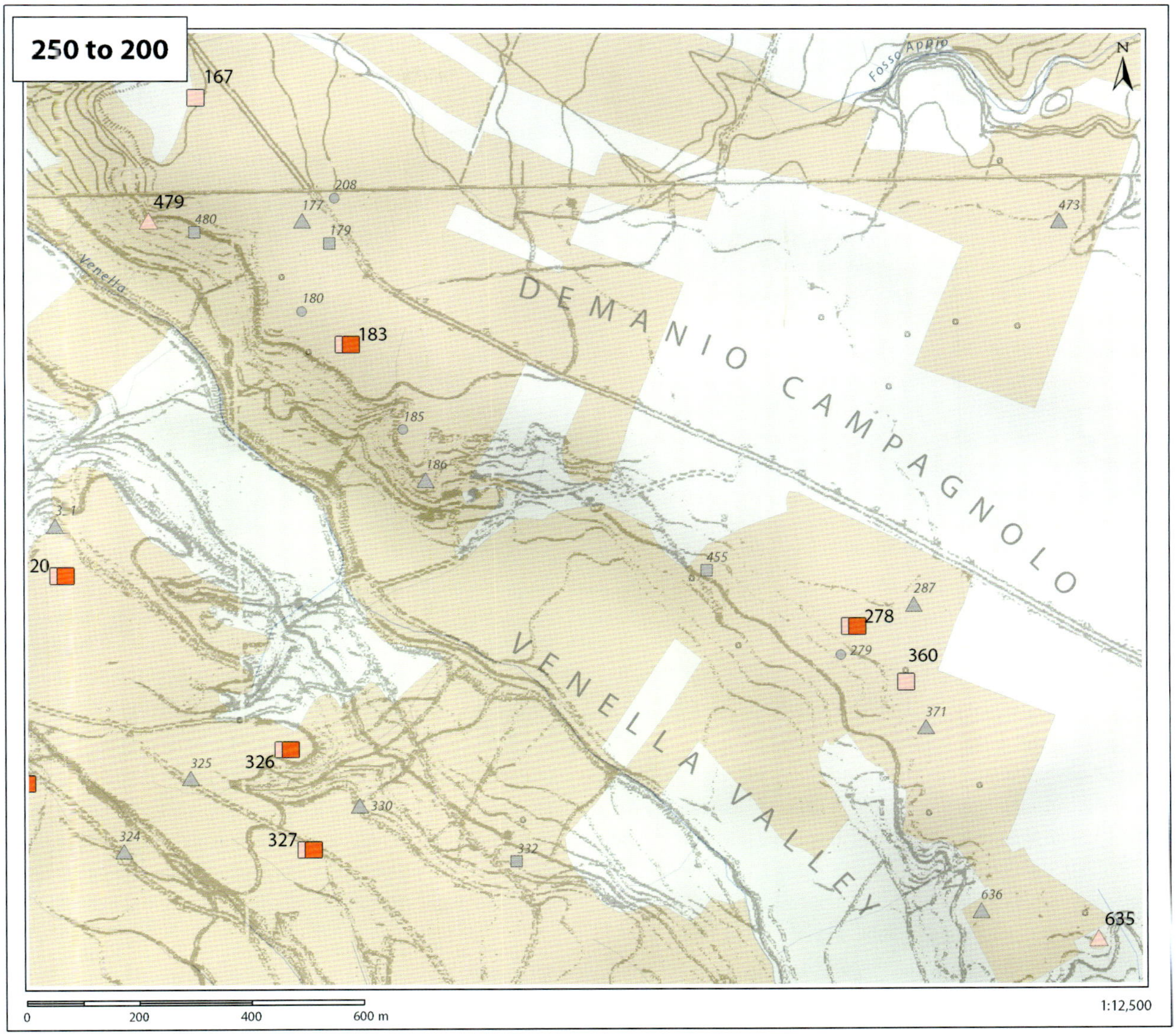

Figure 26.7 Venella in **250** and **200**. See Fig. 26.1 for map key. (JT)

were actually higher, for example at Sant'Angelo. The fact remains, however, that there were only half as many farmhouses as there were a century earlier. We may conclude that the crisis of **250** affected most areas of the chora without discrimination—the plowland of the Central Plateau and the narrow valleys of the interior in almost equal measure.

Far From the City *(Fig. 26.1)*

Campagnolo. Of the five farmhouses of **300**, there was not a single survivor in **250**, despite the fact that site 46F had been the second most significant farmhouse in the study area in **300** (FMCE value 4.0). This locality, beyond the Acqua Fetente to the west, was abandoned and only briefly revisited with the return to significance of 47F in **100**. The one necropolis, 48N, from the 5th century continued in use in **250**.

Bufalara (*Fig. 26.1; Appendix C3*). Only two of the six farmhouses of the previous period survived in **250**. There were no dateable necropoleis in this (or any) period. The two survivors were site 488F, a farmhouse that was occupied in **600** and almost continuously thereafter, and 309F, which was ex novo in **300**. No farmhouse in Bufalara was significant after **250**.

Scarciullo (*Fig. 26.11, Appendix C3*). Seven of 14 farmhouses that were there in **300** disappeared from the Scarciullo plateau in **250** along with both of the necropoleis. This loss of 50% of existing farmhouses

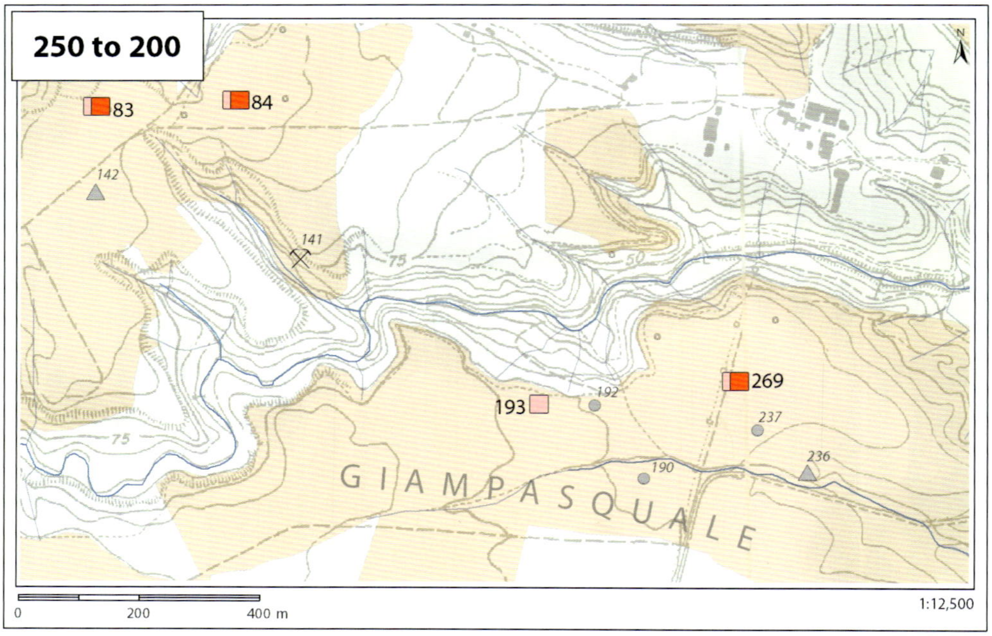

Figure 26.8 Giampasquale in **250** and **200**. See Fig. 26.1 for map key. (JT)

is slightly less than that of the other sizeable localities Far from the City. All of the survivors are located either along the Gaudella valley (376F, 375F, 365F), at the head, or along drainages that join the Venella or flow directly to the Basento. All farmhouses are on sites with long histories, going back at least as far as **400**, except for 376F, which was ex novo in **300**. All of the new sites in **300** (those that lined up, bisecting the plateau), sites 15F, 16F, and 17F (new in **350**), vanish completely and forever in **250**. The line formed by these sites was interpreted as an indication of a division of the plateau into possibly equal-sized plots, in an analogy with the system of division lines in other parts of the chora at an earlier date in the 5th century BC.[12] The disappearance of the sites might indicate that the division was no longer valued or enforced in **250**. This ought to be associated with the disappearance of farmhouses in close proximity to division lines in Avinella and Lago del Lupo in **250**.

The retreat of farmhouses to the valley's edge and the natural drainages, instead, seems to reinforce the observation made about the relatively higher rate of survival of sites in Avinella and the Venella, where the majority line the natural drainages as they had since the earliest settlements of the chora. Only one (354F) of the seven farmhouses in Scarciullo in **250** was of above average significance. It was located perhaps along another axis, defined by 376F (ex novo in **300**) and 60F, dividing the secondary plateau to the east of the main one.

A large percentage—five of seven or 71%—of the farmhouses that survived to **250** remained in use in **200**. This and San Marco (see below) are the only areas in this region where farmhouses continued in use. In this respect Scarciullo and San Marco in **200** resemble Avinella and the Venella, Lago del Lupo, and Cogno del Prete, all of which show an unusual resiliency in **200**. Scarciullo, a small version of the Central Plateau, realized its potential late in the 4th century BC, perhaps because of the introduction of improved wells (see Ch. 25). Avinella and Venella, as always well supplied by springs and streams, had access to the plateaus behind and the valleys below to raise a diversity of crops as well as animals.

San Marco. Half of the four significant farmhouses the survey documented in this locality date to **450** or before. The two that survived to **250**, and subsequently to **200**, were ex novo in **350** and **300**. They lie along the banks of the Acqua Fetente, and in contrast to the earlier, more exposed sites stand back from the Bradano on the easily defended heights about 200 m apart.[13] One of the two survivors, 83F, had a far below average FMCE value in **300**, but above average in **250** and **200**, while site 84F, with a substantially higher value in **300**, had the dubious distinction of being the least significant site in the study area in **250** and **200**.

AVS. The two surviving sites in portions of localities Far from the City are 494F in the upper Venella and 312F in Avinella, also high up the Venella, in perhaps the most remote location in the chora.

[12] See Ch. 25, 824.

[13] As noted earlier, this location also attracted a village in the Medieval period (Ch. 34, 1166).

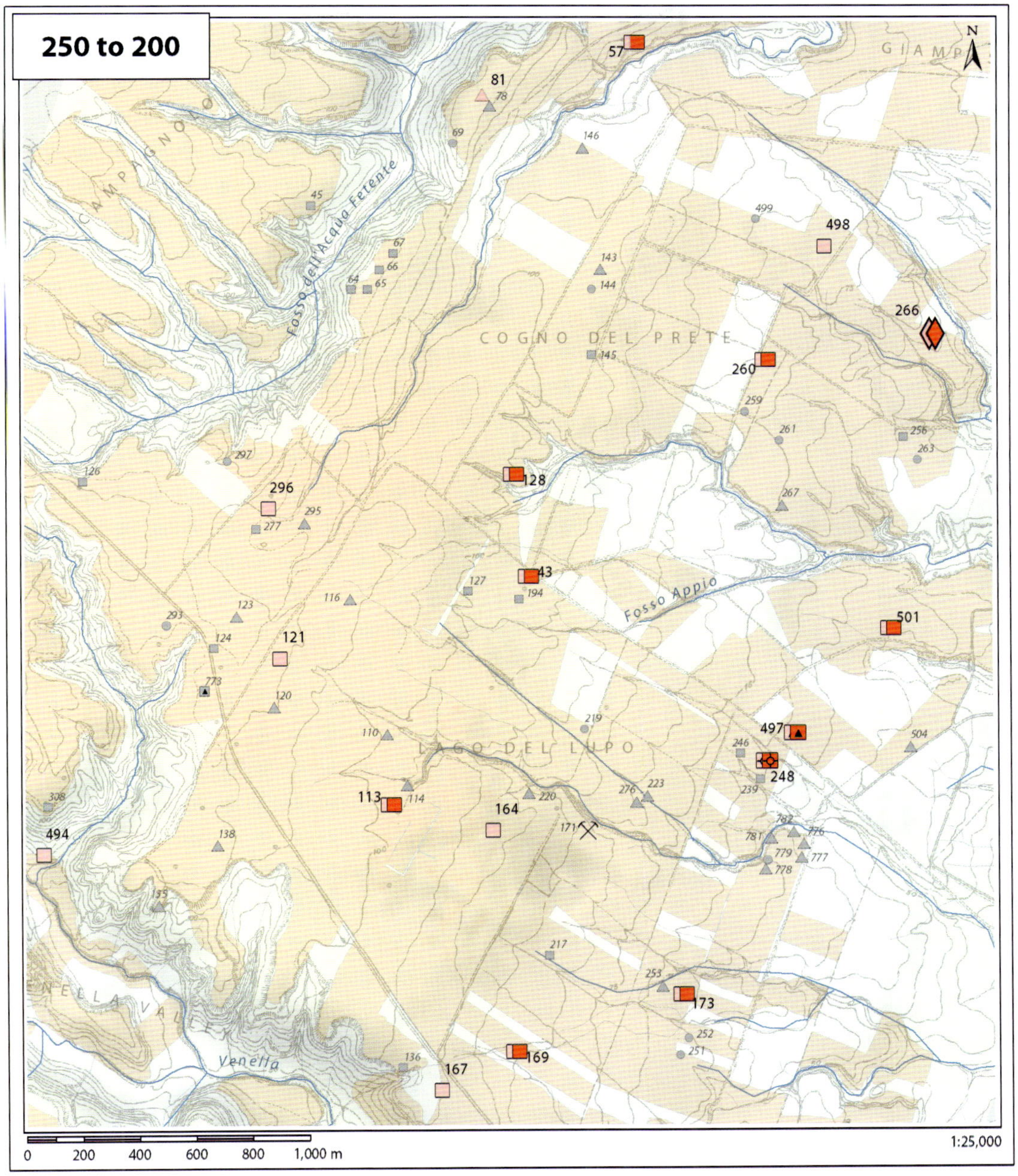

Figure 26.9 Central Plateau in **250** and **200**. See Fig. 26.1 for map key. (JT)

Close to the City

The region Close to the City was clearly the most seriously affected by the crisis of the 3rd century BC, losing all but four of the 18 farmhouses in existence in 300—a loss of 78%.

Pizzica. Of the five farmhouses active in **300**, only one farmhouse, 730F/T, the most significant, continued in use in **250**. It was joined by a farmhouse on site 734F/T that was first occupied in **550** and not again until **250**, and by the only ex novo farmhouse in the chora in **250**, site 745F, which continued, alone of all the farmhouses Close to the City, to be significant in **200**. Not until **150** did the percentage of new farmhouses in the locality return to its historic levels.

Pantanello. All farmhouses were abandoned for a century, but two saw new life in **150** and **100**, in the locality closest to the asty.[14]

A Synoptic View of the Chora in 250 and 200

The most obvious indication that the chora under-

[14] See Ch. 27, 895.

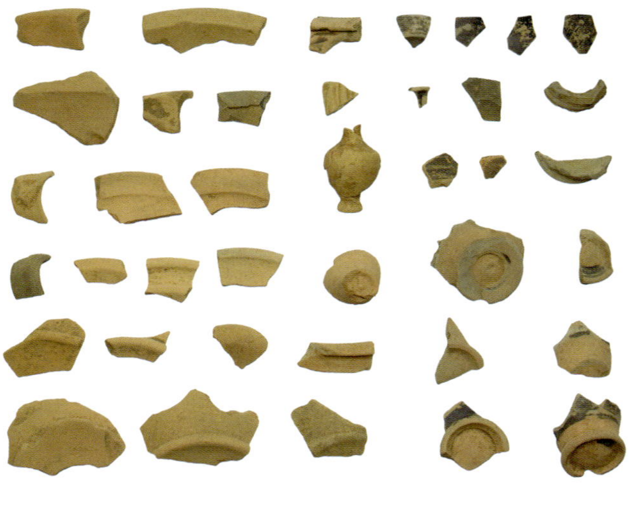

Figure 26.10 Assemblage from site 81N, Cogno del Prete. (MLG)

went enormous change in the period 275–225 BC is the precipitous drop in the number of farmhouses and necropoleis. The dating of the pottery, especially the black gloss, of the late 4th and early 3rd centuries BC, is not entirely independent of the great political events. The failure of Pyrrhos and loss of independence of the Greek poleis used to be taken as the end date for the production of much black gloss and other fine wares. This is no longer generally accepted,[15] and the chronology of the Archaic fine wares and black-gloss pottery adopted here[16] extends the production much longer than previously thought. There were other equally obvious signs of decline, most of them also dependent on the pottery evidence, such as the diminished average of FMCE values of the farmhouses (*Appendix C7*).

The fact that there is only one ex novo farmhouse in the study area in this period (*Appendix C6*) is indicative, as is the great decrease of burials in the chora with respect to active farmhouses. The increasing distances between farmhouses might appear obvious given the greatly decreased numbers, but the contraction of the settlement did not affect all areas in the same way. The region Close to the City, for example, was the most seriously affected in this period. Other factors that reflect the changing pattern of settlement and are not quite so obvious are the large-scale shift in the distance of farmhouses from the city (*Appendix C9a*) and the changing distribution of farmhouses within the four regions. The increased distance between farmhouses and the division lines suggests that these may no longer have been in use as before to drain fields and divide properties.

The number of farmhouses in the study area fell from its maximum of 142 in **300** to 55 in **250**, a decrease of -61%. This sudden decline is comparable in its severity to the practical desertion of the chora that took place in **500**. It was accompanied by decline in the significance of the farmhouses that survived. It fell from 2.57 in **300** to 1.99 in **250**, a decrease of -23%, which exceeds that of **500** (2.14 in **550** to 1.77 in **500**, or -17%). Though the significance of farmhouses in all areas was affected, some areas suffered more than others. These differences were discussed in the detailed description of the villages and localities above.

Significant farmhouses were constructed on new sites in the study area in every period with the exception of **200** and **50**. In **250** there was only one ex novo farmhouse, 745F in Pizzica. It was the most significant site in the chora in **250** and throughout its century-long existence. This singular event stands in sharp contrast to the preceding three centuries, when the rate of the creation of new farmhouses, while steadily slowing (from 70% in **550** to 23% in **350**) as the chora filled up, dipped below 23% only once, in **500**, when, amidst the desolation, a single newly significant farmhouse (site 726F/T) appeared. All 30 farmhouses occupied in **200**, with the exception of 745F, were on sites that had been previously significant for two centuries or more.

Fundamental changes in the nature of the settlement are signaled by the plunge in the number of necropoleis from 36 in **300** to ten in **250**, a decrease of -72%. Though this is very dramatic, a telling fact is that activity in these ten surviving necropoleis, as indicated by the EAW values, was rapidly slowing. At the larger rural necropoleis like 401N in Avinella and 471N in Giardinetto, less than a tenth of the amount of dateable pottery was being deposited. At 415N (the Eastern necropolis at Pantanello),[17] the latest burials were probably no later than the first decades of the 3rd century BC, though ongoing studies of Gnathia and black-gloss pottery will no doubt affect their dat-

[15] Lippolis 1994b. See also Carter 1998f, 223–35.

[16] See Ch. 6. Because of ongoing research and modifications at both ends (the early production in the 6th c. and the latest in the 2nd c.), greater flexibility in the date ranges of all pottery may be incorporated in the GIS, as discussed in Ch. 4.

[17] Carter 1998b, 40–41.

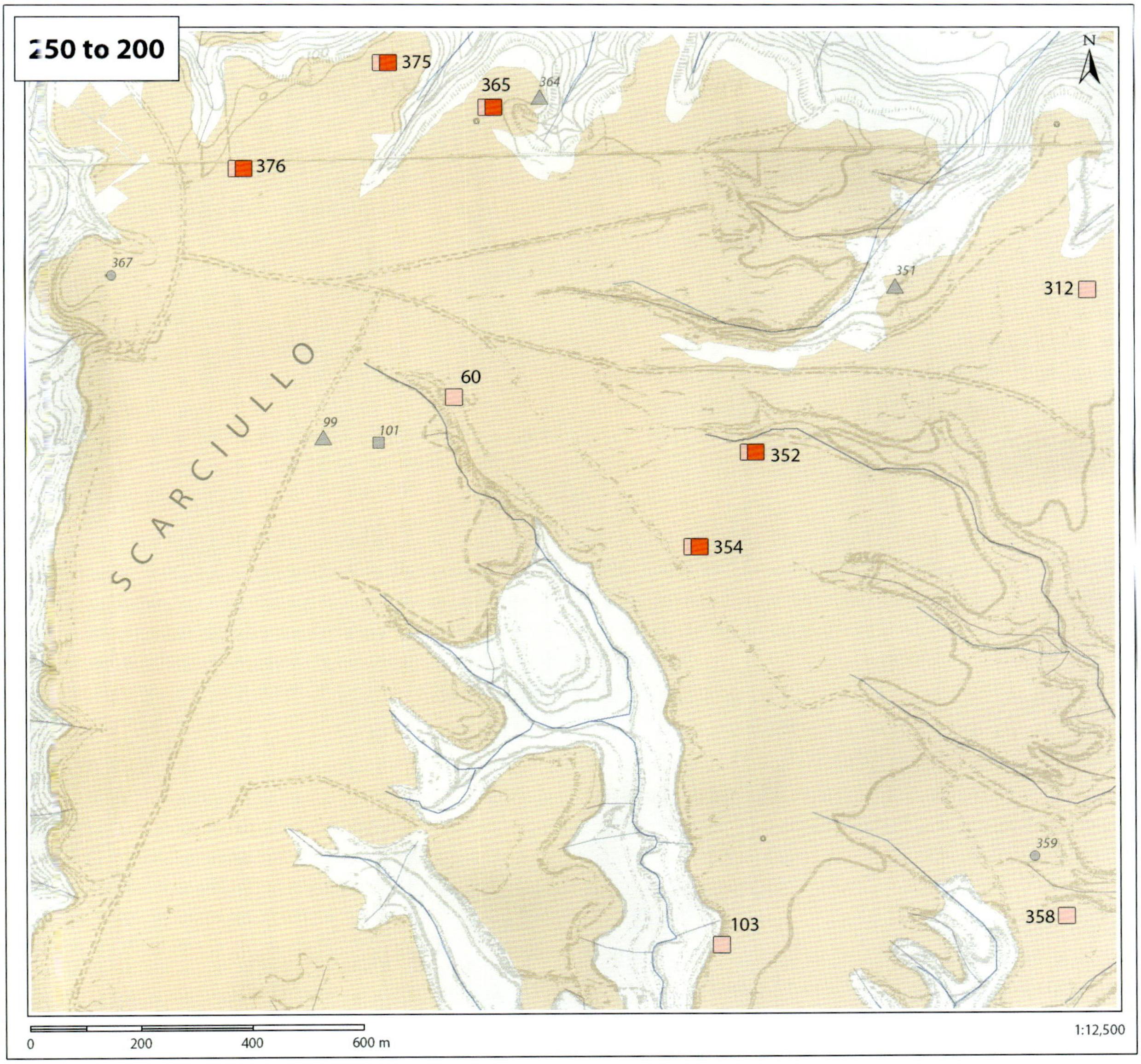

Figure 26.11 Scarciullo in **250** and **200**. See Fig. 26.1 for map key. (JT)

ing, and bring some of it down into the 3rd century BC. Of the ten necropoleis active in **250**, all have much less dateable fine pottery in **250** than they had in **300** (*Table 19.6*).[18] Most affected were the larger necropoleis.[19] In any case there were no necropoleis with an EAW ≥ 0.45 in **200** and very few later. Necropoleis and observable signs of burial practically disappear from the chora in **200**. There is a single exception (439N) in Pantanello. Either the occupants of the farmhouses are burying elsewhere or they are disposing of the dead in ways that have left no permanent traces in the archaeological record in large areas of the chora. A fair number of burials of the 3rd century BC have been documented in the areas of the urban necropolis, for example in Casa Ricotta.[20] Recent excavations, particularly in the area known as the Svincolo, where the Metaponto-Matera road SS 175 intersects with coastal SS 106, will probably add substantially to that number (see *Map 12*, Vol. IV).[21]

As the number of farmhouses dropped, the space

---

[18] In the Pantanello necropolis an attempt has been made to re-date tomb 71 to the late 3rd c. BC. Jackson 2004.

[19] Three of them are very little frequented indeed, to judge by EAW values, but long-lived (48N in Cogno del Prete and 638N and 732N in Pizzica).

[20] Rocchietti 2003, 70, Table 4.

[21] See discussion of the urban necropoleis in Ch. 25, 843–44.

between them increased. The average distance between a farmhouse and its nearest neighbor was 274 m in **300**, and 379 m in **250** (*Appendix C12*). As noted above, the thinning out of the settlement was not uniform, and was particularly noticeable in those areas of the chora where the greatest densities of farmhouses had been formed in **350** and **300**, those around the preserved sections of division lines in Lago del Lupo (see *Fig. 25.22*), but also in Avinella. By **200** the distance between nearest neighbors had practically doubled.

The elevation of a farmhouse or necropolis is related both to its distance from the city, given the slope of the marine terraces downward to the coast and the city, and to its location in a geomorphological unit, whether in the lower down on the fluvial terraces of the valley bottoms, the Pleistocene marine terraces, or the valley walls between the two. As noted earlier in the study,[22] there is a general trend towards increased elevation for the farmhouses from **600**, where the average is 57 masl (*Appendix C10*). It reached 77 masl in **250** and its maximum in **200** at 83 masl (after which it abruptly fell to an even lower level than in **600**). It might be imagined that the number of farmhouses on the marine terraces Far from the City, and hence higher, had increased (*Appendix C11a*). In fact, the number of sites on these terraces did increase from 74% to 80% between **300** and **250**—and by **200**, to 97%! At the same time, while the percentage of the settlements on the valley walls remained the same in **250**, that of the alluvial terraces dropped from 11% in **300** to 6% in **250** and 0 in **200** (before rebounding in **150**). In effect, what was happening was the abandonment of the major river valleys and a withdrawal towards the interior. This was explored in greater detail in the discussions of the individual areas of the chora, as well as possible motivations for this shift.

Despite the high elevation and location in the interior, the farmhouses Far from the City decreased in the century between 275 and 175 BC (*Appendix C9a*). There is a steady decrease, in fact, in the number of farmhouses between 9 and 11 km from the city, and a concomitant increase in those between 6 and 8 km, while those between 3 and 5 km, represented by a much smaller sample, continually decrease.

[22] Ch. 19, 638–39.

Figure 26.12 Proposed reconstruction of the farmhouse at the Pantanello sanctuary. (ALP/EH)

Excavated Farmhouses in the Study Area, the Chora, and the Interior

Farmhouse at Pantanello. Among the excavated farmhouses in the chora are two of unusual historical importance for this period, if our tentative conclusions about them are correct. The farmhouse at Pantanello was revealed as such in 1977 during the excavation of the Pantanello sanctuary (*Fig. 26.12*).[23] Work continued in 1978, but was only completed after more than a decade's interruption, in 1990 and 1991.[24] The structure is on high ground, well above the 6th- to 4th-century BC sanctuary structures.[25] The foundation course, constructed of fragments of roof tile mixed with conglomerate and limestone blocks, lies directly on top of the larger, solid conglomerate stones of the 5th-century BC oikos, which were opportunistically appropriated by the farmhouse's builders as a support for the foundation of the northern row of its rooms.

The plan (*Fig. 26.13*), in its arrangement of rooms and absence of a courtyard, closely resembles that of Fattoria Fabrizio.[26] It is a type of plan that is encountered early on in the chora at Cogno del Pero,[27] but also in indigenous contexts.[28] The rooms vary in

[23] Soundings in 1975 uncovered a small portion of the southwest corner of the structure, but not enough to identify its function. For a preliminary discussion of the sanctuary as a whole, see Carter 2006a, 163–73.

[24] In a preliminary study (Carter 1994a, 196–98), the "farmhouse" was interpreted as an example of a Lucanian sanctuary. This building, and the sanctuary over which it lies, is the subject of a forthcoming study in which the full evidence for the dates and function of all the buildings will be presented.

[25] For a plan and view of the "oikos," see Carter 1994a, Ills. 11 and 17. For a plan of the site as a whole, see Carter 1994a, Ill. 7.15.

[26] Carter 2006a, Fig. 4.3.

[27] See Ch. 21, 695, and Carter 2006a, 140.

[28] E.g., Tolve, phase 1. Russo Tagliente 1992, 173–75, Fig. 105.

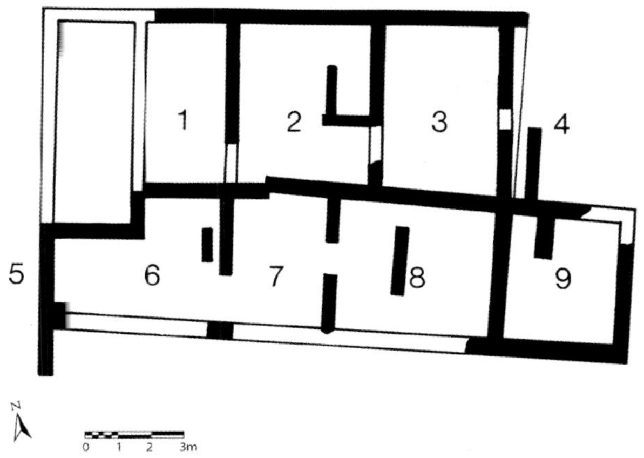

Figure 26.13 Plan of the farmhouse, Pantanello sanctuary, early 3rd century BC.

size with no obvious pattern, representing a cobbling together of walls from several periods, especially in the southwestern corner. The northwestern corner was demolished by the intrusive kiln deposit (see below) of the mid-2nd century BC. The southern wall had largely eroded down the slope, over the Archaic sanctuary. The probable maximum dimensions of the Pantanello farmhouse are 11 x 18 m, with a footprint of approximately 162 m^2. The pottery and other objects from the occupation level of the farmhouse are typical material of the late 4th and early 3rd century BC, and the assemblage was remarkable only for the almost total absence of objects related to a cult. For example, there were fragments of a terracotta doll, but not a single one of the satyr maenad plaques (e.g., *Fig. 25.38*) of the type found by the hundreds in the votive deposit by the sacred spring, just a few feet away, at a lower level. A revealing discovery in the farmhouse was the articulated skeleton of a young male found in the northwest corner of room 2, under the fallen tile roof. To all appearances the farmhouse was occupied, and underwent modification before it was hurriedly abandoned. The death of one of its occupants is hardly likely to have been an isolated act of violence, since to leave a body unburied would indicate that there were either no survivors or that for some compelling reason they were prevented from burying their dead. We may only speculate that this was connected in some way with the presence of Pyrrhos and the Romans in the area. The dating of the pottery in use would appear to favor Pyrrhos,[29] but a date in the Hannibalic period—the other historically documented hostilities in the region in the 3rd century—is also possible. In 214–213 BC he obtained grain from Metaponto before Metaponto had gone over to him. The ferocity of Hannibal's foraging was recounted by Livy.[30]

Farmhouse at San Biagio. A recent discovery in the Sanctuary of Artemis at San Biagio not only provides support for the interpretation of the site at Pantanello, it makes a similar statement about the nature of the occupation of the chora in the mid-3rd century BC. As described by the excavator, Antonio De Siena, the sanctuary ceased to function as such in the first decades of the 3rd century BC,[31] the same period in which evidence for worship runs out at Pantanello. Among the latest artifacts at San Biagio that refer to its use as a sanctuary are the antefixes of the same type employed in the area near the stage buildings of the theater in the early 3rd century BC.[32] A structure identified by De Siena as a farmhouse was erected beside the sanctuary buildings. The similarities with the Pantanello farmhouse are evident in the plan and the contents of the rooms. The San Biagio farmhouse is large and fits neatly into a rectangle measuring 13 x 19 m, with a total area of 247 m^2. The longer axis runs almost perfectly north-south. There is a courtyard on the east side. The rooms on the east, according to the excavator, served as storerooms, and the numerous burned patches, either hearths or caused by braziers, suggested that this might be a multi-family establishment (*Fig. 26.14*). In fact there are eight rooms not counting the courtyard, but this is not more than the possible eight at Pantanello, and the probable eight (and possible ten) at Fattoria Fabrizio, or eight at Fattoria Stefan (see above). The contents at San Biagio resemble those at Pantanello—fragments of storage amphorae, cooking ware, and loom weights, but they also include female statuettes of an earlier period.

The conclusion that De Siena draws about the transformation of the Greek sanctuary at San Biagio applies equally well to Pantanello and together they make a strong argument that sometime in the first third of the 3rd century sacred land had been appropriated for building by people for whom the tradition and beliefs of the population resident there previously

[29] Studies of the coin hoards buried in the surroundings of Metaponto and the chora show them to be concentrated in the early 3rd c. BC. See Stazio 1974a, 103–05.

[30] Livy 24.20.15. Cf. Brunt 1971, 272.

[31] De Siena 2005, 443–47.

[32] Tempesta 2005.

Figure 26.14 Farmhouse at San Biagio. (EH after De Siena 2005, Fig. 2)

had no importance and was accorded no respect.[33] The cultural identity of the chora had changed and the authority that had maintained important places of worship and congregation in the chora was either seriously weakened or nowhere to be found. The asty, as will be clear from the discussion below, was simultaneously undergoing similar tumultuous change.

Farmhouse at Montalbano (*Fig. 26.15*). In the 3rd century BC a new type of farmhouse, clearly different from those familiar in the chora of Metaponto, appeared in other parts of Lucania.[34] A very complete example was excavated at Montalbano,[35] some 20 km to the southwest of Metaponto on the right bank of the Cavone, not far from the luxurious 2nd-century BC villa at Termitito.[36] The building is sited on the first hills that rise above the alluvial plain in gently rolling countryside adapted to agriculture. Another outstanding excavated example is the villa at Tolve,[37] not far from Potenza, deep in the interior.

The farmhouse, built over a preexisting settlement, measured 26 x 34 m, but the plan is not a rectangle like that of San Biagio. There is an extra, shorter row of rooms on the east side. Its relatively enormous size (803 m^2) immediately set it off from the 4th-century BC farmhouses of the chora, including the 3rd-century farmhouses at Pantanello and San Biagio, none of which have a footprint of more than 300 m^2, and are mostly much smaller. Around a central atrium, a feature it shares with the villa at Tolve, are arranged a total of 17 rooms of varying sizes and shapes. Rainwater from the impluvium was channeled to the side of the house. Along the south end is a two-room complex identified by De Siena as a kitchen and bath. If this is correct it would be, with the bath at Tolve, among the earliest signs of modest luxury known in a rural dwelling in this part of the world. Cooking in farmhouses of the chora took place in a larger room, as at Fattoria Fabrizio and Fattoria Stefan.[38] In three of five rooms on the west side, pithoi for storing grain or other produce were interred in the floor. The smallest of these rooms, perhaps because its contents were more precious and required greater security, opened into the middle room, and not onto the atrium space. The two large rooms in the northern sector were used as stalls to house animals, and there was a grinding stone (*macina*) in the larger one. De Siena surmises from the multiple hearths (*piani di cottura*) that multiple families dwelt here, as he thought was the case at San Biagio. Here the considerably greater size and number of rooms makes that more likely.

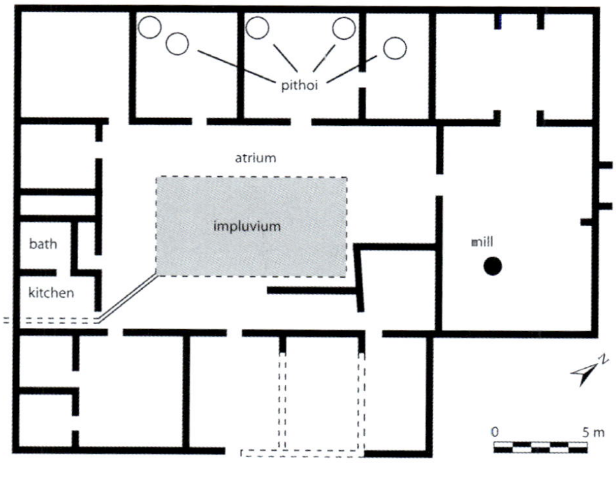

Figure 26.15 Plan of Montalbano farmhouse. (After De Siena 2005, Fig. 3)

Rooms that were clearly intended for storage (those with pithoi) account for about an eighth of the total space in this farmhouse, and there may have been much more for this purpose as the function of many rooms remains unclear; the stalls and work area on

[33] It is not likely to have been the same group that appropriated burial plots in the Pantanello necropolis in the late 4th c.. They showed both familiarity and some respect. See Ch. 25, 843, and Carter 1998f, 222–36.

[34] See Ch. 25, 862–63.

[35] De Siena 2005, 446–48, with a plan and brief description.

[36] See Ch. 27, 912–13.

[37] A. Russo 1993b. See also Ch. 25, 863.

[38] Carter 2006a, 139, 143.

the north occupy roughly a quarter of the farmhouse, a space about equal to the atrium. The paved bath, kitchen, and a possible pantry occupy only 5% of the total that includes the partially roofed atrium. De Siena interprets this clear emphasis on production and storage as an indication that this farm is the embodiment of a new direction in farming, anticipating the so-called Catonian model, the ideal profit-making establishment of the early 2nd century BC described by Cato in the *De agricultura*. The model specializes in the production of olives and wine and made use of low cost labor.[39] The very great size of the Montalbano farmhouse in comparison with the 4th-century BC farmhouses of the Metapontine chora implies a corresponding increase in the size of the land under cultivation. The villa at Tolve in its 3rd-century BC phase is in many ways comparable but with a greater emphasis on luxury. It has been interpreted as a rural residence, similar to urban examples at Roccagloriosa occupied by members of the Lucanian elite, and to pre-Roman Pompeii. The lifetime of these structures at Montalbano and Tolve appears to have been confined to the 3rd century BC. Tolve may possibly have survived the century and the conclusion of the Second Punic War, but that is not certain.[40]

Conclusions. We have no way of knowing with certainty the ethnicity of the occupants of the chora in any period, much less in the mid-3rd century BC, with such enormous political and social changes that took place in this relatively brief period. We can only make inferences from a fairly small body of relevant facts which may not be representative. One thing I think we can be sure of is that the fields were being cultivated. It would require truly exceptional circumstances to force prime grain growing fields out of production, and very few areas of the Mediterranean are as well adapted to growing cereals as the Metapontino. The dramatic decrease in the number of individual farm units in the chora in the 3rd century BC meant opportunity for others, though it may not have been immediately seized. Farmhouses like that at Montalbano, as De Siena suggests, paved the way. Still, there is no evidence that this model was adopted in the chora at this time. The ex novo site 745F (see above) may be one of a few exceptional Metapontine examples that could have fit the requirements. Another unexcavated exceptionally large farmhouse in the chora between the Basento and Cavone is 829F.[41] Essentially the organization of the chora remained the same as it had been in the preceding century, and all sites in the study area but 745F had been previously occupied. This situation was to change again, rather suddenly, in the next century.

As early as the 5th century BC some Lucanians had begun to settle in the Metapontino.[42] The revival of the chora in the 4th century BC, as we have argued,[43] was propelled by a fresh infusion of population from the interior of Lucanians well-familiar with the pattern of life in the chora. The proportion of farmers of Lucanian origin in the chora in the 3rd century BC increased and the newcomers, to judge by the farmhouses at Pantanello and San Biagio, were neither as acculturated nor as respectful of tradition as their predecessors had been.[44] The people who occupied the chora in the mid-3rd century BC, and continued to live on the earlier sites, differed in at least one other important respect from the 4th- and early-3rd-century BC population. Burials in rural necropoleis dropped dramatically and virtually ceased at Pantanello.[45]

In the first half of the 3rd century BC, many of the fortified Lucanian centers that had appeared for the first time less than a century earlier, in the mid-4th century BC,[46] had been abandoned, as occurred at Pomarico Vecchio near Metaponto, but a number of sanctuaries and farmhouses continued much as before, though diminished in number. The fortified centers were not necessarily the most important part of the settlement, as a study by Isayev maintains.[47] Their organization of the territory was cantonal, not polis-centered. This was true of other areas as well, for example, on the Tyrrhenian side of Lucania around Roccagloriosa, where an intensive survey and care-

[39] Toynbee 1965, II.296–312. Cf. also Merlo 2003, Ch. 4.

[40] A. Russo, 1993b, 42, with references. See above, Ch. 25, 863.

[41] D. Davis 2003, 78–80. Some of this site's plan has been revealed in a gradiometer survey of the site in 2003; see above, Ch. 25, n. 176.

[42] See Ch. 24, 806–07.

[43] See Ch. 25, 865–66.

[44] The question of who in this period was striking coins in the Metapontine tradition and for what purpose is intriguing. Sometime, perhaps earlier in the 3rd c. rather than later, Metapontine type coins with the ethnic ΛΟΥΚΑ began to be minted (Taliercio Mensitieri 1989, 43–47; 1999; E.S.G. Robinson 1964, 49–52), presumably but not necessarily at Metaponto. Coins with the monogram ΛΥΚ on coins of the Carthaginian standard weight (shekel) appear at the end of the century (Johnston 1990, 17–18; E.S.G. Robinson 1964, 49–52).

[45] See Ch. 25, 842–43.

[46] Ibid, 861.

[47] Isayev 2001, 115.

ful study of the pottery has been carried out.[48] The wealthiest do not seem to have spent time or money in the cities. Their major expenditures, rather, were on sanctuaries and tombs, according to Isayev.

For reasons that are made very clear in the discussion of the castrum,[49] it and the area of the city would almost certainly not have been under Lucanian control even if the Lucanians had had an interest in controlling it. Their business was in the fields; the Romans', in the port and castrum that protected it (see below). The result, we may speculate, was a relatively peaceful coexistence.

The probability that the Metapontino may have been under Roman control after Pyrrhos, however, cannot be excluded, despite appearances to the contrary. Toynbee asserts without proof that it was already part of the territory of Grumentum, and had become ager Romanus.[50] The real discontinuity in the settlement,[51] however, occurred in the Metapontino in the 2nd century, when it was much more likely that it became ager Romanus (Ch. 27).

This and any reconstruction of the 3rd century BC at Metaponto will remain extremely provisional until the publication of excavated sites, like the farmhouses and the castrum, are complete, until the residential quarter of the city is better known, and the debated questions of the coinage and of the chronology of pottery in this century, especially the Gnathia and black gloss, have been resolved.

The Asty in the 3rd Century BC

The changes that took place in the urban center of the polis in the 3rd century BC, like those recorded in the chora in **250** and **200**, were wide-ranging and to a large extent irreversible. The period considered here was bracketed by the campaigns of Pyrrhos and Hannibal in Italy, and no small part of the action took place around and even in Metaponto. The Metapontines, as all the Greek poleis of Magna Grecia, had been increasingly exposed to political and economic developments, whose origins and effects were beyond their control. This had already become apparent in the second half of the preceding century.[52] Few poleis came through the 3rd century BC relatively unscathed, neighboring Herakleia being apparently a rare exception. The question, much debated by modern historians, is the degree to which the society of the poleis was affected by these events. Less is known about Metaponto than the others, but evidence from various sources, literary, numismatic, and archaeological, seems to coalesce to form a possible reconstruction of this period.

The Castrum and the Port

Pyrrhos, with a large force including elephants, arrived in Italy in 281 BC, and in 280 BC achieved a costly victory over the Romans at Herakleia. Despite the fact, as noted,[53] that he would have marched across her chora to reach the battlefield, Metaponto does not, unsurprisingly, appear in our historical sources. By 275 BC he had withdrawn, and Rome's power in Magna Grecia was unopposed.

Metaponto moves to center stage sixty years later. Hannibal arrived in Italy, descending from the Alps, with an even larger force and many more elephants in 217 BC, and reached southern Italy the same year. Metaponto, with its ever productive fields, became a base for the Carthaginian army. Here we are supplied with comparatively abundant information by Livy, Polybios, and Appian. Hannibal fed his army with Metapontine grain,[54] and he occupied the city from 212 to 207 BC. The Romans earlier garrisoned at Metaponto went to the aid of those in Taras, leaving behind only a small force which was then betrayed by the Metapontines.[55] When Hannibal withdrew five years later to Bruttium to the south, he took with him the entire population of Metaponto,[56] surely including the citizens that he had recently imported from the northern Apulian town of Herdonia.[57]

Did the Metapontines support Pyrrhos and were they punished by the Romans, as were the other Greek poleis except friendly Herakleia? The numismatic evidence indicates that Metaponto did not

[48] Fracchia 2001.

[49] See discussion below, 889.

[50] See Ch. 27, 918 n. 104. Toynbee supports his assertion with an opaque reference to Tenney Frank.

[51] See Ch. 27, 893.

[52] See Ch. 25, 860–68.

[53] Ibid, 867–68.

[54] Livy 24.20. See also above discussion.

[55] Livy 25.15.6; Polybios 8.34.1; Appian, *Hannibal* 7.35.

[56] Livy 27.15.13.

[57] Livy 27.1.14.

rally around Pyrrhos to the degree that other Greek cities did.[58] Some of the coin displays are Pyrrhic symbols,[59] but the fact that the Metapontines ceased to mint silver coins during Pyrrhos' stay must indicate that their support ceased early on in his campaign. Even if they had had the will, they may not have had the means. The tribute owed to Kleonymos (see Ch. 25), if it was indeed paid, may have bled them dry, and there may well have been other factors contributing to the Metapontines' apparent inability to participate with Taras and her other allies in opposing Rome.

There are a number of strong indications, besides the cessation of the silver coinage, of a lack of continuity in the normal functioning of the city in the years immediately post-Pyrrhos. The excavators have noted the progressive failure to maintain public buildings. The efforts that continued to be made in the 4th century BC ceased altogether in the 3rd. Major building projects of the preceding century, the theater and the stoa on the eastern edge of the agora, as noted, were left incomplete.[60] The kerameikos ceased production after the first two decades of the 3rd century BC. A healthy output of smaller-sized amphorae and table wares of good quality continued throughout the 3rd century BC at other centers in and around the asty, even after work stopped in the kerameikos.[61] One development, however, overshadowed all of these and represents a violent interruption of the developments of the urban plan up to this point—the castrum.

A rectangular feature enclosing an area of nearly 14 ha within the circuit of the city walls was apparent in aerial photographs of the 1940s and 50s (*Fig. 26.16*). It was interpreted as the agger and fossatum of a castrum or military camp by Adamesteanu and Lo Porto.[62] They connected it with the passage of Livy, cited above, that mentions Roman troops at Metaponto in 212 BC. Excavations followed in 1975, 1976, and 1977, and are still largely unpublished, though preliminary reports provide some facts.[63] They were limited to a sounding at the southwest corner, and to a more extensive area in the west center of the structure overlying the intersection of the north-south Plateia IV and the major east-west Plateia A, also visible in the photographs (*Fig. 26.16*). Its basically rectangular outline was truncated at the southeast corner, probably by a branch of the Basento (*Fig. 26.17*).[64] The north and south sides were aligned roughly with the underlying grid of streets. The west cut across them at right angles, and the east, toward the sea, was found to be parallel to the west, and thus forms a regular polyhedron. The central NE-SW axis of the castrum corresponds to the stenopos just to the south of Plateia A. The complete west side measures ca. 338 m (1,142 Roman feet); the longer northern side ca. 430 m. The ratio of width to length is 4:5.

Figure 26.16 Aerial photo of Metaponto. (Aerofototeca Neg. No. 30305, GAI)

The sounding at the northwest corner revealed that the construction of the castrum followed the basic design features of the Roman military camp as described by Polybios.[65] The outer limit was defined by a fossatum that measured 7.56 m wide, behind which arose the agger construction made with the earth removed from the fossatum. A feature on top of the model Polybian camp was a wooden palisade on top of the agger, which is mentioned in the discussion of the Metapontine castrum, but not in D'Andria's initial report.[66] The ideal proportion of length to breadth, according to Polybios and Hyginus, was 3:2. Surviving examples of early Roman temporary military camps

[58] See Ch. 25, 868.

[59] Johnston 1990, 57.

[60] See Ch. 25, 859.

[61] L. Giardino 2005, 410.

[62] De Siena 1990b and 1993, for discussion and bibliography; more recently, De Siena 2005.

[63] D'Andria 1976, D'Andria 1977, and L. Giardino 1978a.

[64] De Siena 1990b, 305.

[65] Polybios 6.22–46.

[66] D'Andria 1976, 540.

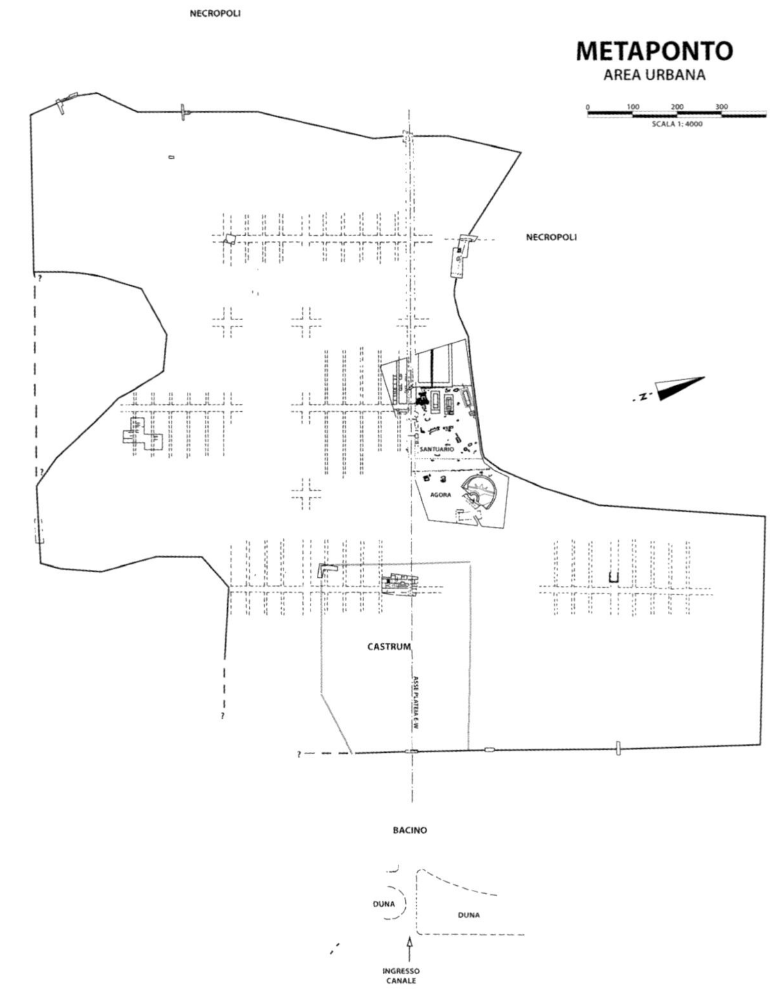

Figure 26.17 Plan of Metaponto with castrum and port (De Siena 1990, Pl. 6).

are few, and in their outline and proportions they do not conform to the ideal. Good parallels for the castrum at Metaponto are to be found, as we shall see, but not among these temporary camps.

Inside this military perimeter the castrum made use of existing buildings. The late-4th-century BC stoa that framed the agora on the east (see *Fig. 25.37*) was converted. Its colonnades at one point were walled up and storerooms created. The passage of Plateia A through the stoa was not blocked, and it is likely that Plateia A formed a sort of off-center decumanus for the castrum, connecting it to the port and to the chora. The castrum rose in an area of the asty that was among the earliest to be occupied.[67] It may have been chosen in both periods because it was one of the highest and driest areas of the city. It is also close to the sea, and to the rivers.

Research has shown that a port was constructed between the eastern limit of the castrum and the coastline, some 500 m away (*Fig. 26.17*).[68] The port was reached from the sea by means of a narrow canal that was on axis with Plateia A and with the western gate (Porta Settembrini) leading into the chora. It cut through an area of the urban necropolis on the dune to the east of the city, with monumental tombs dated to the late 4th century BC.

The date of the castrum has been much discussed. We can be sure that it, or something like it, was in existence in Hannibal's time, but when was it built? D'Andria observed that the fossatum surrounding it cut through buildings of the 3rd century BC,[69] which provides a rather broad terminus post quem. Giardino's study of the amphorae found in the storeroom, which she places in the first phase of the castrum, concludes that they date to the 3rd century BC, but with the majority belonging to the first half of the century.[70] Ciaceri supposed that Metaponto, like the poleis of Thourioi, Croton, Locri, and Rhegion that turned against Rome after the battle of Herakleia, would also have received a garrison.[71] Herakleia, which returned to the Roman fold in 277 BC, apparently did not suffer this fate. We cannot be sure, however, that Metaponto offended the Romans at this time. The first mention of a Roman *praesidium* at Metaponto is in the early years of the Hannibalic War, in 212 BC. The structure that we have been calling a castrum is not a praesidium, though it may have housed one. Neither is it a castrum of the sort described by Polybios (or Hyginus or Vegetius), that is, a marching camp. If indeed it were built in the first half of the 3rd century BC, and was still in use in the late 3rd century to house the praesidium betrayed to Hannibal, it was either intended as a semi-permanent structure, or de facto became one. Marching camps, as Johnson notes,[72] could become permanent camps,

67 The sounding produced pottery of the immediately pre-Greek period in the lowest strata. D'Andria 1976. See Ch. 20, 672–73,

68 De Siena 1990b.
69 D'Andria 1976, 541.
70 L. Giardino 2005, 406–11.
71 *Storia della Magna Grecia* III, 88.
72 A. Johnson 1983.

Figure 26.18 Map of Roman colonies of the 3rd century BC. (JT)

but if this were a marching camp it did not have the shape or proportion of those described by the authors cited above, for which a 3:2 ratio of length to breadth seems to have been the rule. Nor was it like the surveyed examples of the 2nd to 1st centuries BC Roman Republican military camps in Spain.[73]

What our castrum does resemble in many ways are the forts at Ostia, Pyrgi, and Minturnae that the Romans founded as *coloniae marittimae*.[74] These are the examples which have left substantial physical remains, but there were many others dating as early as the 4th and 3rd centuries BC—such as Sinuessa near Minturnae, and Castrum Novum at Santa Marinella near Santa Severa (Pyrgi), both forming pairs, separated from their counterparts by 10 km or less (*Fig. 26.18*).

The similarities between the Metapontine castrum and the so-called castrum at Ostia,[75] the earliest of these, dating early in the second half of the 4th century BC, are striking. Both are sited close to a river and the sea, rectangular in shape, and planned around

[73] A. Johnson 1983, 222–27. The oldest known to the author, Renebla, Numantia, in Spain, has been dated to 195 BC and thought to have been built by Cato the Elder.

[74] A stimulating and provocative article by Crawford (2003) calls attention to these sites and asks why more attention has not been paid to them in discussions of the castrum at Metaponto.

[75] Meiggs 1973, Ch. 3; the earliest phase is called the "castrum."

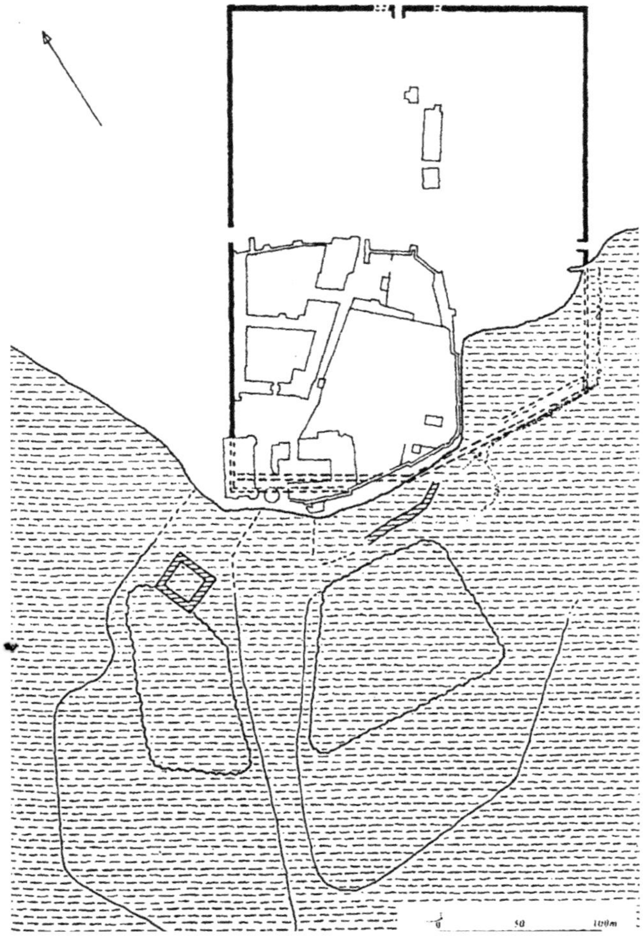

Figure 26.19 Pyrgi. Plan of the fort. (After Oleson 1974, Figs. 1 and 3)

an approximate east-west and a north-south axis (decumanus and cardo). Both were probably an extension of the organization of space into the interior beyond the coast and close to a source of wood for shipbuilding. Neither was mentioned in the literary sources.[76]

In the 4th and early 3rd centuries BC *coloniae marittimae* were the first line of defense against a seaborn invasion of Italy. The Romans began seriously to develop a fleet in the 3rd century BC. The need to cut the Carthaginian supply line to Sicily in the First Punic War proved a major stimulus. The castrum at Metaponto would hardly have been necessary to keep this defeated and rather helpless former enemy from mischief, nor to defend it from a Lucanian threat. That would have been a possible reason for having a praesidium at Metaponto before the Pyrrhic War, but afterward the Lucanians and the Bruttians had been controlled and were incorporated into the Roman commonwealth. We hear of Roman praesidia of considerable size in the rebellious poleis in the 270s BC, but not again until the time of the Second Punic War. In the interim, there were no local threats that would have required a fortress the size of the castrum at Metaponto. In terms of surface area (about 14 ha) it was large enough to house comfortably an entire legion.[77]

In shape and proportion the castrum most resembles that at Pyrgi (*Fig. 26.19*),[78] which was constructed at an unknown date early in the 3rd century BC.[79] Both are rectangles with the narrow end facing the sea. In both cases a corner is truncated—at Metaponto the corner between the river and the coastline; at Pyrgi, the southeast corner fronting the sea. It would appear from the plan, which is based on an aerial photograph,[80] that the modification was caused by the presence of a narrow channel separated from the sea by a dune or breaker.[81] The ratio of the length and breadth of the walls at Pyrgi is 4:5, which is identical to that at Metaponto. The walls of Ostia instead are wider on the side facing the sea and the ratio is 3:2.[82]

In his discussion of the planning of Ostia, Brandt suggests that there may have been a relation between the planning of the town or castrum and the area of the land to be distributed among the colonists and he entertains the possibility that the traces of land division in the division in the Imperial period of Ostia may not reflect centuriation but an older type of division, *per strigas*.[83] We do not know whether the castrum at Metaponto was related to the forming of a colony. If so, like Ostia, it is not mentioned in the literary sources, but if it were so, and presuming that there was no Lucanian opposition, the colonists would have found a ready-made division of the land, of the per strigas variety, a relic of the polis in **300**. The decumanus of the castrum of Metapontum, Plateia A, continued through the Western gate, according to De Siena's re-

[76] Meiggs 1973, 18.

[77] Estimated to be 4,200–5,000 strong in the 3rd c. BC. The surface area of the Polybian camp to accommodate two legions was 36 ha: A. Johnson 1983, 27.

[78] R. Brandt 1985, 65–87.

[79] R. Brandt 1985, 67. Probably after 273 BC, when the territory became Roman and Cosa was founded. In the same territory *Castrum Novum* was found in 264 BC.

[80] R. Brandt 1985, Fig. 19.

[81] Oleson 1974.

[82] R. Brandt 1985, 32, Fig. 4. Brandt observes that a 10 iugera town with a ratio of 4:5 would have been easier to lay out.

[83] R. Brandt 1985, 50.

construction, into the center of the chora along division line L13.[84] If land were distributed to new colonists in the 3rd century BC, a reasonable place to look for new settlement would be along the decumanus. Can it be pure coincidence that very near division line L13 (at 500 and 200 m respectively) lie two sites that spring suddenly to life in **250**, in the midst of a general decline in the population? These are the unique ex novo and quasi ex novo sites in this period, site 745F, with the highest FMCE value (3.00) in the chora in both this and the next period (3.09)—after which it disappears—and the revived site 734F/T, which was previously significant in **550**, three centuries earlier.

The castrum at Metaponto was more than five times as extensive as those Ostia, Minturnae (founded in 296 BC), and Pyrgi, all of which cover about 2.5 ha. What might have justified creating a castrum of such relatively enormous size? With the elimination of Pyrrhos, and the subjugation of the Greeks, Samnites, and Lucanians, Rome could now turn her attention to the growing threat of Carthage.[85] A fortress to protect one of the richest sources of grain on the peninsula would have made good sense. In fact, Hannibal's later reliance on Metapontine grain, as noted earlier, confirms its strategic importance. And this may very well have been the reason for the Roman praesidium there, even before the outbreak of the Second Punic War, but there may have been another strong motive. Could the creation of this castrum not have been prompted by the realization at the outbreak of the First Punic War in 264 BC that the contest could not be decided by land forces alone?[86] That a fleet had to be constructed, and that would best be done near the sources of basic materials by a work force experienced in shipbuilding? These could be found nearby, at Taras, still the leading maritime city in Italy even after its defeat. The *socii navales*, after all, which Taras became in defeat, were required to furnish ships and seamen.[87]

The forests were the major source of strategically vital material that needed to be protected. In the choice of the site for the colony at Ostia Meiggs mentions that the stand of coastal pine trees offered a ready source of material for shipbuilding, but adds that when Polybios refers to the building of large fleets he does not tell us where the ships were made. The pines of Ostia are lovely, but the best wood for shipbuilding available in the western Mediterranean, as Meiggs in his study of timber informs us, citing Theophrastus, was southern Italy.[88] Vast, mixed forests covered the Alpine plateau of the Sila, with hardwoods and conifers. No less impressive were the forests of interior Lucania.[89] Streams which become torrents in spring, down which logs could be floated to the sea from the Sila, flow past Croton (the Neto) and Thourioi (the Crati). Past Metaponto flow two major rivers. The Bradano and Basento drop nearly 1000 m over their 100-km course from deep in the Lucanian forest to the coast. They, too, could overflow their banks violently, flooding the valleys, as we have remarked on numerous occasions,[90] but according to our sources, as noted earlier, the Basento, at least as late as the Medieval period, was navigable for some distance upstream. The most useful trees for shipbuilding were the firs,[91] and to a lesser extent the oak. Both are major components of the Lucanian forests, and were in ancient times. Pine was preserved also, and its primary importance was as a source of pitch. From the paleobotanical research at Metaponto we know that fir was used to make planks,[92] and the oak appears to have been a major component of the forests near Metaponto, together with maple and pine.[93]

The castrum located at the mouth of the two rivers provided that protection and could have accommodated a large work force engaged in the construction of ships. The smaller amphorae and table ware production that continued after the closure of the kerameikos may, at least in part, have been destined for such a group. We have not a shred of evidence to indicate that this was in fact going on in Metaponto in the 3rd century BC, but then the port itself has only recently been identified, and it and the surrounding dunes may hold the evidence of shipsheds of the period. None of the dates proposed for the creation of the castrum has greater probability, in my estimation, than the fourth

[84] The slight difference in orientation of Plateia A and the division line is negligible and falls within the margin of error in assigning a direction to the broad and blurry anomalies that mark the division lines in the aerial photographs.

[85] Meiggs 1973, 24 n. 7, 25 n. 1.

[86] Meiggs 1982, 140–04.

[87] Ciaceri 1931.

[88] Meiggs 1973, 25.

[89] Brunt 1971, 359. For Roman interest in the Sila, besides Toynbee, see La Torre 2011.

[90] See Ch. 22, 736–37.

[91] Meiggs 1983, 118.

[92] Carter 2006a, Fig. 1.18 (typo for plank, not "plant"). Cf. *Territory of Metaponto* (35) for a photograph of the plank.

[93] Carter 2006a. For a pollen diagram, see Fig. 1.20.

decade of the 3rd century BC. Meiggs maintains that the most important stage in the build up of the Roman navy came in 267 BC with the creation of additional quaestors to equip and keep in repair a fleet.[94]

The castrum no doubt continued to play an important role throughout the period between the First Punic War and the Second, when it fell under Carthaginian control for five years. After Hannibal's evacuation and withdrawal to Bruttium, and the final defeat of Carthage in Africa in 203 BC, its importance to the Romans would have declined considerably. The excavator, Giardino, reports that the deposit of amphorae was destroyed at the end of the 3rd century by a fire (set by Hannibal in retreat?), and never again utilized.

Hannibalic Metaponto

While the presence of Hannibal in Metaponto is something of a parenthesis, it does mark a turning point, and was, ironically, the best documented episode in the history of the polis—though at this time the appropriateness of the term "polis" is debatable. With Pyrrhos, Metaponto had ceased to strike silver coins, one of the hallmarks of an autonomous city. The coinages of Taras and other poleis continued at least for a while after the departure of Pyrrhos, e.g. Thourioi, Herakleia, Velia, but not at Metaponto, Croton, or Locri, whose silver coinage stopped soon after. About the half-century preceding Hannibal's arrival, as we have seen, there is much room for conjecture. To what extent and in what form can Metaponto be said to have existed? For a while after Hannibal, if the sources are to be believed, no one lived in Metaponto.

Metaponto was not among the first of the former autonomous poleis that, together with the Lucanians, Bruttians, and others, went over to the Carthaginian in the aftermath of Hannibal's overwhelming victory over the Romans at Cannae in 215 BC. The list includes all those who sooner or later turned against the Romans.[95] The Roman garrison at Taras, with its fleet, the most powerful of the rebellious Greeks, was betrayed to Hannibal by the democratic elements of the population only in 212 BC, three years after Cannae. The Metapontines massacred the small Roman garrison that had stayed behind in 212 BC,[96] after which Metaponto became a base for Hannibal, and an especially important one from 210 to 207 BC. The chora, as noted, had already become an object of his attention as early as 214 BC,[97] and he supplied his winter camp at Salpia with grain from Metaponto and Herakleia. This is ample testimony to the productivity of these cities at this time. Metaponto, Thourioi, and Herakleia were on the Carthaginian side from 212 BC. The closeness of Metaponto and Thourioi, born already in the 5th century BC, was remarkably durable. After a naval battle in Tarentine waters in which the Tarentines were victorious, Metapontines and Thourians captured the Roman ships.[98] In that same year Hannibal transferred the potentially rebellious population of Herdonia in northeast Apulia en masse to Metaponto and Thourioi.[99]

After the Romans had retaken Taras, having used Hannibal's tactics[100]—"the Romans also have their Hannibal," said Hannibal—he withdrew to the Punic praesidium at Metaponto, and from there he attempted to arrange the ambush of the Roman commander at Taras. The plot, as reported, was hatched with the collaboration of the leaders of Metaponto, the *principes civitatis*, but the augural chickens did not cooperate and the unfavorable auspices saved the Roman commander Livius's life.[101] This episode and the original massacre of the Roman praesidium illustrate the unpopularity of Roman rule and show that there was a united front at Metaponto, with the principes, not just the people in opposition. The principes may have been particularly glad to be rid of the Roman praefectus, but this is putting a lot of weight on a literal interpretation of commonplace terms like principes. And civitas might just have been a handy term for referring to a site that had been a famous Greek polis; it need not imply that all the components of a true polis were in place in the late 3rd century BC on the site, and that there was a fully functioning government. Principes might simply have referred to the strong men of the moment, not the nobility of a fully functioning society of the sort that had once existed.

There is some confusion about the events of 207 BC, and Metaponto's role may well have been less prominent than the sources seem to suggest.[102] Hannibal

94 Livy 9.30.4.

95 Livy 22.61.11–13.

96 Livy 25.15.6.

97 Livy 24.15–16.

98 Livy 26.39.18.

99 Livy 27.1.4.

100 Livy 27.16.13.

101 Livy 27.16.13–10.

102 Giannotta 1980, 57.

probably sent Hanno, who had been in charge of the praesidium, into Bruttium to recruit reinforcements. The announcement of Hasdrubal's defeat at Metaurus, accompanied by his head, was more likely to have been made at the *Castra Hannibalis* at Concesium than at Metaponto. It was, however, from Metaponto that Hannibal withdrew from Lucania into Bruttium in 207 BC, taking with him the whole citizen body (*civitatem universam*), the Metapontines' *excitos sedibus suis*. Are the transplanted citizens of Herdonia (who had recently been exciti from their homes) intended by this phrase or were they just a part of the larger, uprooted crowd?[103] He took with him also the Lucani under his jurisdiction (*qui suae dicionis erant*). Hannibal made good use of Metaponto. It offered a steady supply of food for the troops, and a convenient location between his recruitment base in Bruttium and the battlefields of Apulia and Lucania. If nothing else his attention shows that Metaponto continued to be quite agriculturally productive, though he left it a desert. The most important and vital feature of the 3rd-century BC site, the castrum, quite soon after Zama in 203 BC, ceased to have a military value. The habitation of the 2nd century BC, as we shall see, was a feeble re-creation of pre-Pyrrhic Metaponto.

103 Livy 27.51.13.

27
The Last Two Centuries BC

Joseph Coleman Carter

The Chora in 150–50

The organization in the chora between **150** and **50**, which represents nearly a third of the total span considered in this volume, was distinctly different from that in the preceding four centuries, as will become apparent in this discussion. The number of farmhouses fell well below the new low reached in **200**. Though very much thinned out, the pattern of villages and groupings of sites found in the previous centuries is still discernible in its general outline in **200**. By **150**, some of the previously most prosperous and thriving localities, especially certain of the Early Village Clusters and those Close to the City, had been totally and definitively abandoned.

The most immediately apparent change is that the farmhouse sites of **150** are always isolated from their nearest neighbors and surrounded by large expanses of open land. This is especially obvious in the center of the study area and on the north side. What clusters of sites there are, are very small—we should call them hamlets or vici rather than villages—and they occur mostly on the south side of the study area (*Fig. 27.1*). A nearly complete lack of evidence for burial in the chora raises the possibility that the land was being explored by a population that did not reside there year-round or disposed of their dead in a manner that has left almost no trace. Individual sites are found at lower elevations on fluvial terraces and along valley walls as well as on marine terraces, where the majority of sites had been concentrated from the 6th century BC onwards. On the whole sites are closer to the coast, the port, and what remained of the former urban center.

For the first time in nearly two centuries, there is a high percentage (32%) of completely new sites in **150** (see *Appendix C6*). The average significance of the chora as a whole was impressively greater in **150**—in fact, it surpassed the average for **300**, though only a seventh of the number of sites were occupied (see *Appendices C1a* and *b, C7*).

All of the significant changes together would indicate that a new beginning was made in the 2nd century BC. The historical sources are virtually silent about the status of the territory of Metaponto after the defection of the city and its abandonment by Hannibal. We must rely entirely on inferences from the archaeological evidence—especially the survey results. It is reasonable to assume that Metaponto was treated by the Romans as they treated that of other traitors, and that it became at least temporarily *ager publicus* in the late 3rd and early 2nd centuries BC. The distribution of sites and their characteristics, as we shall see, reveals a pattern that is consistent with that conclusion. These phenomena are best explained if we identify the new occupants or landlords as Romans or others operating under the Roman system.

The Distribution of Sites in the Study Area, 150–50

This chapter begins, as have all that deal with the settlement pattern and its development over time, with a detailed description of the spatial distribution of the sites. In the changed world of the study area in these last two centuries, it would be cumbersome and serve little purpose to frame the discussion in terms of the regions and localities that organized the analysis of the patterns in the earlier periods. The Early Village Clusters, for example, have all but disappeared. Instead, this analysis of a numerically much diminished population is organized geographically in larger areas: the major river valleys and the interior valleys, divided as before into 50-year periods.

The limitations of our evidence in this (as in every period) need to be kept in mind. Though representative, the coverage was far from complete. One of the biggest gaps, of particular relevance in the 2nd and 1st centuries BC, occurs in the interior valley of Fosso Appio.[1] The coverage of valley floors and edges was generally lacking.[2]

[1] Many were noted by the surveyors in the late 1960s. These are indicated on *Maps 4–16* in Vol. IV (see map key).

[2] Sites have nevertheless been found there. One very prominent example from this period is of a site not recorded by the survey, the large villa rustica known as "Destra Basento," discovered during the excavation for the SNAM Pipeline. See Excavated sites, below.

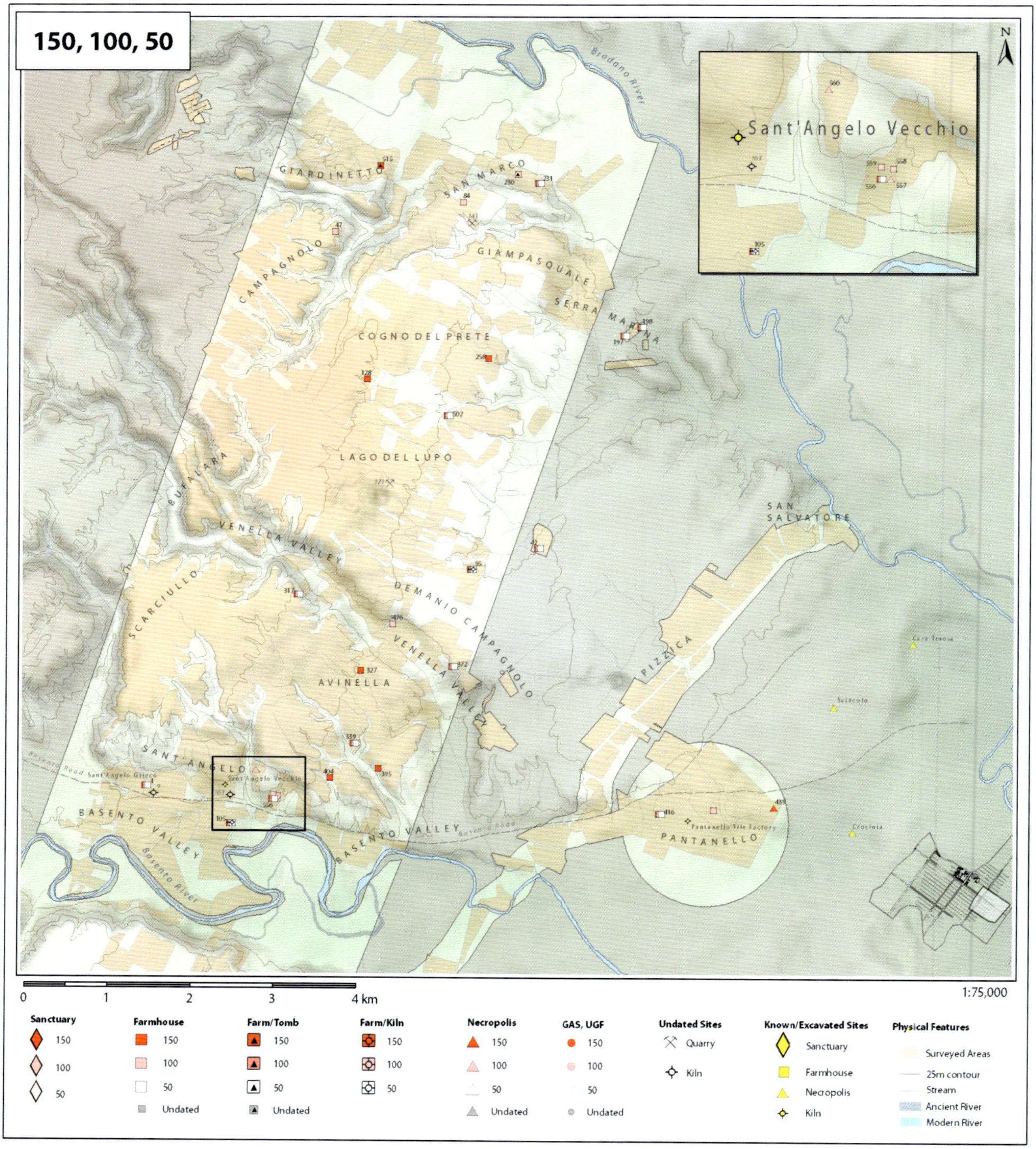

Figure 27.1 The study area in **150, 100, 50**. (JT)

The buried landscape of the period 150 to 50 was perhaps no more populous than the exposed one on higher ground, but the fact that such a high percentage of sites of this period line the edge of the valley does suggest that the settlers of this period found the valleys, with their easy access to water for animals and clay for ceramic production, particularly attractive. In this analysis of the settlement pattern, which is more complex and revelatory than the low numbers of sites might at first suggest, we shall consider excavated as well as surveyed sites.

The Mid-2nd Century BC (150).

Major river valleys (Fig. 27.1; Map 14, Vol. IV*).* Along

the Basento two sites in Sant'Angelo, the excavated site Sant'Angelo Grieco (1F) and 105F/K, were occupied continuously between **150** and **50** and both continued in the Early Imperial period. These sites were 1.1 km apart and in a thoroughly surveyed area, so we may be sure that this was their true spacing. Site 105F/K with 556F and the excavated kilns at Sant'Angelo Vecchio (SAV) form a nearly equilateral triangle with sides half a km long, through which ran the Fosso Sant'Angelo. The kilns at SAV were active in the period **150** to **100**, if not longer. Pottery was being produced at 105F/K, though the precise period cannot be known, and this was the case also at Sant'Angelo Grieco, where a kiln of still undetermined date was excavated.[3] The easternmost site of the triangular configuration was joined in **50** by two others to form a small hamlet (see Hamlets below).

The Basento valley to the east of Sant'Angelo was surely populated at this time, although the point where the Conca and Venella join the Basento valley was not covered by the survey (because it was excessively disturbed).[4] Survey coverage picks up again along the Basento valley's edge at Pantanello. It covers a stretch that runs from a point about 4.5 km from the city's walls to within 1.5 km of them. Two sites, 416F and the tile factory at Pantanello,[5] a large building with kilns perhaps associated with a villa rustica (the working part of a larger one, probably) are separated by slightly less than 0.5 km. The presence of water, in the form of an all weather spring,[6] once the focus of a major Greek rural sanctuary (see *Fig. 20.33*),[7] and deposits of good clay, were the resources being obviously exploited by the tile factory. Both sites were continuously occupied from **150** to **50**.[8]

The picture of the Bradano side of the study area is notably less complete than that on Basento, since survey of the valley floor was not productive and excavation did not deal with any site of the period under consideration here.[9] Sites of the middle of the 2nd century BC identified by the survey are more widely spaced. Site 515F in once densely populated Giardinetto lies nearly 2 km upstream from site 231F in San Marco, which in turn is 2 km above the pair of sites 197F and 198F in Serra Marina. Of the six sites documented in the very small area of Serra Marina (0.61 km^2 or 2.3% of the study area) there were two farmhouses (197F and 198F) in **150**, and only one site (197F) was significant before **150**. Such a close spacing of sites and the regular intervals between small groups of them is unusual in **150**, but became more common in **100**.

Interior valleys. The two sites along the Venella that were continuously occupied from **150 to 50**, sites 317F and 372F, like those along the Bradano, are 2 km apart; and here because the survey coverage is virtually complete we may be quite certain that there were no intervening sites. It is precisely 2 km as the crow flies between 372F and the farmhouse at San Biagio at the mouth of the Venella. It would appear that these sites were placed on opposite banks to control equally long stretches of an inland waterway and pasture. In **150** the mid-point between the two, on the west bank, was occupied by 327F, which survived only through this period. In **100**, on the opposite bank, at the mid-point between 317F and 372F, another, 476F, was occupied, again for no more than half a century.

A triangular constellation of sites (404F, 339F, 395F) between 400 and 600 m apart, similar to that at the mouth of the Fosso Sant'Angelo, occupied the bluffs on either side of the Conca drainage. It existed only in **150**, as two of the sites, 404F and 395F,[10] were at last abandoned after four centuries in **100**.

Two major architecturally impressive sites,[11] 42F and 95F/K in Demanio Campagnolo near the center of the chora, were both utilized throughout **150** to **50**, and were probably not the only ones along the Fosso Appio, which, as noted above, was only very selectively surveyed. Site 42F, in fact, lies just outside the original transect. The Appio's gently sloping banks, in contrast to those of the Venella and Conca, offer wide vistas. To the north and west of 95F/K are the near-

[3] Sant'Angelo Grieco will be the subject of a future study.

[4] A Late Roman farmhouse at San Biagio was excavated on a previously occupied site. It was situated so as to overlook both the Venella and the Basento valleys. It is the subject of another volume in this series.

[5] The subject of a future volume in this series.

6 At the heart of the sanctuary active from the 6th to the early 3rd c. BC.

[7] Subject of a future volume in this series.

[8] And on the site of the tile factory, a kiln continued to operate in the Early Imperial period and perhaps later. See Carter 1983a, 421–22, 447, for a preliminary publication of the factory.

[9] Excavated sites on the Bradano include three Greek farmhouses excavated in the 1960s (unpublished) in the Saldone locality, the Saldone necropolis dating to the 5th and 4th c. BC (*Necropoleis*, 429–42), a Bronze Age site at San Marco (see Ch. 17 above), and the major indigenous—and later Greek—and Lucanian-controlled hilltop site of Cozzo Presepe (just outside the study area to the northwest).

[10] 395F was again occupied in the Early Imperial period.

[11] See Gazetteer, 42F and 95F/K.

level fields of the Central Plateau. That area is best adapted to plow-based agriculture but could well have been used in this period as pasture land.[12] This pair of prominent sites is separated from site 502F in the Central Plateau by slightly less than 2 km. Site 502F was the only site there to have been active from **150** to **50**. In **150** it is part of a triangular configuration, which includes sites 128F and 258F. Neither of these sites survived beyond **150**.

A typical pattern in **150** is one of sites of long duration, usually separated by about 1 km. These are often associated with two sites of short duration, located about 0.5 km away. This is the case at Sant'Angelo, the Conca valley, and the Central Plateau. The components of these groups of three are too widely separated to imagine that they constituted a hamlet. The individual sites probably operated independently, each controlling a certain area. By **100**, in the Conca and Central Plateau at least, only one remained in these groups where three sites had been before. The impression that the multi-period sites were larger and gobbled up the satellites around them is supported by the evidence of the FMCE values of the components of these triangular configurations. Those of the multi-period sites are consistently higher (*Appendix B*).

Late 2nd to early 1st century BC **(100)** (*Fig. 27.1*). Between 125 and 75 BC the number of sites occupied hardly changed at all. As noted above, several concentrations of sites, the triangular groups in the Conca valley in Avinella along the Basento and that in the Central Plateau around the Fosso Solfegna in Cogno del Prete and Lago del Lupo, contracted and one dominant site survived. Elsewhere new sites whose existence was also brief appeared along the major valleys, at Sant'Angelo, and San Marco along interior valleys, the Acqua Fetente and the Venella.

New additions 558F and 559F near the Basento are ex novo or practically so and appear beside the very significant 556F.[13] The three sites were very close together and constitute one of four possible very small hamlets which are the outstanding developments of this period. The only other ex novo site in **100** was site 428F, located also along the Basento at Pantanello.[14] It may have been the residence of the operator of the tile factory some 250 m upwind, and separated by a slight dip.

On the Bradano side, site 230F/T[15] joined 231F to form another close grouping similar to the one that existed at Serra Marina already in **150**[16] (sites 197F and 198F). The two sites 47F[17] and 84F, separated by 1.5 km, evince a renewed interest in the valley of the Acqua Fetente after a century and a half of apparent neglect.

The FMCE values of sites of this period were the highest on the average of any period in the history of the chora (*Appendix C7*). The individual sites that stand out among the rest are 339F in Avinella, 502F in Lago del Lupo, 556F in Sant'Angelo, and sites 42F and 95F/K in DBSBS. Apart from the last two they are distant one from another, but seem to favor the interior over the river valleys and all are reasonably close to the city. Sites Far from the City and along the river valleys, however, were not among the least important settlements.[18] Though FMCE values are not necessarily always an infallible guide to the relative importance of sites, it is reassuring to note that the two sites with large scatters and architecturally impressive surface remains, 42F and 95F/K, are also among the very few with FMCE scores higher than 4.0 in **100** (*Appendix B*).

The middle of the 1st century BC **(50)** (*Fig. 27.1*). In the preceding period there had been two ex novo sites (559F and 428F), neither of which survived in this period, in contrast to the six ex novo sites of **150**, all of which were active until **50**. In **50** there were no new farmhouses, and the numbers of previously existing ones had fallen from 20 to 13; the recorded necropoleis, from two to zero. The largest losses were experienced in the localities Far from the City. In **100** there had been eight significant farmhouses, the largest number in any of the regions.[19] In **50** that number

[12] To the east and south of 42F, the almost equally vast expanse of arable terrace land in Pizzica was probably utilized for olive culture and as pasture at some time before **150**, as we know from paleobotanical investigations. Mercuri 2009.

[13] 558F had been significant in **300**.

[14] 428F was also occupied in the Early Imperial period.

[15] A very significant site in **450**.

[16] A modern village of Serra Marina (see below). The sites are listed as belonging to DBSBS.

[17] Possibly, with 49F, a 2nd magnitude farmhouse in **100**, part of a small hamlet.

[18] FMCE values could not be calculated for Sant'Angelo Vecchio (SAV) and the tile factory at Pantanello.

[19] Six Close to the City; 5 in the Early Village Clusters; and 3 in the Central Plateau.

had fallen to three. The areas least visibly affected by the decline were the hamlets at Sant'Angelo and Serra Marina, equidistant at 7.5 km from the settlement on the coast, along the Basento and Bradano valleys.

The average FMCE value for the study area declined with respect to the preceding period, but was slightly greater than in **150** (*Appendix C7*). In the sites continuously occupied, the increase between **150** and **100** is probably due in large measure to the relatively higher quantities of grey ware, on which the chronology of this period is based. Local production of Metapontine Greco-Italic and collared Metapontine amphorae at Pantanello and probably also in the urban center continued in this period,[20] though these are poorly represented in the survey assemblages (see *Fig. 275*; *Appendix C39a*). They would have served local wine and possibly olive growers. Evidence from other sources, such as pollen, and evidence for cattle breeding, indicates that the agricultural economy was enjoying a modest revival in **150–50**, especially around **100**.[21] Agriculture would not necessarily have been excluded by large scale pastoralism, which, when reintroduced in the 2nd century BC, appears to have grown in importance throughout the 1st century BC.[22] The larger drops in FMCE values between **100** and **50** occur among sites that were ex novo in **150** (*Appendix B*). Two sites, 339F and 502F, out of four ex novo sites had absorbed smaller neighbors. Half of those whose FMCE values were least affected, and had declined less than 20%, were at some point centers of ceramic production, for example 1F (the excavated Sant'Angelo Grieco) and 105F/K, and they continued to be occupied in the Early Imperial period. This was also the case at 95F/K, also a ceramic production site, the Pantanello tile factory, and its near neighbor 428F.

Hamlets. Clustering of sites takes place in all periods of the chora. The Early Village Clusters and those of the Central Plateau were constant features of the settlement pattern. The clustering that took place in the last centuries BC was ephemeral, and it seems to have been more limited with a more precise economic scope.

The pair of two significant farmhouses at Serra Marina (197F and 198F, separated by just 200 m) stood out in sharp contrast to the other looser concentration of sites at Sant'Angelo, Avinella, and the Central Plateau mentioned above. There the distance separating the components was never less than 400 m and usually closer to 500 m. In **100** three more such close groupings of sites appear at Sant'Angelo along the Basento, at San Marco along the Bradano, and possibly around site 95F/K along the Appio. All of them lie along roads, the Basento Road on the Basento side, the road or *tratturo* that led into the interior along the Bradano,[23] and the road corresponding to division line L15 that appears to have continued to serve Pizzica and the Central Plateau in the last two centuries BC.

A notable aspect of these new clusters is that they included a center of ceramic production or they were reasonably close to one. The cluster at Sant'Angelo was 600 m to the east of the kiln at Sant'Angelo Vecchio and 600 m from site 105F/K. Though not identified in the survey, the San Marco cluster may also have had a pottery production center. The clusters at San Marco and Serra Marina lie very close to the limits of the surveyed areas on the Bradano. There is a possibility that other components of these clusters could lie hidden under the alluvial fill of the valley. The lifespan of these new clusters in **100** was very short. They had all disappeared before **50**.

Sant'Angelo. The most unusual of the four clusters was certainly that at Sant'Angelo. Here the distance separating the three component farmhouses dating to **100** is about 50 m. This is a fraction of the distance between farmhouses in the other hamlet-like clusters. Like 230F/T at San Marco, it may have had associated burials.[24] Necropoleis are very rare after **300** and **250**, but there is no doubt that the one at site 560N, 400 m to the north on the plateau above the Sant'Angelo cluster, and opposite the kilns at Sant'Angelo Vecchio, was in use in **100** and continued to be used in the Imperial period.

The survey in the area where these farmhouses are located was very restricted and it is possible that the settlement was more extensive. It had attracted settlers in earlier periods, and all of the farmhouses and necropoleis had black-gloss pottery of **400** and **350**. Site 556F, the predominant member of the clus-

[20] See Ch. 12, 462–64.

[21] Carter 1994b, 185–87.

[22] Bökönyi and Gál 2010, 54–59.

[23] Nava 2003a, 666.

[24] There was a doubt that the one piece of grey ware recovered from 557N might have washed down from the farmhouses (see Gazetteer).

Figure 27.2 Sant'Angelo, site 162, near 105F/K. A number of straight, tile-lined drainage channels were visible on the surface. They may well have played some role in pottery production. (CDA)

ter, had an FMCE value of 3.30 in **100**.[25] The very limited space around the structures suggests that the occupants in **100**, and earlier as well, may have been engaged in non-agricultural activity. The proximity to two centers of ceramic production (Sant'Angelo Vecchio and 105F/K) is suggestive. Site 105F/K, as noted earlier, was a very extensive site with a dense scatter measuring 6,300 m^2 with little pottery. Hence it may have been primarily a site of pottery production as opposed to residences in the way that farmhouses were.[26] Two of the three components of the cluster were rather modest in terms of FMCE value and area of surface scatter. Sites 558F and 559F had FMCE values of 1.43 and 1.78 in **100**, and areas of 625 and 1600 m^2 respectively (both with moderately dense scatters).[27] 556F had an FMCE value of 3.30 and a heavy scatter measuring 1000 m^2. Taken together, these facts suggest a scenario in which site 556F served as the residence of a manager or *vilicus*, with subordinates housed nearby, who was responsible for a factory producing roof tiles and pottery on the bank of the Basento at site 105F/K (*Fig. 27.2*). Of course, a number of other scenarios are possible, but if 105F/K was linked to another site, it would likely have been 556F, not 1F or 4F, which, at 1.1 km, were twice as far away.

Serra Marina, San Marco, and Fosso Appio. This pair of sites (197F and 198F) appeared for the first time in Serra Marina in **150**; a second pair emerged in San Marco, when site 231F, ex novo in **150**, was joined by 230F/T. The two pairs of sites, 2 km apart, may have been part of a larger array of sites placed at regular intervals along a main thoroughfare. Serra Marina is located at the mouth of the Appio valley and so may have served as a junction point for transport of agricultural products from the valley along the main road from the interior toward the coast, much as does its 20th-century successor, Serra Marina, a modern village created during the agrarian reform of the 1950s. San Marco, 2 km farther along that road toward the interior, may have served a similar purpose. It is placed just about equidistant from the Fosso dell'Acqua Fetente and the Fosso Appio. In contrast to the situation at Serra Marina some six sites in San Marco were found in close proximity to sites 231F and 230F/T. Unfortunately the majority are so damaged that neither their function nor date can be known. They include: 232N, 408N, 233UGF, and 234GAS, and 409GAS (all undated). If these sites ever constituted a hamlet it was a fairly loosely organized one.

More problematic is the collection of sites around 95F/K on the western slope of the valley of the Fosso Appio. Site 539F, a second magnitude site in **150** and **100**, was 200 m distant; the undated necropolis 473N, about 50 m away. Between 200 and 300 m away were 475F, a much earlier unidentified Greek feature, and two farmhouses 538F and 540F of the Imperial period. Here, as with all of the other potential hamlets, the survey coverage of the surroundings was unfortunately limited and the conclusions must consequently remain very tentative.

Areas of control. The farmhouses of the last two centuries BC, as shown above, are with few exceptions widely separated from their neighbors. We have recognized close groupings as nascent, rural hamlets or vici, all of which were newly formed in **100**. The others stand proudly alone, with distances ranging between 0.45 km and 1.90 km separating them from their nearest neighbor. The distance in **150** between individual farmhouses is approximately 1025 m on the average (*Appendix C12*). In **100**, the distance between farmhouses and hamlets, considered as a single unit, is

[25] It was significant in **300**. The necropolis 557N was active from **400** to **300** and the farmhouse 558F had been a second magnitude farmhouse in **550**, **400**, **350**, and **250**, though significant only in **300** and **100**.

[26] It was significant from **400** to **250**, and again in the last two centuries BC and in the Early Imperial period as well.

[27] See Fracchia 2001.

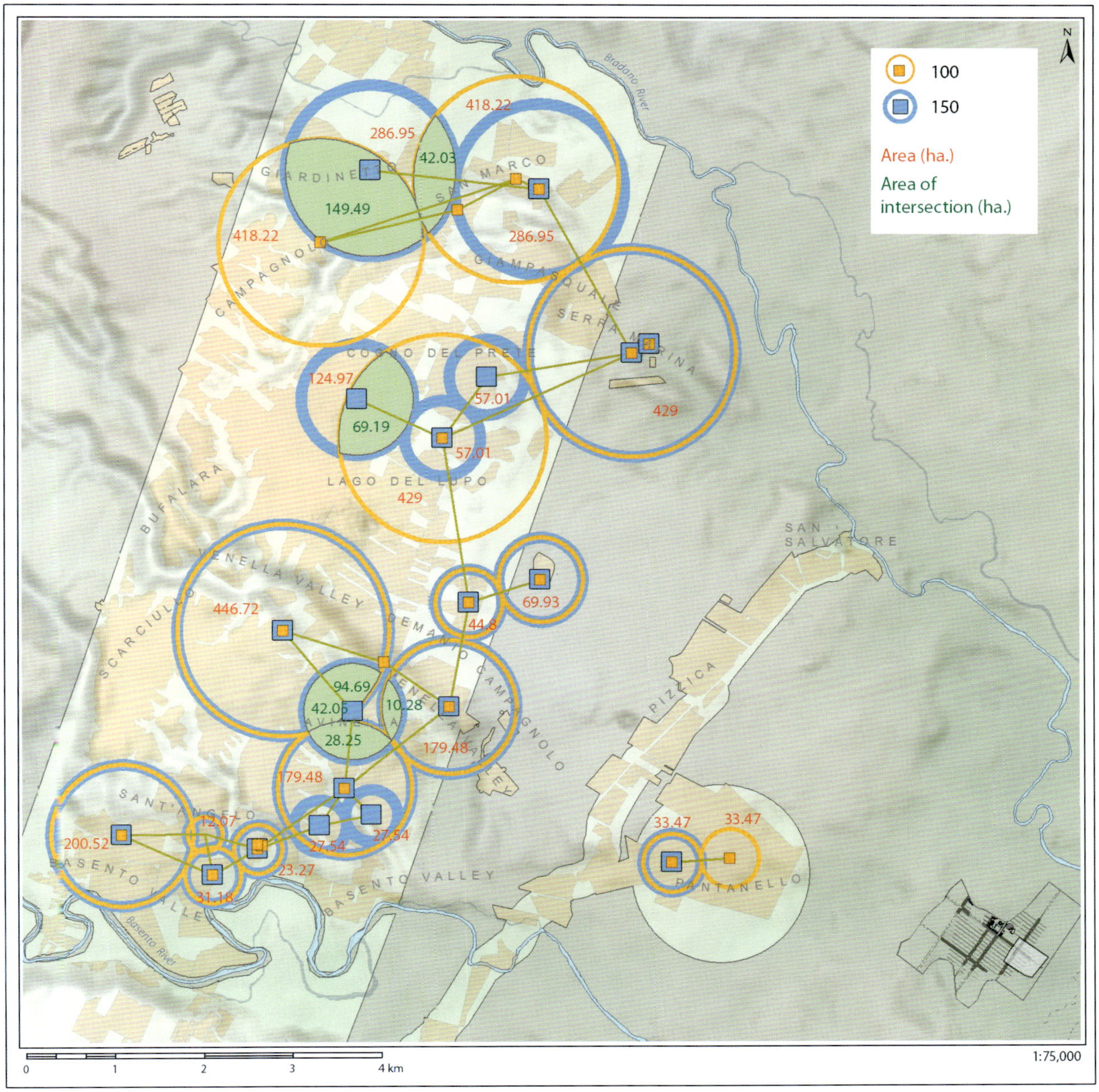

Figure 27.3 Hypothetical spheres of control, **150** and **100**. (JT)

approximately 1100 m.[28] The increased space between sites in the open countryside in **100** suggests that the land around sites also increased in that period. This is borne out only by a hypothetical calculation of the area of land that might have been controlled by individual sites in **150** and **100**. For convenience, circles with the site at the center were drawn to suggest these spheres of control (*Fig. 27.3*).[29]

Circles drawn around sites in existence in **150** have been colored blue, those that came into being in **100** are orange, and those that existed in both periods have bicolored circles. The radii of the circles were limited, so that the circles do not overlap. There was only one instance where that was not possible: the circle around 327F.[30]

[28] The discrepancy between this figure and that in the tables, 804 m (*Appendix C12*), is explained by the fact that in the tables the distance between all farmhouses including those in hamlets were averaged.

[29] A similar result might have been attained using polygons.

[30] The radius was taken as half of the average distance between three surrounding sites: 372F, 317F, and 339F. Sites 404F and 395F are wholly contained within the circle around 339F.

In cases of overlaps, the appropriate subtraction was made,[31] e.g., the entire areas of 404F and 395F were subtracted from the area of 339F in **150**, and the portion of 327F which was active in **150** but not later was subtracted from the areas in **150** of farmhouses 317F, 372F, and 339F. The areas and adjusted areas are recorded in Table 27.1. These calculations are intended to approximate the maximum area that a farmhouse may have controlled. Of course, the territory under a landlord's control may have been smaller or irregularly shaped, but the construct was designed to give an idea of the order of magnitude of a holding. What is surprising about it is that though there is considerable variation, with some very large areas and others quite small, the average is 136 ha in **150** and 222 ha in **100**.[32] The actual values are probably too high, but they make it possible to attempt to link the archaeological situation, the actual distribution of sites in this period, with what is known of Roman agrarian history and legislation in the period.

The amount of land that an individual citizen could possess was limited by the Licinian Law *De Modo Agrorum* to 500 *jugera* (about 125 ha). The date of the law has been disputed. Rathbone has argued for an early-3rd-century date.[33] In any case, it was well-known and apparently respected when Cato delivered his oration *Pro Rhodiensis* in 167 BC.[34] It has been taken to apply to the ager publicus, but probably limited private ownership of land in general. The limit was raised to 1000 jugera for a Roman citizen and two sons in the Gracchan legislation of 133 BC.[35]

The analysis of the maximum areas of control around farmhouse sites in the study area of the Metapontine chora in **150** appears to indicate that a third of the areas exceeded the 500 jugera as a maximum. They may, of course, have been considerably smaller. The analysis of the maximum areas of control, and the average values of the land surrounding farmhouses in the territory of Metaponto in **150**, shows that only 6 out of 18 areas of control exceeded the 500 jugera maximum limit; in **100** that number rose to 8 out of 13 and the average area of control to 885 jugera (*Table 27.1*). Could it be that the local situation at Metaponto roughly reflects the historically attested increased limit, permitted by the Gracchan legislation? Perhaps, but in any case, the division of the land at Metaponto has undergone a fundamental change that is best explained if we identify the new occupants or landlords as Romans, or others operating under the Roman system, and if the Metapontino became *ager publicus populi Romani* in the 2nd century BC, and that like most Roman land in conquered territories, it quickly became private property.[36]

Site	150		100	
	Area (ha)	Area (jugera)	Area (ha)	Area (iugera)
1F (SAG)	201	804	201	804
105F/K	31	124	31	124
SAV	12	48	12	48
556F	23	92	23	92
404F	28	112	—	—
395F	28	112	—	—
339F	96	384	179	716
327F	95	380	—	—
372	169	676	179	716
317F	404	1,616	447	1,780
95F/K	45	180	45	180
42F	70	280	70	280
502F	57	228	429	1,716
258F	57	228	—	—
128F	125	500	—	—
197F	429	1,716	429	1,716
230F	—	—	418	1,672
231F	287	1,148	—	—
515F	286	1,114	—	—
47F	—	—	418	1,672
Average	**136**	**544**	**222**	**885**

Table 27.1 Hypothetical size of areas around farmhouses, **150** and **100**.

An Overview of the Study Area in the Last Two Centuries BC

Between **200** and **150** the number of occupied farmhouses fell from 30 to 19, a decrease of 37% (see *Ap-*

[31] The entire areas of 404F and 395F were subtracted from the area of 339F in **150**, and the portion of 327F which was active in **150** but not later was subtracted from the areas in **150** of farmhouses 317F, 372F, and 339F.

[32] The equivalence in Roman land measurement of 544 jugera and 885 jugera, respectively (*Table 27.1*).

[33] Rathbone 2003, 146.

[34] Toynbee 1965, 554–61. Rathbone 2003, 144.

[35] Rathbone 2003, 161.

[36] Rathbone 2003, 141. Rathbone offers an incisive comparison of the major ways the Greek and Roman systems of landholding differ.

pendix C1a). What sets the fifty years between 175 and 125 BC off in terms of numbers of farmhouses is the sudden increase in ex novo farmhouses (see *Appendix C6*). The count of new farmhouses had been gradually declining but was still a healthy number, as the chora expanded throughout the 5th and 4th centuries BC, and then, abruptly, there was only one new farmhouse in **250** and none in **200**. In **150** the six new farmhouses represent 32% of the total active farmhouses, and two more were added in **100**. Though the numbers are low the rate of growth is impressive, like that of an earlier era. Any illusion of a revival, like that at the time of Alexander the Molossian, however, is belied by the other changes that were taking place. The political and military events of the Hannibalic War and their consequences had irretrievably altered the situation of the chora in **150**, and for the future.

Beside the sudden increase in new establishments on previously unoccupied sites, the other statistic that is most indicative of the changed nature of the settlement is the increase in the FMCE values of the sites occupied in **150** (see *Appendices C7, C38*). The average jumped from 1.85 in **200** to 2.14 in **150** and 2.63 in **100**, the highest in the entire history of the chora—higher even, though not by much, than the 2.57 of **300**. The obvious difference between these two periods is that the average of **300** is derived from the FMCE value of 142 separate farmhouses, while that of **100** is based on just twenty. Slightly more than half of the values of individual sites are below the average for the period; some are considerably below it; and others considerably above it. In terms of the distribution of high, low, and intermediate FMCE values, **100** resembles none of the previous periods. There are, in all periods considered together, only four FMCE values greater than or equal to 4.0. Two of them are found in **300**, the other two, in **100**.[37] Both of these periods have very significant sites, but in most other respects, they are quite dissimilar. A statistical comparison with the distribution in **100** and **300** reveals some of the rather important differences.[38] Considering all values in all periods, **300** has the highest percentage, by far, of sites with values in the highest range (2.60–4.335). In **150–50**, there are just slightly more sites in that range than would be expected. It is a far more egalitarian distribution than **300**, and far more than intuitively we might have expected.

Fewer sites—and there were far fewer sites in **150** than **300** or even **250**—does not necessarily mean more widely separated neighbors, but this is the case, and the distances between nearest neighbors makes this clear (*Appendix C12*). In **150** there are a number of individual sites that are more or less isolated, but also a number that cluster together in a generous, uncrowded way: at Sant'Angelo and Avinella on the Basento side of the chora, where two Early Village Clusters had once existed, but also at San Marco and Serra Marina on the Bradano where there had been no clusters before. In all these cases sites that were ex novo in **150** help to create the new clusters or hamlets.

The average distance between nearest neighbors increased relatively gradually after **300**. The real break comes in **150**, when the increase is over a hundred percent, as opposed to less than 40% in the two immediately preceding periods. There is a slight decline in **100** and by **50** the average distance is nearly 1.3 km, the greatest separation between neighbors in the history of the chora down to this time, surpassing even the figure for **500**, when the chora was practically deserted.

The average elevation of farmhouses rose continuously and almost without interruption for four centuries, from the days of the earliest settlements in the chora, and it reached a peak (83 masl) in **200** (**500** was exceptional also in this respect). The drop in the average from **200** to **150** therefore was quite a dramatic reversal. It decreased by 37% to 52 masl, and continued to trend downward in the 1st century BC (see *Appendix C10*).

Elevation, as noted earlier, is related both to the geomorphological unit in which a site is located and its distance from the city. The percentage of significant farmhouses located on the Pleistocene marine terraces of the Central Plateau and other regions also rose steadily from **600** onward, reaching a maximum of 97% in **200** (*Appendix C11a*). Only one (site 404F in Avinella) of the 30 sites in the chora at this time was located on a valley wall, not on a marine ter-

[37] In **300**: 296F Lago del Lupo (4.32), 46F Campagnolo (4.0). In **100**: 42F Demanio (4.17), 95F/K Demanio (4.02).

[38] The FMCE values in all periods were compared using analyses of variants (Anova) and chi-square tests (see *Appendix D*). The chi-square (chi-square test 1 in *Appendix D*; see the cross-tabulations, and standard residuals) indicates that the distribution in the period **150–50** is closer to those of the four other periods analyzed than that of **300**, which has fewer than expected lower and medium values and many more high values than **150–50**. Of all six periods, the distribution of low, medium, and high values in **100** varies least from what would be expected based on the assumption that the distribution is equal across all cells of the cross-tabulation.

race. This upward trend of average elevation above sea level was reversed abruptly in **150**. Only slightly more than half of the 19 significant farmhouse sites in **150** could be found on marine terraces. More than a quarter were on valley walls and the remaining ones on the alluvial T_2–T_4 terraces along the river valleys and one on the mass movement deposit in Giardinetto.

With the high percentage on valley walls, the settlement in **150**, and even more so in **100**, resembles that of the very first years in **600** of the Greek settlement of the chora. This cannot be entirely coincidental, since there was a definite shift away from the plow-based agriculture in the last centuries BC toward pastoralism, which was, as we have seen, an important component of the economy of the early settlers. A difference of course would be the almost certainly greater scale of the herds in these later phases, which would probably have included together with sheep and goats a higher proportion of cattle and horses. The valley provided protection, forage, and access to water and the terraces, extensive grazing land. Farmhouses disappeared from the Central Plateau because there was little need for them.

In **200**, the percentages of significant farmhouses in the areas Close to the City (3–5 km from the former asty) and Far from the City (9–11 km from the former asty) were lower than they had ever been before (*Appendices C9a* and *b*).[39] In the last two centuries BC (**150–50**), the percentages of farmhouses in these two areas remained low, while that of the farmhouses located between 6 and 8 km from the site of the former asty, the middle range between those Close to the City and Far from the City, rose steadily and was greatest in the last half of the 1st century BC. Already by **200**, the number of sites Close to the City was reduced to just one, and by far the largest number lay in the middle range. Sites Far from the City, from virtual parity with those in the middle range in **300** and **250**, were proportionately halved in **200**, and reduced further in **150**. The motive for the partial abandonment of this region was perhaps because the Central Plateau was better for the livestock. The virtual disappearance of sites Close to the City reflected the decline of the asty.

Assemblages, 100

Assemblages from sites with a best date in **100** differ from those with a best date in **600**, **550**, **450**, or **300** in an important aspect that requires special consideration and necessary corrections in the raw data. In the earlier periods, the individual farmhouses chosen for analysis had the largest concentrations of dateable pottery, the black gloss, in the period under consideration. There was often some residual pottery from an earlier fifty year period and some from a later period, but essentially the chosen sites were as close to single period sites as was possible in a survey situation. With farmhouses whose best date (an average of the date range of all dateable fine wares) is **100**, we are dealing with two main types of fine wares:[40] the black gloss (BG) and the grey ware (GW), dated to the 2nd and 1st centuries BC (see Ch. 7), which are concentrated in periods for the most part widely separated in time.[41]

The farmhouses with the largest amounts of pottery on the surface were in all cases occupied also in the 4th century and earlier as well as in the 2nd century. These we shall refer to as a multi-period. The group of sites occupied for the first time in **100** in the period **150–50** was, as we have noted, unusually large. These ex novo farmhouses generally have dense scatters of roof tiles and other building materials covering relatively large areas and comparatively sparse pottery remains on the surface. The analyses of assemblages from these farmhouses by ware class, and especially by functional groups, are less robust than those of farmhouses belonging also to an earlier period. One of the consequences is that disappointingly little may be said about the activities carried out at farmhouses that begin in this period.

The residual material on most of the multi-period sites with a best date of **100** is often quite considerable. Site 372F, for example, has 30 fragments of black-gloss pottery (and one red-figure sherd), but only 19 fragments of grey ware. The average of the date ranges of this pottery results in a best date of 98 BC, only because a precise date range could not be assigned to a large percentage of the fragmentary black-gloss pottery. For the purpose of this analysis, in order to give a more realistic picture of the nature of these sites in the last two centuries BC,

[39] It should be recalled that the much lower percentages of farmhouses in the 3–5 km range from the city (*Appendix C9a*) is due to the fact that the survey sample in that area is much lower than for areas farther out. The area Close to the City accounts for only 15% of the total coverage.

[40] Limited quantities of figured pottery including Gnathia style (Gna) pottery belong to the earlier period.

[41] The black gloss belongs to the period **550** to **200**, and the grey ware to **150** to **50**.

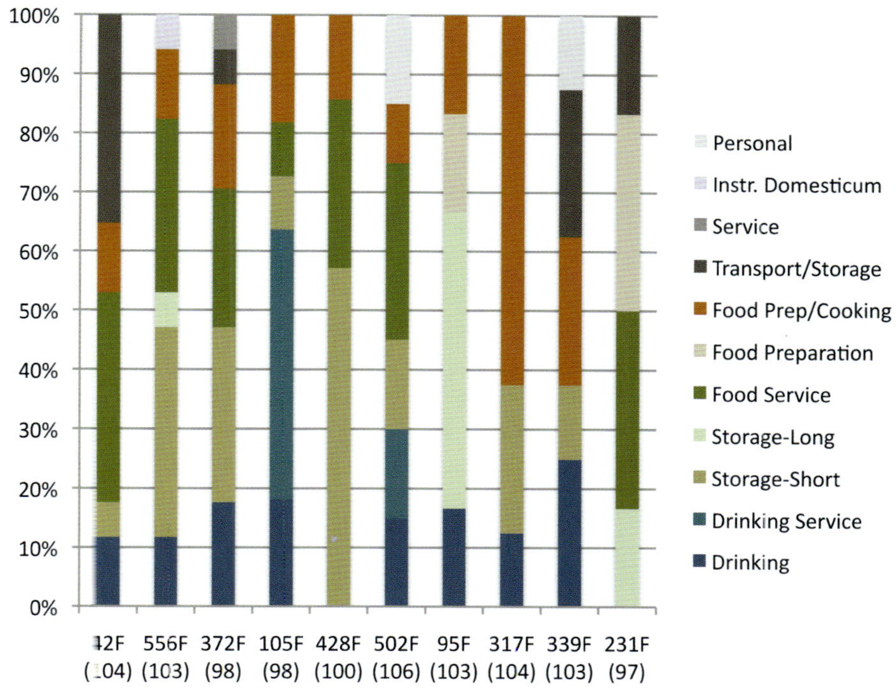

Figure 27.4 Farmhouse assemblages in **100**: functional groups—counts and percentages.

the residual black gloss dateable before **150–50** and undateable black-gloss pottery has been excluded from the assemblages. The black gloss actually dated to this period has been retained. The quantities of undateable pottery (BW, PW, CW, undated amphorae) have been reduced by a factor determined by the ratio of black gloss to grey ware. In the case of 372F, this ratio is 31:19, and so forth. Only two sites of this group, 556F and 42F, had proportionately more grey ware than 4th- and 3rd-century BC black gloss and other fine wares.

These changes result only in very approximate estimates of the quantities of the undateable plain/banded ware, cooking ware, and amphorae at the sites in this period, but it is the best that can be done within our limitations. Naturally we cannot know which particular fragments of plain ware, for example, were on the site in **100**, and therefore the analysis of functional groups in **100** is only possible at sites where grey ware (GW) is the predominant dateable pottery—these are sites 556F and 42F (*Fig. 27.4* and *Appendix C39a*). In the case of the ex novo sites, we can be reasonably sure that the undateable pottery is representative of the period, but only site 502F had a sufficient quantity present to make at least a reliable functional groups analysis possible.

It is fortunate that the one excavated farmhouse of the period, Sant'Angelo Grieco, was also identified as survey site 1F, weeks before it was excavated. This farmhouse, as many others, was occupied with interruptions from the 6th century BC to the 3rd, but is unusual for its continued existence in the last two centuries BC and into the 1st century AD. It provided a rare opportunity to explore the relationship between a surface assemblage and an excavated one—especially welcome, because there are relatively few known farmhouses in this period.

Functional groups in 100 (*Fig. 27.4; Appendix C39a*). The main problem that the assemblages of **100** pose is the small body of material. This is especially serious for analysis by functional group. In the case of the ex novo farmhouses of **100**, there are, with the exception of site 502F, fewer than ten fragments whose form and therefore function could be determined. It is hardly an assemblage at all. (For the period **300**, for example, there is an average of 84 fragments with identifiable forms per site.) We have included in this analysis the multi-period sites 556F and 42F which have a preponderance of precisely-dated pottery, and hence a best date in **100**. The ratios of functional groups from these two sites, however, inevitably include the residual material from the earlier periods on the sites, even after the adjustment described above, and so they must be viewed with caution.

The functional groups from ex novo sites are indeed very spare. That from 502F, the fullest assemblage with 19 fragments identifiable as to form, gives us the clearest idea of an undiluted functional group assemblage of **100**. Clear differences emerge when it is compared with the averages of functional groups in earlier periods (*Appendix C17*). There is a lower proportion of vessels associated with drinking service. There are lower than expected percentages of short-term storage vessels and cooking ware compared to those in **300**, but higher than expected food service vessels and vessels devoted to personal hygiene. Because there are so few fragments involved, these imbalances could, for example, be the result of having included two undated fragments of black-gloss pottery.[42] If they are removed from the analysis, in fact, the percentage of short-term storage and cooking ware from site 502F rises, along with the already unexpectedly high percentage of personal vessels—based on a grey ware bowl, lid, and unguentarium. Of the three vessels only the unguentarium is clearly related to personal hygiene. This should suffice to illustrate the problems of interpreting these small assemblages!

The situation is better in the case of the multi-period sites, such as 556F, with more than twice the number of identifiable forms (*Appendix C39a*). The percentages compare much more closely with the average for **300** (cf. *Fig. 25.29* and *Appendix C24a*). The percentage of vessels connected with drinking service is high again, and for the same reason. For the purpose of this analysis, only the vessels dated precisely before the period **150–50** have been excluded. Two large black-gloss vessels of uncertain date, but possibly belonging to the period **150–50** were, however, included.[43] Vessels corresponding to the drinking service function are absent from eight of the ten sites in this table. The percentage of short-term storage ware at sites 556F and 428F is considerably higher than the average for **300** or any of the other individual farmhouses included in the analysis of functional groups in **100**. Long-term storage is represented by a single vessel fragment at site 556F, by one at site 231F, and by three at site 95F, the latter two being ex novo sites in **150–50**.

Food service is well represented at both multi-period sites, in percentages comparable to the average in **300**. The average number of cooking ware fragments in **100** is higher than that in **300**, though again the average is high because of the large number at site 317F, where the majority of the eight identifiable fragments (out of 17) were of cooking ware—so the percentage, as a result, is misleading. It should be noted, however, that cooking ware and drinking vessels occur at all farmhouses of **100** but site 231F.

[42] In this case it is owed to two black-gloss fragments from a krater and a larger vessel, both undated but probably belonging to the period **150–50**, as is the black-gloss dish (type D28) which has been assigned a date **200–120** (see assemblage table for site 502F in Vol. IV).

[43] See assemblage table for site 556F in Vol. IV.

As these observations show, a strictly quantitative analysis of functional groups (and as we shall see, of ware class proportions) for the assemblages of **100** could lead to false conclusions. The best the percentages can do is indicate very general trends. The percentage of drinking vessels, as indicated by the averages, declined after **600/550**, remained steady through **450** and **300**, and fell again in **100** (*Appendix C17*). The percentage for all but one of the individual sites—the ex novo site 339F—was below that for **300**. Short-term storage declined by 10% relative to the percentages for all other periods where it very regularly made up a quarter of the entire assemblage. Food service, represented at four of the seven sites with more pottery, also declined. It, too, constituted about a quarter of the assemblages of previous periods except for the earliest, where it made up about an eighth of the total. Cooking ware, as noted above, allowing for the anomalous site 317F, was at a level comparable to earlier periods.

Ware class proportions in 100 (*Fig. 27.5, Appendix C39b*).[44] The most obvious difference, of course, between this period and the previous ones is in the dateable fine ware, the black gloss, and the grey ware; these account for almost 60% of the total fragments of pottery recovered from the sites. Black gloss of the 2nd and 1st centuries BC had not completely disappeared from farmhouse assemblages in **100**. Black gloss dated to the period **150–50** is very slight as compared to the grey ware. Only half of the ten farmhouses analyzed here have some black gloss.

The quantities of both banded ware (BW) and plain ware (PW) make up a much smaller portion of the total in **100**. In **300**, their combined averages were nearly half of the total assemblage. In **100**, together they are less than 30%. Cooking ware has declined slightly. It is only slightly more common than it was in the earliest occupation of the chora. The percentage of the assemblage represented by amphorae increased to its highest level. That change is small, but it is significant.

[44] For this analysis the black-gloss pottery and other wares definitely dated before the period **150–50** have been removed from the assemblages and the quantities of the undated pottery from the five multi-period sites reduced by a constant factor equal to the ratio between the grey ware and the pre-2nd-c. BC fine ware from the sites. This was done, as in the functional group analysis above, so that the different ware class totals from the multi-period farmhouses are comparable to those from the ex novo farmhouses.

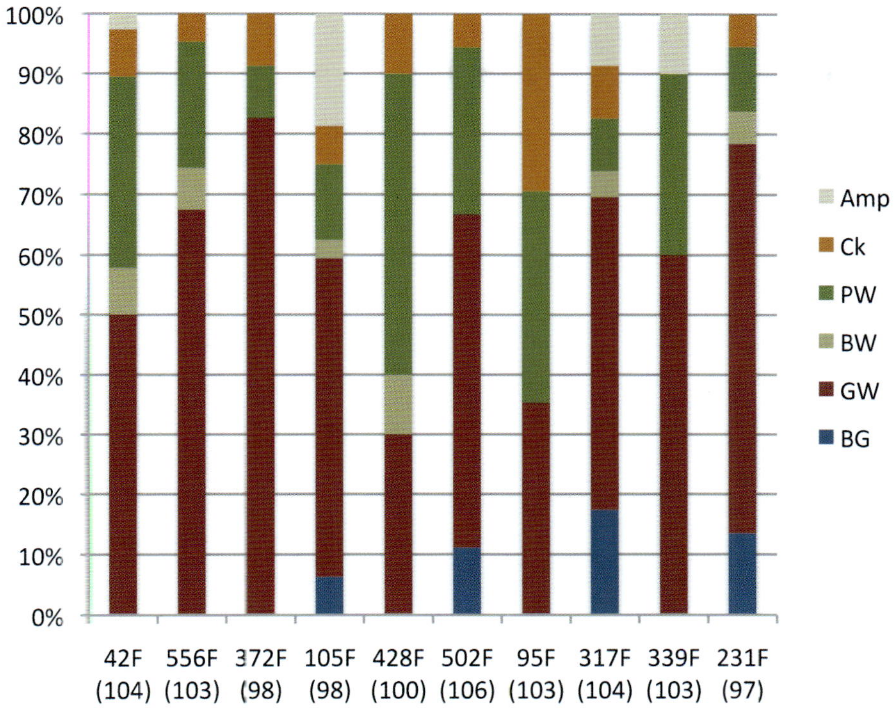

Figure 27.5 Farmhouse assemblages in **100**: ware classes—counts and percentages.

have been dated to the last two centuries BC. While there was a significant earlier presence, the phase of the farmhouse that spanned the 2nd and 1st centuries BC was one of great economic activity (see Amphorae, below).

A comparison of the assemblage consisting of 67 fragments collected from the surface of site 1F with the nearly 4,000 from the excavated farmhouse Sant'Angelo Grieco reveals considerable discrepancies both in the percentages of ware classes and functional groups. The most significant differences are in the dateable black gloss, grey ware, and the amphorae. The quantity of black gloss from the excavation is 14% of the total, compared to 40% from the surface finds; the excavated grey ware is 19%, which is relatively closer to the survey result of 11%. Only one amphora fragment was recovered from the surface, while 575 or 14% of the total excavated pottery consisted of amphorae. On the basis of the surface finds of this site alone, the site would be judged to be relatively more important in the 5th and 4th centuries than it actually was, compared to the 2nd and 1st centuries.

How do the assemblages from the survey and excavation compare if we consider just the pottery from the 2nd and 1st centuries BC? That comparison is made (see *Table 27.2*) by eliminating the black gloss from the counts of both.[46] Further, we can eliminate 133 amphorae that date before the 2nd century. The final major adjustment is made by dividing the undateable pottery, such as plain/banded ware, and cooking ware, using the proportion between grey ware and black gloss, which is 19% vs. 14% for Sant'Angelo Grieco and 11% vs. 40% for the pottery from the surface. This results in the proportions recorded in the "corrected" table. A similar operation was performed to divide the pottery classified by functional group.

The percentages of ware classes and functional groups in the corrected tables for site 1F do not com-

Excavated vs. surveyed site assemblages: site 1F and Sant'Angelo Grieco (*Table 27.2*). As was the case in the study of the assemblages from surveyed farmhouse sites in **600/550**, especially **300**, we have an opportunity to compare the results from the surveyed farmhouses of **100** with those of an excavated farmhouse occupied in the same period (*Figs. 27.8* and *27.9*). The comparison will help to show how accurately a surface collection reveals the reality beneath.[45]

The farmhouse known as Sant'Angelo Grieco was built on a site previously inhabited in the 6th, 5th, and 4th centuries BC. As will be noted below, the 2nd- and 1st-century building incorporated parts of the earlier structure and there was a considerable amount of pottery from all the previous periods, especially from the 4th century BC.

The total raw assemblages. The total assemblage consists of 3,994 fragments. The black gloss and other fine wares (e.g., red figure) of the 4th century and earlier periods constitute 15% of the total assemblages. The grey ware and Roman fine ware (e.g., Arretine) make up 20%. In addition to the household pottery, there is a very large quantity of amphorae, which makes up 14% of the total. The majority of the amphorae (77%)

[45] The preliminary data from Sant'Angelo Grieco was made available by Eloisa Vittoria. The study of this site continues.

[46] Only two black-gloss fragments from Sant'Angelo Grieco out of the total of 533 have a portion of their date range in this period.

Ware class	Survey average*	Site 1 adjusted	SAG adjusted
Total fragments		40	1665
AFW	0	0	46/3%
BG	5%	0	2/< 1%
GW	54%	7/18%	317/19%
BW	4%	8/20%	70/4%
PW	24%	22/55%	589/35%
Ck	9%	2/5%	20/12%
Amp	4%	1/2%	442/27%
Min	< 1%	0	0

Functional group	Survey average**	Site 1 adjusted	SAG adjusted
Total fragments***		13	1080
Drinking	13%	0	124/11%
Drinking service	6%	0	10/1%
Storage-short	19%	4/30%	112/10%
Storage-long	7%	1/8%	7/1%
Food service	19%	6/46%	74/7%
Food prep.	5%	0	29/3%
Food prep./Cooking	19%	1/8%	268/25%
Transport/Storage	8%	1/8%	375/35%
Service	< 1%	0	19/2%
Lamp	< 1%	0	16/1%
Personal	3%	0	46/4%
Votive/Ritual	0	0	0
Known		13/32%	1080/65%

Table 27.2 Comparison assemblages of site 1F and Sant'Angelo Grieco (SAG)—functional groups and ware classes *Appendix C18. **Appendix C17. ***Total identifiable forms.

pare closely to the average for survey sites. The major discrepancy is that between the grey ware from site 1F, 18% of the total, and the average from the survey, 54%. In the ten sites included in the average, grey ware is more than 50% of the total at all but two sites. The percentage of grey ware, while very low compared to that of the other survey sites, is in fact almost identical to that in the assemblage of the excavated Sant'Angelo Grieco. The high level of grey ware in the survey average results in lower percentages for the other components of the assemblages. The combined plain/banded wares total is just 28%. At site 1F, plain/banded wares make up 75% of the total. The percentages of amphorae in the survey average (4%) and at site 1F (2%) are quite low compared to Sant'Angelo Grieco (27%). In this three-way comparison, it is clear that the survey assemblages are biased in favor of the fine dateable wares—site 1F is an exception—and that the coarse wares and amphorae are underrepresented in surface collections.

The most obvious and significant discrepancy between the assemblages of the surface and excavated collection is the very high percentage of amphorae fragments from the excavated site. Even considering the relatively larger number of fragments that amphorae produce compared to other types of vessels, it is still impressive that a quarter of the fragments from the sites were from amphorae.

The fact that the site 1F assemblage appears in some respects closer to that of Sant'Angelo Grieco in its latest phase than the survey averages suggests that site 1F/Sant'Angelo Grieco might be a different kind of site, or at least one with a greater emphasis on production—agricultural, ceramic, or both.

Some clear trends in the survey assemblages of **100** emerge (*Appendix C17*): there is a perceptible drop in drinking vessels. While short-term domestic storage also declined slightly, the proportion of vessels devoted to the serving, preparation, and cooking of food (43% vs. 40% in **300**) remained about the same.

The drop in the percentage of vessels associated with drinking among the surveyed farmhouses of **100** with respect to previous periods has been noted. Short-term domestic storage declined in **100**. The percentages of cooking ware among surveyed sites of **100** are perhaps higher than in previous periods, but were lower than that from the excavated site at Sant'Angelo Grieco (19% vs. 25%). As noted above, the percentage of cooking ware is much higher at excavated sites than at surveyed sites.[47] The extreme case of imbalance between survey results and those from excavation, however, occurs with the amphorae. In the case of the farmhouses whose principal period of occupation is

[47] Exaggeratedly so at the atypical 6th-c. single room house on the hill at Sant'Angelo Vecchio.

the 4th century BC,[48] excavated amphora fragments outnumber those collected from the surface by a factor of three or four times, but amphorae in **100** are clearly a special case. In comparing survey assemblages with those of excavated farmhouses in the periods **600/550**, **300**, and **100** (*Appendix C36*), we note that black-gloss and grey ware are overrepresented, and that amphorae are very seriously underrepresented in surface collections by a factor of as great as six times. In all these respects, the assemblages of **100** are generally consistent with the experience of the preceding periods.

A full discussion of amphorae should await the final publication of the excavated farmhouses, but some preliminary observations can be made now, not only about the overall numbers but also about the quantification of individual types from Sant'Angelo Grieco. As should be expected from a site that was occupied with interruptions during the preceding four centuries, there is a considerable percentage (23%) of amphora fragments dating before the last two centuries BC, but Sant'Angelo Grieco stands out among the other excavated farmhouses, all dating before the period with amphorae, making up nearly 15% of the total pottery (cf. Fattoria Fabrizio with 5.4%, Sant'Angelo Vecchio with 8.8%, and Fattoria Stefan with 3.5%, provisionally).

The actual 2nd- and 1st-century BC amphora types and their numerical weight are quite revealing. The best represented are the Metapontine collared amphorae and the Lamboglia type 2 (see Ch. 11). The minimum number of individual analysis (MNI) has reduced the 158 fragments of collared amphorae to four and the 154 Lamboglia type 2 to five. Allowing for the necessary approximation of the method, it is clear from the quantity of amphorae present that the farmhouse did not serve as a warehouse for large-scale agricultural production, but rather as a residence, as the plan indicates. The Lamboglia type 2, which is generally—though not exclusively—thought to have contained oil, was not locally produced, which would suggest that these farmers imported the oil they used. The collared amphorae (*Fig. 27.6*), thought to be the continuation of the locally produced Greco-Italic type, but under the influence of the Dressel types, were locally produced, and are thought to be containers for wine.[49] The farmhouse could be seen to consume its own wine, and perhaps have produced a surplus for export. Local centers for production of collared amphorae have been discovered in the chora at Pantanello and close to the site of the much-reduced asty.[50] The distribution of this type, as revealed by the survey (*Fig. 27.7*), was quite limited. It included the major center, site 42F, and two lesser sites (339F and 193F) on the Basento and Bradano sides of the chora, respectively. Keith Swift opines that the oil was probably produced in North Africa or Spain,[51] and the owner of the land (not necessarily the occupant of the farmhouse), if he were a capitalist following Cato's model of farming as a profitable investment, would have wanted to maximize his profits by investing in a single crop with proven success, as he may have diversified his investment elsewhere with other crops

Figure 27.6 Metapontine type collared amphora from the tile factory at Pantanello, 2nd century BC (h 92 cm). (CW)

48 Fattoria Fabrizio, Fattoria Stefan, and Sant'Angelo Vecchio.

49 Vandermersch 1994.

50 De Siena and Giardino 1994, 201; L. Giardino 2005, 424.

51 K. Swift, pers. comm., May 2010.

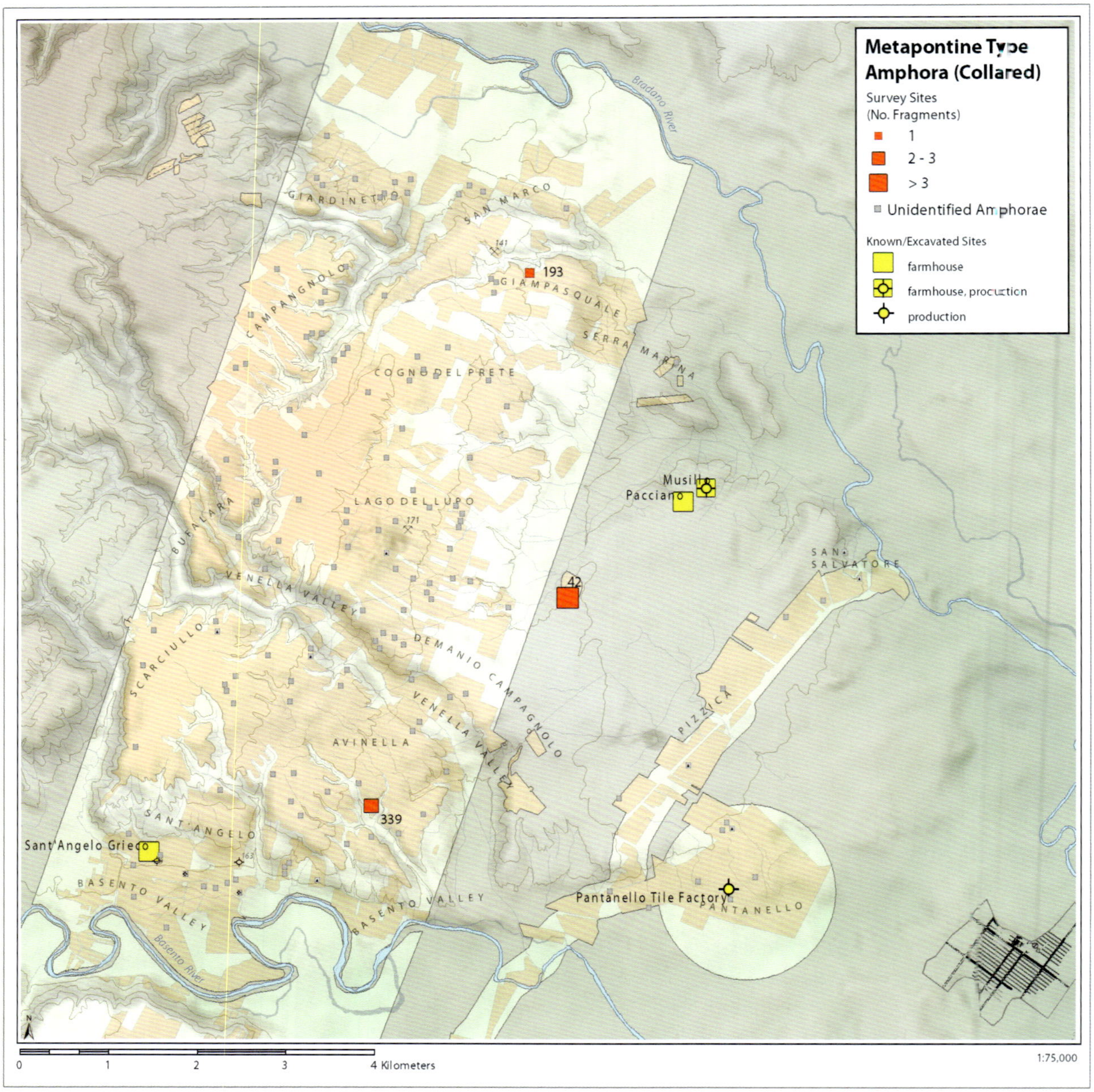

Figure 27.7 Map of Metapontine-type amphorae. (JT)

and with animal husbandry. By the 2nd century BC, the Metapontino had been subsumed into the pan-Mediterranean economy.

Excavated Structures in the Study Area

New types of structures appeared in the chora in the post-Hannibalic era, adapted to the expanded range and increased scale of economic activities. The traditional single family farmhouse did not disappear altogether, but was modified to meet the new demands. Villas of varying size and complexity arose, as did industrial structures that may have stood alone or been part of larger establishments. There were relatively modest but still substantial farmhouses, as well as a range of subsidiary structures that have left slight remains. Examples of a number of these structures dating to the last centuries BC were found in the study area and fully excavated. They are the subjects of preliminary reports,[52] and as with the farmhouses cited in earlier chapters, this

[52] Carter 2006a, 146–48, Sant'Angelo Grieco (Nuovo), and 150, Pantanello tile factory, with bibliography. Cf. the Republican-period kilns at Sant'Angelo Vecchio in Edlund 1979.

Figure 27.8 Farmhouse at Sant'Angelo Grieco, reconstruction. (ALP)

treatment will be limited to general statements that attempt to place the excavated sites in the context of the survey and populate the landscape with more than site numbers.

The Farmhouses.

Sant'Angelo Grieco. Many if not all of the new structures of the 2nd century BC will have been built on the site of previously existing farmhouses. This is true of the villae (see below) and of the surveyed sites. The farmhouse at Sant'Angelo Grieco illustrates this very well (*Figs. 27.8* and *27.9*). Nothing remains of the 6th-century BC structure on the site. The still extant foundations of the 4th-century BC farmhouse were utilized in the 2nd- to 1st-century BC structure in the western wing, which included the kitchen and storage. The eastern half with a dining room was added ex novo. The footprint covering some 260 m^2 is smaller than the exceptionally large late 4th- to 3rd-century BC farmhouse Fattoria Stefan (320 m^2), and the Late Roman house of the *vilicus* at San Biagio (310 m^2).[53] Like other buildings of this and the later Roman period, creative use was made of *spolia* from earlier Greek structures. Among the building materials at Sant'Angelo Grieco was an impressive Doric capital

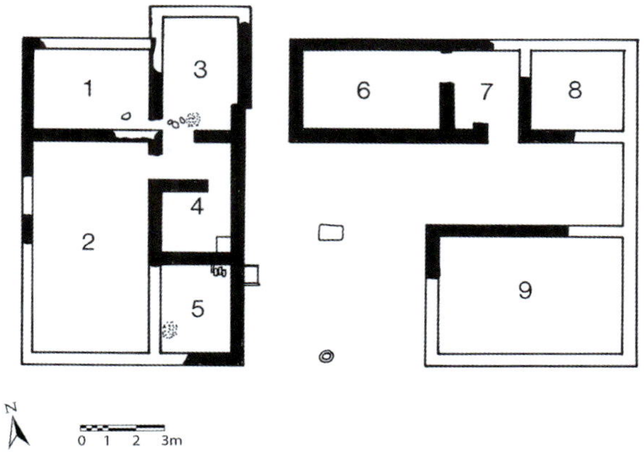

Figure 27.9 Farmhouse at Sant'Angelo Grieco, plan. (ALP)

of the 5th century BC that served as a base for the stairway to the second floor. Another fine limestone capital of Hellenistic date may have crowned a column supporting the second story balcony that looked out over the Basento. The plan of this building with its south-facing open courtyard is unlike any of the farmhouses of the 4th century BC or earlier, but it does bear a superficial resemblance at least to the arrangement of spaces in the tile factory at Pantanello, discussed below.

The contents of the house included quite a lot of fine grey ware pottery, but the floors were dirt paved and if there was plaster on the walls it has left no

[53] See Carter 2006a, 134–51, Table 4A, for dimensions and bibliography of all the structures mentioned here.

trace. The occupants of Sant'Angelo on the new social and economic scale of the 2nd century BC were probably of modest means. In their appropriation of earlier buildings one suspects that their motive was not an eye for art, but rather a need for good stone as building material. We can imagine that they were either tenant farmers, perhaps descendants of the earlier inhabitants of the chora, or *vilici* like the occupants of the Late Roman farmhouse at San Biagio, a medium-sized farmhouse.[54]

Figure 27.10 Reconstruction of Roman tile factory at Pantanello. (BP)

Pantanello tile factory.[55] As an instance of the opportunistic reuse and complete transformation of a site, there could be no better example than the tile factory at Pantanello (*Fig. 27.10*). At the base of a slight rise overlooking the Basento at Pantanello, and the edge of the valley, was a sanctuary beside a spring from the end of the 7th century to the early 3rd century BC (Chs. 20 and 23), after which time the lower portion of it was briefly and partially covered by a farmhouse in the mid-3rd century (Ch. 26). As a result of recent excavation and study (2008), it is now certain that the hilltop was occupied in the 4th century BC by a monumental structure whose sizeable ashlar blocks were reused in the foundations of the factory building.[56] This is another example of the opportunistic re-appropriation of elements of a former Greek sanctuary, as at Sant'Angelo Grieco, but on a much larger scale.

The factory itself consists of three wings arranged around a rectangular courtyard that faces and slopes to the south, as did Sant'Angelo Grieco's. The footprint including the open courtyard is ca. 900 m^2, plus 120 m^2 of an extension of the west wing that included the kilns. There was no indication from the plan or the material excavated from this building that it was lived in. The extensive tile-covered space may well have been entirely used as drying sheds for the roof tiles, amphorae and other pottery that was fired in the large and small kilns (*Fig. 27.11*). The deposit, by contrast, yielded much evidence for occupation nearby, as well as production. It included large quantities of fine grey ware pottery and mold-made "Megarian" bowls, as well as a range of domestic pottery, amphorae, and tiles, including misfired pieces. Evidence for the active production of pottery were the kiln furniture, bellows mouths, and a great mass of metal slag that shows that iron working went on there. It also contained in the upper levels substantial quantities of fine, imported Arretine pottery, which points to a degree of wealth and luxury in the late 1st century and Early Imperial period that is not evident in the factory itself. Numerous animal bones mixed throughout provide firm evidence for animal husbandry and breeding in this period. This has suggested the strong possibility that the factory was one detached element of a *pars rustica* of a quite substantial villa. If it existed, it has disappeared leaving no obvious trace. Possibly the structure on site 428F (significant in **100**) which is some 100 m distant may have been the dwelling of the manager of the factory—a potential analogy for the possible factories at 105F/K and 556F (see Hamlets, above).

The pair of kilns excavated at Sant'Angelo Vecchio belong to the same period as those at Pantanello. They were superimposed on earlier contexts, of farmhouses of the 4th century BC and of kilns of that period producing votive terracottas.[57] The construction technique employed for the kilns is the same as that at Pantanello, but their designs are entirely different. Again, there is no hard evidence for a residence of the period close by.

[54] Gualtieri 2009, 348 and n. 21.

[55] A full study of this site will appear in this series.

[56] For a preliminary study of the tile factory, see Carter 1983a.

[57] Edlund 1979. The kilns will be included in the future study of Sant'Angelo Vecchio in its entirety.

Figure 27.11 The kilns at Pantanello. (JCC)

The villas. Since Kahrstedt's pioneering study,[58] excavations have significantly altered our understanding of the economic situation in the Roman period along a stretch of coastal Italy between Nuova Siri and Ginosa, including the chora of Metaponto. The upper end of the social and economic spectrum is well illustrated now by the discovery of the princely and prestigious villa at Termitito, completely excavated in the 1980s and by other less thoroughly investigated structures of similar dimensions and importance. De Siena has divided them into two classes; Type 1 is characterized by its exceptional size, division into residential quarters (*pars domestica*), and the rest devoted to agricultural and artisan activities (*pars rustica*). The construction techniques (paved floors, plastered and painted walls), materials, and the amenities (extensive facilities for bathing) indicate a tenor of life comparable, in the case of Termitito, to that of the highest Italic and Roman elites.

The modest constructions of lesser dimensions, using river cobbles for foundations and walls of mudbrick, which continue the old tradition of the Metapontine farmhouses of earlier times, are grouped together as Type 2. Within a 20-km radius of Metaponto there are currently four examples of Type 1: Termitito, Incoronata La Cappella, Bivio Franchi, and a villa on the boundary between Montescaglioso and Ginosa; and three of Type 2: Destra Basento (Soprintendenza site 77) and, in the territory of Herakleia, two more: proprietà De Leo near Scanzano, and Recoleta between the Cavone and Agri. In all cases excavation was limited.[59]

None of the three rural structures excavated in the study area fits neatly into De Siena's two types. Sant'Angelo Grieco, as noted, is small and built using traditional materials—a possible Type 3?—and

[58] Kahrstedt 1960.

[59] For discussions of those in the jurisdiction of the Soprintendenza della Basilicata, see De Siena and Giardino 1994, L. Giardino 2005, and De Siena 2005. With reference to the villa near Ginosa under the Soprintendenza della Puglia, see De Siena 2005, 453 n. 20, and *Museo Taranto* II.1, 30–1.

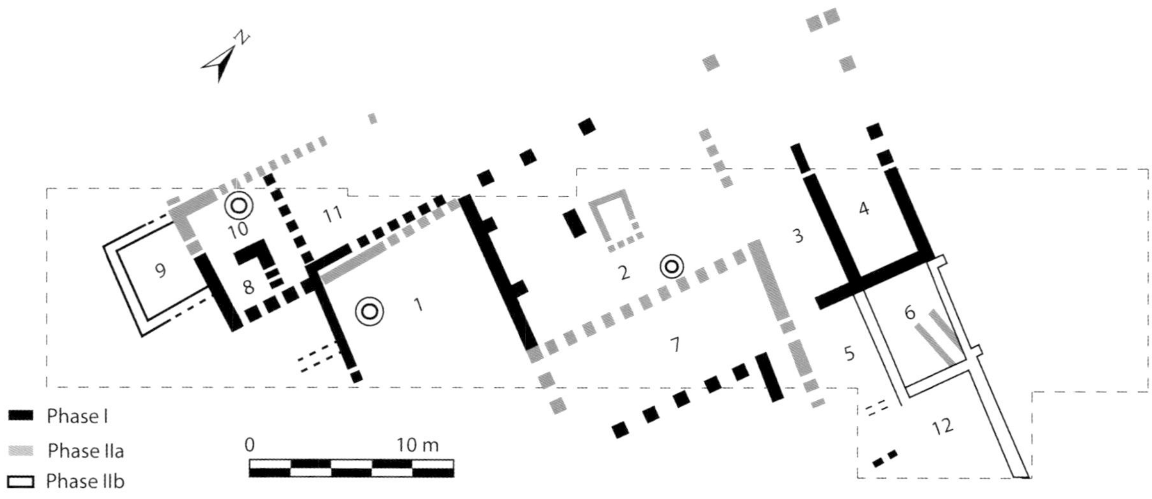

Figure 27.12 Partial plan of Villa Destra Basento. (After De Siena 2005, Fig. 4)

the tile factory at Pantanello was perhaps part of the productive side (*pars rustica*) of a large villa of Type 1. The best candidate among the survey sites for Type 1 status would be site 42F (see Gazetteer) with its abundant evidence of monumental architecture—also possibly recycled remnants of a Greek sanctuary or a colonnaded villa. Only excavation will tell. In this section the two best examples of De Siena's Types 2 and 1 will be briefly discussed.

Destra Basento. This portion of a larger structure—some 525 m^2 have been excavated—is the most recent addition to the monuments of the 2nd century BC in the chora of Metaponto. It was discovered and excavated as a part of the archaeological mitigation for the SNAM pipeline (*Fig. 27.12*).[60] The site lies along the edge of the valley at the foot of the marine terrace just to the south of the plateau of Incoronata greca.[61] This is one of the many sites which once must have filled the valley. It rose over an earlier 5th- to 4th-century BC farmhouse or possibly a rural sanctuary, in an area that had serious drainage problems. The spring that caused them then (and still does) was captured and channeled for domestic use.[62] The building consists of several sections using different construction techniques and functions that represent at least three separate phases. The central section consists of three spaces with thick walls and a tile roof. A square-paved impluvium collected the rain water from the roof (no. 2 on the plan, *Fig. 27.12*). This part was probably two storeys high, and was the domestic quarter. Food was stored here, and women worked at their looms, as the loom weights, fragments of commercial amphorae, and grinding stones (*macine*) testify. The date in the 2nd and 1st centuries BC is based on pottery—grey ware and black gloss—and lamps, unguentaria, and amphorae of the Metaponto type, found also in the castrum and at Pantanello. This nucleus was destroyed by fire early in the 1st century BC—a possible victim of the Social War or Spartacus?—and not rebuilt. The atrium opened out onto a great courtyard with peristyle. The other sectors of the building had thinner walls and were possibly not tile roofed.

The villa at Termitito. The villa at Termitito, exceeding 3,000 m^2, covers most of the small hilltop terrace.[63] It enjoys panoramic views. On a clear day it is possible to see as far as Capo Leuca at the tip of the Salentine peninsula. This location had previously attracted an important Bronze Age settlement,[64] and it was more or less continuously occupied from that time forward (*Fig. 27.13*). The unified plan filling a rectangle ca. 45 x 95 m is among the larger Republican-era Roman villas in Italy. Its earliest phase has been dated by the excavators to the 2nd century BC, a date well before the better-known examples at Settefinestre near Cosa in Tuscany or the two villae

[60] Nava 2003a, 673–75.
[61] Ch. 17, Fig. 17.8.
[62] De Siena 2001, 449–51.
[63] De Siena 2005, 453–56.
[64] See Ch. 17, 576.

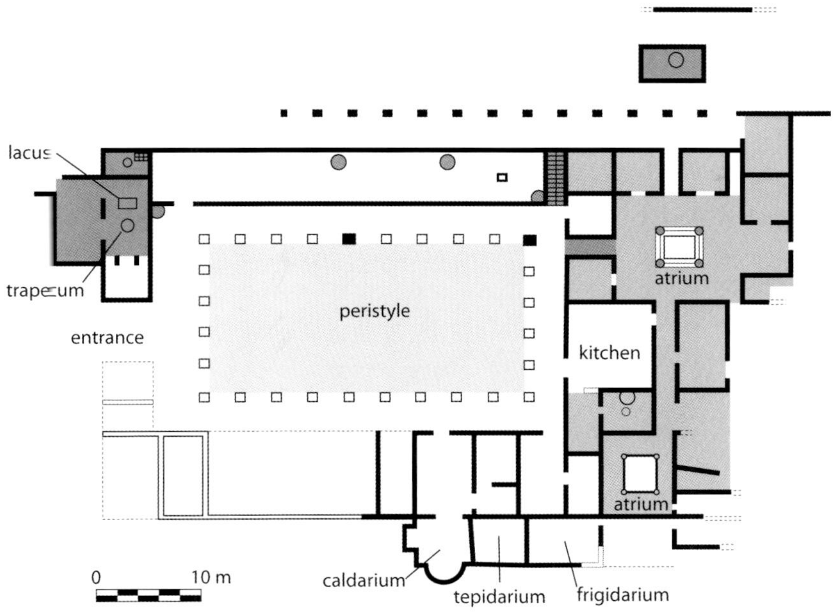

Figure 27.13 Plan for villa at Termitito. (After De Siena 2005, Fig. 5)

at Francolise in Campania,[65] which date early in the 1st century BC. The whole southeastern end contains the *pars domestica*. There are two atria with impluvia separated by the kitchen. The larger, at the southeast corner, would have been occupied by the dominus and his family. The plan is that of a typical Hellenistic Roman townhouse with fauces opening onto an enormous peristyle courtyard (measuring ca. 22 x 36 m), cubicula, a tablinum, and triclinium.[66] A long portico with an unobstructed view to the southeast was entered directly by a passageway leading from the larger atrium. Around the smaller atrium is an apartment for guests and adjacent to it a very elaborate bath complex with hypocaust heated caldarium and the full complement of related rooms, six in all.

The *pars rustica* opened off the northwest end of the great courtyard. Here were found the olive press (*trapetum*) and the settling tank (*lacus*). A narrow space that ran the whole length of the courtyard but was closed off from it contained buried pithoi for storage of oil and wine and perhaps spaces for slaves. More precious produce or objects were stored in an underground room placed under the northwest cubiculum of the master's apartment so as to be under his surveillance.[67] The villa was constructed using *opus caementicum*, one of the first appearances of this technique in this area of Italy. The floors of the atrium and other residential areas were paved in *opus signinum*. Mosaic rosettes embellished the bath. Similar decoration has been identified in residences in Herakleia and in Taras, but was certainly not typical of the countryside. An interesting touch was the terracotta antefixes with female heads that were executed for this structure. They recall those of the sanctuaries of the asty and chora of another era. The owners of Termitito with their taste for the antique style were in an economic position—in contrast to the owners of Sant'Angelo Grieco (see above) and the Late Roman farmhouse at San Biagio[68]—to commission original reproductions, not simply to appropriate spolia from nearby ruins. De Siena notes that the family of the Licinii, whose association with Herakleia and Cicero (see below) is documented, are among the most likely of the Roman nobility to have occupied this site.[69]

Cattle and Horse Breeding and Sheep Farming

More than 40 years have passed since the appearance of Toynbee's *Hannibal's Legacy* (1965). His theses, built exclusively on the ancient sources, were that southern Italy was widely depopulated in the aftermath of the Second Punic War in the 2nd century BC, that urban life had practically disappeared, and that the agricultural economy of the coastal areas and the interior had been largely replaced by vast latifundia and pastoralism conducted using slave labor. These assertions have recently come in for perhaps excessive criticism and have been shown to be oversimplifications of a complex situation. This is the result of his-

[65] Potter 1987, Ch. 5, "Villas, Farms, and the Countryside," with bibliography.

[66] Cf., the Villa of the Mysteries at Pompeii or the House of the Stags at Herculaneum, which were also entered through the courtyard.

[67] De Siena 2005, 456.

[68] Carter 2006a, 142, Fig. 4.10.

[69] De Siena 2005, 456. If a roof tile with the stamp AUF from late Republican Sant'Angelo Vecchio, can be traced to Publius Aufidius Pontianus mentioned by Varro, it would be another sign of senatorial or equestrian interest in the territory of Metaponto. Francesca Silvestrelli,pers. comm., June 2011.

torical studies,[70] but also and in large part the result of archaeological investigation focused on the Roman period in the three principal regions affected: Apulia, Basilicata, and Calabria.

The present study establishes that, in the Metapontino, a great change did indeed take place in the 2nd century BC. There are few small farmhouses, in our study area at least. They had begun to disappear in the 3rd century BC and by the end of it there were only a handful left. In their place we find a few larger establishments—some very large indeed, as revealed by excavation (e.g., at Termitito) in the wider area between the Bradano and the Agri. Among the survey sites in the study area appear, for the first time in over a century, new farmhouses or villa structures on virgin sites with much higher average FMCE values than those of the previous century. The wide spacing of the farmhouses in the study area indicates that they controlled relatively large surrounding areas, but they could hardly qualify as latifundia on the scale of those that are known from historical sources to have existed under the Empire in Sicily and some areas of Italy. Rather they suggest holdings of no more than 500 jugera (125 ha) in the 2nd century, possibly larger in the 1st century BC. An establishment of this size could have provided winter pasture for moderate-sized flocks of sheep or herds of cattle and horses. Certainly the owner or manager of a property this size would have made use of its agricultural potential in the summer months benefitting from the manure laid down in fallow fields during the winter.

Too much, perhaps, has been made of the isolated passage of Varro mentioning *Metapontinos saltus* (see below), if it leads us to imagine that pastoralism was the only source of profit for the new owners of the chora, or that transhumance meant necessarily the long range sort with bases in Metapontine fields and central Italy.[71] The interior of Lucania with its forest and highland pastures was quite sufficient then, as in the recent past, to sustain herds year round. The occasional ancient sources on animal husbandry in Apulia and Lucania make it clear it was an important activity in the last century of the Republic.[72] The wool of neighboring Taras was famous in antiquity. Support for this in the case of Metaponto comes from archaeozoological research on sizeable deposits of faunal remains from well-dated sites. These do not contribute to the debate on long distance versus more local transhumance, but they do indicate that the Metapontine farmers were engaged in transhumance and it establishes that they were engaged in conscious breeding of horses and cattle for size. The faunal assemblage also includes the shepherd's essential helpers—his dogs.

The study by Bökönyi and Gál shows that the remains of Metapontine dogs found together with the cattle, sheep, and horses in the Pantanello midden were all of large size, either as a result of conscious selection or of local domestication of wolves.[73] They form a single group, with none of the smaller luxury dogs that are found elsewhere in Imperial times. The Metapontine horses were among the largest in the Roman world (*Fig. 27.14*), standing 138.5 m at the withers, some 13 cm higher on the average than the eastern, Scythian horses, the best horses known before the Imperial period. Conscious breeding was most apparent and successful with cattle with an average withers height of 133.5 cm; they surpass in height the average Roman cattle (124–125 cm) by almost 10 cm. Bökönyi remarks that they are almost as large as wild cattle (*aurochsen*) and like them they were longhorns. Bökönyi notes that according to Aristotle the cattle of Molossos are longhorns and Pliny remarks on the excellent cattle of Epiros. Bökönyi wondered whether the large cattle were specific to southern Italy in the Roman period, or if they occurred in the Greek period and were possibly imported. If this later possibility had indeed been the case, Epiros would be a good candidate, and a good occasion would be the period of intense interaction with Epiros in the 4th and 3rd centuries BC, when, as noted earlier,[74] the large horse was also introduced.

What purpose did these large cattle serve? Elsewhere I have argued that they were bred to be more powerful tractors and draft animals for agriculture in the Roman period,[75] and that still seems to me the best explanation. Bökönyi's suggestion that meat and milk were more important than traction and hides seems just the reverse of what would appear to be the most obvious use. Small cattle are as good a source

[70] E.g., *Modalità insediative*, reviewed by Fentress 2005. Crawford 2005. Most recently, *Agricoltura e scambi*.

[71] Cf. Crawford 2005.

[72] E.g., Strabo 6.3.9 and Varro *RR 2.10.11*.

[73] Bökönyi and Gál 2010, 37–38, Ch. 2 on animal husbandry in Roman Metaponto.

[74] Carter 1998h.

[75] Carter 1994b, 186–87.

Figure 27.14 Graphic reconstruction of the size difference between the smallest and the largest Roman period horse found in the Pantanello Kiln Deposit. (After Bökönyi and Gál 2010, Fig. 2.4)

of protein as larger ones. The evidence of cut marks on the long bones suggests that the midden contained the remains of animals who were either eaten or had died and been disposed of after skinning. There are more complete long bones of cattle and horses than of any other species, and they are unlikely to have been table scrap. The evidence of the kiln deposit strongly suggests that the owners of the villa with which the Pantanello tile factory and kiln deposit were associated were engaged in animal breeding, agriculture, and transhumance.

The Asty in the Last Two Centuries BC

The picture of the urban center of Metaponto in this period, as in all others, emerges primarily from the archaeological evidence, but the documentary sources can add important details and color. From the 3rd century BC onward these sources for Metaponto, as for all of southern Italy, are Roman and this has shaped the discussion. The questions of the status of Metaponto vis-à-vis the Romans from the time of Pyrrhos and before, of Hannibal, and the aftermath of the Second Punic War, of the Social Wars, and Spartacus, and finally of the civil war, may never receive definitive answers. The documentary sources that refer to Metaponto specifically are, as ever, short and sparse. Arguments from silence are always suspect. In attempting to understand this period in the history of the Metapontino, however, we need to make use of all available evidence.

The Written Sources

Metaponto. Did Metaponto after Hannibal, Giannotta wonders,[76] recover the status of *civitas foederata* that she supposes it enjoyed in the period after Pyrrhos and preceding Hannibal? There is no positive evidence that Metaponto was ever a civitas foederata or ally of the Romans in the 3rd, much less the 2nd century BC. We cannot conclude from the fact that Metaponto's name did not appear together with those of Taras, Petelia, Locri, and her close neighbor Herakleia as receiving a visit from the *theorodokoi*, sacred ambassadors from Greece in 198 or 194 BC,[77] that Metaponto was no longer a Greek polis, but it is likely at least that she was not able to participate in the festivals at Delphi, where in better times her "golden harvest" had attracted much notice. Thourioi, whose past relations with Rome were similar to Metaponto's, was also absent from the list.

Polybios writing in the mid-2nd century BC lists Metaponto among the poleis of the Ionian coast (*Fig. 27.15*), but we cannot be sure to what period he is referring. It would seem improbable that Metaponto could return to the status she had before Pyrrhos. The Romans treated harshly the allies of Pyrrhos after 272 BC and even more so those cities which had gone over to Hannibal, including those who did so against their will.[78] They were hardest hit by confiscations.[79] Treachery against a Roman garrison was punished by massacre. Metaponto should have been considered among the worst traitors, and therefore received

[76] Giannotta 1980, 60.

[77] Sartori 1967, 87, with references.

[78] Toynbee 1965, II: 19–23.

[79] Brunt 1971, 278.

Figure 27.15 Map of 2nd-century BC Roman colonies and poleis. (JT)

the maximum punishment. Hannibal, who was no more kindly to defectors, foresaw this when he took the population of Metaponto, which included many whom he had recently transferred from Herdonia, with him, as he did at Atella in Campania. He apparently did not raze to the ground what remained of the asty, as he is reported to have done at Atella.[80]

Early in the century, between 81 and 79 BC probably, Cicero stopped with his friend Gaius Calpurnius Piso at Metaponto, and would not go to his host until he had first seen the place where Pythagoras had lived and died.[81] If this was in fact the Temple of Demeter on the stenopos of the Muses,[82] then Cicero was probably staying in the countryside as he was accustomed to do. Good candidates would have been any of a number of the luxurious villae, like Termitito.

If, in consequence of the Social War (90-87 BC) Metaponto was offered municipal status, the event went unrecorded. Again, it would seem unlikely. Herakleia, always favored by Rome, had its status altered from civitas foederata to *municipium*, at this time,

[80] Toynbee 1965, II: 19.

[81] Cicero, *De finibus* 5.2.4.

[82] Vallet 1974.

though apparently not all that willingly.[83] Herakleia's tabularium, as recalled by Cicero in the *Pro Archia*,[84] was destroyed by fire and Metaponto likewise suffered damage either during the Social War or the city's occupation by Spartacus, according to Giardino.[85] Metaponto, according to Florus (2.8.5), was destroyed with terrible slaughter, as were Thourioi, Nola, and Nuceria. Not content with the recruits from villages (*vici*) and villae, Orosius recorded that the rebel slave found large numbers of willing fighters in the areas around Metaponto and Consentia (Cosenza).[86]

Consentia was the capital of the Bruttii and dominated routes into the interior of the Sila, a vast area of alpine forest (see Ch. 26) and pastures, ideally suited to transhumance grazing of large slave-tended flocks. Metaponto for the same reason is likely to have had in the surroundings a large slave population. In addition to the herds, slaves were required for the large farms and villae that began to populate the countryside in the 2nd century BC. The difficult terrain of Lucania and Bruttium must have been an attraction for Spartacus and his army for the same reason that the Roman legions opposing Pyrrhos and Hannibal avoided it. It was ideally suited for guerilla warfare.

A much quoted passage from Varro's treatise on farming,[87] composed about 37 BC, has recently been reinterpreted with a textual emendation. The anecdote, used to illustrate the loyalty of dogs to their shepherd masters and not to the sheep, recounts how herds and dogs recently purchased in Umbria, having been left in *Metapontinos saltus et Heracleae emporium* together with the sheep, return home to Umbria on their own. Crawford reinterprets the defective text by reading *oram*, not *emporium*, and asks for any evidence of economic activity and social structure in the "deurbanized" areas of the south.[88] Recent work by Giardino has produced evidence of a small cottage wool industry and of the presence of shepherds in depressed but still inhabited Herakleia.[89] Metaponto, and Herakleia too, were winter pasture for long distance transhumance. The definition of a saltus by Aelius Gallus, cited by Crawford, leaves no doubt about its function. A saltus is "where there are forests and pastures with houses also to enable their exploitation; if some small part is cultivated to feed the shepherds or watchmen, that fact does not take away the name of *saltus*."[90]

This picture of a saltus is easily imaginable in the landscape of Metaponto in the 2nd and 1st centuries BC on the basis of our knowledge of the plant and animal remains.[91] The strong representation of wild animals, deer especially, in the faunal record, attests to the presence of gallery forests close by. Deer and wild cattle would have been in competition for pasture land with the domesticated animals which were being bred at Pantanello and elsewhere in the territory.[92]

Writing in the Augustan period, Strabo, who drew heavily on earlier sources and rarely if ever from firsthand experience, treats Metaponto as if it were an inhabited polis, but gives no clue as to its present status or importance.[93] His sources are referring to the place that existed in the 1st century BC, and probably in the later part of the century.[94] Metaponto is mentioned by name in connection with a meeting between Mark Antony and Caesar in the period of the triumvirate that took place μεταξὺ Μεταποντίου καὶ Τάραντος.[95] It was a place remembered for its ancient glory, and also, no doubt, for its recent history as a stronghold of Hannibal and Spartacus. This is the sum of the documentary evidence illuminating directly or indirectly the history of Metaponto proper in the last two centuries BC.

Other poleis. More is known about the fate of her neighbors and former allies, and it may help, in trying to understand this eventful period, to consider briefly several of these.

Taras' failed attempt to regain full independence by switching to the Carthaginian side in 212 BC (three years after Cannae) was dealt with harshly. She lost her freedom and an important part of her territory became *ager publicus populi Romani*, and was probably distributed to the pro-Roman Tarentines at

[83] Sartori 1967, 88–91.

[84] Cicero, *Pro Archia* 4.8.

[85] L. Giardino 2005, 424.

[86] Orosius 5.24.2.

[87] Varro, *Res rustica* 2.9.6.

[88] Crawford 2003, 19.

[89] L. Giardino 2005.

[90] Crawford 2003, 19, quotes Aelius Gallus (Festus 392L). Translation by Crawford.

[91] Carter 2002 (2003), 499–509. Bökönyi and Gál 2010, 14.

[92] See Ch. 19, 623–25.

[93] Strabo 6.1.15.

[94] Giannotta 1980, 63; Kahrstedt 1960, 1. Strabo is relating to a situation ca. 100 BC, but Brunt (1971, 354) suggests that it is a half-century later.

[95] Appian, *Bellum Civile* 5.392; Giannotta 1980, 65.

the expense of those who opposed Roman rule.[96] In 122 BC, a Gracchan colony, Neptunia, was established in Tarentine territory. Wuilleumier considered Taras a "dead city" by the 2nd century BC, but this is an exaggeration. She continued to enjoy limited prosperity resulting from the wool business, agriculture, and trade, though the rapid growth of Brundisium as a port eclipsed much of her former importance.[97] After Hannibal, Taras lost an estimated 30,000 inhabitants and much treasure. Her coinage ceased. Works of art were paraded in a triumph except for her "angry gods," which Fabius wisely left to the Tarentines. She was required to give up ships and destroy her walls, though these conditions may not have been strictly enforced.[98] Taras was governed by a Roman *praetor* for a while before regaining her status as a civitas foederata about 180 BC.

The treatment of Croton and of Thourioi, whose course was, as we have seen, in many ways parallel to Metaponto's, offers perhaps closer analogies for the likely treatment of Metaponto. They are mentioned more often by the historians than Metaponto, but again, because they were not by themselves of very great interest to Rome, the passages about them, too, are quite limited. Thourioi, like Taras but unlike Croton, continued to be an ally of Rome, on a part of its formerly vast territory, until it was enfranchised as a result of the Social War, in 91–90 BC.[99] Part of her territory was expropriated for a Latin colony, Copia, in 193 BC (see *Fig. 27.15*). Very stingy portions of the land were distributed to the colonists and it is supposed that parts were held back in order to create winter pasture for flocks that spent the summer in the Sila, a large part of which had been previously annexed by Rome in 266 BC[100] in order to provide a secure source of timber for constructing the fleet.[101] The transhumant shepherds presumably worked for Roman or local pro-Roman investors.[102]

The population of Croton was drastically reduced as a result of the Pyrrhic war and numbered only 2000 according to Livy by the time the Carthaginians arrived.[103] The Crotoniates defended the citadel against Hanno and were forced to withdraw to Locri and Croton soon became a Bruttian city. Despite its resistance to the Carthaginians, Croton was treated as an enemy of Rome, its territory seized, and a Roman colony established there in 194 BC.

At the same time colonies were being sent out to the territories of Thourioi and Croton, on the Ionian side, other coastguard colonies including Buxentum, and Valentia on the site of Greek Hipponion were founded at strategic places on the Tyrrhenian Sea (see *Fig. 27.15*). In the first decade of the 2nd century BC the Carthaginian threat was still very real. Rhegion, Locri, and Herakleia, which had been mostly loyal to Rome, did not have to suffer a colony. By this time, Rome had expropriated much of Lucania and Bruttium, along both coasts, including Buxentum and, according to Toynbee and Brunt, down to and including a stretch of the Ionian coast between Metaponto and Herakleia.[104]

These somewhat isolated facts of Roman history will serve as broad guides to the sort of treatment Metaponto might have expected from the victorious Romans. There was no colony on her soil, as far as we know, from the documentary, and only a slight suggestion of one from archaeological sources.[105] The case of Thourioi suggests that not having a colony might have served better the interest of those Romans and their friends who recognized the possibility of the greater wealth to be derived from staking out large farms, and ample pastures. And this is what the distribution of sites in the study area would seem to indicate was the case (see *Fig. 27.1*). The farmhouses which were clearly active can have been the residences of large land owners, or of the *vilici*, whose numbers may have included survivors of the old farming stock of the Metapontine countryside. They may have worked the land as tenants paying a portion of their

[96] This is an inference from Livy 27.35; cf. Toynbee 1964, 117 n. 2.

[97] Brauer 1985; see also Brunt 1971, 370.

[98] Wuilleumier 1939, 167.

[99] Toynbee 1965, II: 662.

[100] Toynbee 1965, II: 120 n. 7, 545–46. Brunt 1971, 278 n. 3.

[101] See Ch. 26, 889.

[102] Toynbee 1965, II: 235–40.

[103] Livy 23.30.6.

[104] Toynbee 1965, II: 117–21, on the treatment of re-subjugated secessionist allied states in southeast Italy. "The annexation of Grumentum carried the new Lucanian Ager Romanus southeastward to the coast of the Gulf of Taranto, since Grumentum had expanded its territory to the coast along a stretch between Metapontum and Heraclea before the completion of the Roman conquest of southeast Italy in 266 BC" (Toynbee 1965, II.120). Toynbee cites Frank 1959 113, who in turn cites Livy 23.37, who records a Roman victory at Grumentum over Hanno; cf. Brunt (1971), 360, who states that the territory of Grumentum reached the southern coast. The expansion of Grumentum in the 3rd c. is asserted by Toynbee and Frank, without proof.

[105] See Ch. 26, 888–89.

Figure 27.16 Map of the asty in **150–50**. (JT, based on Mertens 2006, Fig. 270)

profits to the Roman state directly, or to absentee landlords, who would have been responsible for rent and the *scriptura*,[106] the head tax for livestock. There is documentary evidence for an active agricultural economy in Magna Grecia in the last centuries BC. Huge ranches existed but not to the exclusion of farms.[107] It would have been impossible to run such extensive farms or manage the herds without the large numbers of slaves which became much more available in southern Italy beginning in the 2nd century BC. A slave revolt involving 7,000 was violently put down by the Roman commander at Tarentum in 185 BC.[108] The recruiting success that Spartacus had in areas around Thourioi and Metaponto in the early 1st century BC is ample testimony to the importance of slaves in the *Metapontinos saltus*.[109]

[106] Toynbee 1965, II: 291–92.

[107] Brunt 1971, 364.

[108] Livy 39.29. Cf. Toynbee 1965, II: 320 and n. 3. It has been suggested that the slaves were being persecuted because of their worship of Dionysos, expressly forbidden by the *Senatus consultum de Bacchinalibus* of 186 BC.

[109] Osirius V.24.2.

The asty: archaeological evidence

The abandonment of Metaponto by Hannibal with "the entire population" in tow, as reported by Livy,[110] was probably exaggerated, and in any case it did not last long. Archaeological investigation of the urban center provides evidence of renewed but quite modest building activity in the 2nd century BC, and of some localized destruction and continued contraction in the 1st century (*Fig. 27.16*). Caution has been observed in interpreting these results which have been published only in useful but still preliminary articles and reports. There are naturally ongoing debates about the dating of pottery, coins, and other materials. Some are currently undateable, or only broadly dated, like the Metapontine type amphorae.[111] Further research will no doubt refine and modify current thinking, but the biggest question mark, the residential quarter, can only be resolved by large-scale excavation.

Castrum. Large areas of the asty lay deserted in the 2nd century BC, according to its excavator, De

[110] See Ch. 26, 891.

[111] See Ch. 12, 465.

Siena.[112] Settlement was concentrated around the castrum, by this time devoid of military functions. The basin just beyond the castrum that assured its independence and supply line from the sea ceased to operate. The stoa which served as a storehouse for provisions also lost its function and was broken up into separate units,[113] which Giardino suggests may have been dwellings. Plateia A, which passed through the stoa and was a major axis of the settlement within the castrum, was given a thick pavement in the 2nd century BC.[114] There is evidence in the form of metal slag of artisan activity in the spaces opening onto this street at the point where Plateia IV intersected it.[115] The stoa was destroyed early in the 1st century BC and never rebuilt. Associated with the destruction are amphorae of the Metapontine type, misfired examples of which were found in a depot outside the city to the north,[116] where slag, kiln furniture and molds, molded bowls, and abundant quantities of grey ware were also excavated. Grey ware, the characteristic fine pottery of Metaponto and southern Italy from the early 2nd to the 1st century BC, was finally superseded by Arretine ware.[117] Similar materials have been found in the kiln deposit of the factory at Pantanello.[118] This early-1st-century BC destruction of the stoa by fire, which left its columns blackened, has been connected, reasonably enough, with the presence of Spartacus in the area.[119] A road surface dateable to the second half of the 1st century BC, with a coin of Caesar, brings us down to the end of the last 50-year period covered in this volume.[120]

Agora. The limited new construction that took place in the 2nd century BC in the excavated areas of the city look back to earlier periods in the city's history. The Sanctuary of Apollo in the agora, located in the most prominent spot in the venerable city, the intersection of its two main and still most frequented arteries, Plateiai A and III, was remodeled.[121] It was surrounded by a wall, crowned by a triglyph-metope frieze that was both scenographic and practical. It was just high enough to shield from curious gaze the rites carried out within. A low wall decorated in relief ran along the east edge of the open cloaca that separated the agora from the sanctuary. Inside the closure a new public structure consisting of two interconnecting rooms fronted by a long portico with rooms at either end has been identified by De Siena as an *hestiatorion*, or guest house, for visitors consulting the oracle. The Manteion with the altar of Apollo, statues of the god and of Aristeas, and a well, had been erected in the 5th century BC,[122] as the mysteries variously associated with Pythagoras, Orpheus, and Dionysos swept through southern Italy. These beliefs survived the tumultuous events of the 3rd century BC. Their popularity was perceived as a threat by the Roman government, as evidenced by the *Senatus consultum de Bacchanalibus* of 186 BC. The remodeling of this place of worship is another indication that a segment, at least, of the population of Metaponto—an influential one to judge by the prominence of this shrine—continued the religious practices of earlier times. De Siena theorizes that these may have been survivors of the old aristocracy, whom, he suggests, were now in control of large areas of the chora.[123] To judge by the number of ex novo establishments in the chora in **150**, some of the elite in the countryside could have been newcomers. The stone used in the construction of Apollo's shrine was a sign of the times. Thriftiness in the choice of building materials was dictated by the straitened circumstances of the building's commissioners. It consisted entirely of stone cannibalized from earlier buildings around the site.

There were limited signs of life elsewhere in the agora. The exterior retaining wall of the theater, which had largely collapsed, was not repaired, but there was some modification of the stage building. It is uncertain, however, whether the theater was again in use as a place of public assembly or simply a shelter for squatters.[124]

Sanctuary. By the 2nd century BC the big temples in the sanctuary area were no longer in use. They were not only in disrepair but actually falling down. They, too, were a source of building material. The Ionic capitals from Temple D were reutilized in an L-shaped

[112] De Siena 1993, 115.

[113] L. Giardino 2005, 423.

[114] See the plan of the castrum in Ch. 26, Fig. 26.17.

[115] Giannotta 1980, 61.

[116] Giannotta 1980, 62.

[117] L. Giardino 1980a.

[118] Carter 1983a, 465–90; cf. also Carter 2006a, Fig. 4.22.

[119] Giannotta 1980, 66.

[120] Giannotta 1980, 66.

[121] De Siena 1998, 156–59. Mertens 1999, 285–86.

[122] See Ch. 23, 781–83.

[123] De Siena 1993, 115.

[124] De Siena 1990b, 305.

structure, possibly a portico, of light construction, that occupied the southeast corner of the sanctuary.[125] It would have screened the sanctuary from the southwestern end of the agora and the Sanctuary of Apollo, and from the earlier stoa that ran parallel to Plateia A and replaced the southern temenos wall of the sanctuary.[126]

The Temple of Apollo was robbed to provide a cover for the cloaca and the foundation of Oikos I, which rose on higher ground directly centered between and on axis with the Temples of Hera and Apollo.[127] When a censor stole roof tiles from the Temple of Hera at Croton, the superstitious senators hastened to return them for fear of Hera's wrath.[128] The Metapontines may have been unable to save the temples, but the construction of the small and poor Oikos I between the temples of the Oikoi G and H near the venerable Oikos C would indicate a healthy respect for the traditional gods of the place.

The numerous altars and small structures that crowd together at the eastern edge of the sanctuary, along Plateia III, mostly date to the 3rd and 2nd centuries BC.[129] The western half, with its crumbling monuments of bygone times, was not wholly abandoned. Oikoi G and H were constructed in the 2nd century BC with poor materials and their siting, beside and on axis with Metaponto's earliest surviving temple, Temple C, was not casual. Here, as in the case of the refurbished Manteion, is further evidence that some aspects of the traditional worship survived Hannibal and the reported departure of the Metapontines to Bruttium. Though, as De Siena admits,[130] there is no indication which gods were worshipped here, there is a possibility that these are the humble abodes of Hera and Apollo, the principal gods of the city from the beginning on the sites where their temples first arose. Apollo's worship also continued on the site of the Manteion just to the south, and Hera's perhaps in her temple, the Tavole Palatine (see below).

The southern side of the sanctuary, just to the south of Oikoi G and H, was occupied by a stoa that acted as a frame as early as the early 3rd century BC.[131] In the 2nd century, rooms alternately occupied by wells, surrounded by stone pavements, faced on the Plateia A to the south. At either end and at the center, there were passageways to both street and sanctuary. The necessity for publicly accessible wells may be a sign that the system of terracotta pipes that earlier carried water along the streets was no longer functioning,[132] but it is also an indication of a resident population that made use of the old center of the asty.

The main east-west thoroughfare of the asty continued to be a link between the area of the castrum and the earlier port area at the mouth of the Basento and the chora. It began to be encroached upon by artisans and their workshops, which clustered around the south end of the great stoa and the southeastern edge of the agora. A high proportion of artisan activity in this period was devoted to metalworking, and this was, like the production of amphorae and other pottery, not limited to the asty. Besides producing amphorae and tiles, the factory at Pantanello was engaged also in forging if not smelting iron.[133] Plateia A led out through the abandoned western part of the city, past the defunct kerameikos, now agricultural terrain, and through the western gate of the city wall, which had not been repaired for a century or more, and into the open countryside. The Crucinia necropolis was still in use but burials were few, though grave goods occasionally reflect wealth.[134]

The dramatic changes that were triggered in the early 3rd century BC included the disappearance of cult places in the chora. One may have survived—the architecturally most prepossessing of them all and virtually an extension of the cult of the asty—the extramural temple of Hera, the Tavole Palatine. The evidence is slender. There was a report that a mosaic of the 2nd century BC was removed from the temple in relatively recent times by the Archbishop of Taranto presumably to that city and to a location still to be discovered and confirmed.[135] Unfortunately, in addition to the mosaic, the archaeological levels surrounding the temple were removed as well. If indeed the temple had been left intact and were actually in use in the 2nd century BC, we can only speculate about the reason. Could this have been a rare display of religious

125 De Siena 1993, 117.

126 De Siena 1993, plan following 118; Mertens 2006, Fig. 279.

127 It has been provisionally dated to the second half of the 3rd c. BC.

128 Toynbee 1965, II.631–32, based on Livy 42.1.

129 De Siena, pers. comm., October 2009.

130 De Siena 1993, 119.

131 See Ch. 25, 860.

132 Mertens 2006, Fig. 270.

133 C. Giardino, pers. comm., July 2008.

134 De Siena 1993, 126–35.

135 De Siena 1993, 121 n. 18; cf. Giannotta 1980, 67, n. 76. Apparently it was seen by Kahrstedt (1960).

scruple such as that shown by the Romans for Demeter at Locri or even more to the point, for Hera in her sanctuary on the Lacinian Cape?

Conclusion

The last two centuries BC represent one third of the long span, from ca. 600 BC to the time of Augustus, through which we have traced the changing pattern of the Greek and later settlement in the chora. They stand apart from the preceding four hundred years, which, while not without some quite dramatic vicissitudes, follow a pattern that is fairly constant. The central phenomena of Magna Grecia between Hannibal and Augustus that Toynbee recognized, a new agricultural regime based on huge estates, the latifundia, populated by slaves and sheep, and malaria, he attributed to the aftereffects of the Second Punic War in Italy. We have attempted to show here that while the agricultural regime changed, farming survived, and as the anthropological studies on the necropoleis of the chora presented in the first volume of this series have clearly shown, malaria was not a new thing in the Metapontino.[136] Its indicators appear already, together with those of an inherently more deadly disease, treponematosis, in the heyday of the chora, and probably from the beginning of the Greek settlement of the chora, if not before.

There were elements of the countryside that did not change. The fertility of the land remained, as it does today, very high. The arc of coastal plain between Sybaris and Metaponto is still considered among the potentially best-adapted land in Italy for cereal cultivation, though much of it is occupied today by citrus groves. If anything, the fertility of the land probably increased in the 2nd and 1st centuries BC as the flooding in the river valleys dumped more rich top soil from the interior on the valley floors and coastal plains. Successful agricultural production, however, does not depend on fertility alone.

The forests continued to provide raw materials, lumber and game,[137] and they were exploited more in this period than previously, but still, probably on a small scale. The deforestation that impressed Toynbee in his travels through Magna Grecia at the beginning of the 20th century largely occurred in the second half of the 19th as forests were turned into crossties for the railroads,[138] after which runoff increased dramatically, creating the fiumare, or torrents, that he observed especially in Calabria.

The changes in the settlement pattern that appear in the last centuries BC have been detailed at length above. There are fewer and larger settlements in the study area, mostly isolated and not clustered together except for the rare hamlet. The necropoleis related to farmhouses had virtually disappeared by the 3rd century BC, together with the rural sanctuaries, and the widespread worship of Dionysos, so characteristic of the countryside and city in the later 4th and early 3rd centuries BC. The asty lost many of the features that once made it a city as its area was reduced to little more than that of the castrum. The physical changes were accompanied by intangible losses. With the disappearance of the urban center and of the villages that were a constant feature of the countryside for four centuries went a sense of community among its inhabitants; with the necropoleis, a visceral attachment to the land from generation to generation, with the sanctuaries large and small, a spiritual void that apparently was not filled again until the Medieval period. The continued worship of the ancient cults around the Manteion and in the sanctuary in this period was a survival that had no future.

In this period and from this point onward, the economy was pan-Mediterranean, the laws and government were determined from the outside, a pattern that continued under the Empire and throughout the Middle Ages, under Byzantine, Norman, French, and Spanish rule. By comparison, the productivity, creativity, and independence of the first four centuries appear as a remarkable, unsustainable experiment, a measure and a reminder of what was humanly possible.

136 Henneberg and Henneberg 1998, 527–29.

137 See Bökönyi and Gál 2010, 7–8, and Brunt 1971, 359.

138 Brunt 1971, 358.

Appendix A
EAW of Dated Sites, 600–50

Joseph Coleman Carter

Site	Type	Locality	Geomorph	Elev	Dated Arts.	600	550	500	450
1	F	Sant'Angelo	T3	31	18	0.63	1.36	0.10	3.28
2	F	Sant'Angelo	T3	31	1	0.50	0.50		
3	N	Sant'Angelo	T3	34	0				
4	F	Sant'Angelo	T3	41	4				
5	GAS	Sant'Angelo	T3	35	0				
6	N	Sant'Angelo	T3	32	0				
7	F	Sant'Angelo	T3	36	2	0.27	1.58	0.15	
8	N	Sant'Angelo	T3	36	4	0.13	1.16	0.71	1.95
9	K	Sant'Angelo	T3	34	0				
10	F	Sant'Angelo	T3	37	0				
11	N	Sant'Angelo	T3	27	1			0.01	0.99
12	F/K	Sant'Angelo	T2	33	3	0.26	1.73	0.01	0.03
13	F	Sant'Angelo	T2	37	13	1.49	3.46	0.19	3.16
14	F	Sant'Angelo	T2	38	1			0.01	0.95
15	F	Scarciullo	P.T.	117	2				
16	F	Scarciullo	P.T.	117	4				
17	F	Scarciullo	P.T.	115	3				
18	F	Scarciullo	P.T.	114	2				
35	F	Cogno Prete	P.T.	96	8				0.35
36	F	Cogno Prete	P.T.	96	1				0.35
37	F	Cogno Prete	P.T.	92	1				
38	F	DBSBS	P.T.	51	5				0.35
39	N	DBSBS	V.W.	40	0				
41	N	DBSBS	P.T.	52	2				0.35
42	F	DBSBS	P.T.	54	28				
43	F	Cogno Prete	P.T.	99	40	1.01	0.99		0.82
44	F	Campagnolo	P.T.	104	4				
45	F	Campagnolo	P.T.	101	0				
46	F	Campagnolo	P.T.	114	2				
47	F	Campagnolo	P.T.	100	7				
48	N	Campagnolo	P.T.	97	3			0.01	0.99
49	F	Campagnolo	P.T.	100	3			0.01	1.98
50	N	Campagnolo	P.T.	102	1		0.03	0.94	0.03
51	GAS	Campagnolo	P.T.	104	0				
52	F	Campagnolo	P.T.	104	3				0.35
53	F	Venella	P.T.	52	1	0.13	0.87		
54	F	Venella	P.T.	55	1				0.35
55	F	Cogno Prete	P.T.	108	1				
56	N	Cogno Prete	P.T.	108	1				
57	F	Cogno Prete	P.T.	80	18			0.01	1.02
58	F	Cogno Prete	P.T.	92	9	0.55	4.11	0.33	0.35

Type: **F** = Farmhouse; **N** = Necropolis; **T** = Tomb; **K** = Kiln; **Q** = Quarry; **GAS** = Greek Agrarian Structure; **UGF** = Unidentified Greek Feature; *Geomorph:* **M.M.D.** = Mass Movement Deposit; **P.T.** = Pleistocene Terrace; **T2–T4** = T2–T4 Terrace; **V.W.** = Valley Walls.

400	350	300	250	200	150	100	50	Best Date
1.56	2.37	1.67	0.03	0.08	1.50	3.83	1.50	386
								575
								0
	0.26	2.54	0.70	0.44	0.06			278
								0
								0
								553
0.05								490
								0
								0
0.01								450
0.94	0.03							557
1.34	1.31	1.95	0.10					565
0.04								450
	0.27	1.64	0.09					307
	0.36	2.27	0.87	0.44	0.06			282
0.26	1.25	1.46	0.03					330
	0.03	0.94	0.03	0.01	0.21	0.55	0.21	300
0.93	1.53	3.22	0.61	0.34	0.25	0.55	0.21	322
0.65								418
	0.01	0.99	0.01					300
0.65	0.26	3.60	0.14					300
								0
0.90	0.72	0.03						391
0.14	0.87	2.39	0.60	0.86	5.99	12.54	4.34	104
3.68	8.70	17.21	3.55	1.83	0.70	0.94	0.47	319
0.87	1.72	1.31	0.09					344
								0
	0.05	1.83	0.12					300
	1.02	2.83	0.15	0.02	0.93	1.59	0.43	313
0.01	0.01	1.19	0.79					280
0.01				0.01	0.21	0.55	0.21	450
								500
								0
0.67	0.75	1.20	0.03					353
								557
0.65								418
0.50	0.50							375
	0.03	0.94	0.03					300
1.88	2.57	8.33	1.95	1.41	0.79	0.03		305
0.66	0.99	1.19	0.69	0.11				550

Site	Type	Locality	Geomorph	Elev	Dated Arts.	600	550	500	450
60	F	Scarciullo	P.T.	112	4				0.03
61	N	Scarciullo	P.T.	116	0				
62	F	Sant'Angelo	T2	28	1				
63	N	Sant'Angelo	T3	35	0				
64	F	Cogno Prete	P.T.	108	0				
65	F	Cogno Prete	P.T.	108	0				
66	F	Cogno Prete	P.T.	108	0				
67	F	Cogno Prete	P.T.	108	0				
68	F	Cogno Prete	P.T.	108	5		0.29	0.69	1.08
69	GAS	Cogno Prete	P.T.	103	0				
70	N	Cogno Prete	P.T.	106	18	4.28	7.57	3.11	2.02
78	N	Cogno Prete	P.T.	105	0				
81	N	Cogno Prete	P.T.	106	14		0.50	0.50	0.14
82	F	Cogno Prete	P.T.	106	15				0.24
83	F	San Marco	P.T.	91	5				
84	F	San Marco	P.T.	91	12				0.35
88	F	Cogno Prete	P.T.	102	3	0.13	0.87		0.09
95	F/K	DBSBS	P.T.	54	11				
96	GAS	Campagnolo	P.T.	106	2				
97	F	Campagnolo	P.T.	101	1				
98	N	Campagnolo	P.T.	122	0				
99	N	Scarciullo	P.T.	116	0				
100	N	Scarciullo	P.T.	116	1				
101	F	Scarciullo	P.T.	114	0				
102	F	Campagnolo	P.T.	104	1	0.13	0.87		
103	F	Scarciullo	P.T.	102	19	1.00	1.00	0.96	1.74
104	F	Avinella	P.T.	87	1				
105	F/K	Sant'Angelo	T2	16	34				
106	F	Avinella	P.T.	91	0				
107	N	Avinella	P.T.	92	1	0.50	0.50		
108	N	Avinella	P.T.	88	0				
109	F	Avinella	P.T.	88	1				
110	N	Lago Lupo	P.T.	98	0				
111	F	Lago Lupo	P.T.	98	1				
112	N	Lago Lupo	P.T.	96	1				
113	F	Lago Lupo	P.T.	99	25			0.01	1.65
114	N	Lago Lupo	P.T.	96	0				
115	F	Lago Lupo	P.T.	106	0				
116	N	Lago Lupo	P.T.	106	0				
117	F/T	Lago Lupo	P.T.	109	1				
118	F	Lago Lupo	P.T.	107	1				0.35
119	F	Lago Lupo	P.T.	109	2	0.50	0.50		0.35
120	N	Lago Lupo	P.T.	111	0				
121	F	Lago Lupo	P.T.	111	7			0.50	0.50

400	350	300	250	200	150	100	50	Best Date
1.32	0.64	0.15	1.03	0.76	0.07			384
								0
					0.06	0.44	0.44	75
								0
								0
								0
								0
								0
1.94	0.04	0.95	0.01					433
								0
0.96	0.04							542
								0
0.73	2.57	8.44	1.11					306
3.06	4.36	6.13	0.77	0.39	0.05			339
0.93	0.07	0.95	1.76	1.22	0.06			247
0.79	1.00	5.27	1.12	1.18	0.74	1.09	0.43	264
0.89	0.05	0.94	0.03					300
				0.26	2.86	5.61	2.14	103
	0.07	1.89	0.04					300
	0.93	0.07						350
								0
								0
		0.61	0.38					281
								0
								557
2.20	3.50	7.11	0.97	0.44	0.06			345
		0.36	0.62	0.02				268
1.23	6.21	5.45	1.73	0.51	3.77	10.26	4.53	98
								0
								575
								0
		0.09	0.62	0.28	0.01			241
								0
	0.08	0.88	0.04					304
	0.88	0.12						344
1.78	3.81	12.81	2.70	1.80	0.40	0.03		302
								0
								0
								0
		0.06	0.44	0.44	0.06			225
0.65								418
0.65								418
								0
	0.15	3.68	1.88	0.29				283

Site	Type	Locality	Geomorph	Elev	Dated Arts.	600	550	500	450
122	F	Lago Lupo	P.T.	107	4				
123	N	Lago Lupo	P.T.	112	0				
124	F	Lago Lupo	P.T.	113	0				
125	F	Lago Lupo	P.T.	114	3				
126	F	Lago Lupo	P.T.	114	0				
127	F	Cogno Prete	P.T.	99	0				
128	F	Cogno Prete	P.T.	101	35			0.05	2.07
129	F	Cogno Prete	P.T.	104	1				0.03
130	F	Cogno Prete	P.T.	86	1				
131	F	Lago Lupo	P.T.	111	7	0.63	1.36	0.05	2.79
132	N	Lago Lupo	P.T.	111	1				
133	GAS	Lago Lupo	P.T.	111	0				
134	F	Lago Lupo	P.T.	111	1				
135	N	Lago Lupo	V.W.	102	0				
136	F	Lago Lupo	P.T.	91	0				
137	F	Lago Lupo	P.T.	96	1				
138	N	Lago Lupo	P.T.	111	0				
139	F	Bufalara	P.T.	111	1				
140	N	San Marco	P.T.	87	0				
141	Q	San Marco	V.W.	75	0				
142	N	San Marco	P.T.	92	0				
143	N	Cogno Prete	P.T.	96	0				
144	GAS	Cogno Prete	P.T.	96	0				
145	F	Cogno Prete	P.T.	96	0				
146	N	Cogno Prete	P.T.	92	0				
147	F	Sant'Angelo	T2	18	0				
148	F	Pantanello	T3	18	2				0.03
149	UGF	Sant'Angelo	T2	18	0				
150	F	Sant'Angelo	T2	18	0				
151	UGF	Sant'Angelo	T2	17	0				
152	F	Lago Lupo	P.T.	87	4				0.35
153	F	Sant'Angelo	T2	19	1				
154	F	Sant'Angelo	T2	19	25	2.76	4.23	0.01	0.78
155	F	Sant'Angelo	T2	19	27	3.71	10.51	6.31	3.36
156	F	Sant'Angelo	T2	18	3				
157	F	Sant'Angelo	T2	18	1				0.35
158	F	Sant'Angelo	T2	21	1				0.35
159	N	Sant'Angelo	T2	17	1				0.16
161	F	Sant'Angelo	V.W.	26	0				
162	other	Sant'Angelo	V.W.	27	0				
163	K	Sant'Angelo	V.W.	27	0				
164	F	Lago Lupo	P.T.	91	7				
165	F	Lago Lupo	P.T.	91	1				
166	F	Lago Lupo	P.T.	86	6			0.01	1.95

400	350	300	250	200	150	100	50	Best Date
	0.58	2.84	0.58					300
								0
								0
	0.30	1.67	0.52	0.44	0.06			281
								0
								0
1.97	6.91	15.90	4.08	2.38	1.61			305
0.94	0.03							400
		0.06	0.44	0.44	0.06			225
1.65	0.50							423
		0.06	0.44	0.44	0.06			225
								0
	0.03	0.94	0.03					300
								0
								0
		0.06	0.44	0.44	0.06			225
								0
	0.20	0.80						310
								0
								0
								0
								0
								0
								0
								0
								0
0.94	0.06	0.94	0.03					300
								0
								0
								0
0.65	0.15	1.87	0.47	0.44	0.06			276
	0.12	0.88						306
3.87	2.91	9.08	0.92	0.39	0.05			334
0.73	1.57	0.79						531
	0.43	2.47	0.11					307
0.65								418
0.65								418
0.81	0.03							408
								0
								0
								0
0.14	1.31	3.57	0.98	0.89	0.11			289
		0.06	0.44	0.44	0.06			225
0.61	1.30	2.07	0.06					371

Site	Type	Locality	Geomorph	Elev	Dated Arts.	600	550	500	450
167	F	Venella	P.T.	91	12				
168	GAS	Lago Lupo	P.T.	86	3	0.26	1.73	0.01	0.99
169	F	Lago Lupo	P.T.	81	18				
170	F	Lago Lupo	P.T.	77	10	1.04	4.88	1.07	1.01
171	Q	Lago Lupo	P.T.	81	0				
172	F	Lago Lupo	P.T.	74	4				0.03
173	F	Lago Lupo	P.T.	66	22			0.01	1.66
174	F	Lago Lupo	P.T.	109	8		0.29	0.84	3.59
175	F	Lago Lupo	P.T.	98	2			0.01	1.31
176	N	Venella	P.T.	89	5				0.05
177	N	Venella	P.T.	85	0				
178	F	Venella	P.T.	86	1			0.50	0.50
179	F	Venella	P.T.	82	0				
180	GAS	Venella	P.T.	82	0				
181	F	Venella	P.T.	81	4	0.40	2.45	0.15	
182	F/T	Venella	P.T.	82	12	0.64	1.51	0.98	5.61
183	F	Venella	P.T.	77	23	0.13	0.87		0.89
184	N	Venella	P.T.	76	3				
185	GAS	Venella	V.W.	71	0				
186	N	Venella	V.W.	63	0				
187	F	Venella	P.T.	77	3				0.35
188	N	Venella	P.T.	74	2				
189	UGF	Giardinetto	T4	105	0				
190	UGF	Giampasquale	P.T.	68	0				
191	F/T	Giampasquale	P.T.	72	1				0.03
192	UGF	Giampasquale	P.T.	69	0				
193	F	Giampasquale	P.T.	72	24	0.50	0.50		0.70
194	F	Cogno Prete	P.T.	96	0				
195	F	Venella	P.T.	78	4		0.11	0.72	1.16
196	F	Venella	P.T.	78	1				
197	F	DBSBS	V.W.	26	17				
198	F	DBSBS	V.W.	26	6				
199	UGF	DBSBS	V.W.	21	1	0.14	0.72	0.14	
200	F	Giampasquale	P.T.	75	10	0.63	1.36	0.01	1.56
201	F	Giampasquale	P.T.	75	7			0.01	1.31
207	GAS	Sant'Angelo	T2	30	0				
208	GAS	Venella	P.T.	82	0				
209	F	Venella	P.T.	78	2				
210	F	Lago Lupo	P.T.	70	2				
211	F	Lago Lupo	P.T.	68	11	0.64	1.72	0.64	0.35
212	F	Lago Lupo	P.T.	68	1				
213	F/K	Sant'Angelo	T3	23	14	0.13	0.87	0.01	1.85
214	GAS	Sant'Angelo	T2	22	0				
215	F	Giampasquale	P.T.	81	8	0.39	2.89	0.72	1.99

400	350	300	250	200	150	100	50	Best Date
1.46	1.92	5.55	2.04	0.92	0.11			304
0.01								557
0.14	3.71	10.69	2.13	1.17	0.15			305
	0.01	1.90	0.09					537
								0
0.42	1.50	2.01	0.04					330
2.39	2.28	9.59	3.42	1.87	0.75	0.01		312
2.25	0.38	0.65						439
0.69								433
2.52	1.93	0.50						370
								0
								475
								0
								0
	0.35	0.65						557
3.25								453
2.08	4.71	10.76	2.23	1.00	0.28	0.03		317
0.02	1.42	1.53	0.03					324
								0
								0
1.79	0.85							392
0.28	1.43	0.28						350
								0
								0
0.94	0.03							400
								0
4.77	4.25	10.05	2.27	0.59	0.30	0.06		327
								0
0.51	0.79	0.69	0.01					405
0.14	0.72	0.14						350
	0.66	5.13	0.53	0.69	3.14	5.04	1.72	300
	0.29	0.69	0.01	0.05	1.00	2.90	1.00	100
								550
1.10	2.70	2.58	0.06					360
0.97	1.70	2.11	0.76	0.14				350
								0
								0
	0.35	0.79	0.72	0.14				284
0.02	0.92	1.03	0.03					324
1.29	2.00	4.18	0.17					331
0.21	0.67	0.11						355
2.02	0.89	5.29	1.80	0.64	0.50			335
								0
0.01	0.29	0.75	0.46	0.44	0.06			508

Site	Type	Locality	Geomorph	Elev	Dated Arts.	600	550	500	450
216	F/T	Lago Lupo	P.T.	87	4				0.04
217	F	Lago Lupo	P.T.	82	0				
218	F	Lago Lupo	P.T.	75	1				
219	GAS	Lago Lupo	P.T.	86	0				
220	N	Lago Lupo	P.T.	86	0				
221	F	Lago Lupo	P.T.	86	5	0.13	0.87		
222	F	Lago Lupo	P.T.	79	4	0.01	0.95	0.04	
223	N	Lago Lupo	P.T.	77	0				
224	F/T	Lago Lupo	P.T.	65	5	0.26	1.73	0.05	0.95
225	F	San Marco	V.W.	57	2				0.35
226	F	San Marco	V.W.	40	0				
227	UGF	San Marco	V.W.	40	0				
228	F	San Marco	V.W.	42	3	1.50	1.50		
229	N	Venella	P.T.	70	1			0.50	0.50
230	F/T	San Marco	V.W.	28	5			0.01	1.02
231	F	San Marco	V.W.	35	6				
232	N	San Marco	V.W.	40	0				
233	UGF	San Marco	V.W.	39	0				
234	GAS	San Marco	V.W.	36	0				
235	F	Giampasquale	P.T.	81	7				
236	N	Giampasquale	P.T.	61	0				
237	UGF	Giampasquale	P.T.	66	0				
238	F	Giampasquale	P.T.	78	4	0.27	1.58	0.19	0.95
239	F	Lago Lupo	P.T.	69	0				
240	N	AVS	V.W.	25	1				0.18
241	F/T	AVS	V.W.	27	8	0.50	0.50		0.71
242	GAS	DBSBS	V.W.	21	0				
243	UGF	DBSBS	V.W.	19	2	0.26	1.73	0.01	
244	F	Lago Lupo	P.T.	66	2	0.13	0.87		0.02
245	F	Lago Lupo	P.T.	67	2	0.13	0.87		
246	F	Lago Lupo	P.T.	71	0				
247	F	Lago Lupo	P.T.	71	2				
248	F/K	Lago Lupo	P.T.	70	15	0.50	0.50		
249	F	Lago Lupo	P.T.	68	8				
250	F	Lago Lupo	P.T.	70	8			0.01	1.38
251	GAS	Lago Lupo	P.T.	67	0				
252	GAS	Lago Lupo	P.T.	66	0				
253	N	Lago Lupo	P.T.	67	0				
254	F	Lago Lupo	P.T.	67	6	0.89	3.10	0.02	0.99
255	F	Cogno Prete	P.T.	84	1				
256	F	Cogno Prete	P.T.	63	0				
257	F	Cogno Prete	P.T.	71	2				1.05
258	F	Cogno Prete	P.T.	61	18	0.76	2.23	0.01	
259	GAS	Cogno Prete	P.T.	76	0				

400	350	300	250	200	150	100	50	Best Date
0.87	1.00	2.02	0.06					335
								0
	0.20	0.80						310
								0
								0
0.14	1.30	2.44	0.12					317
0.38	1.00	1.59	0.03					330
								0
0.03	0.72	0.31	0.44	0.44	0.06			550
0.65	0.03	0.94	0.03					300
								0
								0
								575
								475
0.93	0.93	0.12		0.02	0.43	1.09	0.43	402
				0.06	1.13	3.18	1.51	97
								0
								0
								0
1.01	1.65	4.25	0.09					327
								0
								0
0.01	0.50	0.50						550
								0
0.80	0.02							409
2.72	0.63	1.93	0.50	0.44	0.06			368
								0
								557
0.34	0.56	0.08						550
0.14	0.72	0.14						550
								0
	0.01	0.69	0.29	0.50	0.50			230
	1.47	8.61	2.22	1.02	0.35	0.24	0.08	297
0.50	3.19	3.28	0.03	0.06	0.44	0.44	0.06	325
2.32	2.10	1.19		0.14	0.72	0.14		378
								0
								0
								0
0.01	0.03	0.94	0.03					561
		0.21	0.67	0.11				255
								0
0.95								426
1.02	4.30	6.66	0.87	0.90	1.22	0.03		320
								0

Site	Type	Locality	Geomorph	Elev	Dated Arts.	600	550	500	450
260	F	Cogno Prete	P.T.	78	32			0.02	4.10
261	GAS	Cogno Prete	P.T.	74	0				
262	F	Cogno Prete	P.T.	77	1				
263	GAS	Cogno Prete	P.T.	63	0				
264	GAS	Cogno Prete	P.T.	58	1				
265	N	Cogno Prete	P.T.	59	2	0.13	0.87	0.50	0.50
266	S	Cogno Prete	P.T.	67	33	0.13	1.36	0.51	1.95
267	N	Cogno Prete	P.T.	74	0				
268	N	Cogno Prete	P.T.	64	0				
269	F	Giampasquale	P.T.	67	55	0.50	1.29	1.20	3.44
270	S	Giampasquale	P.T.	67	21	4.95	13.00	2.01	0.03
271	UGF	Giampasquale	P.T.	67	0				
272	F/T	DBSBS	P.T.	32	0				
273	UGF	DBSBS	P.T.	58	1				
274	F	Lago Lupo	P.T.	67	5	0.26	2.02	0.83	1.38
275	F	Lago Lupo	P.T.	61	3	0.13	0.87		
276	N	Lago Lupo	P.T.	79	0				
277	F	Lago Lupo	P.T.	107	0				
278	F	Venella	P.T.	63	10				
279	GAS	Venella	P.T.	57	0				
280	F	Lago Lupo	P.T.	104	1				
281	F	Lago Lupo	P.T.	97	2				
282	F	Cogno Prete	P.T.	96	2				0.35
283	F	Cogno Prete	P.T.	96	2	0.13	0.87		0.04
284	F	Sant'Angelo	T2	24	0				
285	F/T	Sant'Angelo	T2	23	23	10.09	8.64	1.30	1.31
286	F	Sant'Angelo	T3	22	7	0.97	4.01	0.02	0.38
287	N	DBSBS	P.T.	64	0				
288	F	Venella	P.T.	45	10			0.50	1.21
289	N	Venella	P.T.	50	6	0.76	2.52	0.79	1.92
290	N	Venella	P.T.	58	0				
291	F	Lago Lupo	P.T.	113	3				0.35
292	F	Lago Lupo	P.T.	112	18			0.04	1.00
293	UGF	Lago Lupo	P.T.	112	0				
294	N	Lago Lupo	P.T.	107	1				
295	N	Lago Lupo	P.T.	107	0				
296	F	Lago Lupo	P.T.	107	19				0.02
297	GAS	Lago Lupo	P.T.	109	0				
298	N	Lago Lupo	P.T.	111	1			0.13	0.87
299	N	Lago Lupo	P.T.	111	3			0.27	2.72
300	N	AVS	P.T.	105	4				0.15
301	F	Campagnolo	P.T.	117	10	0.63	1.36	0.13	3.16
302	F	Campagnolo	P.T.	117	3			0.70	1.30
303	F	Campagnolo	P.T.	121	3			0.01	1.94

400	350	300	250	200	150	100	50	Best Date
3.10	3.11	13.35	4.06	3.21	0.94	0.09		318
								0
0.50	0.50							375
								0
	0.03	0.94	0.03					300
								510
1.33	3.93	18.60	3.88	1.18	0.11			300
								0
								0
5.75	8.74	23.18	6.85	3.55	0.48			315
	0.50	0.50						557
								0
								0
0.38	0.61							369
0.50								496
0.75	1.10	0.14						370
								0
								0
1.00	1.59	4.43	1.26	1.43	0.28			297
								0
	0.03	0.94	0.03					300
0.85	1.12	0.02						372
0.65	0.03	0.94	0.03					300
0.73	0.23							550
								0
0.79	0.72	0.14						577
1.59	0.03							560
								0
3.54	1.19	3.47	0.09					363
0.01								518
								0
0.79	0.72	1.07	0.07					357
2.45	4.69	7.89	0.90	0.52	0.50			332
								0
	0.12	0.85	0.04					306
								0
0.13	1.76	14.03	2.48	0.52	0.05			298
								0
								457
0.01								454
1.81	1.20	0.84						363
2.15	1.50	0.95	0.04					398
0.93	0.07							446
0.05	0.61	0.38						450

Site	Type	Locality	Geomorph	Elev	Dated Arts.	600	550	500	450
304	F	Bufalara	P.T.	122	8	0.13	1.36	0.65	2.48
305	F	Bufalara	P.T.	109	4		0.11	0.68	1.16
306	F	Bufalara	P.T.	106	2				
307	F	Bufalara	P.T.	122	1			0.13	0.87
308	F	Bufalara	P.T.	111	0				
309	F	Bufalara	P.T.	122	8				0.35
310	F	AVS	P.T.	106	10	0.26	2.23	0.53	3.28
311	N	AVS	P.T.	103	0				
312	F	AVS	P.T.	103	19			0.01	2.63
313	N	AVS	P.T.	103	0				
314	N	DBSBS	V.W.	51	0				
315	N	DBSBS	V.W.	51	0				
316	GAS	DBSBS	V.W.	51	1				
317	F	AVS	P.T.	92	6				
318	N	AVS	P.T.	93	1				
319	F/T	AVS	P.T.	97	5			0.50	0.85
320	F	Venella	P.T.	89	48	0.13	1.16	0.75	5.08
321	N	Venella	P.T.	85	0				
322	F	Venella	V.W.	72	7			0.07	3.20
323	F	Venella	P.T.	78	7	1.40	3.45	0.16	0.99
324	N	Avinella	P.T.	80	0				
325	N	Venella	P.T.	78	0				
326	F	Venella	P.T.	76	39			0.14	2.35
327	F	Venella	P.T.	73	17				1.19
328	F	Avinella	P.T.	66	4	0.26	1.73	0.55	1.45
329	F	Venella	P.T.	37	10		0.29	0.69	0.20
330	N	Venella	P.T.	68	0				
331	N	Venella	P.T.	65	0				
332	F	Venella	P.T.	60	0				
333	F	Venella	P.T.	49	1				
334	S	Venella	V.W.	36	21	1.94	11.27	2.25	2.50
335	F	Venella	V.W.	35	3			0.01	1.31
336	F	Venella	V.W.	35	11	1.81	6.84	0.35	1.31
337	N	Venella	P.T.	55	1				
338	N	Venella	P.T.	43	2			0.01	0.99
339	F	Avinella	P.T.	65	13				
340	F	Avinella	P.T.	71	0				
341	F	Avinella	P.T.	75	44				0.75
342	F	Avinella	P.T.	86	37	0.13	0.87	0.05	2.97
343	F	Avinella	P.T.	86	14				
344	F	Avinella	P.T.	91	4	0.14	1.85	0.02	1.98
345	F	Avinella	P.T.	90	8			0.13	0.87
346	N	Avinella	P.T.	95	1				
347	N	Avinella	P.T.	92	2	0.13	0.87		0.35

400	350	300	250	200	150	100	50	Best Date
2.63	0.74							445
0.21	1.43	0.40						468
		0.17	0.94	0.78	0.10			234
								457
								0
0.65	0.54	4.96	1.00	0.44	0.06			302
0.91	0.73	1.94	0.12					390
								0
0.69	3.82	8.31	2.11	0.92	0.50			306
								0
								0
								0
0.50	0.50							375
				0.06	1.57	3.23	1.07	104
0.50	0.50							375
1.15	1.42	1.03	0.05					384
7.77	5.47	19.98	4.17	2.91	0.54	0.03		338
								0
0.65	1.16	1.88	0.04					388
0.01	0.01	0.69	0.29					564
								0
								0
5.95	8.82	16.17	3.19	1.71	0.65			326
1.31	1.36	9.46	1.18	1.44	1.06			316
0.01								504
2.40	2.26	3.06	0.73	0.32	0.04			337
								0
								0
								0
0.14	0.72	0.14						350
0.98	0.14	1.84	0.08					535
0.69	0.03	0.94	0.03					433
0.69								560
0.14	0.72	0.14						350
0.01	0.35	0.65						450
				0.14	3.19	7.11	2.41	103
								0
4.65	9.10	23.66	3.63	1.91	0.27			318
2.69	2.30	16.15	7.79	3.55	0.47	0.01		299
	1.63	8.02	2.10	1.41	0.79	0.03		288
0.01								450
0.89	1.21	3.49	0.92	0.44	0.06			332
0.23	0.73	0.04						362
0.65								550

Site	Type	Locality	Geomorph	Elev	Dated Arts.	600	550	500	450
348	F	Avinella	P.T.	96	10	0.13	1.16	0.70	0.73
349	F	Avinella	P.T.	97	6	1.04	3.66	0.29	
350	F	Avinella	P.T.	99	3				
351	N	AVS	P.T.	100	0				
352	F	Scarciullo	P.T.	108	27	0.50	0.50		1.42
353	F	Scarciullo	P.T.	110	21			0.14	3.97
354	F	Scarciullo	P.T.	108	35				
355	F	Scarciullo	P.T.	107	2				0.35
356	N	Avinella	V.W.	85	4	1.15	2.21	0.65	
357	N	Avinella	V.W.	80	0				
358	F	Avinella	P.T.	92	26				1.59
359	GAS	Avinella	V.W.	92	0				
360	F	Venella	P.T.	61	5				0.35
362	F	Scarciullo	P.T.	108	2	0.88	0.42	0.69	0.01
363	F/T	Scarciullo	P.T.	116	11				0.72
364	N	Scarciullo	P.T.	110	0				
365	F	Scarciullo	P.T.	112	29	0.50	1.29	1.20	2.31
366	F	Scarciullo	P.T.	115	2	0.26	1.73	0.01	
367	GAS	Scarciullo	P.T.	111	0				
368	GAS	Scarciullo	P.T.	112	0				
369	F	Scarciullo	P.T.	116	4				0.71
370	F	Scarciullo	P.T.	115	11				0.35
371	N	Venella	P.T.	63	0				
372	F	Venella	P.T.	62	31	0.28	1.72	0.98	0.92
373	F	Scarciullo	P.T.	116	2				
374	N	Scarciullo	P.T.	117	2	0.50	0.50		
375	F	Scarciullo	P.T.	114	21	0.26	1.73	0.01	1.56
376	F	Scarciullo	P.T.	114	9	0.50	0.50		0.01
377	F	AVS	P.T.	96	1	0.13	0.87		
378	GAS	AVS	T4	72	0				
380	N	AVS	T4	50	0				
382	UGF	AVS	T4	67	0				
383	F	AVS	T4	67	0				
385	N	Sant'Angelo	T3	57	0				
386	GAS	Sant'Angelo	T3	38	0				
389	N	Sant'Angelo	T3	33	0				
390	F	Avinella	P.T.	85	10				1.06
391	N	Avinella	P.T.	90	0				
392	N	Avinella	P.T.	86	2				0.35
393	N	AVS	P.T.	56	0				
394	F	Avinella	P.T.	61	17				0.03
395	F	Avinella	P.T.	58	61	1.40	3.95	1.23	5.58
396	N	Avinella	P.T.	53	1				
397	S	Avinella	P.T.	60	62	8.56	34.82	9.33	6.05

400	350	300	250	200	150	100	50	Best Date
2.85	1.36	2.89	0.19					392
	0.03	0.94	0.03					561
	1.05	1.90	0.05					318
								0
4.10	6.57	10.76	1.47	1.42	0.25	0.01		334
4.46	2.92	9.27	0.23					358
1.28	4.02	18.79	7.76	1.92	0.44	0.56	0.21	294
0.65	0.03	0.94	0.03					300
								556
								0
4.26	4.73	12.84	2.35	0.21				323
								0
0.79	0.72	2.06	1.08					323
								555
1.81	2.70	5.62	0.15					331
								0
1.94	5.32	12.23	2.98	0.96	0.21	0.03		320
								557
								0
								0
1.29	0.06	1.88	0.06					300
0.79	1.67	6.81	0.94	0.39	0.05			311
								0
3.59	2.73	1.74	0.03	0.19	3.93	9.39	4.65	98
	0.85	1.15						321
	0.03	0.94	0.03					300
2.79	3.05	6.22	2.61	1.92	0.81	0.03		319
0.69	0.38	2.14	1.98	1.91	0.85	0.03		239
								557
								0
								0
								0
								0
								0
								0
								0
2.08	1.59	4.23	0.44	0.45	0.14	0.01		350
								0
0.65	0.08	0.88	0.04					300
								0
1.59	3.66	9.08	1.76	0.78	0.10			316
7.55	10.09	19.44	6.42	4.22	1.05	0.02		336
0.50	0.50							375
1.35	0.72	0.76	0.38					539

Site	Type	Locality	Geomorph	Elev	Dated Arts.	600	550	500	450
398	F	Avinella	P.T.	60	1	0.14	0.72	0.14	
399	F/K	AVS	P.T.	60	0				
400	F	AVS	P.T.	62	4	0.63	1.36		
401	N	AVS	P.T.	61	47	0.64	1.22	0.14	1.27
402	F	Avinella	V.W.	70	4				
403	N	Avinella	V.W.	68	0				
404	F	Avinella	V.W.	69	34	0.26	1.73	0.05	0.95
405	N	Avinella	V.W.	68	3				
406	F/T	Avinella	V.W.	66	7	0.76	2.23	0.01	0.35
407	N	Avinella	V.W.	60	1			0.04	0.73
408	N	San Marco	V.W.	39	0				
409	GAS	San Marco	V.W.	45	0				
410	N	San Marco	V.W.	35	5			0.01	1.44
411	F/K	San Marco	V.W.	29	1	0.13	0.87		
412	UGF	Pantanello	V.W.	25	2				
413	F	Pantanello	T3	15	12				0.06
414	F	Pantanello	T3	15	8	0.63	1.36	0.88	1.46
415	N	Pantanello	T3	19	19				0.77
416	F	Pantanello	T3	14	18				0.71
417	F	Pantanello	T2	8	3	0.50	0.50		0.09
418	F	Pantanello	T3	16	1				
419	N	Pantanello	T2	9	0				
420	N	Pantanello	T3	16	0				
421	N	Pantanello	T3	15	2				0.01
422	N	Pantanello	T3	13	0				
423	UGF	Pantanello	T3	18	1				
424	UGF	Pantanello	T3	15	0				
425	F/T	Pantanello	T3	13	1				
426	N	Pantanello	T3	13	0				
427	N	Pantanello	T3	9	0				
428	F	Pantanello	T3	10	5	0.13	0.87		
429	N	Pantanello	T2	6	2				0.04
430	N	Pantanello	T2	6	1				
431	UGF	Pantanello	T2	7	0				
432	F	Pantanello	T2	7	0				
433	UGF	Pantanello	T2	6	0				
434	UGF	Pantanello	T2	6	0				
435	UGF	Pantanello	T2	6	0				
436	UGF	Pantanello	T2	6	0				
437	N	Pantanello	P.T.	13	2				
438	N	Pantanello	P.T.	20	1				
439	N	Pantanello	P.T.	21	9				0.01
440	N	Pantanello	P.T.	20	0				
441	UGF	Pantanello	V.W.	20	0				

400	350	300	250	200	150	100	50	Best Date
								550
								0
	0.50	0.59	0.47	0.39	0.05			566
12.41	20.93	9.60	0.77					353
	0.14	1.77	1.58	0.44	0.06			267
								0
0.43	3.92	18.32	4.69	2.37	1.25	0.01		291
1.43	0.61	0.90	0.05					359
0.69	1.34	1.59	0.03					563
0.23								438
								0
								0
2.54	0.87	0.14						406
								557
	0.38	1.59	0.03					310
3.32	2.82	4.73	0.56	0.44	0.06			344
1.09	1.19	1.37						437
6.03	7.88	3.41	0.67	0.21				358
2.58	1.90	3.86	0.81	0.24	1.72	4.37	1.72	342
0.47	0.42	0.99	0.03					336
	0.03	0.94	0.03					300
								0
								0
0.42	1.33	0.24						355
								0
	0.23	0.73	0.04					312
								0
0.50	0.50							375
								0
								0
0.21	0.67	0.11		0.04	0.64	1.64	0.64	100
0.73	0.73	0.50						356
0.14	0.72	0.14						350
								0
								0
								0
								0
								0
								0
	0.54	1.38	0.08					314
0.20	0.80							360
2.02	3.91	1.68	0.38	0.50	0.50			352
								0
								0

Site	Type	Locality	Geomorph	Elev	Dated Arts.	600	550	500	450
442	F/T	Pantanello	P.T.	21	0				
443	F/T	Pantanello	T2	13	1		0.11	0.67	0.21
444	UGF	Pantanello	T2	17	1				0.07
445	F	Pantanello	P.T.	19	3	0.50	0.50		
446	N	Pantanello	P.T.	16	8				0.35
447	N	Pantanello	T2	14	0				
448	N	Pantanello	P.T.	11	0				
449	F	Pantanello	T2	11	8	0.50	0.50		0.71
450	UGF	Pantanello	T2	8	1				
451	F	Pantanello	P.T.	22	3				0.04
452	F	Pantanello	P.T.	23	1				
453	F	Pantanello	P.T.	25	1				
454	F	Venella	P.T.	53	10	0.13	0.87		0.35
455	F	Venella	P.T.	55	0				
456	F	Giardinetto	M.M.D.	38	2				
457	N	Giardinetto	M.M.D.	50	1				
458	F	Giardinetto	M.M.D.	38	0				
459	F	Giardinetto	M.M.D.	32	2	0.26	1.73	0.01	
460	F	Giardinetto	M.M.D.	40	19				0.03
461	F	Giardinetto	M.M.D.	38	2	0.13	0.87		
462	F	Giardinetto	M.M.D.	34	10	0.76	2.23	0.01	2.08
463	N	Giardinetto	M.M.D.	30	0				
464	N	Giardinetto	M.M.D.	43	3				
465	F	Giardinetto	M.M.D.	54	43	1.39	3.71	0.75	3.78
466	N	Giardinetto	M.M.D.	50	3				0.07
467	N	Giardinetto	M.M.D.	50	0				
468	F	Giardinetto	V.W.	67	3	0.13	0.87		
469	F	Giardinetto	M.M.D.	35	4				
470	F	Giardinetto	M.M.D.	38	1				
471	N	Giardinetto	M.M.D.	41	25			0.01	1.39
472	N	DBSBS	T4	31	0				
473	N	DBSBS	P.T.	61	0				
474	F	DBSBS	P.T.	64	0				
475	UGF	DBSBS	P.T.	63	3			0.01	1.31
476	F	Venella	T2	40	6				
477	S	Venella	T2	36	51	8.14	23.22	4.69	12.07
478	GAS	Venella	T2	35	0				
479	N	Venella	V.W.	70	10				0.71
480	F	Venella	V.W.	70	0				
481	F	Venella	V.W.	65	2				0.35
482	N	AVS	V.W.	55	6			0.88	1.82
484	N	DBSBS	T0	15	0				
485	F	DBSBS	T2	15	3	0.13	1.86	1.00	
486	F	AVS	V.W.	61	4	0.40	2.95	0.65	

400	350	300	250	200	150	100	50	Best Date
								0
								495
0.67	0.26							386
	0.03	1.00	0.37	0.45	0.14	0.01		265
0.93	2.66	3.24	0.80					324
								0
								0
2.09	2.67	1.50	0.03					364
	0.04	0.88	0.08					296
0.73	0.29	1.88	0.06					330
	0.50	0.50						325
	0.20	0.80						310
1.15	1.36	5.72	0.42					322
								0
0.36	0.65	0.96	0.03					335
	0.50	0.50						325
								0
								557
3.13	4.34	9.50	1.73	0.27				324
0.23	0.73	0.04						550
2.91	0.04	1.08	0.52	0.32	0.04			421
								0
0.50	1.50	1.00						342
4.22	11.40	14.75	2.47	0.46	0.06			346
1.30	1.50	0.13						373
								0
0.04	0.96	0.87	0.13					326
0.23	0.79	2.36	0.62					302
	0.50	0.50						325
4.31	9.35	9.40	0.53					339
								0
								0
								0
1.05	0.62	0.02						412
	0.09	3.07	0.75	0.10	0.71	1.05	0.21	290
1.87	1.00							528
								0
1.53	1.84	4.31	1.11	0.44	0.06			331
								0
0.69	0.95	0.01						385
1.30	0.08	1.84	0.08					445
								0
								533
								547

Site	Type	Locality	Geomorph	Elev	Dated Arts.	600	550	500	450
487	F/T	Bufalara	P.T.	115	5	0.50	0.79	0.69	0.02
488	F	Bufalara	P.T.	116	31	1.00	1.00	0.08	4.64
489	F	Bufalara	P.T.	111	5				
490	F	Bufalara	P.T.	97	9				
491	F	AVS	T2	55	5	0.13	0.87		0.03
492	F	AVS	T2	57	2	0.26	1.73	0.01	
493	F	AVS	T2	63	0				
494	F	AVS	V.W.	62	3				
495	N	Cogno Prete	P.T.	80	3				
497	F/T	Lago Lupo	P.T.	68	26			0.17	1.86
498	F	Cogno Prete	P.T.	78	17				0.04
499	GAS	Cogno Prete	P.T.	80	0				
500	UGF	Lago Lupo	P.T.	61	0				
501	F	Lago Lupo	P.T.	61	17	0.13	0.87		
502	F	Lago Lupo	P.T.	72	25				
503	F	Lago Lupo	P.T.	68	19	0.28	1.43	0.29	1.56
504	N	Lago Lupo	P.T.	57	0				
505	F	Avinella	V.W.	62	2	1.01	0.99		
506	F	Avinella	P.T.	80	19				0.31
508	N	Avinella	P.T.	63	0				
509	F/T	Avinella	P.T.	67	4	0.21	1.78	0.01	0.03
510	N	Bufalara	P.T.	105	0				
511	F	Bufalara	P.T.	112	6	0.14	0.72	0.14	0.39
512	N	Avinella	P.T.	84	0				
513	N	Avinella	P.T.	80	0				
514	F	Giardinetto	M.M.D.	48	0				
515	F/T	Giardinetto	M.M.D.	44	9	0.13	0.87	0.14	2.17
516	F	Giardinetto	M.M.D.	46	5			0.13	0.87
517	F	Lago Lupo	P.T.	111	1				
518	F	Giardinetto	M.M.D.	48	4				
519	F	Giardinetto	M.M.D.	41	10	0.27	1.58	0.15	1.77
520	N	Giardinetto	M.M.D.	40	3				
521	F/T/K	Giardinetto	M.M.D.	40	55	0.50	0.50	0.15	7.39
522	N	Giardinetto	M.M.D.	54	0				
523	F/T	Giardinetto	M.M.D.	52	21	2.33	9.34	2.35	1.34
524	N	Giardinetto	M.M.D.	52	5	0.63	1.36	0.55	2.44
525	F	Giardinetto	M.M.D.	42	7				
526	F	Giardinetto	M.M.D.	42	0				
527	N	Giardinetto	M.M.D.	57	0				
528	F	Giardinetto	V.W.	63	11	0.50	0.50		
529	F	Giardinetto	V.W.	60	5		0.29	0.69	0.37
530	F/K	DBSBS	V.W.	30	10	3.53	3.47	0.98	1.00
531	F/T	DBSBS	T4	32	3	0.28	1.43	0.28	
532	F	Avinella	V.W.	62	1				

400	350	300	250	200	150	100	50	Best Date
0.28	0.68	1.97	0.06					313
5.98	6.32	10.99	0.98					358
0.11	1.88	2.14	0.72	0.14				312
0.14	1.81	5.52	0.53	0.50	0.50			312
1.20	1.50	0.56	0.62	0.09				342
								557
								0
	0.04	1.05	1.19	0.65	0.06			257
0.28	1.47	1.22	0.03					334
1.09	4.27	14.15	2.91	1.21	0.29	0.03		303
1.02	3.14	10.08	2.19	0.46	0.06			303
								0
								0
0.28	2.28	8.65	3.64	1.05	0.09			295
				0.64	7.19	12.07	4.37	106
4.87	3.80	5.63	0.77	0.32	0.04			357
								0
								575
0.69	1.59	12.83	0.93	1.63	0.94	0.09		306
								0
1.08	0.75	0.14						550
								0
1.38	1.10	1.86	0.26					344
								0
								0
								0
0.69	0.37	1.59	0.90	0.42	1.43	0.28		438
0.26	1.10	1.67	0.47	0.44	0.06			297
0.01	0.95	0.04						350
1.07	0.84	2.02	0.06					338
3.73	0.53	1.05	0.70	0.21				416
	0.06	2.49	0.44					292
10.30	11.17	21.20	3.20	0.58				354
								0
1.92	2.24	1.42	0.06					509
0.01								502
0.64	1.45	3.36	1.33	0.21				310
								0
								0
0.09	1.79	6.95	1.02	0.14				304
1.15	0.97	1.49	0.04					376
0.01				0.14	0.72	0.14		553
		0.14	0.72	0.14				550
	0.03	0.94	0.03					300

Site	Type	Locality	Geomorph	Elev	Dated Arts.	600	550	500	450
533	N	Avinella	V.W.	66	1				
534	N	Avinella	V.W.	61	0				
535	F	Giardinetto	M.M.D.	55	4	0.27	1.58	0.15	
539	F	DBSBS	P.T.	39	1				
540	UGF	Lago Lupo	P.T.	56	1				
541	N	Lago Lupo	P.T.	56	1				
542	F	Lago Lupo	P.T.	56	8	0.13	0.87		
544	UGF	Lago Lupo	V.W.	40	1	0.13	0.87		
547	N	San Marco	P.T.	60	0				
548	N	Venella	P.T.	40	2				0.36
549	F	Venella	P.T.	30	4				0.71
553	N	AVS	T4	48	1				
554	F	AVS	V.W.	27	1				
555	N	AVS	V.W.	28	0				
556	F	AVS	V.W.	42	32				0.35
557	N	AVS	V.W.	40	3				0.04
558	F	AVS	V.W.	43	7	0.13	0.87		0.36
559	F	AVS	V.W.	46	4				0.35
560	N	AVS	T4	59	1				
622	F	DBSBS	V.W.	55	8	1.13	2.45	1.59	0.83
623	UGF	DBSBS	P.T.	50	1				0.35
624	GAS	DBSBS	P.T.	59	0				
625	UGF	DBSBS	P.T.	53	1				0.04
626	N	DBSBS	P.T.	60	2	0.13	0.87		
627	UGF	DBSBS	P.T.	60	0				
628	UGF	DBSBS	P.T.	62	1				
629	N	DBSBS	P.T.	50	0				
630	N	DBSBS	P.T.	58	0				
631	UGF	DBSBS	P.T.	56	1				
632	UGF	DBSBS	P.T.	58	1				0.03
633	F	Venella	P.T.	61	5				0.03
634	F	Venella	P.T.	64	1				
635	N	Venella	P.T.	60	14	2.14	3.15	0.70	0.01
636	N	Venella	P.T.	60	0				
637	F	Pizzica	P.T.	46	10	0.50	0.50		0.36
638	N	Pizzica	P.T.	29	5			0.29	1.38
639	UGF	Pizzica	P.T.	45	2				
726	F/T	DBSBS	T3	19	7	0.13	0.87	1.00	1.41
727	F	DBSBS	T3	19	6			0.17	1.82
728	F/T	Pizzica	P.T.	43	29	0.13	0.87	0.20	1.56
729	UGF	Pantanello	T3	13	0				
730	F/T	Pizzica	P.T.	43	30	0.50	0.50		0.20
731	N	Pizzica	P.T.	41	2				
732	N	Pizzica	P.T.	40	2				0.14

400	350	300	250	200	150	100	50	Best Date
	0.03	0.94	0.03					300
								0
0.28	1.43	0.28						550
					0.50	0.50		125
	0.35	0.65						318
0.01	0.95	0.04						350
0.93	1.38	4.45	0.24					324
								557
								0
1.00	0.62	0.02						393
2.22	0.10	0.94	0.03					412
	0.35	0.65						318
	0.03	0.94	0.03					300
								0
0.79	1.01	0.83	0.01	0.32	7.07	15.71	5.57	103
0.73	0.73	0.50		0.01	0.21	0.55	0.21	356
0.86	0.71	1.19	0.52	0.35	0.47	1.09	0.43	329
0.67	0.72	0.26		0.02	0.43	1.09	0.43	100
				0.01	0.21	0.55	0.21	100
		0.21	0.67	0.61	0.50			532
0.65								418
								0
0.73	0.23							388
0.14	0.72	0.14						550
								0
		0.09	0.59	0.31	0.01			238
								0
								0
	0.03	0.94	0.03					300
0.92	0.06							400
0.94	0.26	1.85	0.73	0.52	0.35	0.24	0.08	259
		0.06	0.44	0.44	0.06			225
0.14	2.80	4.50	0.56					319
								0
1.79	1.23	4.98	0.64					330
0.46	0.86	0.91	0.58	0.44	0.06			348
	0.47	1.49	0.04					312
2.27	0.32				0.50	0.50		454
0.01	0.72	1.36	0.85	0.78	0.25	0.03		450
5.85	10.55	9.08	0.74					344
								0
4.45	7.36	14.98	1.55	0.39	0.05			330
	0.53	1.44	0.03					313
0.72	0.14	0.50	0.50					337

Site	Type	Locality	Geomorph	Elev	Dated Arts.	600	550	500	450
733	N	Pizzica	V.W.	35	0				
734	F/T	Pizzica	P.T.	41	5	0.63	1.36	0.13	0.88
735	F	Pizzica	P.T.	41	7				0.06
736	N/Q	Pizzica	P.T.	35	6				0.23
737	N	Pizzica	P.T.	38	1				0.04
738	GAS	Pizzica	P.T.	36	0				
739	N	Pizzica	P.T.	41	2				
740	N	Pizzica	P.T.	42	1				
741	F	Pizzica	P.T.	30	6	0.14	1.56	0.30	0.99
742	F/T	DBSBS	P.T.	24	12	0.27	1.58	0.34	1.65
743	UGF	DBSBS	P.T.	23	0				
744	F	Pizzica	P.T.	37	4		0.11	0.87	1.01
745	F	Pizzica	P.T.	42	5				0.35
760	N	Pizzica	P.T.	38	2			0.01	0.95
762	F/T	DBSBS	T2	15	12	0.13	0.87		0.50
763	N	Pizzica	P.T.	45	2				
764	N	Pizzica	P.T.	45	2				
769	UGF	DBSBS	T2	13	0				
771	F	Sant'Angelo	T2	18	1	0.14	0.72	0.14	
772	F/T	Lago Lupo	P.T.	96	3		0.50	0.50	
773	F/T	Lago Lupo	P.T.	112	0				
774	F	DBSBS	T1	15	0				
775	UGF	Sant'Angelo	T2	20	0				
776	N	Lago Lupo	P.T.	57	0				
777	N	Lago Lupo	P.T.	59	0				
778	N	Lago Lupo	P.T.	66	0				
779	UGF	Lago Lupo	P.T.	65	0				
780	UGF	DBSBS	T0	18	0				
781	N	Lago Lupo	P.T.	60	0				
782	N	Lago Lupo	P.T.	59	0				
783	UGF	DBSBS	T0	13	2				
787	UGF	Sant'Angelo	T2	19	0				
791	N	Giardinetto	M.M.D.	40	1				
792	GAS	DBSBS	T3	19	1				
793	UGF	Giardinetto	M.M.D.	53	0				
796	N	Giardinetto	M.M.D.	57	0				
797	N	Avinella	P.T.	62	2			1.00	1.00
798	N	Avinella	P.T.	62	0				
799	UGF	Venella	V.W.	50	1				
800	UGF	Scarciullo	P.T.	114	0				

400	350	300	250	200	150	100	50	Best Date
								0
		0.19	0.99	0.59	0.21	0.03		566
2.30	1.54	2.61	0.49					341
4.19	0.60	0.80	0.18					380
0.95	0.01							400
								0
0.80	0.23	0.94	0.03					347
	0.50	0.50						325
0.15	1.10	1.73	0.03					319
3.10	1.13	2.96	0.81	0.14				369
								0
1.15	0.85							424
0.65		0.63	1.38	1.48	0.48	0.03		227
0.18	0.72	0.14						399
1.65	2.19	3.45	2.29	0.86	0.05			307
0.28	1.43	0.28						350
0.14	1.22	0.64						337
								0
								550
	0.12	0.91	0.48	0.44	0.06			263
								0
								0
								0
								0
								0
								0
								0
								0
								0
								0
	0.08	0.94	0.34	0.43	0.18	0.02		254
								0
0.50	0.50							375
0.14	0.72	0.14						350
								0
								0
								475
								0
	0.02	0.89	0.09					295
								0

Appendix B
Farmhouse Significance (FMCE), 600–50 BC

Joseph Coleman Carter

Site	Type	Locality	Area	Artifacts	Geomorph	Elev	Best Date	600	550
1	F	Sant'Angelo	1943	18	T3	31	386		1.64
4	F	Sant'Angelo	1750	4	T3	41	278		
7	F	Sant'Angelo	300	2	T3	36	553		2.88
12	F/K	Sant'Angelo	900	3	T2	33	557		2.68
13	F	Sant'Angelo	600	13	T2	37	565	1.12	1.53
14	F	Sant'Angelo	375	1	T2	38	450		
15	F	Scarciullo	2250	2	P.T.	117	307		
16	F	Scarciullo	2025	4	P.T.	117	282		
17	F	Scarciullo	2025	3	P.T.	115	330		
35	F	Cogno Prete	2250	8	P.T.	96	322		
37	F	Cogno Prete	1400	1	P.T.	92	300		
38	F	DBSBS	1600	5	P.T.	51	300		
42	F	DBSBS	10000	28	P.T.	54	104		
43	F	Cogno Prete	800	40	P.T.	99	319	1.52	1.49
44	F	Campagnolo	1600	4	P.T.	104	344		
46	F	Campagnolo	3000	2	P.T.	114	300		
47	F	Campagnolo	2500	7	P.T.	100	313		
49	F	Campagnolo	1050	3	P.T.	100	450		
52	F	Campagnolo	1225	3	P.T.	104	353		
57	F	Cogno Prete	2500	18	P.T.	80	305		
58	F	Cogno Prete	50	9	P.T.	92	550		1.89
60	F	Scarciullo	1575	4	P.T.	112	384		
68	F	Cogno Prete	900	5	P.T.	108	433		
82	F	Cogno Prete	625	15	P.T.	106	339		
83	F	San Marco	1600	5	P.T.	91	247		
84	F	San Marco	1600	12	P.T.	91	264		
95	F/K	DBSBS	6400	11	P.T.	54	103		
103	F	Scarciullo	2025	19	P.T.	102	345	1.34	1.31
105	F/K	Sant'Angelo	6300	34	T2	16	98		
113	F	Lago Lupo	750	25	P.T.	99	302		
121	F	Lago Lupo	1600	7	P.T.	111	283		
122	F	Lago Lupo	1800	4	P.T.	107	300		
125	F	Lago Lupo	2100	3	P.T.	114	281		
128	F	Cogno Prete	2700	35	P.T.	101	305		
131	F	Lago Lupo	1600	7	P.T.	111	423		1.76
152	F	Lago Lupo	2025	4	P.T.	87	276		
154	F	Sant'Angelo	1600	25	T2	19	334	1.98	2.08
155	F	Sant'Angelo	4000	27	T2	19	531	2.33	3.00
156	F	Sant'Angelo	2500	3	T2	18	307		
164	F	Lago Lupo	3200	7	P.T.	91	289		
166	F	Lago Lupo	1500	6	P.T.	86	371		

Type: **F** = Farmhouse; **N** = Necropolis; **T** = Tomb; **K** = Kiln; **Q** = Quarry; **GAS** = Greek Agrarian Structure; **UGF** = Unidentified Greek Feature;
Geomorph: **M.M.D.** = Mass Movement Deposit; **P.T.** = Pleistocene Terrace; **T2–T4** = T2–T4 Terrace; **V.W.** = Valley Walls

500	450	400	350	300	250	200	150	100	50
	2.01	1.68	1.82	1.69			1.72	2.15	1.75
				2.88					
	1.47	1.05	1.04	1.17					
	3.11								
				3.48					
				2.93					
			2.53	2.72					
			1.88	2.48					
				3.48					
				3.18					
				2.46			3.25	4.17	3.02
		1.70	2.07	2.65	1.72	1.62			
			2.52	2.23					
				4.00					
			1.78	2.48				2.03	
	2.51								
				2.24					
	1.43	1.58	1.70	2.66	1.61	1.55			
			0.84	0.90					
		2.17			1.97				
	1.78	2.26		1.70					
		1.61	1.85	2.17					
				1.74	2.23	1.95			
			1.08	2.17	1.12	1.17		1.12	
							3.10	4.02	2.87
1.35	1.42	1.49	1.68	2.18	1.31				
		2.00	2.47	2.38	2.06		2.39	3.03	2.58
	1.12	1.13	1.38	2.44	1.26	1.19			
				2.63	1.95				
				3.50					
				3.22					
	2.02	2.00	2.51	3.38	2.26	2.13	2.02		
	2.37	1.89							
				2.62					
		2.04	1.91	2.60					
2.85	2.19		1.97						
				3.36					
			2.16	3.05	2.04				
	2.07		1.76	2.10					

Site	Type	Locality	Area	Artifacts	Geomorph	Elev	Best Date	600	550
167	F	Venella	2200	12	P.T.	91	304		
169	F	Lago Lupo	2800	18	P.T.	81	305		
170	F	Lago Lupo	1000	10	P.T.	77	537	1.30	2.45
172	F	Lago Lupo	700	4	P.T.	74	330		
173	F	Lago Lupo	2100	22	P.T.	66	312		
174	F	Lago Lupo	1000	8	P.T.	109	439		
175	F	Lago Lupo	800	2	P.T.	98	433		
181	F	Venella	225	4	P.T.	81	557		2.28
182	F/T	Venella	875	12	P.T.	82	453		1.64
183	F	Venella	4200	23	P.T.	77	317		
187	F	Venella	2700	3	P.T.	77	392		
193	F	Giampasquale	3200	24	P.T.	72	327		
195	F	Venella	1200	4	P.T.	78	405		
197	F	DBSBS	4200	17	V.W.	26	300		
198	F	DBSBS	150	6	V.W.	26	100		
200	F	Giampasquale	18000	10	P.T.	75	360		2.38
201	F	Giampasquale	300	7	P.T.	75	350		
210	F	Lago Lupo	1000	2	P.T.	70	324		
211	F	Lago Lupo	2400	11	P.T.	68	331		2.17
213	F/K	Sant'Angelo	8000	14	T3	23	335		
215	F	Giampasquale	3500	8	P.T.	81	508		2.44
216	F/T	Lago Lupo	2000	4	P.T.	87	335		
221	F	Lago Lupo	3500	5	P.T.	86	317		
222	F	Lago Lupo	1050	4	P.T.	79	330		1.60
224	F/T	Lago Lupo	4000	5	P.T.	65	550		2.47
228	F	San Marco	1200	3	V.W.	42	575	2.23	2.19
230	F/T	San Marco	12000	5	V.W.	28	402		
231	F	San Marco	1750	6	V.W.	35	97		
235	F	Giampasquale	2000	7	P.T.	81	327		
238	F	Giampasquale	1500	4	P.T.	78	550		1.86
241	F/T	AVS	1800	8	V.W.	27	368		
248	F/K	Lago Lupo	500	15	P.T.	70	297		
249	F	Lago Lupo	6000	8	P.T.	68	325		
250	F	Lago Lupo	4000	8	P.T.	70	378		
254	F	Lago Lupo	2400	6	P.T.	67	561		2.86
257	F	Cogno Prete	1000	2	P.T.	71	426		
258	F	Cogno Prete	3000	18	P.T.	61	320		2.21
260	F	Cogno Prete	3000	32	P.T.	78	318		
269	F	Giampasquale	4000	55	P.T.	67	315		1.88
274	F	Lago Lupo	800	5	P.T.	67	496		2.02
275	F	Lago Lupo	1350	3	P.T.	61	370		
278	F	Venella	2100	10	P.T.	63	297		
281	F	Lago Lupo	700	2	P.T.	97	372		

500	450	400	350	300	250	200	150	100	50
		2.03	2.13	2.94	2.18				
			2.41	3.49	2.18	2.04			
1.33	1.26			1.52					
			1.91	2.24					
	1.90	2.00	1.98	2.92	2.16	1.99			
	2.14	1.64							
	2.55								
1.58	2.60	2.04							
		2.12	2.46	3.20	2.15	2.01			
		2.88							
		2.48	2.41	3.06	2.19				
	1.56								
				2.72			2.52	2.82	2.25
							1.24	2.28	1.26
	2.44	2.30	2.74	2.69					
	1.35	1.21	1.50	1.65					
				2.29					
		2.06	2.24	2.78					
	2.63	2.66		3.29	2.63				
	2.14								
			1.81	2.46					
			2.63	3.21					
			1.63	2.01					
	2.06								
	2.53							2.58	
							1.86	2.94	2.11
		1.79	2.03	3.00					
	1.44								
		2.06		1.77					
			1.63	2.98	1.80	1.58			
			3.11	3.13					
	2.27	2.58	2.50	2.19					
	1.92								
	2.33	2.21							
		2.01	2.53	2.87			2.09		
	2.26	2.15	2.14	3.04	2.27	2.27			
1.94	2.02	2.15	2.31	3.05	2.27	2.14			
	1.67								
			1.88						
		1.59	1.77	2.59	1.68	1.77			
			2.41						

Site	Type	Locality	Area	Artifacts	Geomorph	Elev	Best Date	600	550
285	F/T	Sant'Angelo	2500	23	T2	23	577	3.22	2.74
286	F	Sant'Angelo	2500	7	T3	22	560	1.65	2.77
288	F	Venella	700	10	P.T.	45	363		
291	F	Lago Lupo	1500	3	P.T.	113	357		
292	F	Lago Lupo	3000	18	P.T.	112	332		
296	F	Lago Lupo	6300	19	P.T.	107	298		
301	F	Campagnolo	1050	10	P.T.	117	398		1.59
302	F	Campagnolo	1050	3	P.T.	117	446		
303	F	Campagnolo	450	3	P.T.	121	450		
304	F	Bufalara	1200	8	P.T.	122	445		1.58
305	F	Bufalara	1350	4	P.T.	109	468		
309	F	Bufalara	1400	8	P.T.	122	302		
310	F	AVS	1600	10	P.T.	106	390		1.96
312	F	AVS	2750	19	P.T.	103	306		
317	F	AVS	700	6	P.T.	92	104		
319	F/T	AVS	1400	5	P.T.	97	384		
320	F	Venella	2000	48	P.T.	89	338		1.73
322	F	Venella	1125	7	V.W.	72	388		
323	F	Venella	1000	7	P.T.	78	564	1.53	2.27
326	F	Venella	2000	39	P.T.	76	326		
327	F	Venella	1200	17	P.T.	73	316		
328	F	Avinella	875	4	P.T.	66	504		2.18
329	F	Venella	2250	10	P.T.	37	337		
335	F	Venella	300	3	V.W.	35	433		
336	F	Venella	375	11	V.W.	35	560	1.33	2.54
339	F	Avinella	2000	13	P.T.	65	103		
341	F	Avinella	1800	44	P.T.	75	318		
342	F	Avinella	1600	37	P.T.	86	299		
343	F	Avinella	750	14	P.T.	86	288		
344	F	Avinella	1350	4	P.T.	91	450		2.75
345	F	Avinella	625	8	P.T.	90	332		
348	F	Avinella	2250	10	P.T.	96	392		2.02
349	F	Avinella	375	6	P.T.	97	561	1.20	2.37
350	F	Avinella	225	3	P.T.	99	318		
352	F	Scarciullo	1400	27	P.T.	108	334		
353	F	Scarciullo	2700	21	P.T.	110	358		
354	F	Scarciullo	1400	35	P.T.	108	294		
358	F	Avinella	875	26	P.T.	92	323		
360	F	Venella	1350	5	P.T.	61	323		
363	F/T	Scarciullo	1925	11	P.T.	116	331		
365	F	Scarciullo	2475	29	P.T.	112	320		1.63
366	F	Scarciullo	1000	2	P.T.	115	557		3.20
369	F	Scarciullo	600	4	P.T.	116	300		

500	450	400	350	300	250	200	150	100	50
1.86	1.80								
		1.85							
	1.23	1.89	1.22	1.84					
				1.74					
	2.02	2.27	2.64	3.14					
			2.52	4.32	2.65				
	2.11	1.82	1.63	1.47					
	2.38								
	2.60								
	1.96	2.01							
	1.93		2.11						
				3.19	1.88				
	2.27			1.86					
	1.97		2.14	2.79	1.89				
							1.60	2.41	1.36
		1.78	1.92	1.71					
	2.02	2.19	2.01	2.88	1.97	1.95			
	2.47		1.67	1.93					
	1.33								
	1.85	2.14	2.35	2.87	1.94	1.85			
	1.45	1.47	1.47	2.83	1.45	1.54	1.46		
	1.99								
		2.37	2.33	2.53					
	1.76								
	1.16								
							2.31	3.41	2.10
		1.71	2.02	2.96	1.66	1.57			
	1.83	1.80	1.76	2.84	2.29	1.99			
			1.49	2.79	1.61	1.50			
	2.84								
			1.54	2.31					
		2.49	2.07	2.48					
			1.75	2.52					
	1.45	1.75	2.01	2.43	1.46	1.49			
	2.36	2.42	2.19	3.02					
		1.65	1.88	3.10	2.29	1.77			
	1.11	1.40	1.43	2.25	1.20				
				2.38	1.87				
		1.86	2.08	2.81					
1.68	1.75	1.70	2.04	2.69	1.83	1.63			
		1.77		2.16					

Site	Type	Locality	Area	Artifacts	Geomorph	Elev	Best Date	600	550
370	F	Scarciullo	1400	11	P.T.	115	311		
372	F	Venella	2700	31	P.T.	62	98		1.73
373	F	Scarciullo	900	2	P.T.	116	321		
375	F	Scarciullo	900	21	P.T.	114	319		1.43
376	F	Scarciullo	2000	9	P.T.	114	239		
390	F	Avinella	1800	10	P.T.	85	350		
394	F	Avinella	3000	17	P.T.	61	316		
395	F	Avinella	1943	61	P.T.	58	336	1.76	1.86
400	F	AVS	900	4	P.T.	62	566		1.60
402	F	Avinella	300	4	V.W.	70	267		
404	F	Avinella	2000	34	V.W.	69	291		1.56
406	F/T	Avinella	2000	7	V.W.	66	563		1.78
413	F	Pantanello	1050	12	T3	15	344		
414	F	Pantanello	2000	8	T3	15	437		1.93
416	F	Pantanello	3000	18	T3	14	342		
417	F	Pantanello	800	3	T2	8	336		
428	F	Pantanello	2400	5	T3	10	100		
445	F	Pantanello	2200	3	P.T.	19	265		
449	F	Pantanello	1600	8	T2	11	364		
451	F	Pantanello	1200	3	P.T.	22	330		
454	F	Venella	2400	10	P.T.	53	322		
456	F	Giardinetto	3900	2	M.M.D.	38	335		
459	F	Giardinetto	1400	2	M.M.D.	32	557		3.36
460	F	Giardinetto	3900	19	M.M.D.	40	324		
462	F	Giardinetto	1200	10	M.M.D.	34	421		1.61
465	F	Giardinetto	5200	43	M.M.D.	54	346	2.14	2.28
468	F	Giardinetto	750	3	V.W.	67	326		
469	F	Giardinetto	1000	4	M.M.D.	35	302		
476	F	Venella	1943	6	T2	40	290		
481	F	Venella	1400	2	V.W.	65	385		
485	F	DBSBS	1200	3	T2	15	533		2.73
486	F	AVS	375	4	V.W.	61	547		3.07
487	F/T	Bufalara	2000	5	P.T.	115	313		
488	F	Bufalara	1000	31	P.T.	116	358	1.33	1.30
489	F	Bufalara	1000	5	P.T.	111	312		
490	F	Bufalara	900	9	P.T.	97	312		
491	F	AVS	500	5	T2	55	342		
492	F	AVS	750	2	T2	57	557		3.57
494	F	AVS	1500	3	V.W.	62	257		
497	F/T	Lago Lupo	4200	26	P.T.	68	303		
498	F	Cogno Prete	1400	17	P.T.	78	303		
501	F	Lago Lupo	3600	17	P.T.	61	295		
502	F	Lago Lupo	2400	25	P.T.	72	106		

500	450	400	350	300	250	200	150	100	50
			1.71	2.99					
1.70		1.93	1.83	1.72			2.12	2.69	2.31
				2.41					
	1.41	1.58	1.61	2.02	1.57	1.52			
				2.04	2.02	2.05			
	1.66	1.94	1.80	2.51					
		1.87	2.21	3.08	1.91				
1.77	1.97	2.07	2.18	2.60	2.05	2.02	1.74		
				2.01	1.90				
	1.49		1.76	3.00	1.88	1.70	1.57		
			1.42	1.52					
		1.89	1.76	2.19					
	1.98	1.83	1.86	1.92					
		2.04	1.92	2.22			1.95	2.40	1.99
				2.03					
								2.41	
				2.31					
		1.80	1.99	1.58					
				2.92					
		1.80	1.86	3.05					
				3.04					
		2.54	2.73	3.51	2.32				
	1.58	1.80		1.28					
	2.29	2.32	2.82	2.99	2.20				
			1.75						
				2.63					
				2.81				1.89	
			2.38						
1.98									
				2.45					
	1.68	1.81	1.82	2.23	1.31				
			2.23	2.36					
			1.76	2.93					
		1.82	2.00						
				2.17	2.32				
	2.32	2.22	2.60	3.70	2.46	2.28			
		1.46	1.84	3.01	1.69				
			2.38	3.48	2.66	2.20			
							2.97	3.70	2.47

Site	Type	Locality	Area	Artifacts	Geomorph	Elev	Best Date	600	550
503	F	Lago Lupo	3000	19	P.T.	68	357		1.83
505	F	Avinella	400	2	V.W.	62	575	2.10	2.04
506	F	Avinella	2475	19	P.T.	80	306		
509	F/T	Avinella	1000	4	P.T.	67	550		2.42
511	F	Bufalara	1400	6	P.T.	112	344		
515	F/T	Giardinetto	4800	9	M.M.D.	44	438		
516	F	Giardinetto	3150	5	M.M.D.	46	297		
517	F	Lago Lupo	900	1	P.T.	111	350		
518	F	Giardinetto	500	4	M.M.D.	48	338		
519	F	Giardinetto	1200	10	M.M.D.	41	416		1.45
521	F/T/K	Giardinetto	3500	55	M.M.D.	40	354		
523	F/T	Giardinetto	3000	21	M.M.D.	52	509	2.00	2.93
525	F	Giardinetto	3600	7	M.M.D.	42	310		
528	F	Giardinetto	875	11	V.W.	63	304		
529	F	Giardinetto	1600	5	V.W.	60	376		
530	F/K	DBSBS	625	10	V.W.	30	553	2.08	1.96
531	F/T	DBSBS	875	3	T4	32	550		2.26
535	F	Giardinetto	1600	4	M.M.D.	55	550		2.14
542	F	Lago Lupo	1800	8	P.T.	56	324		
549	F	Venella	1350	4	P.T.	30	412		
556	F	AVS	1000	32	V.W.	42	103		
558	F	AVS	625	7	V.W.	43	329		
559	F	AVS	1600	4	V.W.	46	100		
622	F	DBSBS	450	8	V.W.	55	532	0.86	1.30
633	F	Venella	150	5	P.T.	61	259		
637	F	Pizzica	3600	10	P.T.	46	330		
726	F/T	DBSBS	42000	7	T3	19	454		
727	F	DBSBS	22500	6	T3	19	450		
728	F/T	Pizzica	11200	29	P.T.	43	344		
730	F/T	Pizzica	13000	30	P.T.	43	330		
734	F/T	Pizzica	7800	5	P.T.	41	566		2.21
735	F	Pizzica	5600	7	P.T.	41	341		
741	F	Pizzica	7000	6	P.T.	30	319		2.43
742	F/T	DBSBS	6000	12	P.T.	24	369		2.19
744	F	Pizzica	2600	4	P.T.	37	424		
745	F	Pizzica	10800	5	P.T.	42	227		
762	F/T	DBSBS	13000	12	T2	15	307		

500	450	400	350	300	250	200	150	100	50
	1.86	2.40	2.21	2.48					
			1.98	3.61		2.05			
		1.94							
		1.91	1.78	2.11					
	2.69			2.48			2.49		
			2.21	2.51					
			3.41						
		1.85		2.50					
	1.51	2.07		1.29					
	2.43	2.59	2.61	3.11	2.17				
2.06	1.80	1.87	1.91	1.79					
			2.29	3.03	2.26				
			1.63	2.91	1.45				
		1.93	1.83	2.10					
1.23	1.19								
			2.03						
			1.73	2.81					
		2.51							
			1.32				2.31	3.30	2.19
				1.45				1.43	
								1.78	
1.08									
				1.47					
		2.27	2.09	3.20					
2.43	2.53	2.85							
	2.56			2.34					
	2.42	2.85	3.31	3.12					
		2.70	2.98	3.66	2.42				
					2.02				
		2.64	2.34	2.74					
	2.19		2.23	2.50					
	2.22	2.61	2.06	2.55					
	1.94	2.03							
					3.00	3.09			
		2.39	2.51	2.79	2.55				

Appendix C
Basic Graphs and Tables

Joseph Coleman Carter

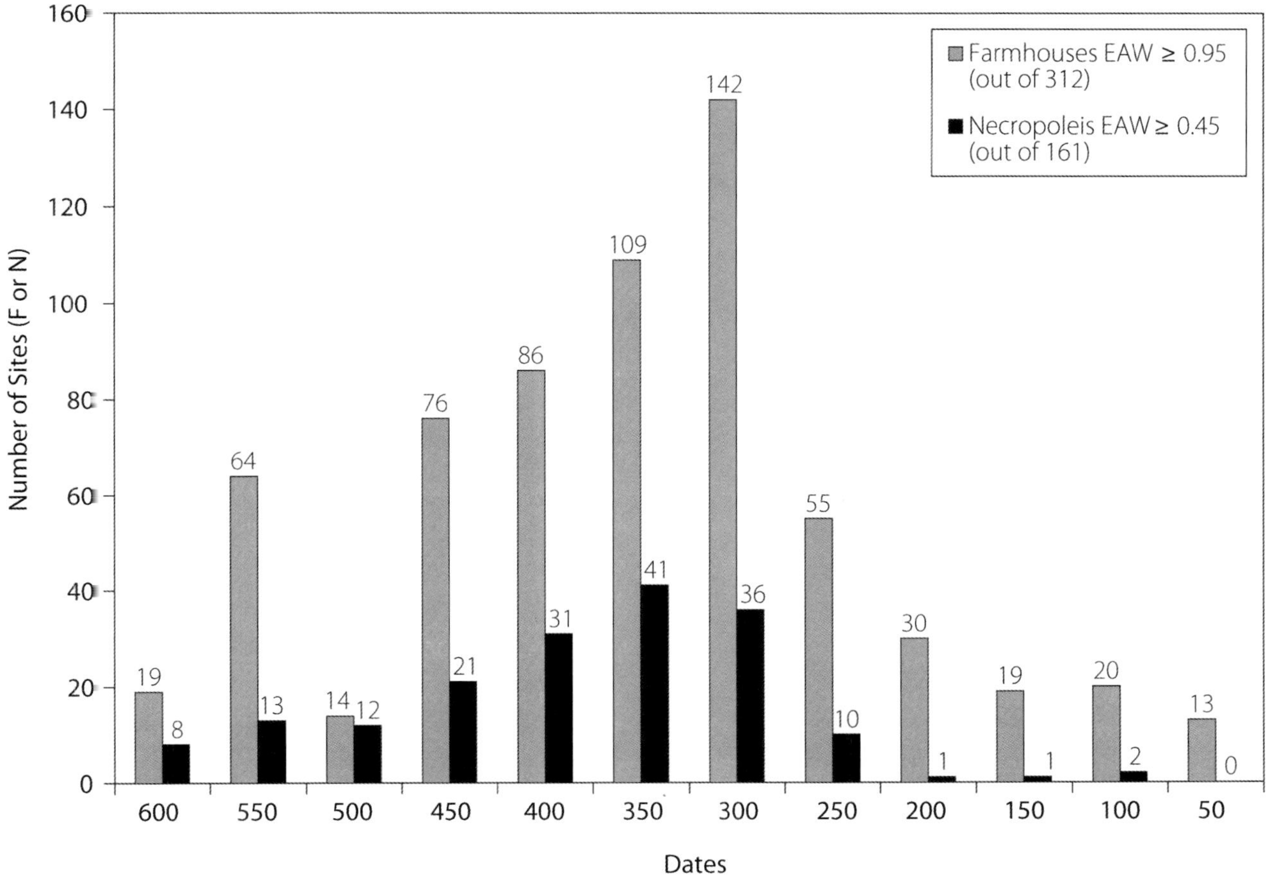

C1a Counts of farmhouses and necropoleis by period.

Farmhouses	600	550	500	450	400	350	300	250	200	150	100	50
EAW ≥ 0.95	19	64	14	76	86	109	142	55	30	19	20	13
EAW ≥ 0.45	51	108	41	98	148	155	180	103	58	47	27	15
(Out of 312)												
Necropoleis	**600**	**550**	**500**	**450**	**400**	**350**	**300**	**250**	**200**	**150**	**100**	**50**
EAW ≥ 0.95	3	7	2	15	15	19	16	2	0	0	0	0
EAW ≥ 0.45	8	13	12	21	31	41	36	10	1	1	2	0
(Out of 161)												

C1b Counts of farmhouses and necropoleis by period.

Site Type	Total	EAW ≥ 0.95	EAW ≥ 0.45	Undated (EAW = 0)	Dated
Farmhouses	312	207	263	43	269
Farmhouses without tombs (F)	282	185	237	40	242
Farmhouses with tombs (F/T)	30	22	26	3	27
Necropoleis	161	40	73	87	74
Sanctuaries	5	5	5	0	5
Greek Agrarian Structures (GAS)	35	2	5	30	5
Unidentified Greek Features (UGF)	47	4	18	29	18
Other	1	0	0	1	0
Kilns	2	0	0	2	0
Quarries	3	0	0	3	0
Totals	**566**	**258**	**364**	**195**	**371**

C2 Types of sites.

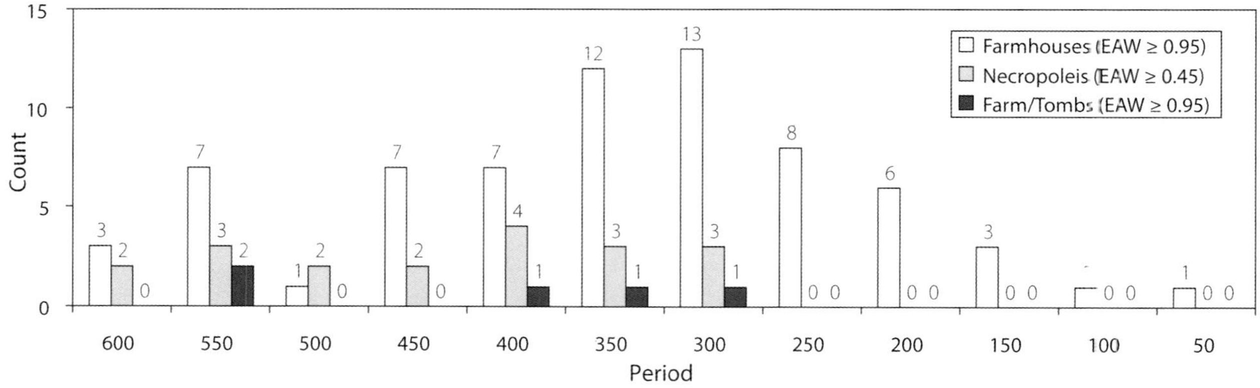

C3 Avinella Farmhouses, necropoleis, and farmhouse/tombs, **600–50**.

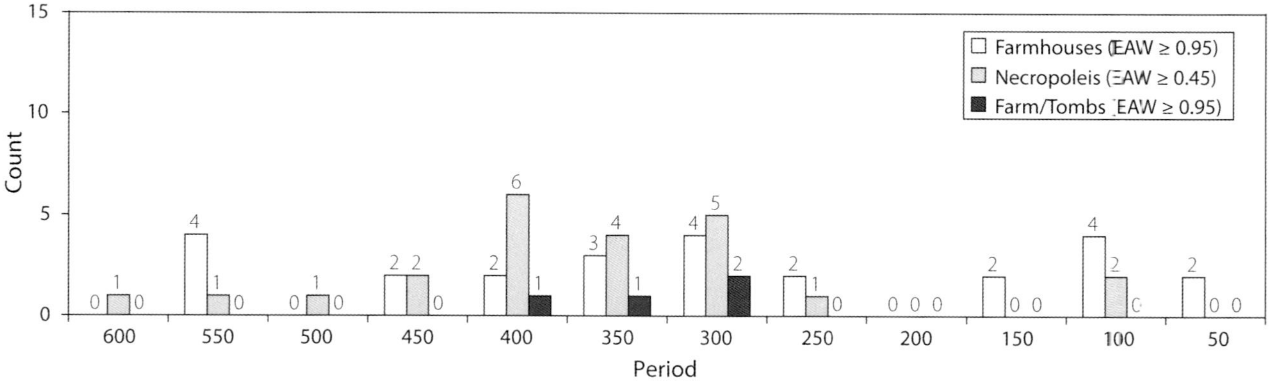

C3 AVS Farmhouses, necropoleis, and farmhouse/tombs, **600–50**.

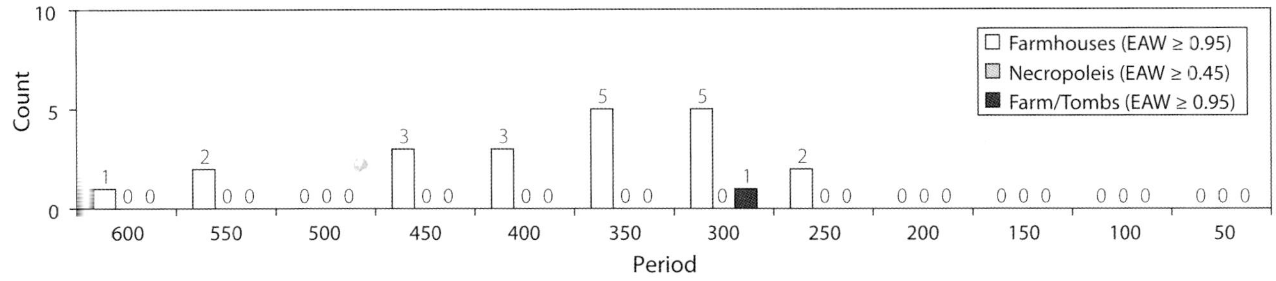

C3 Bufalara Farmhouses, necropoleis, and farmhouse/tombs, **600–50**.

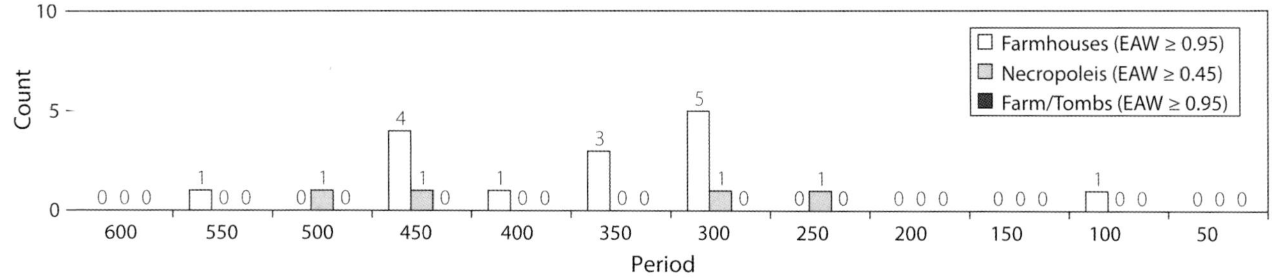

C3 Campagnolo Farmhouses, necropoleis, and farmhouse/tombs, **600–50**.

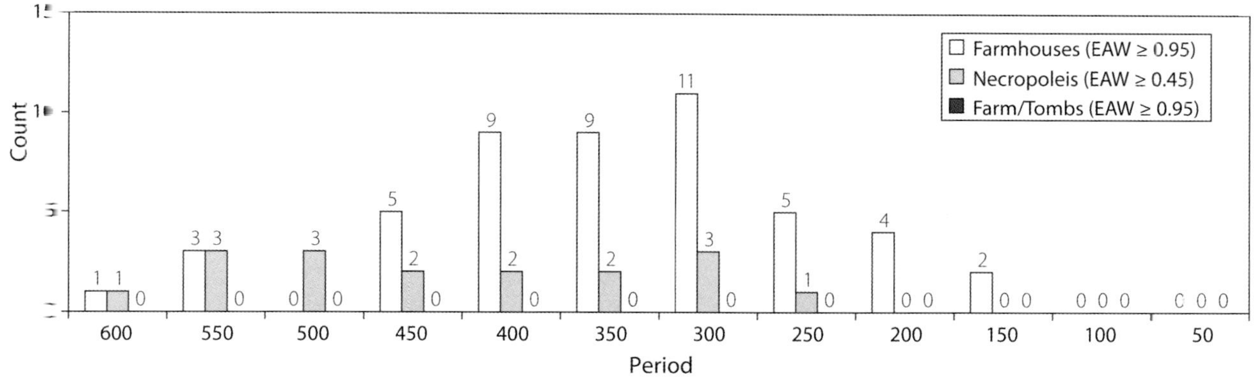

C3 Cogno del Prete Farmhouses, necropoleis, and farmhouse/tombs, **600–50**.

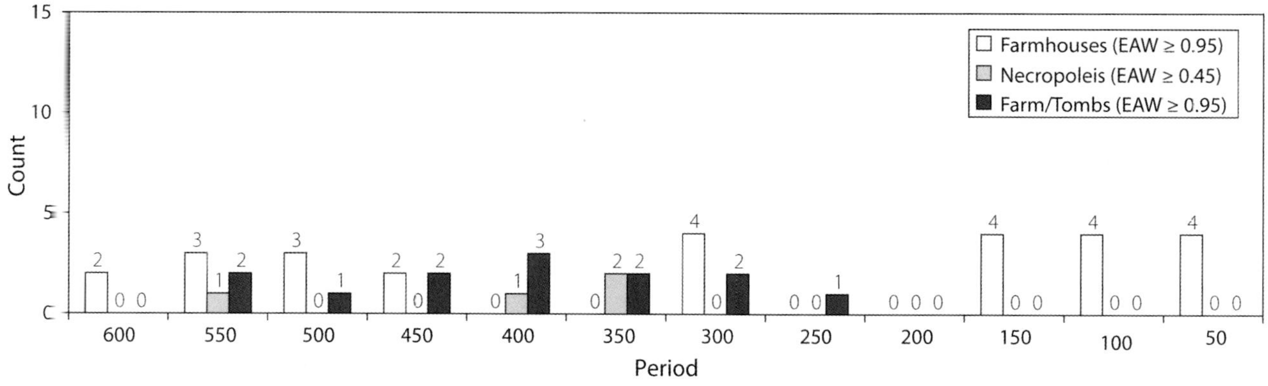

C3 DBSBS Farmhouses, necropoleis, and farmhouse/tombs, **600–50**.

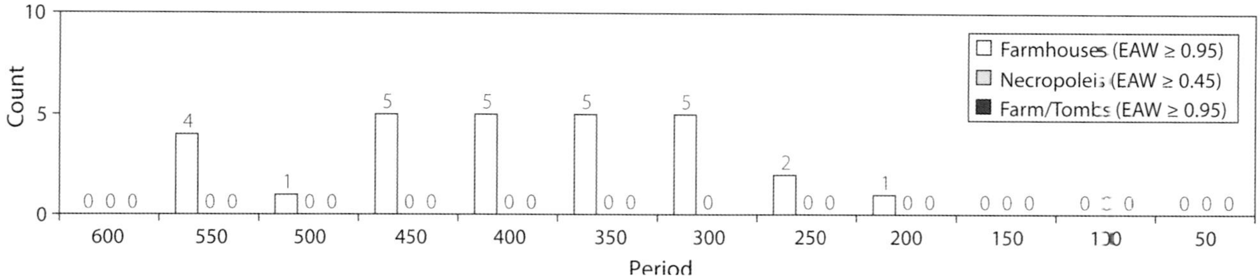

C3 Giampasquale Farmhouses, necropoleis, and farmhouse/tombs, **600–50**.

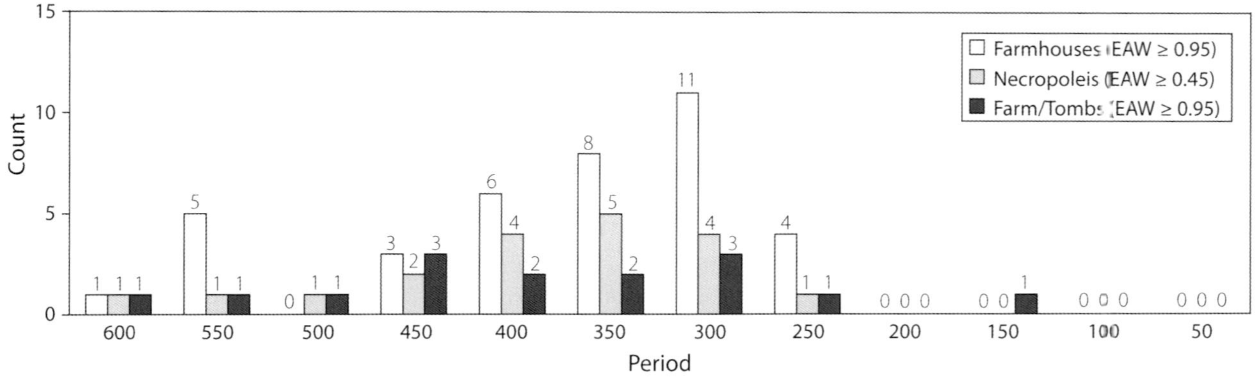

C3 Giardinetto Farmhouses, necropoleis, and farmhouse/tombs, **600–50**.

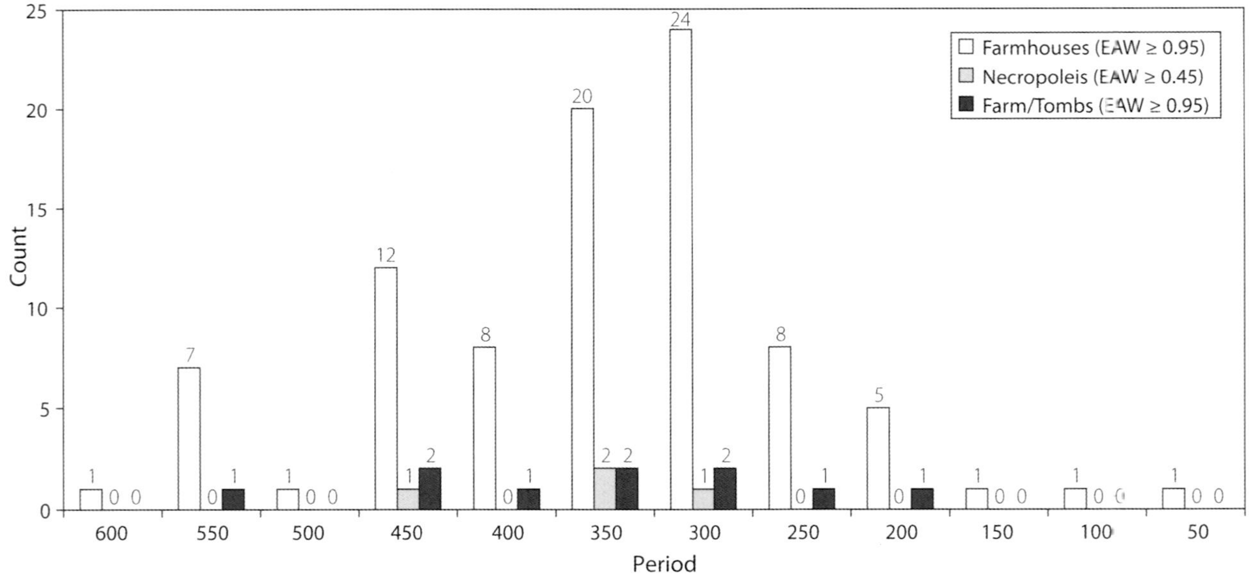

C3 Lago del Lupo Farmhouses, necropoleis, and farmhouse/tombs, **600–50**.

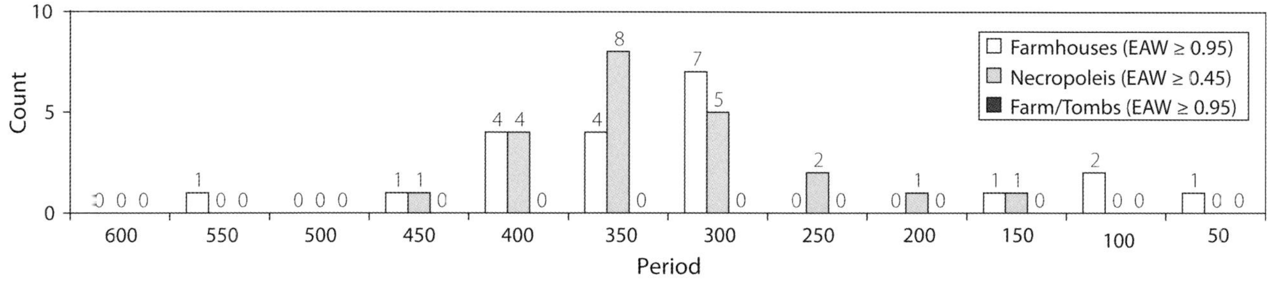

C3 Pantanello Farmhouses, necropoleis, and farmhouse/tombs, **600–50**.

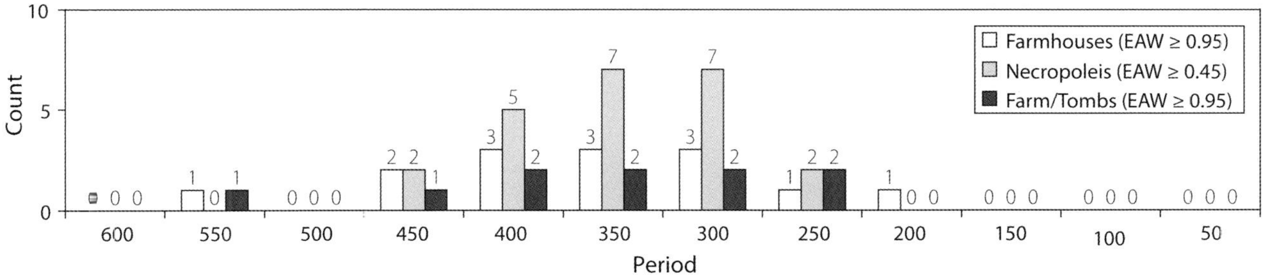

C3 Pizzica Farmhouses, necropoleis, and farmhouse/tombs, **600–50**.

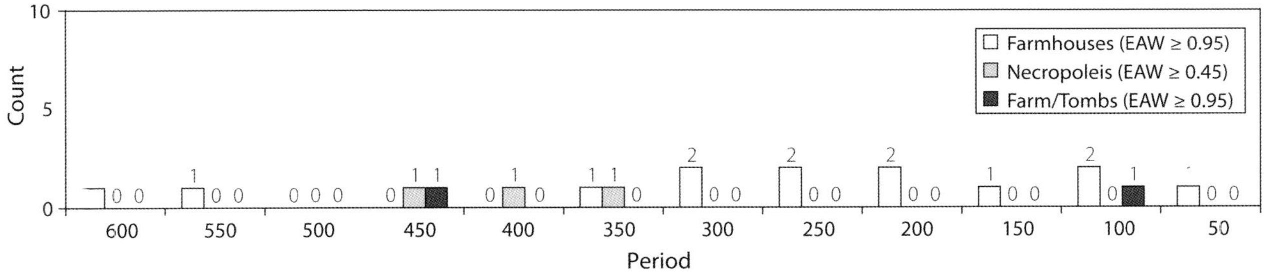

C3 San Marco Farmhouses, necropoleis, and farmhouse/tombs, **600–50**.

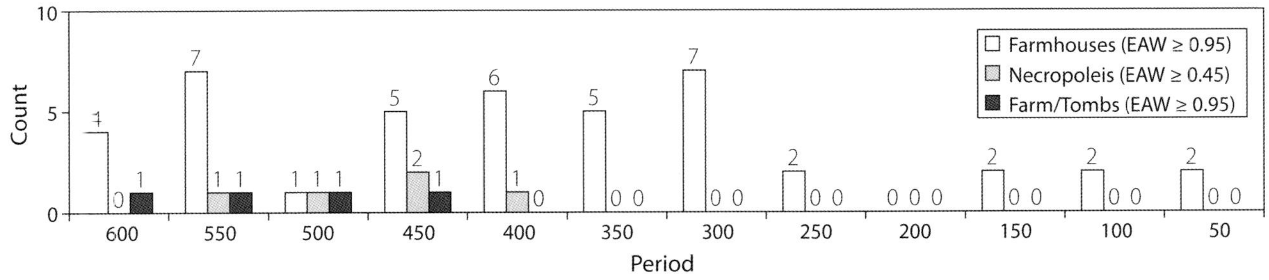

C3 Sant'Angelo Farmhouses, necropoleis, and farmhouse/tombs, **600–50**.

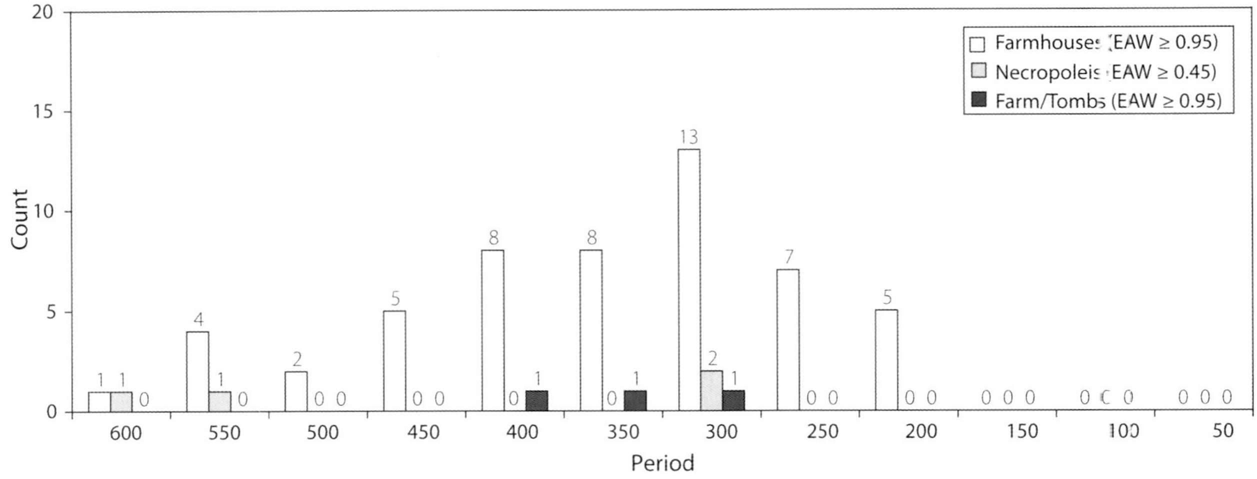

C3 Scarciullo Farmhouses, necropoleis, and farmhouse/tombs, **600–50**.

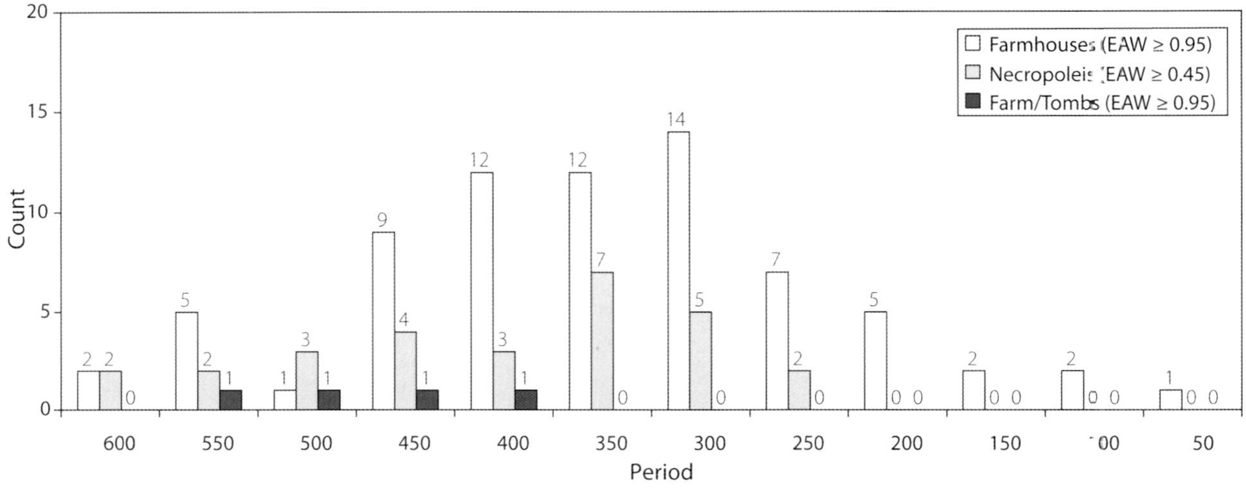

C3 Venella Farmhouses, necropoleis, and farmhouse/tombs, **600–50**.

Localities (% of study area)	600	550	500	450	400	350	300	250	200	150	100	50
Early Clusters (22%)	*12/80 (15%)*	*33/80 (41%)*	*7/80 (9%)*	*34/80 (43%)*	*40/80 (50%)*	*45/80 (58%)*	*54/80 (68%)*	*24/80 (30%)*	*12/80 (15%)*	*8/80 (10%)*	*5/80 (6%)*	*4/80 (5%)*
Giardinetto (3.3%)	2/17 (12%)	6/17 (35%)	1/17 (6%)	6/17 (35%)	8/17 (47%)	10/17 (59%)	14/17 (82%)	5/17 (29%)	0/17 (0%)	1/17 (6%)	0/17 (0%)	0/17 (0%)
Sant'Angelo (3.7%)	5/13 (38%)	8/13 (62%)	2/13 (15%)	6/13 (46%)	6/13 (46%)	5/13 (38%)	7/13 (54%)	2/13 (15%)	0/13 (0%)	2/13 (15%)	2/13 (15%)	2/13 (15%)
Avinella (7.5%)	3/20 (15%)	9/20 (45%)	1/20 (5%)	7/20 (35%)	8/20 (40%)	13/20 (65%)	14/20 (70%)	8/20 (40%)	6/20 (30%)	3/20 (15%)	1/20 (5%)	1/20 (5%)
lower (4.7%)	2/20 (10%)	6/20 (30%)	1/20 (5%)	4/20 (20%)	4/20 (20%)	6/20 (30%)	7/20 (35%)	4/20 (20%)	3/20 (15%)	3/20 (15%)	1/20 (5%)	1/20 (5%)
upper (2.8%)	1/20 (5%)	3/20 (15%)	0/20 (0%)	3/20 (15%)	4/20 (20%)	7/20 (35%)	7/20 (35%)	4/20 (20%)	3/20 (15%)	0/20 (0%)	0/20 (0%)	0/20 (0%)
Venella (5.8%)	2/23 (9%)	6/23 (26%)	2/23 (9%)	10/23 (43%)	13/23 (57%)	12/23 (52%)	14/23 (61%)	7/23 (30%)	5/23 (22%)	2/23 (9%)	2/23 (9%)	1/23 (4%)
west (2.5%)	2/23 (9%)	3/23 (13%)	0/23 (0%)	7/23 (30%)	5/23 (22%)	5/23 (22%)	5/23 (22%)	3/23 (13%)	3/23 (13%)	1/23 (4%)	0/23 (0%)	0/23 (0%)
east (3.3%)	0/23 (0%)	3/23 (13%)	2/23 (9%)	3/23 (13%)	8/23 (35%)	7/23 (30%)	9/23 (39%)	4/23 (17%)	2/23 (9%)	1/23 (4%)	2/23 (9%)	1/23 (4%)
Giampasquale (2.3%)	0/7 (0%)	4/7 (57%)	1/7 (14%)	5/7 (71%)	5/7 (71%)	5/7 (71%)	5/7 (71%)	2/7 (29%)	1/7 (14%)	0/7 (0%)	0/7 (0%)	0/7 (0%)
Central Plateau (25.6%)	*2/48 (4%)*	*11/48 (23%)*	*1/48 (2%)*	*19/48 (40%)*	*18/48 (38%)*	*31/48 (65%)*	*37/48 (77%)*	*14/48 (29%)*	*10/48 (21%)*	*3/48 (6%)*	*1/48 (2%)*	*1/48 (2%)*
Lago Lupo ([illegible]5.3%)	1/36 (3%)	8/36 (22%)	1/36 (3%)	14/36 (39%)	9/36 (25%)	22/36 (61%)	26/36 (72%)	9/36 (25%)	6/36 (17%)	1/36 (3%)	1/36 (3%)	1/36 (3%)
Cogno Preti (10.3%)	1/12 (8%)	3/12 (25%)	0/12 (0%)	5/12 (42%)	9/12 (75%)	9/12 (75%)	11/12 (92%)	5/12 (42%)	4/12 (33%)	2/12 (17%)	0/12 (0%)	0/12 (0%)
Far from City (26.6%)	*3/50 (6%)*	*12/50 (24%)*	*2/50 (4%)*	*15/50 (30%)*	*16/50 (32%)*	*22/50 (44%)*	*33/50 (66%)*	*13/50 (26%)*	*7/50 (14%)*	*3/50 (6%)*	*8/50 (16%)*	*3/50 (6%)*
Campagnolo (4.9%)	0/8 (0%)	1/8 (13%)	0/8 (0%)	4/8 (50%)	1/8 (13%)	3/8 (38%)	5/8 (63%)	0/8 (0%)	0/8 (0%)	0/8 (0%)	1/8 (13%)	0/8 (0%)
Bufalara (2.[illegible]%)	1/8 (13%)	2/8 (25%)	0/8 (0%)	3/8 (38%)	3/8 (38%)	5/8 (63%)	6/8 (75%)	2/8 (25%)	0/8 (0%)	0/8 (0%)	0/8 (0%)	0/8 (0%)
Scarciullo ([illegible].0%)	1/16 (6%)	4/16 (25%)	2/16 (13%)	5/16 (31%)	9/16 (56%)	9/16 (56%)	14/16 (88%)	7/16 (44%)	5/16 (31%)	0/16 (0%)	0/16 (0%)	0/16 (0%)
San Marco (3.4%)	1/5 (20%)	1/5 (20%)	0/5 (0%)	1/5 (20%)	0/5 (0%)	1/5 (20%)	2/5 (40%)	2/5 (40%)	2/5 (40%)	1/5 (20%)	3/5 (60%)	1/5 (20%)
AVS (6.4%)	0/13 (0%)	4/13 (31%)	0/13 (0%)	2/13 (15%)	3/13 (23%)	4/13 (31%)	6/13 (46%)	2/13 (15%)	0/13 (0%)	2/13 (15%)	4/13 (31%)	2/13 (15%)
Close to City (25.3%)	*2/29 (7%)*	*8/29 (28%)*	*4/29 (14%)*	*8/29 (28%)*	*12/29 (41%)*	*11/29 (38%)*	*18/29 (62%)*	*4/29 (14%)*	*1/29 (3%)*	*5/29 (17%)*	*6/29 (21%)*	*5/29 (17%)*
Pizzica (5.8%)	0/8 (0%)	2/8 (25%)	0/8 (0%)	3/8 (38%)	5/8 (63%)	5/8 (63%)	5/8 (63%)	3/8 (38%)	1/8 (13%)	0/8 (0%)	0/8 (0%)	0/8 (0%)
Pantanello (7.4%)	0/8 (0%)	1/8 (13%)	0/8 (0%)	1/8 (13%)	4/8 (50%)	4/8 (50%)	7/8 (88%)	0/8 (0%)	0/8 (0%)	1/8 (13%)	2/8 (25%)	1/8 (13%)
DBSBS (12.[illegible]%)	2/13 (15%)	5/13 (38%)	4/13 (31%)	4/13 (31%)	3/13 (23%)	2/13 (15%)	6/13 (46%)	1/13 (8%)	0/13 (0%)	4/13 (31%)	4/13 (31%)	4/13 (31%)

C4 Rates of occupation, **600–50**, significant farmhouses.*

*Percentage of total number of significant sites occupied in the individual localities of the four major regions of the study area in all periods.

Localities (% of study area)	600	550	500	450	400	350	300	250	200	150	100	50
Early Clusters (22.6%)	*12/19 (63%)*	*33/64 (52%)*	*7/14 (50%)*	*34/76 (45%)*	*40/86 (47%)*	*45/109 (41%)*	*54/142 (38%)*	*24/55 (44%)*	*12/30 (40%)*	*8/19 (42%)*	*5/20 (25%)*	*4/13 (31%)*
Giardinetto (3.3%)	2/19 (11%)	6/64 (9%)	1/14 (7%)	6/76 (8%)	8/86 (9%)	10/109 (9%)	14/142 (10%)	5/55 (9%)	0/30 (0%)	1/19 (5%)	0/20 (0%)	0/13 (0%)
Sant'Angelo (3.7%)	5/19 (26%)	8/64 (13%)	2/14 (14%)	6/76 (8%)	6/86 (7%)	5/109 (5%)	7/142 (5%)	2/55 (4%)	0/30 (0%)	2/19 (11%)	2/20 (10%)	2/13 (15%)
Avinella (7.5%)	3/19 (16%)	9/64 (14%)	1/14 (7%)	7/76 (9%)	8/86 (10%)	13/109 (11%)	14/142 (10%)	8/55 (14%)	6/30 (20%)	3/19 (15%)	1/20 (5%)	1/13 (8%)
lower (4.7%)	2/19 (11%)	6/64 (9%)	1/14 (7%)	4/76 (5%)	4/86 (5%)	6/109 (5%)	7/142 (5%)	4/55 (7%)	2/30 (7%)	2/19 ([illegible]1%)	0/20 (0%)	0/13 (0%)
upper (2.8%)	1/19 (5%)	3/64 (5%)	0/14 (0%)	3/76 (4%)	4/86 (5%)	7/109 (6%)	7/142 (5%)	4/55 (7%)	4/30 (13%)	1/19 (5%)	1/20 (5%)	1/13 (8%)
Venella (5.8%)	2/19 (11%)	6/64 (9%)	2/14 (14%)	10/76 (13%)	13/86 (15%)	12/109 (11%)	14/142 (10%)	7/55 (13%)	5/30 (17%)	2/19 (11%)	2/20 (10%)	1/13 (8%)
west (2.5%)	2/19 (11%)	3/64 (5%)	0/14 (0%)	7/76 (9%)	5/86 (6%)	5/109 (5%)	5/142 (4%)	3/55 (6%)	2/30 (7%)	1/19 (5%)	0/20 (0%)	0/13 (0%)
east (3.3%)	0/19 (0%)	3/64 (5%)	2/14 (14%)	3/76 (4%)	8/86 (9%)	7/109 (6%)	9/142 (6%)	4/55 (7%)	3/30 (10%)	1/19 (5%)	2/20 (10%)	1/13 (8%)
Giampasquale (2.3%)	0/19 (0%)	4/64 (6%)	1/14 (7%)	5/76 (7%)	5/86 (6%)	5/109 (5%)	5/142 (3%)	2/55 (4%)	1/30 (3%)	0/19 (0%)	0/20 (0%)	0/13 (0%)
Central Plateau (25.6%)	*2/19 (11%)*	*11/64 (17%)*	*1/14 (7%)*	*19/76 (25%)*	*18/86 (20%)*	*31/109 (28%)*	*37/142 (26%)*	*14/55 (25%)*	*10/30 (33%)*	*3/19 (15%)*	*1/20 (5%)*	*1/13 (8%)*
Lago Lupo (15.3%)	1/19 (5%)	8/64 (13%)	1/14 (7%)	14/76 (18%)	9/86 (10%)	22/109 (20%)	26/142 (18%)	9/55 (16%)	6/30 (20%)	1/19 (5%)	1/20 (5%)	1/13 (8%)
Cogno Prete (10.3%)	1/19 (5%)	3/64 (5%)	0/14 (0%)	5/76 (7%)	9/86 (10%)	9/109 (8%)	11/142 (8%)	5/55 (9%)	4/30 (13%)	2/19 (11%)	0/20 (0%)	0/13 (0%)
Far from City (26.6%)	*3/19 (16%)*	*12/64 (19%)*	*2/14 (14%)*	*15/76 (20%)*	*16/86 (19%)*	*22/109 (20%)*	*33/142 (23%)*	*13/55 (24%)*	*7/30 (23%)*	*3/19 (16%)*	*8/20 (40%)*	*3/13 (23%)*
Campagnolo (4.9%)	0/19 (0%)	1/64 (2%)	0/14 (0%)	4/76 (5%)	1/86 (1%)	3/109 (3%)	5/142 (4%)	0/55 (0%)	0/30 (0%)	0/19 (0%)	1/20 (5%)	0/13 (0%)
Bufalara (2.9%)	1/19 (5%)	2/64 (3%)	0/14 (0%)	3/76 (4%)	3/86 (4%)	5/109 (5%)	6/142 (4%)	2/55 (4%)	0/30 (0%)	0/19 (0%)	0/20 (0%)	0/13 (0%)
Scarciullo (9.0%)	1/19 (5%)	4/64 (6%)	2/14 (14%)	5/76 (7%)	9/86 (10%)	9/109 (8%)	14/142 (10%)	7/55 (12%)	5/30 (17%)	0/19 (0%)	0/20 (0%)	0/13 (0%)
San Marco (3.4%)	1/19 (5%)	1/64 (2%)	0/14 (0%)	1/76 (1%)	0/86 (0%)	1/109 (1%)	2/142 (1%)	2/55 (4%)	2/30 (7%)	1/19 (5%)	3/20 (15%)	1/13 (8%)
AVS (6.4%)	0/19 (0%)	4/64 (6%)	0/14 (0%)	2/76 (3%)	3/86 (4%)	4/109 (3%)	6/142 (4%)	2/55 (4%)	0/30 (0%)	2/19 (11%)	4/20 (20%)	2/13 (15%)
Close to City (25.3%)	*2/19 (11%)*	*8/64 (13%)*	*4/14 (29%)*	*8/76 (10%)*	*12/86 (14%)*	*11/109 (11%)*	*18/142 (13%)*	*4/55 (7%)*	*1/30 (3%)*	*5/19 (26%)*	*6/20 (30%)*	*5/13 (38%)*
Pizzica (5.8%)	0/19 (0%)	2/64 (3%)	0/14 (0%)	3/76 (4%)	5/86 (6%)	5/109 (5%)	5/142 (4%)	3/55 (6%)	1/30 (3%)	0/19 (0%)	0/20 (0%)	0/13 (0%)
Pantanello (7.4%)	0/19 (0%)	1/64 (2%)	0/14 (0%)	1/76 (1%)	4/86 (5%)	4/109 (4%)	7/142 (5%)	0/55 (0%)	0/30 (0%)	1/19 (5%)	2/20 (10%)	1/13 (8%)
DBSBS (12.1%)	2/19 (11%)	5/64 (8%)	4/14 (29%)	4/76 (5%)	3/86 (3%)	2/109 (2%)	6/142 (4%)	1/55 (1%)	0/30 (0%)	4/19 (2[illegible]%)	4/20 (20%)	4/13 (30%)

C5 Significant farmhouses as percentages of study area totals, **600–50**.

Localities (% of study area)	600	550	500	450	400	350	300	250	200	150	100	50
Early Clusters (22.6%)	*12*	*21/33 (64%)*	*0/7 (0%)*	*14/34 (41%)*	*15/40 (38%)*	*9/45 (20%)*	*8/54 (15%)*	*0/24 (0%)*	*0/12 (0%)*	*1/8 (13%)*	*0/5 (0%)*	*0/4 (0%)*
Giardinetto (3.3%)	2	4/6 (67%)	0/1 (0%)	2/6 (33%)	3/8 (38%)	4/10 (40%)	2/14 (14%)	0/5 (0%)	0/0 (0%)	0/1 (0%)	0/0 (0%)	0/0 (0%)
Sant'Angelo (3.7%)	5	3/8 (38%)	0/2 (0%)	2/6 (33%)	1/6 (17%)	0/5 (0%)	2/7 (29%)	0/2 (0%)	0/0 (0%)	0/2 (0%)	0/2 (0%)	0/2 (0%)
Avinella (7.5%)	3	6/9 (67%)	0/1 (0%)	3/7 (43%)	2/8 (25%)	4/13 (31%)	1/14 (7%)	0/8 (0%)	0/6 (0%)	1/3 (33%)	0/1 (0%)	0/1 (0%)
lower (4.7%)	2	4/6 (67%)	0/1 (0%)	1/7 (14%)	1/8 (12%)	1/13 (8%)	1/14 (7%)	0/4 (0%)	0/3 (0%)	1/3 (33%)	0/1 (0%)	0/1 (0%)
upper (2.8%)	1	2/3 (67%)	0/0 (0%)	2/7 (29%)	1/8 (12%)	3/13 (23%)	0/14 (0%)	0/4 (0%)	0/3 (0%)	0/0 (0%)	0/0 (0%)	0/0 (0%)
Venella (5.8%)	2	4/6 (67%)	0/2 (0%)	6/10 (60%)	7/13 (54%)	1/12 (8%)	3/14 (21%)	0/7 (0%)	0/5 (0%)	0/2 (0%)	0/2 (0%)	0/1 (0%)
west (2.5%)	2	1/3 (33%)	0/0 (0%)	4/10 (40%)	2/13 (15%)	0/12 (0%)	0/14 (0%)	0/3 (0%)	0/3 (0%)	0/1 (0%)	0/0 (0%)	0/0 (0%)
east (3.3%)	0	3/3 (100%)	0/2 (0%)	2/10 (20%)	5/13 (39%)	1/12 (8%)	3/14 (21%)	0/4 (0%)	0/2 (0%)	0/1 (0%)	0/2 (0%)	0/1 (0%)
Giampasquale (2.3%)	0	4/4 (100%)	0/1 (0%)	1/5 (20%)	2/5 (40%)	0/5 (0%)	0/5 (0%)	0/2 (0%)	0/1 (0%)	0/0 (0%)	0/0 (0%)	0/0 (0%)
Central Plateau (25.[illegible]%)	*2*	*9/11 (82%)*	*0/1 (0%)*	*13/19 (68%)*	*2/18 (11%)*	*14/31 (45%)*	*7/37 (19%)*	*0/14 (0%)*	*0/10 (0%)*	*1/3 (33%)*	*0/1 (0%)*	*0/1 (0%)*
Lago Lupo (15.3%)	1	7/8 (88%)	0/1 (0%)	8/14 (57%)	0/9 (0%)	13/22 (59%)	6/25 (23%)	0/9 (0%)	0/6 (0%)	1/1 (100%)	0/1 (0%)	0/1 (0%)
Cogno Prete (10.3%)	1	2/3 (67%)	0/0 (0%)	5/5 (100%)	2/9 (22%)	1/9 (11%)	1/11 (9%)	0/5 (0%)	0/4 (0%)	0/2 (0%)	0/0 (0%)	0/0 (0%)
Far from City (26.5%)	*3*	*9/12 (75%)*	*0/2 (0%)*	*8/15 (53%)*	*8/16 (50%)*	*8/22 (36%)*	*11/33 (33%)*	*0/13 (0%)*	*0/7 (0%)*	*2/3 (67%)*	*1/8 (13%)*	*0/3 (0%)*
Campagnolo (4.9%)	0	1/1 (100%)	0/0 (0%)	3/4 (75%)	0/1 (0%)	2/3 (67%)	2/5 (40%)	0/0 (0%)	0/0 (0%)	0/0 (0%)	0/1 (0%)	0/0 (0%)
Bufalara ([illegible].9%)	1	1/2 (50%)	0/0 (0%)	1/3 (33%)	1/3 (33%)	2/5 (40%)	2/6 (33%)	0/2 (0%)	0/0 (0%)	0/0 (0%)	0/0 (0%)	0/0 (0%)
Scarciullo (9.0%)	1	3/4 (75%)	0/2 (0%)	2/5 (40%)	4/9 (44%)	2/9 (22%)	4/14 (29%)	0/7 (0%)	0/5 (0%)	0/0 (0%)	0/0 (0%)	0/0 (0%)
San Marco (3.4%)	1	0/1 (0%)	0/0 (0%)	1/1 (100%)	0/0 (0%)	1/1 (100%)	1/2 (50%)	0/2 (0%)	0/2 (0%)	1/1 (100%)	0/3 (0%)	0/1 (0%)
AVS (6.4%)	0	4/4 (100%)	0/0 (0%)	1/2 (50%)	3/3 (100%)	1/4 (25%)	2/6 (33%)	0/2 (0%)	0/0 (0%)	1/2 (50)%	1/4 (25%)	0/2 (0%)
Close to City (25.3%)	*2*	*6/8 (75%)*	*1/4 (25%)*	*3/8 (38%)*	*7/12 (58%)*	*0/11 (0%)*	*6/18 (33%)*	*1/4 (25%)*	*0/1 (0%)*	*2/5 (40%)*	*1/6 (17%)*	*0/5 (0%)*
Pizzica (5.8%)	0	2/2 (100%)	0/0 (0%)	2/3 (67%)	3/5 (60%)	0/5 (0%)	0/5 (0%)	1/3 (33%)	0/1 (0%)	0/0 (0%)	0/0 (0%)	0/0 (0%)
Pantanello (7.4%)	0	1/1 (100%)	0/0 (0%)	0/1 (0%)	3/4 (75%)	0/4 (0%)	3/7 (43%)	0/0 (0%)	0/0 (0%)	0/1 (0%)	1/2 (50%)	0/1 (0%)
DBSBS (1[illegible].1%)	2	3/5 (60%)	1/4 (25%)	1/4 (25%)	1/3 (33%)	0/2 (0%)	3/6 (50%)	0/1 (0%)	0/0 (0%)	2/4 (50%)	0/4 (0%)	0/4 (0%)
Chora as a whole	19 (100%)	45/64 (70%)	1/14 (7%)	38/76 (50%)	32/86 (37%)	31/109 (28%)	32/142 (23%)	1/55 (2%)	0/30 (0%)	6/19 (32%)	2/20 (10%)	0/13 (0%)

C6 Ex novo: newly significant farmhouses as percentages of significant site totals, **600–50**.

	600	550	500	450	400	350	300	250	200	150	100	50
Early Village Clusters												
Giardinetto	2.07 (2)	2.29 (6)	2.06 (1)	2.05 (6)	2.12 (8)	2.18 (10)	2.51 (14)	2.08 (5)	—	2.49 (1)	—	—
Sant'Angelo	2.07 (5)	2.42 (8)	2.35 (2)	2.20 (6)	1.88 (6)	1.84 (5)	2.48 (7)	2.34 (2)	—	2.06 (2)	2.59 (2)	2.17 (2)
Avinella	1.69 (3)	2.11 (9)	1.77 (1)	1.84 (7)	1.90 (8)	1.80 (13)	2.60 (14)	1.81 (8)	1.80 (6)	1.87 (3)	3.41 (1)	2.10 (1)
lower	1.93 (2)	1.98 (6)	1.77 (1)	1.78 (4)	1.96 (4)	1.89 (6)	2.62 (7)	1.93 (4)	1.92 (3)	1.87 (3)	3.41 (1)	2.10 (1)
upper	1.20 (1)	2.38 (3)	—	1.93 (3)	1.85 (4)	1.72 (7)	2.59 (7)	1.69 (4)	1.69 (3)	—	—	—
Venella	1.43 (2)	2.03 (6)	1.64 (2)	1.74 (10)	2.08 (13)	1.96 (12)	2.50 (14)	1.89 (7)	1.82 (5)	1.79 (2)	2.29 (2)	2.31 (1)
west	1.43 (2)	2.18 (3)	—	1.72 (7)	2.14 (5)	1.97 (5)	2.61 (5)	1.79 (3)	1.78 (3)	1.46 (1)	—	—
east	—	1.88 (3)	1.64 (2)	1.80 (3)	2.04 (8)	1.95 (7)	2.44 (9)	1.97 (4)	1.89 (2)	2.12 (1)	2.29 (2)	2.31 (1)
Giampasquale	—	2.14 (4)	1.94 (1)	1.88 (5)	1.99 (5)	2.20 (5)	2.69 (5)	2.23 (2)	2.14 (1)	—	—	—
Weighted avg.	**1.86 (12)**	**2.21 (33)**	**1.96 (7)**	**1.92 (34)**	**2.01 (40)**	**1.98 (45)**	**2.55 (54)**	**1.97 (24)**	**1.84 (12)**	**1.97 (8)**	**2.64 (5)**	**2.19 (4)**
Central Plateau												
Lago Lupo	1.30 (1)	2.14 (8)	1.33 (1)	1.97 (14)	2.02 (9)	2.22 (22)	2.79 (26)	2.13 (9)	1.88 (6)	2.97 (1)	3.70 (1)	2.47 (1)
Cogno Prete	1.52 (1)	1.86 (3)	—	1.96 (5)	1.89 (9)	1.93 (9)	2.58 (11)	1.91 (5)	1.89 (4)	2.06 (2)	—	—
Weighted avg.	**1.41 (2)**	**2.07 (11)**	**1.33 (1)**	**1.97 (19)**	**1.95 (18)**	**2.14 (31)**	**2.72 (37)**	**2.05 (14)**	**1.88 (10)**	**2.36 (3)**	**3.70 (1)**	**2.47 (1)**
Far from City												
Campagnolo	—	1.59 (1)	—	2.40 (4)	1.82 (1)	1.97 (3)	2.48 (5)	—	—	—	2.03 (1)	—
Bufalara	1.33 (1)	1.44 (2)	—	1.86 (3)	1.91 (3)	1.94 (5)	2.54 (6)	1.59 (2)	—	—	—	—
Scarciullo	1.34 (1)	1.89 (4)	1.52 (2)	1.68 (5)	1.82 (9)	1.97 (9)	2.64 (14)	1.78 (7)	1.69 (5)	—	—	—
San Marco	2.23 (1)	2.19 (1)	—	2.53 (1)	—	1.08 (1)	1.96 (2)	1.68 (2)	1.56 (2)	1.86 (1)	2.22 (3)	2.11 (1)
AVS	—	2.55 (4)	—	2.12 (2)	1.89 (3)	1.84 (4)	1.96 (6)	2.11 (2)	—	1.96 (2)	2.23 (4)	1.77 (2)
Weighted avg.	**1.63 (3)**	**2.03 (12)**	**1.52 (2)**	**2.02 (15)**	**1.85 (16)**	**1.90 (22)**	**2.43 (33)**	**1.78 (13)**	**1.65 (7)**	**1.93 (3)**	**2.20 (8)**	**1.88 (3)**
Close to City												
Pizzica	—	2.32 (2)	—	2.18 (3)	2.50 (5)	2.59 (5)	3.04 (5)	2.48 (3)	3.09 (1)	—	—	—
Pantanello	—	1.93 (1)	—	1.98 (1)	1.89 (4)	1.88 (4)	2.17 (7)	—	—	1.95 (1)	2.40 (2)	1.99 (1)
DBSBS	1.47 (2)	2.09 (5)	1.68 (4)	2.12 (4)	2.62 (3)	2.29 (2)	2.67 (6)	2.55 (1)	—	2.53 (4)	3.32 (4)	2.35 (4)
Weighted avg.	**1.47 (2)**	**2.13 (8)**	**1.68 (4)**	**2.13 (8)**	**2.32 (12)**	**2.28 (11)**	**2.58 (18)**	**2.50 (4)**	**3.09 (1)**	**2.41 (5)**	**3.02 (6)**	**2.28 (5)**
Study area average	1.74 (19)	2.14 (64)	1.77 (14)	1.97 (76)	2.01 (86)	2.04 (109)	2.57 (142)	1.99 (55)	1.85 (30)	2.14 (19)	2.63 (20)	2.17 (13)

C7 Average FMCE values, **600–50**.

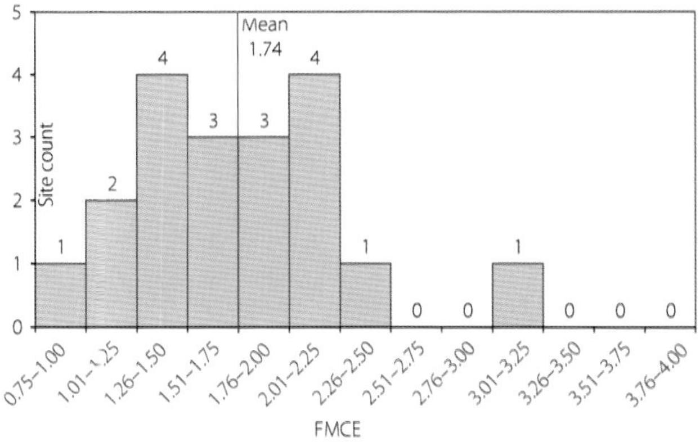

C8 Farmhouse FMCE distribution: **600**.

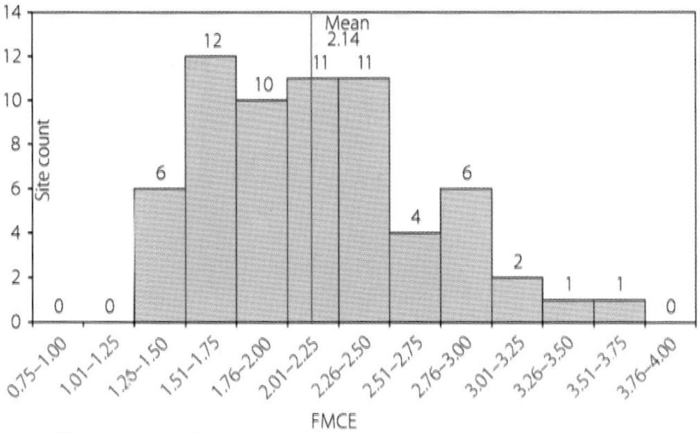

C8 Farmhouse FMCE distribution: **550**.

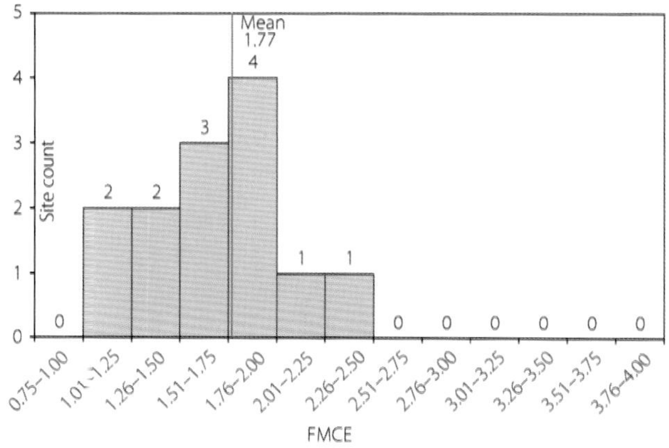

C8 Farmhouse FMCE distribution: **500**.

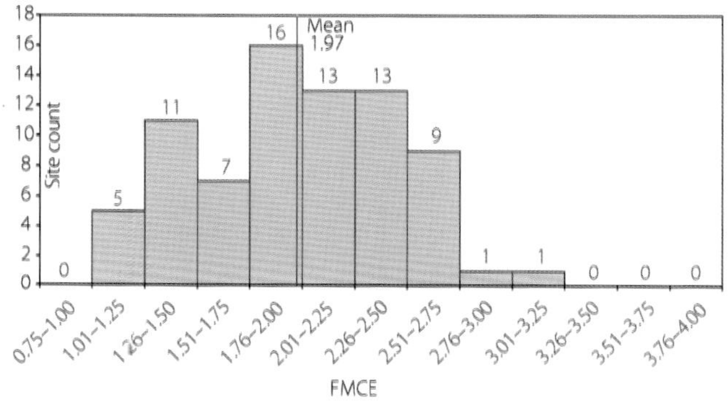

C8 Farmhouse FMCE distribution: **450**.

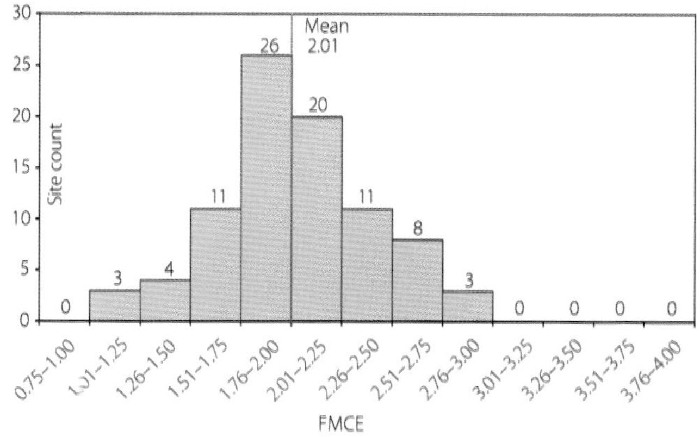

C8 Farmhouse FMCE distribution: **400**.

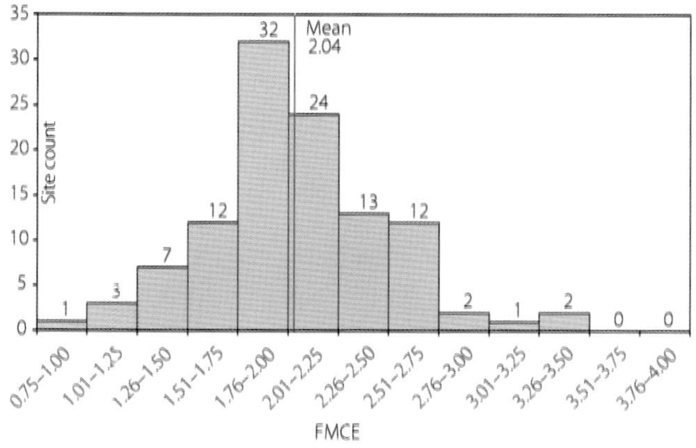

C8 Farmhouse FMCE distribution: **350**.

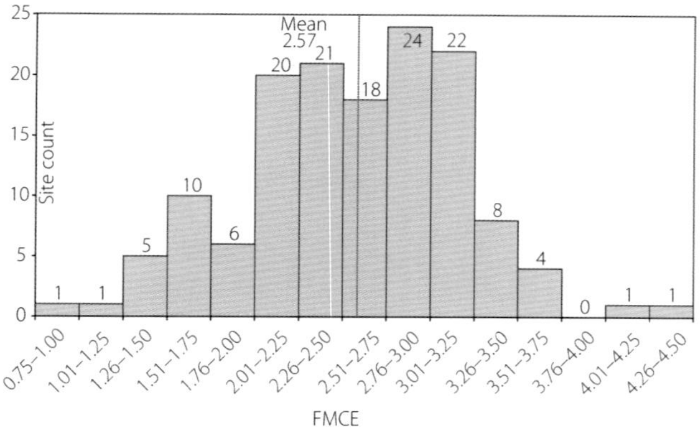

C8 Farmhouse FMCE distribution: **300**.

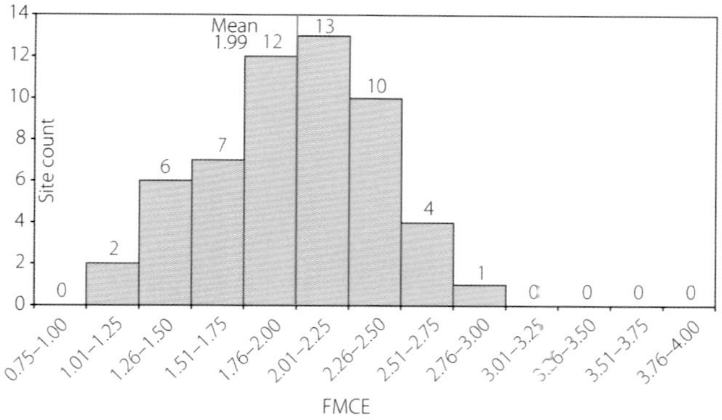

C8 Farmhouse FMCE distribution: **250**.

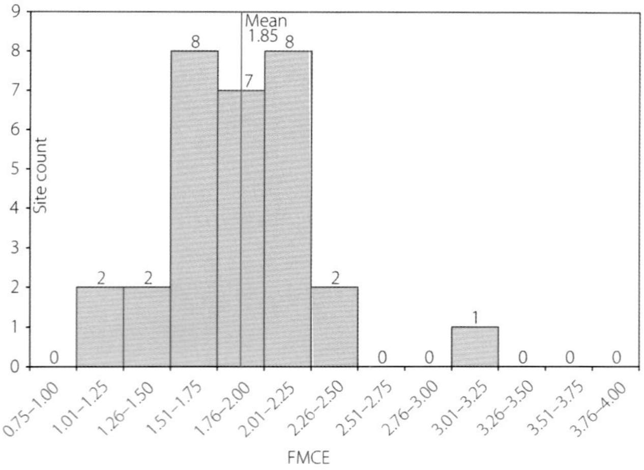

C8 Farmhouse FMCE distribution: **200**.

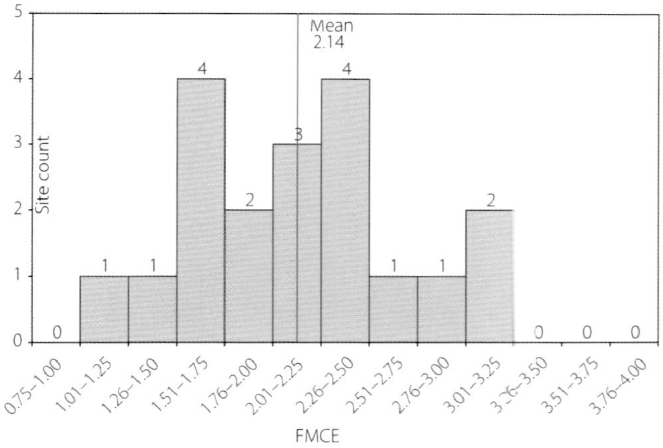

C8 Farmhouse FMCE distribution: **150**.

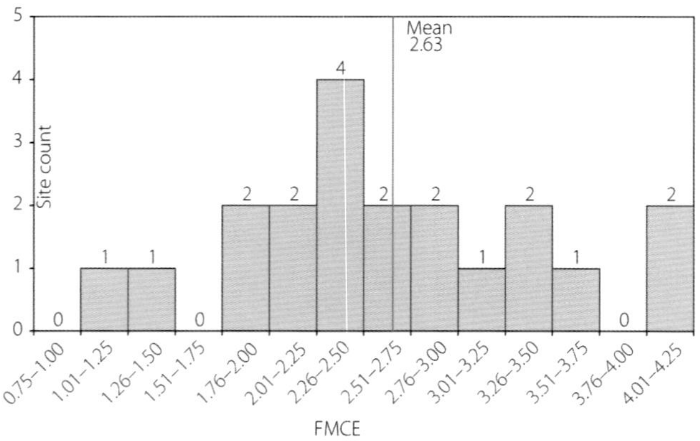

C8 Farmhouse FMCE distribution: **100**.

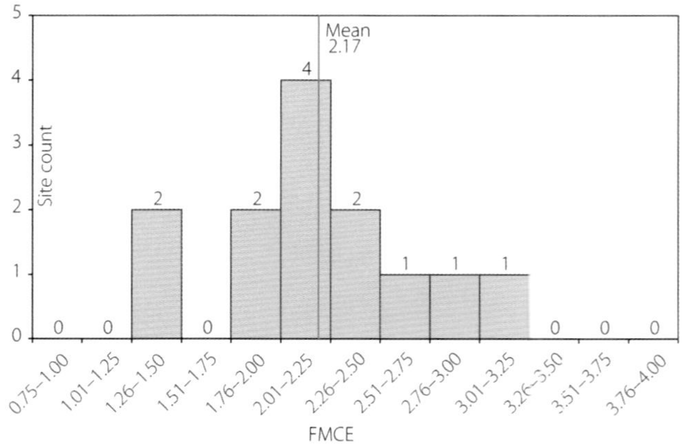

C8 Farmhouse FMCE distribution: **50**.

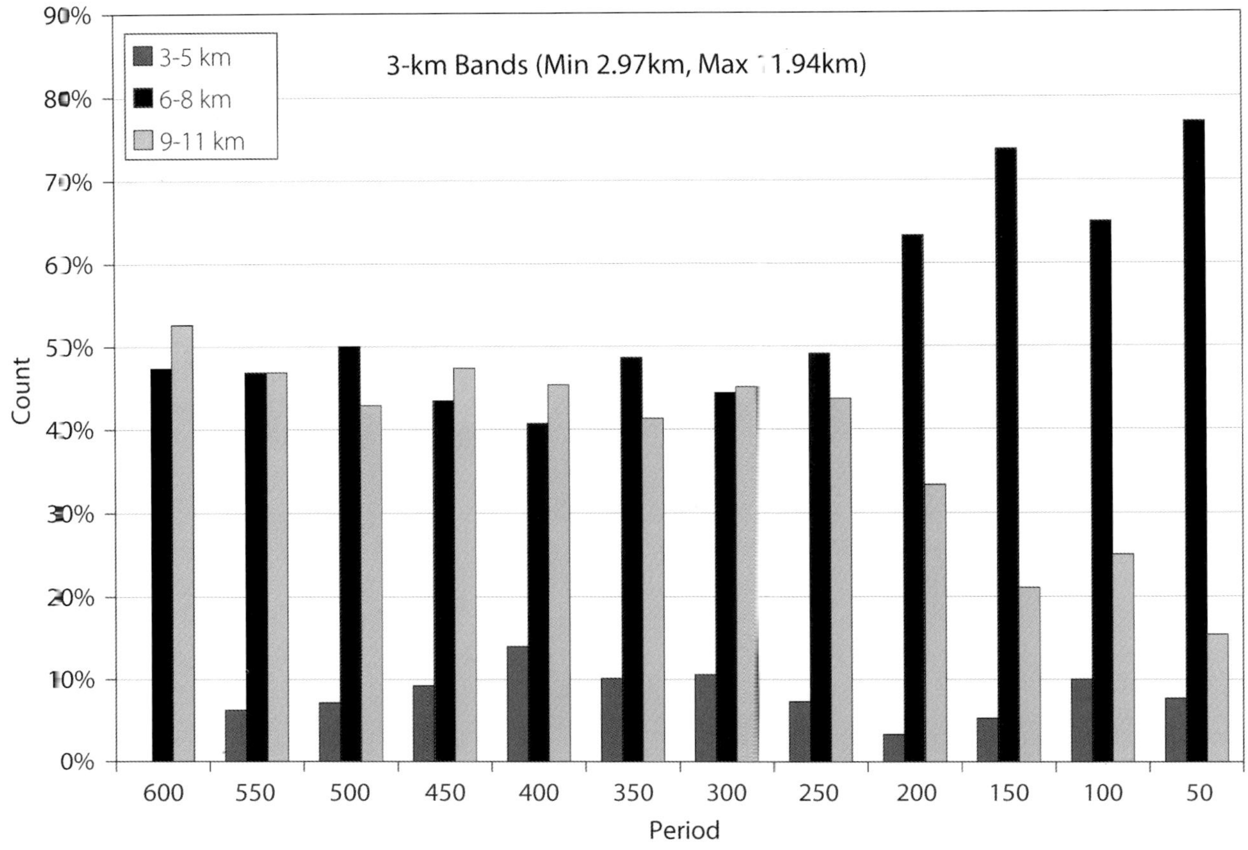

C9a Distances from asty to significant farmhouses, **600–50** (percentages).

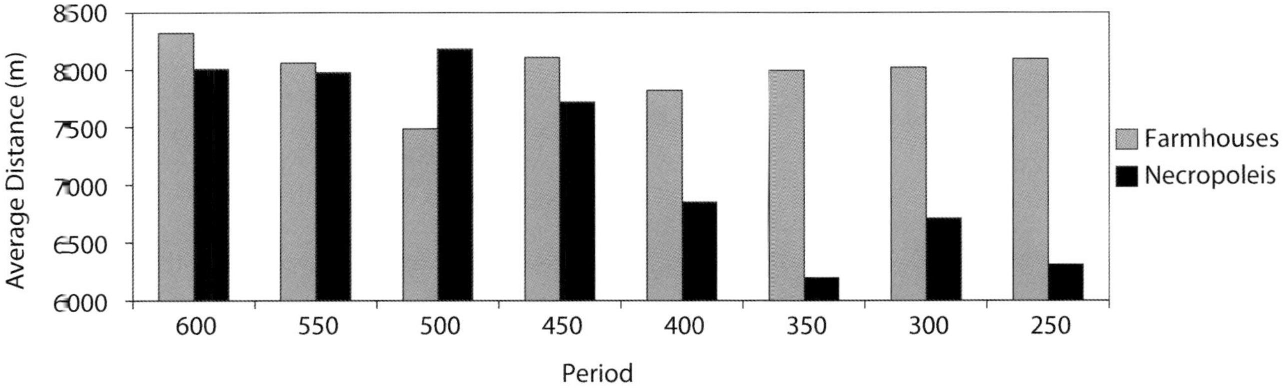

C9b Distances from asty to significant farmhouses and necropoleis, **600–50** (averages).

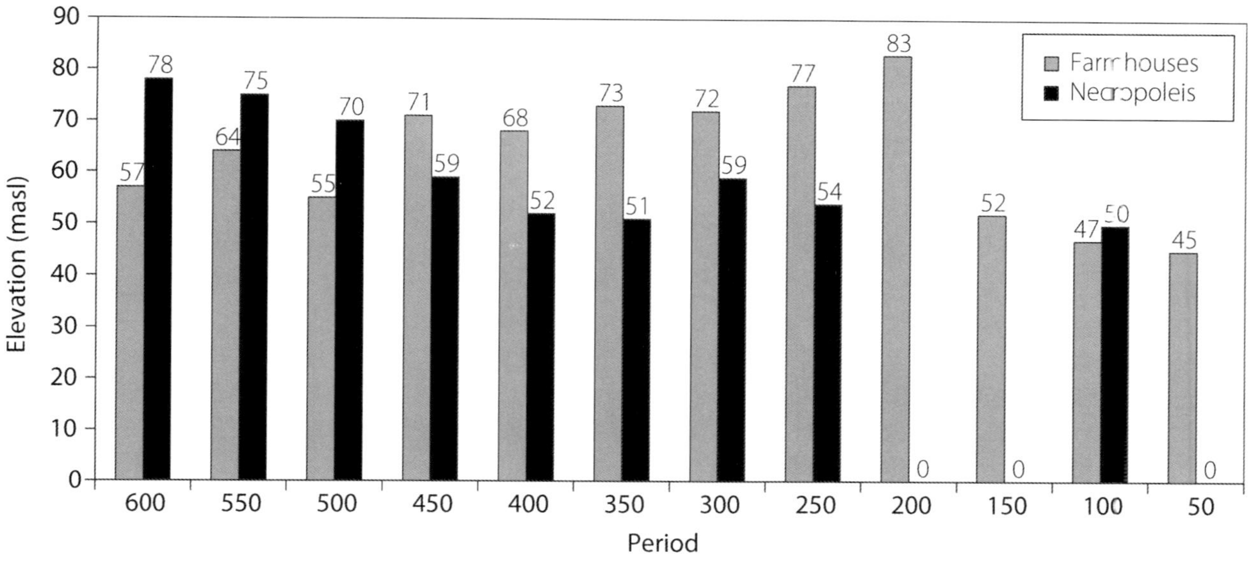

C10 Average elevations, **600–50**, farmhouses (EAW ≥ 0.95) and necropoleis (EAW ≥ 0.45).

Geomorphological unit	% of study area	600	550	500	450	400	350	300	250	200	150	100	50
Pleistocene marine terrace	56	7 (37%)	38 (59%)	7 (50%)	55 (72%)	64 (75%)	83 (76%)	105 (74%)	44 (80%)	29 (97%)	10 (53%)	8 (40%)	6 (46%)
T_0 flood plain + T_1–T_4 alluvial terraces	19	5 (26%)	12 (19%)	4 (29%)	9 (12%)	13 (15%)	11 (10%)	15 (11%)	3 (6%)	—	3 (16%)	5 (25%)	3 (23%)
Valley walls	24	5 (26%)	8 (13%)	2 (14%)	6 (8%)	2 (2%)	8 (7%)	10 (7%)	4 (7%)	1 (3%)	5 (26%)	7 (35%)	4 (31%)
Mass movement deposit	1	2 (11%)	6 (9%)	1 (7%)	6 (8%)	7 (8%)	7 (7%)	12 (8%)	4 (7%)	—	1 (5%)	—	—
Totals		**19**	**64**	**14**	**76**	**86**	**109**	**142**	**55**	**30**	**19**	**20**	**13**

C11a Counts and percentages in geomorphological units, **600–50**, significant farmhouses (EAW ≥ 0.95).

	Area (km^2)	Percent of study area	No. of sites	Density
Alluvial terraces (T_0–T_2)	5.55	15%	51	(9.2)
Alluvial terraces (T_3–T_4)	1.35	4%	47	32.6
Valley walls	8.7	24%	68	7.8
Mass movement deposits	0.5	1%	32	64
Marine terraces	20.7	56%	367	17.7
Totals	**36.8**	**100%**	**509**	**16.3**

C11b Observed density of sites in geomorphological units.

PointFile	MinNNDist	MeanNNDist	MaxNNDist
All Farmhouses	30.0	182.4	861.5
Farmhouses **600**	30.0	761.5	1893.4
Farmhouses **550**	30.0	341.4	1151.6
Farmhouses **500**	421.4	1035.4	2474.9
Farmhouses **450**	72.1	519.4	1861.2
Farmhouses **400**	54.8	342.4	1260.0
Farmhouses **350**	30.0	363.5	7766.5
Farmhouses **300**	30.0	273.6	1605.1
Farmhouses **250**	92.2	378.5	2081.8
Farmhouses **200**	140.7	506.7	1853.2
Farmhouses **150**	222.7	1025.0	3036.2
Farmhouses **100**	50.0	804.0	1934.0
Farmhouses **50**	222.7	1281.7	3036.2
All Dated Necropoleis	50.0	247.3	844.3
All Necropoleis	22.4	217.9	811.0
Necropoleis **600**	402.7	938.9	1677.1
Necropoleis **550**	50.0	799.6	1991.1
Necropoleis **500**	50.0	902.0	1991.1
Necropoleis **450**	141.4	641.1	2010.6
Necropoleis **400**	50.0	545.9	2327.0
Necropoleis **350**	70.7	518.9	1538.7
Necropoleis **300**	107.3	440.7	1175.5
Necropoleis **250**	500.9	1044.1	2120.0

C12 Distances between each site and its nearest neighbor: minimum, mean, maximum. ArcGIS Spatial Analyst Tool: *Calculate Distance Band from Neighbor Count.* This tool calculates the distance for all features in the input feature class to their *n*th neighbor (specified in the Neighbor's parameter). Then it examines the distances and returns three values: the maximum distance, the minimum distance, and the average distance (in meters). Peter Dana, 2009.

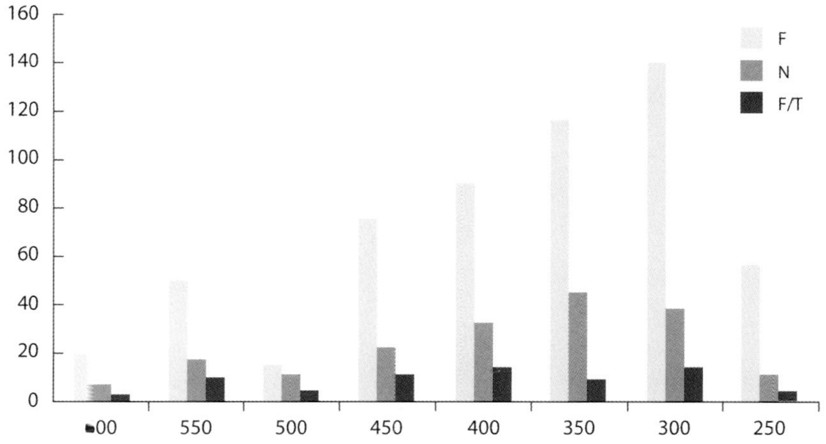

Period	Burials (Necs, F/Ts)	Farm-houses	Ratio
600	10 (8, 2)	19	1:2
550	21 (13, 8)	64	1:3
500	16 (12, 4)	14	1:1
450	31 (21, 10)	76	1:2.5
400	43 (31, 12)	86	1:2
350	50 (41, 9)	109	1:2
300	50 (36, 14)	142	1:3
250	15 (10, 5)	55	1:3.5

C13 Ratios of total burials to farmhouses, **600–50**.

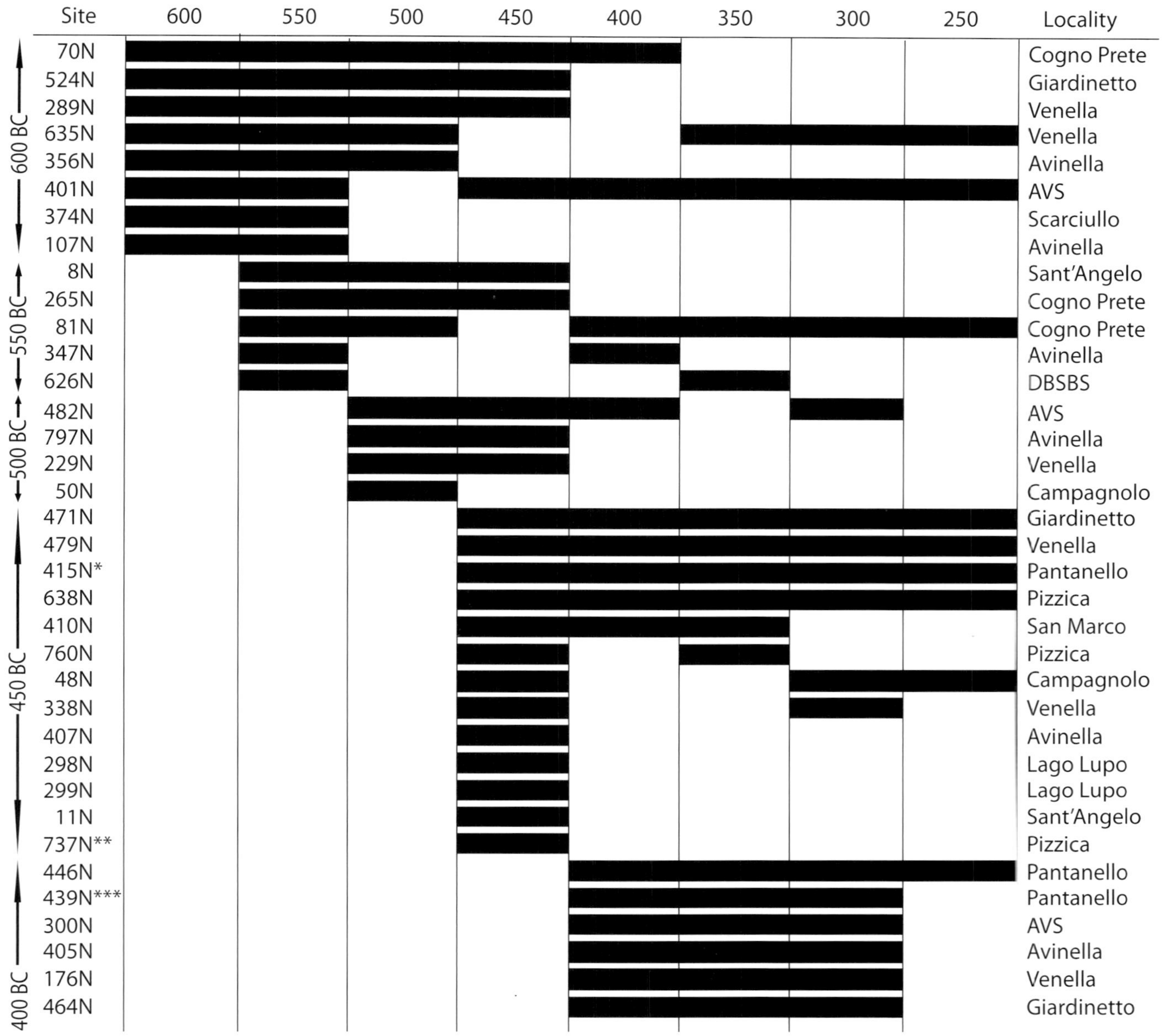

* 415N Pantanello, excavated. ** 737N Pizzica, excavated. See Ch. 8. *** 439N Pantanello, active 200–150 BC.

C14 Duration of necropoleis.

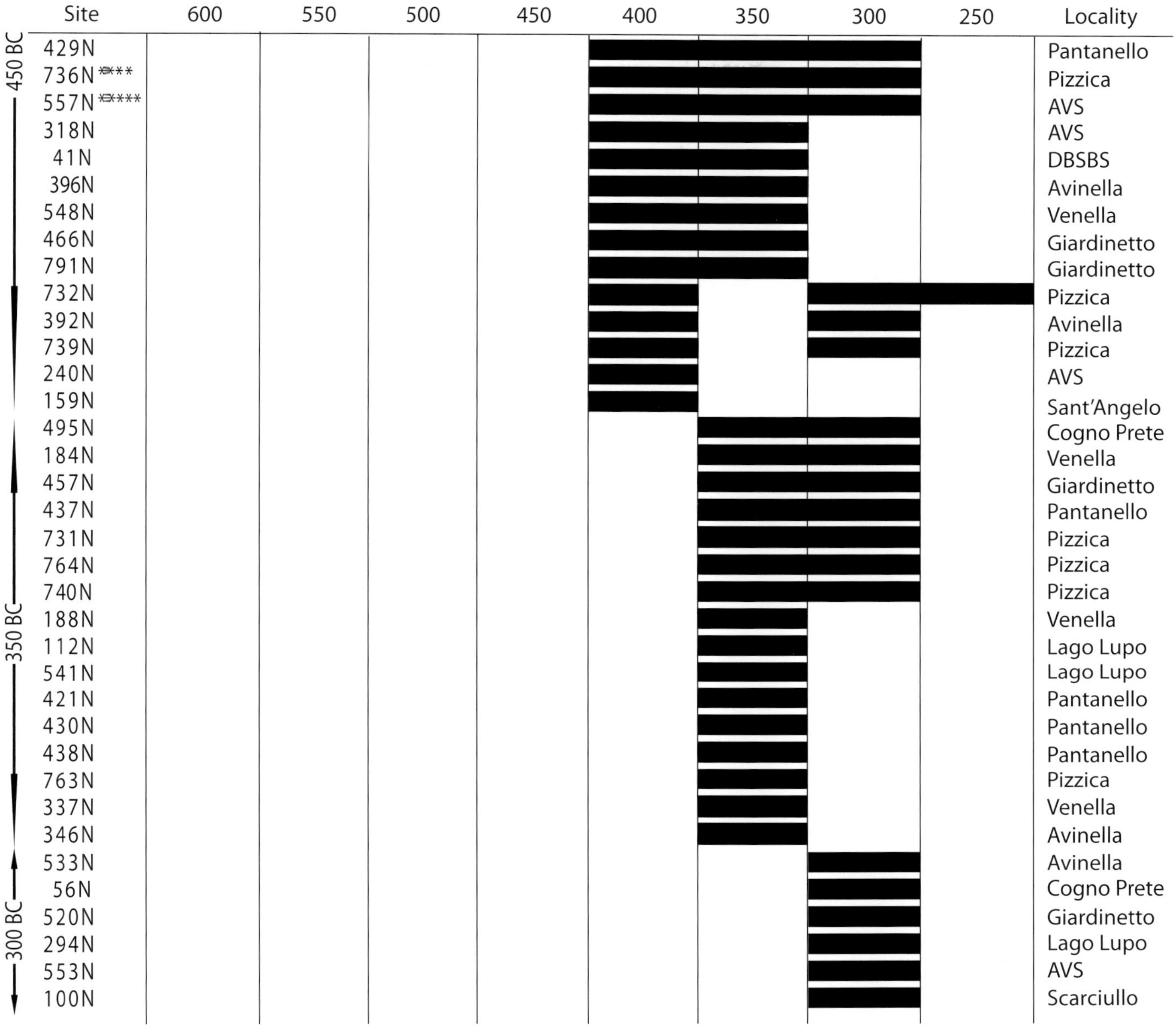

**** 736N Pizzica, excavated.

***** 557N, active 100 BC.

C14 *cont.* Duration of necropoleis.

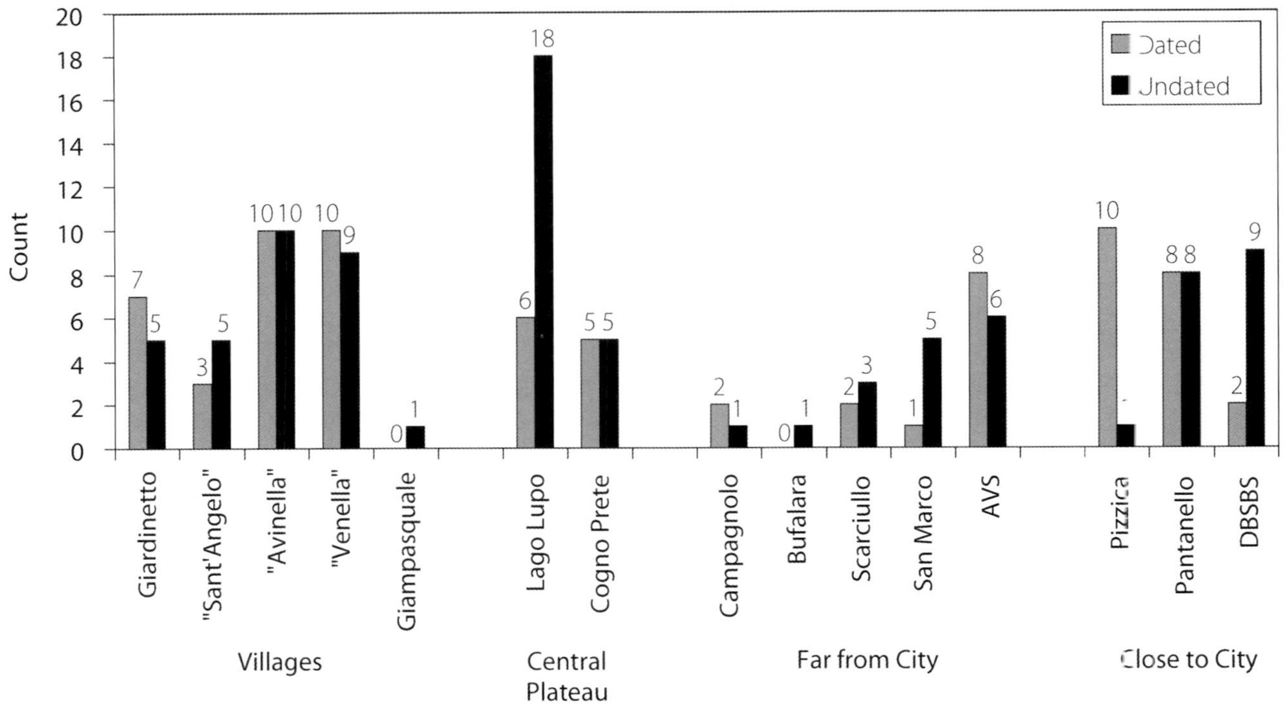

C15 Necropoleis, dated and undated, by region and locality.

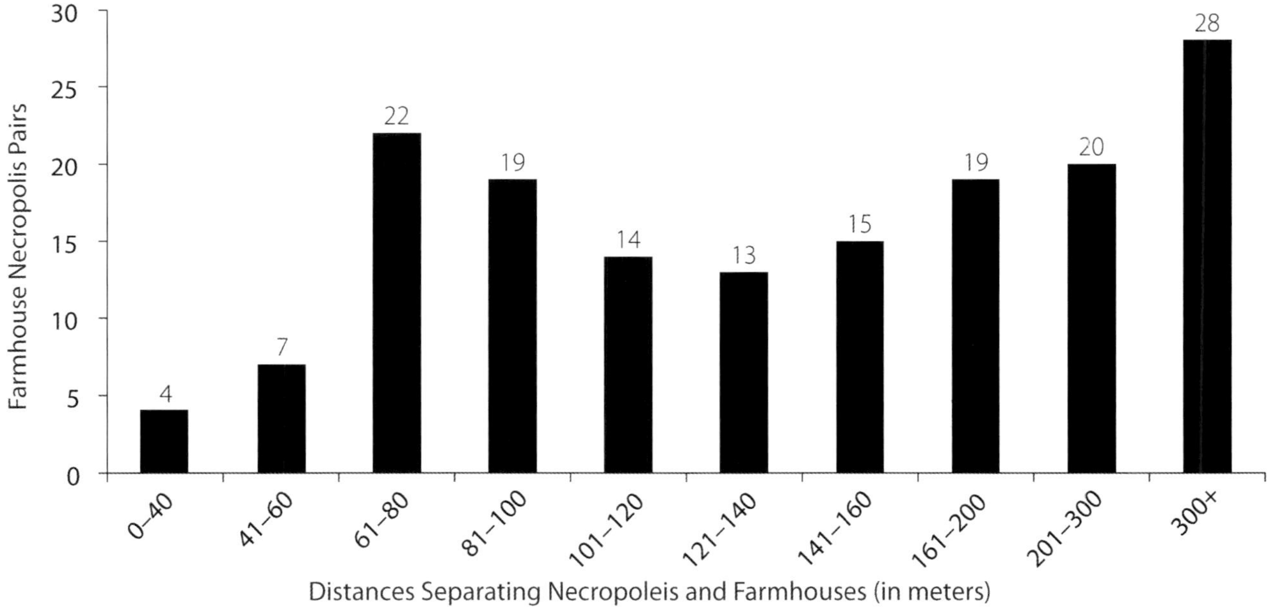

C16 Nearest neighbor distances: all necropoleis to all farmhouses (without burials).

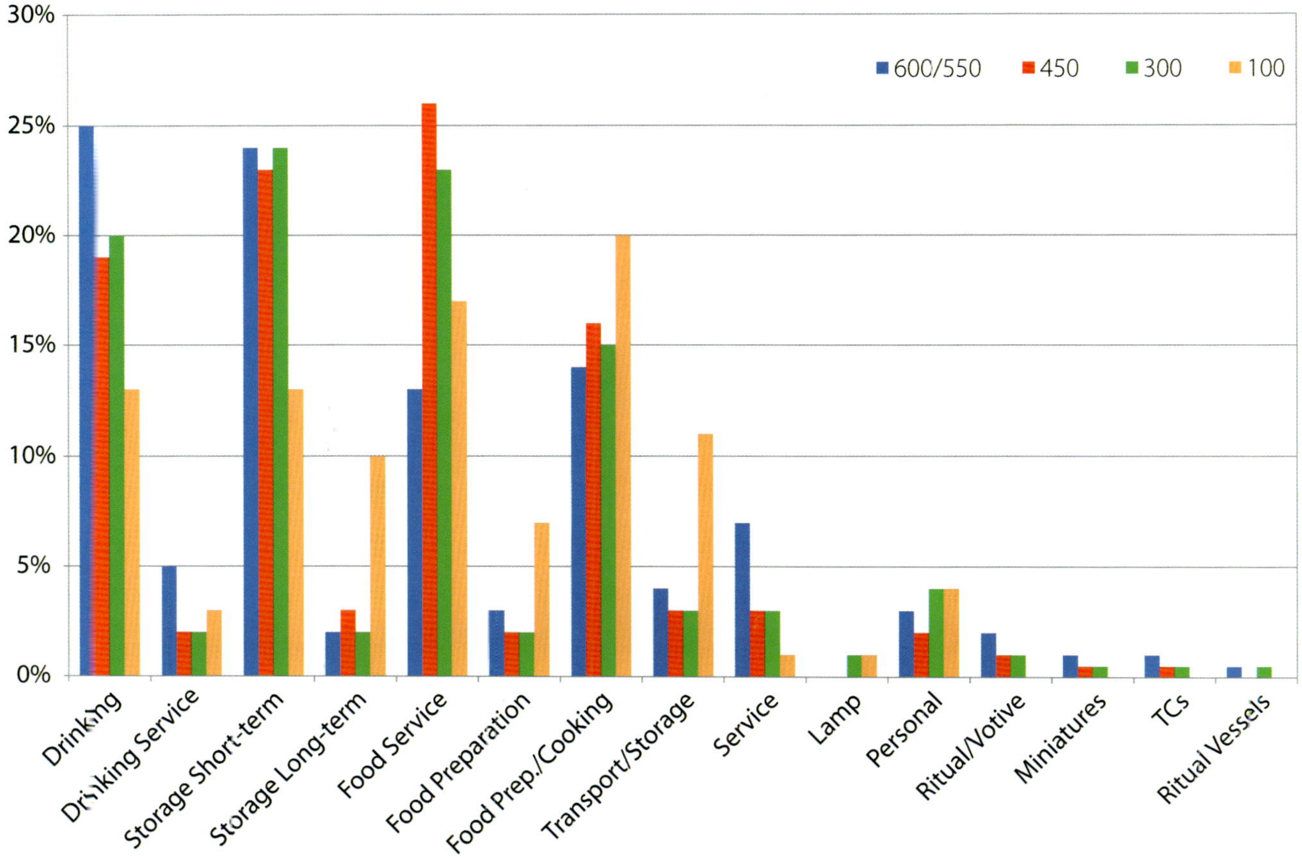

	600/550	450	300	100*
Drinking	25%	19%	20%	13%
Drinking Service	5%	2%	2%	6%
Storage Short-term	24%	23%	24%	19%
Storage Long-term	2%	3%	2%	**7%
Food Service	13%	26%	23%	19%
Food Preparation	3%	2%	2%	5%
Food Prep./Cooking	14%	16%	15%	**19%
Transport/Storage Amphorae	4%	3%	3%	8%
Service	7%	3%	3%	(< 1%)
Lamp			1%	(< 1%)
Personal	3%	2%	4%	3%
Ritual/Votive	2%	1%	1%	0
Miniatures	1%	(< 1%)	(< 1%)	0
TCs	1%	(< 1%)	(< 1%)	0
Ritual Vessels	(< 1%)		(< 1%)	0

* Functional groups analysis based on ex-novo sites and sites 556F and 142F.
** Unusually high values at single sites.

C17 Functional-group averages, **600–100,** surveyed farmhouses.

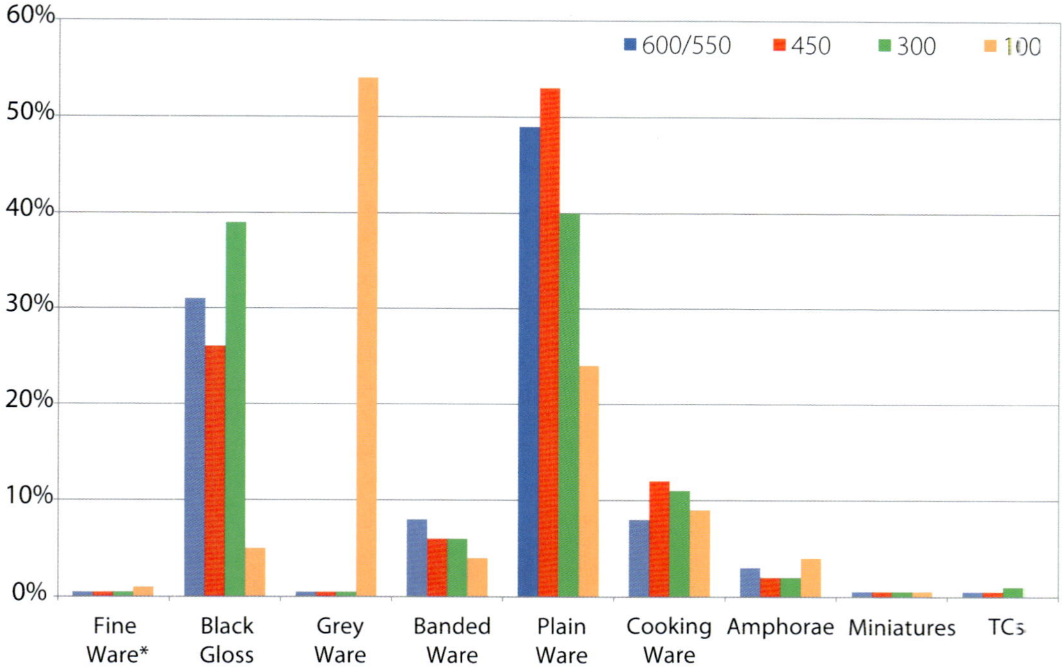

	600/550	450	300	100
FW	<1%	<1%	<1%	0
BG	31%	26%	39%	5%
GW	0	<1%	<1%	54%
BW	8%	6%	6%	4%
PW	49%	53%	40%	24%
CW	8%	12%	11%	9%
Am	3%	2%	2%	4%
Min	<1%	<1%	<1%	<1%
TC	<1%	<1%	1%	0

C18 Ware class averages, **600–100**, surveyed farmhouses.

* Figured ware: **550, 450, 300**; molded ware: **100.**

	285F/T	228F	13F	323F	349F	336F	286F	366F	459F	530F	170F	1[illegible]5F
Best date	577	575	565	564	561	560	560	557	557	553	537	531
Total fragments	188	47	108	131	87	163	182	61	31	61	60	201
Drinking	29 (27%)	5 (22%)	15 (25%)	18 (19%)	14 (22%)	18 (21%)	22 (21%)	8 (20%)	7 (23%)	17 (42%)	11 (28%)	4[illegible] (32%)
Drinking service	3 (3%)	2 (9%)	3 (5%)	2 (2%)	3 (5%)	5 (6%)	4 (4%)	1 (3%)	1 (3%)	3 (7%)	2 (5%)	7 (5%)
Storage-short	30 (27%)	5 (22%)	11 (19%)	34 (36%)	13 (21%)	18 (22%)	26 (25%)	12 (30%)	7 (23%)	7 (17%)	5 (13%)	38 (29%)
Storage-long	4 (4%)		1 (2%)	1 (1%)			4 (4%)		3 (10%)	1 (2%)		4 (3%)
Food service	7 (6%)		15 (25%)	12 (13%)	4 (6%)	12 (14%)	13 (13%)	8 (20%)	3 (10%)	4 (10%)	10 (26%)	1[illegible] (13%)
Food prep.	3 (3%)	4 (17%)	2 (3%)	3 (3%)	3 (5%)		2 (2%)		1 (3%)			1 (1%)
Food prep./Cooking	18 (16%)	6 (26%)	6 (10%)	7 (7%)	10 (17%)	10 (12%)	13 (13%)	5 (13%)	1 (3%)	6 (15%)	8 (21%)	13 (10%)
Transport/Storage (amphorae)	5 (5%)		4 (7%)	5 (6%)	4 (6%)	4 (5%)	1 (1%)		4 (13%)	1 (2%)		2 (2%)
Service	6 (6%)		1 (2%)	7 (7%)	9 (14%)	13 (15%)	16 (15%)	3 (8%)	2 (6%)	2 (5%)		
Personal	2 (2%)		1 (2%)	4 (4%)	2 (3%)	4 (5%)	3 (3%)	1 (3%)	1 (3%)		2 (5%)	2 (2%)
Votive/Ritual												
Miniatures				2 (2%)	1 (1%)			1 (3%)				1 (1%)
TC figurines		1 (4%)									1 (2%)	2 (2%)
Other vessels	1 (1%)											
Unknown	80 (43%)	24 (51%)	49 (45%)	36 (27%)	24 (28%)	79 (48%)	78 (43%)	21 (34%)	1 (3%)	20 (33%)	21 (35%)	7[illegible] (36%)
Known	108	23	59	95	63	84	104	40	30	41	39	128

C19a Functional groups, **600/550**, farmhouses.

Site (best date)	285F/T (577 BC)	228F (575 BC)	13F (565 BC)	323F (564 BC)	349F (561 BC)	336F (560 BC)	286F (560 BC)	366F (557 BC)	459F (557 BC)	530F (553 BC)	170F (537 BC)	1[illegible]5F (53[illegible] BC)
Total fragments	188	47	108	131	87	163	182	61	31	61	60	201
AFW			1 (1%)									
BG	47 (25%)	9 (19%)	45 (42%)	29 (22%)	30 (35%)	41 (25%)	48 (26%)	14 (23%)	12 (39%)	26 (43%)	22 (37%)	81 (40%)
BW	12 (6%)		5 (5%)	14 (11%)	6 (7%)	21 (13%)	22 (12%)	3 (5%)	2 (7%)	5 (8%)	5 (8%)	2[illegible] (11%)
PW	106 (56%)	31 (66%)	47 (44%)	74 (57%)	36 (41%)	87 (53%)	98 (54%)	37 (61%)	12 (39%)	23 (38%)	24 (40%)	79 (39%)
Ck	18 (10%)	6 (13%)	6 (6%)	7 (5%)	10 (12%)	10 (6%)	13 (7%)	5 (8%)	1 (3%)	6 (10%)	8 (13%)	1[illegible] (7%)
Amp	5 (3%)		4 (4%)	5 (4%)	4 (5%)	4 (3%)	1 (1%)	1 (2%)	4 (13%)	1 (2%)		[illegible] (1%)
Min				2 (2%)	1 (1%)			1 (2%)				[illegible] (1%)
TC		1 (2%)									1 (2%)	[illegible] (1%)

C19b Ware classes, **600/550**, farmhouses.

Site	**270S**	**397S**	**334S**	**477S**
Best date	557 BC	539 BC	535 BC	528 BC
Total Fragments	193	425	209	315
Drinking	37 (26%)	95 (27%)	39 (27%)	67 (29%)
Drinking Service	13 (9%)	11 (3%)	5 (3%)	7 (3%)
Storage-Short	33 (23%)	51 (15%)	14 (10%)	35 (15%)
Storage-Long	1 (1%)	1 percent?	—	—
Food Service	7 (5%)	34 (10%)	13 (9%)	38 (17%)
Food Prep.	3 (2%)	10 (3%)	2 (2%)	5 (2%)
Food Prep./Cooking	15 (11%)	11 (3%)	13 (9%)	14 (6%)
Transport/Storage (Amphorae)	3 (2%)	3 (1%)	—	1
Service	10 (7%)	32 (9%)	25 (17%)	12 (5%)
Personal	3 (2%)	3 (1%)	4 (3%)	4 (2%)
Votive/Ritual	[12%]	[28%]	[20%]	[21%]
Miniatures	14 (10%)	85 (25%)	24 (17%)	10 (4%)
TC figurines	3 (2%)	10 (3%)	4 (3%)	38 (17%)
Other vessels				
Unknown	50 (26%)	77 (18%)	66 (32%)	83 (26%)
Known	143	348	143	232

C20 Functional groups, **600/550,** sanctuaries.

Functional Group	**Farmhouse average, 600/550**	**SAV, House on Hill**
Drinking	25%	21%
Drinking Service	5%	<1%
Short Storage	24%	7%
Long Storage	2%	<1%
Food Service	13%	1%
Food Preparation	3%	4%
Food Prep./Cooking	14%	54%
Transport/Storage	4%	4%
Service	5%	2%
Lamp	0	0
Personal	3%	1%
Ritual/Votive	2%	0
Miniatures	(1%)	0
Terracottas	(1%)	0
Ritual Vessels	(<1%)	0

C21 Functional groups, surveyed farmhouses average and excavated farmhouse SAV House on Hill.

	SAV Hill	FF	SAV	FS	SAG
Total Fragments	**677**	**4345**	**919**	**9413**	**3994**
Greek FW	<1%	<1%	2%	1%	1%
Roman FW	0	0	0	0	1%
BG	25%	18%	22%	24%	14%
GW	0	0	0	<1%	19%
BW	4%	7%	18%	8%	4%
PW	46%	40%	26%	37%	35%
Ck	22%	27%	23%	23%	12%
Amp	2%	5%	9%	3%	14%
Min	0	1%	0	1%	<1%
TC	0	1%	0	2%	<1%

C22 Ware classes, **600/550–100**, excavated farmhouses.

Site	**515F/T**	**344F**	**335F**	**182F/T**	**175F**	**414F**	**68F**	**304F**	**305F**	**303F**	**302F**	**49F**	**257F**
Best date BC	438	450	433	453	433	437	433	445	468	450	446	450	426
Total fragments	78	85	30	210	31	42	49	113	53	39	43	30	41
Drinking	9 (16%)	16 (28%)	6 (28%)	14 (8%)	6 (30%)	12 (34%)	4 (9%)	14 (18%)	5 (11%)	5 (18%)	5 (18%)	4 (14%)	3 (11%)
Drink. service		2 (3%)		5 (3%)	1 (5%)	3 (9%)		1 (1%)	3 (7%)				
Storage-short	25 (43%)	10 (18%)	2 (10%)	44 (24%)	3 (15%)	9 (25%)	17 (38%)	16 (20%)	15 (34%)	5 (18%)	5 (18%)	5 (18%)	6 (21%)
Storage-long	1 (2%)			5 (3%)	1 (5%)		2 (4%)	2 (3%)			2 (7%)	4 (14%)	
Food Service	10 (17%)	12 (21%)	4 (19%)	61 (34%)	5 (25%)	3 (8%)	13 (29%)	22 (28%)	12 (27%)	11 (38%)	4 (14%)	11 (39%)	11 (39%)
Food prep.	2 (4%)	4 (7%)		11 (6%)		1 (3%)		4 (5%)	2 (5%)				
Food prep./cooking	6 (10%)	6 (11%)	7 (33%)	28 (15%)	2 (10%)	2 (5%)	5 (11%)	12 (15%)	7 (16%)	4 (14%)	9 (32%)	3 (11%)	6 (21%)
Transport/storage (amphorae)	1 (2%)	2 (3%)		3 (2%)	2 (10%)	1 (3%)	3 (7%)	3 (4%)				1 (4%)	
Service	1 (2%)	3 (5%)	2 (10%)	6 (3%)		2 (6%)		2 (3%)		1 (4%)	2 (7%)		1 (4%)
Personal		1 (2%)		1 (1%)		2 (6%)	1 (2%)	2 (3%)		1 (4%)			1 (4%)
Votive/ritual													
Miniatures		1 (2%)		1 (1%)						1 (4%)			
TC figurines	2 (4%)										1 (4%)		
Other vessel													
Unknown	21 (30%)	28 (33%)	9 (30%)	31 (15%)	11 (35%)	7 (20%)	4 (8%)	35 (31%)	9 (17%)	11 (28%)	15 (35%)	2 (7%)	13 (32%)
Known	57	57	21	179	20	35	45	78	44	28	28	28	28

C23a Functional groups, **450**, farmhouses.

Site	**515F/T**	**344F**	**335F**	**182F/T**	**175F**	**414F**	**68F**	**304F**	**305F**	**303F**	**302F**	**49F**	**257F**
Best date	438	450	433	453	433	437	433	445	468	450	446	450	426
Total fragments	78	85	30	210	31	42	49	113	53	39	43	30	41
AFW						2 (5%)	1 (2%)	1 (1%)					
BG	23 (30%)	35 (41%)	7 (23%)	43 (21%)	9 (29%)	21 (50%)	7 (14%)	47 (42%)	11 (21%)	8 (21%)	6 (14%)	6 (20%)	5 (12%)
GW												1 (3%)	
BW	10 (13%)	2 (2%)	2 (7%)	26 (12%)	1 (3%)	6 (14%)	2 (4%)	4 (4%)	4 (8%)	1 (3%)	2 (5%)	1 (3%)	2 (5%)
PW	36 (46%)	39 (36%)	14 (47%)	109 (52%)	17 (55%)	10 (24%)	31 (63%)	46 (41%)	31 (59%)	25 (64%)	25 (58%)	18 (60%)	28 (58%)
Ck	6 (8%)	6 (7%)	7 (23%)	28 (13%)	2 (7%)	2 (5%)	5 (10%)	12 (11%)	7 (13%)	4 (10%)	9 (21%)	3 (10%)	6 (15%)
Amp	1 (1%)	2 (2%)	3 (1%)	2 (7%)	1 (2%)	3 (6%)	3 (3%)					1 (3%)	
Min		1 (1%)	1 (1%)							1 (3%)			
TC	2 (3%)										1 (2%)		

C23b Ware classes, **450**, farmhouses.

Site	417F	451F	637F	730F/T	249F	460F	454F	43F	741F	341F	183F	490F	370F
Best Date	336	330	330	330	325	324	322	319	319	318	317	312	311
Total fragments	43	50	55	286	71	79	98	349	72	271	165	61	112
Drinking	7 (19%)	6 (15%)	9 (22%)	27 (21%)	3 (5%)	18 (24%)	21 (22%)	25 (10%)	8 (22%)	46 (26%)	19 (18%)	6 (13%)	18 (21%)
Drinking service				7 (5%)	5 (9%)	3 (4%)		3 (1%)		6 (4%)	2 (2%)	1 (2%)	3 (3%)
Storage-short	5 (14%	11 (27%)	15 (37%)	47 (36%)	18 (31%)	17 (23%)	21 (22%)	83 (32%)	13 (35%)	30 (17%)	15 (14%)	5 (13%)	23 (28%)
Storage-long		2 (5%)	1 (2%)		2 (3%)	5 (6%)	3 (3%)	6 (2%)			2 (2%)	3 (7%)	
Food service	9 (24%)	20 (48%)	10 (24%)	33 (25%)	9 (15%)	17 (23%)	14 (14%)	87 (33%)	10 (27%)	28 (16%)	28 (26%)	9 (20%)	24 (28%)
Food prep.				1 (1%)	4 (7%)	2 (3%)		6 (2%)	3 (8%)	2 (1%)	2 (2%)	1 (2%)	1 (1%)
Food prep./Cooking	7 (19%)	2 (5%)	4 (10%)	4 (3%)	11 (19%)	4 (5%)	20 (21%)	32 (12%)	2 (5%)	34 (19%)	22 (20%)	15 (35%)	10 (12%)
Transport/Storage (amphorae)				1 (1%)	2 (3%)	2 (3%)	4 (4%)	4 (2%)		1 (1%)	7 (6%)	1 (2%)	3 (3%)
Service	3 (8%)		2 (5%)	3 (2%)	2 (3%)	1 (1%)	8 (8%)	5 (2%)		18 (10%)	4 (4%)	2 (4%)	1 (1%)
Instr. domesticum					1 (2%)	2 (3%)				2 (1%)			
Personal	4 (11%)			3 (2%)	2 (3%)	4 (5%)	6 (6%)	9 (3%)	1 (3%)	9 (5%)	4 (4%)	1 (2%)	3 (3%)
Votive/Ritual													
Miniatures				1 (2%)				1 (0.5%)			1 (1%)		
TC figurines	2 (5%)							1 (0.5%)			1 (1%)		
Other vessels (louterion etc.)				(louterion) 1 (1%) (thymiaterion) 1 (1%)							(louterion) 1 (1%)		
Unknown	6 (14%)	9 (18%)	14 (25%)	156 (55%)	12 (20%)	4 (5%)	1 (1%)	87 (25%)	35 (49%)	95 (35%)	57 (35%)	15 (25%)	26 (23%)
Known	37	41	41	130	59	75	97	262	37	176	108	46	86

C24a Functional groups, **300**, farmhouses.

Site	417F	451F	637F	730F/T	249F	460F	454F	43F	741F	341F	183F	490F	370F
Best date	336	330	330	330	325	324	322	319	319	318	317	312	311
Total fragments	43	50	55	286	71	79	98	349	72	271	165	61	112
AFW				2 (1%)	3 (4%)	1 (1%)	1 (1%)	3 (1%)		1 (0%)	2 (1%)		1 (1%)
BG	15 (35%)	9 (18%)	16 (29%)	157 (55%)	14 (20%)	33 (42%)	38 (39%)	129 (37%)	20 (28%)	132 (49%)	54 (33%)	26 (43%)	42 (38%)
GW													
BW	7 (16%)	1 (2%)	6 (11%)	16 (6%)	4 (6%)	9 (11%)	6 (6%)	40 (12%)	3 (4%)	6 (2%)	16 (10%)	2 (3%)	4 (4%)
PW	12 (28%)	38 (76%)	29 (53%)	104 (36%)	37 (52%)	30 (38%)	29 (30%)	139 (40%)	47 (65%)	97 (36%)	62 (38%)	16 (26%)	52 (46%)
Ck	7 (16%)	2 (4%)	4 (7%)	4 (1%)	11 (16%)	4 (5%)	20 (20%)	32 (9%)	2 (3%)	34 (13%)	22 (13%)	16 (26%)	10 (9%)
Amp				1 (0%)	2 (3%)	2 (3%)	4 (4%)	4 (1%)		1 (0%)	7 (4%)	1 (2%)	3 (3%)
Min				2 (1%)				1 (0%)			1 (1%)		
TC	2 (5%)							1 (0%)			1 (1%)		

C24b Ware classes, **300**, farmhouses.

		FW	BG	GW	BW	PW	Ck	Amp	Min	TC
Excavated	FF	0%	18%	0	7%	40%	27%	5%	1%	1%
	SAV	2%	27%	0	18%	26%	23%	9%	0	0
	FS	1%	4%	<1%	8%	37%	23%	3%	<1%	2%
Surveyed	506F	0	39%	0	4%	26%	14%	1%	1%	16%
	169F	0	39%	0	10%	39%	11%	1%	1%	0
	128F	2%	56%	0	4%	24%	11%	2%	0	2%
	497F/T	5%	49%	0	6%	22%	15%	4%	0	0
	354F	0	58%	1%	2%	27%	12%	0	1%	1%
	417F	0	35%	0	16%	28%	16%	0	0	5%
	730F/T	1%	55%	0	6%	36%	1%	0	1%	0
	460F	1%	42%	0	11%	38%	5%	3%	0	0
	183F	1%	33%	0	10%	38%	13%	4%	1%	1%
	741F	0	28%	0	4%	65%	3%	0	0	0
	637F	0	29%	0	11%	53%	7%	0	0	0
	528F	0	49%	0	16%	28%	7%	0	0	0
	490F	0	43%	0	3%	26%	26%	2%	0	0
	454F	1%	39%	0	6%	30%	20%	4%	0	0
	415F	0	18%	0	2%	76%	4%	0	0	0
	418F	0	29%	0	0	48%	19%	5%	0	0
	370F	1%	38%	0	4%	46%	9%	3%	0	0
	341F	0	49%	0	2%	36%	13%	0	0	0
	249F	4%	20%	0	6%	52%	16%	3%	0	0
	121F	0	39%	0	5%	49%	5%	2%	0	0
	43F	1%	37%	0	12%	40%	9%	1%	0	0
	113F	0	38%	0	7%	37%	18%	0	0	0
	16F	0	46%	0	7%	39%	7%	0	0	2%
	4F	0	17%	0	0	67%	6%	11%	0	0

C25 Ware classes, **300**, excavated and surveyed farmhouses.

Localities (% of study area)	450	400	350	300
Early Clusters (22.6%)	5/10 (50%)	9/12 (75%)	5/15 (33%)	2/12 (17%)
Giardinetto (3.3&)	1/2 (50%)	3/4 (75%)	1/5 (20%)	1/4 (25%)
Sant'Angelo (3.7%)	1/2 (50%)	1/1 (100%)	0/0 (0%)	0/0 (0%)
Avinella (7.5%)	1/2 (50%)	3/4 (75%)	1/3 (33%)	1/3 (33%)
lower (4.7%)	1/2 (50%)	3/4 (75%)	0/3 (0%)	1/3 (33%)
upper (2.8%)	0/2 (0%)	0/4 (0%)	1/3 (33%)	0/3 (0%)
Venella (5.8%)	2/4 (50%)	2/3 (67%)	3/7 (43%)	0/5 (0%)
west (2.5%)	1/4 (25%)	1/3 (33%)	1/7 (14%)	0/5 (0%)
east (3.3%)	1/4 (25%)	1/3 (33%)	2/7 (29%)	0/5 (0%)
Giampasquale (2.3%)	0/0 (0%)	0/0 (0%)	0/0 (0%)	0/0 (0%)
Central Plateau (25.6%)	2/4 (50%)	0/2 (0%)	3/4 (75%)	2/4 (50%)
Lago Lupo (15.3%)	2/2 (100%)	0/0 (0%)	2/2 (100%)	1/1 (100%)
Cogno Prete (10.3%)	0/2 (0%)	0/2 (0%)	1/2 (50%)	1/3 (33%)
Far from City (26.6%)	2/4 (50%)	4/7 (57%)	0/5 (0%)	2/8 (25%)
Campagnolo (4.9%)	1/1 (100%)	0/0 (0%)	0/0 (0%)	0/1 (0%)
Bufalara (2.9%)	0/0 (0%)	0/0 (0%)	0/0 (0%)	0/0 (0%)
Scarciullo (9.0%)	0/0 (0%)	0/0 (0%)	0/0 (0%)	1/2 (50%)
San Marco (3.4%)	1/1 (100%)	0/1 (0%)	0/1 (0%)	0/0 (0%)
AVS (6.4%)	0/2 (0%)	4/6 (67%)	0/4 (0%)	1/5 (20%)
Close to City (25.3%)	3/3 (100%)	8/10 (80%)	8/17 (47%)	0/12 (0%)
Pizzica (5.8%)	2/2 (100%)	4/5 (80%)	4/7 (57%)	0/7 (0%)
Pantanello (7.4%)	1/1 (100%)	3/4 (75%)	4/8 (50%)	0/5 (0%)
DBSBS (12.1%)	0/0 (0%)	1/1 (100%)	0/2 (0%)	0/0 (0%)
Chora as a whole	**12/21 (57%)**	**21/31 (68%)**	**16/41 (39%)**	**6/36 (17%)**

C26 Ex novo: new necropoleis as a percentage of the total, **450–300**.

Nucleus	600	550	500	450	400	350	300	Total Burials
Crossroads necropolis								
Division line road L21								
N1	0	0	0	0	3	0	1	4
N2	0	0	1	2	1	1	0	5
N3	0	0	0	2	12	5	8	27
N4	0	0	0	5	1	6	5	17
N5	0	0	2	4	8	2	2	18
N6	0	0	1	1	2	1	32	37
Subtotal	0	0	4	14	27	15	48	108
Basento Road (west)								
N7	2	0	2	1	1	0	0	6
N11	1	0	2	12	10	2	8	35
N17	0	0	0	3	4	1	1	9
N10	0	0	3	5	9	0	1	18
N8	2	2	8	7	4	3	4	30
N9	0	0	0	10	10	11	2	33
N16	0	0	0	8	0	0	1	9
N18	2	0	0	1	6	0	1	10
Subtotal	7	2	15	47	44	17	18	150
Crossroads total	**7**	**2**	**19**	**61**	**71**	**32**	**66**	**258**
Eastern necropolis								
Basento Road (east)								
N12	0	0	0	0	2	3	1	6
N13	0	0	0	7	5	1	0	13
N15	0	1	0	2	6	5	2	16
N14	0	0	0	0	2	3	5	10
Subtotals	0	1	0	9	15	12	8	45
Basento Road total	**7**	**3**	**15**	**56**	**59**	**29**	**26**	**195**
Pantanello necropoleis total	**7**	**3**	**19**	**70**	**86**	**44**	**74**	**303**

C27 Chronology of burials at Pantanello Crossroads and Eastern necropoleis.

Site	482N	338N	299N	11N	298N	229N	407N	797N
Total fragments	70	18	61	15	17	8	2	4
Drinking	6 (12%)	3 (27%)	10 (21%)	1 (11%)	5 (56%)		2 (100%)	1 (25%)
Drinking service	1 (2%)							
Storage-short	16 (32%)	4 (37%)	14 (30%)	1 (11%)	2 (22%)	3 (43%)		1 (25%)
Storage-long		1 (9%)				1 (14%)		
Food service	11 (22%)	1 (9%)	15 (32%)	1 (11%)	1 (11%)			
Food prep.			2 (4%)					
Food prep./Cooking	6 (12%)	2 (18%)	6 (13%)	1 (11%)	1 (11%)	2 (29%)		
Transport/Storage (amphorae)				4 (45%)				
Service	10 (20%)							
Personal				1 (11%)		1 (14%)		2 (50%)
Votive/Ritual								
Miniatures								
TC figurines								
Other vessels								
Unknown	20 (28%)	7 (39%)	14 (23%)	6 (40%)	8 (47%)	1 (14%)	0 (0%)	0 (0%)
Known	50	11	47	9	9	7	2	4

C28 Functional groups, **450**, necropoleis.

Locality	Site no.	Cat. no.	Site type	Ware	Shape	Date	Attribution/comment
Giardinetto	524		N	BF			
	462	08	F	RF	type C skyphos	450–400	Metaponto
	465	09	F	RF	type C skyphos	400–350	
		11	F	RF	pelike	340–320	Late Apulian
	521	02	F/T/K	RF	kantharoid vase	375–350	Xenon group
		06	F/T/K	Gna	skyphos	350–300	
	457	01	N	RF	krater	350–300	
	471	04	N	RF	ps.-pa amph.	400–375	Lucanian (Dolon Painter?)
		12	N	RF	pelike	350–300	
		08	N	RF	hydria	4th c.	
		07	N	RF	ps.-pa amph.	350–300	misfired
		06	N	RF	pelike	330–300	
		06rev01	N	RF	pelike	330–300	
		03	N	RF	hydria	330–300	Late Apulian
		05	N	RF	calyx-krater	ca. 330	
	464	04	N	RF	volute-krater	340–320	
		02	N	Gna	bell-krater	325–300	
	535		F	RF	1 frag.		
	470		F	RF	1 frag.		
	466		N	RF	1 frag.		
	460		F	Gna	1 frag.		
	461		F	RF	1 frag.		
	469		F	RF	1 frag.		
Sant'Angelo	553	01	N		squat lekythos	340–300	Late Apulian
	240		N		1 frag.		
	557		N	RF	1 frag.		
	1		F	RF	1 frag.		
	13		F	RF	1 frag.		
	556		F	RF	1 frag.		
Avinella	401	08	N	RF	pelike	450–420	Lucanian (Pisticci Painter)
		12	N	RF	ps.-pa amph.	450–400	
		01	N	RF	hydria	425–390	Lucanian (Sisyphus Painter)
		16	N	RF	pelike	400–375	
		13	N	RF	pelike	400–350	
		02rev90	N	RF	pelike	400–350	Apulian
		15	N	RF	pelike	350–325	
		29	N	RF	volute-krater	350–325	Late Apulian
		24	N	RF	volute-krater	350–325	Late Apulian
		01	N	RF	volute-krater	350–325	
		02-03	N	RF	volute-krater	350–325	
	401*		N	RF	krater	350–300	*= Plot2001.046.09
	395	02rev01	F	RF	krater	450–400	Lucanian
		01	F	RF	calyx-krater	400–350	
		08rev90	F	RF	krater	420–400	Lucanian
		21	F	Gna	skyphos	340–330	
	300	01	N	RF	pelike	425–375	Lucanian
	394		F	Gna	skyphos	350–300	
		09	F	Gna	1 frag.		
	342	29	F	Gna	skyphos	end 4th c.	
	326	26	F	Gna	type C bell-krater	350–325	
	341	28	F	Gna	type A skyphos	end 4th c.	or beginning 3rd c.
	312	03	F	BFW	skyphos		Metaponto
		04	F	BFW	skyphos		Metaponto

C29 Figured vases from farmhouses, farmhouses with burials, and necropoleis, catalogued and uncatalogued.

Locality	Site no.	Cat. no.	Site type	Ware	Shape	Date	Attribution/comment
Giampasquale	200	01	F	RF	krater	425–400	Lucanian
	269	21	F	Gna	skyphos	3rd c.	
Demanio C.	176	01	N	RF	ps.-pa amph.	350–325	Middle Apulian
	184		N	RF	ps.-pa amph.	350–300	
	635	07	N	Gna	bombylios	300–250	
	635		N	RF	3 frags.	end 4th c.	
	454	02	F	RF	lekanis lid		
	626		N	RF	1 frag.		
	38		F	Gna	1 frag.		
	183		F	RF	2 frags.		
	278		F	RF	1 frag.		
	372		F	RF	1 frag.		
Lago del Lupo	503	10	F	RF	type C skyphos	375–325	Metaponto
			F	Gna	1 frag.		
	250	02	F	RF	ps.-pa amph.	350–300	Late Apulian
		01	F	Gna	type C bell-krater	350–325	
	497	12-20	F/T	RF	hydria	350–300	
	118	02*	F	RF	volute-krater	350–300	*Plot1999.118.02; Late Apulian
	542	04	F	RF	squat lekythos	320–310	
	152		F	Gna	1 frag.		
	172		F	RF	2 frags.		
	210		F	RF	1 frag.		
	248		F	RF	1 frag.		
	249		F	RF	2 frags.		
			F	Gna	1 frag.		
	296		F	Gna	1 frag.		
Cogno del Prete	265	1	N	BF	lekythos	500–480	
	57	01	F	RF	patera	375–300	Metaponto — Darius Painter
	43	137	F	RF	lekanis lid	350–300	
			F	Gna	1 frag.		
	266	45	S	XG	lekane?	4th c.	
	128	19	F	Gna	skyphos	3rd c.	
			F	RF	1 frag.		
	81		N	Gna	1 frag.		
	260		F	RF	1 frag.		
Campagnolo	50	155	N	BF	hydria/neck amph.	510–480	
	97		F	RF	1 frag.		
Bufalara	309	04	F	RF	squat lekythos	350–300	Late Apulian
	304		F	RF	1 frag.		
	488		F	RF	2 frags.		
			F	BFW	1 frag.		
	511		F	RF	1 frag.		
Scarciullo	365	22	F	RF	volute-krater	375–325	Lykourgos Painter
	103	13	F	RF	dish	340–320	Late Apulian
	375	06	F	Gna	lebes gamikos	300–250	
			F	RF	1 frag.		
	363		F/T	RF	1 frag.		
	352		F	Gna	1 frag.		
	353		F	RF	1 frag.		
	370		F	RF	1 frag.		

C29 *cont.* Figured vases from farmhouses, farmhouses with burials, and necropoleis, catalogued and uncatalogued.

Locality	Site no.	Cat. no.	Site type	Ware	Shape	Date	Attribution/Comment
San Marco	410	03	N	RF	type C skyphos	350–300	Metaponto
Basento V.	154	22	F	RF	pelike		
	727	01	F	Gna	type B bell-krater	370–320	Metaponto
	726	01*	F/T	RF	krater	350–300	*Plot1999.073
	159		N	RF	1 frag.		
	153		F	Gna	1 frag.		
Pizzica	728	B03	F/T	RF	volute-krater	410–350	
		01rev00	F/T	RF	krater	400–350	
		B01	F/T	RF	volute-krater	350–325	
		A01	F/T	RF	krater	350–300	
	734	01	F/T	RF	calyx-krater	400–375	Lucanian/Apulian
	764	02	N	RF	hydria	400–350	
	731		N	RF	1 frag.		
	736		N	RF	6 frags.		excavated (see Ch. 18)
	737		N	RF	1 frag.		excavated (see Ch. 18)
	739		N	RF	1 frag.		
	760		N	RF	1 frag.		
	763		N	RF	1 frag.		
Pizzica	730		F	RF	2 frags.		
	762		F	RF	1 frag.		
Pantanello	425	01	F/T	RF	calyx-krater		
	415	04	N	RF	krater		
		07	N	RF	ps.-pa amph.		
		03	N	RF	column-krater		
	413	02	F	RF	pelike		
			F	Gna	1 frag.		
	430		N	RF	1 frag.		
	437		N	RF	1 frag.		
	438		N	RF	1 frag.		
	439		N	RF	3 frags.		
			N	Gna	1 frag.		
	416		F	RF	1 frag.		
	414		F	RF	2 frags.		
	429		N	RF	1 frag.		

C29 *cont.* Figured vases from farmhouses, farmhouses with burials, and necropoleis, catalogued and uncatalogued.

Localities	Newly active farmhouses		Newly active necropoleis	
Villages	**14**		**6**	
Giardinetto	5	468, 516, 525, 528, 535	1	464
Sant'Angelo	1	155	0	
Avinella	6		1	
lower	2	404, 406	0	
upper	4	343, 345, 350, 506	1	346
Venella	2		4	
west	1	322	1	337
east	1	481	3	184,188, 635
Giampasquale	0		0	
Central Plateau	**17**		**2**	
Lago del Lupo	15	160, 164, 169, 172, 216, 221, 222, 248, 249, 275, 281, 296, 501, 517, 542,	1	541
Cogno del Prete	2	35, 58	1	495
Far from City	**10**		**0**	
Campagnolo	2	44, 47	0	
Bufalara	3	305, 489, 490	0	
Scarciullo	2	17, 370	0	
San Marco	1	84	0	
AVS	2	312, 556	0	
Close to City	**1**		**9**	
Pizzica	1	741	4	731, 740, 763, 764
Pantanello	0		4	421, 430, 437, 438
DBSBS	0		1	626

Localities	Abandoned Farmhouses		Abandoned Necropoleis	
Villages	**9**		**2**	
Giardinetto	3	462, 518, 519	0	
Sant'Angelo	2	213, 286	1	159
Avinella	1		1	
lower	1	509	0	
upper	0		1	347
Venella	3		0	
west	1	549	0	
east	2	183, 187	0	
Giampasquale	0		0	
Central Plateau	**4**		**1**	
Lago Lupo	2	131, 174	0	
Cogno Prete	2	68, 257	1	70
Far from City	**4**		**2**	
Campagnolo	0		0	
Bufalara	1	304	0	
Scarciullo	2	60, 369	0	
San Marco	0		0	
AVS	1	241	2	240, 482
Close to city	**2**		**3**	
Pizzica	1	744	3	732, 737, 739
Pantanello	0		0	
DBSBS	1	726	0	

C30 Transition tables, **400–350**.

Localities	Ex-novo farmhouses 350		Ex-novo necropoleis 350	
Villages	**9**		**5**	
Giardinetto	4	468, 516, 525, 528	1	457
Sant'Angelo	0		0	
Avinella	4		1	
lower	0		0	
upper	4	343, 345, 350, 506	1	346
Venella	1		3	
west	0		1	337
east	1	481	2	184, 188
Giampasquale	0		0	
Central Plateau	**14**		**2**	
Lago Lupo	13	164, 169, 172, 216, 221, 248, 249, 275, 281, 296, 501, 517, 542	1	541
Cogno Prete	1	35	1	495
Far from City	**8**		**0**	
Campagnolo	2	44, 47	0	
Bufalara	2	489, 490	0	
Scarciullo	2	17, 370	0	
San Marco	1	34	0	
AVS	1	556	0	
Close to City	**0**		**8**	
Pizzica	0		4	731, 740, 763, 764
Pantanello	0		4	421, 430, 437, 438
DBSBS	0		0	

Localities	Net gain/loss farmhouses		Net gain/loss necropoleis	
Villages	**+5**		**+4**	
Giardinetto	+2		+1	
Sant'Angelo	-1		-1	
Avinella	+5		0	
lower	+1		0	
upper	+4		0	
Venella	-1		+4	
west	0		+1	
east	-1		+3	
Giampasquale	0		0	
Central Plateau	**+13**		**+1**	
Lago Lupo	+13		+1	
Cogno Prete	0		0	
Far from City	**+6**		**-2**	
Campagnolo	+2		0	
Bufalara	+2		0	
Scarciullo	0		0	
San Marco	+1		0	
AVS	+1		-2	
Close to City	**-1**		**+6**	
Pizzica	0		+1	763, 764, 731, 740
Pantanello	0		+4	437, 438, 430, 421
DBSBS	-1		+1	

C30 *cont.* Transition tables, **400–350**.

Localities	Ex-novo farmhouses 350		Ex-novo necropoleis 350	
Villages	**9**		**5**	
Giardinetto	4	468, 516, 525, 528	1	457
Sant'Angelo	0		0	
Avinella	4		1	
lower	0		0	
upper	4	343, 345, 350, 506	1	346
Venella	1		3	
west	0		1	337
east	1	481	2	184, 188
Giampasquale	0		0	
Central Plateau	**14**		**2**	
Lago Lupo	13	164, 169, 172, 216, 221, 248, 249, 275, 281, 296, 501, 517, 542	1	541
Cogno Prete	1	35	1	495
Far from City	**8**		**0**	
Campagnolo	2	44, 47	0	
Bufalara	2	489, 490	0	
Scarciullo	2	17, 370	0	
San Marco	1	84	0	
AVS	1	556	0	
Close to City	**0**		**8**	
Pizzica	0		4	731, 740, 763, 764
Pantanello	0		4	421, 430, 437, 438
DBSBS	0		0	

Localities	Net gain/loss farmhouses		Net gain/loss necropoleis	
Villages	**+5**		**+4**	
Giardinetto	+2		+1	
Sant'Angelo	-1		-1	
Avinella	+5		0	
lower	+1		0	
upper	+4		0	
Venella	-1		+4	
west	0		+1	
east	-1		+3	
Giampasquale	0		0	
Central Plateau	**+13**		**+1**	
Lago Lupo	+13		+1	
Cogno Prete	0		0	
Far from City	**+6**		**-2**	
Campagnolo	+2		0	
Bufalara	+2		0	
Scarciullo	0		0	
San Marco	+1		0	
AVS	+1		-2	
Close to City	**-1**		**+6**	
Pizzica	0		+1	763, 764, 731, 740
Pantanello	0		+4	437, 438, 430, 421
DBSBS	-1		+1	

C31 Transition tables, **350–300**.

Localities	Newly active farmhouses		Newly active necropoleis	
Villages	**13**		**5**	
Giardinetto	6	456, 462, 469, 516, 518, 519	1	520
Sant'Angelo	3	4, 213, 156	0	
Avinella	1		2	
lower	1	402	1	533
upper	0		1	392
Venella	3		1	
west	0		1	338
east	3	360, 476, 633	0	
Giampasquale	0		0	
Central Plateau	**9**		**2**	
Lago Lupo	7	121, 122, 125, 152, 170, 210, 291	1	294
Cogno Prete	2	37, 68	1	56
Far from City	**14**		**5**	
Campagnolo	2	46, 52	1	48
Bufalara	2	309, 487	0	
Scarciullo	5	15, 16, 369, 373, 376	2	374, 100
San Marco	1	83	0	
AVS	4	241, 310, 494, 558	2	432, 553
Close to City	**7**		**2**	
Pizzica	0		2	732, 739
Pantanello	3	417, 445, 451	0	
DBSBS	4	38, 42, 197, 727	0	

Localities	Abandoned Farmhouses		Abandoned Necropoleis	
Villages	**4**		**7**	
Giardinetto	2	468, 535	2	466, 791
Sant'Angelo	1	155	0	
Avinella	0		2	
lower	0		1	396
upper	0		1	346
Venella	1		3	
west	0		2	337, 548
east	1	481	1	188
Giampasquale	0		0	
Central Plateau	**3**		**2**	
Lago Lupo	3	275, 281, 517	2	112, 541
Cogno Prete	0		0	
Far from City	**3**		**2**	
Campagnolo	0		0	
Bufalara	1	305	0	
Scarciullo	0		0	
San Marco	0		1	410
AVS	2	491, 556	1	318
Close to City	**0**		**7**	
Pizzica	0		2	760, 763
Pantanello	0		3	421, 430, 438
DBSBS	0		2	41, 626

C31 *cont.* Transition tables, **350–300**.

Site	446N	184N	437N	731N	81N	533N	520N
Best date	324	324	314	313	306	300	290
Total fragments	51	33	50	41	62	15	30
Drinking	3 (10%)	4 (27%)	9 (25%)	2 (13%)	11 (22%)	1 (10%)	4 (17%)
Drinking service	11 (35%)	2 (13%)	2 (6%)	3 (19%)	1 (2%)		2 (9%)
Storage-short	7 (23%)	1 (6%)	11 (31%)	5 (31%)	13 (26%)	3 (30%)	
Storage-long			1 (3%)	1 (6%)	3 (7%)		8 (35%)
Food service	6 (19%)	4 (27%)	8 (23%)	2 (13%)	12 (24%)	1 (10%)	1 (4%)
Food prep.							
Food prep./Cooking	3 (10%)	4 (27%)	2 (6%)	1 (6%)	5 (10%)	3 (30%)	5 (22%)
Transport/Storage amphorae	1 (3%)			2 (13%)		1 (10%)	
Service					1 (2%)		2 (9%)
Instr. domesticum					1 (2%)		
Personal			1 (3%)		3 (6%)	1 (10%)	
Ritual/Votive			1 (3%)				1 (4%)
Unknown	20 (39%)	8 (55%)	15 (30%)	25 (61%)	12 (19%)	5 (33%)	7 (23%)
Known	31	15	35	16	50	10	23

C32 Functional groups, **300**, necropoleis.

Functional Group	Form	Type
Drinking		
Drink Service		
Short-term Storage	Dinos	Dn2a, Dn5d
Short-term Storage	Jar	Ja1a, Ja1b
Short-term Storage	Jug	J16a, J18+, J14, J15, J19a
Long-term Storage	Deep Basin(/Basin)	DBas/Bas 7, DBas 4
Long-term Storage	Pithos	Pi1a, Pi7a, Pi8b2
Long-term Storage		Pi3a/b
Food Service	Dish	D22, D22a+, D24/27
Food Service	Bowl	B5, B2, B3, B25b+, B26a, B26b++
Food Service	Flanged Lekane	FLne2, FLne1
Food Service	Lekane/Basin	Lne/Bas1
Food Service	Lekane	Lne5a, Lne12/13a
Food Service	Plate/Lekane	Pl/Lne4d2
Food Service	Saltcellar	SA3a+++
Food Service	Small Bowl	SB6++, SB1+
Food Preparation	Mortar	Mor13a+, (Mor5), Mor12b1
Food Preparation	Deep Basin/Basin	DBas/Bas8a
Food Preparation	Basin	Bas1b
Cooking	Casserole	CC1a+
Cooking	Casserole	CG1a++, CG2a, CG3+
Cooking	Pot/Jar	P1a++, P1b, P2
Cooking	Pan	PD1a, PD4
Cooking	Necked Pot/Jar	PJ1+++, PJb
Cooking	Lid	Li1+++
Cooking	Base Pressed Disk	BD3+
Service	Cup	CO6a, CO1

C33a Ceramic forms and types,**300**.

Functional Group	Form	Type
Drinking	Kylix+	
Drink Service		
Short-term Storage	Dinos	Dn6a
Short-term Storage	Jug	J1d2, J1e, J7b, J8, J17, J21a
Long-term Storage	Pithos	Pi4a
Food Service	Flanged Lekane	FLne1
Food Service	Lekane/Basin	Lne/Bas2+, Lne/Bas1, Lne/Bas3a
Food Service	Lekane	Lne1/4b, Lne2b, Lne8a+, Lne9b+, Lne9/10, Lne2b, Lne6b
Food Service	Carinated Lekane	CLne2b, CLne3a
Food Preparation	(Deep Basin/)Basin	Bas2, DBas/Bas8c
Food Preparation	Mortar	Mor5a, Mor5b, Mor7a, Mor7b, Mor12c
Cooking	Necked Pot/Jar	PJ2, PJ3+, PJ2/PJ3
Cooking	Casserole	CC1e+
Cooking	Pan	PD5
Personal	Lebes Gamikos	LG1b

C33b Ceramic forms and types, **450**.

	300							450						
Site no.	**169F**	**354F**	**121F**	**16F**	**490F**	**249F**	**Total**	**182F**	**344F**	**302F**	**175F**	**14F**	**304F**	**Total**
Total identifiable fragments	**117**	**106**	**52**	**26**	**46**	**59**		**179**	**57**	**28**	**20**	**31**	**78**	
Drinking	X	X	X	X	X	X	6/6	X	X	X	X	X	X	6/6
Cup (C)	X	X	X	X	X	X	6/6	X	X	X	X	X	X	6/6
Skyphos (Sk)	X	X	X	X	X	X	6/6	X	X	X	X		X	5/6
Cup-skyphos (CSk)	X			X			2/6			X			X	2/6
Drink/Service	X	X	X		X	X	5/6	X	X		X	X		4/6
Hydria (H)		X					1/6	X	X					2/6
Amphora (Amp)						X	1/6	X						1/6
Pelike (P)						X	1/6	X						1/6
Krater (Kr)	X		X				2/6						X	1/6
Short-Storage	X	X	X	X	X	X	6/6	X	X	X	X	X	X	6/6
Jug (J)	X	X	X	X	X	X	6/6	X	X	X		X		4/6
J4a	X						1/6	X						1/6
J5a	X						1/6	X						1/6
J5b		X		X			2/6						X	1/6
J7	X						1/6	X						1/6
J12				X			1/6	X				X		2/6
J14a	X	X	X				3/6	X		X	X			3/6
J14b				X			1/6	X						1/6
Closed (Cl) Cl2	X		X			X	3/6				X	X		2/6
Dinos (Dn)	X					X	2/6	X					X	2/6
Dn6c	X						1/6	X						1/6
Long-Storage	X	X			X	X	4/6	X		X	X		X	4/6
Basin (Bas)		X					1/6			X				1/6
DBas1		X					1/6			X				1/6
Pithos (Pi)	X	X			X	X	4/6	X		X	X		X	4/6
Pi1b		X					1/6	X						1/6
Food Service	X	X	X	X	X	X	6/6	X	X	X	X	X	X	6/6
Lekane (Lne)	X	X	X	X	X	X	6/6	X	X	X	X	X	X	6/6
Carkinated Lekane (CLne)	X		X				2/6	X	X			X	X	4/6
CLne6a	X						1/6	X						1/6
CLne7			X				1/6		X					1/6
Lekane Basin (Lne/Bas)			X				1/6	X				X	X	3/6
Lne/Bas1b1			X				1/6	X				X		2/6
Plate Lekane (Pl/Lne)		X	X				2/6		X	X			X	3/6
Pl/Lne4b3		X	X				2/6		X	X			X	3/6
Lne4b	X		X				2/6					X		1/6
Lne4c	X						1/6	X						1/6

C34 Ceramic forms and types, **300**, **450**.

	300							450						
Site no.	**169F**	**354F**	**121F**	**16F**	**490F**	**249F**	**Total**	**182F**	**344F**	**302F**	**175F**	**14F**	**304F**	**Total**
Total identifiable fragments	**117**	**106**	**52**	**26**	**46**	**59**		**179**	**57**	**28**	**20**	**31**	**78**	
Lne5	X						1/6				X			1/6
Lne9a	X		X		X		3/6	X					X	2/6
Lne10			X				1/6	X	X					2/6
Lne11	X				X		2/6	X				X	X	3/6
Lne12	X			X		X	3/6	X						1/6
Lne12/13	X						1/6	X						1/6
Lne13	X				X	X	3/6	X						1/6
Lne13a	X						1/6	X			X			2/6
Plate (Pl)	X						1/6		X				X	2/6
Pl1	X					X	2/6		X				X	2/6
Small Bowl (SB)	X	X	X				3/6		X			X		2/6
Bowl (B)					X	X	2/6	X						1/6
Food Prep	X	X			X	X	4/6	X	X				X	3/6
Basin (Bas)		X			X	X	3/6	X	X				X	3/6
Bas7b						X	1/6	X					X	2/6
Lekane (Lne)						X	1/6		X					1/6
Mortar (Mor)	X	X				X	3/6	X					X	2/6
Food Prep/Cook	X	X	X	X	X	X	6/6	X	X	X	X	X	X	6/6
BD2/BD3	X						1/6	X						1/6
Transport/Storage	X		X		X	X	4/6	X	X		X		X	4/6
Instrumentum Domesticum (Lamp)		X	X			X	3/6							0/6
Service	X	X			X	X	4/6	X	X	X	X		X	5/6
One-handler (O)		X			X	X	3/6	X	X	X			X	4/6
Small Bowl/One-handler (SB/O)						X	1/6		X	X	X			3/6
SB/O2						X	1/6				X			1/6
Personal	X	X		X	X	X	5/6	X	X				X	3/6
Lebes Gamikos (LG)	X	X				X	3/6	X					X	2/6
Lekythos (L)	X	X					2/6						X	1/6
Unguentarium (U)				X	X		2/6		X					1/6
Votive/Ritual	X	X		X			3/6	X	X			X		3/6
Louterion (Lou)	X						1/6					X		1/6
Lou2	X						1/6					X		1/6
TC Figurine		X		X			2/6		X					1/6
Miniature (Min)	X	X					2/6	X	X					2/6

C34 *cont.* Ceramic forms and types, **300**, **450**.

Date/ Site no.	Pot/ Jug	Jug/ Jar/ Pot	Jar/Jug	Pot	Casserole	Lopas	Casserole/ Lopas	Base Pressed Disk	Round/ Flat Base	Lid	Lid/ Plate	Dish/ Pan
600/550												
285F/T		PJ3(2)			CC1a		CL1					
530F		PJ3(5)										
323F*		PJ3(2)										
336F			J4		CG2a							
286F		PJ3(4)	J4				CL3					
155F								BD2/ BD3(2)				
450												
182F/T*		PJ2, PJ3(2)					CL1, CL2(5)	BD2/ BD3(2)				
49F*				P1a			CL2/CG1b					
302F*		PJ3/PJ4					CL1					
301F			J4		CC1d		CL1					PD2
322F*	PJ1	PJ3			CG1a, CC1c					Li1		
400												
304F					CC1e(2)					Li1		PD5
519F			J4		CG2a			BD2				
549F*		PJ2					CL1					
350												
329F		PJ3(2)		P1c, P2(2)	CC1b, CG3		Kl?	BD2/BD3		Li1		PD2
326F	PJ1(2)		J4	P1a, P1b	CC1a(2), CG2a		Kl?			Li2		
300												
358F	PJ1	PJ2	J1	P2	CC1c, CC1b(3)			BD2/BD3	BR	Li1	DC1(2)	
454F*			J4	P1a(3), P1b, P2(4)	CG2a, CG3			BD2(3)		Li1 Li1a		
183F				P1(a), P2	CG3			BD		Li1a Li1		PD1a, PD3
394F		PJ3		P1a, P1b, P2, P3	CC1c				BF	Li1a		
490F	PJ1			P1b	CC1a, CG2a(2), CG3					Li1(2)		
506F*	PJ1		Roman?	P1a	CG1a/cover, CG3			BD2/ BD3(2)	BF	Li	DC1	
167F*	PJ1	PJ2(2), PJ3(2)	J4	P1d, P2(2)	CC1a, CC1b			BD2/BD3	BR(2), BF(2)	Li1		
498F				P1a(3), P1b, P2	CC1a(2), CC1b(2), CG2a			BD2	BR			PD1a
113F	PJ1	PJ2	Ja3, J4	P1a(4), P1b, P2(2), P2a, P3, P4	CG3(4)			BD2/ BD3(3)	BR	Li1		
342F	PJ1(2)			P1a(2), P1b, P1c, P1a/ P1c, P2	CC1a, CC1b(2), CC1c(2), CG1a, CG3(4)						DC1	PD2

C35 Cooking ware forms, **600–100**, farmhouses. (*Indicates a site closely dated to the period by black-gloss pottery.)

Date Site No.	Pot/ Jug	Jug/ Jar/ Pot	Jar/Jug	Pot	Casserole	Lopas	Casserole/ Lopas	Base Pressed Disk	Round/ Flat Base	Lid	Lid/ Plate	Dish/ Pan
296F*			Ja1a	P1d, P2	CC1, CC1a, CG2a(3), CG2a, CG3		CL1	BD	BF			
501F			J1	P1a, P2(3)	CG1a, CG2a			BD2(2)				PD2
354F*	PJ1			P1a(2), P2(3)	CC1a, CG1a, CG3			BD3		Li1(2)		PD4(2), PD1a
404F				P1a, P2(3)	CC1a, CC1b, CG2, CG3			BD2, BD3(2)	BF	Li1b		PD2, PD3(2)
250												
104F				P1a	CG1b				BF	Li1		PD3
306F	PJ1				CC1b							
745F					CG1a, CG2a							
100												
317F			Ja1a, J2a									
105F			votive						BF		DC1	

C35 *cont.* Cooking ware forms, **600–100**, farmhouses. (*Indicates a site closely dated to the period by black-gloss pottery.)

	600/550		450	300		100	
	Survey	**SAV, House on Hill**	**Survey**	**Survey**	**FF, SAV, FS**	**Survey**	**SAG**
AFW	<1%	<1%	<1%	<1%	1%	0	3%
BG	31%	25%	26%	39%	22%	5%	0
GW	0	0	<1%	<1%	<1%	54%	19%
BW	8%	4%	6%	6%	11%	4%	4%
PW	49%	46%	53%	40%	34%	24%	35%
Ck	8%	22%	12%	11%	25%	9%	12%
Amp	3%	2%	2%	2%	6%	4%	27%
Min	<1%	0	<1%	<1%	<1%	<1%	0
TC	<1%	0	<1%	1%	<1%	0	0

C36 Ware class percentage averages, **600–100**, surveyed and excavated farmhouses.

Functional groups	Form	Type	No.
Oil vessels	Unguentarium	U2	1
		U3b	1
Covered containers	Lebes gamikos	rb	1
Drinking vessels	Cup	Unknown	2
		Op3	1
	Skyphos	C-S20/28	2
		C-S67	2
		Unknown	1
		C-S65	1
		PSk1-4	1
		A-S72	1
Large containers (liquids)	Hydria	H1b	1
Dishes	Dish	Unknown	1
		D24/27	1
		D25a	1
	Plate	Pl1	1
Other vessels, objects	Lamp	Unknown	1
	Saltcellar	SA3a	1
	Small bowl/one-handler	SB/)2	1
Tomb markers	Large vessel	Unknown	1
	Pithos	Unknown	3
Rites at tombs	Dinos	Unknown	1
	Jug	Cl2	3
		J18	1
		Unknown	4
		Vertical	2
		H3a	1
	Lekane	Lne	1
		Lne12	2
		Unknown	1
		Pl/Lne4b2	1
		Lne4b	1
		CLne3a	1
	Unknown	Unknown	5
Total			**50**

C37 Ceramic vessels from site 81N, form, type, no.

Locality	Pre-250	250	200	Ex-novo 150	150	Ex-novo 100	100	50	Post-50
Early Village Clusters									
Giardinetto	515F				2.49				
Sant'Angelo	1F				1.72		2.15	1.75	Early Imperial
	105F/K	2.06			2.39		3.03	2.58	Early Imperial
Avinella				339F	2.31		3.41	2.10	
	395F	2.05	2.02		1.74				Early Imperial
	404F	1.88	1.70		1.57				
Venella	327F	1.45	1.54		1.46				
	372F				2.12		2.69	2.31	Imperial, Late Antique, Medieval
	476F						1.89		
Central Plateau									
Lago Lupo				502F	2.97		3.70	2.47	
Cogno Prete	128F	2.26	2.13		2.02				
	258F				2.09				
Far from City									
Campagnolo	47F						2.03		
San Marco	84F	1.12	1.17				1.12		Medieval
	230F						2.58		Imperial, Late Antique
				231F	1.86		2.94	2.11	Imperial, Late Antique, Medieval
AVS (Avinella)				317F	1.60		2.41	1.36	
(Sant'Angelo)	556F				2.31		3.30	2.19	
(Sant'Angelo)	558F						1.43		
						559F	1.78		
Close to City									
Pantanello	416F				1.95		2.40	1.99	
						428F	2.41		Early Imperial, Imperial, Late Antique
DBSBS (Demanio)	42F				3.25		4.17	3.02	
(Demanio)				95F/K	3.10		4.02	2.87	Early Imperial
(Serra Marina)	197F				2.52		2.82	2.25	
(Serra Marina)				198F	1.24		2.28	1.26	

C38 FMCE values, **150–50**, farmhouses.

	Multi-period farmhouses (adjusted*)					Ex-novo farmhouses					Average
Site no.	**42F**	**556F**	**372F**	**105F**	**428F**	**502F**	**95F**	**317F**	**339F**	**231F**	
Best date	104	103	98	98	100	106	103	104	103	97	
Total fragments	32	43	38	23	10	37	18	17	23	10	25
Drinking	2 (11%)	2 (13%)	3 (18%)	2 (18%)	0	3 (16%)	1 (17%)	1 (13%)	2 (25%)	0	13%
Drinking service	0	0	0	5 (46%)	0	3 (16%)	0	0	0	0	6%
Storage-short	1 (5%)	6 (37%)	5 (29%)	1 (9%)	4 (57%)	3 (16%)	0	2 (25%)	1 (13%)	0	19%
Storage-long	0	1 (6%)	0	0	0	0	3 (50%)	0	0	1 (17%)	7%
Food service	6 (34%)	5 (31%)	4 (23%)	1 (9%)	2 (29%)	6 (32%)	0	0	0	2 (33%)	19%
Food prep.	0	0	0	0	0	0	1 (17%)	0	0	2 (33%)	5%
Food prep./Cooking	2 (11%)	2 (13%)	3 (18%)	2 (18%)	1 (14%)	2 (11%)	1 (17%)	5 (63%)	2 (25%)	0	19%
Transport/Storage	6 (34%)	0	1 (6%)	0	0	0	0	0	2 (25%)	1 (17%)	8%
Service	0	0	1 (6%)	0	0	0	0	0	0	0	<1%
Lamps	0	1 (5%)	0	0	0	0	0	0	0	0	<1%
Personal	0	0	0	0	0	3 (16%)	0	0	1 (13%)	0	3%
Votive/Ritual											0
Miniatures											
Known Forms	17 (53%)	16 (37%)	17 (45%)	11 (48%)	7 (70%)	19 (51%)	6 (33%)	8 (44%)	8 (35%)	6 (60%)	

C39a Functional groups, **100**, multi-period and ex novo farmhouses.

	Multi-period farmhouses (adjusted*)					Ex-novo farmhouses					Average
Site no.	**372F**	**556F**	**105F**	**42F**	**428F**	**95F/K**	**317F**	**339F**	**231F**	**502F**	
Best date	98	103	98	104	100	103	104	103	97	106	
Total fragments	38	43	23	32	10	18	17	23	10	37	25
BG	0	0	0	2 (6%)	0	2 (11%)	0	4 (17%)	0	5 (14%)	5%
GW	19 (50%)	29 (67%)	19 (82%)	17 (53%)	3 (30%)	10 (55%)	6 (35%)	12 (52%)	6 (60%)	24 (65%)	54%
BW	3 (8%)	3 (7%)	0	1 (3%)	1 (10%)	0	0	1 (4%)	0	2 (5%)	4%
PW	12 (31%)	9 (21%)	2 (9%)	4 (13%)	5 (50%)	5 (28%)	6 (35%)	2 (9%)	3 (30%)	4 ([illegible]%)	24%
Ck	3 (8%)	2 (5%)	2 (9%)	2 (6%)	1 (10%)	1 (5%)	5 (30%)	2 (9%)	0	2 (5%)	9%
Amp	1 (3%)	0	0	6 (19%)	0	0	0	2 (9%)	1 (10%)	0	4%
Min	0	0	0	0	0	0	0	0	0	0	<1%

C39b Ware classes, **100**, multi-period farmhouses and ex novo farmhouses.

Appendix D
Exploratory Statistical Analyses

Alberto Prieto

Several methods of quantitative analysis were employed to extract additional information about historical patterns from the large mass of data obtained from the Metaponto field survey.[1] The methods selected were strictly exploratory in nature. All data manipulation was performed with Microsoft Excel, and all statistical analyses were performed with v.18 of IBM SPSS Statistics (formerly PASW).

Mixed Model Analyses

The Metaponto survey Final Multiple Criteria Evaluation (FMCE), which constitutes a single numerical summary of several ratio- and ordinal-scale variables associated with the individual colonial farmhouse scatters, is the primary metric for determining the relative importance of individual farmhouses, both across historical periods and within a single period. A set of frequency histograms of the FMCE values in the 50-year periods (*Fig. D1*) reveals that the spread, shape, and position of the distributions appear to vary considerably from period to period.

A mixed model analysis was performed initially to test for differences between farmhouse locations and time periods in their FMCE values. The analyses were performed on the 207 dated farmhouses, using all 647 FMCE values calculated for the standard 12 50-year datebins from **600** to **50** BC. The FMCE value was the dependent variable, while the broad locality category of the site (Locality[2]) and the 50-year

[1] The author wishes to thank the Division of Statistics and Scientific Computation of the University's College of Natural Sciences for the invaluable help and guidance provided throughout the analytical process.

[2] Early Village Clusters, Central Plateau, Far from City, and Close to City.

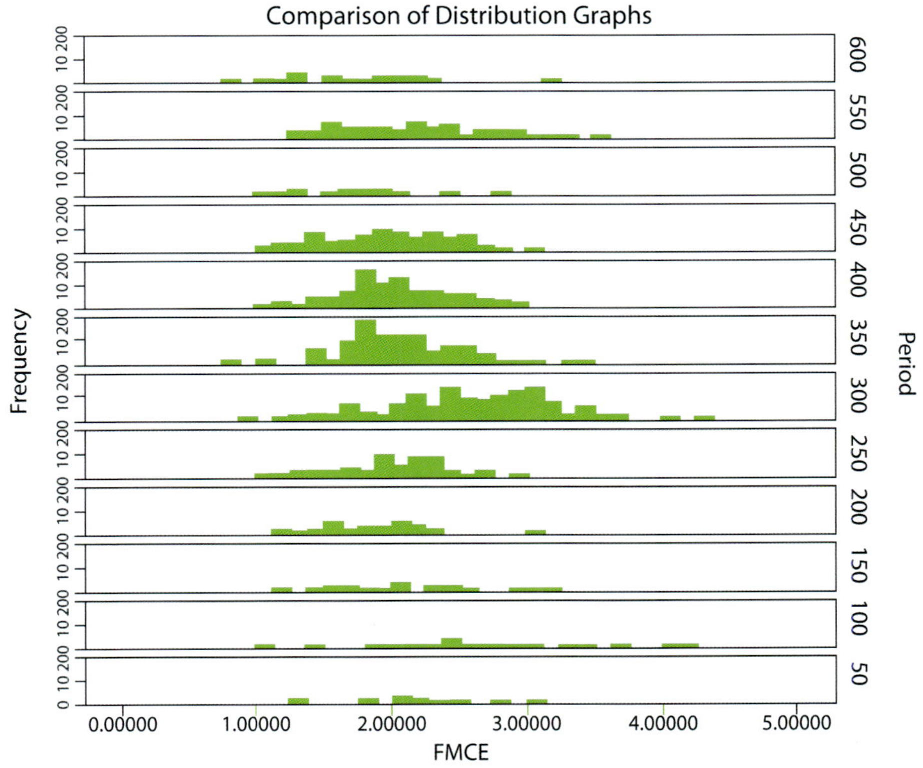

Figure D1 Frequency histograms of FMCE values by 50-year periods, **600–50**.

Source	Numerator df	Denominator df	F	p
Intercept	1	301.438	2685.309	0.000
Locality	3	290.542	2.147	0.094
Period	11	468.842	20.976	0.000
Locality * Period	33	451.347	1.043	0.405

Table D2 Type III tests of fixed effects in mixed model analyses on FMCE.

Test No.	F	df1	df2	p
1: 600 & 450	1.112	1	93	0.294
2: 550 & 500	0.815	1	76	0.369
3: 550 & 450	1.882	1	138	0.172
4: 550 & 300	1.000	1	204	0.319
5: 550 & 100	5.187	1	82	0.025
6: 500 & 450	0.041	1	88	0.841
7: 450 & 300	6.338	1	216	0.013
8: 450 & 100	11.685	1	94	0.001
9: 300 & 100	2.966	1	160	0.087

Table D3 Levene's Tests of Equality of Error variances.

period of the FMCE value (Period) were the independent variables. The results, presented in Table D2, demonstrate that the differences in the FMCE values between periods are significant ($p < 0.001$), while the differences in FMCE values between broad locality categories (which approximately represent varying distances from the urban center) are not significant ($p = 0.094$). Therefore a site's significance appears to have been related more closely to the period(s) of its occupation and use than to its geographic location.

One-Way Analysis of Variance (ANOVA)

The histograms above show different dispersions of FMCE values in different periods. In order to determine which periods differ in their range of FMCE values, the Levene's statistic, which tests for homogeneity among the variances, was calculated as part of the ANOVA of FMCE values by chronological period. The test was run on pairs of 50-year periods that saw large changes in the quantity of active farmhouses:

Table D3 reports that significant results ($p \leq 0.05$) were obtained for test nos. 5 (periods 550 vs. 100), 7 (periods 450 vs. 300), and 8 (periods 450 vs. 100), indicating that the spread of FMCE values in 100 was significantly larger than in 450 and 550, and that the spread of FMCE values in 300 was significantly larger than in 450.

Given the large number of tests performed using a small set of data categories (see below), the Bonferroni adjustment[3] was calculated as a more restrictive indicator of significance:

$$\frac{\alpha}{n} = \frac{0.05}{18} = 0.003$$

where α is the desired level of significance and n is the number of tests performed. Consequently, in order to be considered statistically significant, any given result (p) had to be less than or equal to 0.003, rather than the nominal 0.05. The only Levene's result that passes this test is no. 8, so that the only significant difference is between periods 450 and 100.

Chi-Square Tabulations

A total of 18 chi-square tests were performed using the following categories of data pertaining to all types of sites:

- Chronological period (century or half-century)
- FMCE value (farmhouses only)
- Ceramic functional group
- Ceramic ware class
- Site type

One of the 18 tests was subsequently found to be of doubtful utility in the research. Therefore in the following section only the 17 useful tests are described, with their results.

Chi-square 1

Research question: Is there a relationship between categories of FMCE values for dated farmhouses and categories of time periods?

Setup: The FMCE values were organized as three categories (*Table D4*)[4]; 11 of the 12 50-year chronological periods were organized as six categories comprising from one to three periods (the period 500 was excluded, leaving out 14 of the 647 FMCE values).

Result: There is a significant relationship between the FMCE categories and the chronological categories ($\alpha 2 = 133.484$, df = 10, $p < 0.001$). See Table D4.

Discussion: In the earliest periods (600, 550, and 450) the lowest FMCE values are much more numerous than expected, and in 450 the highest FMCE values are fewer than expected. In 400–350 the middle

3 Holland and DiPonzio Copenhaver 1988.

4 Although half of the FMCE value is derived from date-dependent criteria (see Ch. 4), the 647 FMCE values are technically not fully independent, since they belong to a pool of 207 farmhouses. The test was performed with the intention of serving simply as a broad indicator, with limited interpretive scope.

			FMCE Category			Total
			0.82–1.59	1.60–2.59	2.60–4.35	
Period	600–550	Count	21.0	48.0	14.0	83.0
		Expected count	12.7	53.8	16.5	83.0
		Std. residual	2.3	-0.8	-0.6	
	450	Count	19.0	52.0	5.0	76.0
		Expected count	11.6	49.2	15.1	76.0
		Std. residual	2.2	0.4	-2.6	
	400–350	Count	22.0	156.0	17.0	195.0
		Expected count	29.9	126.3	38.8	195.0
		Std. residual	-1.4	2.6	-3.5	
	300	Count	10.0	61.0	71.0	142.0
		Expected count	21.8	92.0	28.3	142.0
		Std. residual	-2.5	-3.2	8.0	
	250–200	Count	18.0	62.0	5.0	85.0
		Expected count	13.0	55.1	16.9	85.0
		Std. residual	1.4	0.9	-2.9	
	150–50	Count	7.0	31	14	52
		Expected count	8.0	33.7	10.4	52.0
		Std. residual	-0.3	-0.5	1.1	
Totals		**Count**	**97.0**	**410.0**	**126.0**	**633.0**
		Expected count	**97.0**	**410.0**	**126.0**	**633.0**

Table D4 Tabulation of FMCE and chronological categories.

FMCE values are much more numerous than expected, while the highest values are significantly less numerous. In 300 the lowest and middle FMCE values are much fewer (the middle values more so), but the highest values are very significantly more numerous than expected. In 250–200 the highest values are once again much less numerous than expected.

Chi-square 2

Research question: Is there a relationship between the functional categories in the fineware (black-gloss and figured wares) ceramic repertoire and time periods?

Setup: The 2728 fineware fragments that could be dated and represented the six most common functional groups were organized as three functional categories combining related functions. The full suite of 12 50-year periods was employed as the chronological framework.

Result: There is a significant relationship between fineware functional categories and time periods ($\alpha 2$ = 586.919, df = 22, $p < 0.001$). See Table D5.

Discussion: Vessels related to drinking were generally more numerous than expected in the periods from 600 to 400 (significantly in 550 and 400), while vessels pertaining to service at the table were much fewer. By 300 this pattern was reversed, so that service vessels became much more numerous than drinking vessels. There were fewer lamps than expected in the earliest periods, becoming more numerous than expected by 450 and remaining so through approximately 200.

Chi-square 3

Research question: Is there a relationship between the ceramic functional categories, including all ceramic classes, and site type for farmhouses and necropoleis?

Setup: The 16,910 ceramic fragments of all ware classes that were recovered at farmhouses or necropoleis and could be identified by function were organized as seven functional categories representing either unique functions or combinations of related functions.

Result: There is a significant relationship between ceramic functional categories and site type for farm-

			Fine ware functional category			Total
			Drinking + Drinking Service	Food Service + Service	Lamp + Personal	
Period	600	Count	23.0	0.0	0.0	23.0
		Expected Count	17.4	4.2	1.4	23.0
		Std. Residual	1.4	-2.1	-1.2	
	550	Count	370.0	0	1.0	371.0
		Expected Count	279.9	68.4	22.7	371.0
		Std. Residual	5.4	-8.3	-4.6	
	500	Count	79.0	0.0	3.0	82.0
		Expected Count	61.9	15.1	5.0	82.0
		Std. Residual	2.2	-3.9	-0.9	
	450	Count	167.0	1.0	22.0	190.0
		Expected Count	143.3	35.0	11.6	190.0
		Std. Residual	2.0	-5.7	3.0	
	400	Count	305.0	6.0	14.0	325.0
		Expected Count	245.2	59.9	19.9	325.0
		Std. Residual	3.8	-7.0	-1.3	
	350	Count	286.0	40.0	36.0	362.0
		Expected Count	273.1	66.7	22.2	362.0
		Std. Residual	0.8	-3.3	2.9	
	300	Count	663.0	317.0	48.0	1028.0
		Expected Count	775.5	189.5	62.9	1028.0
		Std. Residual	-4.0	9.3	-1.9	
	250	Count	66.0	67.0	20.0	153.0
		Expected Count	115.4	28.2	9.4	153.0
		Std. Residual	-4.6	7.3	3.5	
	200	Count	72.0	25.0	17.0	114.0
		Expected Count	86.0	21.0	7.0	114.0
		Std. Residual	-1.5	0.9	3.8	
	150	Count	10.0	26.0	1.0	37.0
		Expected Count	27.9	6.8	2.3	37.0
		Std. Residual	-3.4	7.3	-0.8	
	100	Count	15.0	16.0	3.0	34.0
		Expected Count	25.6	6.3	2.1	34.0
		Std. Residual	-2.1	3.9	0.6	
	50	Count	2.0	5.0	2.0	9.0
		Expected Count	6.8	1.7	.6	9.0
		Std. Residual	-1.8	2.6	2.0	
Totals		**Count**	**2058.0**	**503.0**	**167.0**	**2728.0**
		Expected Count	**2058.0**	**503.0**	**167.0**	**2728.0**

Table D5 Tabulation of fine ware functional categories and periods.

			Site type		Total
			Farmhouse	Necropolis	
Functional category (all types)	Drinking + Drinking Service	Count	3328.0	503.0	3831.0
		Expected count	3473.5	357.5	3831.0
		Std. residual	-2.5	7.7	
	Storage (long + short)	Count	4153.0	378.0	4531.0
		Expected count	4108.2	422.8	4531.0
		Std. residual	0.7	-2.2	
	Food Service	Count	3322.0	294.0	3616.0
		Expected count	3278.6	337.4	3616.0
		Std. residual	0.8	-2.4	
	Food Preparation + Cooking	Count	2683.0	214.0	2897.0
		Expected count	2626.7	270.3	2897.0
		Std. residual	1.1	-3.4	
	Transport + Storage	Count	442.0	69.0	511.0
		Expected count	463.3	47.7	511.0
		Std. residual	-1.0	3.1	
	Service + Personal	Count	1091.0	101.0	1192.0
		Expected count	1080.8	111.2	1192.0
		Std. residual	0.3	-1.0	
	Votive	Count	313.0	19.0	332.0
		Expected count	301.0	31.0	332.0
		Std. residual	0.7	-2.2	
Totals		**Count**	**15332.0**	**1578.0**	**16910.0**
		Expected count	**15332.0**	**1578.0**	**16910.0**

Table D6 Tabulation of functional categories of all ceramic ware classes for farmhouses and necropoleis.

houses and necropoleis ($\alpha 2$ = 106.327, df = 6, $p < 0.001$). See Table D6.

Discussion: Farmhouses show no significant pattern in the functional distribution, apart from many fewer than expected drinking-related vessels. The necropoleis are marked by several interesting features: significantly more than expected drinking-related and transport/storage vessels, and many fewer than expected vessels in all of the remaining categories except the service/personal.

Chi-square 4

Research question: Is there a relationship between the categories of fine ceramic ware classes from all types of sites, considered as a corpus, and time periods?

Setup. The 2728 fineware fragments used in Chi-square 2 (see above) were organized as two categories, (1) black gloss and (2) all figured wares. The full suite of 12 50-year periods was employed as the chronological framework.

Result: There is a significant relationship between the fineware categories and time periods ($\alpha 2$ = 259.601, df = 11, $p < 0.001$). See Table D7.

Discussion: The figured fine wares are fewer than expected, and often significantly so, in all periods except for 400 and 350; in this latter period the figured finewares are significantly more numerous than expected. The black-gloss pottery is represented consistently, apart from 350, when it is significantly less numerous than expected, probably a consequence of the greater circulation of figured wares at the same time.

Chi-square 5

Research question: Is there a relationship between the functional and ware class categories in the fineware ceramic repertoire?

Setup: The 2728 fine ware fragments used in Chi-square 2 (see above) were organized as two ware class categories, (1) black gloss and (2) all figured wares, us-

			Site type		Total
			Farmhouse	Necropolis	
Period	600	Count	0.0	23.0	23.0
		Expected Count	1.8	21.2	23.0
		Std. Residual	-1.3	0.4	
	550	Count	0.0	371.0	371.0
		Expected Count	28.4	342.6	371.0
		Std. Residual	-5.3	1.5	
	500	Count	1.0	81.0	82.0
		Expected Count	6.3	75.7	82.0
		Std. Residual	-2.1	0.6	
	450	Count	6.0	184.0	190.0
		Expected Count	14.6	175.4	190.0
		Std. Residual	-2.2	0.6	
	400	Count	37.0	288.0	325.0
		Expected Count	24.9	300.1	325.0
		Std. Residual	2.4	-0.7	
	350	Count	96.0	266.0	362.0
		Expected Count	27.7	334.3	362.0
		Std. Residual	13.0	-3.7	
	300	Count	68.0	960.0	1028.0
		Expected Count	78.8	949.2	1028.0
		Std. Residual	-1.2	0.3	
	250	Count	1.0	152.0	153.0
		Expected Count	11.7	141.3	153.0
		Std. Residual	-3.1	0.9	
	200	Count	0.0	114.0	114.0
		Expected Count	8.7	105.3	114.0
		Std. Residual	-3.0	0.9	
	150	Count	0.0	37.0	37.0
		Expected Count	2.8	34.2	37.0
		Std. Residual	-1.7	0.5	
	100	Count	0.0	34.0	34.0
		Expected Count	2.6	31.4	34.0
		Std. Residual	-1.6	0.5	
	50	Count	0.0	9.0	9.0
		Expected Count	0.7	8.3	9.0
		Std. Residual	-0.8	0.2	
Totals		Count	209.0	2519.0	2728.0
		Expected Count	209.0	2519.0	2728.0

Table D7 Tabulation of two fineware categories and period.

ing three functional categories combining related functions.

Result: There is a significant relationship between functional and ware class categories in the fine wares ($\alpha 2 = 46.319$, df = 2, $p < 0.001$). See Table D8.

Discussion: The black-gloss pottery shows no clear pattern. Figured drinking and personal vessels, as well as lamps, are more numerous than expected, while figured service vessels are significantly fewer than expected, numbering only 3 of 503 fragments.

Chi-square 6

Research question: Is there a relationship between the ceramic functional groups within farmhouse assemblages, including all ware classes, and time periods?

Setup: In order to test whole farmhouse assemblages, most of which consist primarily of plain ceramic artifacts that were created in different historical periods but cannot be dated with certainty, this and the following seven chi-square analyses (through Chi-square 13) depend on a fundamental assumption made purely for the sake of argument: in any given assemblage, all of the undated artifacts date to the same moment of occupation as the Best Date calculated for the site during the analysis of its fine ware artifacts.[5]

In this test, four periods were selected based on their perceived importance in the history of settlement, and each of these periods was populated with sites characterized by (1) a Best Date in the same period and (2) a predominance of fine ware artifacts dating to that period. The periods selected were 600–550,[6] 450,[7] 300,[8] and 100.[9] The total number of artifacts considered was 4107.

Result: There is a significant relationship between functional groups within farmhouse assemblages and time periods ($\alpha 2 = 191.963$, df = 33, $p < 0.001$). See Table D9.

Discussion: In 600-550 significantly more vessels than expected were used for drinking and

[5] See Dana in this volume.

[6] Sites 7, 12, 13, 155, 170, 181, 228, 285, 286, 323, 336, 349, 366, 459, 485, 492, 505, and 530.

[7] Sites 14, 49, 68, 175, 182, 257, 302–305, 335, 344, 414, and 515.

[8] Sites 4, 15, 16, 37, 38, 43, 46, 113, 121, 125, 128, 169, 183, 249, 341, 354, 370, 373, 417, 418, 445, 454, 456, 457, 460, 469, 487, 490, 497, 506, 528, 637, 730, and 741.

[9] Sites 42, 95, 105, 231, 317, 339, 372, 428, 502, and 556.

			Site type		Total
			Farmhouse	Necropolis	
Fine ware functional category	Drinking + Drinking Service	Count	185.0	1873.0	2058.0
		Expected count	157.7	1900.3	2058.0
		Std. residual	2.2	-0.6	
	Food + Food service	Count	3.0	500.0	503.0
		Expected count	38.5	464.5	503.0
		Std. residual	-5.7	1.6	
	Lamp + Personal	Count	21.0	146.0	167.0
		Expected count	12.8	154.2	167.0
		Std. residual	2.3	-0.7	
Totals		Count	209.0	2519.0	2728.0
		Expected count	209.0	2519.0	2728.0

Table D8 Tabulation of two ware class and three functional categories for finewares.

service, while food-service vessels were significantly fewer than expected. In 450 food-service vessels were significantly more numerous than expected, while personal vessels were much fewer than expected. In 300 there was no pattern in the distribution of vessels, apart from a much higher-than-expected count of votive/ritual vessels. In 100 drinking-service vessels were present in significantly greater quantity than expected.

Chi-square 7

Research question: Is there a relationship between the classes and categories of ceramic wares within farmhouse assemblages and time periods?

Setup: For the underlying assumption of this test, see Chi-square 6 immediately above. This test also used the same structure by four periods and the same sites as the preceding test. The total number of artifacts considered was 4107.

Result: There is a significant relationship between categories and classes of ceramic wares in farmhouse assemblages and time periods ($\alpha 2$ = 87.412, df = 15, $p < 0.001$). See Table D10.

Discussion: The combined periods 600 and 500 were marked by a much greater-than-expected quantity of banded ware. In 450 the fine wares were significantly fewer than expected, while the plain wares were significantly more than expected. In 300 there were relatively fewer amphorae and relatively more votive vessels. In 100 the amphorae were somewhat more numerous than expected, along with the fine wares, while the plain wares were fewer than expected.

Chi-square 8

Research question: Is there a relationship between the ceramic functional groups and individual site for farmhouses in the periods 600 and 550?

Setup: For the underlying assumption of this test, see Chi-square 6 above. The population of sites was restricted to farmhouses with a minimum of 25 total fragments (12 sites[10]) in order to reduce the number of cells with less than the necessary minimum count; the number of insufficient cells was further reduced by merging some of the functional groups into larger categories. The total number of artifacts considered was 810.

Result: There is a significant relationship between the ceramic functional groups and individual site for farmhouses in the periods 600 and 550 ($\alpha 2$ = 64.799, df = 33, $p < 0.001$). See Table D11.

Discussion: This test was designed as a crude metric of the homogeneity of a group of farmhouses that share a Best Date derived from a preponderance of fine wares dating to the 6th century. In other words, are farmhouses dated closely to 600 and 550 similar to each other in terms of the functional character of their ceramic assemblages? The highly significant result indicates that the selected farmhouses represent very different types of functional assemblages, a situation that in turn indicates a degree of flexibility in the domestic organization of Archaic Metapontine farming families. (The interpretation, however, must be tempered by the assumption of chronological uniformity among non-fine wares that underlies this test.)

[10] Sites 13, 155, 170, 181, 285, 286, 323, 336, 349, 366, 459, and 530.

			Period				Total
			600–500	450	300	100	
Functional group	Drinking service	Count	38.0	15.0	66.0	25.0	144.0
		Expected count	32.8	23.8	78.6	8.8	144.0
		Std. residual	0.9	-1.8	-1.4	5.5	
	Drinking	Count	227.0	106.0	402.0	45.0	780.0
		Expected count	177.8	129.0	425.8	47.5	780.0
		Std. residual	3.7	-2.0	-1.2	-0.4	
	Food preparation	Count	27.0	24.0	38.0	6.0	95
		Expected count	21.7	15.7	51.9	5.8	95.0
		Std. residual	1.1	2.1	-1.9	0.1	
	Food preparation/ cooking	Count	124.0	101.0	347.0	35.0	607.0
		Expected count	138.3	100.4	331.4	36.9	607.0
		Std. residual	-1.2	0.1	0.9	-0.3	
	Food service	Count	119.0	188.0	515.0	49.0	871.0
		Expected count	198.5	144.0	475.5	53.0	871.0
		Std. residual	-5.6	3.7	1.8	-0.6	
	Lamp	Count	0.0	0.0	18.0	2.0	20.0
		Expected count	4.6	3.3	10.9	1.2	20.0
		Std. residual	-2.1	-1.8	2.1	0.7	
	Long-term storage	Count	19.0	17.0	42.0	6.0	84.0
		Expected count	19.1	13.9	45.9	5.1	84.0
		Std. residual	0.0	0.8	-0.6	0.4	
	Personal	Count	24.0	9.0	80.0	9.0	122.0
		Expected count	27.8	20.2	66.6	7.4	122.0
		Std. residual	-0.7	-2.5	1.6	0.6	
	Service	Count	65.0	21.0	81.0	5.0	172.0
		Expected count	39.2	28.4	93.9	10.5	172.0
		Std. residual	4.1	-1.4	-1.3	-1.7	
	Short-term storage	Count	243.0	165.0	541.0	52.0	1001.0
		Expected count	228.1	165.5	546.4	60.9	1001.0
		Std. residual	1.0	0.0	-0.2	-1.1	
	Transport/ storage	Count	40.0	26.0	54.0	14.0	134.0
		Expected count	30.5	22.2	73.2	8.2	134.0
		Std. residual	1.7	0.8	-2.2	2.0	
	Votive/ritual	Count	10.0	7.0	58.0	2.0	77.0
		Expected count	17.5	12.7	42.0	4.7	77.0
		Std. residual	-1.8	-1.6	2.5	-1.2	
Totals		Count	936.0	679.0	2242.0	250.0	4107.0
		Expected count	936.0	679.0	2242.0	250.0	4107.0

Table D9 Tabulation of functional groups and four periods.

Ware class or category			Period: 600–500	450	300	100	Total
Ware class or category	Amphorae	Count	40	26	54	14	134
		Expected count	30.5	22.2	73.2	8.2	134.0
		Std. residual	1.7	.8	-2.2	2.0	
	Banded ware	Count	107	58	182	21	368
		Expected count	83.9	60.8	200.9	22.4	368.0
		Std. residual	2.5	-.4	-1.3	-.3	
	Course ware	Count	124	101	347	35	607
		Expected count	138.3	100.4	331.4	36.9	607.0
		Std. residual	-1.2	.1	.9	-.3	
	Fine wares	Count	308	149	736	101	1294
		Expected count	294.9	213.9	706.4	78.8	1294.0
		Std. residual	.8	-4.4	1.1	2.5	
	Plain ware	Count	348	339	872	77	1636
		Expected count	372.9	270.5	893.1	99.6	1636.0
		Std. residual	-1.3	4.2	-.7	-2.3	
	Votive	Count	9	6	51	2	68
		Expected count	15.5	11.2	37.1	4.1	68.0
		Std. residual	-1.7	-1.6	2.3	-1.1	
Totals		**Count**	**936**	**679**	**2242**	**250**	**4107**
		Expected count	**936.0**	**679.0**	**2242.0**	**250.0**	**4107.0**

Table D10 Tabulation of classes and categories of ceramic wares and four periods.

Chi-square 9

Research question: Is there a relationship between the ceramic functional groups and individual site for farmhouses in the period 450?

Setup: For the underlying assumption of this test, see Chi-square 6 above. The population of sites was restricted to farmhouses with a minimum of 25 total fragments (12 sites[11]) in order to reduce the number of cells with less than the necessary minimum count; the number of insufficient cells was further reduced by merging some of the functional groups into larger categories. The total number of artifacts considered was 622.

Result: There is a significant relationship between the ceramic functional groups and individual site for farmhouses in the period 450 ($\alpha2 = 68.604$, df = 33, $p < 0.001$). See Table D12.

Discussion: This test was designed as a crude metric of the homogeneity of a group of farmhouses that share a Best Date derived from a preponderance of fine wares dating to 450. In other words, are farmhouses dated closely to 450 similar to each other in terms of the functional character of their ceramic assemblages? The highly significant result indicates that the selected farmhouses represent very different types of functional assemblages, a situation that in turn indicates a degree of flexibility in the domestic organization of Classical Metapontine farming families. (The interpretation, however, must be tempered by the assumption of chronological uniformity among non-fine wares that underlies this test.)

Chi-square 10

Research question: Is there a relationship between the ceramic functional groups and individual site for farmhouses in the period 300?

Setup: For the underlying assumption of this test, see Chi-square 6 above. The population of sites was restricted to farmhouses with a minimum of 25 total fragments (24 sites[12]) in order to reduce the number of cells with less than the necessary minimum count; the number of insufficient cells was further reduced by merging some of the functional groups into larger categories. The total

[11] Sites 14, 49, 68, 182, 257, 302-305, 344, 414, and 515.

[12] Sites 43, 113, 121, 128, 169, 183, 249, 341, 354, 370, 373, 417, 445, 454, 456, 460, 487, 490, 497, 506, 528, 637, 730, and 741.

			Functional categories (simplified)				Total
			Drinking	Food preparation	Food service	Transport/ storage	
Site	13	Count	18	8	16	16	58
		Expected count	17.3	9.1	12.2	19.4	58.0
		Std. residual	.2	-.4	1.1	-.8	
	155	Count	48	14	17	44	123
		Expected count	36.6	19.3	26.0	41.2	123.0
		Std. residual	1.9	-1.2	-1.8	.4	
	170	Count	13	8	10	5	36
		Expected count	10.7	5.6	7.6	12.0	36.0
		Std. residual	.7	1.0	.9	-2.0	
	181	Count	7	15	7	21	50
		Expected count	14.9	7.8	10.6	16.7	50.0
		Std. residual	-2.0	2.6	-1.1	1.0	
	285	Count	32	21	13	39	105
		Expected count	31.2	16.5	22.2	35.1	105.0
		Std. residual	.1	1.1	-1.9	.7	
	286	Count	26	15	29	31	101
		Expected count	30.1	15.8	21.3	33.8	101.0
		Std. residual	-.7	-.2	1.7	-.5	
	323	Count	20	10	19	40	89
		Expected count	26.5	14.0	18.8	29.8	89.0
		Std. residual	-1.3	-1.1	.0	1.9	
	336	Count	23	10	25	22	80
		Expected count	23.8	12.5	16.9	26.8	80.0
		Std. residual	-.2	-.7	2.0	-.9	
	349	Count	17	13	13	17	60
		Expected count	17.9	9.4	12.7	20.1	60.0
		Std. residual	-.2	1.2	.1	-.7	
	366	Count	9	5	11	13	38
		Expected count	11.3	6.0	8.0	12.7	38.0
		Std. residual	-.7	-.4	1.1	.1	
	459	Count	8	2	5	14	29
		Expected count	8.6	4.5	6.1	9.7	29.0
		Std. residual	-.2	-1.2	-.5	1.4	
	530	Count	20	6	6	9	41
		Expected count	12.2	6.4	8.7	13.7	41.0
		Std. residual	2.2	-.2	-.9	-1.3	
Totals		Count	241	127	171	271	810
		Expected count	241.0	127.0	171.0	271.0	810.0

Table D11 Tabulation of individual sites in **600–550** and functional categories.

number of artifacts considered was 1,922.
Result: There is a significant relationship between the ceramic functional groups and individual site for farmhouses in the period 300 ($\alpha 2$ = 173.342, df = 69, $p < 0.001$). See Table D13.
Discussion: This test was designed as a crude metric of the homogeneity of a group of farmhouses that share a Best Date derived from a preponderance of fine wares dating to 300. In other words, are farmhouses dated closely to 300 similar to each other in terms of the functional character of their ceramic assemblages? The highly significant result indicates that the selected farmhouses represent very different types of functional assemblages, a situation that in turn indicates a degree of flexibility in the domestic organization of early Hellenistic Metapontine farming families. (The interpretation, however, must be tempered by the assumption of chronological uniformity among nonfine wares that underlies this test.)

Chi-square 11

Research question: Is there a relationship between the ceramic functional groups and site type for farmhouses and sanctuaries in 600 and 550?
Setup: For the assumption underlying this test, see Chi-square 6 above. The relevant list of sites from Chi-square 6 (see n. 6 above) was used to represent the farmhouses active in these periods; all five of the known sanctuaries were also used, since each had an FMCE value in these periods.[13] The test involved 1,864 fragments.
Result: There is a significant relationship between ceramic functional groups and site type for farmhouses and sanctuaries in 600 and 550 ($\alpha 2$ = 34.428, df = 10, $p < 0.001$). See Table D14.
Discussion: Despite the significant result, the symmetrical distribution of the standardized residuals demonstrates that farmhouse and sanctuary assemblages in 600 and 550 were broadly similar in their functional assemblages, with the most extreme difference in short-term storage.

Chi-square 12

Research question: Is there a relationship between the ceramic functional groups and site type for farmhouses and necropoleis in 300?
Setup: For the assumption underlying this test, see Chi-square 6 above. The relevant list of sites from Chi-square 6 (see n. 8 above) was used to represent the farmhouses active in this period; the necropoleis were represented by those with EAW ≥ 0.45 in the period.[14] The test involved 3,062 fragments.
Result: There is a significant relationship between ceramic functional groups and site type for farmhouses and necropoleis in 300 ($\alpha 2$ = 228.498, df = 11, $p < 0.001$). See Table D15.
Discussion: In 300 farmhouse assemblages had significantly fewer vessels related to drinking service, while necropoleis had significantly more drinking-service vessels than expected. Necropoleis also had much fewer food preparation/cooking and votive/ritual vessels relative to farmhouses.

Chi-square 13

Research question: Is there a relationship in the occupation of sites between the Prehistoric and Roman periods?
Setup: The occupation histories of all 595 sites considered in this volume were organized as two categories, "Prehistoric" and "Roman," with a simple "yes/no" indicator (1 or 0) in each field.
Result: There is a significant relationship in the occupation of sites between the Prehistoric and Roman periods ($\alpha 2$ = 18.827, df = 1, $p < 0.001$). See Table D16.
Discussion: Sites occupied during the Roman period tended to be occupied during prehistoric times.

Chi-square 14

Research question: Is there a relationship between the ceramic functional groups and individual site for farmhouses in the period 100?
Setup: For the underlying assumption of this test, see Chi-square 6 above. The population of sites was restricted to farmhouses with a minimum of 25 total fragments (5 sites[15]) in order to reduce the number of cells with less than the necessary minimum count; the number of insufficient cells was further reduced by merging some of the functional groups into larger categories. The total number of artifacts considered was 191.
Result: There is not a significant relationship between the ceramic functional groups and individual site for farmhouses in the period 100 ($\alpha 2$ = 19.131, df = 12, p = 0.085). See Table D17.
Discussion: This test was designed as a crude metric of the homogeneity of a group of farmhouses that share a Best Date derived from a preponderance of fine wares dating to 100. In other words, are farmhouses dated

[13] Sites 266, 270, 334, 397, and 477.

[14] Sites 266, 270, 334, 397, and 477.

[15] Sites 42, 105, 372, 428, and 556.

closely to 100 similar to each other in terms of the functional character of their ceramic assemblages? The lack of a significant result indicates that the selected farmhouses represent very similar types of functional assemblages, a situation that in turn indicates a lesser degree of flexibility in the domestic organization of Late Hellenistic Metapontine farming families compared with previous periods (see Chi-squares 8, 9, and 10 above). (The interpretation, however, must be tempered by the assumption of chronological uniformity among non-fine wares that underlies this test, as well as the very small sample of farmhouses available for consideration.)

Chi-square 15

Research question: Is there a relationship in the occupation of sites between the Prehistoric and Greek periods?
Setup: The occupation histories of all 595 sites considered in this volume were organized as two categories, "Prehistoric" and "Greek," with a simple "yes/no" indicator (1 or 0) in each field.
Result: There is not a significant relationship in the occupation of sites between the Prehistoric and Greek periods ($\alpha 2$ = 0.214, df = 1, p = 0.644). See Table D18.
Discussion: Sites occupied in the Greek period were not more likely to have been occupied in prehistoric times than sites not occupied in the Greek period.

Chi-square 16

Research question: Is there a relationship in the occupation of sites between the Prehistoric and Medieval periods?
Setup: The occupation histories of all 595 sites considered in this volume were organized as two categories, "Prehistoric" and "Medieval," with a simple "yes/no" indicator (1 or 0) in each field.
Result: There is not a significant relationship in the occupation of sites between the Prehistoric and Medieval periods ($\alpha 2$ = 1.052, df = 1, p = 0.305). See Table D19.
Discussion: Sites occupied in the Medieval period were not more likely to have been occupied in prehistoric times than sites not occupied in the Medieval period.

Chi-square 17

Research question: Is there a relationship in the occupation of sites between the Roman and Medieval periods?
Setup: The occupation histories of all 595 sites considered in this volume were organized as two categories, "Roman" and "Medieval," with a simple "yes/no" indicator (1 or 0) in each field.
Result: There is not a significant relationship in the occupation of sites between the Roman and Medieval periods ($\alpha 2$ = 3.215, df = 1, p = 0.073). See Table D20.
Discussion: Sites occupied in the Medieval period were not more likely to have been occupied in Roman times than sites not occupied in the Medieval period.

			Functional categories (simplified)				Total
			Drinking	Food preparation	Food service	Transport/ storage	
Site	14	Count	3	4	10	13	**30**
		Expected count	5.2	5.6	9.5	9.6	**30.0**
		Std. residual	-1.0	-.7	.1	1.1	
	49	Count	4	3	11	10	28
		Expected count	4.9	5.2	8.9	9.0	28.0
		Std. residual	-.4	-1.0	.7	.3	
	68	Count	4	5	13	22	44
		Expected count	7.6	8.2	14.0	14.1	44.0
		Std. residual	-1.3	-1.1	-.3	2.1	
	182	Count	19	39	67	52	**177**
		Expected count	30.7	33.0	56.3	56.9	**177.0**
		Std. residual	-2.1	1.0	1.4	-.7	
	257	Count	3	6	12	6	**27**
		Expected count	4.7	5.0	8.6	8.7	**27.0**
		Std. residual	-.8	.4	1.2	-.9	
	302	Count	5	9	6	7	**27**
		Expected count	4.7	5.0	8.6	8.7	**27.0**
		Std. residual	.1	1.8	-.9	-.6	
	303	Count	5	4	12	5	**26**
		Expected count	4.5	4.8	8.3	8.4	**26.0**
		Std. residual	.2	-.4	1.3	-1.2	
	304	Count	15	16	24	21	**76**
		Expected count	13.2	14.2	24.2	24.4	**76.0**
		Std. residual	.5	.5	.0	-.7	
	305	Count	8	9	12	15	**44**
		Expected count	7.6	8.2	14.0	14.1	**44.0**
		Std. residual	.1	.3	-.5	.2	
	344	Count	18	10	15	12	**55**
		Expected count	9.5	10.3	17.5	17.7	**55.0**
		Std. residual	2.7	-.1	-.6	-1.4	
	414	Count	15	3	5	10	**33**
		Expected count	5.7	6.2	10.5	10.6	**33.0**
		Std. residual	3.9	-1.3	-1.7	-.2	
	515	Count	9	8	11	27	55
		Expected count	9.5	10.3	17.5	17.7	55.0
		Std. residual	-.2	-.7	-1.6	2.2	
Totals		**Count**	**108**	**116**	**198**	**200**	**622**
		Expected count	**108.0**	**116.0**	**198.0**	**200.0**	**622.0**

Table D12 Tabulation of individual sites in **450** and functional categories.

			Functional categories (simplified)				Total
			Drinking	Food preparation	Food service	Transport/storage	
Sites	43	Count	28	38	92	93	251
		Expected count	55.6	47.0	73.1	75.2	251.0
		Std. residual	-3.7	-1.3	2.2	2.0	
	113	Count	19	22	15	17	73
		Expected count	16.2	13.7	21.3	21.9	73.0
		Std. residual	.7	2.3	-1.4	-1.0	
	121	Count	9	4	20	18	51
		Expected count	11.3	9.6	14.9	15.3	51.0
		Std. residual	-.7	-1.8	1.3	.7	
	128	Count	25	14	16	11	66
		Expected count	14.6	12.4	19.2	19.8	66.0
		Std. residual	2.7	.5	-.7	-2.0	
	169	Count	16	17	37	41	111
		Expected count	24.6	20.8	32.3	33.3	111.0
		Std. residual	-1.7	-.8	.8	1.3	
	183	Count	21	24	32	24	101
		Expected count	22.4	18.9	29.4	30.3	101.0
		Std. residual	-.3	1.2	.5	-1.1	
	249	Count	8	15	11	22	56
		Expected count	12.4	10.5	16.3	16.8	56.0
		Std. residual	-1.3	1.4	-1.3	1.3	
	341	Count	52	36	46	31	165
		Expected count	36.6	30.9	48.1	49.4	165.0
		Std. residual	2.6	.9	-.3	-2.6	
	354	Count	23	29	24	24	100
		Expected count	22.2	18.7	29.1	30.0	100.0
		Std. residual	.2	2.4	-1.0	-1.1	
	370	Count	21	11	25	26	83
		Expected count	18.4	15.5	24.2	24.9	83.0
		Std. residual	.6	-1.2	.2	.2	
	373	Count	6	5	15	11	37
		Expected count	8.2	6.9	10.8	11.1	37.0
		Std. residual	-.8	-.7	1.3	.0	
	417	Count	7	7	12	5	31
		Expected count	6.9	5.8	9.0	9.3	31.0
		Std. residual	.0	.5	1.0	-1.4	
	445	Count	8	6	4	11	29
		Expected count	6.4	5.4	8.4	8.7	29.0
		Std. residual	.6	.2	-1.5	.8	
	454	Count	21	20	22	28	91
		Expected count	20.2	17.0	26.5	27.3	91.0
		Std. residual	.2	.7	-.9	.1	

Table D13 Tabulation of individual sites in **300** and functional categories.

			Functional categories (simplified)				Total
			Drinking	Food preparation	Food service	Transport/storage	
	456	Count	7	2	3	14	26
		Expected count	5.8	4.9	7.6	7.8	26.0
		Std. residual	.5	-1.3	-1.7	2.2	
	460	Count	21	6	18	24	69
		Expected count	15.3	12.9	20.1	20.7	69.0
		Std. residual	1.5	-1.9	-.5	.7	
Sites	487	Count	10	7	21	19	57
		Expected count	12.6	10.7	16.6	17.1	57.0
		Std. residual	-.7	-1.1	1.1	.5	
	490	Count	7	17	11	10	45
		Expected count	10.0	8.4	13.1	13.5	45.0
		Std. residual	-.9	3.0	-.6	-.9	
	497	Count	32	28	21	31	112
		Expected count	24.8	21.0	32.6	33.6	112.0
		Std. residual	1.4	1.5	-2.0	-.4	
	506	Count	23	31	46	26	126
		Expected count	27.9	23.6	36.7	37.8	126.0
		Std. residual	-.9	1.5	1.5	-1.9	
	528	Count	11	7	11	13	42
		Expected count	9.3	7.9	12.2	12.6	42.0
		Std. residual	.6	-.3	-.4	.1	
	637	Count	9	4	12	16	41
		Expected count	9.1	7.7	11.9	12.3	41.0
		Std. residual	.0	-1.3	.0	1.1	
	730	Count	34	5	36	48	123
		Expected count	27.3	23.0	35.8	36.9	123.0
		Std. residual	1.3	-3.8	.0	1.8	
	741	Count	8	5	10	13	36
		Expected count	8.0	6.7	10.5	10.8	36.0
		Std. residual	.0	-.7	-.2	.7	
Totals		**Count**	**426**	**360**	**560**	**576**	**1922**
		Expected count	**426.0**	**360.0**	**560.0**	**576.0**	**1922.0**

Table D13 *cont.* Tabulation of individual sites in **300** and functional categories.

			Site type		Total
			Farmhouse	Sanctuary	
Functional group	Drinking/service	Count	38	45	**83**
		Expected count	41.2	41.8	**83.0**
		Std. residual	-.5	.5	
	Drinking	Count	227	282	**509**
		Expected count	252.9	256.1	**509.0**
		Std. residual	-1.6	1.6	
	Food preparation	Count	27	25	**52**
		Expected count	25.8	26.2	**52.0**
		Std. residual	.2	-.2	
	Food preparation/ cooking	Count	124	101	**225**
		Expected count	111.8	113.2	**225.0**
		Std. residual	1.2	-1.1	
	Food service	Count	119	140	**259**
		Expected count	128.7	130.3	**259.0**
		Std. residual	-.9	.8	
	Lamp	Count	0	5	**5**
		Expected count	2.5	2.5	**5.0**
		Std. residual	-1.6	1.6	
	Long-term storage	Count	19	7	**26**
		Expected count	12.9	13.1	**26.0**
		Std. residual	1.7	-1.7	
	Personal	Count	24	21	**45**
		Expected count	22.4	22.6	**45.0**
		Std. residual	.3	-.3	
	Service	Count	65	91	**156**
		Expected count	77.5	78.5	**156.0**
		Std. residual	-1.4	1.4	
	Short-term storage	Count	243	182	**425**
		Expected count	211.1	213.9	**425.0**
		Std. residual	2.2	-2.2	
	Transport/storage	Count	40	39	**79**
		Expected count	39.2	39.8	**79.0**
		Std. residual	.1	-.1	
Totals		Count	926	938	1864
		Expected count	926.0	938.0	1864.0

Table D14 Tabulation of ceramic functional groups for farmhouses and sanctuaries in **600–550**.

			Site type		Total
			Farmhouse	Necropolis	
Functional group	Drinking/service	Count	66	145	211
		Expected count	154.5	56.5	211.0
		Std. residual	-7.1	11.8	
	Drinking	Count	402	144	546
		Expected count	399.8	146.2	546.0
		Std. residual	.1	-.2	
	Food preparation	Count	38	11	49
		Expected count	35.9	13.1	49.0
		Std. residual	.4	-.6	
	Food preparation/ cooking	Count	347	85	432
		Expected count	316.3	115.7	432.0
		Std. residual	1.7	-2.9	
	Food service	Count	515	148	663
		Expected count	485.4	177.6	663.0
		Std. residual	1.3	-2.2	
	Lamp	Count	18	3	21
		Expected count	15.4	5.6	21.0
		Std. residual	.7	-1.1	
	Long-term storage	Count	42	12	54
		Expected count	39.5	14.5	54.0
		Std. residual	.4	-.6	
	Personal	Count	80	34	114
		Expected count	83.5	30.5	114.0
		Std. residual	-.4	.6	
	Service	Count	81	28	109
		Expected count	79.8	29.2	109.0
		Std. residual	.1	-.2	
	Short-term storage	Count	541	169	710
		Expected count	519.9	190.1	710.0
		Std. residual	.9	-1.5	
	Transport/ storage	Count	54	34	88
		Expected count	64.4	23.6	88.0
		Std. residual	-1.3	2.1	
	Votive/ritual	Count	58	7	65
		Expected count	47.6	17.4	65.0
		Std. residual	1.5	-2.5	
Totals		**Count**	**2242**	**820**	**3062**
		Expected count	**2242.0**	**820.0**	**3062.0**

Table D15 Tabulation of ceramic functional groups for farmhouses and necropoleis in **300**.

			Roman		Total
			No	Yes	
Prehistoric	No	Count	440	17	457
		Expected count	429.3	27.7	457.0
		Std. residual	.5	-2.0	
	Yes	Count	119	19	138
		Expected count	129.7	8.3	138.0
		Std. residual	-.9	3.7	
Totals		Count	559	36	595
		Expected count	559.0	36.0	595.0

Table D16 Tabulation of Prehistoric and Roman occupation.

			Period(s)				Total
			Drinking	Food preparation	Food service	Transport/ storage	
Sites	105	Count	19	7	7	4	37
		Expected count	11.2	5.4	8.9	11.4	37.0
		Std. residual	2.3	.7	-.6	-2.2	
	372	Count	17	10	13	18	58
		Expected count	17.6	8.5	14.0	17.9	58.0
		Std. residual	-.1	.5	-.3	.0	
	42	Count	5	3	8	9	25
		Expected count	7.6	3.7	6.0	7.7	25.0
		Std. residual	-.9	-.3	.8	.5	
	428	Count	8	5	6	15	34
		Expected count	10.3	5.0	8.2	10.5	34.0
		Std. residual	-.7	.0	-.8	1.4	
	556	Count	9	3	12	13	37
		Expected count	11.2	5.4	8.9	11.4	37.0
			-.7	-1.0	1.0	.5	
Totals		Count	58	28	46	59	191
		Expected count	58.0	28.0	46.0	59.0	191.0

Table D17 Tabulation of individual sites in **100** and functional categories.

			Greek		Total
			No	Yes	
Prehistoric	No	Count	22	435	457
		Expected count	23.0	434.0	457.0
		Std. residual	-.2	.1	
	Yes	Count	8	130	138
		Expected count	7.0	131.0	138.0
		Std. residual	.4	-.1	
Totals		Count	30	565	595
		Expected count	30.0	565.0	595.0

Table D18 Tabulation of Prehistoric and Greek occupation.

			Medieval		Total
			No	Yes	
Prehistoric	No	Count	437	20	457
		Expected count	434.7	22.3	457.0
		Std. residual	.1	-.5	
	Yes	Count	129	9	138
		Expected count	131.3	6.7	138.0
		Std. residual	-.2	.9	
Totals		Count	566	29	595
		Expected count	566.0	29.0	595.0

Table D19 Tabulation of Prehistoric and Medieval occupation.

			Medieval		Total
			No	Yes	
Roman	No	Count	534	25	559
		Expected count	531.8	27.2	559.0
		Std. residual	.1	-.4	
	Yes	Count	32	4	36
		Expected count	34.2	1.8	36.0
		Std. residual	-.4	1.7	
Totals		Count	566	29	595
		Expected count	566.0	29.0	595.0

Table D20 Tabulation of Roman and Medieval occupation.